AF323428

Electro-Rheological Fluids and Magneto-Rheological Suspensions

Electro-Rheological Fluids and Magneto-Rheological Suspensions

Proceedings of the 12th International Conference
Philadelphia, USA, 16 – 20 August 2010

editor

Rongjia Tao

Temple University, USA

NEW JERSEY · LONDON · SINGAPORE · BEIJING · SHANGHAI · HONG KONG · TAIPEI · CHENNAI

Published by

World Scientific Publishing Co. Pte. Ltd.

5 Toh Tuck Link, Singapore 596224

USA office: 27 Warren Street, Suite 401-402, Hackensack, NJ 07601

UK office: 57 Shelton Street, Covent Garden, London WC2H 9HE

British Library Cataloguing-in-Publication Data
A catalogue record for this book is available from the British Library.

ELECTRO-RHEOLOGICAL FLUIDS AND MAGNETO-RHEOLOGICAL SUSPENSIONS
Proceedings of the 12th International Conference

ISBN-13 978-981-4340-22-9
ISBN-10 981-4340-22-7

Printed in Singapore by World Scientific Printers.

PREFACE

The Twelfth International Conference on Electrorheological (ER) Fluids and Magnetorheological (MR) Suspensions was held at an interesting time: The world economy was still slowly recovering from the worst financial crisis, but the positive impacts of the ERMR research on energy, automobile, transportation and other industries are very notable.

Over one hundred scientists and engineers in multidisciplinary areas from 21 countries came to Philadelphia, Pennsylvania, USA, August 16 to 20, 2010 to explore the state-of-art technology and identify the thrust areas for ER and MR fluids. The Organization Committee received 144 papers. Most of them were presented at the conference. Highly productive and scientifically stimulating discussions followed the presentations. Since the last ERMR Conference in 2008, significant progress in the areas of energy application, materials science, and physical mechanism has been made. The Conference was very successful in spite of the fact that many foreign scientists and engineers were unable to come to present their papers at the Conference because they did not get their entry visa to USA on time.

The conference was sponsored by Temple University, Lord Corporation, BASF Chemical Company, Anton Paar GmbH, and Strem Chemicals, Inc. The organizing committee truly appreciates the support from these corporations and institution.

My special appreciation is also due to Dr. Hong Tang, Dr. Xiaojun Xu, Mr. Enpeng Du, and Mrs. Karen Woods-Wilson for their superb support in organizing the conference.

Rongjia Tao, Ph.D.
Temple University
November 2010

In Memory of Dr. Frank E. Filisko

Professor Frank E. Filisko
(1942-2008)

On November 11, 2008, after a two-year battle with chronic lymphocytic leukemia, Dr. Frank Edward Filisko, passed away. The ERMR Community is deeply saddened by the loss of our friend and productive colleague.

Frank was born and raised in Loraine, OH on January 29, 1942. He attended Colgate University, Hamilton, NY as an undergraduate where he studied physics and played football, receiving an All-American honorable mention as a fullback. He graduated with a BA in physics and math in 1964.

Frank received an MS degree in solid state physics in 1966 from Purdue University. He returned to Ohio to attend Case Western Reserve University to study polymer physics. He received his Ph.D. from CWRU in 1969 but stayed on as a post-doctoral fellow.

In 1970, Frank joined the University of Michigan Department of Chemical and Metallurgical Engineering as an Assistant Professor. When the Department split into separate chemical and metallurgical branches, Frank stayed with Metallurgical Engineering, which is now known as the Department of Materials Science and Engineering. He was promoted to Professor in 1984.

Frank's research accomplishments in the area of ER fluids were remarkable. He was the inventor and co-inventor of 4 US patents related to ER fluids. On 9/24/1996, there was an article in New York Times to discuss ER fluids. The article specifically mentioned Frank's ER fluid invention, as "cheap and fairly stable." Frank also authored and co-authored about 100 peer-reviewed publications. The most popular ones were in the area of ER fluids. Some of them, such as his two papers, "Dynamic mechanical studies of electrorheological materials - moderate frequencies" and "An intrinsic mechanism for the activity of aluminosilicate based electrorheological materials", published in 1990 and 1991, are still widely cited today.

Frank also worked in thermodynamic measurements on polymers and collaborated with the School of Dentistry. He always liked to work directly with his students. He felt that as a mentor he would be most successful when he was in the lab working alongside them, experiencing their successes, and maybe having to periodically fix their instruments.

Frank loved teaching. He excelled at teaching the introductory materials science and engineering course. He also taught a polymers course, kinetics course, senior design course, and during his last term of teaching, a junior-level lab. He was quite popular with students.

Frank left behind his contributions to the area of ER fluids. As we gathered here to disseminate the advance of ER technology, we remember and honor him.

CONTENTS

III. Physical Mechanism

IV. Properties and Microstructures

I. APPLICATIONS

ELECTRORHEOLOGY FOR EFFICIENT ENERGY PRODUCTION AND CONSERVATION

R. TAO

Department of Physics, Temple University, Philadelphia, PA 19122, USA

Recently, we developed a theory and new technology, which utilizes a pulsed electric field to change the rheology of complex fluid to reduce its viscosity, while keeping the temperature unchanged. The method is energy-efficient, universal, and applicable to all complex fluids with suspended particles in nano-meters, sub-micrometers, or micrometers. We have applied this technology to crude oil and refinery fuels. While the applications are still developing, the results are very impressive, indicating that electrorheology can play a very important role in energy production, transportation, and conservation.

1. Introduction

At present, most of our energy comes from liquid fuels. The viscosity plays a very important role in liquid fuel production, transportation, and conservation. For example, reducing the viscosity of crude oil is the key for oil extraction and its transportation from off-shore via deep water pipelines. Currently, the dominant method to reduce viscosity is to raise the oil temperature, which not only requires much energy, but also impacts the environment. The application of DRA (drug reduction agent) is not only costly, but also raises concerns at the refinery. Recently, based on the basic physics of viscosity, we developed a theory and new technology, which utilizes electric or magnetic field to change the rheology of complex fluid to reduce its viscosity. The method is energy-efficient, universal, and applicable to all complex fluids with suspended particles in nano-meters, sub-micrometers, or micrometers. We have applied this technology to crude oil and refinery fuels. The results are very significant.

Einstein first studied a dilute liquid suspension of non- interacting uniform spheres in a base liquid of viscosity η_0 and found the effective viscosity η as follows [1],

$$\eta = \eta_0 (1 + 2.5\phi) . \tag{1}$$

Equation (1) is good for a very small particle volume fraction ϕ. For high ϕ, we must consider the maximum volume fraction, ϕ_m available for adding particles. For liquid suspensions, ϕ_m is about 0.64, the maximum value for

4

random packing. Let us consider adding $d\phi$ spheres to a liquid suspension of volume fraction ϕ. As the net available volume fraction to add spheres is only $1-\phi/\phi_m$ now, the increase of viscosity would be $d\eta/\eta = 2.5 d\phi/(1-\phi/\phi_m)$. Integrating this equation gives us an expression to estimate the viscosity at high ϕ

$$\eta/\eta_0 = (1-\phi/\phi_m)^{-2.5\phi_m}. \tag{2}$$

Krieger-Dougherty introduced the intrinsic viscosity $[\eta]$ for particles of different shapes [2],

$$\eta/\eta_0 = (1-\phi/\phi_m)^{-[\eta]\phi_m}, \tag{3}$$

which enables us to estimate the viscosity for particles of any shape by choosing a suitable $[\eta]$.

When ϕ is unchanged, the most widely used method to reduce viscosity η is to reduce η_0, such as raising the temperature. On the other hand, Eq.(3) suggests that there is another method: if we change the rheology of the suspension to increase the value of ϕ_m and lower $[\eta]$, we will reduce the viscosity η. The physics is clear: the effective viscosity depends on how much freedom the suspended particles have in the suspension. A high ϕ_m and low $[\eta]$ mean high freedom for the suspended particles, which leads to lower dissipation of energy and lower viscosity [3].

The following three mechanisms contribute to the viscosity reduction [4-8]: (1) increase the polydispersity; (2) increase the average size of suspended particles; (3) aggregate the particles into clusters with their shapes streamlined.

The value of ϕ_m increases with increasing polydispersity. Let us consider a suspension of binary particle-size distribution as an example. For random close

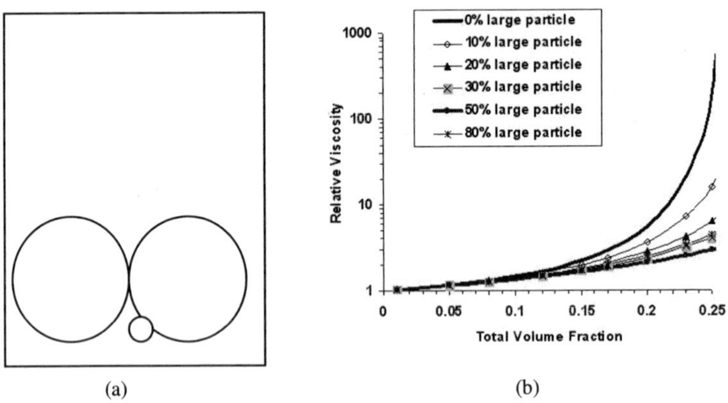

(a)   (b)

Fig. 1 (a) With increasing polydispersity, the maximum volume fraction for adding particles, ϕ_m, is increased. (b) Shown is the viscosity of a binary suspension composed of large particles (diameter 0.5 μm) and small particles (diameter 0.1 μm).

packing, we can let the big ones pack first and add the small ones fill the gaps between the large particles. Therefore, ϕ_m for a suspension of binary particle-size distribution is higher than that for a suspension of mono-disperse distribution. As shown in Fig.1, when the ratio of the large particles to small particles increases, the viscosity is reduced significantly with increasing ϕ_m.

At a high ϕ, the particle size distribution has a strong effect on the viscosity. In liquid suspensions there is always a short-range repulsive force between the suspended particles, which prevents them from touching each other. Let the repulsive force range be λ. If the maximum volume fraction for randomly packing spheres of diameter D without the repulsive force is ϕ_{m0}, the effective maximum volume fraction for randomly packing spheres of diameter D with the short-range repulsive force is [4]

$$\phi_m = \phi_{m0}[D/(D+\lambda)]^3 = \phi_{m0}/(1+\lambda/D)^3 .)$$ (4)

It can now be inferred that as the diameter of the particles D increases, ϕ_m increases and the viscosity decreases (Fig.2).The above analysis indicates that aggregating small particles into large ones in a liquid suspension will reduce the effective viscosity. This viscosity reduction is independent of the temperature. The aggregation of small particles can be realized with electrorheology or magnetorheology. If the suspended particles and the base liquid have different dielectric constants, we can use an electric field to polarize the particles and aggregate them into large ones. If the suspended particles and the base liquid have different magnetic permeability, we can use a magnetic field. Recent experiments indicate that an electric field is universal for all kinds of hydrocarbon fuels, much better than a magnetic field which only works for some paraffin-based crude oil.

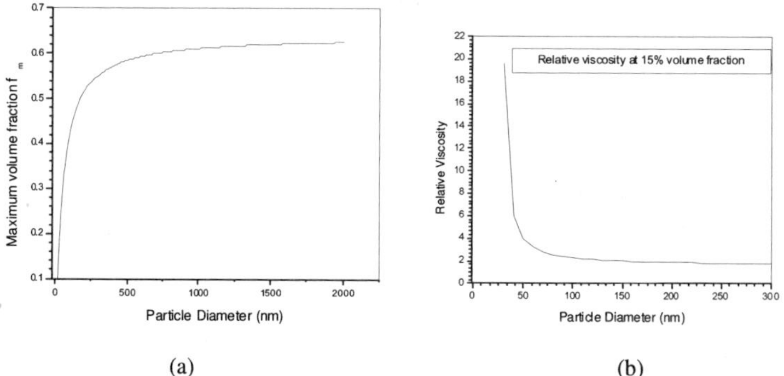

Fig. 2 (a) The effective maximum volume fraction is increased with particle diameter. (b) The effective viscosity is reduced as the particle diameter increases.

Our technology is illustrated in Fig.3. The fuel flows from left to right along a pipe. Initially at the left, the viscosity is high. When the fuel passes a strong local electric field, the suspended particles are polarized by the electric field. The induced dipolar interaction forces the nanoscale particles to aggregate into micrometer-size short chains or ellipsoids. They are of streamline shape with low $[\eta]$ since he electric field is parallel to the flow direction. The viscosity is further reduced by the Serge-Silibergerg effect: large aggregated particles migrate toward the center of the pipe, where the shear rate is the lowest [9]. For illustration, the local electric field is produced by meshes in Fig.3. Actually, the local electric field can also be produced by other methods without the meshes.

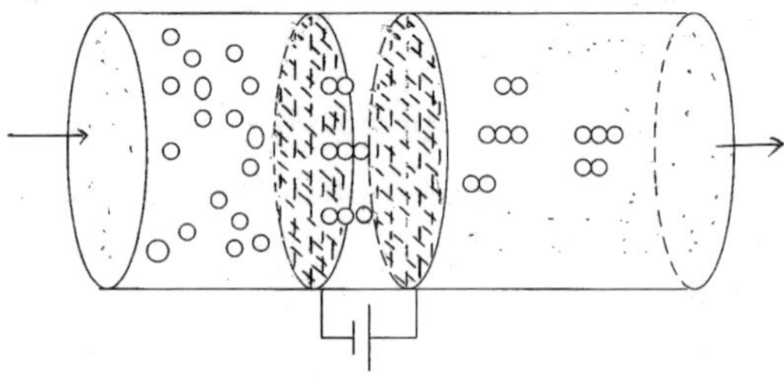

Fig. 3 As the fuel flow passes a strong local electric field, the suspended particles aggregate along the field direction, and the viscosity is reduced.

2. Application to Crude Oil

The typical viscosity reduction pattern for crude oil is shown in Fig.4. This oil sample has viscosity about 1150 cp. After it passes the electric field of 1 kV/mm for 5 seconds, the viscosity is down to 570 cp, and gradually moves up afterwards. However, after about 5 hours, the viscosity is still below 750 cp. Our recent experiment also shows that when the temperature is below 0^0C and the water content inside the oil becomes ice, the viscosity reduction can be as high as above 75%.

This technology is extremely energy-efficient. It does not heat the oil and only aggregates the suspended particles, requiring about 0.01 kW-h per barrel to reduce the viscosity by 30-50%. In comparison with heating, to reach the same level of viscosity reduction, this technology requires less than 1% of the energy needed for heating. With this technology, the oil compounds remain the same, friendly to the refinery process. Moreover, it only takes several seconds to have the viscosity reduced, while heating takes at least several minutes to complete. This approach is applicable to both off-shore pipelines and on-shore pipelines, increasing the flow rate and the capacity of pipelines. It also reduces the footprint required for pumping and has potential for oil extraction.

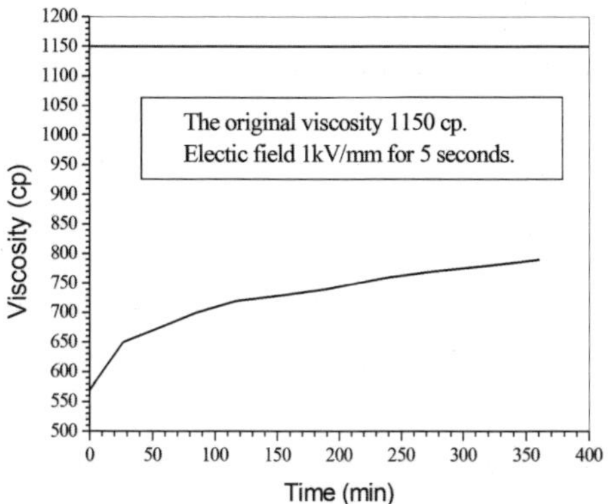

Fig. 4 A crude oil sample has viscosity 1150 cp. After the oil passes an electric field of 1kV/mm for 5 seconds, the viscosity is down to 570 cp, and then slowly rises.

A recent experiment at the NIST Center for Neutron Research, on the NG7 SANS beam line, clearly reveals that the microstructure of crude oil changes under an electric field [12]. As shown in Fig.4, the nanoscale particles were originally randomly distributed in the crude oil sample. In a strong electric field, these particles align to form short chains along the field direction. This SANS experiment supports the notion that electrorheology can change the microstructure of crude oil to reduce its viscosity.

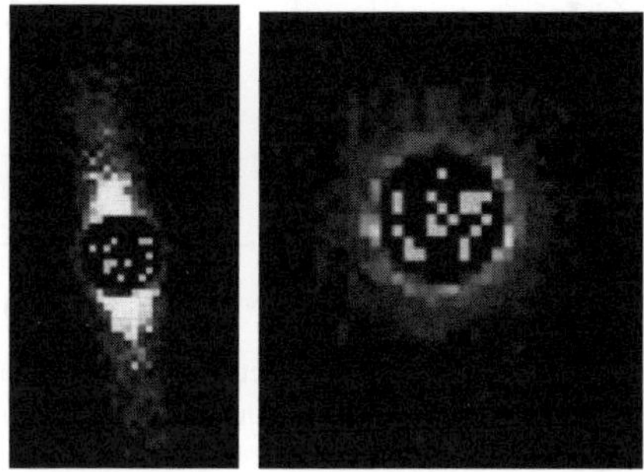

Fig. 5 Small angle neutron scattering (SANS) from crude oil. With no electric field (right), scattering is isotropic. Under an electric field of 750 V/mm (left), the scattering reveals short chains of nanoscale particles aligned along the field direction. Black, blue, red & yellow symbols indicate increasing SANS intensity, in this order.

3. Diesel Fuel Injection

We have also applied electrorheology to reduce the viscosity of diesel fuel just before the fuel atomization [9,10]. Diesel is made of many different molecules and can be regarded as a liquid suspension if we take the large molecules as suspended particles and the base liquid as made of small molecules. Under a strong electric field, the induced dipolar interaction makes the large molecules and suspended nanoparticles aggregate into small clusters. This change reduces the diesel's viscosity. As shown in Figure 6, when the viscosity reduction is just made before the fuel injection, much smaller fuel droplets are produced in the fuel injection. The influence of fuel's viscosity on the atomization can be illustrated by the Ohnesorge number [11], which is defined as

$$Oh = \eta / \sqrt{\rho D \sigma} , \tag{5}$$

where D is the droplet's diameter and η, ρ, and σ are the fuel's viscosity, density, and surface tension, respectively. Typically, there is a universal critical value Oh_c for the Ohnesorge number. As $Oh < Oh_c$, the droplet splits. As $Oh > Oh_c$, the droplet does not split. Therefore, as the viscosity η is reduced, the droplet size is also reduced.

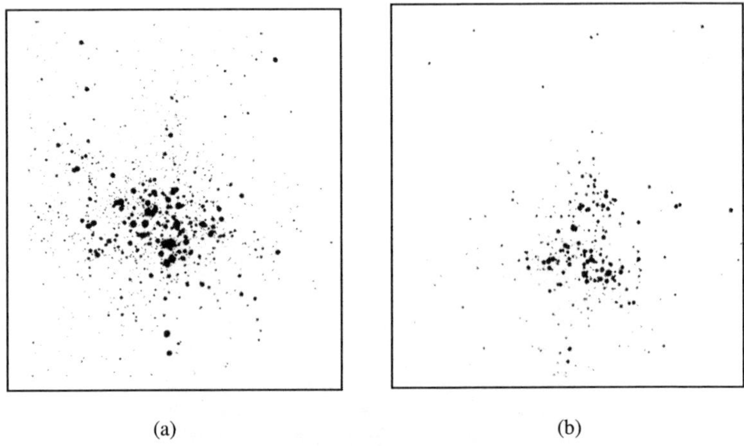

(a)   (b)

Fig. 6 When the electric field is turned on, the injected diesel fuel droplet size is significantly reduced as a result of the viscosity reduction. (a) Without electric field applied (Left). (b) Electric field E=1 kV/mm applied (right).

The fuel efficiency of internal combustion engines depends on the combustion speed and timing. The burning fuel in internal combustion engines releases the heat, expanding the gas and pushing the piston to do work. The combustion speed and timing is very crucial here. If the combustion is slow, some heat released at the time when the piston is near to the bottom-center (BC) crank position will not do work for the engine; then the combustion cannot be efficient. From thermal dynamics modeling, I have found that with the same engine, the fuel efficiency for a slow combustion process, constant pressure combustion process, is 38% while for the fast combustion process the fuel efficiency could reach 52.5%. The improvement can be as high as 38.2%.

As burning starts at the fuel droplet surface, with small fuel droplets, the fuel can mix with air much better, the combustion goes faster and cleaner, and the heat is released on time to push the piston to do work. Both lab tests and road tests show that our technology improves the fuel efficiency significantly. A continuous road test for six months showed that on the highway, our device increased the diesel mileage from 32 miles per gallon (mpg) to 38 mpg. In city driving, the improvement of fuel mileage was averaged at 12-15%. The lab tests with dynamometers confirmed the road tests, and in fact, had a better result than that of the road tests. For example, at low fuel consumption with a fixed fuel consumption rate the average power output increased from 0.3677 hp to 0.4428 hp after the device was turned on. This indicates that the power output was improved by about 20.4% at the same fuel consumption rate. Our recent lab tests with a dynamometer at Stanadyne Corporation further confirm the above results.

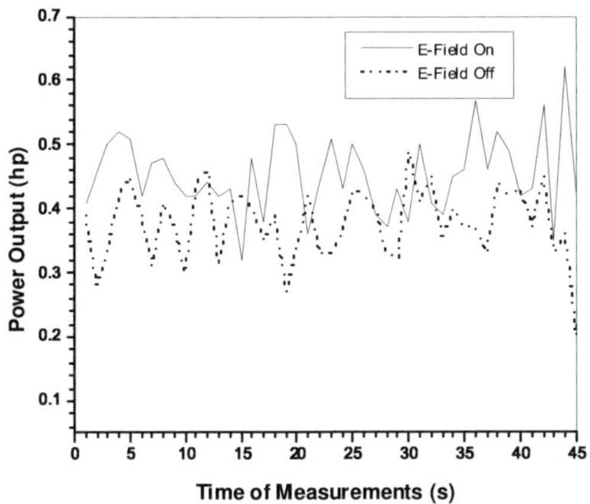

Fig. 6 The lab test of Mercedes-Benz 300D with a dynamometer. At the same fuel consumption, the average power output was increased by 20% when the electric field was turned on.

4. Concluding Remarks

While this technology continues to develop and some large scale tests are on the way, our results clearly show that it is a game changer for energy production, transportation, and conservation. We are looking forward to a bright future in the energy sector.

Acknowledgement

This work was supported in part by STWA and RAND. We are also grateful to the National Institute of Standards and Technology and the US Department of Commerce in providing the neutron research facilities used in this work.

References

1. A. Einstein, Ann. Phsik [4] **17**, 549 (1905); **19**, 289, 371 (1906).
2. I. M. Krieger and T. J. Dougherty, Tans. Soc. Rheol. 3, 137-152 (1959).
3. R. Tao and X. Xu, Energy & Fuels, 20, 2046-2051 (2006).
4. R. Tao, International J. of Modern Physics B, V. 21, N28&29, 4767 (2007).
5. F. L. Saunders, *Jounal of Colloid Science* **16**(1), 13 (1961).
6. K. H. Sweeny and R. D. Geckler, *Jounal of Applied Physics* **25**(9), 1135 (1954).
7. D. G. Thomas, *Journal of Colloid Science* **20**(3), 267 (1965).
8. G. Serge & A. Silibergerg, *J. Fluid Mech.* **14**, 86 (1951).
9. R. Tao, K. Hunag, H. Tang, et al, Energy & Fuels, 22, 3785-3788 (2008).
10. R. Tao, K. Hunag, H. Tang, et al, Energy & Fuels, 23, 3339-3342 (2009).
11. A. H. Lefebevre, *Atomization and Spray* (pp. 27-78). New York: Hemisphere Publishing (1989).
12. 30m SANS: "The 30 m Small-Angle Neutron Scattering Instruments at the National Institute of Standards and Technology", Glinka CJ, et al. *J Appl. Cryst.* **31**(3), 430 (1998).

ADAPTIVE ENERGY ABSORBERS FOR DROP-INDUCED SHOCK MITIGATION

NORMAN M. WERELEY, YOUNG-TAI CHOI and HARINDER J. SINGH

Department of Aerospace Engineering, University of Maryland
College Park, Maryland 20742, USA

This study addresses the nondimensional analysis of adaptive magnetorheological energy absorbers (MREAs) for drop-induced shock mitigation. The governing equation of motion of a single degree of freedom system with an MREA was derived. The Bingham number was defined and its effect on the system response was examined. A comprehensive nondimensional analysis was conducted using nondimensional stroke, velocity and acceleration, where Bingham number and time constant were key parameters. An optimal Bingham number-based on drop velocity, payload mass, and passive damping-minimized the drop-induced shock loads transmitted to the payload by utilizing maximum damper stroke.

1. Introduction

Magnetorheological energy absorbers (MREAs) are receiving great attention in shock mitigation and crashworthiness of systems including vehicles, high-speed boats and helicopters. Severe injuries to operators and crew have been resulted from high levels of shock loads during harsh operating conditions and crash landings [1,2]. The load transmissions can be significantly reduced by employing passive energy absorbers for shock mitigation. However, passive energy absorbers cannot satisfy stringent shock mitigation requirements because of their pre-designed fixed load stroke profiles which render them unsuitable for varying shock conditions. Therefore, MREAs are being considered for the implementation of adaptive shock mitigation [3,4]. MREAs have attractive features, such as rapidly adjustable damping in response to an applied current input without resorting to mechanical moving parts. Furthermore, MREA power consumption is low and, unlike active feedback control, does not produce instabilities like control spillover.

The problem under consideration is based on a single degree of freedom system with a payload falling at a prescribed drop velocity. The payload is isolated from the impact by using an MREA. A key goal is to exploit available stroke completely during the impact such that maximum shock attenuation is achieved. A larger MREA damper force stops the payload early before the MREA reaches its maximum stroke causing larger payload decelerations. On the other hand, smaller damper force can cause the MREA to bottom out, thus producing a severe end-stop impact. Therefore, to maximize shock mitigation performance, the damper force should be appropriately controlled. The performance of MREAs for drop-induced shock mitigation was analyzed by varying the stroking force. The governing equation of motion of a system with an MREA was theoretically derived. A nondimensional analysis was carried out by defining nondimensional stroke, velocity and acceleration. The Bingham number and time constant were of utmost importance in effectively controlling the response of system.

2. Magnetorheological Energy Absorbers (MREAs)

The configuration of MREAs for drop-induced shock mitigation is shown in Figure 1 with the payload mass, m subjected to initial drop velocity, v_0. The total damper stroke available before the impact is S.

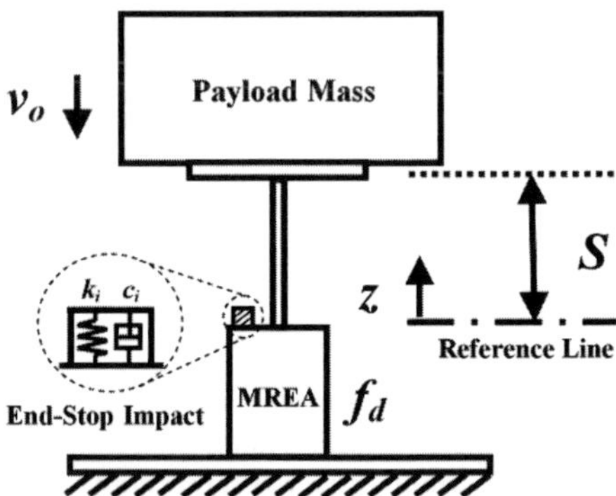

Figure 1. Configuration of magnetorheological energy absorbers (MREAs) for drop-induced shock mitigation.

The governing equation of motion for the system is given by

$$m\ddot{z}(t) = -f_d - mg \tag{1}$$

where

$$f_d = c\dot{z}(t) + f_y sign\{\dot{z}(t)\} \tag{2}$$

with the initial conditions given by

$$z(0) = S \text{ and } \dot{z}(0) = -v_0 \tag{3}$$

Here, f_d is the damper force of the MREA, c is the passive damping coefficient of the MREA, f_y is the yield force of the MREA due to MR effect and g is the acceleration due to gravity.

The governing equation, Eq. (1) can be rewritten in terms of the velocity, $\dot{z}(t) = v(t)$.

$$\dot{v}(t) = -\frac{c}{m}v(t) - \frac{f_y}{m}sign\{v(t)\} - g \tag{4}$$

Integrating Eq. (4) and using the initial condition given by Eq. (3) yields the velocity of the payload.

$$v(t) = -v_0\left\{\left[1 - \frac{f_y sign\{v(t)\}}{cv_0} - \frac{mg}{cv_0}\right]e^{-\frac{t}{\tau_v}} + \frac{f_y sign\{v(t)\}}{cv_0} + \frac{\tau_v g}{v_0}\right\} \tag{5}$$

The Bingham number, Bi, is defined as the ratio of the MR yield force to the passive viscous damping force. The time constant, τ_v, is defined as the ratio of the payload mass to the passive damping coefficient.

$$Bi = \frac{f_y}{cv_0} \; ; \tau_v = \frac{m}{c} \tag{6}$$

Note that, since $v(t)$ is negative during the downstroke of the MREA, the *Signum* function attains a value of -1. Using the Bingham number in Eq. (6), the velocity of the MREA in Eq. (5) can be rewritten as follows

$$v(t) = -v_0\left\{[1 + Bi - R_v]e^{-\frac{t}{\tau_v}} - Bi + R_v\right\} \tag{7}$$

with nondimensional variable, R_v given as

$$R_v = \frac{\tau_v g}{v_0} \tag{8}$$

By integrating Eq. (7) again and using the initial condition given by Eq. (3), we obtain the displacement given as

$$z(t) = S\left[R_s(1 + Bi - R_v)\left(e^{-\frac{t}{\tau_v}} - 1\right) + R_s(Bi - R_v)\frac{t}{\tau_v} + 1\right] \tag{9}$$

with nondimensional variable, R_s given as

$$R_s = \frac{\tau_v v_0}{S} \tag{10}$$

The acceleration of the MREA is obtained by differentiating Eq. (7).

$$a(t) = \frac{v_0}{\tau_v}[1 + Bi - R_v]e^{-\frac{t}{\tau_v}} \tag{11}$$

2.1. *Nondimensional analysis*

Variables such as nondimensional displacement, velocity, acceleration and time (nondimensional quantities are denoted as ($\bar{\cdot}$)) are given by Eq. (12).

$$\bar{z}(\bar{t}) = \frac{z(t)}{S} \; ; \bar{v}(\bar{t}) = \frac{v(t)}{v_0} \; ; \bar{a}(\bar{t}) = \frac{a(t)\tau_v}{v_0} \; ; \bar{t} = \frac{t}{\tau_v} \tag{12}$$

Using Eqs. (7), (9) and (11) and the parameters mentioned in Eq. (12) we obtain

$$\bar{v}(\bar{t}) = -[1 + Bi - R_v]e^{-\bar{t}} + Bi - R_v \tag{13}$$

$$\bar{z}(\bar{t}) = R_s(1 + Bi - R_v)(e^{-\bar{t}} - 1) + R_s(Bi - R_v)\bar{t} + 1 \tag{14}$$

$$\bar{a}(\bar{t}) = [1 + Bi - R_v]e^{-\bar{t}} \tag{15}$$

In order to mitigate maximum shock, the payload should utilize the maximum available damper stroke. Therefore, it is desirable to implement a controlled MREA load such that smooth landing is attained (i.e. payload attains zero velocity at the completion of damper stroke).

The optimal Bingham number corresponding to the nondimensional velocity, $Bi_{o,\bar{v}}$ can be obtained by equating $\bar{v}(\bar{t})=0$ as given by Eq. (13)

$$Bi_{o,\bar{v}} = R_v - \frac{1}{1 - e^{\bar{t}}} \tag{16}$$

The optimal Bingham number, when terminal condition for nondimensional displacement is imposed, $Bi_{o,\bar{z}}$, can be obtained by equating $\bar{z}(\bar{t}) = 0$ as given by Eq. (14).

$$Bi_{o,\bar{z}} = R_v - \frac{1 + e^{\bar{t}}\left(\frac{1}{R_s} - 1\right)}{1 + e^{\bar{t}}(\bar{t} - 1)} \tag{17}$$

The nondimensional time, $\bar{t}_o$ is calculated by equating Eqs. (16) and (17).

$$Bi_{o,\bar{v}}(\bar{t}_o) = Bi_{o,\bar{z}}(\bar{t}_o) \tag{18}$$

From Eq.(18), we obtain the nondimensional time, $\bar{t}_o$ given by

$$\bar{t}_o = \frac{1}{R_s} - 1 - W\left[e^{\left(\frac{1}{R_s} - 1\right)}\left(\frac{1}{R_s} - 1\right)\right] \tag{19}$$

where $W[\cdot]$ is the Lambert W Function [5] also known as product log. Therefore, from Eqs. (16) and (19), the optimal Bingham number, Bi_o, is given by

$$Bi_o = R_v - \frac{1}{1 - e^{\left\{\frac{1}{R_s} - 1 - W\left[e^{\left(\frac{1}{R_s} - 1\right)}\left(\frac{1}{R_s} - 1\right)\right]\right\}}} \tag{20}$$

The optimal Bingham number can also be obtained by using Eqs. (17) and (19).

$$Bi_o = R_v - \frac{1 + e^{\left\{\frac{1}{R_s} - 1 - W\left[e^{\left(\frac{1}{R_s} - 1\right)}\left(\frac{1}{R_s} - 1\right)\right]\right\}}\left(\frac{1}{R_s} - 1\right)}{1 + e^{\left\{\frac{1}{R_s} - 1 - W\left[e^{\left(\frac{1}{R_s} - 1\right)}\left(\frac{1}{R_s} - 1\right)\right]\right\}}\left(\frac{1}{R_s} - 2 - W\left[e^{\left(\frac{1}{R_s} - 1\right)}\left(\frac{1}{R_s} - 1\right)\right]\right)} \tag{21}$$

The end-stop impact between the payload mass and the MREA at the stroke limit is formulated by modeling a spring-damper-mass system as shown in Figure 1. The solution is of the form [6,7]

$$z^+ = \frac{v^-}{\omega_d} e^{-\zeta \omega t} sin(\omega_d t) + z^- \tag{22}$$

$$v^+ = e^{-\zeta \omega t} \left(\frac{-\zeta \omega v^-}{\omega_d} sin(\omega_d t) + v^- cos(\omega_d t) \right) \tag{23}$$

$$a^+ = e^{-\zeta \omega t} \left[\left(\frac{\zeta^2 \omega^2 v^-}{\omega_d} - v^- \omega_d \right) sin(\omega_d t) - 2\zeta \omega v^- cos(\omega_d t) \right] \tag{24}$$

where

$$\zeta = \frac{c_i}{2m\omega} \;\; ; \;\; \omega = \sqrt{\frac{k_i}{m}} \;\; ; \;\; \omega_d = \omega \sqrt{1 - \zeta^2} \tag{25}$$

where, c_i and k_i are the passive damping and spring constant for the end-stop model described in Figure 1. The terms z^- and v^- are the respective displacement and velocity of the payload at the maximum stroke before the end-stop impact. On the other hand, z^+, v^+, and a^+ are the displacement, velocity and acceleration of the payload at the maximum stroke after the end-stop impact. For the simulation, the parameters considered are $k_i = 3000$ kN/m and $\zeta = 0.4$.

3. Results and Discussion

3.1. *Optimal Bingham number evaluation*

Figure 2a presents the optimal Bingham numbers for the nondimensional displacement and velocity with the nondimensional time, $\bar{t}$. From the figure, the optimal Bingham number for nondimensional velocity, $Bi_{o,\bar{v}}$ coincides at time, $\bar{t}_o$, with the optimal Bingham number for nondimensional displacement, $Bi_{o,\bar{z}}$, which means all conditions of smooth landing are satisfied at that point.

The optimal Bingham number, Bi_o, variation with the time constant, τ_v for the available damper stroke of $S=5$ inches is shown in Figure 2b. It is observed that the optimal Bingham number, Bi_o, increases with the time constant, τ_v, which means that higher MREA load is required for optimum drop-induced shock mitigation. This is due to the fact that with the increase in payload mass the total energy of the system also increases and, therefore, more energy has to be dissipated by the MREA requiring a higher MR yield force. A similar trend is observed when the drop velocity is increased for a given payload mass. For low time constant values at lower drop velocity, the optimal Bingham number, Bi_o, is small indicating that good shock mitigation can be achieved by using only passive damping force without MR yield force.

3.2. *Response of MREAs*

The nondimensional response of the MREA for drop-induced shock mitigation is shown in Figure 3. The optimal Bingham number is found to be, Bi_o=0.69 for maximum stroke of $S = 5$ inches, time constant of $\tau_v = 0.06$ s, and the drop velocity, $v_0 = 5$ m/s. It is observed from the figure that the payload utilizes the damper stroke completely without experiencing end-stop impact for optimal Bingham number, Bi_o. In addition, the nondimensional acceleration is smaller for optimum Bingham number, Bi_o, compared to any other case. When the Bingham number is smaller than the optimum value, the payload completes the available stroke with non-zero nondimensional velocity causing end-stop impact, further, resulting in large load transmissions to the payload. For higher Bingham number than optimum, the payload is unable to complete the damper stroke because of high resistance by the MREA. Subsequently, there is increase in nondimensional acceleration which may result in payload damage.

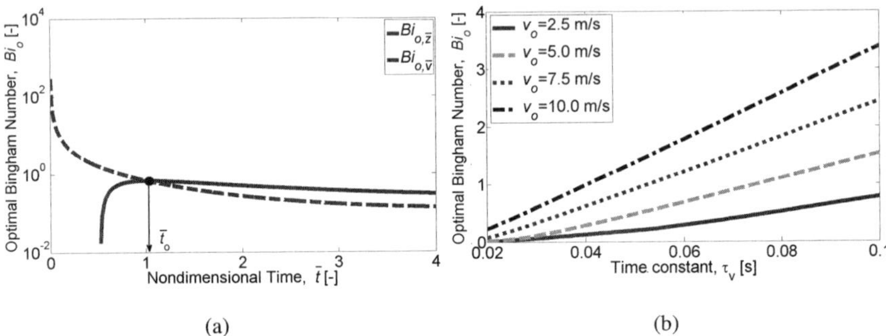

(a)   (b)

Figure 2. The optimal Bingham numbers for (a) nondimensional displacement and velocity with nondimensional time and (b) different time constants and drop velocities.

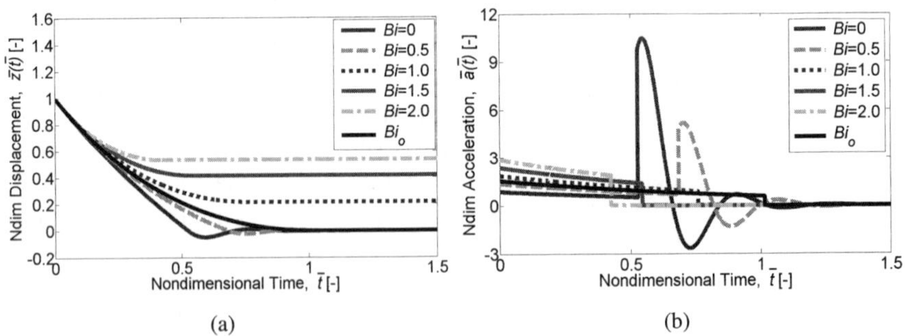

(a)   (b)

Figure 3. The nondimensional response of an MREA for drop-induced shock mitigation at drop velocity of v_o=5 m/s and time constant, τ_v=0.06 s.

4. Conclusions

The nondimensional analysis of MREAs for drop-induced shock mitigation was theoretically analyzed. It was observed that the optimal Bingham number increased as the time constant, τ_v increased. In addition, with increase in drop velocity, similar trend of increase in optimal Bingham number was observed. In other words, if the shock condition becomes severe, higher MR yield force is required for better shock mitigation. It was also shown that, at the Bingham number other than optimum, there was either an end-stop impact or incomplete utilization of damper stroke and both circumstances resulted in large load transmissions.

Acknowledgments

The authors acknowledge support for this research under a contract from The United States Navy Air Warfare Center at Patuxent River (Mr. William Glass as Technical Monitor).

References

1. G.J. Hiemenz, Y.-T. Choi, and N.M. Wereley, "Semi-active control of vertical strolling helicopter crew seat for enhanced crashworthiness," *Journal of Aircraft* **44**(3), 1031-1034, 2007.
2. S.P. Desjardins, "The evolution of energy absorption systems for crashworthy helicopter seats," *Journal of the American Helicopter Society* **51**(2), 150-163, 2006.
3. D. Woo, S.-B. Choi, Y.-T. Choi, and N.M. Wereley, "Frontal crash mitigation using MR impact damper for controllable bumper," *Journal of Intelligent Material Systems and Structures* **18**(2) 1211-1215, 2007.
4. M. Mao, W. Hu, Y.-T. Choi, and N.M. Wereley, "A magnetorheological damper with bifold valves for shock and vibration mitigation," *Journal of Intelligent Material Systems and Structures* **18**(2) 1227-1232, 2007.
5. R.M. Corless, G.H. Gonnet, D.E.G. Hare, D.J. Jeffrey, and D.E. Knuth, "On the Lambert W function," *Advances in Computational Mathematics* **5**, 329-359, 1996.
6. S.M.M. Jafri, *Modeling of impact dynamics of a tennis ball with a flat surface*, Master Thesis, Texas A & M University, 2004.
7. W.T. Thomson, *Theory of vibration with applications*, Prentice-Hall International, Inc., New Jersey, 1988.

MAGNETORHEOLOGICAL DAMPERS WITH HYBRID MAGNETIC CIRCUITS

HOLGER BÖSE* and JOHANNES EHRLICH

*Fraunhofer-Institut für Silicatforschung ISC, Neunerplatz 2
D-97082 Würzburg, Germany
E-mail: boese@isc.fraunhofer.de

Novel concepts for the magnetic circuit in magnetorheological dampers have been proven. In contrast to the known magnetic circuits where the magnetic field for the control of the magnetorheological fluid is generated by the coil of an electromagnet, hybrid magnetic circuits consisting of at least one permanent or hard magnet and an electromagnet are used in the new approaches. Three different technical configurations are distinguished: 1. The electromagnet is combined with two permanent magnets, whose magnetization cannot be modified even by strong magnetic fields of the electromagnet. The main advantage is the improved fail-safe behaviour of the damper in case of a power failure. 2. The electromagnet is combined with a hard magnet, whose magnetization can be modified by the electromagnet. This configuration leads to high energy efficiency, because electric power is only required in short pulses for the switching of the hard magnet. 3. All three types of magnetic field sources, permanent, hard and electromagnet are combined in the magnetic circuit, which gives the highest flexibility of the magnetic field generation and the damping control at the expense of a relatively large effort. Demonstrators for magnetorheological dampers with all three magnetic circuits were constructed and their performances were tested. The results of the investigations are described in this paper.

1. Introduction

Magnetorheological (MR) dampers are semi-active devices, which exploit the outstanding capabilities of MR fluids to change their rheological properties in a magnetic field [1, 2]. Various applications of MR dampers, like automotive shock absorbers, vibration reducers for driver seats in trucks and buses as well as the mitigation of vibration excitation in cable-stayed bridges have already been commercialized [3]. In the past, much work has been devoted to the design of various types of dampers, the development of suitable MR fluids and to different control strategies. However, less attention has been paid to the magnetic circuit and the generation of the magnetic field.

In the MR dampers which are state-of-the-art, the magnetic field is generated by the electric current in a coil and guided by a ferromagnetic material to the MR fluid. The corresponding magnetic circuit includes the electromagnet, the flux-guiding material and the gap which is filled with the MR fluid. This configuration suffers from some drawbacks in terms of fail-safe behavior and energy efficiency. The first problem is, that in case of an electric power failure the magnetic field is not maintained and the damping force drops to its lowest level. The second drawback means, that electric power is permanently consumed, even for the maintenance of a constant magnetic field.

In this work, novel concepts of magnetic circuits are introduced, in which the magnetic field is basically generated by permanent or hard magnets and modified by the additional electromagnet. The magnetic circuits are called hybrid due to the superposition of different magnetic sources. Three cases are distinguished, which differ in the type of the permanent or hard magnet used in the magnetic circuit. The three concepts offer different solutions to the drawbacks identified.

In all concepts, we start from the usual configuration of a conventional MR damper where the entire magnetic circuit is integrated in the piston of the damper. This allows a more compact damper design in comparison to other designs like an MR fluid flow through a bypass.

2. Electromagnet and Permanent Magnets

2.1. *Design of the magnetic circuit*

The first hybrid magnetic circuit contains an electromagnet with a coil in the center of the piston (fine crossing lines in Fig. 1). In addition, two permanent magnets at the top and the bottom of the piston are introduced (dark disks in Fig. 1). These three sources of the magnetic field define three magnetic sub-circuits, whose magnetic fluxes can superpose in the active zones of the annular MR fluid gap between the coil and the permanent magnets. Fig. 1 also shows the three relevant working modes of the magnetic circuit. In the left scheme, the coil is not activated and the magnetic field is generated merely by the permanent magnets. The mid scheme in Fig. 1 depicts a strengthening of the magnetic field of the permanent magnets by the electromagnet. Finally, in the right scheme the coil current is reversed, which leads to a weakening of the magnetic fields of the permanent magnets and the electromagnet.

The components in the hybrid magnetic circuit were designed such, that the magnetic field of the permanent magnets can nearly be canceled by the electromagnet. Furthermore, a very high magnetic flux density of 830 mT in the active MR fluid gap can be achieved in case of the strengthening of the magnetic fields. Fig. 2 shows the results of the simulation of the magnetic flux density for the three working modes described. The magnetic flux density was calculated along the line of the MR fluid gap between the two permanent magnets (see Fig. 2, left). The length of this line is 125 mm which is due to the length of the coil.

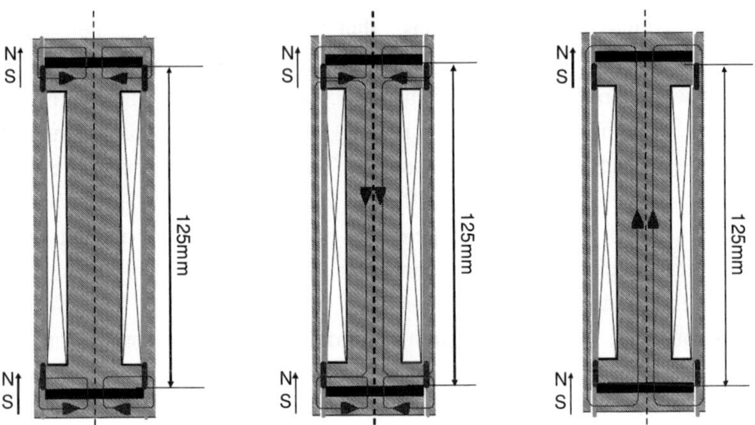

Figure 1. Scheme of the magnetic flux in the MR damper piston with electromagnet and permanent magnets in the three different working modes: magnetic field generated only by the permanent magnets (left), strengthened by the electromagnet (mid) and weakened by the electromagnet (right)

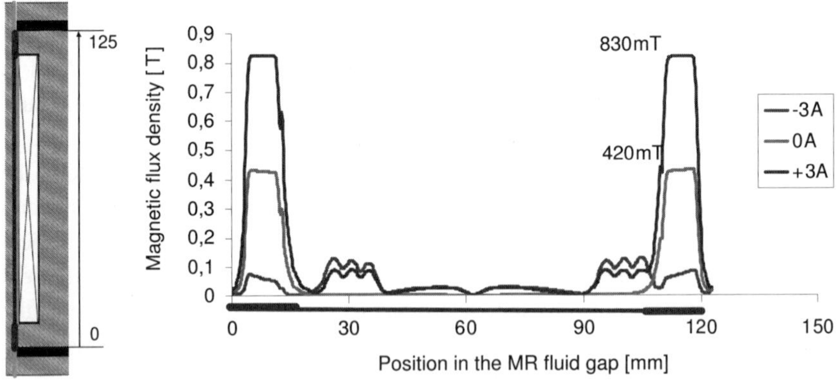

Figure 2. Scheme of the MR fluid gap in the damper piston with electromagnet and permanent magnets with a length of 125 mm (left) and simulated magnetic flux density along this line for the three working modes shown in Figure 2 (right)

Another result of the simulation is the flux density of 420 mT in the powerless working mode, in which no current is applied to the coil. This flux density determines the damping force in case of a power-failure. Finally, in case of weakening of the magnetic field by the two magnetic sources, the magnetic flux density in the active MR gap is less than 100 mT.

2.2. *Demonstrator*

A demonstrator of the MR damper was designed and constructed in order to evaluate the benefit of the hybrid magnetic circuit [4]. Fig. 3 shows the piston and the complete damper, which contains a gas cushion in order to balance the volume change due to the motion of the piston rod.

Figure 3. Demonstrator of the MR damper with electromagnet and permanent magnets: piston (left) and complete damper (right)

The MR damper was filled with an MR fluid having an iron particle concentration of 36 vol.%. The iron particle concentration determines the achievable shear stress and the corresponding damping force at a given magnetic flux density [5]. The demonstrator was mounted in a mechanical testing machine and the performance was investigated.

2.3. *Results*

Fig. 4 shows the measured damping force versus the displacement of the damper piston at a frequency of 1 Hz and different currents in the coil. As already indicated in the magnetic flux simulation, the damping force without any coil current is in a medium range of about ± 1.5 kN. This figure gives the damping force in case of an electric power failure and demonstrates the high fail-safe stability of the MR damper.

If a positive current of +3 A is applied to the coil, which strengthens the magnetic field of the permanent magnets, the corresponding damping force is increased to nearly ± 4 kN. However, if the coil current is reversed to -3 A, the damping force is lowered to ca. ± 0.5 kN (see Fig. 4). These results confirm the intended behavior of the damper and show the wide spread of the achievable damping force.

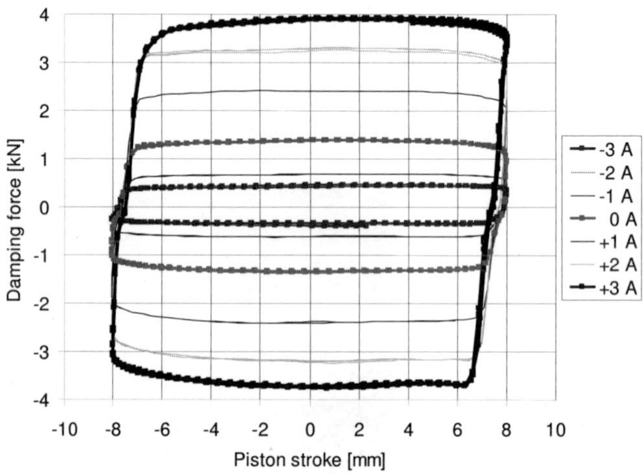

Figure 4. Experimental results of force measurements on the damper with electromagnet and permanent magnets on a mechanical testing machine, the MR fluid contains 36 vol.% iron particles

3. Electromagnet and Switchable Hard Magnet

3.1. *Design of the magnetic circuit*

The purpose of this second type of hybrid magnetic circuit is also, to generate a magnetic field in the active MR fluid gap without any supply of electric energy, by using a hard magnet, like for the damper type discussed before. However, here the magnetization of the hard magnet can be changed by the electromagnet. The benefit of this configuration is, that different magnetic flux densities and corresponding damping forces can be maintained without permanent electric energy consumption [6]. Electric energy supply is required only for the switching of the magnetization of the hard magnet. As the most suitable material for the hard magnet, an AlNiCo alloy was selected, because it combines high saturation magnetization with low coercive field strength.

The task to switch the magnetization of the hard magnet requires another design of the magnetic circuit system than in the first damper type. In this type, the hard magnet is located in the same circuit as the coil. Fig. 5 left depicts the scheme of the magnetic circuit and the flux lines generated by the two magnetic field sources. Fig. 5 right shows also the simulated magnetic flux density along the line of the MR fluid gap over a distance of nearly 40 mm, which is the length of the damper piston in this configuration.

The maximum achievable magnetic flux density in the active MR fluid gap was calculated as ca. 0.6 T (see Fig. 5, right). However, the flux density can be continuously lowered down to zero by a lower magnetization or even a complete demagnetization of the AlNiCo magnet.

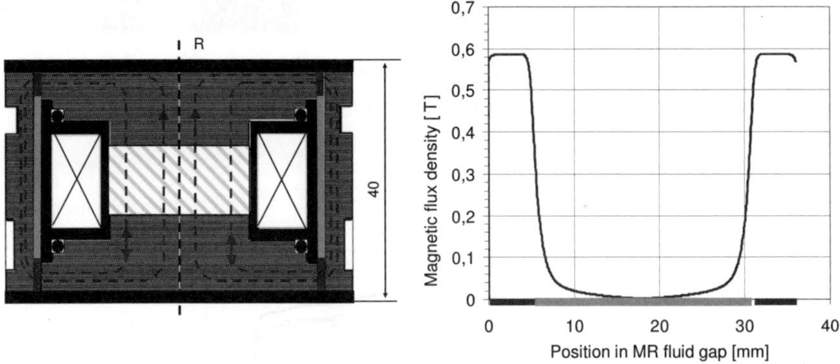

Figure 5. Scheme of the MR fluid gap in the damper piston with electromagnet and switchable hard magnet and with a length of 40 mm, magnetic flux lines included (left) and simulated magnetic flux density along a line through the gap in the state of full magnetization of the switchable hard magnet (right)

3.2. *Demonstrator*

A demonstrator for an MR damper with a piston containing a magnetic circuit with the configuration shown above was constructed and tested. Fig. 6 shows the damper piston as well as the complete damper. Due to the compact design with a single magnetic circuit, the piston is considerably shorter than that of the damper with two permanent magnets.

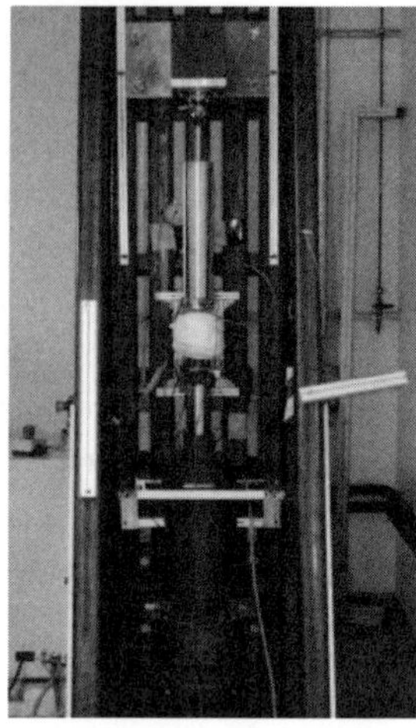

Figure 6. Demonstrator of the MR damper with electromagnet and switchable hard magnet: piston (left) and complete damper in the mechanical test set-up (right)

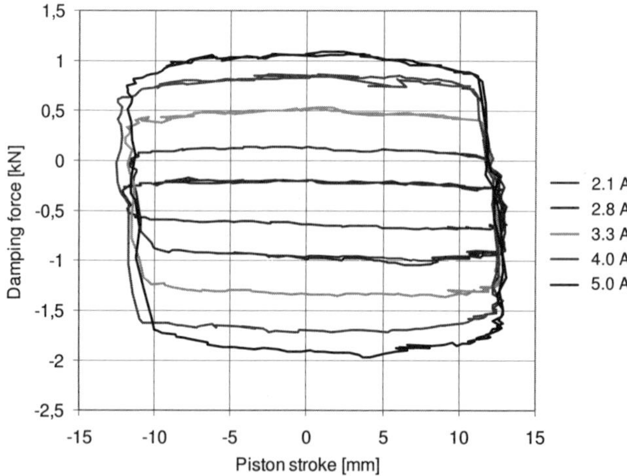

Figure 7. Experimental results of force measurements on the damper with electromagnet and switchable hard magnet, the MR fluid contains 25 vol.% iron particles

3.3. *Results*

The tests of this damper were performed with a self-designed mechanical test set-up, where the motion is generated by a servo motor with a linear gear transmission. In the investigations, the magnetization of the AlNiCo magnet was varied by short pulses of the coil current up to 5 A.

The results of the measurements in the mechanical test set-up are depicted in Fig. 7. Upon rising current in the coil, the magnetization of the AlNiCo magnet and the corresponding damping force are enhanced. The non-symmetric damping force is caused by the air cushion of the damper.

4. Electromagnet and Permanent Magnets and Switchable Hard Magnet

4.1. *Design of the magnetic circuit*

In the third configuration, all three sources of the magnetic field, i.e. the electromagnet, the permanent magnets and the switchable hard magnet are combined in the piston. The design of the magnetic circuit is shown in Fig. 8.

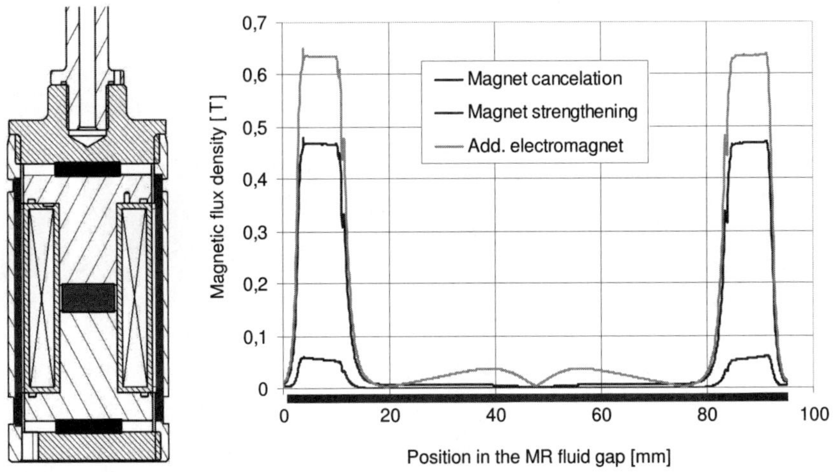

Figure 8. Scheme of the MR fluid gap in the damper piston with electromagnet, permanent and switchable hard magnets with a length of 95 mm (left) and simulated magnetic flux density along a line through the gap for the three states of cancelation of the permanent and switchable hard magnets, strengthening of the magnets and additional application of the electromagnet (right)

As in the first configuration, the magnetic circuit consists of three sub-circuits, where the mid sub-circuit contains the coil and the AlNiCo magnet and the permanent magnets are located in the bottom and top sub-circuits. The magnetic flux density was calculated along the line of the MR fluid gap (see Fig. 8, left). Three operation states are distinguished in Fig. 8, right. If the magnetic fields of the permanent magnets and the switchable hard magnet nearly cancel each other, the flux density in the active MR fluid gap is about 60 mT. The reverse polarization of the AlNiCo magnet causes a strengthening of the fields and leads to a flux density of ca. 570 mT. These operation states and all intermediate states can be held without permanent supply of electric energy. Furthermore, the largest flux density achieved by this way, can even be increased by the additional field of the electromagnet to nearly 630 mT.

4.2. *Demonstrator*

Also for this configuration with all three magnetic sources, a demonstrator of the MR damper was manufactured, mounted and tested. Fig. 9 shows the piston with the integrated magnetic circuit system, which is longer than for the configuration considered before, as well as the complete damper.

Figure 9. Demonstrator of the MR damper with electromagnet, permanent and switchable hard magnets: piston (left), complete damper (mid) and integration in the mechanical test set-up (right)

4.3. *Results*

The investigations of the MR damper with the three magnetic sources were performed on the same self-designed experimental test set-up as those with the MR damper with just the electromagnet and AlNiCo magnet. Measurements were conducted for different magnetizations of the AlNiCo magnet without an additional permanent coil current.

Fig. 10 reveals the results of the force measurements. The inner curve named Cancelation shows the state, in which the magnetic fields of the permanent magnets and the switchable hard magnet nearly cancel each other, which causes the lowest damping force. In the following, the magnetization of the AlNiCo magnet was changed stepwisely by rising short pulses of the coil current. This leads to stepwisely increased flux densities and corresponding damping forces. At a short pulse of 2.5 A, a maximum damping force of 3.3 kN was achieved.

The investigations were performed with an MR fluid which contains only 25 vol.% iron particles. Higher damping forces in the same MR damper can be generated with an MR fluid with a higher concentration of iron particles.

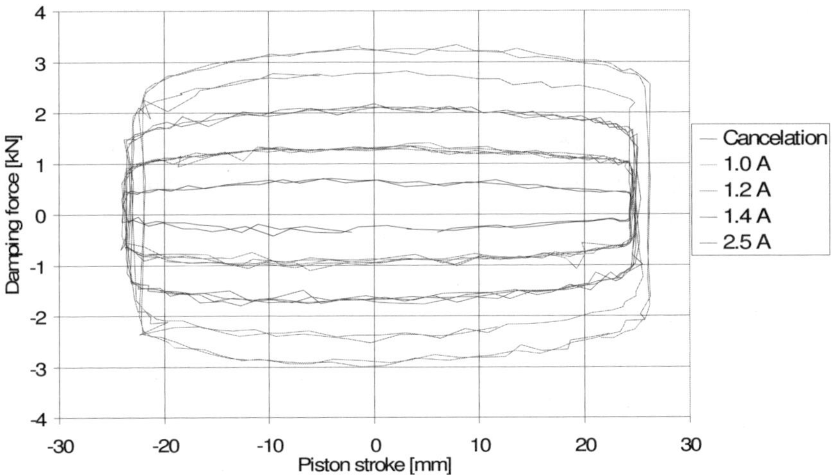

Figure 10. Experimental results of force measurements on the damper with electromagnet, permanent and switchable hard magnets, the MR fluid contains 25 vol.% iron particles

5. Conclusions

Three different types of new MR dampers with hybrid magnetic circuit systems containing an electromagnet and additional permanent or hard magnets were introduced. These approaches result in a high variability of the damping force and offer far-reaching perspectives for MR dampers with improved fail-safe behaviour and increased energy efficiency as well.

The configuration of the MR damper with an electromagnet and permanent magnets gives an improved fail-safe stability, because the permanent magnets generate a medium damping force in case of an electric power failure. This feature is important for shock absorbers, which operate in a stable state even without electric power, but are additionally capable to generate also softer damping for the sake of comfort.

The second MR damper configuration with an electromagnet and a switchable hard AlNiCo magnet has the advantage of increased energy efficiency since the magnetization of the AlNiCo magnet can be changed by short pulses of the coil current. For the switching process an electric energy of less than 1 J is required. This type of MR damper is especially efficient for applications in which the damping force is not continuously changed. An example is a vehicle which transports very different loads, where the damping force can be adapted after loading or unloading.

Finally, the MR damper configuration with an electromagnet, permanent magnets and a switchable hard AlNiCo magnet merges the benefits of the former configurations at the expense of an increased effort. All types of new MR dampers with hybrid magnetic circuit systems have high potentials for various applications.

Acknowledgments

Financial support for this work from the European Community and from the Bavarian State Ministry for economy, infrastructure, traffic and technology is gratefully acknowledged.

The authors thank Jiri Vrbata and Carolin Weis for their support in the damper construction and the performance of investigations.

References

[1] U. Lange, S. Vassileva, L. Zipser: Controllable magnetorheological dampers for shock and vibration. Proceedings of Actuator 2002 – 8th International Conference on New Actuators (2002) 339-342

[2] M. Mao, W. Hu, Y.-T. Choi, N. M. Wereley: A magnetorheological damper with bifold valves for shock and vibration mitigation, J. Intelligent Mater. Syst. Struct. 18 (2007) 1227-1232

[3] http://www.lord.com/Products-and-Solutions/Magneto-Rheological-%28MR%29/Automotive-Suspensions.xml

[4] J. Ehrlich, H. Böse: Novel magnetorheological damper with outstanding fail-safe characteristics. Proceedings of Actuator 2008 – 11th International Conference on New Actuators (2008) 495-498

[5] H. Böse, J. Ehrlich: Performance of magnetorheological fluids in a novel damper with excellent fail-safe behavior. J. Intelligent Material Systems and Structures 10 (2009) doi:10.1177/1045389X09351760

[6] J. Ehrlich, J. Vrbata, H. Böse: Novel magnetorheological damper with improved energy efficiency. Proceedings of Actuator 2010 – 12th International Conference on New Actuators (2010) 545-548

MAGNETORHEOLOGICAL FLUIDS IN HIGH PRECISION FINISHING

WILLIAM KORDONSKI

QED Technologies International, University Avenue 1040, Rochester NY 14607 USA

A new concept of material removal based on the principle of conservation of momentum is applied to analyze Magnetorheological Finishing (MRF®) and Magnetorheological Jet (MR Jet®) processes widely used in precision optics fabrication. According to this concept, a load for surface indentation by abrasive particles is provided at their interaction near the wall with heavier basic particles, which fluctuate (due to collision) in the shear flow of concentrated binary suspension. The model is in good qualitative agreement with experimental results.

1. Introduction

Projection lenses for advanced lithography used in manufacturing of integrated circuits with nanometer features as well as optics for lasers, airborne surveillance, weapon systems, medical devices, digital photography and mirrors for space telescopes are examples of modern optical applications which rely on leading-edge production technologies, especially, the ones delivering high precision aspherical and free form surfaces. The most challenging step in fabrication of such complex surfaces is polishing, particularly, so-called sub-aperture polishing based on zonal material removal. This process requires precision control of position and velocity of the polishing zone. Currently, it is provided by sophisticated contour-controlled precision CNC machines, which execute finishing algorithms according to the prescription. Full advantage of the deterministic nature of CNC machining can only be taken if a sub-aperture polishing tool instantly adapts (conforms) to the local surface and its removal function is well characterized and stable. Commonly used mechanical tools with air pressure or an elastic cushion behind the polishing pad do not provide the required level of adaptability and stability [1].

Liquid substances by its nature can easily conform to any surface and attempts were made to utilize this unique property in controlled material removal including polishing [2, 3]. Based on the 10+ years of experience in the development and study of magnetorheological (MR) fluids and their applications, the use of this liquid smart material for precision finishing was proposed in the late 1980s in Belarus and then two different MR fluid-based finishing methods were further developed and commercialized in the USA and now known as Magnetorheological Finishing® (MRF®) and MR Jet®

Finishing. [4, 5]. Scientific aspects of these technologies are scarcely covered [6, 7, 8, 9]. A new approach to analyze the mechanism of material removal in MRF and MR Jet is under consideration in this paper.

2. Magnetorheological Finishing (MRF)

Schematically, the MRF polishing interface is shown in Fig. 1a. A convex lens is installed at some fixed distance from a moving wall, so that the lens surface and the wall form *a converging gap*. An electromagnet, placed below the moving wall, generates a non-uniform magnetic field in the vicinity of the gap. The magnetic field gradient is normal to the wall. The MR polishing fluid is delivered to the moving wall just above the electromagnet pole pieces to form a polishing ribbon. As the ribbon moves in the field it acquires plastic Bingham properties [10] and the top layer of the ribbon is saturated with abrasive due to levitation of non-magnetic abrasive particles in response to the magnetic field gradient. Thereafter the ribbon, which is pressed against the wall by the magnetic field gradient, is drugged through the gap resulting in material removal over the lens contact zone. This area is designated as the "polishing spot". Two images of the polishing zone are shown in Fig.1. The first one, shown in Fig. 1b, is the high speed photography of the contact zone between a thin stationary meniscus lens and moving rigid wall. An interoferogram of the lens surface after polishing under identical conditions is shown in Fig. 1c. A comparison of those images shows that material removal does occur within the boundaries of the contact zone. The rate of material removal can be controlled by the magnetic field, geometrical parameters of the interface and wall velocity. The polishing process employs a computer program to determine a CNC machine schedule for varying the velocity (dwell time) and the position of the rotating workpiece through the polishing spot. Because of its conformability and subaperture nature, this polishing tool may finish complex surface shapes like

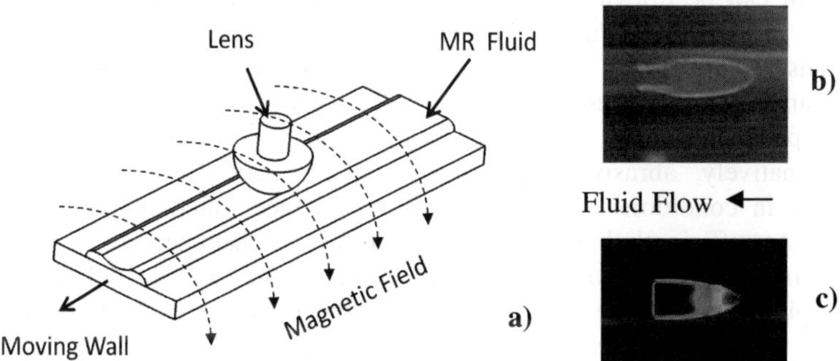

Figure 1. Schematic of Magnetorheological Finishing

aspheres having constantly changing local curvature. A fundamental advantage of MRF over competing technologies is that the polishing tool does not wear, since the recirculating fluid is continuously monitored and maintained. Polishing debris and heat are continuously removed. The technique requires no dedicated tooling or special setup.

3. Modeling of Material Removal in MRF

Most commonly, polishing is carried out by pressing an elastic pad with embedded abrasive particles against moving surface to be polished. According to Preston [11], the removal rate in this case is proportional to the applied pressure and pad relative velocity as well as it depends on properties of polishing interface. Among such properties an important ones are mechanical properties (like elasticity) of the polishing pad which transmits load forces to the abrasive particle. Taking into account that MR fluid in magnetic field stiffens and acquires essential elastic properties it is not unreasonable to suggest that such magnetized material can be considered as a moving polishing pad similar to conventional synthetic polishing tools. To evaluate credibility of this hypothesis, appropriate mechanical properties of typical MR polishing fluid were measured with Anton Paar MCR 301 magneto-rheometer at magnetic field strength and field-shear orientation corresponding to MRF [12]. Measurements were done for sample internal magnetic field strength of 150kA/m and oscillation frequency of 1.592 Hz.

Obtained MR fluid modulus of ~ 0.5 MPa is two orders of magnitude lower than the elastic modulus of conventional pads (~ 50 MPa), which at all conditions equal (like pad velocity) can (theoretically) deliver removal rate the same as MRF (for example, ~ 3 micron/min on fused silica glass). It is reasonable to suggest that particles load, which would be sufficient to support removal rates demonstrated by MRF (3 microns/min and higher) cannot be generated by such much softer analog of a conventional pad. The similar conclusion can be drawn from the measurements with Anton Paar magneto-rheometer of the MR fluid 1st normal stress difference taken at the same conditions as above. The actual wall normal stress of "N1"/4 ~ 5kPa is much lower than the normal stress generated with conventional pads (~100-150 kPa at a typical pad pressure of 70kPa and asperities density of ~ 0.5).

Alternatively, abrasive particles can be energized by a fluid flow, in particular, in conditions of the shear flow of concentrated mixture of solid particles. At sufficiently high shear rates such flow is characterized by intensive particles interaction and collision between them and the surface. In the case of a binary (bimodal) mixture and according to the principle of conservation of momentum larger particles may supply considerable load for smaller particles. When such event takes place near the wall it may result in effective surface indentation by the smaller particle, especially if such particle possesses appropriate mechanical properties. As applied to polishing, this conceptual model suggests that lager or basic particles energized by shear flow provide an indentation load for smaller abrasive particles to penetrate the surface and remove material. Such mechanism of material removal is invoked below to analyze MR polishing process which involves shear flow of a high concentrated mixture (~50 vol.%) of relatively large magnetic particles (microns) and much smaller abrasive particles (tenth of nanometer). As the starting point for the problem modeling and particle force evaluation an assumption is made that particles dynamics in considered case is similar to the general features of granular shear flow described elsewhere [13, 14, 15]. This approach allows evaluation of the surface stress and particle load using constitutive relations accepted for the granular flow, particularly, dependence of the wall normal stress on the shear rate.   It was found that as the solid concentration increases up to 0.7, keeping all other parameters constant, the stress is nearly proportional to the square of the shear rate, then goes down through sharp transition and finally becomes independent of the shear rate at high solid concentration. For relatively moderate concentrations the wall normal stress and consequently particle force take the form of

$$\tau_{22} = K \cdot \rho_p \cdot d_p^2 \cdot \dot{\gamma}^2 \tag{1}$$

$$G_p = K \cdot \frac{\pi}{4} \cdot \rho_p \cdot d_p^4 \cdot \dot{\gamma}^2 \tag{2}$$

where ρ_p is the density of particle, d_p is the diameter of particle and $\dot{\gamma}$ is the shear rate. The dimensionless coefficient K takes into account other flow parameters such as concentration, mechanical properties of particles, carrier fluid damping properties, flow geometry, etc.

According to (2), the problem of evaluation of the particle force is mainly reduced to determining of the flow shear rate at the surface of interest. In the following analysis the shear rate is obtained by numerical modeling of the particular shear flow taking into account rheological properties of the media. The shear rate is then used for calculation of basic particle's force assuming that this force is a load for the abrasive particle.

In what follows, an analysis is restricted to qualitative comparison of experimental removal rate profiles with calculated profiles of surface loading by particles. In doing so, the particle force will be determined for two different methods of supplying mechanical energy to the polishing interface: MR fluid flow through converging gap as applied to MRF and MR fluid jet flow.

In the case of MRF, the effective shear rate was determined by modeling of the Bingham flow in the geometry similar to the one depicted in Fig. 1 using commercially available computational fluid dynamics (CFD) package [16]. To simplify the task and adopt some software limitations the model gap was formed by a cylinder rather than a spherical surface. Other parameters were the same as in experiments (surface radius of curvature, wall velocity, gap thickness, plunging depth and fluid rheological properties).   The three-dimensional solution was found using the free surface volume of fluid (VOF) method and Perzyna hypothesis for effective viscosity of Bingham plastics [17]

$$\mu = Min \begin{cases} A \cdot \mu_\infty \\ \mu_\infty + \dfrac{\tau_0}{\dot{\gamma}} \end{cases} \tag{3}$$

Here A is an arbitrary, dimensionless multiplier supplied by the user (typically, $A = 10^3 - 10^5$), μ_∞ is the viscosity in the limit of very large strain (the fully plastic limit), τ_0 is the yield stress (in the case under consideration depends on magnetic field strength and magnetic particles concentration) and $\dot{\gamma}$ is the shear rate. Rheological parameters required by (3) were obtained with Anton Paar Magneto Rheometer MSR 301. As it would be expected [6, 18, 19], modeling reveals formation of a core of an un-sheared material attached to the moving wall. This fact is illustrated by the shear stress distribution across the gap shown in Fig. 2 for the MR fluid with the yield stress of 20kPa. The shear stress is lower than the yield stress in the core domain and exceeds the yield point in the thin zone near the surface of the lens. Also, the velocity profile is essentially flat in the core region. Thus, an initial interface with a large gap of 2 mm is effectively transformed into the new one with much smaller gap of ~0.2 mm resulting in associated significant increase in the effective shear rate.

3.1. *Model verification*

The shear rate in the sheared zone was determined and used in calculation with (2) of a particle force distribution along the center line of flow. Thereafter,

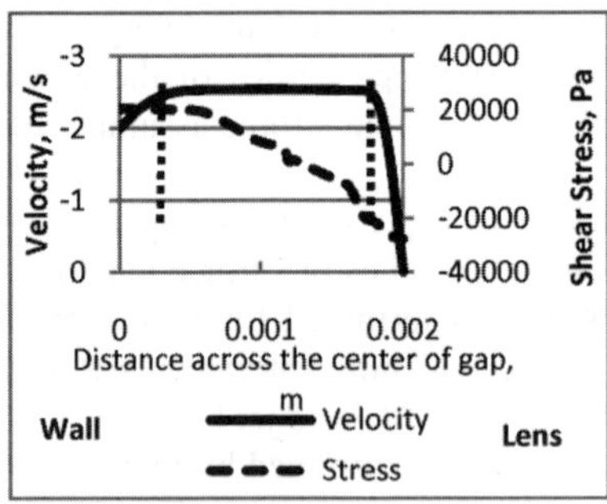

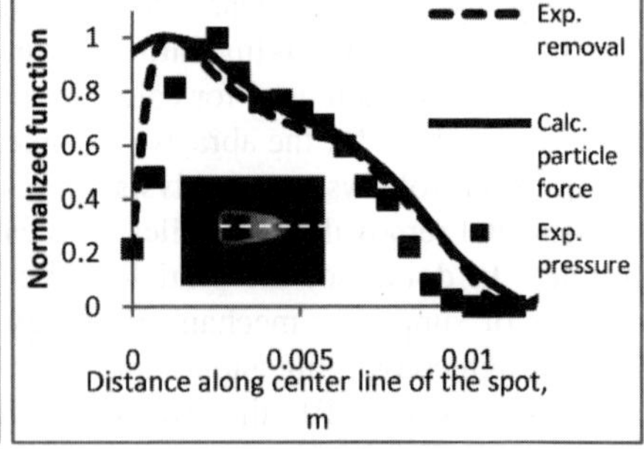

Figure 2. Shear stress distribution and velocity profile across the gap

Figure 3. Calculated and experimental removal rate profiles

normalized values were plotted along with normalized experimental removal rate profiles. Experimental removal functions (spots) were taken on flat parts with the moving wall in the form of the wheel with radius equal to the radius of the cylinder used in the modeling. As an example, results for fluid with the yield stress of 16 kPa are shown in Fig. 3. One can see that correlation between experimental and calculated profile is reasonably good and counts in favor of hydrodynamic (shear flow) mode of material removal. At the same time, the peak of experimental pressure distribution, taken with pressure sensitive film "TekScan" attached to lens surface, is significantly shifted towards the flow, which points to the fact that generated by plastic flow hydrostatic pressure as well as fluid normal stress do not contribute to material removal in the contact zone. The model also adequately predicts such known MRF regularity as an increase of removal rate with magnetic field and concentration of magnetic particles due to appropriate increase in the fluid yield stress. According to (2), the abrasive particle load is very sensitive to the size of basic particle, which should result in an increase of removal rate with the size of basic particle. As experimental results show this prediction was also born out. MRF spots were taken on FS glass with two fluids composed with magnetic particles of sizes 1um and 4 um but at different magnetic fields to make equal fluids' yield stress. Removal rate of 4.45 micron/min which corresponds to the fluid with larger magnetic particles is higher compare to the one (1.77 micron/min) obtained with fluid composed with 1 micron particles at other conditions are equal. Appropriate field strength was determined with magneto rheological measurements done with Anton Parr magneto-rheometer at low shear rate.

Some quantitative model evaluation can be done using the Hertzian theory of surface penetration, which is generally accepted approach in modeling of material removal [20]. In the case of spherical indenter the generated over a contact area tensile stress is given by:

$$\sigma_p = \frac{(1 - 2 \cdot \vartheta_M) \cdot L_p}{2\pi r_c^2} \tag{4}$$

Here r_c is the contact radius, and L_p is the particle load (contact force). The contact radius is given by:

$$r_c = \left[\left(\frac{3}{4} \cdot L_p \cdot r_a \right) \cdot k_E \right]^{1/3} \tag{5}$$

and

$$k_E = \left(\frac{1 - \vartheta_M^2}{E_M} + \frac{1 - \vartheta_p^2}{E_p} \right) \tag{6}$$

where r_a is the radius of abrasive particle ϑ_M and E_M are the Poisson's ratio and Young's modulus for the material (glass) and ϑ_p and E_p are Poisson's ratio and Young's modulus for the abrasive particle.

The Hertzian theory also predicts the depth of penetration in the form of

$$h_t = \left(\frac{9}{16}\right)^{1/3} \cdot \left(\frac{L_p}{E_r}\right)^{2/3} \cdot \left(\frac{1}{r_a}\right)^{1/3} \tag{7}$$

where $E_r = {}^1/_{k_E}$ is the reduced elastic modulus.

In order to evaluate a magnitude of surface contact stress generated by an abrasive particle in addition to the shear rate it is necessary to have a grasp of the size of a basic particle, which are, most likely, some aggregates of original magnetic particles. The size of this fluid sub-structure depends on a ratio between restoring (magnetic) and destroying (hydrodynamic) forces acting on the aggregate. The ratio is known as the Mason number [21]

$$M = \frac{\mu_0 \cdot \kappa_a \cdot H^2}{\eta_0 \cdot \dot{\gamma}} \tag{8}$$

where μ_0 is the magnetic permiability of vacuum, κ_a is the aggregate susceptibility, H is magnetic field strength, η_0 is fluid dynamic viscosity and $\dot{\gamma}$ is the shear rate.

As it was shown in [21] the aggregate size, particularly its aspect ratio (or length of particles chain), decreases as the Mason number decreases. At relatively high shear rates of $\sim 10^4$ 1/s, $H =150$ kA/m and $\kappa_a = 5$, which are characteristic for the case under consideration, the Mason number of 14 predicts the aspect ratio of $\sim$ 1-2 suggesting that the size of aggregate is small. To evaluate the surface tensile stress and depth of penetration an assumption was made that aggregate consists of 4 spherical particles of 4 microns in diameter. Such aggregate may be considered as an ellipsoid with aspect ratio of 1.5. Calculations, which results are shown in Fig. 7, were performed for cerium oxide abrasive particles size of 100 nm and fused silica glass. According to [13, 14] for particles concentration of $\sim$ 0.5 coefficient K can be taken as ~1. Calculated by this way tensile stress in the range of hundreds of MPa is comparable to the ultimate tensile strength for glass (33 MPa) and even some harder materials. Taking into account that ultimate tensile strength is a limit state of tensile stress that leads to tensile failure in the manner of ductile failure or in the manner of brittle failure the predicted values of stress are quite sufficient to result in observed material removal giving some quantitative support to the model. Another approach in model verification can be comparison of the penetration depth calculated with (7) with values of experimental surface roughness. Obtained peak value of a few angstroms surprisingly very well corresponds to actual experimental results obtained in MRF on glasses.

4. MR Jet

Achieving a stable removal function with a liquid jet for high precision finishing is problematic due to fundamental property of a fluid jet to lose its coherence as the jet exits a nozzle. The jet instability results from combination of abruptly imposed longitudinal and lateral pressure gradients, surface tension forces, and aerodynamic disturbances. A method of jet stabilization has been proposed, developed and demonstrated whereby the round jet of magnetorheological fluid is magnetized by an axial magnetic field when it flows out of the nozzle [5, 7]. Such local magnetic field induces longitudinal fibrillation and high apparent viscosity within the portion of the jet that is adjacent to the nozzle resulting in suppression of all of the most dangerous initial disturbances. In MR Jet finishing material removal occurs when the coherent liquid column of slurry comprising magnetic and abrasive particles impinges the surface and spreads in the form of radial laminar flow over the surface. Typical material removal pattern in the polishing spot for stable and unstable jets are shown in Fig 4a and 4b respectively.

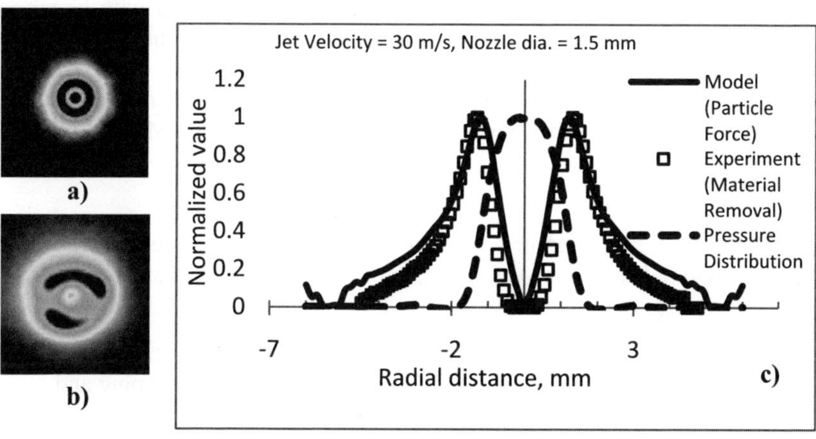

Figure 4. Material removal function generated by MR Jet

Methodically, qualitative analysis of the mechanism of removal in MR Jet finishing was performed in the same manner as for discussed above MRF. Due to the fact that the MR fluid is not affected by the magnetic field at the impingement zone, it can be considered as a Newtonian fluid. This is validated by appropriate rheological measurements. The computed velocity gradient (shear rate) along the radius and (2) were used to calculate the particle force distribution with radius. Assuming that particles in this case do not interact magnetically, a single magnetic particle was considered as the basic particle. An example of such calculations for particle of 4 microns in size is shown in Fig. 4c. Reasonable correlation is observed between normalized experimental removal rate profile and calculated radial distribution of the abrasive particle load generated by the shear flow. It is worth to notice, that again, as in the case of MRF, there is no correlation of removal rate with the distribution of pressure. With the knowledge of the abrasive particle load it is possible to evaluate the surface tensile stress generated by the abrasive particle. Calculated tensile stress in the range up to 1000 MPa is higher than obtained above for MRF. This difference looks reasonable taking into account that the removal rate of MR jet at velocity of 30m/s is also much higher. The same can be said about the penetration depth calculated with (7). Obtained peak value of several nanometers is significantly higher than calculated for MRF. To some extent it looks reasonable taking into account the fact that "MR jet roughness" is always higher than achieved with MRF. On the other hand, accepted in modeling simplifications may result in overrated values of particle force obtained with (2).

5. Summary

A concept of material removal in MRF based on the principle of momentum conservation is proposed wherein a load for surface nano-indentation by abrasive particles is provided at their interaction near the wall with larger and heavier basic particles, which fluctuate due to collision in the shear flow of concentrated binary suspension. Regularity of the granular shear flow and numerical simulation are used in modeling. The model is in good qualitative and quantitative agreement with experimental results for MRF and MR Jet finishing.

Acknowledgments

The author appreciates contribution from Sergei Gorodkin, Justin Tracy, Bob James, Arpad Sekeres, Jerry Carapella, Paul Dumas and Bill Messner.

References

1. Editors, Marinescu I., Uhlmann E, Doi T., Handbook of Lapping and Polishing, CRC Press, Taylor & Francis Group, (2006)
2. Mori, Y., et al., *J. of Japan Soc. of Precision Eng.,* **10(1)** 24 (1988)
3. Momber, A. and Kovacevic R. Principles of Abrasive Water Jet Machining Springer, New York (1999)
4. W. Kordonsky, et al., US Patent # 5,449,313 (1995)
5. W. Kordonski, D. Golini and S.Hogan. US Patent # 5,971,835, (1999).
6. W. Kordonski and S. Jacobs, *Int. J. of Modern Phys. B* **10: 23&24** 2837 (1996)
7. W. Kordonski, A. Shorey, and M. Tricard, *J. of Fluid Eng.,* **128**, 20 (2006)
8. Miao, C., Lambropoulos J., and Jacobs S, *Appl. Optics,* **49**, 1951 (2010)
9. Dai Yifan, et al., *Appl. Optics,* **49**, 298 (2010)
10. W. Kordonski, *J. of Intell. Material Systems and Structures,* **4:1** 65 (1999)
11. F. Preston, *J. Soc. Glass. Tech.* **11** 214 (1927)

12. W. Kordonski and S. Gorodkin, *J. of Phys.: Conf. Series,* **149**, 012064 (2009)
13. Haley H. Shen. Granular shear flows-constitutive relations and   internal structures. 15th ASCE Engineering Mechanics Conference, Columbia University, New York, NY. (2002)
14. A. Karion and M. Hunt. *Powder Technology,* **109** (1-3) 145 (2000)
15. W. Losert, et al., *Physical Review Letters,* **85**, 1428 (2000)
16. Storm/CFD2000. *www.adaptive-research.com*
17. Perzyna, P, *Fundamental Advances in Applied Mechanics,* **9**, 343 (1966)
18. K. Gertzos, et al., *Tribology Int.,* **41** 1190 (2008)
19. J.A. Tichy, *J. Rheol.*; 35(4):47 (1991).
20. L.M.Cook, *J. of Non-Crystalline Solids,* **120** 152 (1990)
21. Z. P. Shulman, et al., *Int. J. of Multiphase Flow,* **12** 935 (1986)

DEVELOPMENT OF A COMPACT BRAILLE DISPLAY USING DIAPHRAGM ACTUATORS CONTROLLED BY ER VALVES

TEPPEI TSUJITA and MASAMI NAKANO

*Institute of Fluid Science, Tohoku University, 2-1-1 Katahira, Aoba-ku
Sendai, Miyagi, 980-8577, Japan*

KEISUKE YOSHIDA

*Graduate School of Engineering, Tohoku University, 2-1-1 Katahira, Aoba-ku
Sendai, Miyagi, 980-8577, Japan*

Electrorheological (ER) suspensions, which behaves like a Bingham fluid having yield stress, is believed to be capable of applying to various mechanical systems since it can be rapidly change in reversible manner by applying electric field. In this research, a Braille display system driven by micro-diaphragm ER actuators which can be driven by relatively low supply pressure is developed. Since the diaphragm actuator uses a polyurethane film of 0.02 mm thickness and very low stiffness, it can be driven by about 240 kPa supply pressure. By using the actuator, experiments for displaying six tactile pins and step response of pin's height are performed.

1. Introduction

Electrorheological (ER) fluids are believed to be capable of applying to various mechanical systems since it can be rapidly change in reversible manner by applying electric field [1]. The electric field can be applied by two electrode plates which can be manufactured in small size. Therefore, ER suspensions suit for MEMS (Micro Electro Mechanical Systems) applications. Nakano et al. developed three-port micro ER valves [2] by using ER suspensions which behaves like a Bingham fluid having yield stress.

In order to support visually-impaired person, a Braille display system using the micro ER valves is developed in this research. Various Braille display systems have been developed until now. For example, Braille cells driven by piezoelectric actuators are commercially available [3]. However, the system is not compact since movable actuator displacement is quite small and mechanisms for amplifying the displacement are required. By using the ER valve, Braille display system is integrated in small size and has high reliability since the valve consists of no moving components. The detail of the developed Braille display system is presented in this paper.

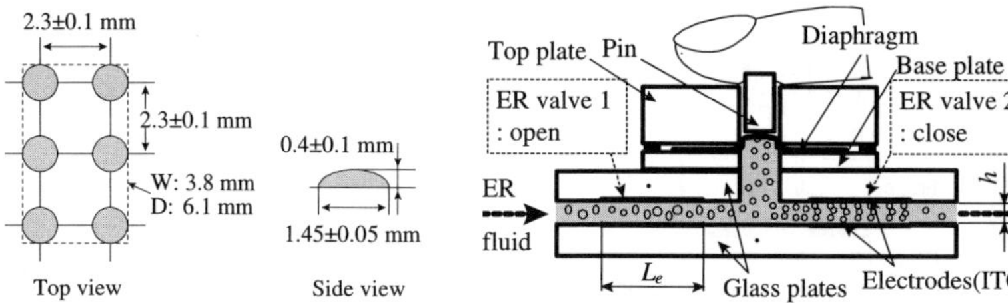

Figure 1. Specification of Braille.　　Figure 2. Mechanism of a micro-diaphragm actuator.

2.　Design of a Micro-Diaphragm ER Actuator

The developed system displays Braille by changing each height of 6 dots. Figure 1 shows a recommended specification of Braille dots defined by JIS (Japanese Industrial Standards). 6 tactile pins must be allocated in a 3.8 mm × 6.1 mm rectangle. Therefore, channels must be integrated in narrow space. To change height of the pins by small actuators, a metal bellows actuator was developed [4]. However, the actuator requires relatively high supply pressure (533 kPa) because its expandable part is made from metal. In order to reduce the supply pressure, a micro-diaphragm actuator using a polyurethane film of 0.02 mm thickness and very low stiffness is developed. This actuator controlled by a pair of micro ER valves pushes up the dot as shown in Figure 2. ER fluid flows through a slit channel between two electrodes of an ER valve. Dispersed particles of the suspension are polarized and form many clusters under an electric field, and changing the electric field strength applied to the electrodes can control pressure drop across the channel. The pressurized ER fluid pushes up a diaphragm when the valve 1 and 2 are respectively opened and closed. On the other hand, the diaphragm shrinks by its elastic restoring force when the valve 1 and 2 are closed and opened, respectively. The diaphragm actuates a tactile pin on it and exerts force on a finger.

To design the ER valves, an experimental apparatus for measuring relationship between display force and supply pressure is developed. A polyurethane film is sandwiched between a top plate and a base as shown in Figure 3. Pressurized silicone oil expands the polyurethane film along a top plate's hole and the polyurethane film pushes up a tactile pin. In order to limit the height of the pin, the pin and the hole have stepped sections. Top and bottom diameters of the pin are respectively 1.5 mm and 1.9 mm. Detail of the mechanism is described in Section 3. Display force is measured by a force gauge and supply pressure is measure by a pressure sensor. The force gauge's probe is adjusted 0.5 mm above the top plate.

42

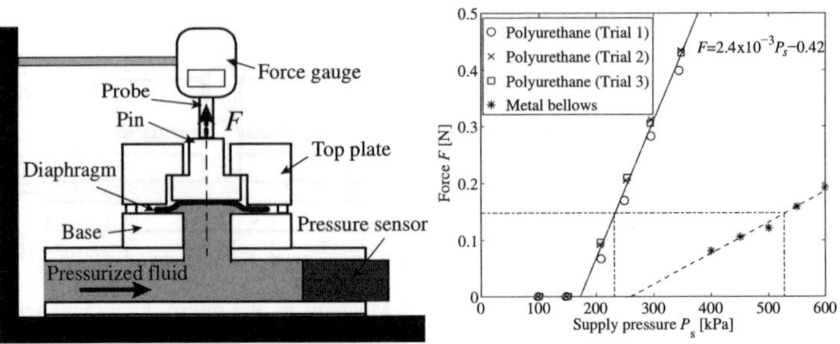

Figure 3. Measurement of display force acting on a tactile pin.

Figure 4. Relationship between supply pressure P_s and display force F.

Figure 4 shows relationship between supply pressure P_s kPa and display force F N. The solid and dotted lines express approximated lines of the diaphragm and the metal bellows, respectively. The solid line can be expressed by the following equation:

$$F = 2.4\times10^{-3} P_s - 0.42 \; . \tag{1}$$

The right hand first term of Eq. (1) indicates linear relationship between supply pressure and contact area between the film and the pin. Pin's bottom area is 3.6×10^{-6} m^2 since its bottom diameter is 1.9 mm. Therefore, the expanded film contacts with the pin partially and the contact area subjected to pressurized fluid is 2.4×10^{-6} m^2. The right hand second term expresses apparent non-linear effect of the polyurethane film. The conceivable reason is the pin cannot contact with the force gauge's probe under 180 kPa. From Eq. (1), the supply pressure which exerts required display force of 0.15 N [4] is 238 kPa. By using the diaphragm, the pressure can be reduced from 533 kPa to 238 kPa. The length of an electrode L_e is calculated by the following equation [5]:

$$L_e = \left(\frac{h}{2\tau_{ER}} \right) \Delta P_{ER} , \tag{2}$$

where h, τ_{ER}, and ΔP_{ER} are gap height between electrodes, shear yield stress and pressure drop of an ER valve, respectively. Figure 2 shows definitions of L_e and h. τ_{ER} depends on fluid property of ER suspension and electric field strength applied to it. Therefore, the relationship between square of electric field strength E^2 and τ_{ER} is measured by using a constant flow rate rheometer for pressure flow mode [6] as shown in Figure 5. In this experiment, ER suspension containing 20 vol.% sulfonated polymer particles (Nippon Shokubai Co., Ltd.) of 5 μm in average diameter is used. E^2 and τ_{ER} have correlative relationship. The solid line express an approximated line $\tau_{ER}=0.3271E^2$. When the height h is 0.3 mm,

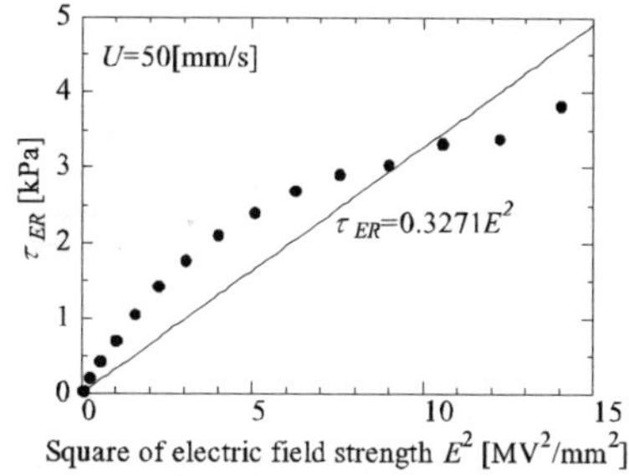

Figure 5. Relationship between yield stress and square of electric field strength in pressure flow mode.

electric field strength E is 2.5 kV/mm and ΔP_{ER} =238 kPa, the electrode length L_e is 17.5 mm.

3. Development of a Braille Display System

In order to integrate the actuators in narrow space, ER valves and flow channels are three-dimensionally fabricated by photolithography. Figure 6 shows an exploded view of the actuator. 12 valves are allocated parallel in a lower channel

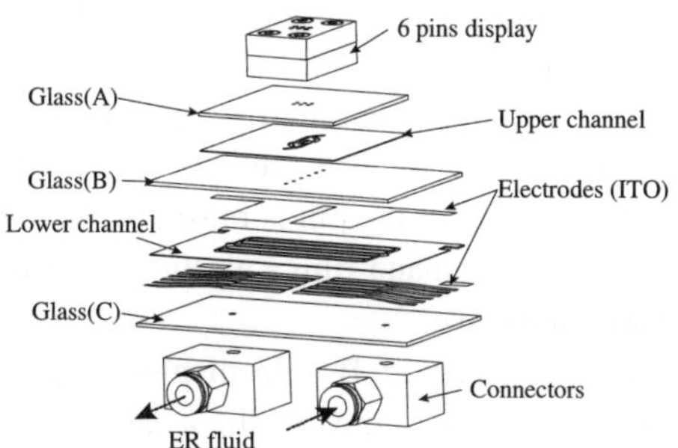

Figure 6. Exploded view of a 6 units actuator.

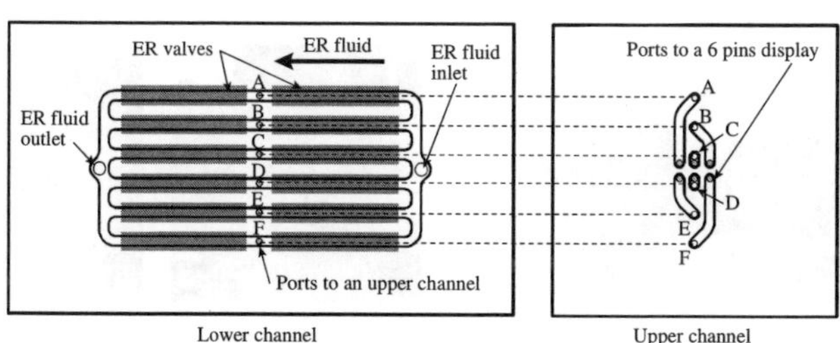

Figure 7. Schematics of channels and ER valves.

44

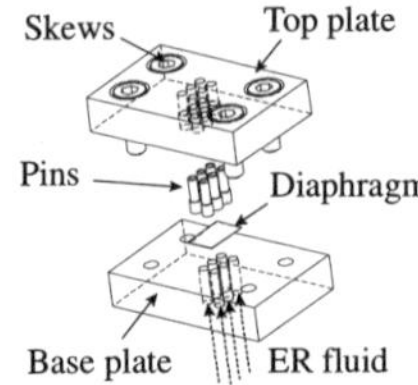

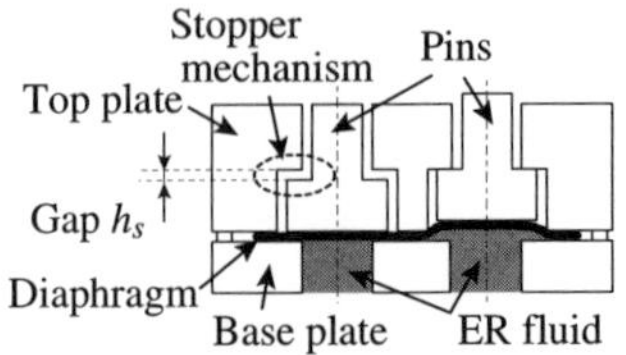

Figure 8. Six pins display part.   Figure 9. Mechanism of pin's height limitation.

as shown in Figure 7. The length of an electrode is 20 mm that includes the estimated length in Section 2 and margin. ER valves and rectangular channels are formed by ITO (Indium Tin Oxide) electrode film and photo resist. Height and width of a channel are 0.3mm and 1.5 mm, respectively. ER fluid is supplied from the inlet and flows an upper channels through ports A~F. The upper channels guide the fluid to 6 pins display part.

On the top of the actuator, 6 pins display part is mounted. As shown in Figure 8, two plates sandwich a film and 6 pins inserted in the top plate's holes. In Figure 2, since the pins can move freely in vertical direction, height of a pin exceeds 0.5 mm when a finger is not put on the display. In order to limit the height, a stopper mechanism is designed as shown in Figure 9. A pin and a hole have stepped sections and there is 1.1 mm gap between the sections. The gap h_s is designed considering the displacement 0.5 mm and margin for plastic deformation of the film which is measured experimentally. When the pin is pushed up 1.1 mm in height, the stepped sections collides with each other and the height is limited. And the top of the pin is raised 0.5 mm above the top plate. To pull down the pin by polyurethane film's elastic restoring force, the pin and the film are attached with adhesive.

Figure 10 shows the developed Braille display. The size of the 6 units actuator is 90 mm × 45 mm × 31 mm (width × depth × height). ER fluid is pressurized by 230 kPa argon gas and high voltage applied to electrodes is controlled by a computer. Figure 11 shows a snapshot of controlling each height.

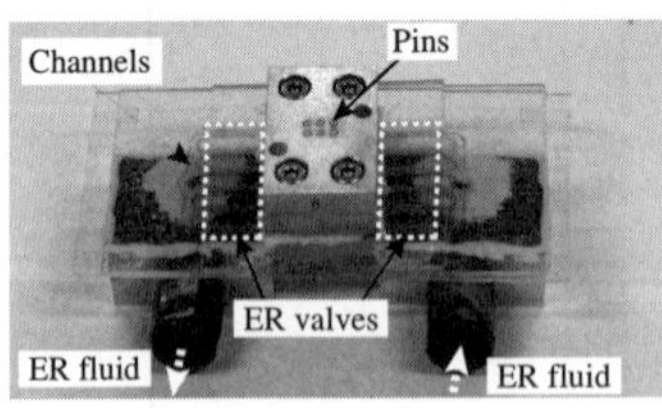

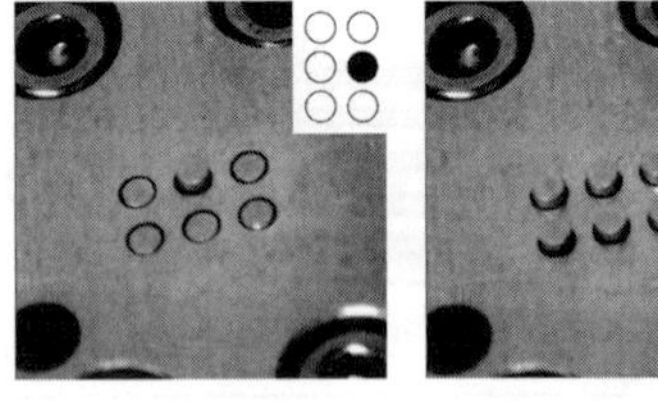

Figure 10. Photograph of a 6 units actuator.   Figure 11. Each height of pins is changed by ER valves control.

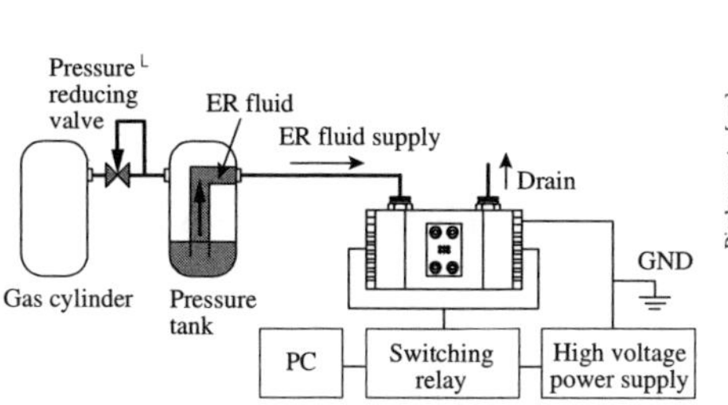

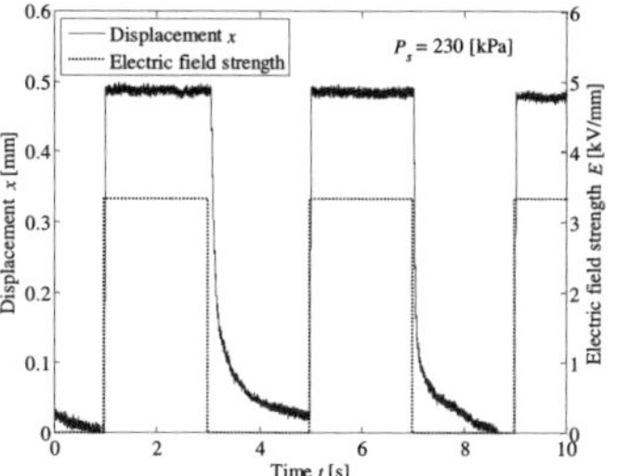

Figure 12. Experimental system for a Braille display system.

Figure 13. Time history of actuator displacement.

4. Evaluation of a Micro-Diaphragm ER Actuator

For evaluating movable length of the actuator, displacement of the top of the pin is measured by a laser displacement sensor. Figure 12 shows an experimental system for the developed Braille display system. The height of the pin is changed at two seconds interval by a computer program. Supply pressure P_s and applied electric field strength E are 230 kPa and 3.3 kV/mm, respectively. Figure 13 shows time history of actuator displacement. The maximum height is about 0.5 mm and the height is changed from 0 mm to 0.5 mm within 0.1 s. The transition from 0.5 mm to 0 mm is relatively slow since the pin is pulled by film's elastic restoring force. But still, the pin is returned under 0.1 mm in height within 0.5 s.

5. Concluding Remarks

A micro-diaphragm ER actuator which can be driven relatively low supply pressure is developed by using a polyurethane film and Braille display using the actuators is performed to display six tactile pins. Displacement and response of the actuator are evaluated and the developed actuator appropriates Braille display system. In order to develop a wearable device, however, the size of the actuator is not sufficient. Smaller actuators and a pump for supplying ER fluid will be developed in the next step.

References

1. W. M. Winslow, *J. Appl. Phys.* **20**, 1137(1949)
2. M. Nakano, T. Katou, A. Satou, K. Miyata and K. Matsushita, *J. Intel. Mat. Syst. Str.* **13**, 503(2002)
3. http://www.kgs-america.com/braille_datasheet.html
4. T. Tsujita, M. Kobayashi and M. Nakano, *Proc. 14th Int. Symp. Appl. Electromag. Mech.*, 637(2009)
5. A. J. Simmonds, *IEE Proc. D* **138**, 400(1991)
6. M. Nakano and T. Yonekawa, *T. Jpn. Soc. Mech. Eng.* **61**, 166(2005)

ELECTRORHEOLOGY IMPROVES TRANSPORTATION OF CRUDE OIL

HONG TANG, KE HUANG and R. TAO

Department of Physics, Temple University
Philadelphia, PA 19122, USA

A system to simulate crude oil flow in pipeline is established in our lab. The system measures the crude oil flow rate with or without the electric field treatment. With this system, it is demonstrated that the crude oil flow rate is increased by about 20% after a n electric field of 12000V/cm is applied in the direction parallel to the flow direction. The electric energy for this electric field treatment is very small and oil temperature is almost unchanged. The electrorheology induced thinning effect will have significant application in crude oil pipeline transportation.

1. Introduction

At present fossil fuels are still the main energy sources for our society. According to a survey from BP Company, the total energy consumption of the world in year 2008 is about 474×10^{18} J [1]. Among the total consumption, 88% came from the fossil fuels, including coal, natural gas, and crude oil. The crude oil and its products account for about 35% of the total energy consumption. Crude oils are usually exploited at rural areas, such as deserts, sea shore, deep water, etc., and transported by pipelines or tank ships. The Trans-Alaska Pipeline System, including 11 pump stations, several hundred miles of feeder pipelines, and the Valdez Marine Terminal, has about 800 miles of the main pipeline with diameter of 48 inches to transport crude oil from Prudhoe Bay to Valdez, Alaska. In current pipeline systems, it usually uses heating method or adding the drag-reducing agent (DRA) to reduce the viscosity of crude oil to improve the transportation. However, the heating method is very energy-consuming and raises concerns about its environment impacts. The DRA addition is not only expensive, but also raises concerns at refineries. Based on the concepts of Electrorheology, we have proposed a novel technology to reduce the viscosity of crude oils by pulsed strong electric field. This technology consumes tiny electric energy and is very fast and efficient [2-4]. In this paper, we report our tube system to test the flow rate of crude oil in the tube systems. The results show that the flow rate of crude oil can be increased by more than 10% after the electric field is applied upon.

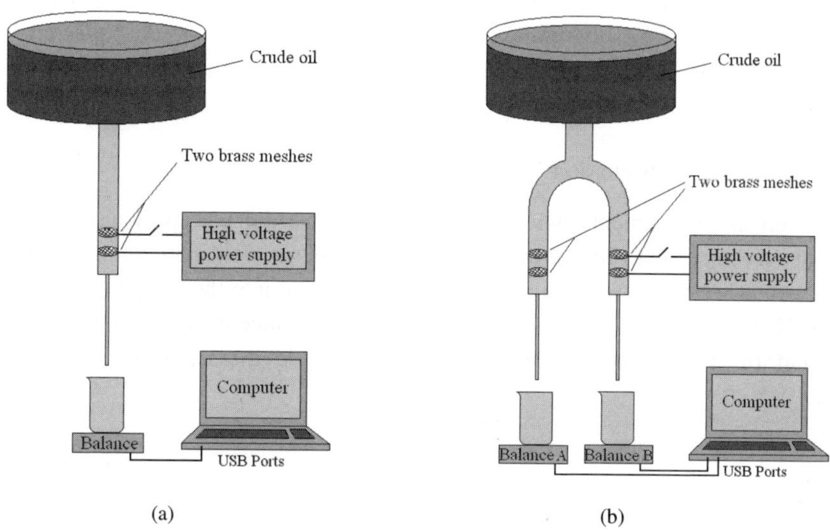

Fig. 1. Schematic diagraphs of experiment setup. (a) single tube system and (b) two tube system.

2. Experiments

In order to simulate the flow of crude oil in pipelines, we established several tube systems in our lab to conduct experiments. Figure 1 (a) shows a single tube system. At the top there is a cylinder reservoir with a diameter of 20cm and height of 15cm, which is used to store crude oil. A middle glass tube is connected with the reservoir through a glass valve. The inner diameter of the middle tube is 2.26cm and its length is about 50cm. There are two brass meshes glued close to the bottom of the middle tube. Between the two meshes it is a glass spacer, which has a same diameter size of the middle glass tube and also glued together with epoxy glue. The height of the spacer is 2.5cm. The outside areas of the meshes and spacer are sealed with thermo-molten plastic glue to prevent electric discharge. The size of hole on the meshes is 0.26cm in diameter with an average open area ratio of 48%. When a high voltage is applied on the two mesh electrodes, a strong local electric field is produced between the two meshes, which is parallel to the flow direction. A fine tube is connected to the bottom of the middle tube. The inner diameter of the fine tube is around 0.37cm. The length of the fine tube is about 27cm. The crude oil used in the experiment is API21. At room temperature, its density of mass is 0.92g/ml. When the glass valve is open, under the gravity the crude oil flows down through the middle tube, the brass meshes, and the fine tube and finally collected by a beaker. The beaker is placed on a microbalance, which is connected to a computer. With the software of Labview, the computer records the mass of crude oil collected in the beaker as a function of time. Hence, the flow rate as a function of time can be obtained easily.

3. Results and Discussion

We run the experiments twice. In the first run, we did not apply any electric field. In the second run, we did not apply any electric field initially, but at 400 seconds when there were about 50g crude oil flowed into the beaker, we turned the DC high voltage 30KV on the mesh electrodes. Since the two electrodes had a gap of 2.5cm, the applied electric field inside the two electrodes was 1200V/mm.

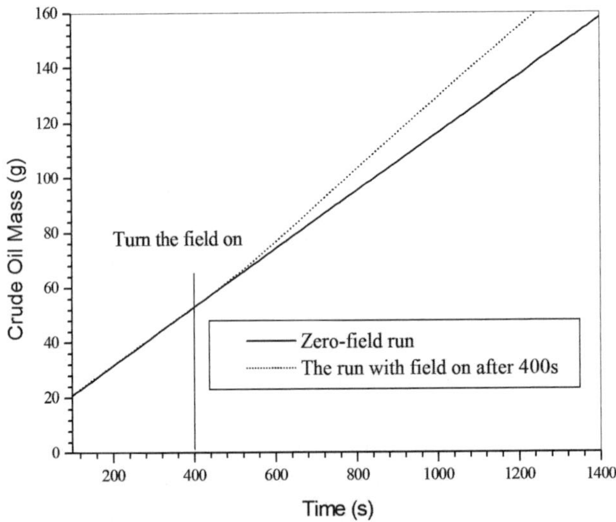

Fig. 2. Collected crude oil mass as a function of elapsed time for both cases, without electric field and with electric field applied.

The result is shown in Fig. 2. The slope of the curve is the flow rate. It is clear from the figure that before the electric field was applied, the two curves were almost identical, implying that both runs had almost the same flow rate in absence of electric field. However, shortly after the electric field was applied, the curve of the second run with field applied began to have a much higher flow rate than the curve of the first run. The delay in the flow rate enhancement indicated that when the electric field was turned on, there was still some untreated crude oil inside the fine tube. Only after the treated crude oil reached the fine tube, the flow rate was increased.

Figure 3 shows the flow rates of API21 crude oil as a function of time for zero-field run and non-zero-field run. The baseline, the zero-field flow rate is almost a constant, at 0.107g/s. After the electric field is turned on, the flow rate climbs up to 0.135g/s; afterwards, the flow rate remains at 0.13g/s. The improvement is between 21.6% to 26.2%.

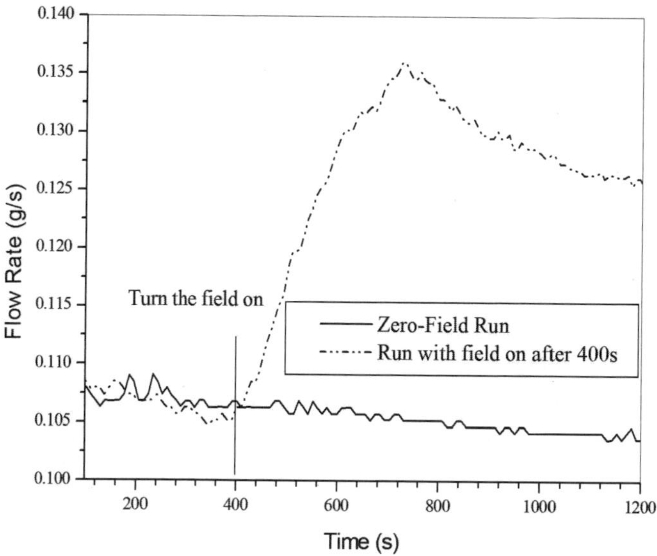

Fig. 3. Flow rates of API 21 crude oil for the two runs. The flow rate goes up quickly after the electric field is turned on.

Because our fine tube has very small diameter, 0.37cm, and is long enough, the flow inside the fine tube must be laminar. With the Navier-Stokes equation for impressible flow,

$$\rho[\partial\vec{u}/\partial t + (\vec{u}\cdot\nabla)\vec{u}] = \rho\vec{g} + \eta\nabla^2\vec{u} = 0 \qquad (1)$$

where $\vec{g}$ is the gravity and $\rho = 0.92$ g/cm^3. From Eq. (1), we have the flow velocity

$$\vec{u} = -\rho\vec{g}(r^2 - R^2)/(4\eta),\tag{2}$$

where $R = 0.185$ cm, the radius of the fine tube. The mass flow rate is given by

$$Q = \rho\int_0^R 2\pi rudr = \pi\rho gR^4/(8\mu).\tag{3}$$

We use the kinetic viscosity, $\mu = \eta/\rho$, in Eq. (3). Therefore, from the oil flow rate, we can calculate the kinetic viscosity,

$$\mu = \pi\rho gR^4/(8Q).\tag{4}$$

With Eq. (4), the crude oil viscosity is 387.6 cSt before the electric field is applied. After the electric field 1200V/mm is applied, the viscosity is down to 307.2 cSt, reduced by 21%.

The Reynolds number is $\mathrm{Re} = 2Ru/\mu$. If we use the speed at the tube center, $u_c = gR^2/(4\mu) = 2Q/(\rho\pi R^2)$, then,

$$\mathrm{Re} = 4Q/(\pi R\rho\mu).\tag{5}$$

For our experiment, the Reynolds number is very small, about 0.33. Therefore, the laminar flow assumption is justified.

It should be pointed out that during our experiment, we monitored the electric current, which was very low and alwayus in the order of μA's. To compare our technology with heating, we note that as the temperature is increased by 1^0C, our API 21 oil sample has its viscosity reduced by 4.5%. Therefore, to have similar viscosity reduction by 21% with heating, we need to raise the oil temperature about 5^0C, requiring energy about 100 times the electric energy needed for our treatment. Therefore, our technology is much energy efficient [4].

We also established a two tube system to check the zero-field and non-zero-field flow rate simultaneously. Figure 1 (b) shows the schematic setup of the two tube system. An up side down Y shape connector splits the flow of crude oil into two almost identical paths: two identical medium size tubes with two identical brass mesh capacitors and two identical fine tubes. Two microbalances are connected to the computers to measure the flow rates for the left and right tubes respectively. In the first run, both sides have no electric field applied and the oil flow rates in both sides are almost the same: 0.09324g/s for the left side and 0.09638 g/s for the right side, respectively. In the second run, the electric field of 1.5kV/mm is applied on the right side while the left side remains electric field free. The oil flow rate of the right side increased to 0.11358 g/s while the left side has the oil flow rate 0.09311g/s, slightly down from the first run, indicating that the crude oil is not completely uniform. With this adjustment, the applied electric field increases the flow rate by 18%.

It is well known that crude oil is a complex fluid. The base liquid, consisting of small molecules for gasoline and diesel, has very low viscosity. There are many tiny particles dispersed in the base liquid, such as paraffin, asphalt, sulfur particle, etc. In paraffin-base crude oil, at low temperature the high viscosity is mainly due to the suspended paraffin particles. In asphalt-base crude oil, at room temperature the asphalt in the crude oil absorbs moisture and solidifies into asphaltene particles, which raise the effective viscosity of crude oil. Generally, the effective viscosity of a liquid suspension depends on the viscosity of the base liquid, the volume fraction and the size distribution of the suspended particles. The apparent viscosity of the suspension can be expressed as [5,6]

$$\eta = \eta_0 (1 - \phi / \phi_m)^{-[\eta]\phi_m} , \qquad (6)$$

where $[\eta]$ is called Krieger-Dougherty intrinsic viscosity and equals to 2.5 for spherical particle, η_0 is the viscosity of base liquid, ϕ is the volume fraction of the dispersed particles, ϕ_m is the maximum volume fraction of particle for a given particle size or a particle size distribution. As can be seen, the apparent viscosity decreases as ϕ_m is increases and $[\eta]$ is reduced.

Dr. Tao first presented a theory and tests about reducing the viscosity of liquid suspensions with pulsed electric field or magnetic field [4]. Usually there is a mismatch in dielectric constant or magnetic permeability between the suspended particles and the base liquid. Once an electric field or magnetic field is applied to the suspension, the particles are polarized with induced dipoles. The dipolar interaction is attractive to aggregate these particles to large ones. This process will not alter ϕ, but makes ϕ_m increased as a result of increase of average particle size and polydispersity.

In addition, the streamline shape of the aggregated large particles is also important for the viscosity reduction and increase of flow rate of crude oil. When the direction of electric field is parallel to the flow of crude oil, the aggregated particles under the electric field are of short chain or ellipsoid. The length along the electric field direction of the aggregated particles is larger than the one vertical to the electric field direction. From a fluid mechanical view point, this kind of shape is streamlined, having low intrinsic viscosity [η] in

Eq. (6). Therefore, the effective viscosity is further reduced. Hence the effective apparent viscosity η is reduced. In short, proper application of eletrorheology or magneto-rheology can reduce the viscosity of liquid suspensions [4].

4. Conclusion

It is well known that laminar flow occurs when the Reynolds number Re<2300 and turbulent flow occurs when Re>4000. In the interval between 2300 and 4000, laminar and turbulent flows are possible ('transition' flows), depending on other factors, such as pipe roughness and flow uniformity [7]. For pipelines made of h 12inch diameter or 6 inch diameter pipes, the crude oil flow is mostly laminar. For pipelines made of 24 inch diameter pipes, the crude oil flow could be transitional or turbulent. For example, if our API 21 is transported via pipeline at a typical speed of 3 miles per hour, the Reynolds number will be 1319.6 on 12in diameter pipeline, indicating that the flow will be laminar. The Reynolds number will be 2639.2 for 24in diameter pipeline, indicating that the flow will be transitional.

In our lab system, the Reynolds number is very small. Therefore, the flow is always laminar. It is conceivable that our lab results can be extended to the laminar flow situation in pipelines. As for the turbulent flow, to extend our lab results to pipeline with large diameter requires some field tests on actual pipelines. However, it is clear that our technology will reduce the effective viscosity of the crude oil. The turbulence may break the aggregated short chains quicker than laminar flow, shortening the duration of viscosity reduction. This implies that we have to retreat the crude oil more for turbulent flow than that for laminar flow. On the other hand, it has no doubt that this new technology will play very important role in the future crude oil transportation.

Acknowledgement

This work was supported in part by STWA and RAND.

References

1. www.bp.com, BP statistical review of world energy 2009.
2. R. Tao, K. Huang, H. Tang, and D. Bell, *Energy Fuels* **22**, 3785 (2008).
3. R. Tao, *Int. J. Mod. Phys. B* **21 (28&29)**, 4767 (2007).
4. R. Tao, X. Xu, *Energy Fuels*, **20**, 2046 (2006).
5. I. M. Krieger, T. J. Dougherty, *Trans. Soc. Rheol.*, **3**, 137 (1959).
6. W. B. Russel, D. A. Saville, and W. R. Schowalter, *Colloidal Dispersion*; Cambridge University Press: Cambridge, U.K., pp 456-503 (1991).
7. J. P Holman *Heat transfer*, McGraw-Hill, 2002, p.207

REDUCING THE VISCOSITY OF DIESEL FUEL WITH ELECTROREHOLOGICAL EFFECT

ENPENG DU[1], H. TANG[1], K. HUANG[1,2] and R. TAO[1]

[1]Dept. of Physics, Temple University, Philadelphia, PA 19122
[2]Dept. of Radiation Oncology, University of Michigan, Ann Arbor, MI, 48109

Improving engine efficiency and reducing pollutant emissions are extremely important. Here we report our finding, using electrorheology to reduce the viscosity of diesel fuel. Diesel is made of many different molecules, 75% small molecules and 25% large molecules. In addition, it contains other nanoscale particles, such as sulfur. Therefore, diesel can be regarded as a liquid suspension. Under a strong electric field, the large molecules aggregate into small clusters, yielding a lower viscosity. This viscosity reduction leads to finer mist in fuel atomization, improving the combustion and engine efficiency.

1. Introduction

It is more and more urgent and important now to improve engine efficiency and reduce pollutant emissions. While some progress in this area has been made since fuel injection technology was developed, internal combustion engines are still a long way from being clean and efficient [1].

As combustion starts at the interface between fuel and air, reducing the size of fuel droplets would increase the total surface area. With small fuel droplets, the fuel can mix with air much better, the combustion goes faster and cleaner, and the heat is released on time to push the engine to do work. This improves the engine efficiency [2,3].

In last 10-20 years, vehicles with internal combustion engines have improved their fuel mileage more than 30%. Most of them come from improving the combustion with increasing fuel pressure or new fuel injectors. The higher fuel pressure produces smaller fuel droplets. The discussions about future engine for efficient and clean combustion are focused on ultra-dilute mixtures at extremely high pressure to produce much finer mist of fuel for combustion [1-3].

Unfortunately, the new combustion technology with extreme pressure is still under development and not applicable to current engines as they cannot sustain such high pressure. Here we report our finding, using electrorheology to reduce the viscosity of diesel fuel. This viscosity reduction improves the fuel injection,

54

leading to finer mist in fuel atomization, improving the combustion and engine efficiency.

2. Method

Diesel is composed of about 75% saturated hydrocarbons, and 25% aromatic hydrocarbons, ranging approximately from $C_{10}H_{20}$ to $C_{15}H_{28}$. In addition, it contains other nanoscale particles, such as sulfur. If we treat the small molecules in diesel as base liquid and the rest 25% large molecules and nanoscale particles as suspended particles, diesel can be regarded as a liquid suspension. With such a liquid suspension, we can use an electric field to reduce its viscosity [4,5].

For diesel with many nanoscale sulfur particles, we can easily reduce its viscosity with an electric field. Our first test used diesel with sulfur, 2000 ppm. As shown in Fig.1, when an electric field of 1000V/mm was applied for 2 seconds, the diesel viscosity was down from 4.6cp to 4.18cp, 10% reduction. Then the viscosity slowly went up. In about 60 minutes, the viscosity came to 4.4cp, still 4.4% below the original value.

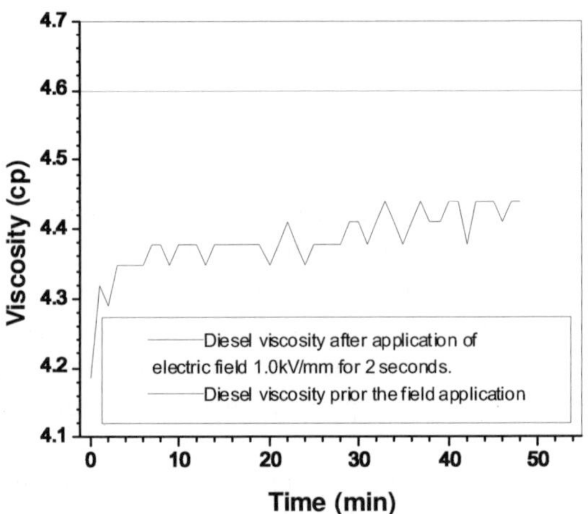

Fig. 1. Diesel with high level sulfur can have its viscosity easily reduced by an electric field.

The physical mechanism of this viscosity reduction is illustrated in Fig.2. The diesel fuel flows from left to right. As it passes the electric field, the suspended particles are polarized and aggregate into short chains, leading to a lower viscosity. The following three mechanisms contribute to the viscosity reduction: (1) As the suspended particles are aggregated , the polydispersity is increaded, leading to the viscosity reduction; (2) As the average suspended particle size is increased, the effective viscosity is reduced [6-8]; (3) As the aggregated particles and clusters have their shape streamlined, the viscosity is reduced [4,5].

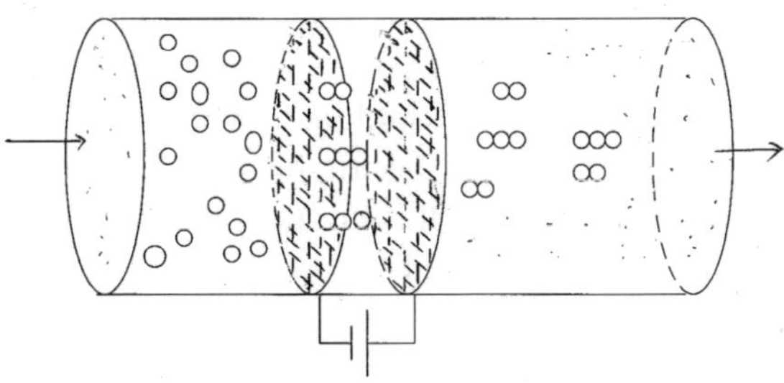

Fig. 2. As the diesel passes the electric field, the suspended particles aggregate into short chains, leading to a lower viscosity.

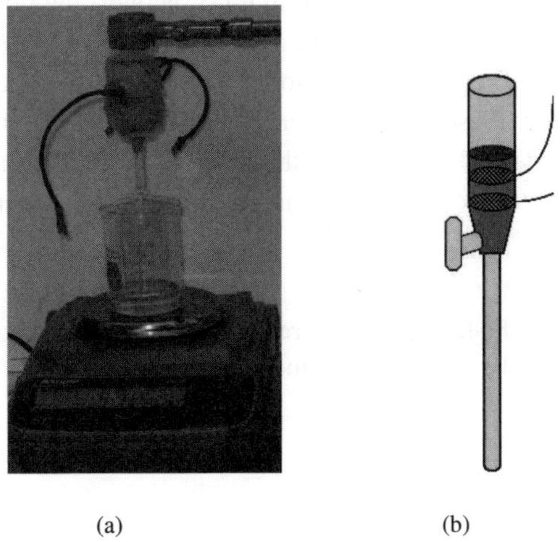

(a)   (b)

Fig. 3. (a) The device we made to determine the viscosity reduction for ultra low sulfur diesel. (b) The outline of the main part of the device. There are two metallic meshes, serving as two electrodes. A high voltage can be applied on the two meshes to produce an electric field in the flow direction.

Since sulfur particles have much higher dielectric constant than the base liquid, the effect of electric field is strong and the viscosity reduction of high sulfur diesel is relatively easy.

However, with ultra low sulfur diesel (ULSD), we have found that the viscosity reduction is much more difficult and requires much high electric field.

In order to do the test, we made a device (Fig.3). Inside a container, we placed two metallic meshes to serve as electrodes. During the test, the device is placed vertically. Then we close the flow switch first and fill the container with ULSD to cover the upper electrode mesh. After the diesel sits in the container still, we open the switch to let diesel flow down along the capillary tube and measure the flow rate. Afterwards, we refill the container with the same diesel. Before open the switch, we apply an electric field for a selected time interval. Then we open the switch and measure the flow rate of the treated diesel.

3. Results

With this method, we found that the required electric field for ULSD is much stronger. For diesel with high sulfur level, application of an electric field of 1kV/mm for 2 seconds can produce significant viscosity reduction. However, with ULSD, we have to apply much higher electric field. As shown in Fig.4, when we apply an electric field of 2kV/mm for 5 seconds, the flow rate increased about 30, indicating that the viscosity of treated diesel is reduced by about 23%. However, this viscosity reduction does not last very long. In about 4 minutes, the flow rate of treated diesel and untreated diesel become almost the same, indicating that the viscosity returns to the original value. For diesel with high sulfur level, the viscosity reduction can last for more than 60 minutes.

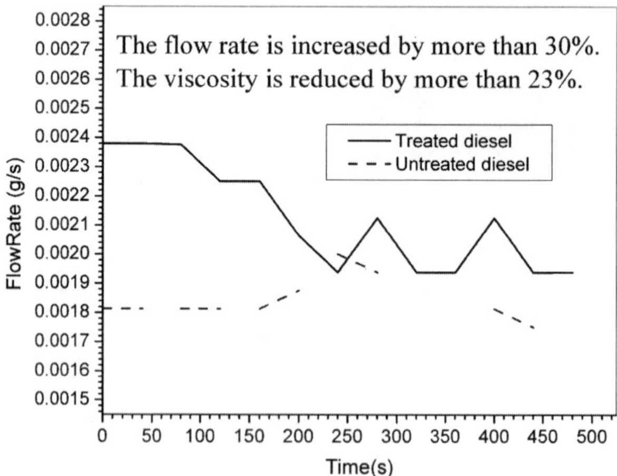

Fig. 4. At E=2kV/mm for 5 seconds, the diesel viscosity is reduced by 23%, but it only lasts for about 240 seconds.

This phenomenon is easy to understand. In ULSD, the electric field forces the large molecules to aggregate. Because the large molecules are much smaller than nanoscale sulfur particles, the required electric field is much stronger. Also because the large molecules are much lighter than the sulfur particles, they are much easier to be kicked out from aggregated chains by the thermal vibrations. Therefore, the viscosity reduction for ULSD does not last very long. As soon as the aggregated clusters are disassembled, the viscosity reduction is disappeared.

On the other hand, even with such short time of viscosity reduction, we can still use it to improve the engine efficiency. Our fuel injection device is illustrated in Fig. 5. Because the device is just next to the fuel injector, the injected diesel still has reduced viscosity, leading to finer mist [2].

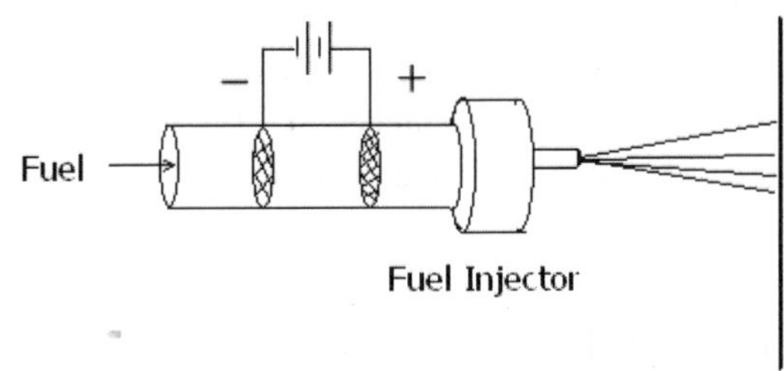

Fig. 5. Our viscosity reduction device is just next to the fuel injector.

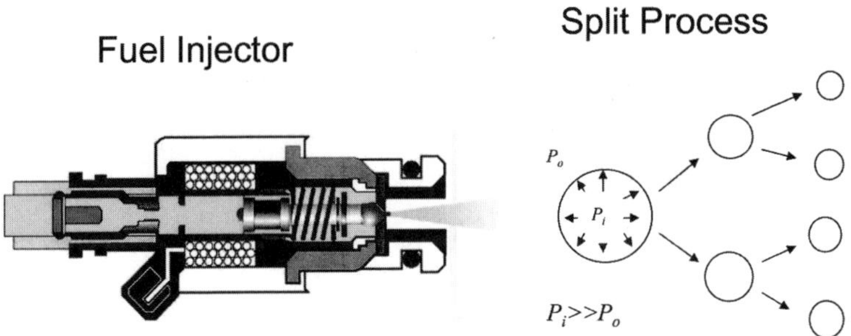

Fig. 6. The emitted droplets from a fuel injector are split to become smaller and smaller.

Reducing the fuel viscosity improves the fuel atomization. As shown in Fig. 6, the injected fuel has a pressure higher than that in the combustion chamber. The droplets are thus split, becoming smaller and smaller after they are emitted from the nozzle. If the droplets are allowed to reach the equilibrium, their radius is given by

58

$$a = 2\sigma / \Delta P , \tag{1}$$

where σ is the fuel's surface tension and $\Delta P = P_i - P_o$ is the pressure difference between the fuel's inside pressure P_i and the pressure outside the fuel P_o. However, in reality, the fuel droplets can never reach equilibrium because the viscosity acts against any deformation of the droplets. The following fact illustrates this issue: Diesel fuel has a surface tension around 10 dyne/cm. When $\Delta P = 2$ bar, a should be about 0.1μm from Eq. (1). However, in the fuel atomization, most diesel droplets are much bigger than 0.1μm.

The influence of fuel's viscosity on the atomization can be illustrated by the Ohnesorge number, which is defined as

$$Oh = \eta / \sqrt{\rho D \sigma} , \tag{2}$$

where D is the droplet's diameter and η, ρ, and σ are the fuel's viscosity, density, and surface tension, respectively. Typically, there is a universal critical value Oh_c close to 0.1 for the Ohnesorge number. As $Oh < Oh_c$, the droplet splits under high pressure. As $Oh > Oh_c$, the droplet does not split even it is under high pressure. Therefore, the average droplet size can be estimated as

$$D = \eta^2 / (\rho \sigma Oh_c) . \tag{3}$$

As the viscosity η is reduced, the average droplet size is reduced dramatically. For example, when the viscosity is reduced by 23%, the average droplet size can be reduced by 41%.

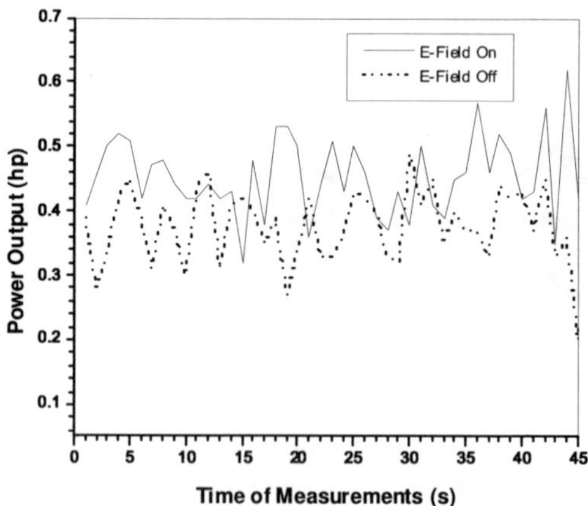

Fig. 7. The lab test of Mercede-Benz 300D with a Dynamometer. The average power output was originally about 0.368hp and increased to 0.443hp after the device was turned on.

Both lab tests and road tests with Mercede-Benz 300D show that our technology improves the fuel efficiency significantly. A continuous road test for six months showed that on the highway, our device increased the diesel mileage from 32 miles per gallon (mpg) to 38 mpg. In city driving, the improvement of fuel mileage was averaged at 12-15%. The lab tests with dynamometers confirmed the road tests, and in fact, had a better result than that of the road tests. For example, at low fuel consumption with a fixed fuel consumption rate the average power output increased from 0.3677 hp to 0.4428 hp after the device was turned on (Fig.7). This indicates that the power output was improved by about 20.4% at the same fuel consumption rate. Our recent lab tests with a dynamometer at Stanadyne Corporation further confirm the above results.

Acknowledgement

This work was supported in part by STWA.

References

1. http://www.sc.doe.gov/bes/reports/abstracts.html
2. R. Tao, K. Huang, H. Tang, and D. Bell, *Energy & Fuels* **22**, 3785-3788 (2008).
3. R. Tao, K. Huang, H. Tang, and D. Bell, *Energy & Fuels* **23**, 3339-3342 (2009).
4. R. Tao & X. Xu, *Energy & Fuels* **20**, 2046-2051 (2006).
5. R. Tao, *Intern. J. of Modern Physics B* **21**, N28&29, 4767-4773 (2007).
6. F. L. Saunders, *Journal of Colloid Science* **16**(1), 13 (1961).
7. K. H. Sweeny and R. D. Geckler, *Journal of Applied Physics* **25**(9), 1135 (1954).
8. D. G. Thomas, *Journal of Colloid Science* **20**(3), 267 (1965).

ELECTRORHEOLOGY IMPROVES E85-ENGINE PERFORMANCE AND EFFICIENCY

K. HUANG[1,2], ENPENG DU[1], H. TANG[1], and R. TAO[1]

[1]*Dept. of Physics, Temple University, Philadelphia, PA 19122*

[2]*Dept. of Radiation Oncology, University of Michigan, Ann Arbor, MI, 48109*

E85 is an alternative fuel with 85% ethanol and 15% gasoline. However, it is widely reported that E85 vehicles have difficulties to start in winter and have poor performance. Here we report that with proper application of electrorheology, we can solve these issues. E85 vehicles all have port injected engines. The fuel is injected into cylinders as droplets. Before the ignition, the fuel evaporates. Because E85 is more viscous than gasoline, the injected E85 droplet size is not small. Especially, in winter the cold weather makes the viscosity even higher, leading the E85 droplets even bigger. Since evaporation starts from the droplet surfaces, large droplets are difficult to be evaporated before the ignition comes. When there are no enough fuel vapors, the engine cannot start. To solve this problem, we introduce a small device just before the fuel injection, which produces a strong electric field to reduce the fuel viscosity, leading to much smaller fuel droplets in atomization. The evaporation is much faster and the engine is easier to start. As the small fuel droplets produced by our device make the combustion fast and timely, engine efficiency and performance are also expected to be improved.

1. Introduction

E85 fuel is a mixture of 85% ethanol and 15% gasoline. Ethanol has excellent fuel property. Its combustion in an internal combustion engine yields many of the products of incomplete combustion produced by gasoline and significantly larger amounts of formaldehyde and related species such as acetaldehyde. This leads to a significantly larger photochemical reactivity that generates much more ground level ozone [1]. The Clean Fuels Report makes a comparison of fuel emissions and shows that ethanol exhaust generates 2.14 times as much ozone as does gasoline exhaust [2,3]. However, the efficiency of E85 engines remains lower in comparison with gasoline engines. The difficulty to start the E85 engine in cold winter is a big issue.

All current E85 vehicles use port injected engines. The fuel is first injected into cylinders as droplets. Before the ignition, the fuel evaporates. Most port injected engines have the fuel pressure around 30-70 psi. Because E85 is more viscous than gasoline, the injected E85 droplet size is not small. During cold winter, the injected fuel droplet size is even bigger because the fuel's viscosity is higher as the temperature is low. As evaporation starts from the droplet surfaces, the bigger the droplets are, the smaller the total surface areas for evaporation would be. Therefore, for big fuel droplets, it takes much time to evaporate at low temperature. When there is no enough fuel evaporated at ignition, the engine cannot start. The well-known recipe for this problem is to inject much smaller droplets into the engine cylinders. Then the evaporation is much faster because the small fuel droplets mixed with air better and the total surface areas for evaporation is also dramatically increased. The engine is easier to start and feels more powerful with small fuel droplets at the starting. After the engine is started, the metal surfaces are warmed enough to help evaporate the gasoline and the engine may operate fairly well without hesitation. On the other hand, continuous injection of finer fuel droplets into the engine will increase the fuel efficiency because the combustion is faster and produces heat on time to push the pistol to do work with small fuel droplets [4,5].

To solve this problem to start E85 engine in cold weather, we introduce a small electrorheological (ER) device just before the fuel injection, which produces a strong electric field to reduce the fuel viscosity, leading to much smaller fuel droplets in atomization. The evaporation is thus much faster and the engine is easier to start. This electrorheological device is similar to our new technology, which is used to improve diesel engine efficiency [4]. We also expect that this ER device will improve E85 engine efficiency.

In this paper, we report our studies about the distribution of droplet size for E85 fuel without and with our ER device at temperature 0 °C.

2. Theory and Experiment

E85 fuel is a liquid suspension. Therefore, we can use electric field to aggregate the suspended particles or emulsions to reduce the viscosity. The physical mechanism of this viscosity reduction is illustrated in Fig.1. The E85 fuel flows from left to right. As it passes the electric field, the suspended particles and emulsions are polarized and aggregate into short chains, leading to a lower viscosity. The following three mechanisms contribute to the viscosity reduction [6-10]: (1) As the suspended particles (or emulsions) are aggregated, the polydispersity is increased, leading to the viscosity reduction; (2) As the average suspended particle (or emulsion) size is increased, the effective viscosity is reduced; (3) As the aggregated particles (or emulsions) and clusters have their shape streamlined, the viscosity is reduced.

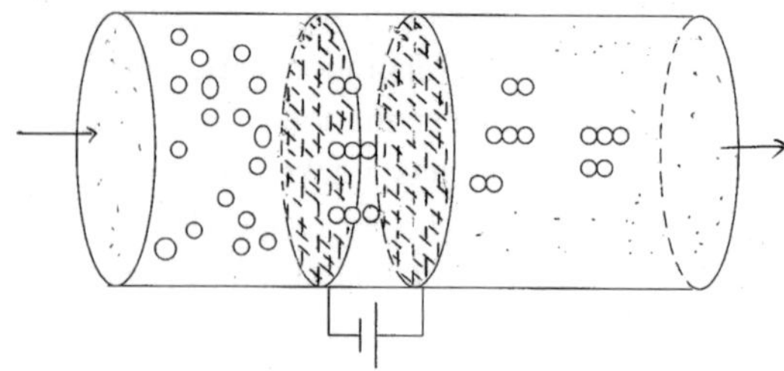

Figure 1. As E85 fuel passes the electric field, the suspended particles and emulsions aggregate into short chains, leading to a lower viscosity.

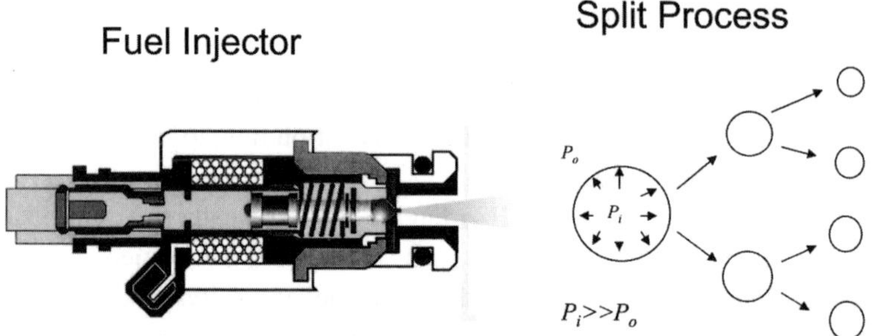

Figure 2. The emitted fuel droplets from a fuel injector are split to become smaller and smaller.

Reducing the fuel viscosity improves the fuel atomization. As shown in Fig. 2, the injected fuel has a pressure higher than that in the combustion chamber. The droplets are thus split, becoming smaller and smaller after they are emitted from the nozzle. If the droplets are allowed to reach the equilibrium, their radius is given by

$$a = 2\sigma / \Delta P , \qquad (1)$$

where σ is the fuel's surface tension and $\Delta P = P_i - P_o$ is the pressure difference between the fuel's inside pressure P_i and the pressure outside the fuel P_o. However, in reality, the fuel droplets can never reach equilibrium because the viscosity acts against any deformation of the droplets.

The influence of fuel's viscosity on the atomization can be further illustrated by the Ohnesorge number [11], which is defined as

$$Oh = \eta / \sqrt{\rho D \sigma} , \qquad (2)$$

where D is the droplet's diameter and η, ρ, and σ are the fuel's viscosity, density, and surface tension, respectively. Typically, there is a universal critical value Oh_c close to 0.1 for the Ohnesorge number. As $Oh < Oh_c$, the droplet splits under high pressure. As $Oh > Oh_c$, the droplet does not split even it is under high pressure. Therefore, the average droplet size can be estimated as

$$D = \eta^2 / (\rho \sigma O h_c) . \tag{3}$$

As the viscosity η is reduced, the average droplet size is reduced dramatically. For example, when the viscosity is reduced by 20%, the average droplet size can be reduced by 36%.

In the experiment, our ER device is used just before the fuel atomization. The device and its schematic are shown in Figure 3. Two mesh grids are inside a guard ring that is made of insulating material. The mesh grids are mounted perpendicular to the direction of flow and are separated by a spacer. Two leads are soldered to the meshes for connection to a high voltage power supply. The whole device is sealed and covered by metal which enable it to withstand high pressure. It has openings on both ends with female threads for the connection to brass tubes for fuel flow. The sample flows through the meshes where the field is applied. Our applied field direction is opposite to the flow direction to utilize the possible electrostatic effect [4]. Electric field strength and the duration for the E85 fuel to pass the electric field are the two important parameters. During the experiment, we vary these two parameters to search for the best condition.

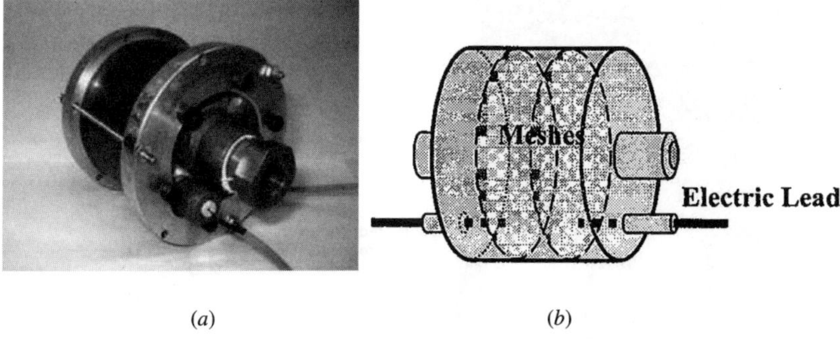

(a)   (b)

Figure 3. *a*. Field Treatment Device and *b*. Schematic of the Device.

Pressurized fuel container is connected to the ER device and supplies pressurized fuel. The container and its schematic are shown in Figure 4. Pressurized air coming from the nitrogen gas cylinder pushes the fuel in the container out through the fuel pipe into the field treatment device. The other end of the field treatment device is connected to a fuel injector in an adapter. The fuel injector responds to an electric pulse and injects out the fuel droplets. The duration of the ejection is controlled by the electric pulse. A homemade circuit including a 555 Timer and a RC circuit is used as the controller. The duration of the pulse is changeable, in the order milliseconds, by choosing different combinations of a resistor and a capacitor.

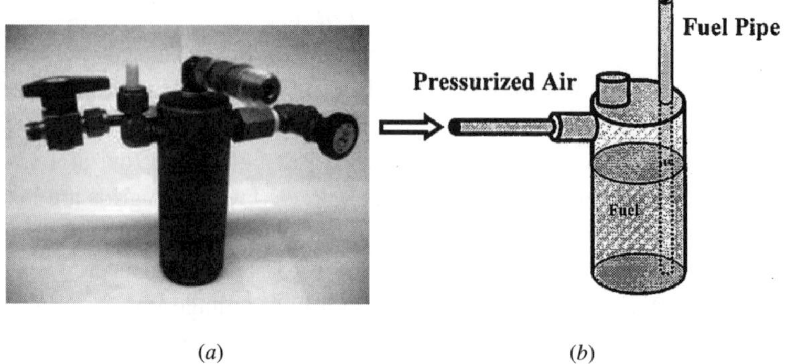

(a)   (b)

Figure 4. *a*. Pressurized Fuel Container and *b*. Schematic of the Container.

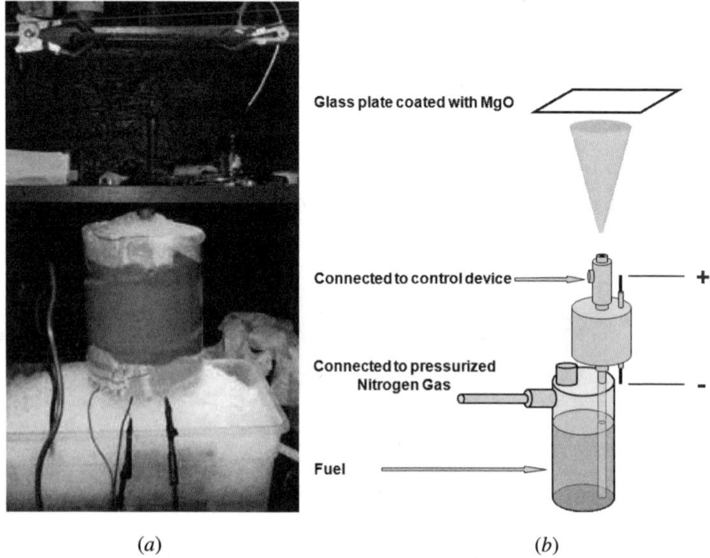

(a)   (b)

Figure 5. *a*. Setup for E85 Spray Experiment and *b*. the Schematic Diagram.

The setup and the schematic diagram are illustrated in Figure 5.

All the components in the spray experiment for E85 were buried in ice-water mixture at temperature around 0 °C. This is to simulate the condition where E85 engines are having difficulty to start in winter. We select the applied electric field on our ER device and the duration for E85 fuel to pass the electric field.

The E85 was sprayed out from the fuel injector upward onto the glass plate coated with magnesium oxide (MgO). The MgO was coated to the glass plate with fume of MgO from burning magnesium ribbons under the glass plate. White fumes of MgO are emitted while the magnesium is burning. If the burning ribbon is held below the glass plate, the plate is coated with MgO. The fumes of Magnesium Oxide are composed of tiny MgO particles with diameter at the micrometer level or less. During the spray, the fuel droplets knock MgO particles off the glass plate and leave small dots on the plate.

3. Results

In our experiment, we found that E85 is different from diesel or gasoline: E85 is easy to absorb water. As a result, the applied electric field cannot be as high as that for diesel. In our E85 tests, we used the electric field around 150V/mm to 200V/mm, but made the duration around 20 seconds. In comparison, for ultra low sulfur diesel fuel, the applied electric field is around 2000V/mm and the duration is about 5 seconds.

On the other hand, the effect for E85 is very strong. Figures 6(a) and 6(b) are two typical plates. It is clear that when our ER device is turned on, the ejected E85 droplet size is dramatically reduced. Figure 6(b) was produced with electric field 200V/mm for 20 seconds.

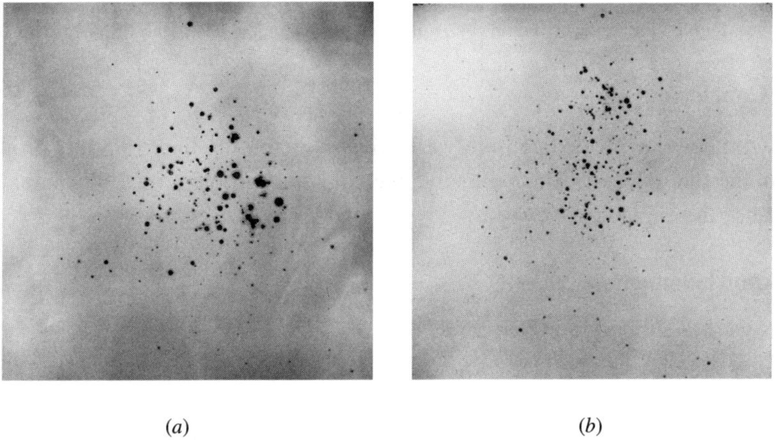

(a)   (b)

Figure 6. *a*. The E85 droplets when there was no electric field applied. *b*. The E85 droplets are much smaller when our ER device was turned on.

To make statistical average, we used ten plates to collect E85 droplets sprayed without and with our ER device each. These plates were scanned with high definition scanner at a resolution of 9600 dpi. Afterwards, we used ImageJ to analyze the droplet size distributions.

66

The particle distributions for the E85 spray without and with our ER device are plotted in Figure 7. The applied electric field was 150 V/mm and the duration was 20s. It is clear that when the electric field was applied, the number of droplets with radius of 30 ± 5µm increased from 2.9% to 7.7%, a factor of 2.7 times. The number of droplets with radius of (55 ± 5) µm increased from 5.1% to 9.7%, a factor of 1.9 times. It is clear that much smaller droplets were produced when our ER device was turned on.

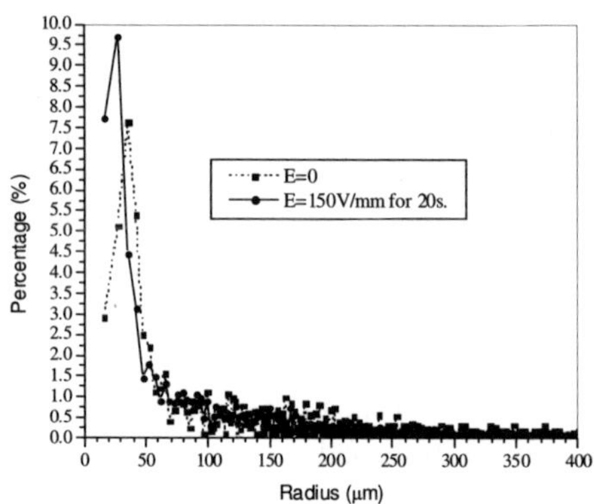

Figure 7. E85 injected droplet size distributions without and with our ER device.

4. Conclusion

In summary, our ER device is very effective in increasing finer E85 fuel droplets from the fuel injector. This will solve the cold-start problem for E85 engine and enhance the engine efficiency.

Acknowledgment

This work is supported in part by STWA.

References

1. California Air Resources Board, Definition of a Low Emission Motor Vehicle in Compliance with the Mandates of Health and Safety Code Section 39037.05, second release, October 1989
2. Lowi, A. and Carter, W.P.L.; A Method for Evaluating the Atmospheric Ozone Impact of Actual Vehicle emissions, S.A.E. Technical Paper, Warrendale, PA; March 1990
3. Jones, T.T.M. The Clean Fuels Report: A Quantitative Comparison Of Motor Fuels, Related Pollution and Technologies (2008)

4. R. Tao, K. Huang, H. Tang, and D. Bell, *Energy & Fuels*, **22**, 3785 (2008).

5. R. Tao, K. Huang, H. Tang, and D. Bell, *Energy & Fuels*, 23, 3339 (2009)

6. F. L. Saunders, *Jounal of Colloid Science* **16**(1), 13 (1961).

7. D. G. Thomas, *Jounal of Colloid Science* **20**(3), 267 (1965).

8. R. G. Larson, *The Structure and Rheology of Complex Fluids* (pp. 263-323). New York: Oxford University Press (1999).

9. R. Tao & X. Xu, *Energy & Fuels,* **2006**, 20**,** 2046-2051.

10. R. Tao, *Intern. J. of Modern Physics B*, **2007**, 21, N28&29, 4767-4773.

11. A. H. Lefebevre, *Atomization and Spray* (pp. 27-78). New York: Hemisphere Publishing (1989).

HIGH SPEED SWITCHING CONTROL OF 1DOF MANIPULATOR USING ER CLUTCH

MASAHIRO YOSHIKAWA

Department of Precision Mechanics, Chuo University, 1-13-27 Kasuga, Bunkyou-ku, Tokyo, 112-8551, JAPAN, m_yoshikawa@bio.mech.chuo-u.ac.jp

KAZUHIKO BOKU

Chuo University, Biomechatronics Laboratory, 1-13-27 Kasuga, Bunkyou-ku, Tokyo, 112-8551, JAPAN

TARO NAKAMURA

Chuo University, Biomechatronics Laboratory, 1-13-27 Kasuga, Bunkyou-ku, Tokyo, 112-8551, JAPAN

Recent advances in robotics have resulted in robots working in close proximity with humans. As a consequence, safety issues are becoming increasingly important, for example when a collision unexpectedly occurs between a robot and a person. To address such a scenario, a 3-DOF soft manipulator has been developed that incorporates an electrorheological (ER) clutch and a pneumatic sensor. The pneumatic sensor is used to decrease the impact force, while the ER clutch also decreases the collision force by making the robotic joint flexible during a collision. In this study, we aim to further improve the safety during collisions between a robot and a person by controlling the electric field applied to the ER clutch after a collision. We add energy dissipation to the robot arm by increasing the friction in the ER clutch. Moreover, we examine recoil of a robot arm just after a collision by reversing a motor. In this paper, we report on several reversal experiments conducted on a 1-link arm that does not undergo collision, and we investigate the energy dissipation effect that can be generated by an 45ER clutch.

1. Introduction

Recently, in the health care industry, robotic equipments are being increasingly used because of the growing extent of automation performed by robots in many settings. It is necessary to assume frequent contact between robots and human beings, when robots share a working environment with people [1]. To improve the safety for people in contact with robotic equipments, a soft manipulator that uses an ER clutch and a pneumatic cushion have been developed for a robotic arm [2], [3], [4]. The soft manipulator detects a collision with the pneumatic cushion, and then the force of the collision is reduced by making an arm joint flexible using the ER clutch. However, this collision force reduction method does not carry out control. It merely reduces the power transferred through the mechanism. Furthermore, in order to make an arm recoil significantly away from a collision, it is not only difficult to make the arm return to its previous position but also there is a danger of a re-collision. Thus, in order to reduce collision power more effectively and to increase the control performance after a collision, high-speed switching control of a manipulator using an ER clutch is considered in this paper.

However, if a high-speed reversal of a robotic arm is carried out at the time of a collision, it is expected that a large inertial load will be applied to a joint. The kinetic energy of the arm can be dissipated by controlling the electric field that is supplied to the ER clutch that acts to alter the frictional torque of the clutch, thereby decreasing the inertial load. In this manner, a high-speed reversal can be performed without supplying a large torque to the motor.

Based upon the above considerations, several reversal experiments in this research are conducted with a 1-link arm, which does not actually collide with any other objects, and we investigate a controller's ability to perform a high-speed reversal of a soft manipulator.

2. Characteristic of an ER Clutch

The ER clutch we adopted and the characteristics of the ER fluid are shown in Fig. 1.

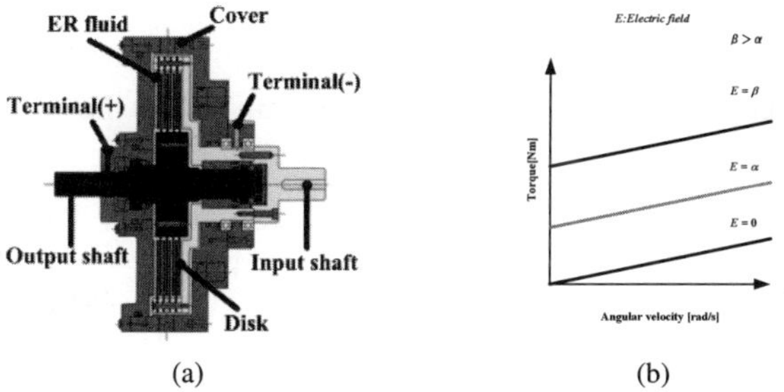

Figure 1. Characteristic of an ER clutch

As shown in Fig. 1(a), the ER fluid [5] is enclosed between disks and the torque can be increased (Fig. 1(b)) by applying an electric field. The torque characteristics can generally be approximated by a quadratic equation as follows:

$$\tau_{ER} = C_{ER}\dot{\theta} + K_{ER1}E^2 + K_{ER2}E \tag{1}$$

where θ is the angular velocity, C_{ER} is the base viscosity of the ER fluid, K_{ER1} is the quadratic torque increase due to the electric field, K_{ER2} is the linear torque increase due to the electric field, and E is the applied electric field. In the experiment described in the next section, the value of each of the coefficients was determined to be as follows: $C_{ER} = 0.013$, $K_{ER1} = 1\times10^{-7}$, $K_{ER2} = 5\times10^{-4}$

3. Experiment

The reversal experiments are conducted with a 1-link arm, and the effect of energy dissipation is observed. The experimental device and experimental system are shown in Fig. 2. The control system is constituted by MATLAB and Simulink interface with dSPACE. The angle of the output shaft is measured by a potentiometer, while that of the input shaft is measured by measuring the pulse of an encoder with a PC. The procedure for reversing the movement of the arm is shown in Fig. 3. First, a desired angle θ_d is given to the 1-link arm and the arm moves from its initial position. When it reaches the dissipation start angle θ_s, a motor is stopped, and the electric field applied to the ER clutch is set to zero kV/mm. The energy of the system at this time is set as U, and this is dissipated over a change in angle of θ_c with the friction energy W of the ER fluid. Next, an electric field is imposed such that the frictional torque gradually increases.

During this process, the relationship between the energy at the time of collision and the frictional torque (and taking into account the characteristics of the ER clutch and the electric field imposed upon the ER clutch at the time of energy dispersion) can be expressed with Eq. (2):

$$E = \frac{-K_{ER2} + \sqrt{K_{ER2}^2 + 4K_{ER1}(\tau_{loss} - C_{ER}(\dot{\theta}_2 - \dot{\theta}_1))}}{2K_{ER1}}, \tag{2}$$

where $\dot{\theta}_1$ is the angular velocity of the input shaft and $\dot{\theta}_2$ is the angular velocity of the output shaft.

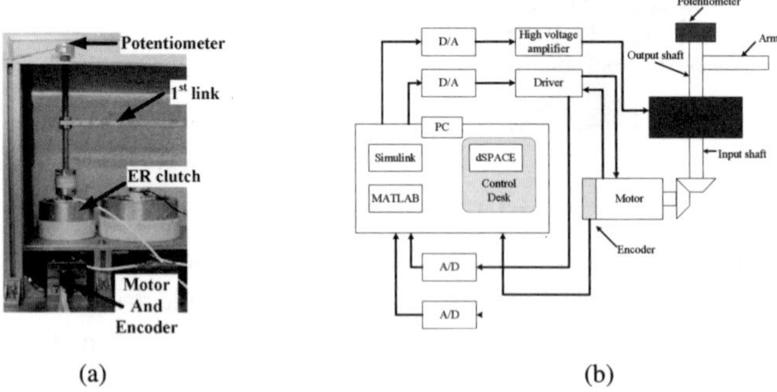

(a)   (b)

Figure 2. Experimental apparatus using the soft manipulator

when the arm reaches the desired angle, the control system changes the electric field imposed on the ER clutch again, while completing the high-speed reversal of the motor.

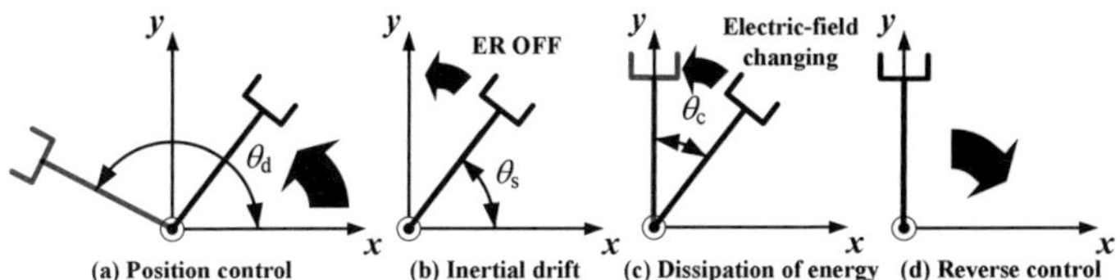

Figure 3. Process of reverse motion control

3.1. *Arm reversal experiments*

The purpose of this experiment is to investigate the effect of high-speed reversal control by controlling the electric field of the ER clutch. In this experiment, the dissipated zone considered is 20° because the energy dissipation effect over this range would be large. The desired angle is set as 180°, and a dissipation start angle θ_s is set so that the arm may start reversal at 90°. Moreover, after reversal, the arm returns to the default position (0°). The following four reversal experiments were run and compared:

Case I: While continuing to supply an electric field, ordinary position control is performed, and after reaching a desired angle, the position control is reversed.

Case II: While continuing to supply an electric field without performing energy dissipation, the arm reverses at 90° before reaching the desired angle.

Case III: A motor is immediately reversed with energy dissipation after reaching the dissipation start angle θ_s.

Case IV: A motor acts as a brake with energy dissipation after reaching the dissipation start angle θ_s, slowing the angular velocity to 0 rad/s, and then the motor is started again.

The response of the angular velocity of the input and output shaft, as well as the motor torque for each of the four cases outlined above is shown in Fig. 4.

For Case I, the dashed line in Fig. 4 indicates the time at which the arm reaches the desired angle of 180° and begins reversal. In this case, since the reversal is started after completing position control, the large magnitude of the motor torque at the time of the reversal is comparable to the motor torque when the drive process started. The maximum torque at the time of the reversal was 2.3 N-m.

For Case II, the dashed line in Fig. 4 indicates the time at which the output shaft reached 90° and started reversal. In this case, it turns out that a larger torque at the time of reversal occurs than at the time the drive process starts. The maximum torque at this time was 2.74 N-m. Note that this value exceeds the torque at the time of reversal in Case I. Moreover, there is a portion of time over which the torque is temporarily stable during reversal. The fact that a condition arose where the power exceeded the torque in the ER fluid shows that a slide has occurred between the input and output shafts. The results of Case II show that a sudden reversal operation resulted in a large load on the motor.

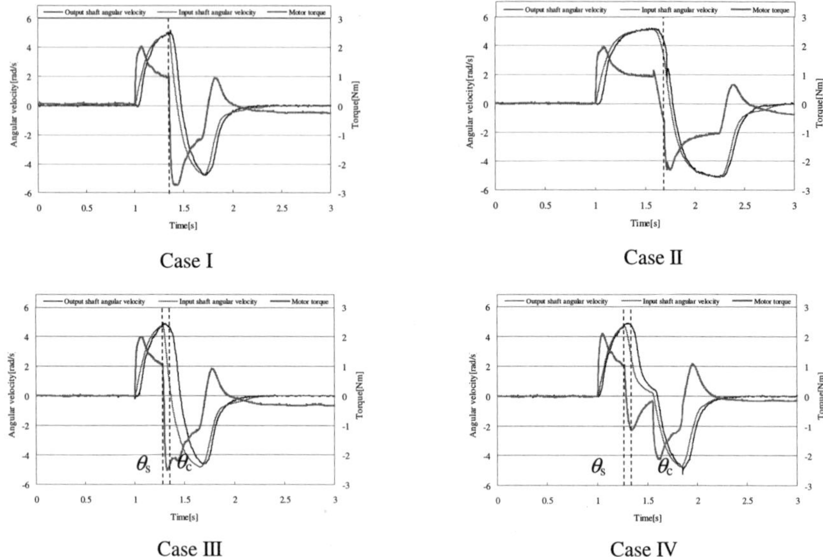

Case I

Case II

Case III

Case IV

Figure 4. Results of the arm reversal experiments

For Case III, the dashed line at θ_s in Fig. 4 indicates the time at which the arm reaches the desired angle and starts the reversal and energy dissipation processes, and the dashed line at θ_c in Fig. 4 indicates the time at which the arm reaches the angle where the dissipation zone ends and the electric field was reintroduced at 3 kV/mm. The maximum torque at the time of the reversal is comparatively small at 2.51 N-m, which is less than the torque measured in Case II. This may be because the torque is reduced by having set the imposition of the electric field on the ER clutch to zero kV/mm for some time during the process. When the end angle of the dissipation zone is reached and the electric field is re-imposed with a value of 3 kV/mm, the torque becomes larger, but then decreases. The time over which all the operation is completed is nearly the same as in Case II.

For case IV, the dashed line at θ_s in Fig. 4 indicates the time at which the arm reaches the desired angle, and the dashed line at θ_c in Fig. 4 indicates the time at which the arm reaches the angle at the end of the dissipation zone, which is where the reversal starts and where the electric field is re-imposed with a value of 3 kV/mm. When the dissipation start angle is reached, the torque increases because the motor is used for braking. In addition, it turns out that the arm cannot be stopped in the dissipation zone; thus, after passing through the dissipation zone, the reversal process is then started. The maximum torque at the time of an inversion is 2.12 N-m, and it reaches a value similar to that reached during the torque that occurs at the time the drive starts. This may be because the braking action of the motor is applied to the input shaft at the time of energy dissipation, so it was not different from the case where the arm reverses after it is already stopped. Note that the torque occurring at the time of reversal for Case IV is less than the torque occurring at the time of reversal for Cases II and III. And compared to Case III, the same time elapses in Case IV until the operation is complete.

The above results show that the load on the motor at the time of the reversal was reduced by carrying out energy dissipation using the ER clutch. Moreover, in Case III, the reversal can be done quickly, by reducing the torque to the motor, and the results show that this is a very effective strategy for performing the reversal in a short time. Moreover, although the time taken to execute Case IV completely was relatively long, the results show that the load torque can be reduced the most in this case. For the future, when a quick arm reversal is given priority, it appears that Case III should be used; however, when reduction of the load torque to the motor is given priority, it appears that Case IV should be used.

4. Conclusion

In this research, the energy dissipation effect generated by controlling the electric field of an ER clutch was confirmed by several experiments. The results showed that a large load torque is generated at the time of a high-speed reversal, but the torque could be reduced by performing energy dissipation.

In future work, we will conduct a collision experiment to further investigate the effect of energy dissipation in robotic arm control.

References

1. K. Boku, Y. Kusaka, Y. Akamatsu, T. Nakamura, Development And the Control Of Three Degrees Of Freedom Soft Manipulators Which Considered A Collision With The Environment That Was Able To Include A Human Being, Proceedings of Robotics Symposia, Vol. 14, pp. 343-348 (March 2009)
2. K. Boku, T. Nakamura, "Development of 3 DOF Soft Manipulator with ER Fluid Clutches," *Journal of Intelligent Material Systems and Structures,* Sage Publication (in press)
3. T. Nakamura, Y. Akamatsu, Y. Kusaka, Development of Soft Manipulator with Variable Rheological Joints and Pneumatic Sensor for Collision with Environment, Journal of Robotics and Mechatronics, Vol. 20, No. 4, pp. 634-640, (2008)
4. K. Boku, T. Nakamura, Development of 3 DOF Soft Manipulator with ER Fluid Clutches, Proceedings of 11th International Conference on Electrorheological Fluids and Magnetorheological Suspensions (ERMR 2008)
5. Winslow W.M., J. Applied Physics, Vol. 20, pp.1137-1140, (1949)

PUMP USING NEMATIC LIQUID CRYSTALLINE FLOW UNDER APPLICATION OF ELECTRIC FIELD

TETSUHIRO TSUKIJI

Department of Engineering and Applied Sciences, Sophia University
7-1 Kioi-cho, Chiyoda-ku, Tokyo 102-8554, Japan

RYO FUCHIMOTO

Graduate School of Science and Technology, Sophia University
7-1 Kioi-cho, Chiyoda-ku, Tokyo 102-8554, Japan

This paper is concerning to a small pump for liquid crystal, which is called liquid crystal pump in the present paper. The mechanism of the induced flow of liquid crystal by application of electric fields on the liquid crystal is used to generate the rotational flow in the present pump. Liquid crystal pump has a channel to change rotational flow into one way flow from inlet to outlet of the pump. Rotational electric fields are used in the present study. The rotational electric fields are generated by circle-electrode structure fabricated on planar surface, which are applied three-phase alternating currents. Using integrated electrode plate, connection of pumps with plate type is easier than cylindrical one. The pressure-flow rate characteristics of the pump were measured. In addition the relation between non-dimensional flow rate and pressure of the pump were obtained. In order to obtain a high pressure and high flow rate, various shapes of electrode and channel were investigated and the measured results were compared each other. Besides connecting pumps with integrated circle electrode plate were designed to get the required pressure and flow rate.

1. Introduction

Research has actively been undertaken to provide a better understanding of pump dynamics. Diaphragm- and micro-syringe pumps are typical examples of a mechanical micro-pump and the other mechanical micro-pumps are developed recently. The advantage of these pumps over conventional ones is that they can be used to pump any liquid or gas. However, they are difficult to micro-fabricate and assemble because they contain many parts, so research is currently being conducted to develop low-noise pumps that use functional fluids and have simplified designs with no sliding parts. Such a micro-pump would have various applications. For example, it could be used in power sources of equipment that supply liquid, cooling systems, micro machines, and supplying fuel to the ultra-micro gas turbine. For this wide range of applications, various micro-pumps are needed to enable its use in any environment and under any conditions. Therefore, various micro-pumps are being designed and are actively being studied. The system with fluid control type would be used widely in micro fields because of the decreased number of parts and the sliding part using the fluid drive by the characteristic of the functional fluid. The properties of functional fluids can be controlled by electric or magnetic fields. Typical fluid pumps use some flow mechanisms, including ion drag [1], electro hydro dynamics (EHD)[2,3], electro-conjugate fluid (ECF) jets [4,5] and electroosmotic flow[6].

In our previous researches, when three-phase alternating current is applied to the cylinder electrode in order to apply the voltage on the liquid crystal, one of authors found rotational flow of liquid crystal in cylinder electrode. And they reported the cylindrical pump that consists of a spiral flow channel wrapped inside cylinder electrode [7]. As a result, the liquid crystal flows in the axial direction of the pump. On the other hands, when three-phase alternating currents apply the planar electrodes which are placed at the bottom of the cylinder, they found rotational flow of liquid crystal on the bottom electrodes. And one of authors suggested the motors using rotational flow of liquid crystal [8].

In this study, we developed plate type pump with planar electrodes and flow channel which change rotational flow into one way flow. This plate type pump is shorter in the axial direction than the previous cylindrical type one. We designed the plate type pump and measured the pressure and the flow rate. Especially the relation between various shapes of electrode and characteristics of the pump is reported in this paper. Beside connection of this plate type pumps is easier than connection of the cylindrical one, because this is able to use integrated electrode plate. So we designed the pump with integrated electrode plate to get high pressure and high flow rate.

The structure of liquid crystal pump is simple and the size can be decreased by further researches. So this pump has the advantage of small source of actuator by changing electric power into fluid power. In addition our pump has a good possibility of the cooling system's source because this pump makes no noise and no mechanical vibration.

2. Properties of Liquid Crystal and Electrodes

The liquid crystal which is used in our study is MLC6650 supplied from MERCK Co. Ltd. This is liquid crystal mixtures mixed with some nematic liquid crystals. The operating temperature is from -44 ℃ to 90 ℃, the kinematic viscosity for 25℃ is 59.7mm^2/s, the dielectric constant of the vertical direction to longitudinal of the liquid crystal molecule ε_{per} for 20℃ is 9×10^{-11}F/m, the dielectric constant of the parallel direction to longitudinal direction of the liquid crystal molecule ε_{para} for 20℃ is 55.6×10^{-11}F/m and the density for 25℃ is 1099kg/m^3.

Electrodes and channels which are used in our research are shown in Fig.1. This electrode is called 'Circle-electrode'. The diameter of the electrode is 6mm and width of it is 0.5mm, and the inter electrode distance is 0.2mm. The channel is along by electrodes, and the width of the flow channel is 1mm.

3. Experiment of Pumps

The principle of this pump is shown in Fig.2. Three-phase alternating currents were used as rotational electric fields to generate the rotational flow of liquid crystal , so liquid crystal flows in the way circumferential direction along the channel .The center hall of this pump is inlet ,and outside hall is outlet because the rotation direction of the field is anticlockwise direction from top view. The

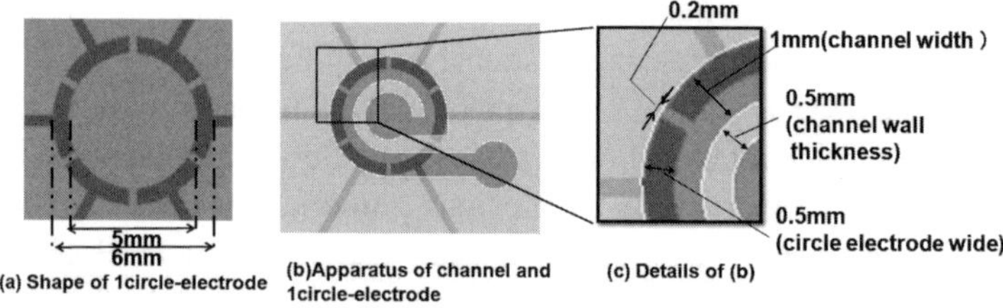

Fig.1 Electrode and channel of pump

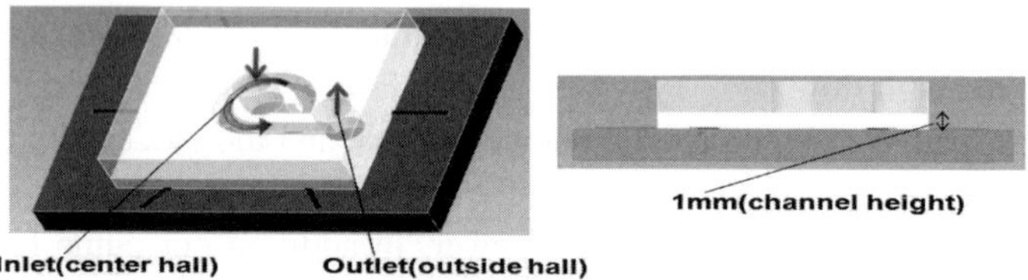

Fig.2 Flow in pump

experimental apparatus for investigating the characteristics of pump is shown in Fig.3. In this experiment, a 50-Hz, 200-V three-phase alternating current is input to a voltage transformer and the voltage means the effect value. An output voltage between 0 and 240 V is generated using this voltage transformer (A). Next, the amplitude of the output voltage is amplified about 15 times using the transformer (B), and the resulting voltage of three-phase alternating current (named 'R', 'S' and 'T') are applied to the electrodes. So the rotating electric field is generated. The resulting voltage is applied to the electrodes of the tested pump(C).

In this experiment, voltages between 0 and 1400 V were applied to the electrodes. The flow rate was measured when the total head, h, was changed. To measure the flow rate, we first took a video of the movement of the free surface of the liquid crystal in a pipe whose inner diameter is 4 mm. The video was taken from the direction vertical to the plate. We then enlarged the images in the video to see the flow clearly. The maximum pressure, p_m, is defined without the flow. The maximum pressure p_m is calculated by $p_m = \rho g h$, where ρ is the density, g is the acceleration due to gravity (9.81 m/s^2), and h is the total head.

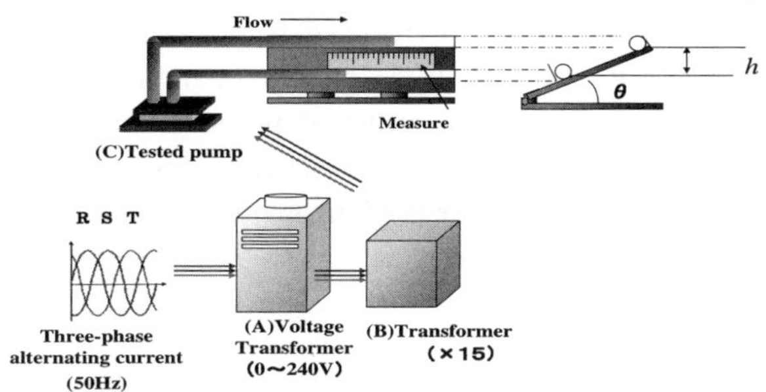

Fig.3 Experimental apparatus for pump characteristics

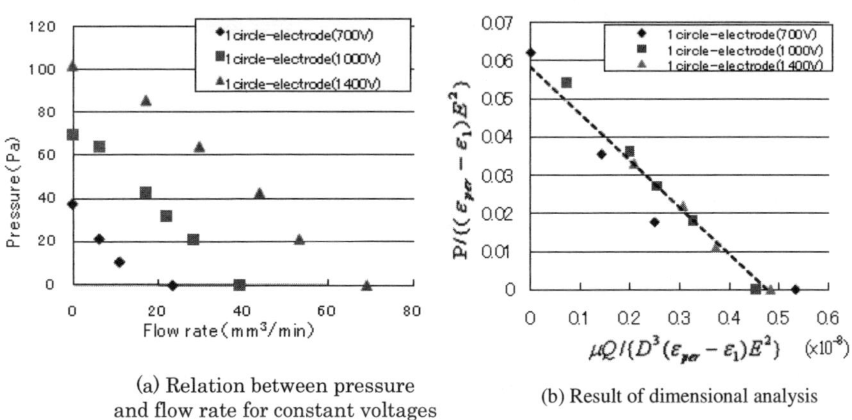

(a) Relation between pressure and flow rate for constant voltages

(b) Result of dimensional analysis

Fig.4 Experimental results

4. Experimental Results

The relation between the flow rate and the pressure are shown in Fig.4(a) for the pump with 1circle-electrode. The pressure is calculated by $p = \rho g H$. The flow rate increased with the voltage when the pressure was the same. Furthermore, the flow rate was a maximum when the pressure was 0, and the pressure dropped off nearly linearly with flow rate with constant voltage.

The results of the non-dimensional analysis of the pump properties are shown in Fig.4(b). The relations between $\mu Q / \{D^3(\varepsilon_{per} - \varepsilon_1)E^2\}$ and $p / \{(\varepsilon_{per} - \varepsilon_1)E^2\}$ derived by the π theorem almost become the same line, where μ is the value of viscosity, Q is the value of flow rate, D is the value of channel length , ε_{per} is dielectric constant of the vertical direction to longitudinal of the liquid crystal molecule, ε_1 is dielectric constant of insulating part (Glass epoxy resin) in the electrode, relative permittivity of 4.7 was used, E is electric field intensity, P is the value of pressure. The pressure-flow rate characteristics of the pump can be guessed from this result when the liquid crystals with different properties are used in the present pump.

In order to increase pressure and flow rate, connecting pumps with integrated circle-electrode plate were shown in Fig.5. Three-phase alternating

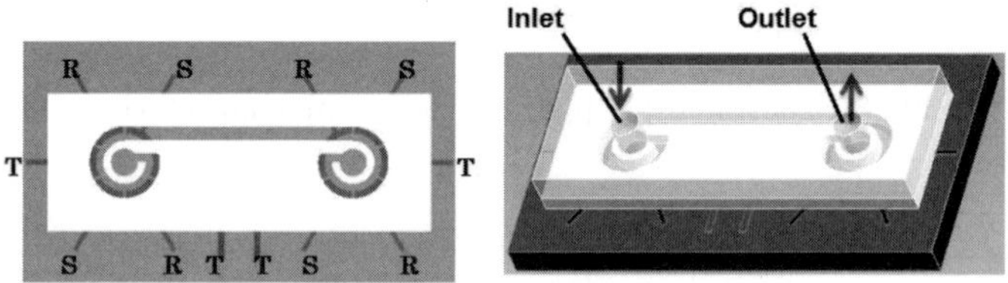

Fig.5 Connected pumps

current applied on the two circle-electrodes respectively as rotations of electric fields are clockwise direction from the top view. So rotations of liquid crystal are same and the liquid crystal flow one way along this channel. The experimental results for this pump is shown in Fig.6 (a) and (b). Using the pump with integrated two circle electrodes, the maximum pressure and pressure-flow rate characteristics are almost twice that of a single pump.

Furthermore, 2 and 3circle-electrodes are designed in order to get more output power. These electrodes are shown in Fig.7, and channel's scroll number fit the electrode. We investigated the relation between the electrode's circle number and pressure- flow rate characteristics of the pump.

Characteristics of the pumps with 1,2,3circle-electrode are shown in Fig.8(a) and (b). As circle number is large, the maximum pressure is increased. But circle number is large, the flow rate is decreased.

5. Conclusions

In the present study, rotation electric fields were applied to the circle-electrodes to add the voltage on the liquid crystal (MLC6650), and flow was induced by the property of the liquid crystal. A pump with the circle-electrode and channel which change rotational flow into one way flow was designed, and the pressure-flow rate characteristics of the pump were measured. We made the following findings.

1. The pressure dropped off nearly linearly with flow rate with constant voltage.

2. For the dimensionless parameters calculated using π theorem, all relations between $\mu Q / \left\{ D^3 \left(\varepsilon_{per} - \varepsilon_1 \right) E^2 \right\}$ and $p / \left\{ \left(\varepsilon_{per} - \varepsilon_1 \right) E^2 \right\}$ almost become the same line.

3. The connecting pumps are able to increase both pressure and flow rate.

4. As a number of circle-electrode is large, the maximum pressure is increased, but the flow rate is decreased.

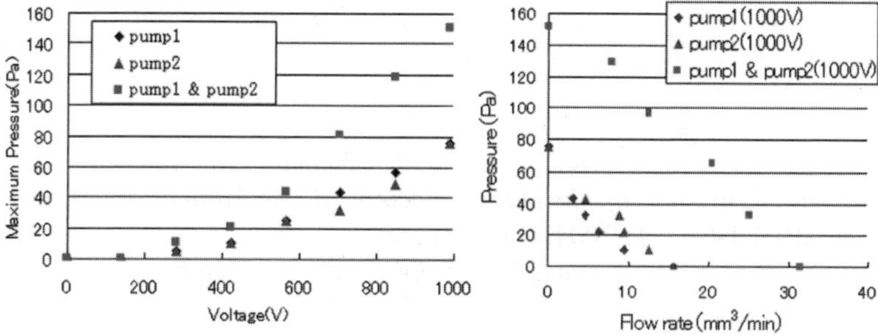

(a) Relationship between voltage and maximum pressure (single and 2 connecting pumps)

(b) Relationship between the flow rate and pressure (single and 2 connecting pumps)

Fig.6 Experimental results for connected pumps

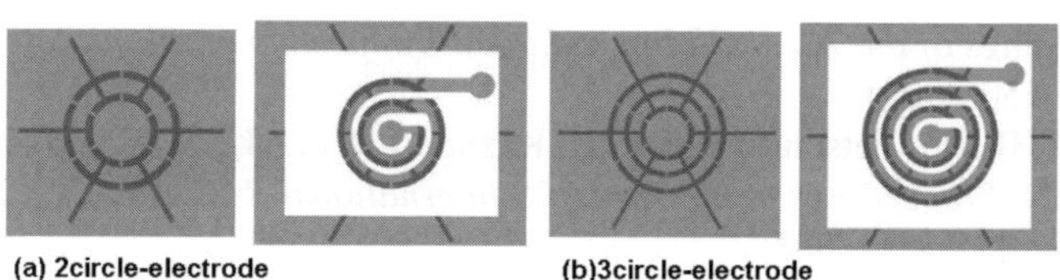

(a) 2circle-electrode   (b)3circle-electrode

Fig.7 Two and three circle-electrodes

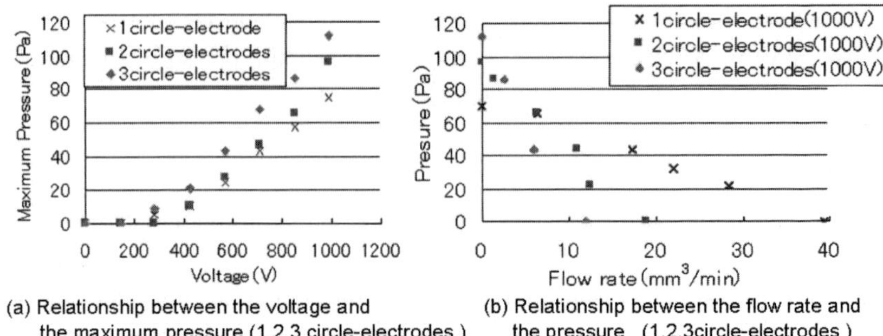

(a) Relationship between the voltage and
the maximum pressure (1,2,3 circle-electrodes)

(b) Relationship between the flow rate and
the pressure   (1,2,3circle-electrodes)

Fig.8 Experimental results for the pumps with 1,2 and 3 circle-electrodes

Acknowledgments

We thank Mr. Sumio Syuuto for his assistance in designing our experimental rig, and MERCK Co., Ltd. for supplying us with the liquid crystals.

References

1. Yamada, H., Hakama, S., Miyashita, T. and Zhang, N., *Proc. Instn Mecha. Engrs., 216 Part C, J.Mechanical Engineering Science*, pp.325-335(2006).
2. Feng, Y. and Seyed-Yagoobi, J., *IEEE Transactions on Industry Applications*, 42-2, No.2, pp.369-377(2006).
3. Kano, I., Mizuochi, K. and Takahashi, I., *Proceedings of the Sixth International Symposium on Fluid Power Tsukuba,* pp.575-579(2005).
4. Yokota, S., Seo, W., Yoshida, K. and Edamura, K., *Robotics and Mechatronics Division of the Japan Society of Mechanical Engineers*, No.04-4, 2P1-L1-61(1-4)(2004) (in Japanese).
5. Sakurai, Y., Kadoi, H., Nagata, T. and Edamura, K., *Transactions of the Japan Society of Mechanical Engineers*, C, 72-715, pp.991-996(2006) (in Japanese).
6. Okazaki, T.,Miyazaki, K. and Tsukamoto, H., *The preprint of the Mechanical Engineering Congress in Japan(MECJ-06)*, No.06-1, No.2, pp.291-294(2006) (in Japanese).
7. Tetsuhiro TSUKIJI and Hiroki SATO,
Proceedings of the 7th JFPS International Symposium on Fluid Power, pp.545-550(2009).
8. Liang CHENG, Tetsuhiro TSUKIJI, Kazunori HAYAKAWA, and Xiao don g RUAN, *Proceedings of the 7th JFPS International Symposium on Fluid P ower*, pp.833-836(2008).

BASIC PERFORMANCE ANALYSIS OF ER GEL IN VACUUM

Y. SHIDA, Y. KAKINUMA and T. AOYAMA

Department of System Design Engineering, Keio University 3-14-1 Hiyoshi, Kouhoku-ku, Yokohama 223-8522, Japan

H. MINAMI

ULVAC, Inc., 2500 Hagisono, Chigasakishi, Kanagawa 253-8543, Japan

H. ANZAI

Fujikura Kasei Co., Ltd, 2-6-15 Shibakouen, Minato-ku, Tokyo 105-0011, Japan

Electro-rheological gel (ER gel) exhibits various adhesive characteristics according to the applied electric field. This characteristic is called the electro-adhesive effect (EA effect). The results of a recent study reveal that the EA effect also occurs in a vacuum. Therefore, it is expected that ER gel can be applied to a chucking or damping device in a vacuum process. However, the characteristics of ER gel in a vacuum have not been sufficiently clarified. The purpose of this study is to experimentally analyze the characteristic of ER gel in a vacuum. The performance of an ERG fixture element used for a silicon wafer in a vacuum is evaluated through shearing tests. The experimental results indicate that the ER gel demonstrates satisfactory performance in a vacuum. Moreover, the outgas released from the ER gel in a vacuum is evaluated, and the influence of the degassing time on the outgas is investigated.

1. Introduction

Electro-rheological gel (ER gel) is a functional elastomer whose adhesion characteristics change on being subjected to an electric field. Its surface adhesion can be changed quickly and reversibly by applying an electric field. It is a composite material consisting of ER particles and silicone gel. ER particles are dispersed uniformly in the gel and suspended elastically. ER gel is placed between plane-parallel electrodes. When an electric field is applied to it, the phenomenon of adhesion between silicone gel and electrode occurs in regions surroundings the ER particles. This phenomenon is called the electro-adhesive effect (EA effect).

The results of a recent study reveal that the EA effect also occurs in a vacuum [1]. Therefore, it is expected that ER gel can be applied to a chucking or damping device [2, 3] in a vacuum process. However, the characteristics of ER gel in a vacuum have not been sufficiently clarified. The purpose of this study is to experimentally analyze the characteristics of ER gel in a vacuum. The performance of the ER gel in a vacuum is evaluated through shearing tests. The outgassing characteristics of the ER gel in a vacuum environment are also examined. The effect of the degassing time on the outgas released from the ER gel is simultaneously investigated.

2. Sample of ER gel for Vacuum Test

2.1. *Fabrication procedure*

ER gel is produced by the hydrosilylation reaction, which is caused by mixing a hydrogen silicone, unsaturated compounds, and ER fluid. The ER fluid consists of silicone oil and ER particles. These particles comprise an organic core and an inorganic shell layer. The average diameter of ER particles is 15 μm. In order to promote the abovementioned gelation reaction, the reactants are heat-treated at 373 K in the presence of a platinum catalyst. After gelation, excess oil is removed from the surface of ER gel. Moreover, to suppress outgassing from ER gel in a vacuum, vacuum degassing is carried out over half an hour under vacuum and high-temperature condition (393 K). The ER gel produced by these processes is of a standard type.

3. ERG Fixture Element

Considering the practical use in a vacuum process, an ERG fixture element with one-sided electrodes is trial-manufactured as shown in Figure 1. Basic performance of ER gel in a vacuum is investigated utilizing this fixture element. In order to produce the EA effect, it is necessary to apply an electric field to ER gel. For applying an electric field, ER gel is sandwiched between plane-parallel electrodes, called both-sided electrodes. If ER gel is used as a fixture element for semiconducting and insulating materials, this type of electrode is not available because the object must be conductive. On the other hand, one-sided electrodes [4] can be used to realize the EA effect for nonconductive materials, whose anode and cathode are arranged alternately on an insulating base plate. When voltage is applied to the one-sided electrodes, arch–like electric lines of force emerge between the anode and cathode, instead of straight lines. The EA effect is produced by the arched electric lines passing through the surface of the ER gel [5].

In this study, a silicon wafer is used as the object. The base of the one-sided electrodes is made of glass epoxy resin, which has large thermal resistance and

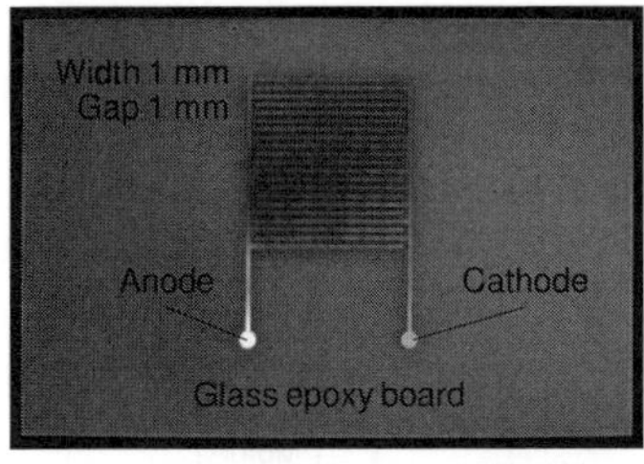

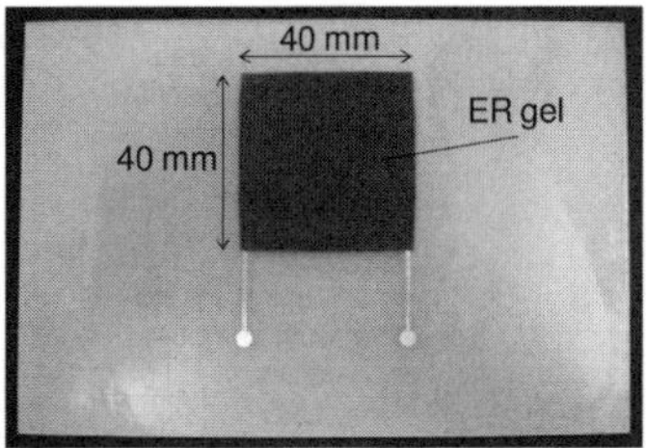

(a) One-sided electrodes (b) ERG fixture element

Figure 1. ERG fixture element

high bending stiffness. The electrode width and gap are 1 mm, and the thickness of the ERG sheet is 0.3 mm.

4. Experimental Setup and Procedure

A system has been developed for evaluating the vacuum characteristics of ER gel. This system can measure the performance of an ERG fixture element in a vacuum via shearing tests in the vacuum chamber (Fig.2). A schematic diagram of the system is shown in Figure 3. A silicon wafer is placed on the ERG fixture and weighted by an aluminum block. This block is attached to the silicon wafer by a heat-resistant adhesive tape in order to carry out the shearing tests under high temperature. The aluminum block and the horizontal movable bar with a load cell are connected by wire. This bar is driven by a motor at a constant speed. When various voltages are applied to the ERG fixture, the maximum fixing force just before the wafer slides is measured by the load cell. A temperature controller equipped with a thermoelectric couple maintains the temperature of the heat stage in the vacuum chamber constant.

In this research, to analyze the characteristics of ER gel, shearing tests are conducted in a vacuum of 4.0×10^{-3} Pa. The temperature of the heat stage is varied from 296 K (room temperature) to 423 K.

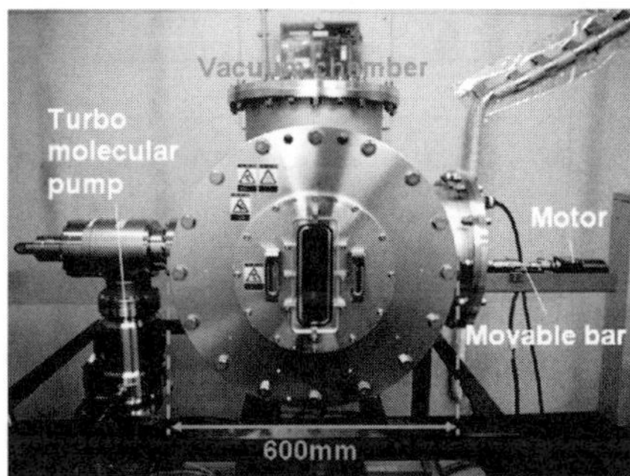

Figure 2. Vacuum chamber

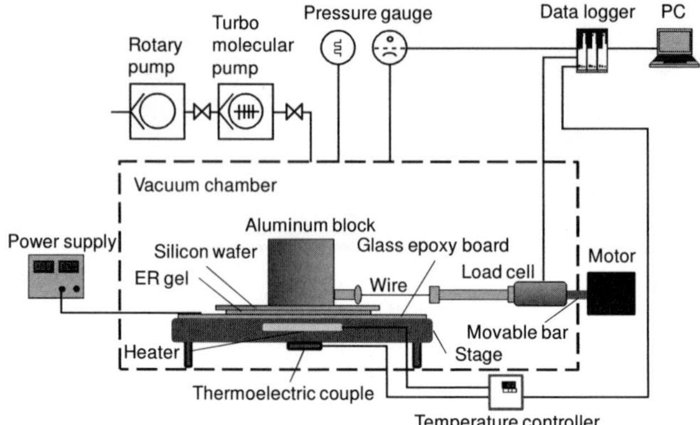

Figure 3. Experimental setup

4.1. *Evaluation of fixing force in a vacuum*

The fixing force generated by the EA effect at vacuum pressure is compared with the one generated at atmospheric pressure. The relation between the horizontal fixing force of the ER gel and the applied voltage is shown in Figure 4. With an increase in the applied voltage, the fixing force increases under both the conditions. This result shows that the EA effect emerges in a vacuum and that the ER gel exhibits satisfactory performance, though the absolute value of the generated fixing force at vacuum pressure is slightly lower than that at atmospheric pressure.

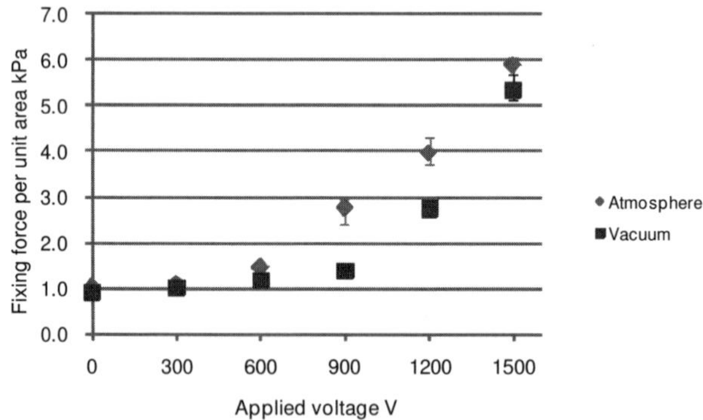

Figure 4. Fixing force in a vacuum

4.2. *Evaluation of fixing fore in a vacuum and high temperature*

In order to investigate the influence of the environmental temperature on the performance of ER gel in a vacuum, the shearing tests are carried out at a

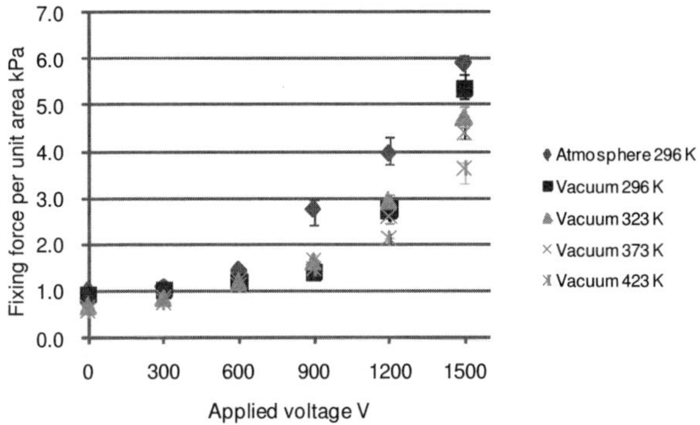

Figure 5. Fixing force in a vacuum and high temperature

temperature ranging from 296 K to 423 K. As shown in Figure 5, the EA effect can be produced in this range of temperature under vacuum conditions. This result indicates that the ER gel exhibits satisfactory performance with regard to fixing the silicon wafer. However, the fixing force tends to decline gradually with an increase in the temperature.

5. Experimental Analysis of Outgas

5.1. *Vacuum degassing of ER gel*

The outgas released from the ER gel in a vacuum is evaluated and the influence of degassing time on the outgas is investigated with a view to develop ER gel appropriate to use in a vacuum because the generation of undesirable gas causes a serious problem related to pollution in a vacuum environment. For analyzing of the outgas, three types of ER gels (20 mm × 20 mm × 0.3 mm) are fabricated under different degassing times (0.5 h, 2.0 h, and 4.0 h) at 10^{-1} Pa and 393 K.

The components of the outgas from these samples are analyzed by means of a quadrupole mass spectrometer (QMS), which can measure the gases present in vacuum environments. This analysis is carried out at a temperature ranging from 296 K to 393 K. Moreover, the relation between the contamination level in a vacuum chamber and the temperature of the ER gel samples is evaluated.

5.2. *Components of outgas*

The gases present in vacuum environments are investigated. For comparison of contamination, the analysis of outgas is also carried out with comparison sample of rubber, which can be available for general use in a vacuum. Figure 6 shows the analytical data of these samples at a temperature of 353 K. Though outgassing from the samples degassed for 0.5 and 2.0 h is not sufficiently suppressed, the sample degassed for 4.0 h does not generate a gas whose molecular mass exceeds 50. This fact indicates that the components having a

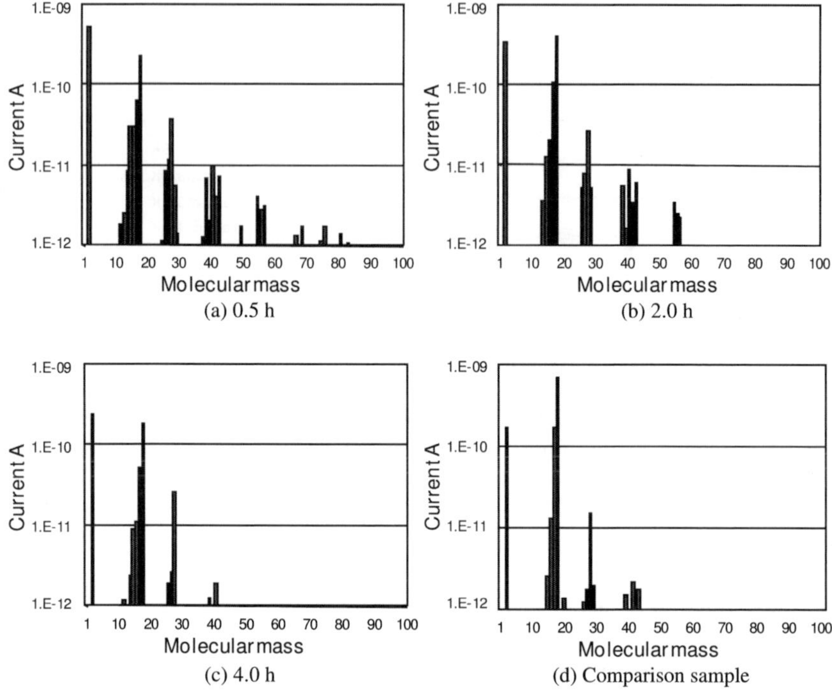

Figure 6. Evaluation of outgas quantity at 353K

high molecular mass, such as silicone, are not generated. It is confirmed that the necessary degassing time is more than 4.0 h.

5.3. *Analysis of contamination level*

The contamination level in a vacuum chamber is defined as the ratio of the total partial pressure of species with a high molecular mass to the total pressure. It is considered that the sum of the partial pressures of species having a high molecular mass is in proportion to the total ion current from species having the molecular mass in the range of 50-100. The formula for the contamination level is given by

Contamination level = (Sum of ion currents from species with a
molecular mass of 50-100) / (Total ion current)   (1)

The relation between the contamination level and the sample temperature is shown in Figure 7. This result indicates that vacuum degassing is effective in suppressing outgassing. From the view point of the contamination level, the sample degassed for 4.0 h is comparable to the rubber sample practically used

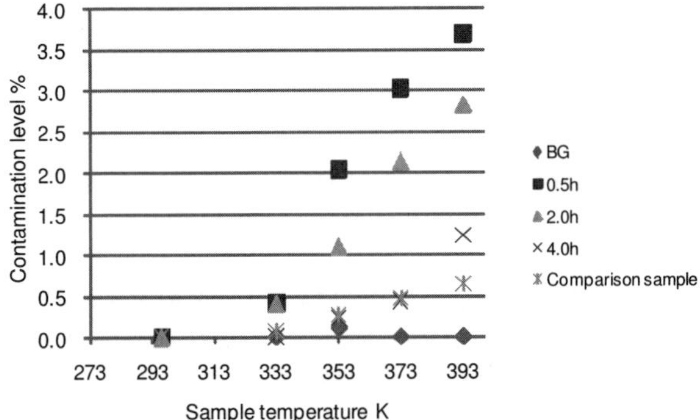

Figure7. Relation between sample temperature and contamination level

under vacuum conditions in a range of environmental temperature from 296 K to 373 K.

6. Conclusion

Characteristics of ER gel in a vacuum and at high temperature are analyzed. In the shearing test, the ER gel demonstrates an satisfactory performance with regard to fixing a silicon wafer in a vacuum and at high temperature, though the EA effect under this condition is slightly weaker than that at atmospheric pressure and room temperature.

Vacuum degassing is effective in suppression of outgas. In particular, a sample degassed for 4.0 h is comparable with a rubber sample in terms of the contamination level in the range of environmental temperature from 296 K to 373 K.

Acknowledgments

This study was supported by a Grant-in-Aid for Scientific Research (B) Japan Society for the Promotion of Science.

References

1. M. Tanaka et al., *J. Adv. Mech. Des. Syst. Manuf.*, **2**, 4, 762 (2008).
2. S. Kanno et al., *J. Vac. Sci. Technol. B.*, **21**, 6, 2371 (2003).
3. S. Kanno et al., *J. Vac. Sci. Technol. B.*, **24**, 1, 216 (2006).
4. N. Takesue et al., *J. Appl. Phys.*, **91**, 3, 1618 (2002).
5. S. Saito et al., *Int. J. Mod. Phys. B.*, **21**, 28 & 29, 4751 (2007).

DEVELOPMENT OF TACTILE DISPLAY DEVICE USING ERG MULTIPLE-DISK CLUTCH

Y. URAKAMI, T. AOYAMA and Y. KAKINUMA

Faculty of Science and Technology, Keio University, 3-14-1
Hiyoshi, Kouhokuku, Kawasakishi, Kanagawa, Japan

H. ANZAI

Fujikura-Kasei Co., Ltd., 2-6-15
Shibakouen, Minato-ku, Tokyo, Japan

Nowadays, due to the widespread use of rehabilitation and virtual reality technology, tactile display systems have attracted the attention of researchers from the medical and amusement fields. A clutch mechanism that employs ER and MR fluids is appropriate for force transmission in a tactile display device, because the ER and MR effects are passive and safe for the operator, as compared with an active actuator using motor. The developed ER gel (ERG) produces approximately twenty times higher shear forces than ER fluid and exhibit stable performance. In this research, an ERG multiple-disk clutch is fabricated and applied to a tactile display system. The experimental results of the control test indicate that the tactile display system with the proposed control method can exhibit the "sense of touch" of various viscous fluids.

1. Introduction

A tactile display system is an interface a between human and a machine, which can show various senses such as touch. Nowadays, due to the widespread use of rehabilitation and virtual reality technology, tactile display systems have attracted attention of researchers from the medical and amusement fields. A clutch mechanism that employs ER and MR fluids is appropriate for force transmission in a tactile display device [1] [2] because the ER and MR effects are passive functionalities. Therefore, its application device is safer for an operator, than the device using an active actuator such as a motor. However, since the devices using these fluids need a sealed mechanism, their structure becomes complicated and the scale of a device tends to enlarge. In a tactile display system, a compact clutch is required as its force transmission mechanism.

Electro-rheological gel (ERG) is a functional material that can change its surface adhesive property via the intensity of the applied electric field. This characteristic phenomenon is called the electro-adhesive effect (EA effect). Applying ERG to a clutch component has several advantages as follows. First, the clutch will be downsized, due to the simplification of its structure. Second, the force transmission performance is stable because the sedimentation of ER particles does not occur. Third, ERG produces twenty times higher shear stress than does ER fluid. From these facts, we apply ERG to a disk clutch mechanism not only for increasing the transmission torque, but also for downsizing the clutch. The main purpose of this study is to develop a tactile display system using an ERG multiple-disk clutch, and evaluate the validity of this system.

2. Outline of ERG

ERG is a composite material that consists of ER particles and silicone gel. In the initial state, the ER particles at the surface are protruded from the silicone gel. The principle of the EA effect is described in Figure 1. ERG is sandwiched between a sliding electrode and a fixed electrode. When the voltage is not applied to the ERG, the sliding electrode will slip on the ERG's surface. This is because the electrode is in contact with slippery ER particles. However, when an electric field is applied to the ERG, the ER particles settle out and the silicone gel rises around them. Due to the contact between the silicone gel and the sliding electrode, adhesive force is exerted. This adhesive force could be controlled by controlling the intensity of the voltage applied across the electrodes. Therefore, ERG is expected to serve as a material that can help create a novel clutch or a dumper mechanism [3].

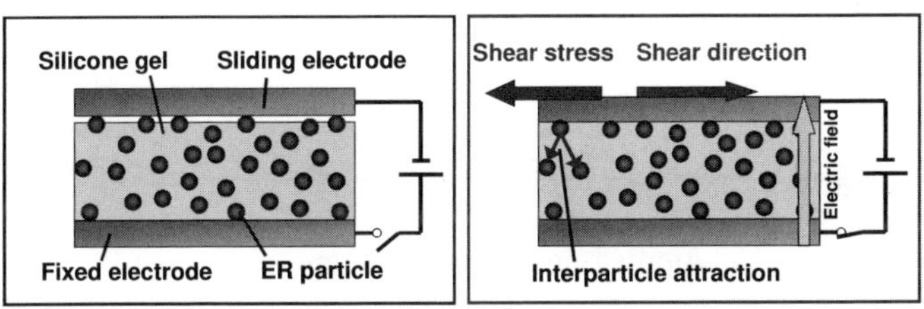

(a) No electric field   (b) Applying electric field

Figure 1. Principle of EA effect

3. Development of ERG Multiple-Disk Clutch

Figure 2 shows the structure and appearance of an ERG multiple-disk clutch. Its structure comprises of rotary and fixed disks. ERGs are pasted on both the sides of fixed disks. The rotary and fixed disks are respectively attached to the shaft and the outer case through key ways and employed as the anode and cathode. The voltage applied across the electrodes can be supplied from the outer case and shaft. The rotary and fixed disks are placed alternately, and the number of ERG layers can be changed from 1 to 6. When a voltage is applied, the ERG adheres to the rotary disks, and then, transmission torque is exerted.

Figure 3 shows the static characteristics of the ERG multiple-disk clutch. The shaft of the clutch is rotated by a driven motor, and its output torque is measured by a torque sensor. The transmission torque is found to increase with the intensity of the applied voltage and the number of ERG layers. Almost no torque is exerted when voltage is not applied.

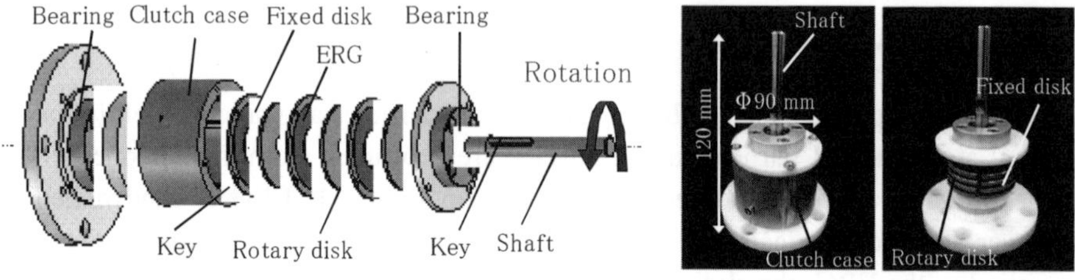

(a)Illustration of Clutch (b) Appearance of Clutch

Figure 2. Structure of ERG multiple-disk clutch

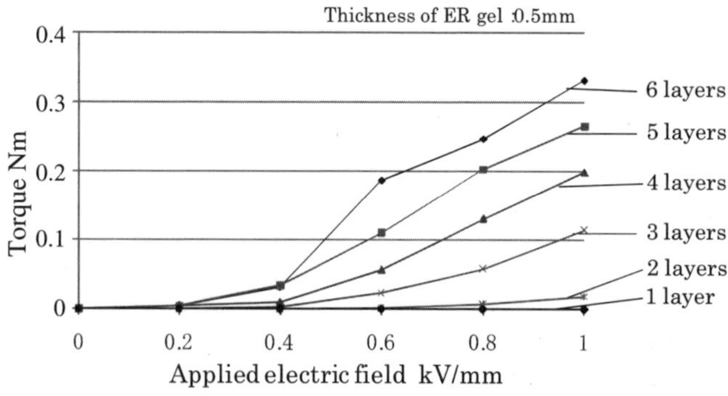

Figure 3. Relation between transmission torque and electric field

4. Dynamic Characteristics of ERG Multiple-Disk Clutch

The ERG multiple-disk clutch is considered as a single-degree-of-freedom rotational system given by Eq. (1).

$$J\ddot{\theta} + C_r\dot{\theta} + K_r\theta = T \tag{1}$$

where J [kg·m^2] denote the inertia moment of rotational parts if the ERG multiple-disk clutch, which is a constant value of $6.29*10^{-6}$ [kgm^2]; K_r [Nm/rad] represents the spring constant; C_r [Nm·s/rad] indicates the viscous damping coefficient; θ [rad] denotes the angle of rotation; and T [Nm] specifies the applied external torque. C_r and K_r can be changed by the applied voltage. In this study, the relation between C_r, K_r, and the voltage are investigated by forced vibration test.

A dynamic characteristic experiment is carried out with the ERG multiple-disk clutch. Figure 4 shows the experimental system. The rotational movement of the motor is converted to $\pm 6°$ shaking movement by a cam mechanism, and vibration with a frequency ranging from 1 to 5 Hz is transmitted to the clutch. Voltage is applied to clutch during the experiment, and transmission torque and angular of shaft are measured by torque sensor and rotary encoder, respectively. Substituting these data in Eq. (1) and solving the equation, C_r and K_r are obtained. Figure 5 shows the relationship of each coefficient to the applied voltage. It indicates that both of the coefficients depend on the applied voltage and angular velocity. C_r and K_r are expressed as a function of the angular velocity and applied voltage and are given by Eq. (2) and (3), respectively. These equations are derived from the gradient of the approximated straight-line plots shown in Figure 5. The magnitude of voltage necessary to output a target torque can be calculated from these equations.

$$C_r = 0.2747 e^{-0.2861 \left(\frac{\omega}{2\pi}\right)} \times V \ \text{[Nm·s/rad]} \tag{2}$$

$$K_r = \left(0.6063 \times \left(\frac{\omega}{2\pi}\right) + 1.335\right) \times V \ \text{[Nm/rad]} \tag{3}$$

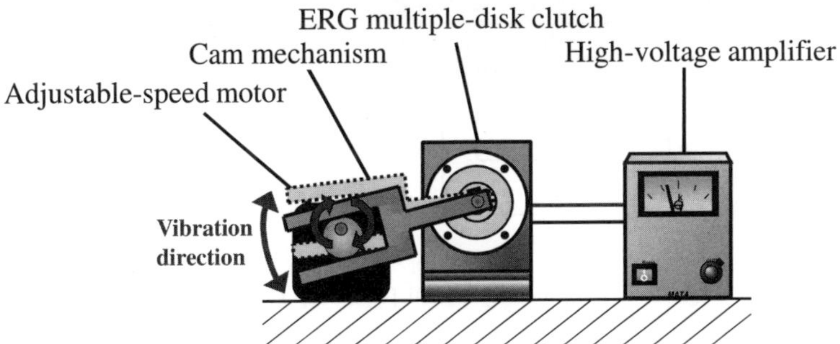

Figure 4. Equipment for measuring the dynamic characteristic of clutch

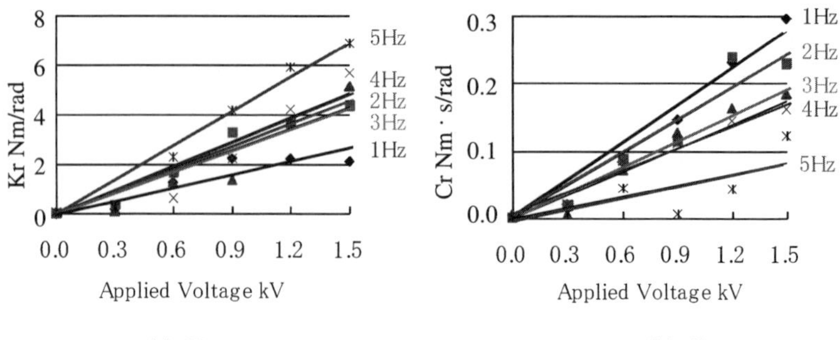

(a) K_r (b) C_r

Figure 5. Variation in K_r and C_r with applied voltage

5. Torque Control System using Open-Loop Control

The transmission torque of a clutch must be controlled freely when it is applied to a tactile device. By using the equation of a dynamical model, the transmission torque can be estimated from the applied voltage and angular velocity. Control The control system developed in this research is designed to change the output torque as a function of the angular velocity $\dot{\theta}$. Figure 6 shows a block diagram of the open-loop control system. Here, the "inverse model" represents the relational expression derived from the formula of the dynamical model. By substituting the displacement angle θ and the torque reference T^{ref} for the inverse model, the voltage command necessary to output a target torque is calculated.

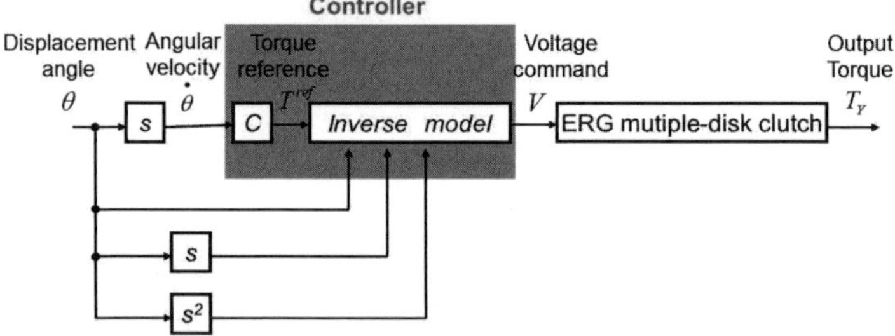

Figure 6. Block diagram of pen-loop control system

When the angular velocity $\dot{\theta}$ is input to the controller, the torque reference T^{ref} is obtained by multiplying the angular velocity $\dot{\theta}$ with the conversion coefficient C (Eq. (4)).

$$T^{ref} = C\dot{\theta} \tag{4}$$

In addition, the displacement angle θ, angular velocity $\dot{\theta}$, and angular acceleration $\ddot{\theta}$ are input to the inverse model and the voltage command is determined. Eventually, the calculated voltage is actually applied to the ERG multiple-disk clutch, and output torque T_o is generated. In this study, we examined and compared two output patterns generated using two values of C (0.02 and 0.05). Figure 7 shows the equipment of torque control system. A handle is set on the shaft of the clutch and can be moved manually. The displacement angle is measured by means of a rotary encoder installed on one side of the clutch. A real-time OS is used as the controller, and the sampling time is set as 1 ms. The calculated voltage is output through a D/A board and is applied to the clutch after it is amplified. In order to confirm the functionality of this system, a torque sensor is set on the clutch to measure the output torque T_o. The torque reference T^{ref} and the output torque T_o are compared.

The amount of controlled torque, when the operator moves the handle freely, is measured. Figure 8 presents the results of this control test. Figure 8(a) shows the results obtained when C is 0.02. The output torque T_o is found to be almost equal to the torque reference T^{ref}. This fact indicates that the force increases in proportion to the angular velocity. On the other hand, Figure 8 (b) shows the results obtained when C is 0.05. This figure indicates that the output torque T_o is higher than that in (a) and indicates that the operator feels the "sense of touch" while pushing through a more viscous fluid than that in the case of (a). These facts indicate that the proposed open-loop control system has sufficient functionality to express the "sense of touch" of various viscous fluids, such as water and oil. However, the figure also reveals that the output torque T_o does not correspond to the command in a low-angular-velocity range. The main cause for this problem is that the equation of the dynamical model is not exact. Hence, the model must be made precise for future research.

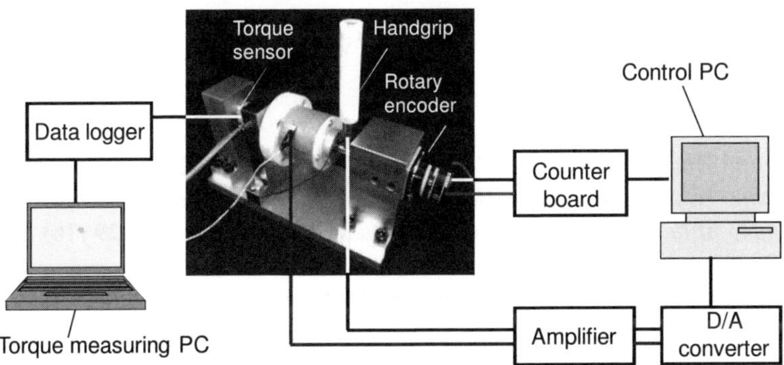

Figure 7. Equipment of torque control system

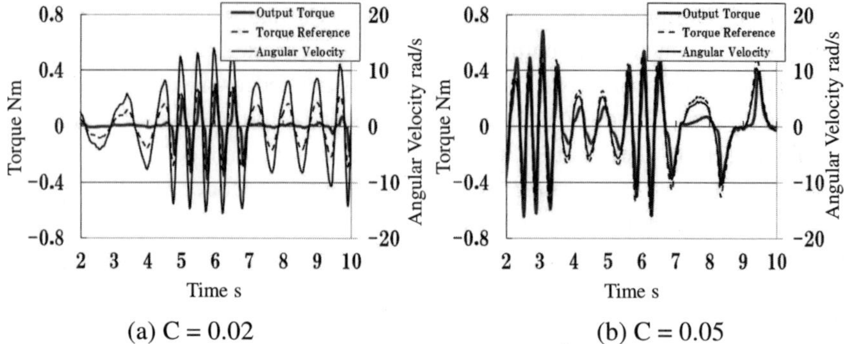

(a) C = 0.02 (b) C = 0.05

Figure 8. Relation between torque reference T^{ref} and output torque T_o

6. Conclusion

In this research, ERG is applied to a multiple-disk clutch. The basic characteristics of this clutch are measured and a model formula is derived. Moreover, an open-loop control test is carried out by using the model formula. Followings are the results obtained in this research.

1. A multiple-disk clutch that employs ERG is developed. The amount of transmission torque increases with the intensity of the applied voltage and the number of ERG layers.
2. The clutch is considered as a single-degree-of-freedom rotational system. From the dynamical characteristic experiment, the relations of the angular velocity and the applied voltage with C_r and K_r are obtained.
3. On the basis of the model formula, an open-loop torque control system is designed. By using this system, the transmission torque is controlled by controlling the angular velocity without using a torque sensor. The operator can feel the "sense of touch" of various viscous fluids.

References

1   Jyunji Hurushou et al, *J. Robotics Soc. Jap* **25(6)**, 867(2007)
2   Tomo Kikuchi et al, *Trans. of Inst. Electrical Engineers of Japan. C, A publication of Electronics, Information and System Society* **129 (10)** 1859 (2009)
3   Yasuhiro Kakinuma, *Trans. Jap. Soc. Mech. Engrs. C* **75(755)**, 2084(2009)

STABILIZING OUTPUT OF ER GEL LINEAR ACTUATOR

K. KOYANAGI

Department of Intelligent Systems Design Engineering, Faculty of Engineering,
Toyama Prefectural University,
Kurokawa 5180, Imizu, Toyama, 939-0398, Japan
E-mail: koyanagi@pu-toyama.ac.jp

Y. KAKINUMA

Faculty of Science and Technology, Keio University,
3-14-1 Hiyoshi, Kohoku, Yokohama, Japan
E-mail: kakinuma@sd.keio.ac.jp

H. ANZAI* and K. SAKURAI

Fujikura Kasei Co., Ltd.,
13-1,Sakurada 5-chome Washimiya-machi Kitakatsushika-Gun, Saitama, Japan
**E-mail: h-anzai@fkkasei.co.jp*

T. YAMAGUCHI and T. OSHIMA[†]

Department of Intelligent Systems Design Engineering, Faculty of Engineering,
Toyama Prefectural University,
Kurokawa 5180, Imizu, Toyama, 939-0398, Japan
[†]E-mail: oshima@pu-toyama.ac.jp

Electro-rheological (ER) gel is a relatively new type of functional material that has been developed. ER gel has a shear stress characteristic on the surface, which increases according to the applied electric field. Introducing the gel into the interface elements of a force transmission system facilitates the realization of functional clutches and brakes. This study aims to develop a linear actuator which is safe, can operate in reverse and can generate a large force using such a functional clutch. Applying an ER gel clutch in a force transmission system results in a decrease of inertia and the mechanical limits of the maximum speed not obstructing high controllability. This contributes to high safety. This paper reports, in particular, on stabilizing the output of the linear actuator.

Keywords: Electro-rheological Gel; Functional Materials; Linear Actuator; Safety; Human-coexistence Robot; Stabilizing Output.

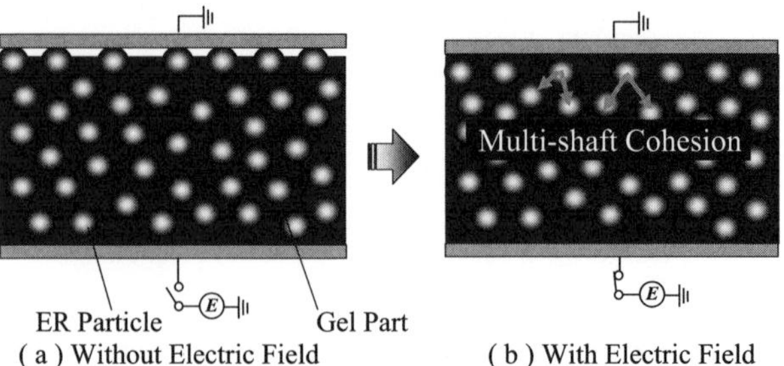

Fig. 1.   Mechanism of ER gel effect.

1. Introduction

Human-coexistence robots manipulated directly by human such as power-assist systems or other welfare application has recently been actively researched. Actuators used in such robots require high safety, backdriveability, large generative force and so on. However, few conventional linear actuators combine such features. We have focused on the safety and the backdriveability that is an ability to operate in reverse, and have created a novel linear actuator applying an ER (Electro-Rheological) gel, which we call the ER Gel Linear Actuator (ERGLA).

The prototype ERGLA, however, had some issues with stability and the magnitude of the output.[1,2] This paper presents the construction and features of the ERGLA and presents modifications of the ER gel clutches used in the ERGLA to stabilize its output force of the ERGLA.

2. ER Gel

ER gels are a functional gel; polymer particles about $20\,\mu$m in diameter are dispersed in an insulating oil, and then a gelling agent is added to that. When an electric field is applied to electrodes sandwiching the ER gel, dielectric polarized particles and the gel component increase the shear stress of the surface of the gel several tenfold as shown in Fig. 1. ER gels which have large ER effects have recently been reported.[3]

For example, an ER gel can be applied to a rotary clutch surface as shown in Fig. 2. While the ER gel is de-energized, the clutch input torque is not transferred because of minimal shear stress. When the ER gel is energized by an electric field, increased shear stress contributes torque transfer, that is, the clutch is en-gaged. The magnitude of the transferred torque can be controlled by the magnitude of the electric field. Devices with ER gels have such advantages against conventional metal friction brakes or powder

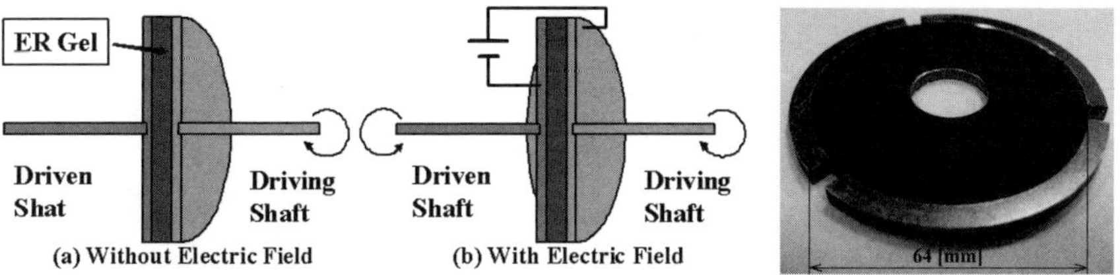

Fig. 2.   Image of an ER gel clutch.   Fig. 3.   An ER gel.

clutches etc. as no burning out, good controllability of the torque, not requiring sealing. Comparing with ER/MR fluid based devices, unnecessary of sealing is big advantage.

Figure 3 shows the ER gel used in this study. The gel is based on semiconducting tin oxide coated polyacrylic particles dispersed in dimethyl silicone oil of 100 cSt as base fluid. The thickness of the gel is about 0.5 mm. The maximum magnitude of the electric field for the gel is under 3.0 kV/mm.

3. The ER Gel Linear Actuator

3.1. *Scheme of ER Gel Linear Actuator[1]*

The ERGLA we aim to develop is shown in Fig. 4 which presents the structure of the center part. The ERGLA unit is attached to a fixed bar and moves on this fixed bar. The main parts of the ERGLA unit are two ER gel drums (ERGDs) including ER gels, a timing belt, guide rollers and a motor. The timing belt is tensioned with the ERGDs. The fixed bar is pinched by the timing belt and guide rollers.

The inside part of the ERGD can be said the rotary type disk clutch. The ERGD consists of the output disc engaging with the outside of the drum, the input disk connecting to the input shaft, and the ER gel as shown in Fig. 5. A sheet of ER gel is pressed on the output disk and sandwiched between the disks. The input shaft is rotated by the motor at a constant speed. Because the shear stress of the ER gel changes according to magnitude of the applied electric field, an arbitrary torque can be transmitted to the output of the ERGD. The advantage of this structure is that it is easy to increase the output force of the ERGLA using multiple ER gels and disks.

The ERGD input parts are rotated in opposite directions at the same rotational speed by one driving motor and reversing gears. Therefore, the ERGLA moves in different directions according to the application of the electric field as follows.

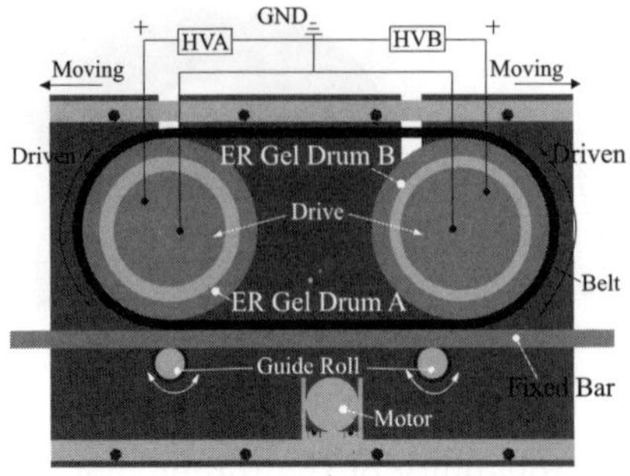

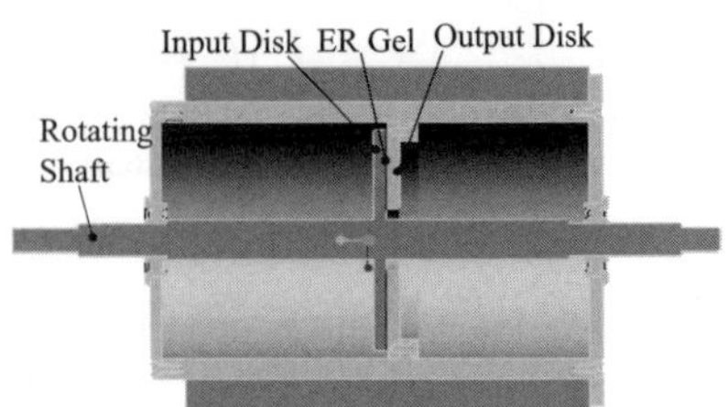

Fig. 4.   Structure of the ERGLA.   Fig. 5.   Design of the ER gel drum.

- <u>No electric field applied</u>: Because the shear stresses at the surfaces of the ER gels are very small, the electrodes slip each other. The torques of the input parts oriented from the motor are not transferred to output parts. Therefore, the ERGLA unit does not generate a driving force. In this case, it is easy to backdrive the ERGLA from the outside.
- <u>An electric field applied to one ER gel drum</u>:
  Since the shear stress at the surface of the drum increases with the magnitude of the electric field, the torque is transmitted to the output of the ERGD. Because the belt is fitted to the fixed bar by the guide rollers, the ERGLA unit moves along the fixed bar, and thus the ERGLA unit generates a driving force.

Essentially, the ERGLA cannot move faster than the speed corresponding to the input motor which rotates at a constant low speed owing to a highly reliable driving method such as battery operation. That is, the maximum speed of the ERGLA can be mechanically governed even if the controller of the ERGLA runs away. It is the reason of the safety.

3.2. *Prototype ERGLA and Basic Experiments*[1]

A photo of the prototype ERGLA is shown in Fig. 6. The design specifications of the ERGLA are $5.32\,\mathrm{kg}$ weight and $0.22 \times 0.26 \times 0.17\,\mathrm{m}^3$ volume. If the shear stress of the ER gel is $4.0\,\mathrm{kPa}$ with a $1.5\,\mathrm{kV/mm}$ applied electric field, the designed maximum torque of the ERGD is $0.55\,\mathrm{Nm}$ and the designed maximum force of the ERGLA is $10\,\mathrm{N}$.

While the mechanism of ER gel effects is considered as being different from that of ER fluids, a shaft set-collar is used to adjust the gap between the electrodes, according to a study on ER fluid actuators.[4]

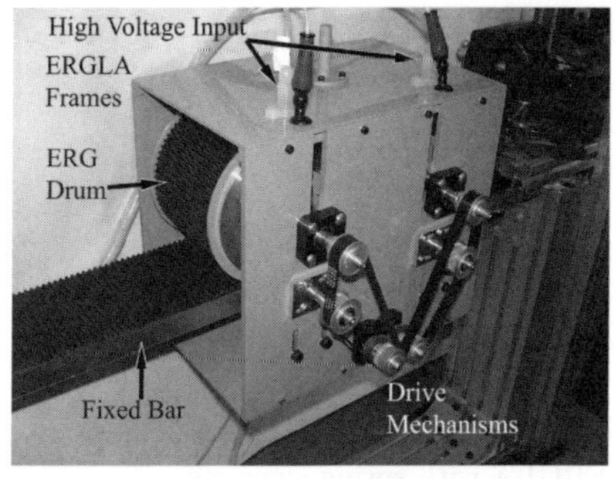

Fig. 6.   Prototype ERGLA.

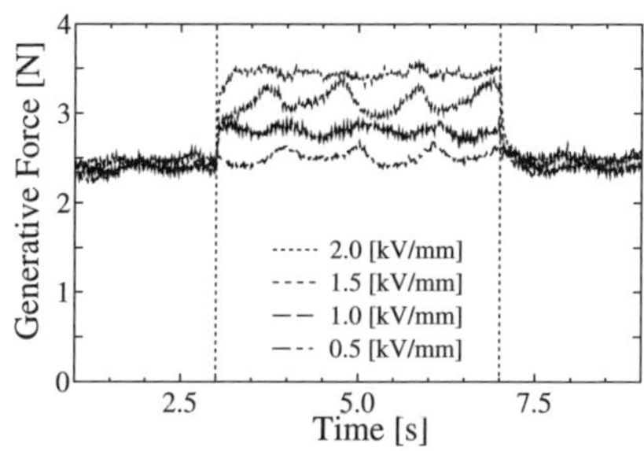

Fig. 7.   Step responses of the prototype ERGLA.

Figure 7 shows experimental results on step responses of the ERGLA. A 60 Hz notch filter to remove noise from the power source and a 200 Hz software low-pass filter were introduced. Each result was averaged from those data of 5 experiments runs for the same condition. The time constant of the ERGLA was about 40 ms. However, the effective thrust generated by the ERGLA was only 1 N even when 2.0 kV/mm electric field was applied. Additionally, responses show a cyclical variation.

One reason of this result is the instability of the generative torques of the ERGDs. It means there would be the instability of the pressure on the ER gels due to a variation of thickness of the gel when energized.

4. Stabilizing Output of the ERGLA

4.1. *Inserting Spring Elements*

We redesigned the ERGD to stabilize and increase its torque. The improved multiple-disk ER gel drum is shown in Fig. 8. The number of clutch disks inside the ER gel drum increased. It was assumed that very little pressure was needed to work the ER gels because the gels were able to move in the normal stress direction. Therefore, clutch disks can slide in the axial direction through three-lines of key ways: input disks are on the input shaft and output disks are on the inside wall of the drum. ER gels are molded on both sides of the output disks. Spring elements were introduced to stabilize the pressure on the ER gels. Springs were set in six positions for the improved ERGD as shown by arrows in Fig. 8 to keep the input and output disks parallel. Since the ERGDs are set to the ERGLA horizontally, only the spring forces act on ER gels as pressure. Pressure of 272 Pa was applied in this case.

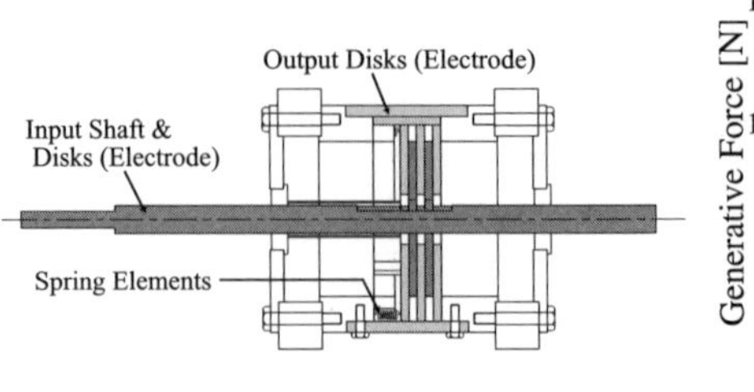

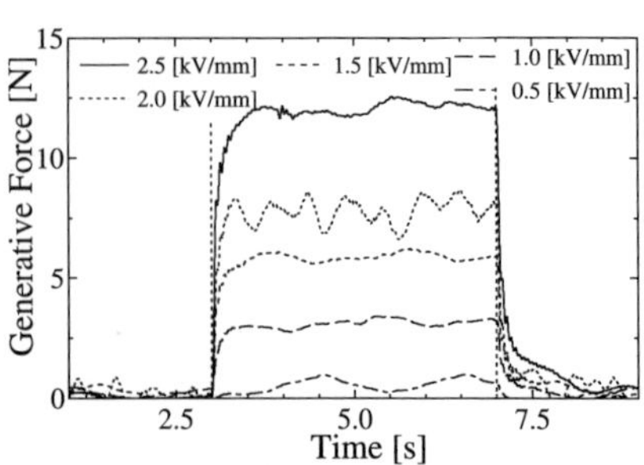

Fig. 8. Structure of improved ERGD inserted spring elements.

Fig. 9. Step responses of the improved ERGLA with spring elements.

Step inputs of electric fields were introduced to investigate the stabilization of the responses, and then the noise reduction filters were introduced as in the previous section. In this section, two ER gel disks were used in each ERGDs. The step responses of the ERGLA are shown in Fig. 9. The time constant of the ERGLA was about 40 ms and smaller than that in the previous section. The figure also shows smaller variances in the generative forces for the improved ERGLA than the variances in the earlier version. This implies that the pressure on the surface of the ER gel was stabilized. The magnitude of the generative force was larger than the initial target of 10 N. Spring elements would contribute to high speed and stable response.

However, a large degree of waviness sometimes occurred, which was cyclical but at a higher frequency than that due to the rotation of the input shafts. Therefore, another method for stabilization was investigated.

4.2. *Inserting Elastomer*

To improve the stability, a cylindrical elastomer was inserted to contact the entire surface of the clutch output disc as shown in Fig. 10. Uniform pressure would be applied to the ER gel by the viscoelastic force of the elastomer for which the contact area to the output disc was larger than that of the spring. For the elastomer, a foamed melamine with 0.12 MPa Young's modulus was adopted.

To verify the stabilization, step response experiments to the applied electric field were carried out. Experimental conditions were the same as in the previous section. Figure 11 shows the responses which were stable and much less variant than those from Fig. 7 or Fig. 9. However, a small amplitude noise was sometimes present. This would be because of unstable pressure on the gel by a slightly uneven thickness of the elastomer, an uneven thickness of the ER gels, fine voids in the gel, dust adhering to the

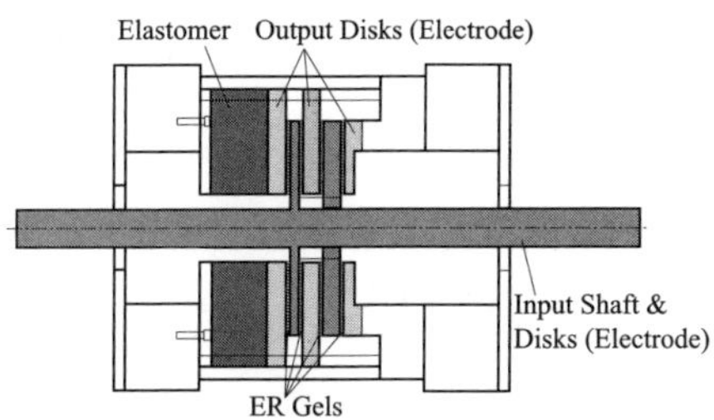

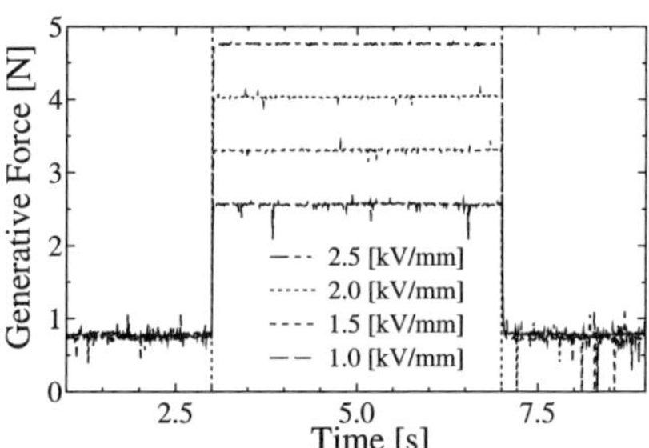

Fig. 10. Structure of improved ERGD inserted the elastomer.

Fig. 11. Step responses of the improved ERGLA with the elastomer.

surface of the gel and so on. This is a topic for future research.

5. Conclusion

This paper presents a novel linear actuator that uses ER gels, its improved design, and basic experiments to investigate its basic characteristics.

- It was confirmed that spring elements stabilized the response.
- It was confirmed that elastomer elements significantly stabilized the response of the ERGLA.
- The specifications of the most recent prototype are a mass of about 4.4 kg and a generative force of more than 12 N. Lightening the ERGLA is still an issue.
- The response time of the force was about 40 ms. It could be improved to around 20–30 ms the same as that of the ER gel.

In future studies, more ER gel disks will be introduced to the ERGLA to improve the force–weight ratio. Controllability of the ERGLA will also be investigated in terms of force control and position control.

References

1. K. Koyanagi, Y. Kakinuma, H. Anzai, K. Sakurai, T. Yamaguchi, T. Oshima, N. Momose and T. Matsuno, Basic structure and prototype of novel linear actuator with electro-rheological gel, in *Proceedings of the 2007 IEEE International Conference on Mechatronics and Automation*, 2007.
2. K. Koyanagi, T. Yamaguchi, Y. Kakinuma, H. Anzai, K. Sakurai and T. Oshima, *Journal of Physics: Conference Series* **149**, p. 012020 (2009).
3. Y. Kakinuma, T. Aoyama, H. Anzai, H. Sakurai, K. Isobe and K. Tanaka, *International Journal of Modern Physics B* **19**, 1339 (2005).
4. K. Koyanagi and J. Furusho, Study on high safety actuator for force display, in *Proceedings of SICE Annual Conference 2002*, 2002.

STUDY OF CORRUGATED DISC ROTARY DAMPER BASED ON INTELLIGENT EFFECT

SHUMEI CHEN[*] and SHI WEI

College of Mechanical Engineering, Fuzhou University, Fujian Province, 350108, China

An experimental device of rotary damper with one corrugated disc and another plate disc has been examined. The narrow gap between two discs is filled with electrorheological fluid (ERF) (or magnetorheological (MR) Liquid,); so electro-structural effect has to be taken into consideration. Computational fluid dynamics (CFD, FLUENT software) is the method used to produce comparative performance data between two types of dampers. Theoretical and numerical studies have been shown that the total output torque increment for corrugated disc rotary damper is significant larger than those of the dual-disc rotary damper at ignoring the impact of electric field strength. The research results help for optima design of a pre prototyping procedure with ER (or MR) fluid damper.

1. Introduction

Now for the research for ERF rotary damper has focused on the disc or cylinder rotary dampers (Chen, Wei, 2006; Chen, Bullough, 2007; Chen, Wei, 2007; Chen, Bullough, 2008. D J Ellam, 2004). Recently, Shi Wei and Li Yang (Li, Huand,2005; Shi, Chen, 2007) present a new prototype of damper, i.e., corrugated rotary disk damper with such a corrugated disk drive instead of the previous flat shape of disk drive. This paper uses FLUENT software to compare corrugated disc ERF damper with dual-disc rotary ERF damper, from which found that the output torque properties differences in both dampers, so that can reduce huge consuming on experiment and benefit to rotary ER damper design optimized.

2. Algorithm and Model of Rotary Damper

2.1. *Model of dual-disc rotary damper*

The geometric structure of dual -disc rotary damper has been shown in Figure 1.

---

[*] Corresponding author. E-mail: smchen@fzu.edu.cn, smch2002@vip.sina.com

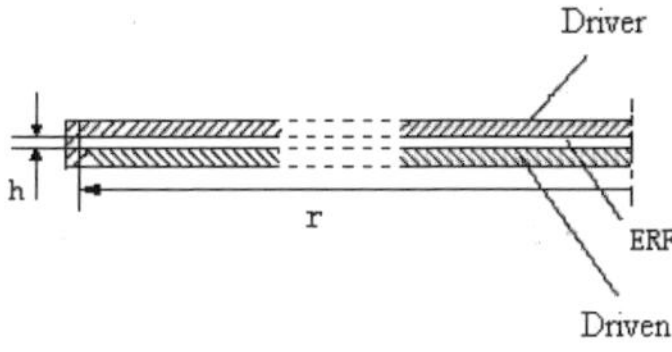

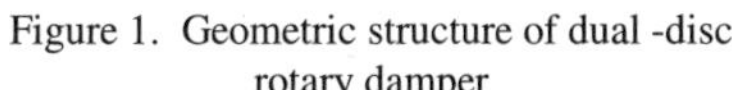

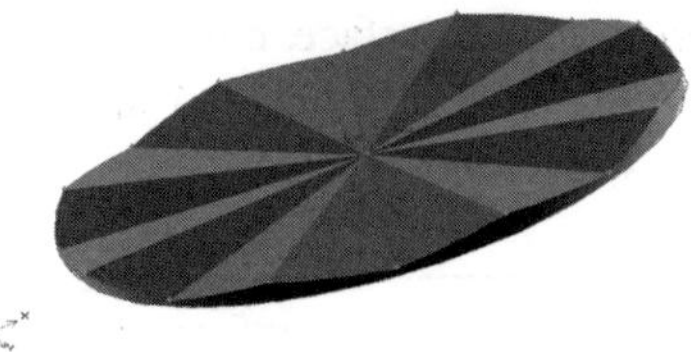

Figure 1. Geometric structure of dual -disc rotary damper

Figure 2. Figuration of corrugated disc

Its transmission torque can be obtained by formula (1).

$$T = \frac{\pi^2 \eta \Delta n r^4}{h} + \frac{2\pi \tau_y r^3}{3}$$

（1）

Where: T is transmission torque, N·m; h is the gap between the both discs of driver and driven, m; η is dynamic viscosity of ERF, Pa·s; τ_y is yield stress induced by electric field, Pa; Δn is rotary speed difference between the driver and driven, rpm; r is radius of disc, m. The second term of equation (1) is 0 when there is no external electric field.

It is sufficient to meet the simulation requirements for using the FLUENT software to solve the case with two-dimensional single-precision because of its typical of axial symmetry. The simulation settings are as follows, the solver chosen is built in FLUENT, use uncoupled solution method and implicit algorithm, spatial attribute is axisymmetric rotary space, time attribute is steady flow and the flow in two rotation surfaces is defined as the laminar flow mode. 50×1525----rectangle grids generated by GAMBIT.

It is important here to link the bi-viscosity model to the FLUENT via user-defined program [Chen, Shu-Mei; Bullough, WA. 2007].

2.2. *Model of corrugated disc rotary damper*

Corrugated disc rotary damper improved from the general dual disc rotary damper, which substituted a corrugated plate for the previous plate driven disc. The corrugated disc rotary damper can increase torque due to the forces of shearing and complex squeeze between both discs. Analyze the pressure distribution of rotary damper and design model of corrugated disc rotary damper according to the theory of fluid dynamics. Model consists of a stationary corrugated disc and a rotary flat disc. Electric (or magnetic) rheological fluids were filled into the gap of both discs. Torque is transferred via shearing and squeeze pressure forces of ERF (or MRF). Figure 2 shows the figuration of corrugated disc.

The corrugated surface consists of 16 same fan-shaped planes. The diagram of corrugated disc expanded along the circumferential direction as shown in Figure 3.

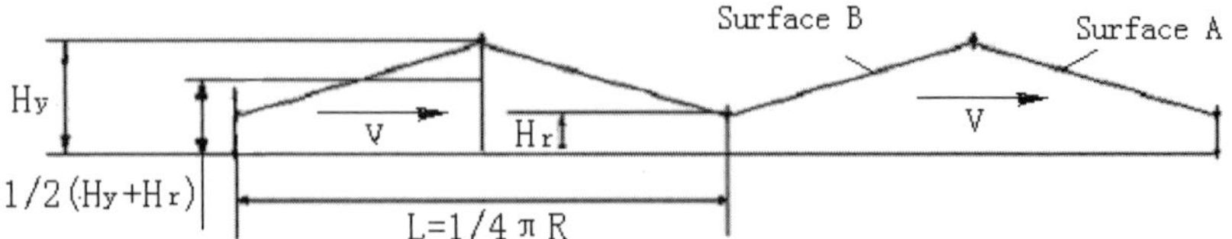

Figure 3. Diagram of corrugated disc expanded along circumferential direction

Where: Hy is the maximum height of corrugated disc circle; Hr is the minimum height of corrugated disc circle; L is wavelength of ripples on the circle; R is the radius of corrugated disc; Surface A is the ripple surface toward to the direction of flowing velocity of ERF (or MRF); and surface B is the ripple surface neighboring surface A.

The sketch of corrugated disc rotary damper is shown in Figure 4. And flat disc is a driver disc, corrugated disc is a driven disc, h is the smallest gap between the discs here.

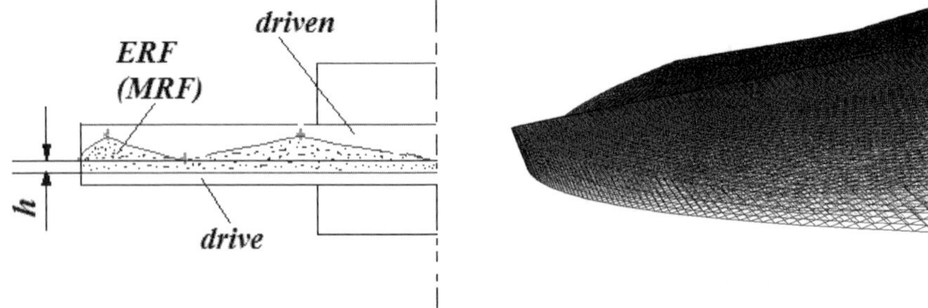

Figure 4. Sketch of corrugated disc rotary damper model

Figure 5. Local grid in flow cavity

The boundary conditions of the model are decided based on the material parameters and the fluid flowing features of the rotating surface.

3. Analysis and Comparison Between Corrugated Disc and Dual-Disc Rotary Dampers

3.1. *Models and boundary conditions*

Ignoring stress and viscosity changes due to different gap of ERF (or MRF), and use FLUENT to simulate the fluid characteristics of corrugated disc rotary damper. Compare corrugated disc rotary damper with dual-disc rotary damper.

Use GAMBIT to generate 3D model of corrugated disc rotary damper, the parameters have been shown in Table 1.

Table 1. Parameters of corrugated disc

Radius of corrugated disc, mm	30
Numbers of fan-shaped surface	16
Amplitude height of waveform: Hy−Hr, mm	1
The minimum distance to flat disc: h, mm	0.5
Material	Al, density: 2719 kg/m³
Temperature	Constant temperature: 300 K
Parameters of ERF	Density: 970 kg/m³ Specific heat: 1880 J/kgK Heat conduction coefficient: 0.12 W/mK
Parameters of bi-viscosity via user-defined program	Yield stress: 5000 Pa Viscosity coefficient at zero electric field: 0.05 Pa.s
Rotary state	Corrugated disc is stationary while the other is rotation with variable rotational speeds

Consider the gap height of adequate grid density, 240,000 cubic grids has been generated from GAMBIT, local grid has been shown in Figure 5.

Consider the economics of calculation by GAMBIT and establish 2D dual-disc model of dual-disc rotary damper, the model geometry has been shown in Figure 1 and the related parameters have been shown in Table 2. The parameters of ERF and the parameters of user-defined program are same as those of corrugated-disc.

Table 2. Parameters of dual-disk

Radius of rotary disc, mm	30
The minimum distance to flat disc: h, m	1
Material	Al, density: 2719 kg/m³
Temperature	Constant temperature: 300 K
Rotary state	A flat disc is stationary while the other is rotation with variable rotational speeds

3.2. *Calculation and analysis of corrugated disc rotary damper*

The residuals are less than 3.466×10^{-4} after 1800 iterations for 3D model of corrugated disc rotary damper. The torque has been shown in Table 3 when the input rotary speed is 250 rpm.

Table 3. Comparison of torques between dual-disc and corrugated disc rotary damper (250 rpm)

	Dual-disc	Corrugated disc
Viscosity torque (N·m)	0.29764	0.27476
Press squeezed torque (N·m)	0	0.01943

Figure 6 is the pressure distribution of flat disc; Figure 8 is the velocity distribution on the flat disc; Figure 7 is pressure distribution of cavity A-A profile (see Figure 6) which is 21.21mm apart from the vertical rotary disc, and Figure 9 is velocity distribution of cavity A-A profile (see Figure 6) which is 21.21mm apart from the vertical rotary disc.

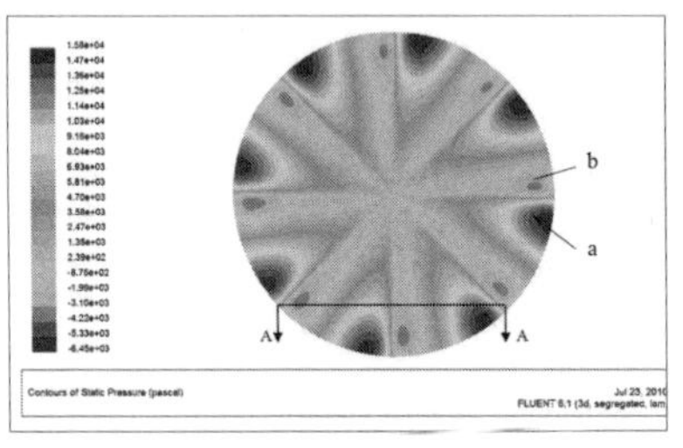

Figure 6. Pressure distribution on the flat disc (250rpm)

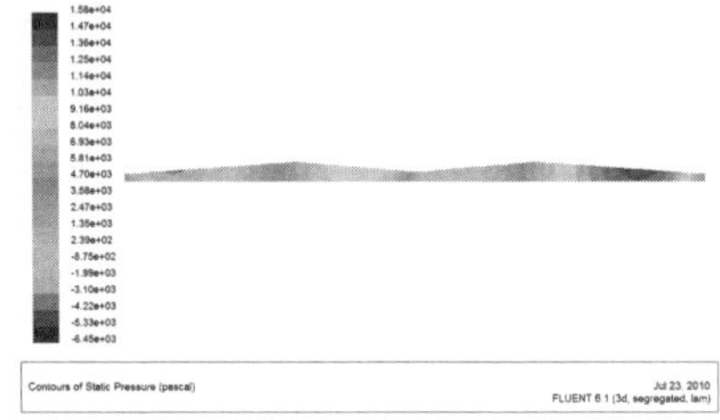

Figure 7. Pressure distribution on A-A cross-section

From Figures 6 and 7 can be found that the greatest positive pressure 15302 Pa is at surface A (point a), the greatest negative pressure 4215 Pa is at the surface B (point b). The distribution area of positive pressure is concentrated in the surface A of corrugated disc and the distribution area of negative pressure is concentrated in the surface B which is adjacent to surface A. Research the cavity of corrugated disc by two fan-shaped planes. The velocity distribution curve is upward off-set on the cavity in Figure 9, which indicates the upward squeezing action of ERF. The static pressure increasing can result in pressure increasing of surface A and pressure decreasing of surface B according to Bernoulli's theorem (2). The reason of press squeezed action by both surfaces of A and B will produce the output torque of corrugated disc 0.01943 N·m and viscous torque 0.27476 N·m.

$$p + \frac{\rho v^2}{2} = C \tag{2}$$

Where p is fluid static pressure, Pa; ρ is fluid density, kg/m^3; v is fluid velocity, m/s; and C is constant. Simulation results show that torque of corrugated disc rotary damper consists of pressure (press squeezed) torque and viscous torque.

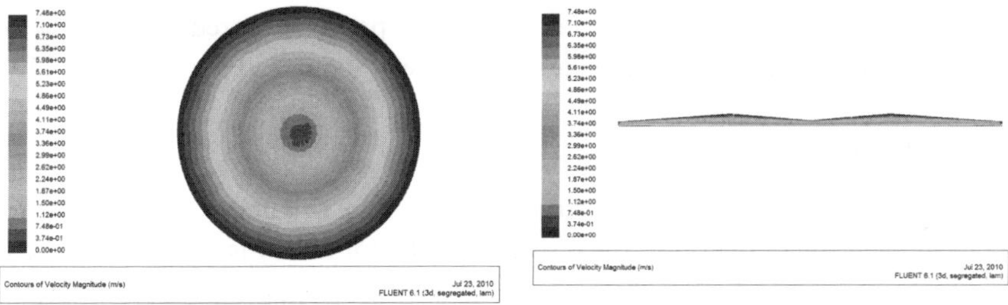

Figure 8. Velocity distribution on the flat disc (250 rpm)

Figure 9. Velocity distribution on A-A cross-section

3.3. *Comparison between corrugated disc and dual-disc rotary dampers*

Calculate the viscous torque and total torque under different rotary speed and a given yield stress 5000Pa. Conclusions are as follows in Table 4, Figure 10 and Figure 11.

Table 4. The results of torque at different rotary speed and a given yield stress (5000 Pa)

Rotary Speed (Ω)	Dual-disc rotary damper				Corrugated disc rotary damper			
	100	250	500	750	100	250	500	750
Viscous Torque (N·m)	0.283	0.287	0.289	0.292	0.2817	0.2916	0.3150	0.3404
Total torque (N·m)	0.283	0.287	0.289	0.292	0.2978	0.310	0.3401	0.3766

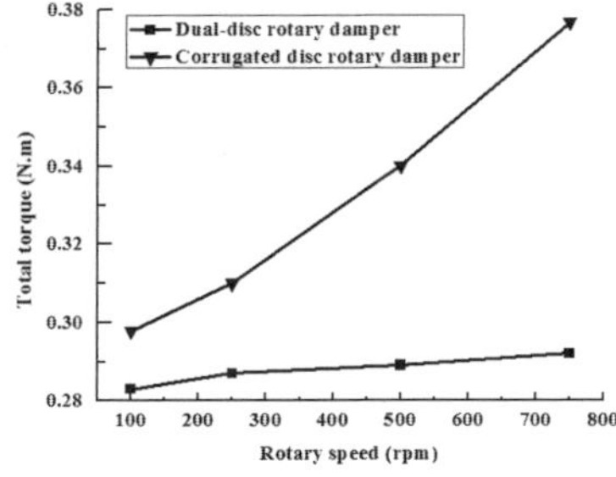

Figure 10. The relationship between total torque and rotary speed at constant yield stress

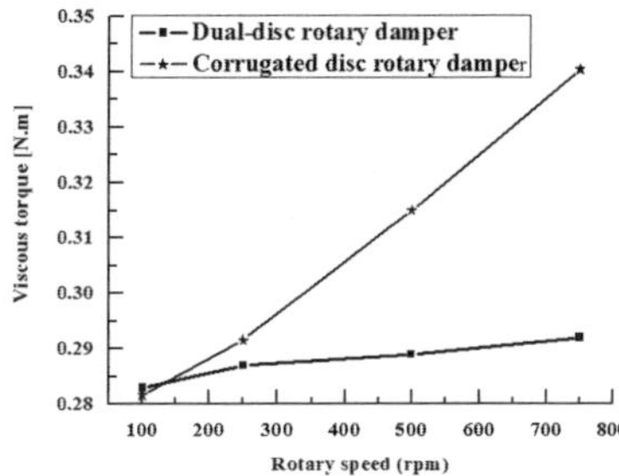

Figure 11. The relationship between viscous torque and rotary speed at constant yield stress

Simulation results of dual-disc damper show that the total torque equals to the viscous torque (Table.4). However, the total torque is not same result for the corrugated plate rotating damper due to two parts: the viscous torque and an additional press squeezed torque.

Enhancing the input rotational speed is only to boost the torque generated by the dynamic viscosity for the double-disc rotary damper; while to boost viscous torque (i.e., total torque) and an additional press squeezed torque for a corrugated disc rotary damper; the result have been shown in Figures 10 and 11.

4. Conclusion

The torque performance of corrugated disc rotary damper is superior to conventional rotary duel-disc shear damper. ERF (or MRF) is squeezed directly on the corrugated surface so that it can generate the output press squeezed torque which exits not in conventional rotary duel-disc shear damper. Simulation shows that the total output torque of corrugated disc rotary damper can be increased 25% than dual-disc rotary damper in 10000rpm.

The total torque and viscous torque of corrugated disc rotary damper are decreasing with the amplitudes height increasing.

Acknowledgements

Financial support from Fujian Provincial Department of Science and Technology of China (Grant No. 2008HZ0002-1) is gratefully acknowledged.

References

1. Chen, Shu-Mei; Wei, Chen-Guan. Experimental study of the rheological behavior of electrorheological fluids. Smart Materials and Structures. 2006, v 15, n 2, p 371-377.
2. Chen, Shu-Mei; Bullough, WA. Examination of Through Flow in a Radial ESF Clutch. Journal of Intelligent Material Systems and Structures. 2007.18,12, 1175-1179.
3. Chen, Shu-Mei; Wei, Chen-Guan. Experimental study on percolation structure of electrorheological suspensions. International Journal of Applied Mechanics and Engineering. 2007, vol.12, No.1, pp.299-308.
4. Chen, Shu-Mei; Bullough, WA. CFD Study of the Flow in a Radial Clutch with a Real Electrorheological Fluid. The 11th International conference on Electrorheological fluids and Magnetorheogical Suspensions. Dresden, Germany. 2008.
5. D J Ellam. The University of Sheffield, UK. PhD thesis. 2004.
6. Shi We, Chen Shu-mei (supervisor). The numerical simulate study of the ER rotary damper. Dissertation to the Academic Degree of Fuzhou University. 2007.6
7. Li Yang, Huand Yijian (supervisor). Study on Electrorheological Transmission Mechanism with Ripple-shaped Disc. Dissertation to the Academic Degree of Huaqiao University. 2005.5

THE STUDY OF AUTO ABS PERFORMANCE AND CONTROL BASED ON ELECTROLOGICAL EFFECT

SHUMEI CHEN[*], KUNQUAN YANG and BO-FENG CHEN

College of Mechanical Engineering, Fuzhou University, Fujian Province, 350108, China

W. A. BULLOUGH

Department of Mechanical Engineering, the University of Sheffield, S1 3JD, UK

The anti-lock brake system (ABS) based on the conventional solenoid valve and hydraulic brakes in car agencies existed shiver during the breaking processing. Brake friction block and brake disc friction led to a sharp temperature rising and resulted in some brake lag phenomenon. In this paper, it has been proposed and developed an auto ABS brake based on electrorheological (ER) effect. The result shows that brake effect based on ER is superior to solenoid valve and hydraulic brake, and also it can be achieved an automobile anti-lock braking processing by adjusting voltage.

1. Introduction

The brake performance is one of the main parameters for an automotive brake. In present, a largest number of disc-type and drum-type brakes are used currently, and the manipulation method is primarily by controlling of pressure of wheel cylinder or brake caliper to complete the wheels anti-locked. Good braking is following the processes of pressurization, decompression and keeps pressure; however, because of the shiver, braking friction block and braking disc friction lead to a sharp temperature rising; and there is a certain hysteresis and complicated configuration for the conventional solenoid valve and hydraulic brake agencies. In this paper, so an electro-rheological fluid with variational inherent characteristics is tried to be introduced skillfully as automobile ABS, and thus to some extent to resolve the existing brake shortcomings. Besides simple configuration with no hardware connections, this ER brake has the advantages with the realization of automatic control, convenient installation and maintenance.

Though ER technology has broad been researched in the mechanical and vehicle engineering, however, with regard to apply brake system only few paper

---

[*]Corresponding author. E-mail: smchen@fzu.edu.cn, smch2002@vip.sina.com

published (S.-B. Choi, J.-Y. Yook, M.-K. Choi, Q.H. Nguyen, Y.S. Lee and M.S. Han. 2007). The study of ABS based on ER effect is an original work.

2. Working Principle and Working Process for an Automobile ABS

The basic working principle for an ABS is to make full use of vehicle's wheel tires, the ground gummy coefficient and the control strategy to obtain a suitable braking force. Slip rate is usually used to measure the degree of wheel slip when it is in braking. Slip rate, 'S', is defined as the speed difference between vehicle and wheel by the percentage of the speed of vehicle, namely:

$$S = \frac{v_V - \omega_W R}{v_V} \times 100\% \tag{1}$$

where v_V is automotive vehicle linear velocity; ω_W is wheel angular velocity; and R is effective radius of wheel (Yazicioglu, Yigit, Samim Unlusoy, Y. 2008). If the ratio of wheel slip is near to a value given, the electronic control device will control fluid pressure regulator to keep a certain pressure. In this way, each ratio of wheel slip is maintained at the desired range (0.1~0.2) to prevent the wheels locked completely [3].

3. The Principle of Auto ABS Based on ER Effect

In this paper, we take the control method of the automobile drive shaft to prevent anti-lock and the braking form based on the ER effect is shown in Figure 1. It consists of two coaxial cylinders A, B and a shaft C; cylinder A is fixed, cylinder B is fixed on the shaft C and rotated together with the shaft C; inner radius of cylinder A is noted as $R_1^{'}$, the outer radius of cylinder B as R_1, the gap spacing between the two cylinders is $h = (R_1^{'} - R_1)$. When it is filled with electrorheological fluid between two coaxial cylinders A, B and imposed on a series of variation electric fields, the axis C generated a variational torques 'T' to block the rotating cylinder B and also braking axis C because of the ER effect electric-induced. To facilitate research, it is assumed that: (1) incompressible fluid; (2) the flow pattern for laminar flow; (3) it is a steady-state exercise, that is, all of the time derivative are zero in the equations of continuity and momentum; (4) excluding edge effects; (5) system isothermal.

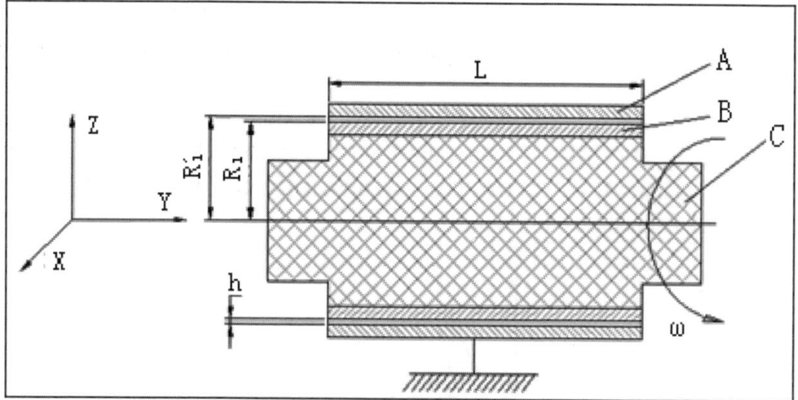

Figure 1. Sketch of cylindrical ER brake

When the C-axis is in rotation, then the angular velocity ω is distribution along the z direction, the shear strain rate $\dot{\gamma}$ is as follows:

$$\dot{\gamma}=\frac{du}{dz}=\frac{d(R_1\omega)}{dz}=R_1\frac{d\omega}{dz}=\frac{R_1\omega}{h} \tag{2}$$

Set electrorheological fluid as a Bingham fluid (Chen, Shumei; Wei, Chenguan. 2007) under an excitated electro field, the fluid constitutive equation is:

$$\tau=\tau_y+\eta_0\cdot\dot{\gamma}=\tau_y+\eta_0\cdot\frac{R_1\omega}{h} \tag{3}$$

where τ_y is yield stress induced by electric field (Pa); η_0 is liquid viscosity (m^2/s); and ω is angular velocity (rad/s)[2][13][14];

$$\tau_y = k\cdot E^\alpha = k\cdot(\frac{U}{h})^\alpha \tag{4}$$

$$F=A\cdot\tau=2\pi R_1 L\cdot\tau=2\pi R_1 L\cdot(\tau_y+\eta_0\frac{R_1\omega}{h}) \tag{5}$$

$$=2\pi R_1 L\tau_y+2\pi R_1 L\eta_0\frac{R_1\omega}{h}$$

Torque 'T' is

$$T = F\cdot R_1 =2\pi Lk(\frac{U}{h})^\alpha R_1^2 + 2\pi L\eta_0\omega\frac{R_1^3}{h} \tag{6}$$

When it is in absence of external electric field, the behavior of electrorheological fluid is Newtonian fluid, or else behaves as Bingham fluid under the action of external electric field. The first item in the formula (6) is representative of torque generated by electric-induced yield stress; and the second one on behalf of torque by fluid viscous. From above, we can see that the braking torque can be controlled by the voltage U. And also we can achieve automatic control for anti-lock brake requirements via different control methods and strategy (Chen, SM; Bullough, WA. 2007).

Generally a simple adhesion coefficient vs. slippage-rate, i.e. μ-S bilinear model and the Pacejka μ-S model are used to describe the character of wheel (Lin, J.-S.; Ting, W.-E. 2007). The μ-S bilinear model has been used in this paper; it has been shown in Figure 2.

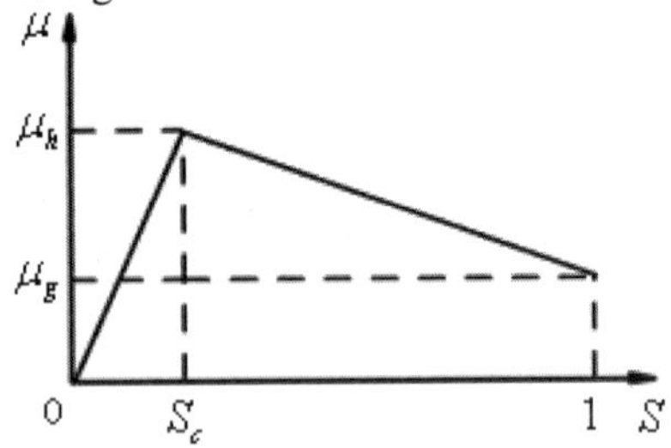

Figure 2. Character curve of adhesion coefficient of a wheel tyre

$$\mu = \mu_h S / S_c \qquad\qquad\qquad S \le S_c \qquad\qquad (7)$$

$$\mu = (\mu_h - \mu_g S_c)/(1-S_c) - (\mu_h - \mu_g S)/(1-S_c) \qquad S > S_c \qquad (8)$$

when $S \le S_c$, it is a stable zone and when the $S > S_c$, it is an unstable zone; S_c is the optimal slippage-rate; S is the slippage-rate of a wheel; μ_g is the adhesion coefficient at 100% slippage-rate; μ_h is the peak value adhesion coefficient (Cheng Jun. 1998).

4. Design of Experimental Unit of Automobile ABS

The experimental unit of automobile ABS is shown in Figure 3.

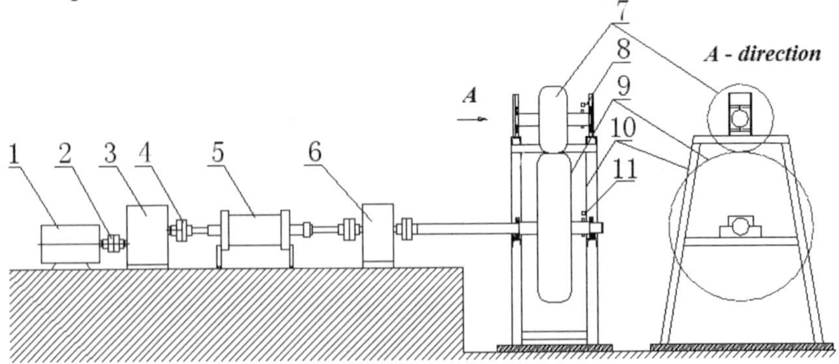

Figure 3. Experimental unit of automobile ABS
1. Motor; 2. Electrical coupling; 3. Reducer 1; 4. Coupling; 5. ER brake; 6. Reducer 2;
7. Round weight; 8. Speed sensor; 9. Wheel; 10. Backstop; 11. Speed sensor

The details are as follows:

(1) When the motor gets power and rotates quickly, motor speed will be controlled in a given value so that the vehicle obtains an initial velocity. Put the initial velocity into a computer. As soon as the computer gets the signal and presses the brake button, the vehicle enters into the braking state.

(2) Turn off the power, analog of vehicle body is still rotating in a high-speed because of the inertial force. Because the body is completely acted on wheels, so the wheel at this time will continue rotate with the analog body in a high-speed. However, the ERF breaking torque is corresponding variety and the speed of the wheel also is changed in an intermittent high-voltage. At this time, the speed deviation between the wheel and vehicle analog will occur.

(3) Two speed-sensors are used to collect the speeds of vehicle and wheel respectively. The data of speed are sent to a computer using the LABVIEW software in order to calculate real-time slip ratio S (see the equation (1)).

(4) Compare the real-time slip ratio S with the ideal slip ratio, the difference is fed back to control the output voltage of the high power, that is, try to control timely the electro field to keep the slip ratio in an ideal value. And then, the computer analyzes and calculates the speed data collected by single-chip to obtain the testing results.

5. Simulation Experiments and Results Analysis

In this paper, take Mazda RX-7 Turbo 2[2] for the test object, with a total weight M = 1585kg, the main reduction ratio: i= 3.91, radius of wheel R = 0.3m. The ERF, which the main components of particles are cellulose, the liquid is silicone oil; the ERF has higher shear strength to meet the basic requirements of the brake system. The core of control system is the PIC16F874A micro-controller, through the MCU I/O port, collecting and computing pulses generated by two speed sensors, and transmitting to the computer through the RS232 serial port, LABVIEW process associated antilock calculation and output the corresponding curves.

Testing research includes two parts.

(1) ER brake performance testing

Using the experimental unit based on the ER effect, pouring ERF (cellulose solid mainly 30% volume fraction) into the ER brake, putting an electro field on the electrorheological fluid by voltage of 2kV, 3kV, 4kV, attain the different speeds for both round weight and the wheels respectively as shown in Figure 4.

114

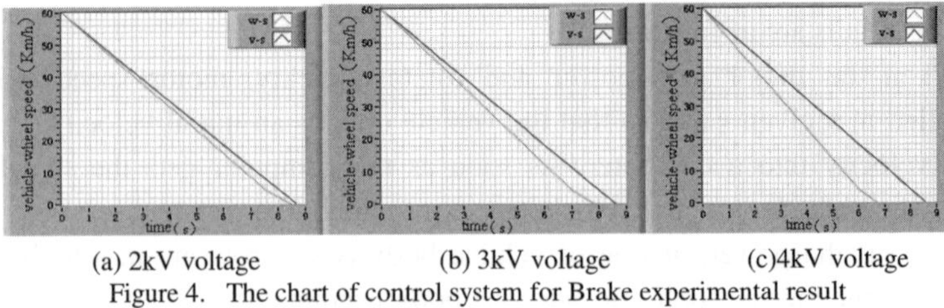

(a) 2kV voltage   (b) 3kV voltage   (c)4kV voltage

Figure 4.   The chart of control system for Brake experimental result

As can be seen from Figure 4, when the voltage on the electrorheological fluid is added to 2kV, the electrorheological fluid effect is happened, but is very weak, at this time the wheel speed and the vehicle speed is near closer. With the increasing of voltage, electrorheological effect is getting more and more obvious; the locked time of wheels is becoming shorter and shorter, when the voltage is added to 4kV, the locked time is 6.8s. It has verified that the braking effect based on electrorheological fluid is obvious and proposed application in engineering.

(2) The simulation testing of ER anti-lock brake performance

In accordance with the design of experimental unit, the results of simulation testing for ER antilock brake performance are shown in Figure 5.

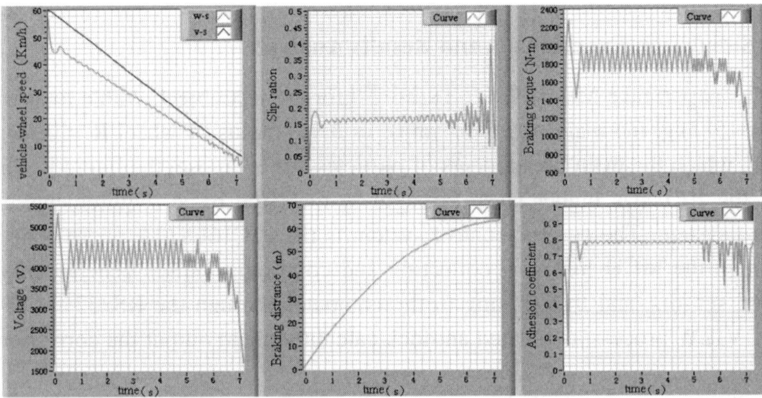

Figure 5.   The curves of anti-lock simulation result of control system

As can be seen from Figure 5, the control system almost meets the basic requirements of automotive ABS braking, i.e., the time reached the minimum braking speed requirements (4km / h) is shorter. In the initial speed of 60km / h, the simulation parameters of control system are shown in Table 1.

Table 1. Simulation parameters of control system

Initial velocity (km/h)	Braking time (s)	Brake distance (m)	Voltage	Slip ratio	Adhesion coefficient
60	7.376	63.24	It is stable in 4281V after 2s	It is stable at 0.16	It is stable in 0.78 after 2s

6. Conclusion

Electro-rheological technology used in automobile anti-lock braking system is an innovation. The results of simulation show that ABS performance based on the ER technology can meet the requirement of anti-lock brake of vehicle. The research findings of this paper can provide further experimental data for the braking effect generated by electro-rheological technology, which is significance for the future development of automotive ABS. However, there are several aspects should be researched emphatically in the future, such as temperature effect on the breaking process, the control method and strategy for shortening control distance and time and the reliability.

Acknowledgements

Financial support from Fujian Provincial Department of Science and Technology of China (Grant No. 2008HZ0002-1) is gratefully acknowledged.

References

1. S.-B. Choi, J.-Y. Yook, M.-K. Choi, Q.H. Nguyen, Y.S. Lee and M.S. Han. Speed Control of DC Motor using Electrorheological Break System. Journal of Intelligent Material Systems and Structures.Dec.2007.1191-1196.
2. Yazicioglu, Yigit, Samim Unlusoy, Y. International Journal of Vehicle Design, v 48, n 3-4, 2008, p 299-315.
3. GUO Konghui; REN Lei. A Unified Semi-Empiri-cal Tire Model with Higher Accuracy and Less Parameters. SAE 99PC-17.
4. Chen, Shumei; Wei, Chenguan. Experimental study on percolation structure of electrorheological suspensions. International Journal of Applied Mechanics and Engineering. 2007, vol.12, No.1, pp.299-308.
5. Chen, SM; Bullough, WA. Examination of throughflow in a radial ESF clutch. Journal of Intelligent Material Systems and Structures. Dec.2007. 1175-1179.
6. Lin, J.-S.; Ting, W.-E. IET Control Theory and Applications, v 1, n 1, 2007, p 343-348
7. Cheng Jun. Analogue Investigation of Different Control Methods of Anti-lock Brake System. Automobile Technology. 1998, 8: 1~7.

SOFT MAGNETORHEOGICAL ELASTOMERS AS ACTIVE MATERIALS IN NOVEL VALVES

HOLGER BÖSE[*], RAMAN RABINRANATH and JOHANNES EHRLICH

*Fraunhofer-Institut für Silicatforschung ISC, Neunerplatz 2,
D-97082 Würzburg, Germany
E-mail: boese@isc.fraunhofer.de

The actuation behavior of soft silicone-based magnetorheological eleastomers (MRE) in magnetic fields of variable strength was investigated. An inhomogeneous magnetic field gives rise to a reversible actuation effect, which is the result of the competition between magnetic and elastic forces in the material. MRE are capable to perform more sophisticated deformations than known rigid actuator materials. In this connection, the actuation behavior of MRE ring-shaped bodies in a valve-type device for the control of an air flow is demonstrated. For this purpose, MRE rings with different hardness were prepared and used in the valve. Additionally, the actuation of anisotropic MRE was compared with that of isotropic samples. The inhomogeneity of the magnetic field at the MRE material which is required for the actuation could be strongly influenced by the shape of the magnetic yoke. In the study, the closing characteristics of the valve with different yoke shapes and MRE materials were evaluated by measuring the dependence of the air flow rate on the magnetic field strength. It is demonstrated that the air flow through the valve can be controlled by the current in the field-generating coil, which yields the base for a new type of magnetic valves.

1. Introduction

Magnetorheological elastomers (MRE) represent a relatively new class of magnetically controllable materials [1]. They are composites of magnetic particles which are embedded in an elastomeric matrix. By applying a magnetic field, the material can strongly and reversibly change its mechanical properties like the Young's and the shear modulus. This magnetorheological effect is reversible and occurs within milliseconds, in analogy to the behavior of magnetorheological fluids (MRF). The magnetic control of the mechanical properties of MRE has attracted increasing attention in recent years [2, 3]. This interest is referred to a multitude of potential applications in various technological branches, like adaptive vibration damping, tunable vibration absorption and haptics.

However, only very few work has been devoted to the actuation capabilities of MRE up to now. If the elastomer matrix of the composite is sufficiently soft and the applied magnetic field is strong enough and inhomogeneous, the MRE material shows a pronounced actuation effect [4]. In an inhomogeneous magnetic field the material is deformed or stretched and changes its shape or geometrical dimensions. When the magnetic field is removed, the MRE relaxes to its original shape, behaving as a magnetically driven shape-memory polymer. The degree of actuation depends on the relative strength of the magnetic particle and the elastic polymer forces. MRE composites which reveal such an actuation effect could also be called magnetoactive polymers (MAP), in analogy to the known electroactive polymers (EAP) which actuate in electric fields.

Soft MRE are capable to perform more complex motions in a magnetic field than it is possible with rigid bodies. In comparison with a voice coil, in which a magnetic body moves with respect to the coil, driven by a magnetic force, an MRE body can be strongly deformed in the field. It was demonstrated before, that deformations of ca. 10 % can be achieved [4]. This magnetomechanical effect could be exploited in a magnetic valve for the control of gas flows. The basic function of an MRE valve was recently shown with a laboratory device containing an MRE ring in the space between two concentric yoke parts of the magnetic circuit leaving an inner air gap [5]. Fig. 1 depicts this laboratory device and the closing of the air gap by the magnetic field.

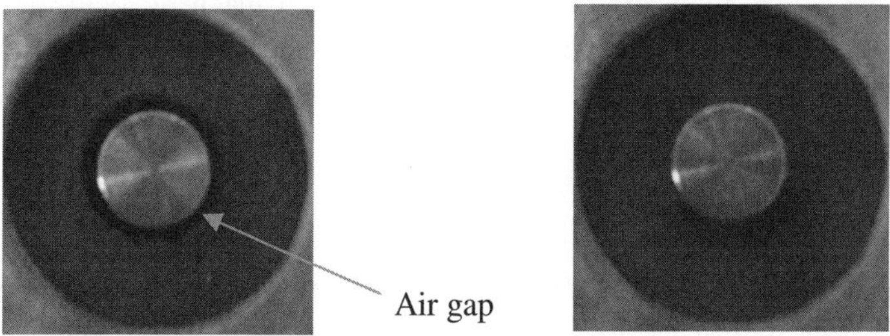

Figure 1. Visible actuation of an MRE in a valve with an inner air gap: top view onto the ring-shaped MRE body in the open (left) and closed state (right)

In this work a more detailed study of the performance of MRE materials in a model valve was conducted. For this purpose, an experimental set-up for an MAP valve with a magnetic circuit with variable yoke shape was used. The magnetic circuit was modularly designed and simulated. Furthermore, MRE rings with different hardness as well as with isotropic and anisotropic particle arrangements were manufactured and investigated in the experimental set-up. The objective of this basic study was to evaluate the actuation performance of

MRE materials in the valve and to investigate the influence of various material and technical parameters.

2. MRE Materials

The MRE used as actuator materials in the valve consisted of carbonyl iron particles with a mean size of about 5 µm dispersed in a silicone elastomer matrix. The concentration of the iron particles in polydimethylsiloxane was 30 vol.%. All MRE composite samples were manufactured in a ring-shaped moulding tool upon a thermal Pt-catalyzed curing process, in which the vinyl end groups of the silicone polymer react with the silane groups of the corresponding crosslinker.

The mechanical hardness of the silicone elastomer can be controlled by the crosslinking density. This crosslinking density was systematically varied in order to study the influence of the MAP hardness on the characteristics of the valve. As a result, MRE actuator rings with different hardness between 55 and 70 on the soft Shore 000 scale were prepared for the investigations in the valve. Furthermore, another series of MRE samples was prepared upon curing in the presence of a magnetic field. For this purpose, a special mould was constructed, which allows the application of a radially oriented magnetic field during the curing process. With this mould, anisotropic ring-shaped MRE samples could be prepared, in which the iron particles are arranged in radially oriented chains.

3. Design of the MRE Valve

The design of the experimental set-up of the valve with the magnetic circuit and the MRE ring is revealed in Fig. 2. The outer yoke around the MRE ring was made exchangeable between flat and edged surface shapes. In the magnetic field the ring-shaped MRE body expands radially and closes the gap around the ring. This expansion blocks the air flow through the gap. Moreover, if the expansion of the ring can be controlled to intermediate positions, the MRE may also be used as a proportional valve.

The experimental set-up for the investigations on the MRE valve is depicted in Fig. 3. A variable pressure source supplies an air flow through the MRE valve. Two pressure sensors detect the air pressure on both sides of the valve. The air flow at the outlet of the valve is determined with a flow meter.

Detailed experiments were performed to study the performance of the various MRE materials in the valve. Starting in the opened state, the current in the coil was stepwisely increased and the air flow rate was measured in order to evaluate the closing characteristics of the MRE valve.

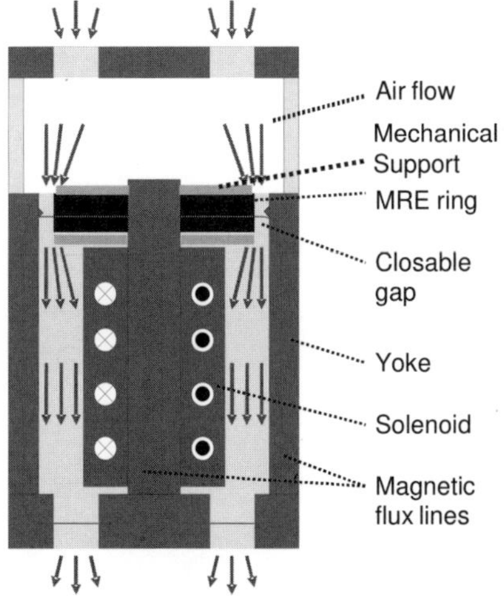

Figure 2. Scheme of the MAP valve

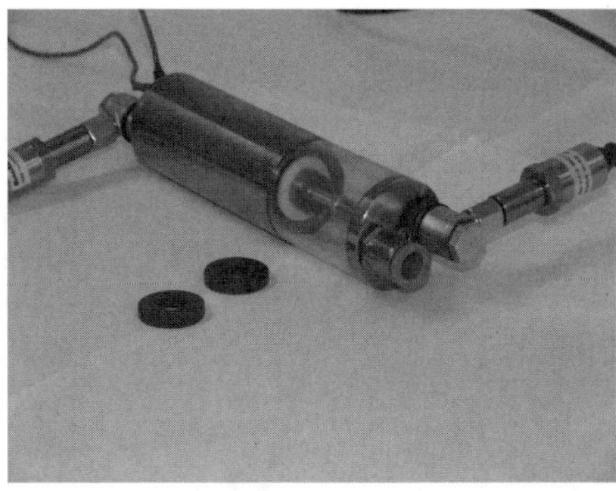

Figure 3. Experimental set-up for the measurements of the MRE. The MRE ring is in the middle of the tube

4. Results

At first, the air flow characteristics of the valve were evaluated with MRE rings of different Shore 000 hardness. Initially, the air flow through the open valve was tuned to a defined value of 0.36 m^3/h and then the coil current was stepwisely increased. With the increasing current, the air flow continuously decreases due to the reduction of the opened gap width of the valve. The measurement was finished when the air flow was below the detection limit of the used flow meter.

The measurements shown in Fig. 4 were performed with the flat outer yoke in the circuit. The increase of the Shore 000 hardness from 55 to 70 tentatively shifts the required current to higher values. However, the softest MRE material with a Shore 000 hardness of 55 is relatively sensitive, can be more easily deformed by the air pressure and does not completely close the valve.

In the next step, the actuation characteristics of MRE samples with isotropic and anisotropic particle arrangements were compared with each other. It becomes apparent in Fig. 5, that the anisotropic MRE ring causes a steeper closing characteristic than the isotropic ring. From this behavior it is concluded,

120

that the isotropic MRE material having the same hardness without applied magnetic field, is easier to control by the coil current.

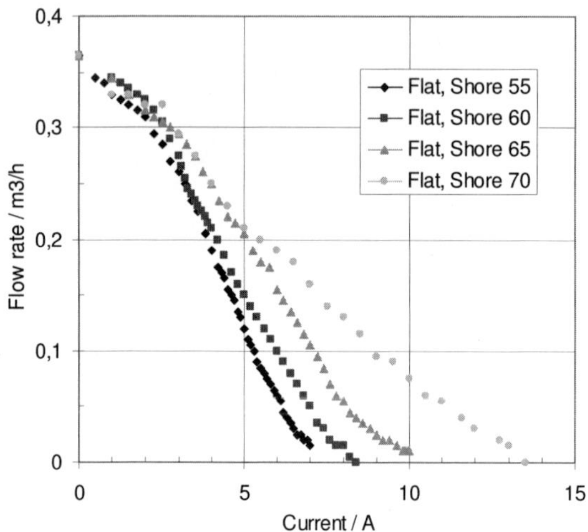

Figure 4. Air flow through the gap between the MRE ring with different Shore hardness and the flat outer yoke at increasing coil current

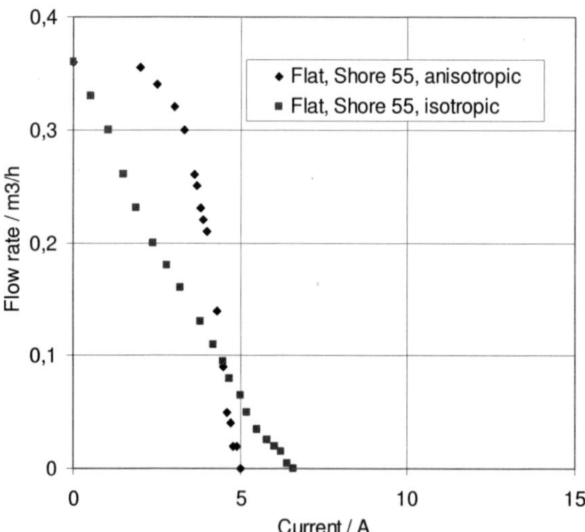

Figure 5. Air flow through the gap between the MRE ring (Shore 000 hardness of 55) with isotropic or anisotropic particle configurations and the flat outer yoke at increasing coil current

In Fig. 6 a remarkable difference between the data for the valve with the flat and with the edged outer yoke is visible. The air flow reduction with the edged outer yoke has a steep characteristic, i.e. the valve closes within a relatively narrow range of the coil current. In contrast to this behavior, the use of the flat outer yoke leads to a more gradual air flow decrease, when the coil current is increased. The gradual decrease of the air flow is more suitable for a precise tuning of the valve. The flat and edged yoke shapes give only extreme configurations, and many intermediate or other yoke shapes are possible to tune the closing characteristics of the valve.

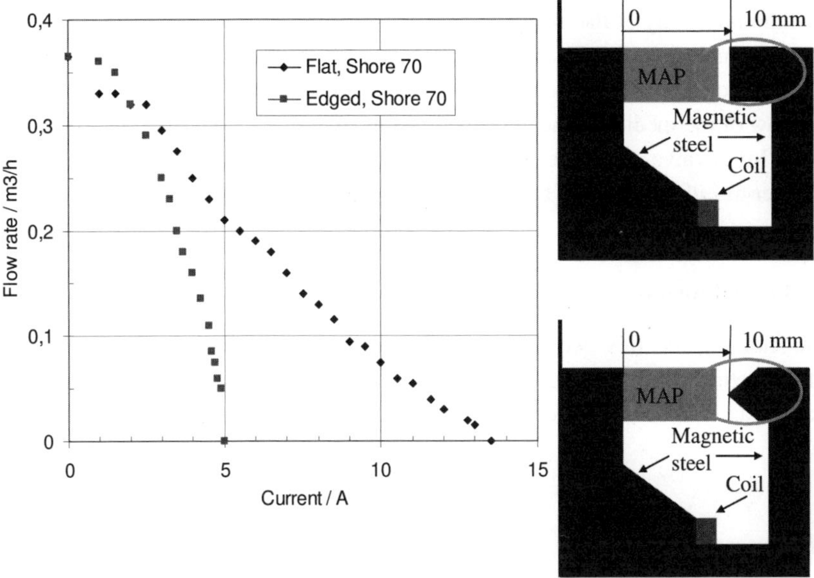

Figure 6. Air flow through the gap between the MRE ring (Shore 000 hardness of 70) and the flat and edged outer yoke, respectively, at increasing coil current (left) and scheme of the flat and edged outer yoke

5. Conclusions

The actuation performance of soft MRE materials prepared with variable properties was investigated. For this purpose, an experimental-set-up for an annular valve was used. The measurements on this MRE valve show, that it can be closed by the magnetic field which forces the MRE material to expand. Furthermore, the air flow through the valve can be controlled by the coil current of the electromagnet.

The results of this study give evidence, that the characteristics of the MRE valve can be influenced by several material and technical quantities. Important parameters, which affect the valve performance, result from the composition and properties of the MRE material, i.e. the hardness of the elastomer matrix of the composite. The hardness can be tuned by the crosslinking density and the use of plasticizers. Another influence on the valve performance is given by the shape of the yoke parts between which the MRE material expands, because it determines the inhomogeneity of the magnetic field.

In conclusion, MRE as a novel class of actuator materials offer a huge potential for new valves, in which more sophisticated motions are able to control the flow of the medium. The described results give first evidence on the behavior of the MRE valve, but more detailed studies and optimization steps are necessary in order to fully exploit the benefit from this new valve technology.

Acknowledgments

The funding of this work by the Bavarian State Ministry for economy, infrastructure, traffic and technology is gratefully acknowledged. The authors thank Daniel Weber for the reliable performance of measurements on the MRE valve.

References

[1] M. R. Jolly, J. D. Carlson, B. C. Munoz, and T. A. Bullions: The magnetoviscoelastic response of elastomer composites consisting of ferrous particles embedded in a polymer matrix. J. Intelligent Mat. Syst. Struct. 7 (1996) 613-622

[2] M. Lokander and B. Stenberg: Performance of isotropic magnetorheological rubber materials. Polymer Testing 22 (2003) 245–251

[3] H. Böse, R. Röder: Magnetorheological elastomers with high variability of their mechanical properties. J. Phys.: Conf. Ser. 149 (2009) 012090

[4] H. Böse: Viscoelastic properties of silicone-based magnetorheological elastomers. Proc. 10th ERMR, World Scientific, (2007) 51-57

[5] H. Böse, J. Ehrlich, P. Löschke, J. Rumpel: Novel valve mechanism based on magnetoactive polymers. Proceedings of Actuator 2010 – 12th International Conference on New Actuators (2010) 876-879

HYSCOM KNEE, A PROSTHETIC KNEE JOINT WITH STANCE AND SWING MOTION CONTROL SYSTEM UTILIZING COMPACT MR FLUID BRAKE

YUICHI HIKICHI

Graduate school of engineering, Tohoku University
2-1-1, Katahira, Aoba-ku, Sendai, Miyagi 980-8577, Japan

MASAMI NAKANO, TEPPEI TSUJITA

Institute of Fluid Science, Tohoku University
2-1-1, Katahira, Aoba-ku, Sendai, Miyagi 980-8577, Japan

HYSCOM knee joint with stance and swing motion control system utilizing a developed compact MR fluid brake has been developed, and the details of the knee design and control system and results of its field walking test on some individual TF amputees are reported.

1. Introduction

In order to support TF(Trans-Femoral) amputees, various above-knee prostheses have been developed and many of them have a passive type knee joint. A TF amputee needs to bend and swing prosthesis's knee joint when he walks using the above-knee prosthesis. However, if the joint is bent incidentally during support phase, there is possibility to fall down because its passive joint cannot support amputee's weight. In addition, the prosthesis needs to support his weight with its knee bent when he ascends stairs. Therefore, a prosthesis which can control its joint torque based on walking situation is required.

Recently, few powered active type prosthesis which can control the knee joint have been developed. For example, POWER KNEE (Össur Co.) [1] is commercially available. However, it is heavy, high energy consumption and expensive. In order to overcome these difficulties, semi-active type prostheses utilizing MR fluid device have been proposed. Kim and Oh developed an above knee prosthesis using MR damper [2]. However, the prosthesis is relatively large and its controller can handle only walking. Therefore, new knee joint with stance and swing motion control system utilizing a compact MR fluid brake is developed. By installing a MR fluid brake to HYSCOM(**HY**draulic **S**tance phase **C**ontrolled by **O**ptional **M**otion) [3], its size is reduced and resistance torque of its knee joint can be controlled voluntarily by a TF amputee. By using the developed prosthesis, a TF amputee is able to ascend and descend stairs and walk slopes with a reciprocating gait. The detail of the developed MR fluid brake, the control scheme and evaluations for field walking are presented.

124

2. Development of a Prosthetic Knee Joint

2.1. *MR fluid rotary brake*

In this research, a compact magnetorheological (MR) fluid rotary brake with multiple disks for an above-knee prosthesis is developed. Figure 1 shows a cross-sectional view of the developed MR fluid rotary brake. The MR fluid brake consists of a shaft, a coil, yokes, brake plates. 11 brake plates are fixed on a housing and 10 brake plates are connected to the shaft. The fixed brake plates and the rotator brake plates are allocated alternately and gap between these plates is 0.15 mm. MR fluid (LORD MRF-132DG) is filled in the gap. The plate is made from a cold rolled steel sheet (SPCC) which is 0.5mm in thickness. The yoke and the shaft are made of steel (SS400) and non magnetic material, respectively.

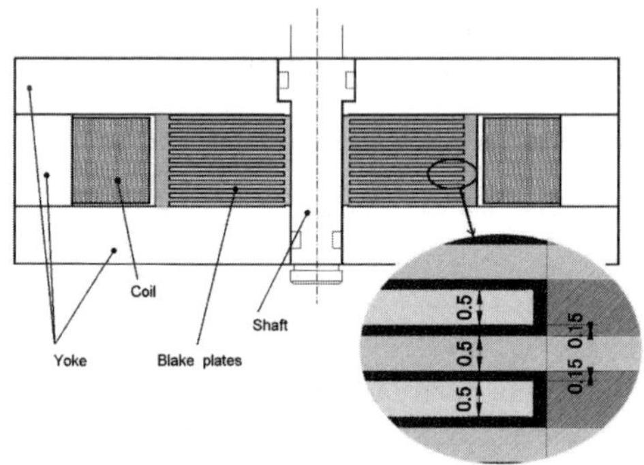

Figure 1. Structure of a MR fluid rotary brake.

In order to design a brake, magnetic flux density was computed by magnetic field analysis software JMAG (JSOL Corp.) as shown in Figure 2. The overall dimensions of a simulation model of the brake are 30.5 mm in thickness, 90 mm in diameter. This simplified simulation model has 9 fixed brake plates and 10 rotator brake plates. The figure shows contour plot of magnetic flux density when coil current I is 1 A and strength of the magnetic field applied the MF fluid is about 1 T. In Figure 3, vectors express directions of magnetic flux. As shown in the figure, the magnetic field is applied to the brake plates vertically. From this simulation result, resistance torque M can be derived by the following equation on assumption that shear yield stress depend on magnetic flux density is dominant comparing with shear stress depend on shear rate.

$$M = 2\pi \int_{R_i}^{R_o} P_s(B(r))r^2 dr , \tag{1}$$

where R_o and R_i are respectively outer and inner diameters of a brake plate. Function $B(r)$ indicates magnetic field strength in radius r which can be obtained by the simulation result and function $P_s(B)$ expresses shear yield stress under magnetic field strength B derived from LORD technical data sheet. From Eq. (1), the designed MR fluid brake can exert 31.5 Nm brake torque at 1 A.

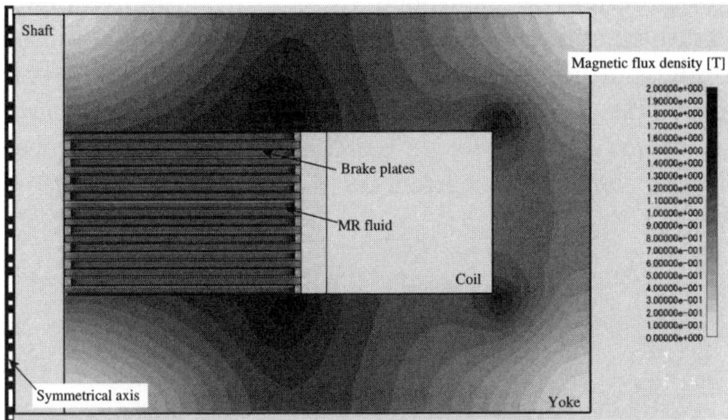

Figure 2. Analytical distribution of magnetic flux density ($I = 1$ A).

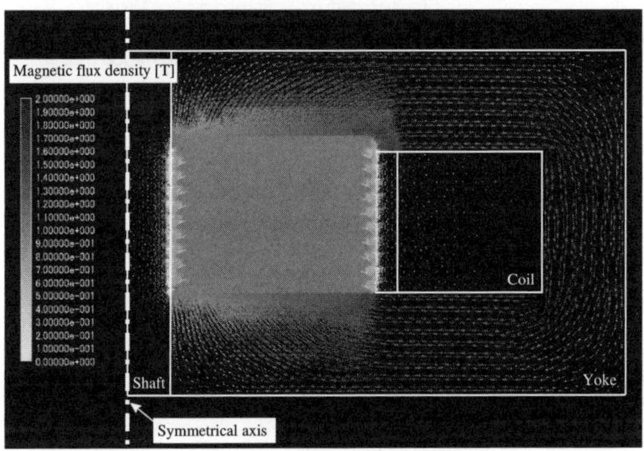

Figure 3. Directions of magnetic flux ($I = 1$ A).

Figure 4 shows a developed MR fluid brake based on the structure shown in Figure 1. The overall dimensions of the developed brake are 31mm in thickness, 94mm in diameter and its weight is about 1.6kg.

Figure 4. Picture of a developed MR fluid brake.

In order to evaluate the developed brake, resistance torque is measured by torque measurement system as shown in Figure 5. The MR fluid brake is driven by a speed controlled motor and resistance torque is measured by a torque transducer. Coil current is controlled by a power supply. Figure 6 shows relationship between coil current and resistance torque. This result indicates that resistance torque can be controlled by coil current and it does not depend on rotational speed. However, the torque is proportional to coil current. Almost 20Nm brake torque is exerted at 0.8A. This experimental result fits the simulation result when coil current is 1 A considering the proportional relationship.

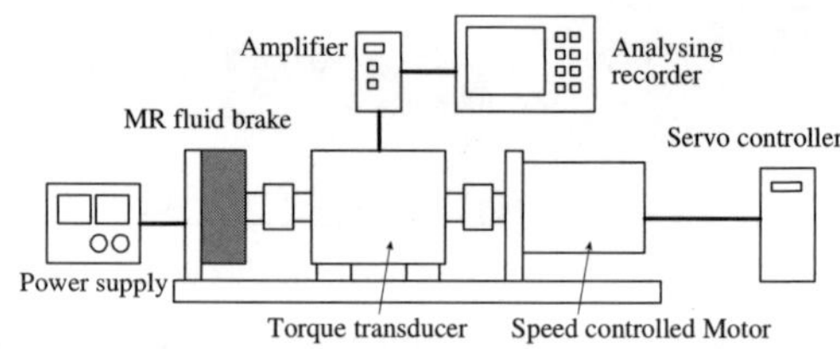

Figure 5. Experimental system for measuring resistance torque.

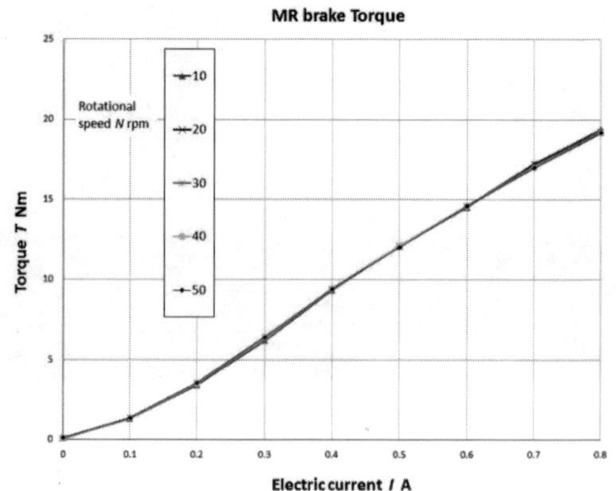

Figure 6. Relationship between electric current and torque.

2.2. *Control methods*

In order to control MR fluid brake voluntarily, four sensors are mounted in this knee joint. The first sensor is put above the knee axis which detects A-P movements, the second one senses knee angular, the third one is put on the lowest part of the knee joint, which measures the ankle moment. The last one measures the knee moment. These data are taken into a laptop computer through a cable and calculated based on a control strategy. Then, adequate voltage is applied to the MR fluid brake to harmonize with individual situations. Figure 7 shows the control system of the knee joint.

The control scheme of the knee joint is as follows. During swing phase; almost MR brake is almost free. In stance phase, stress on the sensor above knee changes resistance to a very high value. When amputees bend their stumps that give signals, the controller directs the MR brake to perform with adequate small torque. When they extend their stumps, it performs with adequate large torque (knee locks).

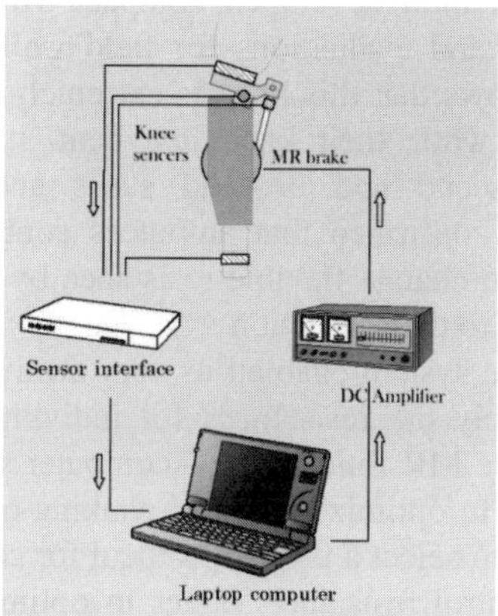

Figure 7. Knee control system.

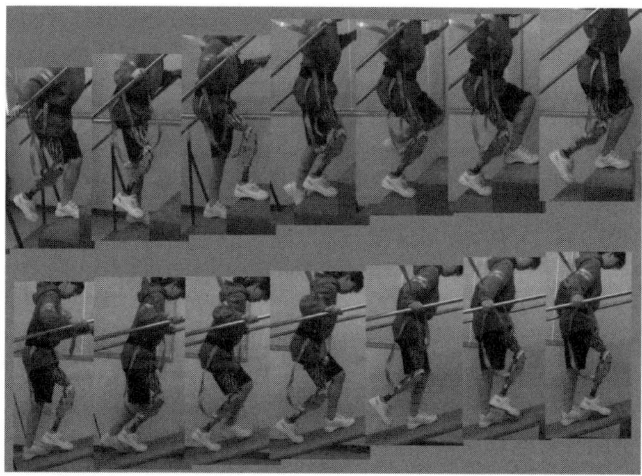

Figure 8. Ascending 20cm height steps of stairs and 20 ° of slopes with a reciprocating gait.

3. Evaluation for Field Walking

Figure 8 shows a TF amputee uses the developed prosthesis to ascent stairs and slopes. This trial evaluation proves that this knee enables TF amputees to ascend and descend two type stairs (15cm and 20cm height steps) and two type slopes (10 ° and 20 ° slopes) with a reciprocating gait.

4. Concluding Remarks

In this paper, the detail of a developed compact MR fluid brake, a control scheme of a knee joint and evaluations for field walking are presented. The evaluation test result shows that this knee is extremely safe as it can lock with TF amputees' intention while their knees are bent. It also proved that a TF amputee is able to ascend and descend stairs and walk slopes with a reciprocating gait. It is confirmed that amputees control the MR fluid brake easily. They were able to change flexible resistance by themselves and the MR brake quickly responded to their intention

This smart MR knee system enabled us very easily to change the program and to let adjust precisely the resistances for individual situations. This trial evaluation has proved the MR fluid brake + computer system has enhanced the adjustability and helped to optimize patients' walking on stairs. It is also found that it is very important to select a proper protocol for collecting all the data and calculating the knee control resistance values to optimize patients' walking. It means that now we have a possibility to construct a good program that makes progress in patients' ADL especially in tough situations such as on stairs, ramps or with obstacles.

References

1. http://www.ossur.com/?PageID=14255
2. J.-H. Kim and J.-H Oh, *Proc. IEEE Int. Conf. Robotics and Automation*, 3686(2001).
3. Y. Hikichi, *Bull. JPN. Soc. Prosthetics & Orthotics* **22**, 147(2006).

A SEMI-ACTIVE CONTROL OF MR MOUNT FOR VIBRATION ISOLATION

W. H. KUO

*Department of Mechatronic Technology, Tungnan University,
Taipei, 22202, Taiwan*

R. PAN[1], J. SHAW[1] and G. LIN[2]

[1]*Institute of Mechatronic Engineering, National Taipei University of Technology,
Taipei, 10608, Taiwan*
[2]*Institute of Mechatronic Engineering, Tungnan University, Taipei, 22202,
Taiwan*

In this paper, developments of a MR mount for vibration attenuation are presented. First, an MR mount was designed, and analysis of the magnetic circuits by using the magneto V64 software was conducted, followed by calculating the corresponding magnetic flux density and the magnetorheological fluid yield shear stress. Then, this MR mount was manufactured and filled with MR fluid (MRF-122EG made by Lord Ltd), and a single degree of freedom system with the MR mount is set up for vibration validation. Utilizing various controllers like the skyhook, PID and self-tuning fuzzy PID controllers, the damping effects of the MR mount are conducted. Experimental results show that it has better damping effectiveness by using the self-tuning fuzzy PID controller than the other controllers for the semi-active control system with the MR mount.

1. Introduction

In order to reduce unwanted vibration of a vibrating system, various types of mounts are adopted either in passive, semi-active or active control manner. They have advantages and disadvantages, depending on their applications. Recently, the adjustable semi-active mounts have been designed by incorporating magneto-rheological (MR) fluids [1-3]. A semi-active control technique can provide real-time dissipation of the system energy, which has proved to provide better performance than the passive control, and which saves energy better than active control. Karnopp et al. [4] used semi-active force generators to vibration control. Lu [5] studied active and semi-active air-spring suspension systems and compared the performances. MR fluids had been studied in many applications of semi-active control. Dyke and Spence [6] proposed the MR dampers for seismic protection.

The PID algorithm is the most popular feedback controller used within the process industries. It has been successfully used for over 60 years. It can provide excellent control performance despite the varied dynamic characteristics of process plant [7]. Fuzzy logic may tune each parameter of PID controller.

Mamdani is the first researcher using fuzzy logic to control steam engine in 1974 [8]. In recent years, fuzzy logic control (FLC) had been applied to control many devices, and many researchers suggested combining the FLC with other controllers. For example, there are fuzzy sliding mode controllers (FSMC), fuzzy PID controller, fuzzy-PI controller etc. Chen and Chang [9] employed an optimal design method to FSMC. Huang and Lin [10] used adaptive fuzzy control with sliding surface to vehicle suspension control. Shaw et al. [11] applied FSMC to control an MR mount for vibration attenuation.

Some researchers study about application of a self-tuning fuzzy PID controller to overcome the appearance of nonlinearities and uncertainties in the vibration systems. The self-tuning fuzzy PID controller is the combination of a classical PID and fuzzy controller. Fuzzy logic is used to tune each parameter of the PID controller [12][13].

In this paper, a self-tuning fuzzy PID controller is developed to improve the performance of the semi-active control system with the MR mount in the attenuation of vibration.

2. MR mount Design

The MR mount includes current coil and iron parts. We wind the wire around the piston to make a simple electromagnet. The design parameters of the vibration isolation system with MR mount contain load mass, displacement, spring constant, yield shear stress of MR fluid and resonance frequency of system. Design structure of MR mount is shown in Figure 1. The simulated magnetic flux density by using the Magneto software is given in Figure 2. The relationship between magnetic flux density and the applied current is illustrated in Figure 3. We use MR fluid made by Lord Ltd (MRF-122EG).

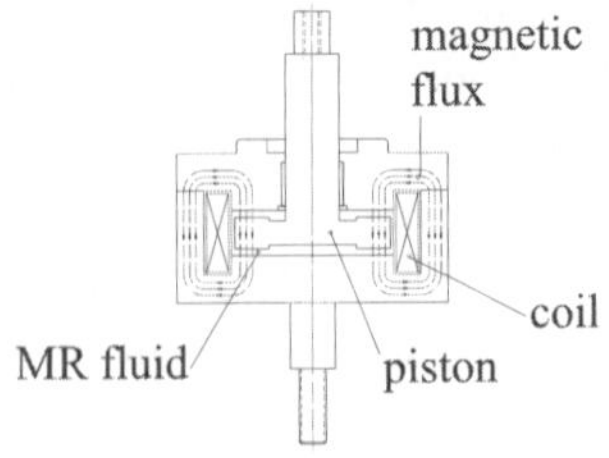

Figure 1. MR mount diagram

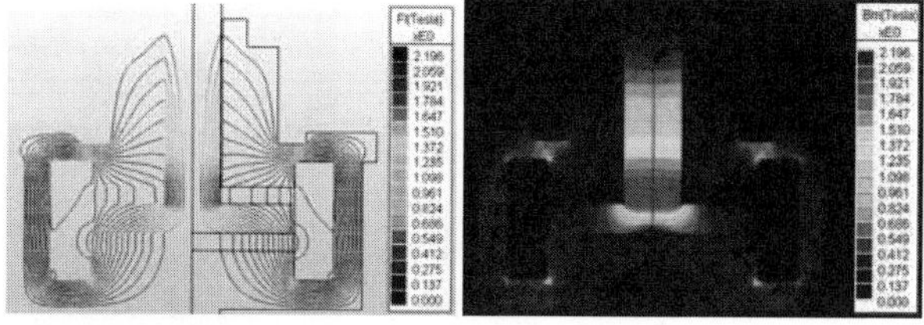

Figure 2.   Simulation of magnetic flux density

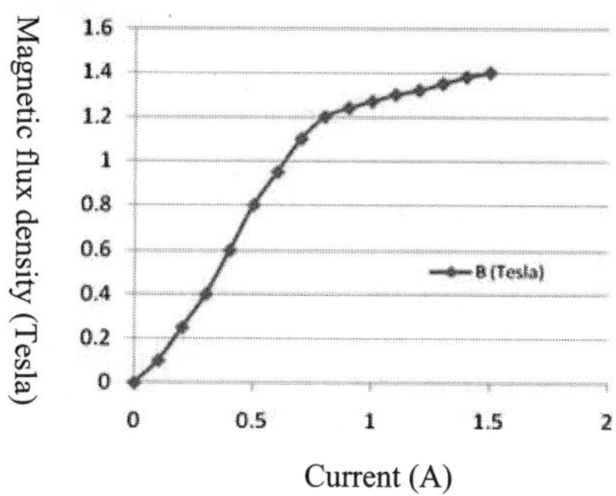

Figure 3. The relationship between magnetic flux density and current

3.   Modeling of MR Mount

The schematic diagram and photo of the one-dimensional MR mount system is shown in Figure 4. The dynamic equation of this vibration system can be derived as [14]

$$m\ddot{x}(t) - A\frac{\eta}{h}[\dot{x}(t) - \dot{y}(t)] + (k + \frac{A_p^2}{C_1 + C_2})[x(t) - y(t)]$$

$$+ \frac{C_2 A_p}{C_1 + C_2}\{(\frac{2n+1}{n})(\frac{2}{Wh^2})(\frac{A}{2} - \frac{C_2 A_p}{C_1 + C_2})|\dot{x}(t) - \dot{y}(t)|\}^n(\frac{2\eta L}{h}) = -F_{MR} \tag{1}$$

It is noticed that A is the flow area, Ap is the piston area of the upper chamber, C_1 and C_2 are the compliances of the upper and lower chamber, h is the gap of the magnetic pole, k is the stiffness of the spring, L is the length of the magnetic pole, m is the mass, n is the flow behavior index of Herschel-Bulkey model, W is the width of the magnetic pole, $x(t)$ and $y(t)$ represent the respective displacements at the mass and base, η is the viscosity of the MR fluid.

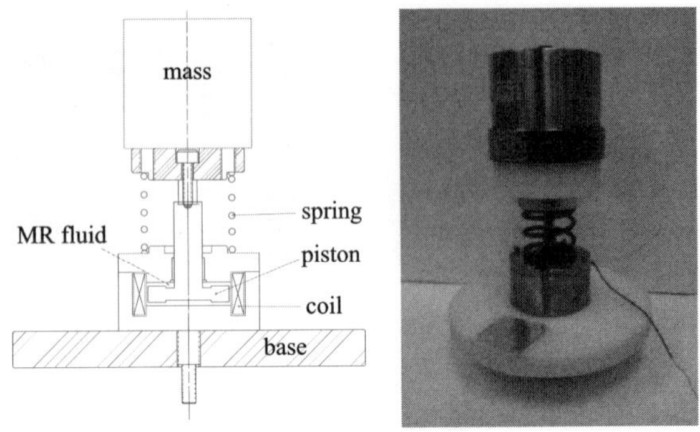

Figure 4. Schematic diagram and photo of MR mount

4. Self-Tuning Fuzzy PID Controller

The self-tuning fuzzy PID control block diagram is shown in Figure 5. The control value of fuzzy controller is utilized to tune parameters of PID controller. The design procedures are as follows.

   (a) First, choose three parameters of PID to set the fuzzy relationship between deviation (e)/deviation rate (ec) and these three parameters.

   (b) Check the deviation (e) and deviation rate (ec) gradually. According to fuzzy control principle, revise these three parameters on-line in order to satisfy the requirement of control parameters in various e and ec.

   (c) According to the rule, the parameters (k_P, k_I and k_D) can self-tune in different e and ec.

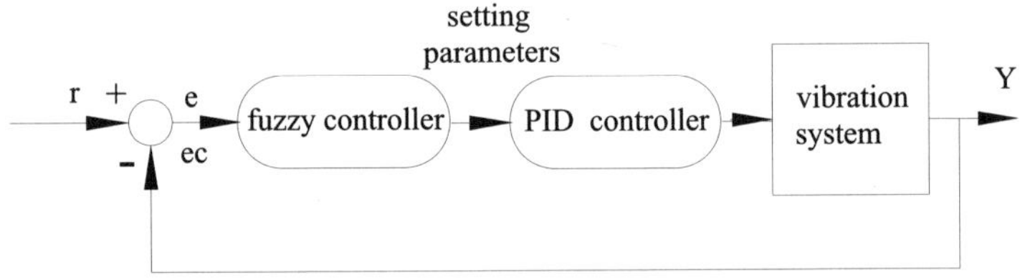

Figure 5.   The fuzzy PID control block diagram

Absolute deviation $|e|$ and absolute deviation rate $|ec|$ are the input variables. Each variable state in this system can be subdivided into a range of small (S), middle(M), and big(B) whose membership functions are depicted in Figure 6. The fuzzy controller has only 5 rules regulating the inputs ($|e|$, $|ec|$)and output (k_P, k_I, k_D). The antecedents of the 5 rules are:

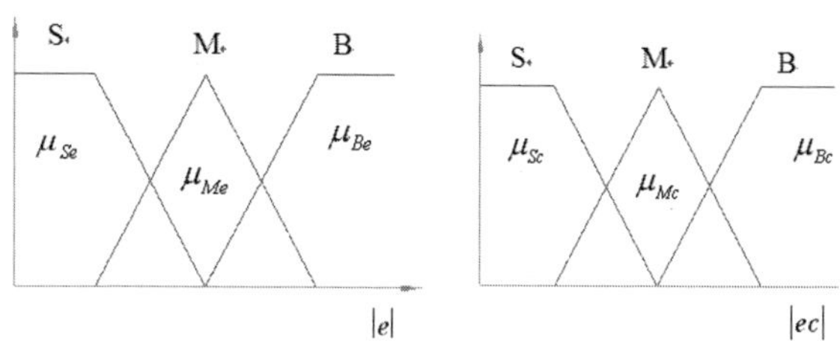

Figure 6. Membership function of deviation $|e|$ and deviation rate $|ec|$

$$(1)\,|e| = B \;;\; (2)\,|e| = M \text{ and } |ec| = B \;;\; (3)\,|e| = M \text{ and } |ec| = M \;;\; (4)\,|e| = M$$
$$\text{and}\,|ec| = S \;;\; (5)\,|e| = S \tag{2}$$

The corresponding firing strength of rule j, $\mu_j(|e|,|ec|)$, can be obtained as

$$(1)\,\mu_1 = \mu_{Be}(|e|); \;(2)\; \mu_2 = \mu_{Me}(|e|) \wedge \mu_{Bc}(|ec|) \;;$$
$$(3)\,\mu_3 = \mu_{Me}(|e|) \wedge \mu_{Mc}(|ec|) \;;\; (4)\,\mu_4 = \mu_{Me}(|e|) \wedge \mu_{Sc}(|ec|) \;;$$
$$(5)\; \mu_5 = \mu_{Se}(|e|) \tag{3}$$

From the measuring values of $|e|$ and $|ec|$, the resulting parameters of the PID controller can be calculated according to

$$k_P = \frac{\sum_{j=1}^{5}\mu_j(|e|,|ec|) \times w_{pj}}{\sum_{j=1}^{5}\mu_j(|e|,|ec|)} \;;\; k_I = \frac{\sum_{j=1}^{5}\mu_j(|e|,|ec|) \times w_{Ij}}{\sum_{j=1}^{5}\mu_j(|e|,|ec|)} \;;\; k_D = \frac{\sum_{j=1}^{5}\mu(|e|,|ec|) \times w_{Dj}}{\sum_{j=1}^{5}\mu_j(|e|,|ec|)} \tag{4}$$

where w_{Pj}, w_{Ij}, w_{Dj} are the weighting value of k_P, k_D, k_I in the different states.

$$(1)\,w_{P1} = w'_{P1}, w_{D1} = 0,\; w_{I1} = 0$$
$$(2)\,w_{P2} = w'_{P2},\; w_{D2} = w'_{D2},\; w_{I2} = 0$$
$$(3)\,w_{P3} = w'_{P3},\; w_{D3} = w'_{D3},\; w_{I3} = 0$$
$$(4)\,w_{P4} = w'_{P4},\; w_{D4} = w'_{D4},\; w_{I4} = 0$$
$$(5)\,w_{P5} = w'_{P5},\; w_{D5} = w'_{D5},\; w_{I5} = w'_{I5} \tag{5}$$

From Equation (4), we can obtain the k_P, k_I, k_D parameters on-line. The PID control value u can be obatined by the discrete formulas as follows.

$$u_n = k_P e_n + k_1 \sum_{i}^{n} e_i + k_D(e_n - e_{n-1}) \tag{6}$$

$$u_{n-1} = k_P e_{n-1} + k_1 \sum_{i}^{n-1} e_i + k_D(e_{n-1} - e_{n-2}) \tag{7}$$

or $\Delta u_n = k_P(e_n - e_{n-1}) + k_1 e_n + k_D(e_n - 2e_{n-1} + e_{n-2})$ $\tag{8}$

134

5. Experimental Results

The experimental setup for studying the vibration attenuation capability of the MR mount is shown in Figure 7. Two displacement sensors are used to detect vibrations at x and y places. The sinusoidal disturbance $y(t)$ with single frequency between 2Hz~30Hz and amplitude at 0.15 *mm* is employed as the base excitation. Vibration amplitude reductions by the skyhook, PID and self-tuning fuzzy PID controllers are illustrated in Figure 8, from which the fuzzy PID controller outperforms other controllers near the resonant frequency.

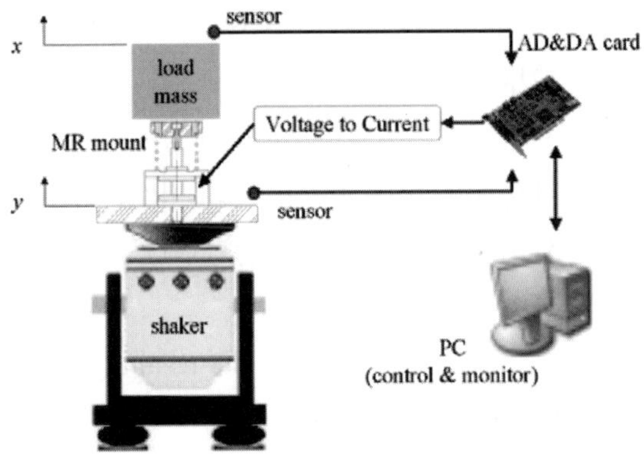

Figure 7.   The experimental setup

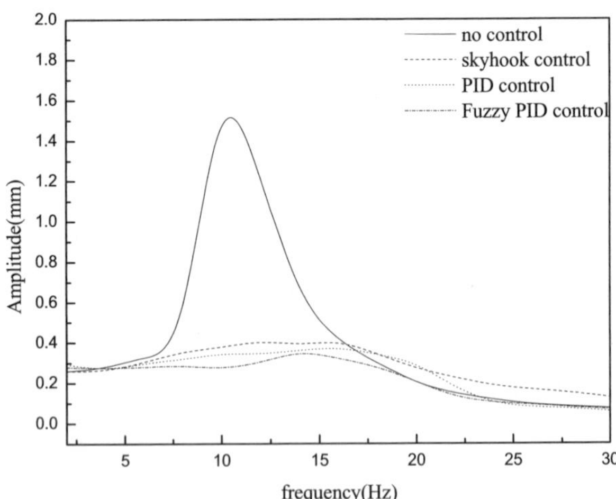

Figure 8.   Amplitude of vibration attenuation for 2Hz~30Hz disturbance

6. Conclusion

In this paper, we design an MR mount and simulate the magnetic field in the MR mount, and use a self-tuning fuzzy PID controller for vibration suppression. Good vibration controls are obtained by fuzzy PID controller as compared with the PID and skyhook controllers. In addition, the MR mount design can avoid the damage from vibration and friction. In the near future, the experiments for vibration suppression of the MR mount with another intelligent controller will be carried out to validate the effectiveness of the constructed MR mount.

References

1. D. Carlson, W. Matthis, and J. R. Toscano, *Proc. of SPIE*, **4332**, 308 (2001).
2. G. Yang, B. F. Spencer, J. D. Carlson and M. K. Sain, *Engr Stru.*, **24/3**, 309 (2002).
3. S. B. Choi, S. R. Hong, K. G. Sung and J. W. Sohn, *Int. J. of Mechanical Sciences* **50,** 559 (2008).
4. D. Karnopp, M. J. Crosby and R. A. Harwood, *J. of Engineering for Industry*, **96/2**, 619 (974).
5. Z. Lu, *J. of the China Railway Society*, **23/ 1**, 33 (2001).
6. S. J. Dyke and B. F. Spencer, *Smart Mat. and Stru.*, **7/ 5**, 693 (1998).
7. K.J. Astrom, *Control System Design*, **Chap. 6**,216(2002).
8. E. H. Mamdani, *Proc. IEE*, **121/12**, 1585 (1974).
9. C. L. Chen and M. H. Chang, *Fuzzy Sets and Systems*, **93/1**, 37(1998).
10. S. J. Huang and W. C. Lin, *IEEE Trans. on Fuzzy Syst.*, **11/4**, 550 (2003).
11. J. Shaw, R. Pan and Y. C. Chang, *World Academy of Science, Engineering and Technology,* **60**,330(2009).
12. Zulfatman and M. F. Rahmat, *Int. J. on Smart Sensing and Intell. Systems*, **2/2**(2009).
13. H.A. Malki, D. Misir, D. Feigenspan, G. Chen, *IEEE trans. On Control Systems Tech.*, **5**, 371(1997).
14. S. B. Choi, H. J. Song, H. H. Lee, S. C. Lim, J. H. Kim and H. J. Choi, *Int. J. of Vehicle Design*, **33/1-3**, 2 (2003).

NONLINEAR DYNAMIC CHARACTERISTIC MODEL OF ARTIFICIAL RUBBER MUSCLE MANIPULATOR USING MR BRAKE

HIROKI TOMORI

Department of Precision Mechanics, Chuo University, 1-13-27 Kasuga, Bunkyou-ku, Tokyo, 112-8551, JAPAN

YUICHIRO MIDORIKAWA

Chuo University, Biomechatronics Laboratory, 1-13-27 Kasuga, Bunkyou-ku, Tokyo, 112-8551, JAPAN

TARO NAKAMURA

Chuo University, Biomechatronics Laboratory, 1-13-27 Kasuga, Bunkyou-ku, Tokyo, 112-8551, JAPAN

An artificial rubber muscle was paid to attention as an actuator in the present study because human was safe for the manipulator to have come in contact with human. However, this actuator occurs easily the vibration, and is late the response because of applying the air pressure. Then, the MR brake that uses the MR fluid with an early response is built into the joint, and controls the vibration. In this paper, we have grasped a manipulator's dynamic characteristics by construction of a model for improvement in the control performance of MR brake. Furthermore, the simulation was performed using the model and efficient breaking of MR brake was examined.

1. Introduction

Recently, there have been advances in the development of partner robots aimed at providing life support and robots aimed at providing medical treatment and care such as power assistance [1]. These robots are expected to be safe to human when these robots contact human.

Then, we developed the straight-fiber-type artificial muscle [2] as actuators. It has a high contraction percentage and high contractive force, and this muscle has a long life compared with past McKibben type rubber artificial muscle [3]-[6]. Moreover, these artificial muscles are lightweight, have high output, and are flexible. However, because the artificial muscle is flexible, it vibrates when there is a high load.

Therefore, we are interested in the MR brake using magnetorheological (MR) fluid [7], because its apparent viscosity can be changed very quickly (in milliseconds) by applying a magnetic field. In this study, by applying MR fluid to the joint part of an artificial muscle manipulator, we aim at suppressing vibration of the manipulator's arm. We used a 1-degree-of-freedome (1-DOF) artificial muscle manipulator that applied MR fluid to a joint.

In this paper, we examine an effect which the MR brake gives to the manipulator by using a manipulator model.

2. The Straight-Fiber-Type Artificial Muscle

The whole artificial muscle figure developed at our laboratory is shown in Figure 1. The form of this artificial muscle is tubular, and the material used is natural-rubber-latex liquid. Moreover, since its structure includes a carbon fiber layer in the direction of an axis, if air pressure is supplied to this artificial muscle, the rubber membrane will expand, but the fiber stops the growth in the direction of an axis so that the membrane is mostly not extended. As a result, the muscle expands only in a radial direction and contracts in the direction of an axis. The contractile force at this time is used as an actuator. This artificial muscle is lightweight, has high output, and is flexible.

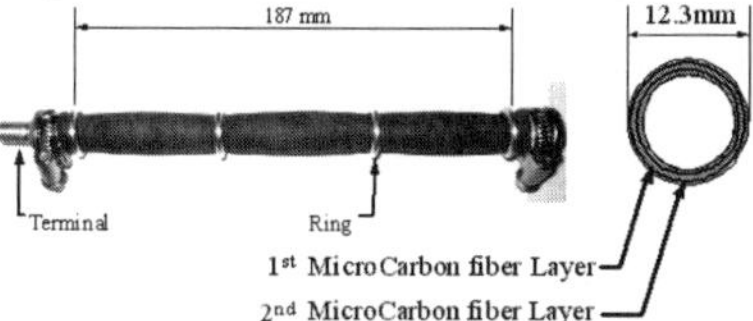

Figure 1. Fiber double-layer type artificial muscle.

3. MR Brakes

We focused on the MR brake as a way to apply MR fluid to the joint. In this study, we used the MRB-2107-3 of the LORD Co. as MR braking equipment. The overall view and the diagrammatic illustration of the MRB-2107-3 are shown in Figure 2. This device generates friction on the surface of the disk as it changes in response to the magnetic field by arranging the MR fluid in the internal disk surroundings. It is possible to apply brakes to a rotation operation according to this mechanism. The MR brake has the following features:

 (i) Since torque density is high, it is small, and it is high output.
 (ii) Since the structure is simple, it is strong.
(iii) The speed of response is in units of several milliseconds.
(iv) It can be changed continuously.
 (v) It is stable.

(vi) It is an energy-saving model.

Figure 2. Fiber double-layer type artificial muscle.

4. 1-DOF Artificial Muscle Manipulator

The overall view of the 1-DOF artificial muscle manipulator that was produced in this study is shown in Fig. 3. The artificial muscle manipulator is arranged in the form where two artificial muscles rival one another, and the contractile force of the artificial muscle is transmitted to the axis of rotation through the pulley. The MR brake is fixed to the 1st link side, and it can apply braking power to the axis of rotation. The manipulator is installed in the load cell to detect the load torque that hangs to the joint and the encoder to detect the joint angle.

Table 1. Parameters of the manipulator.

Mass of 2nd link m_1 [kg]	0.168
Length of 2nd link l_1 [mm]	250
Center of gravity of 2nd link l'_G [mm]	240
Movable range θ [rad]	$-\pi/2 \leqq \theta \leqq \pi/6$
Inertia of 2nd link I_1 [kgm^2]	4.80×10^{-6}
Parallel number of artificial muscles n_p	1

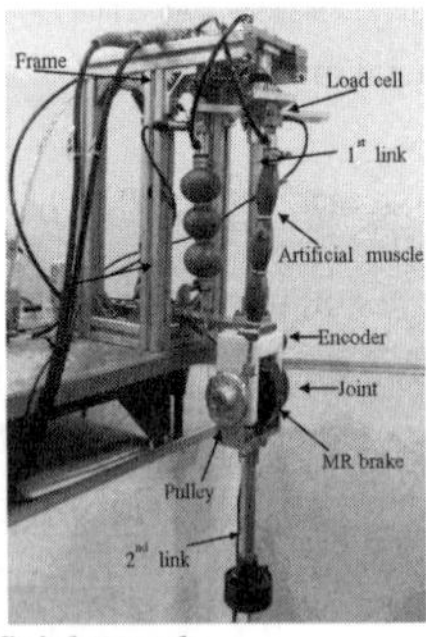

Figure 3. Fiber double-layer type artificial muscle.

5. Manipulator Model

We built a manipulator model while considering the dynamic characteristics of artificial muscles. By building this manipulator model, we were able to approach against control from theory.

The schematic view of this manipulator model is shown in Figure 4. This model consists of three parts. The first part is a mechanical equilibrium model [8] treating the static characteristic of an artificial muscle. It is made easy to use the artificial muscle by linearizing the relation of the amount of pressure, load and contraction. The second part is a model [9] containing the elements related to the dynamic characteristics of the air artificial muscle system, including a pressure valve. The third part is a manipulator's load system model. These models were combined and considered as the manipulator model.

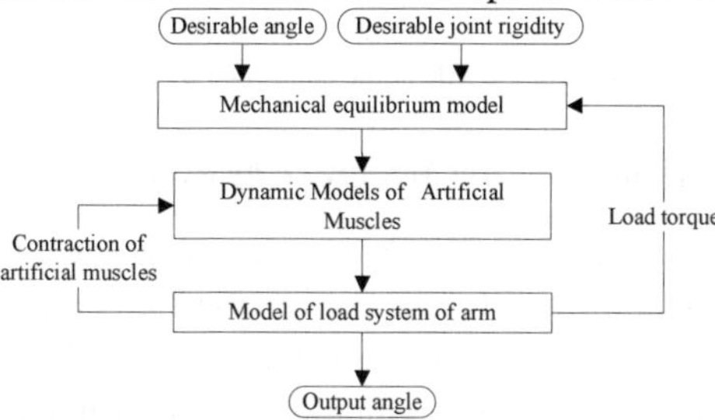

Figure 4. A schematic of the manipulator model.

6. Effect Examination of the MR Brake

6.1. *The Control of the MR Brake*

Until now, we inputted the brake torque value into the MR brake, and had stopped rotation of an arm. However, when stationary error arose, it was observed that MR brake checked compensation in the conventional way, because the MR brake continues outputting brake torque, without being dependent on revolving speed. In this paper, we propose the controller of MR brake for not inducing stationary error. The Propositional technique of control of MR brake is shown Figure 5. In this propositional technique, the command torque of the MR brake depends on the rotary speed of the arm, and the braking torque lowers with an arm becoming slow. By this control, MR brake obtained output characteristics shown in Figure 6.

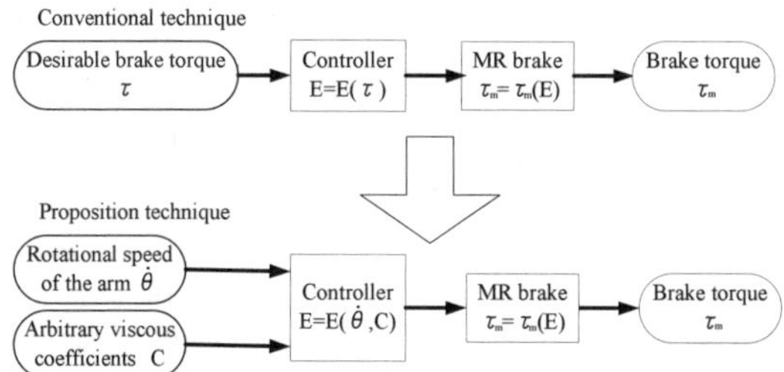

Figure 5. Controller of the MR brake.

140

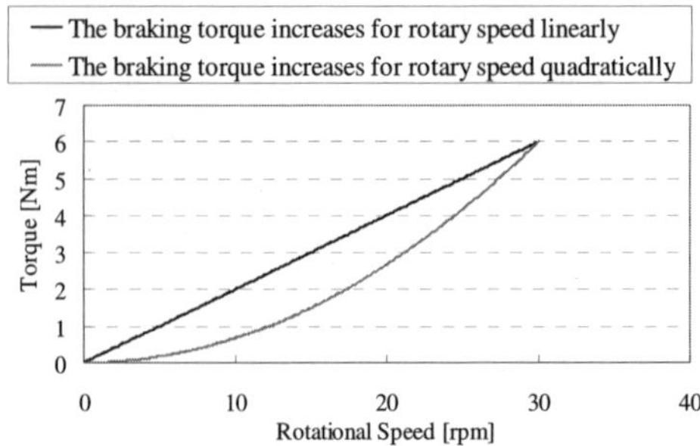

Figure 6. Output characteristics of the MR brake controlled by proposition technique.

6.2. *Effect Examination of the MR brake by using the Model*

We simulated raising load (8.33[N]) to a desirable angle by using manipulator model. The simulation result by the conventional control is shown in Figure 7. And simulation results by the propositional control are shown in Figure 8.

From results of simulations, vibration of the arm was suppressed by MR brake. And stationary error of the angle response decreased by proposition technique. We think that the reduction in the stationary error is an effect of control of the MR brake. When rotation of the arm is slow, generation of a stationary error is prevented by weakening the brake.

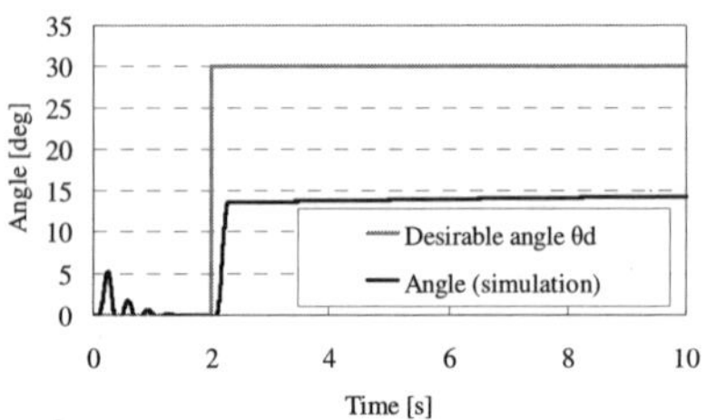

Figure 7. Simulation results by the conventional control.

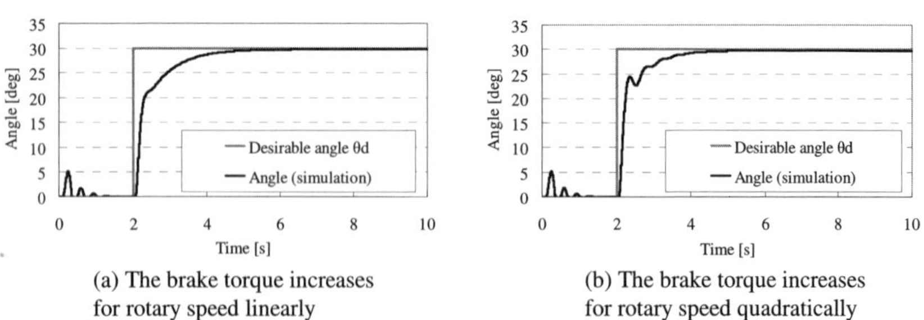

(a) The brake torque increases
for rotary speed linearly

(b) The brake torque increases
for rotary speed quadratically

Figure 8. Simulation results by the proposition control.

7. Conclusion

In this study, we examined the effect of the MR brake by simulation and checked the reduction of vibration and stationary error by the MR brake experimentally. We believe our experimental results show effective control by the MR brake.

As future research, the control in which the MR brake does not interfere superfluously will be designed for flattering to the target angle of the arm. We also hope to further improve the rise time.

References

1. H. Satoh, T. Kawabata, F. Tanaka, and Y. Sankai, "Transferring-Care Assistance with Robot Suit HAL," *Transactions of the Japan Society of Mechanical Engineers, Series C* Vol. 76, pp. 227-235, 2010.
2. V. L. Nickel, M. D. J. Perry, and A. L. Garrett, "Development of useful function in the severely paralysed hand," *Journal of Bone and Joint Surgery* 45A(5), 1963, pp. 933–952.
3. M. M. Gavrilovic and Maric M. R., "Positional servo-mechanism activated by artificial muscles," *Medical and Biological Engineering* 7, 1969, pp. 77–82.
4. G. K. Klute, J. M. Czernieki, and B. Hannaford, "McKibben Artificial Muscles: Pneumatic Actuators with Biomechanical Intelligence," *Proceedings of the IEEE/ASME International Conference on Advanced Intelligent Mechatronics* 1999, pp. 221–226.
5. C. P. Chou and B. Hannaford, "Static and Dynamic Characteristics of McKibben Pneumatic Artificial Muscles," *Proceedings of IEEE International Conference On Robotics and Automation* 1994, pp. 281–286.
6. T. Nakamura, "Experimental Comparisons between McKibben type Artificial Muscles and Straight Fibers Type Artificial Muscles," *SPIE International Conference on Smart Structures, Devices and Systems III*, 2006.
7. B. J. Park, C. W. Park, S. W. Yang, H. M. Kim, and H. J. Choi, "Core-Shell Typed Polymer Coated-Carbonyl Iron Suspension and Their Magnetorheology," *ERMR08*, 2008, pg. 102.
8. T. Nakamura and H. Shinohara, "Derivation of a mathematical model for straight fibers type artificial muscles and application of feed-forward linearization," *11th ROBOTICS Symposia*, pp. 222-227, 2006.
9. H. Maeda, H. Tomori, and T. Nakamura, "Orbit tracking control of 6-DOF lubber artificial muscle manipulator considering nonlinear dynamics model," *15th ROBOTICS Symposia*, pp. 429-435,2010.

PERFORMANCE EVALUATION OF RAILWAY SECONDARY SUSPENSION UTILIZING MAGENTORHEOLOGICAL FLUID DAMPER[*]

SUNG HOON HA, MIN SANG SEONG, HYUNG SUB KIM
and SEUNG-BOK CHOI[†]

Smart Structures and Systems Laboratory, Department of Mechanical Engineering, In-ha University, Incheon, 402-751, Korea

This paper presents to the feasibility for improving the ride quality of railway vehicle equipped with semi-active suspension system using magnetorheological (MR) fluid damper. In order to achieve this goal, a fifteen degree of freedom of railway vehicle model, which includes a car body, bogie frame and wheel-set is proposed to represent lateral, yaw and roll. The MR damper system is incorporated with the governing equation of motion of the railway vehicle which includes secondary suspension. To illustrate the effectiveness of the controlled MR dampers on railway vehicle secondary suspension systems, the sky-hook control law using the velocity feedback is adopted as the system controller. Subsequently, computer simulation of performance evaluation such as vibration control of the car body is performed using Matlab. Various control performance are demonstrated under external excitation by creep force between wheel and rail.

1. Introduction

Recently, the realization of high speed railway vehicle is increasing in the many countries. It has been provided with efficient transportation which is included with passenger and freight railway vehicle. However, the high speed railway vehicle would cause car body vibration which may induce the various problems such as ride stability, ride quality and track abrasion. Thus the vibration control of railway vehicle is necessary for the improvement of ride quality and stability of car body.

In general, the suspension systems of railway vehicle have been proposed depending upon the operation mode: passive, active and semi-active. A passive railway vehicle suspension featuring spring, oil damper and pneumatic damper provides design simplicity, but performance limitations are inevitable in the relatively wide frequency range. The active one generally provides high control

[*] This work was supported by National Research Foundation of Korea (NRF) grant funded by the Korea government (MEST) (No. 2010-0015090).

[†] Corresponding author.

performance in wide frequency range. Therefore, many researchers have been proposed active suspension technology for railway vehicle which is utilized by oil valve and pneumatic actuators [1-2]. However, it requires high power sources as well as sophisticated control algorithm. And active control suspension would need mechanical power into the system, so the stability of the control system needs to be carefully considered. On the other hand, the semi-active suspension offers desirable performance generally enhanced in the active mode without requiring large power sources and expensive hardware. Furthermore, various semi-active suspension system featuring magnetorheological (MR) or electrorheological (ER) fluid have been proposed and successfully applied in the real field, especially in wheel based passenger vehicle suspension system [3-5].

Therefore, in this research, we investigate the feasibility for improving the ride quality of railway vehicle featuring semi-active suspension system using MR fluid damper. Firstly, the MR damper system is designed which is incorporated with the governing equation of motion of the railway vehicle which includes secondary suspension. And then, to illustrate the effectiveness of the controlled MR dampers on railway vehicle secondary suspension systems, the sky-hook control law using the velocity feedback is adopted as the system controller. Subsequently, computer simulation of performance evaluation of vibration control is performed using Matlab.

2. Analytical Model of Railway Vehicle

The governing equations of motion for the railway vehicle with suspension systems can be derived using Newton`s laws. A 15-degree-of-freedom passenger railway vehicle model, shown in Figure 1, is presented to investigate the lateral response on tangent track to random track irregularities. The model degrees of freedom are given in Table 1. The wheel-set of the railway vehicle is assumed to follow the track perfectly in the vertical direction; the wheel-set motion is given by the creep force input. A bogie frame roll is neglected as a degree of freedom. Wheel-rail interaction creep forces are calculated by using Johnson and Vermeulen`s creep theory. The vehicle equations of motion are presented, including the track input terms. The lateral car body response is then investigated for representative alignment and creep force input [6].

The governing equations for the wheel-set can be expressed as follows:

$$m_w \ddot{\delta}_{1,3,7,9} + 2k_{py}(\delta_{1,3,7,9} - \delta_{5,12} - b\delta_{6,12}) + 2F_{y1,2,34} = 0 \qquad (1)$$

$$I_w \ddot{\delta}_{2,4,8,10} + 2k_{px}d_1^2(\delta_{2,4,8,10} - \delta_{6,12}) + 2F_{x1,2,3,4} = 0$$

The governing equations for the bogie frame can be expressed as follows:

144

$$m_f \ddot{\delta}_{5,11} - 2k_{py}(\delta_{1,7} - \delta_{5,11} - b\delta_{6,12}) - 2k_{py}(\delta_{3,9} - \delta_{5,11} + b\delta_{6,12})$$
$$+ 2k_{sy}(\delta_{5,11} - \delta_{13} - h_3\delta_{15} - l\delta_{14}) + 2c_{sy}(\dot{\delta}_{5,11} - \dot{\delta}_{13} - h_3\dot{\delta}_{15} - l\dot{\delta}_{14}) + 2F_{MR} = 0 \qquad (2)$$
$$I_{fy}\ddot{\delta}_{6,12} - b\{2k_{py}(\delta_{1,7} - \delta_{5,11} - b\delta_{6,12})\} + b\{2k_{py}(\delta_{3,9} - \delta_{5,11} + b\delta_{6,12})\} -$$
$$2k_{px}d_1^2(\delta_{2,8} - \delta_{6,12}) - 2k_{px}d_1^2(\delta_{4,10} - \delta_{6,12}) = 0$$

The governing equations for the car body can be expressed as follows:
$$m_c\ddot{\delta}_{13} + 2k_{sy}(\delta_{13} - \delta_5 + h_3\delta_{15}) + 2c_{sy}(\dot{\delta}_{13} - \dot{\delta}_5 + h_3\dot{\delta}_{15}) + 2F_{MR}$$
$$+ 2k_{sy}(\delta_{13} - \delta_{11} + h_3\delta_{15}) + 2c_{sy}(\dot{\delta}_{13} - \dot{\delta}_{11} + h_3\dot{\delta}_{15}) + 2F_{MR} = 0$$
$$I_{cy}\ddot{\delta}_{14} + 2lk_{sy}(\delta_{13} - \delta_5 + h_3\delta_{15}) + 2lc_{sy}(\dot{\delta}_{13} - \dot{\delta}_5 + h_3\dot{\delta}_{15}) + 2lF_{MR} \qquad (3)$$
$$- 2lk_{sy}(\delta_{13} - \delta_{11} + h_3\delta_{15}) - 2lc_{sy}(\dot{\delta}_{13} - \dot{\delta}_{11} + h_3\dot{\delta}_{15}) - 2lF_{MR} = 0$$
$$I_{cr}\ddot{\delta}_{15} + 2h_3k_{sy}(\delta_{13} - \delta_5 + h_3\delta_{15}) + 2h_3c_{sy}(\dot{\delta}_{13} - \dot{\delta}_5 + h_3\dot{\delta}_{15}) + 2h_3F_{MR}$$
$$+ 2h_3k_{sy}(\delta_{13} - \delta_{11} + h_3\delta_{15}) + 2h_3c_{sy}(\dot{\delta}_{13} - \dot{\delta}_{11} + h_3\dot{\delta}_{15}) + 2h_3F_{MR} = 0$$

where, $m_{c,f,w}$ and $I_{c,f,w}$ are the mass and inertia moment of car body, bogie frame, wheel-set, respectively. $k_{p,s}$ and c_s are the stiffness and damping ratio of primary and secondary suspension. l, h are distance of between bogie frame and car body. The MR damping force and creep force are expressed by F_{MR} and $F_{x,y}$.

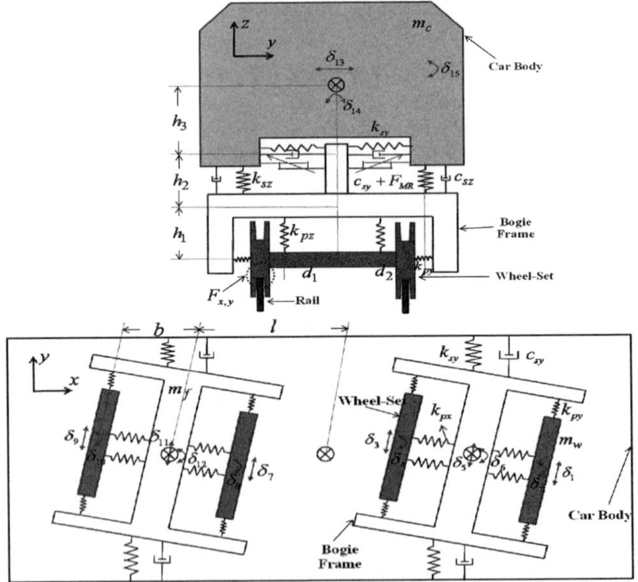

Figure 1. Mechanical model of railway vehicle

Table 1. The degree of freedom of railway vehicle

Element	Lateral	Yaw	Roll
1st Wheel-set	δ_1	δ_2	
2nd Wheel-set	δ_3	δ_4	
3rd Wheel-set	δ_7	δ_8	
4th Wheel-set	δ_9	δ_{10}	
1st Bogie frame	δ_5	δ_6	
2nd Bogie frame	δ_{11}	δ_{12}	
Car body	δ_{13}	δ_{14}	δ_{15}

3. Modeling of MR Damper

In this study, the MR damper which is incorporated with the MR valve and accumulator is considered. The schematic diagram of the MR damper is shown in Figure 2. The outer and inner pistons are combined to form a MR valve structure which divides the MR damper into two chambers: the upper and lower chambers. These chambers are fully filled with the MR fluid. The floating piston incorporated with gas chamber functions as an accumulator to accommodate the piston shaft volume as it enters and leaves the fluid chamber. Using quasi-static behavior of the damper, the damping force can be expressed as follows:

$$F_d = P_2 A_p - P_1(A_p - A_s) \tag{4}$$

where A_p and A_s are the piston and the piston shaft effective cross-sectional areas, respectively. P_1 and P_2 are pressures in the upper and lower chamber of the damper, respectively. The relations between P_1, P_2 and the pressure in the gas chamber, P_a, can be expressed as follows:

$$P_2 = P_a; \quad P_1 = P_a - \Delta P \tag{5}$$

where ΔP is the pressure drops of MR fluid flow through the valve structure. By neglecting minor loss of MR fluid flow, the pressure drops ΔP can be calculated as follows:

$$\Delta P = \frac{6\eta L}{\pi t_d^3 R_d}(A_p - A_s)\dot{x}_p + 2c\frac{L_p}{t_d}\tau_y \tag{6}$$

where τ_y is the yield stress of the MR fluid induced by the applied magnetic field, η is the post-yield viscosity of the MR fluid, L is the length of the inner piston, L_p is the length of the magnetic pole, R_d and t_d are the average radius and gap of the annular duct (the MR valve orifice). The coefficient c depends on the MR flow velocity profile which can be approximately estimated as follows [7]:

$$c = 2.07 + \frac{12Q\eta}{12Q\eta + 0.8\pi R_d t_d^2 \tau_y} \tag{7}$$

where Q is the flow rate of MR fluid flow through the valve orifice. The pressure in the gas chamber can be calculated as follows:

$$P_a = P_0 \left(\frac{V_0}{V_0 + A_s x_p} \right)^{\gamma} \tag{8}$$

where P_0 and V_0 are initial pressure and volume of the accumulator. γ is the coefficient of thermal expansion which is ranging from 1.4 to 1.7 for adiabatic expansion. Using Eq. (4), (5), (6) and piston velocity, the MR damping force can be calculated by

$$F_{MR} = P_a A_s + (A_p - A_s) \frac{2cL_p}{t_d} \tau_y \, \mathrm{sgn}(\dot{\delta}_{5,11,13} - \dot{\delta}_{13,11,5}) \tag{9}$$

The first term in Eq. (9) represents the elastic force from the gas compliance and the last one is the force due to the yield stress of the MR fluid which can be continuously controlled by the magnetic field across the MR fluid duct. In this work, the commercial MR fluid (Lord Cor. 132DG) is used. The induced yield stress of the MR fluid can be approximately estimated by

$$\tau_y = \alpha H^{\beta} \tag{10}$$

Here, H is the applied magnetic field whose unit is A/mm. The α and β are intrinsic values of the MR fluid to be experimentally determined.

4. Evaluation of Control Performance

In order to investigate the performance of the proposed MR damper, among many potential candidates for control algorithms, the semi-active skyhook controller is adopted in this research. It is well-known that the logic of the skyhook is simple and easy to implement in practical field. The desired damping force for the MR damper is set by

$$u = C_{sky} \cdot \dot{\delta}_{5,11,13} \tag{11}$$

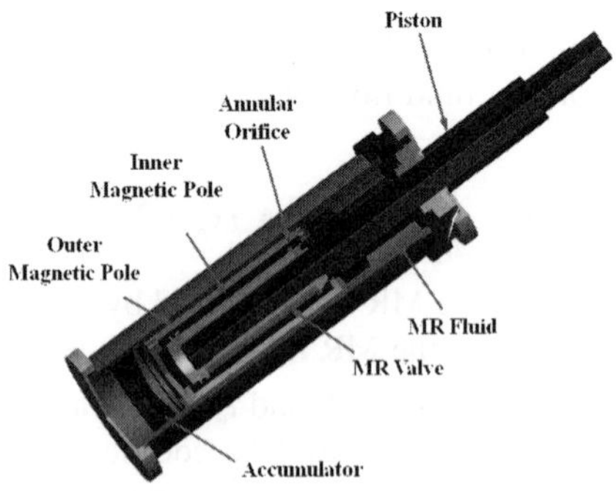

Figure 2. Schematic diagram of MR damper

where C_{sky} is the gain of the skyhook controller. The control gain physically indicates the damping coefficient. On the other hand, the damping force of the MR damper needs to be controlled depending upon the motion of the piston movement. Therefore, the following actuating condition is normally imposed.

$$u = \begin{bmatrix} C_{sky} \cdot \dot{\delta}_{5,11,13}, & for \; \dot{\delta}_{5,11,13}(\dot{\delta}_{5,11,13} - \dot{\delta}_{13,11,5}) > 0 \\ 0, & for \; \dot{\delta}_{5,11,13}(\dot{\delta}_{5,11,13} - \dot{\delta}_{13,11,5}) \leq 0 \end{bmatrix} \qquad (12)$$

The power spectrum density (PSD) of lateral, yaw and roll acceleration of the car body of railway vehicle under random track irregularities are illustrated in Figure 3. The "uncontrolled" case represents the conventional passive system using viscous dampers without MR dampers for the secondary suspension system. For the "controlled" case, the MR dampers are operated in semi-active control mode via the proposed controller.

From Figure 3, it is illustrated that the secondary suspension system integrated with MR dampers is especially effective for reducing the lateral, yaw and roll vibrations of the car body compared to the uncontrolled case. While the

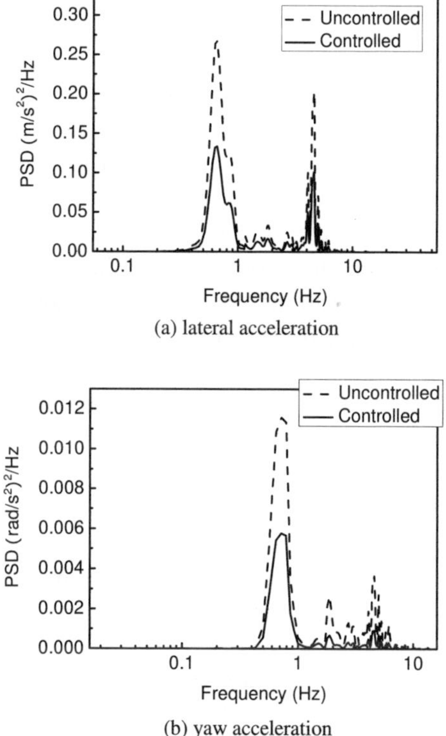

(a) lateral acceleration

(b) yaw acceleration

Figure 3. Power spectrum densities of the car body acceleration of the railway vehicle

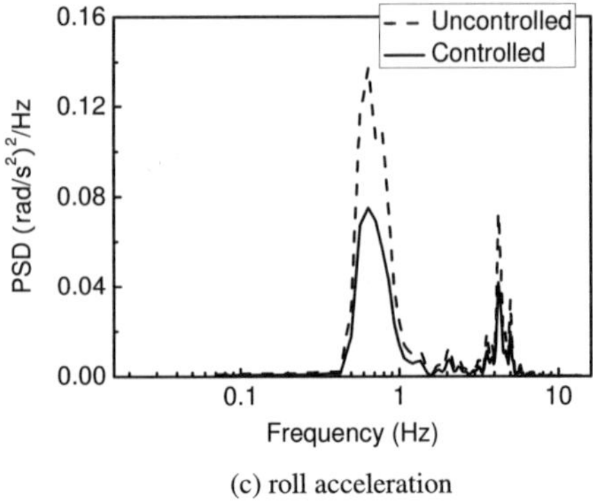

(c) roll acceleration

Figure 3. (*Continued*)

laterally installed MR dampers are effective for controlling the lateral, yaw and roll vibrations of the car body about 24%~30% improvement in terms of vibration reduction.

5. Conclusion

In this work, a semi-active secondary suspension system with MR dampers of railway vehicle has been investigated by considering a lateral response railway vehicle model, which includes three vibration motions such as lateral, yaw and roll of the wheel-set, bogie frame and car body. The governing equations of a fifteen degree-of-freedom railway vehicle model integrated with MR dampers are developed. To illustrate the feasibility and effectiveness of controlled MR dampers on railway vehicle suspension systems, the sky-hook control using the velocity feedback is adopted to the system control algorithm. Subsequently, in order to demonstrate the effectiveness of the MR damper, computer simulation is undertaken showing vibration control performance. It is shown through random irregularity using creep force that ride quality can be substantially improved by employing the proposed MR damper.

References

1.   K. Sasaki, S. Kamoshita and M. Enomoto, *IECON`94*, 2011 (1994).
2.   R. Shimamune and K. Tanifuji, *SICE'95*.1335 (1995).
3.   W. H. Liao and D. H. Wang, *JIMSS.* **14**, 161 (2003).
4.   C. Y. Lai and W. H. Liao, *J VIB CONTROL.* **8**, 527 (2002).
5.   K. G. Sung and S. B. Choi, *P I MECH ENG D-J AUT.* **222**, 2307 (2008).
6.   V. K. Grag, *ACADEMIC PRESS INC.* (1984).
7.   Q. H. Nguyen, S. B. Choi and N. W. Werely. *SMART MATER STRUCT.* **17**, 1 (2008).

EXPERIMENTAL EVALUATION OF MR DAMPER FOR INTEGRATED ISOLATION MOUNT[*]

MIN-SANG SEONG, SEUNG-BOK CHOI[†] and SEUNG-GU LIM

*Smart Structures and Systems Laboratory, Department of Mechanical Engineering,
Inha University, Incheon, 402-751, Korea*

CHEOL-HO KIM

*e-Machining Process Team, Korea Institute of Industrial Technology,
Cheonan, 331-825, Korea*

This paper presents experimental evaluation of a magnetorheological (MR) damper designed for an integrated isolation mount for ultra-precision system. The vibration sources of the ultra-precision system can be classified as two. The one is the environmental vibration from the floor, and the other is the transient vibration occurred from stage moving. The transient vibration occurred from stage moving has serious adverse effect to the process because the vibration scale is quite larger than other vibrations. Therefore in this research, semi-active MR damper, which can control the transient vibration, is adopted. Also the stage needs to be isolated from tiny vibrations from the floor. For this purpose, dry friction of MR damper must be removed. In order to achieve this goal, a new type of MR damper is designed that the friction parts are eliminated and damping force range is optimally increased. Subsequently, the damping force characteristics of MR damper are experimentally evaluated.

1. Introduction

The main stream of ultra-precision machining industries like display and semi-conductor is featured by huge scale, integration and multifunction. Therefore, the accuracy limitation of machining/manufacturing/measuring system is required to be continuously increased [1-3]. The environmental tiny vibrations and transient vibration from structure make noticeable problems to accuracy. For solving these problems, three types of control system have been proposed and successfully implemented; passive, active and semi-active. The passive system featuring rubber mount or air spring is normally used for vibration absorption. This system provides design simplicity and cost-effectiveness. However, it cannot provides

---

[*] This work is supported by National Research Foundation of Korea (NRF) grant funded by the Korea government (MEST) (No. 2010-0015090).

[†] Corresponding author.

150

sufficient performance at these days because the accuracy level of ultra-precision system becomes more important and moving mass on the stage is operated as a huge vibration source. On the other hand, the active control system provides high control performance in wide frequency range. However, this type may require large space and high power source. And the semi-active control system offers a desirable performance without requiring large power source and space. Therefore the vibration and position control systems using active and semi-active actuator are needed to be developed and adopted for ultra-precision system as an integrated module.

Kato et al.[4] proposed an active control of pneumatic vibration isolation tables. They designed a new type of pneumatic pressure regulator for high precision and quick response. Huang et al.[5] proposed an active isolation of a flexible structure from base vibration. They designed an electro-magnetic actuator and controlled the base vibration. Hong et al.[6] proposed electro-rheological fluid mounts for vibration control of a frame structure. They developed an ER fluid mount and controlled the vibration using optimal controller. Wang et al.[7] developed a new magnetorheological fluid-elastomer mount. They designed a new type of MR-elastomer mount and evaluated its performance. Seong et al.[8] proposed a Preisach hysteretic compensator for the damping force control of MR damper. They evaluated the hysteretic behavior of MR damper and formulated the hysteretic model and compensator using Preisach model. The above reviews indicate that many research works about the vibration control of ultra-precision system are carried out. However, the combination and redesign of active and semi-active actuators for the ultra-precision system is rarely researched.

Therefore, in this research, we proposed an integrated isolation mount which consists of the passive air spring, active electromagnetic actuator and semi-active MR damper to obtain superior vibration control performance. As a preliminary work, the semi-active MR damper for the integrated isolation mount is developed in this paper. For this purpose, the integrated isolation mount is introduced and the MR damper is newly designed. After then, MR damper is manufactured and its performance is experimentally evaluated.

2. Integrated Isolation Mount

The integrated isolation mount is consists of three parts: passive air spring, active electromagnetic actuator and semi-active MR damper. Figure 1 shows the ultra-precision machining system (exposure equipment) which has a moving stage with the integrated isolation mount system. As shown in Fig. 1, vibration source can be classified as two parts. The one is environmental vibration from

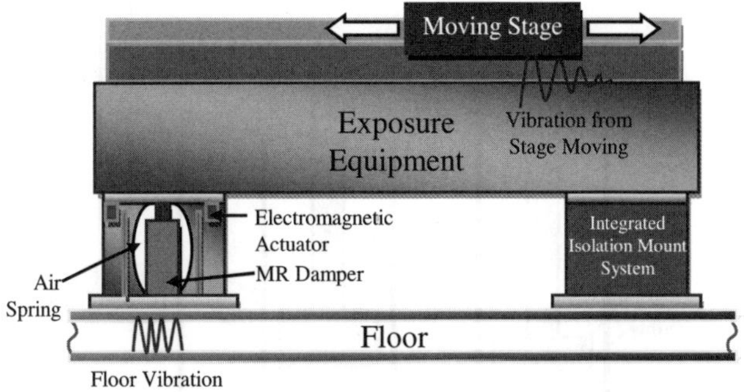

Figure 1. Configuration of the integrated isolation mount.

the floor, and the other is transient vibration occurred from the stage moving. The transient vibration occurred from the stage moving affects seriously to the process because the vibration scale is quite larger than other vibrations. If we try to control this vibration using active actuator, the actuator needs a large power source and space. Therefore, in this research, semi-active MR damper which can control the transient vibration without requiring large power source and space is adopted. On the other hand, micro- or nano-scale tiny vibrations cannot be controlled effectively using semi-active actuator. So the active electromagnetic actuator controls the tiny vibrations such as environmental vibration from the floor. In this case, the stage needs to be isolated from these tiny vibrations. For this purpose, the friction of MR damper must be removed. Therefore, a new design concept of MR damper is proposed in this study.

3. MR Damper

Conventional MR damper has sealing between piston rod and cylinder (housing), and between piston head and cylinder. From these sealing, dry friction is occurred and it makes a vibration transmission from floor to stage which lower the stage accuracy. Therefore, dry friction needs to be removed from the MR damper. Figure 2 shows the configuration of new type MR damper for the integrated isolation mount. Conventional MR damper must consider not only the vertical vibration but also the horizontal shock. Therefore, sealing between piston head and cylinder is required for preventing horizontal impact. However, integrated isolation mount is setup on the high precision system. So the horizontal movement is not considerable factor. From this point, sealing between piston head and cylinder can be removed and the remaining gap can be

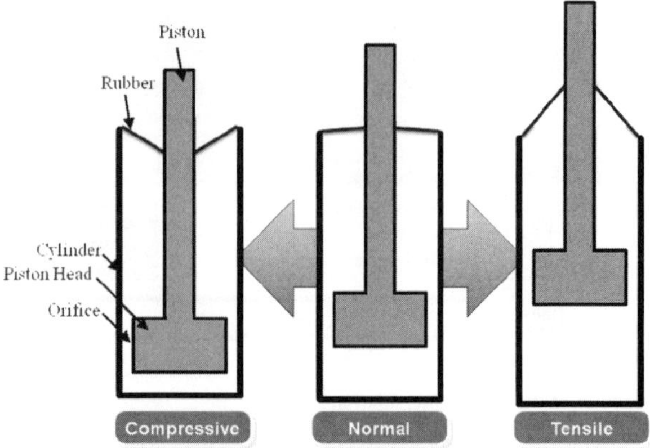

Figure 2. Configuration of the new type MR damper.

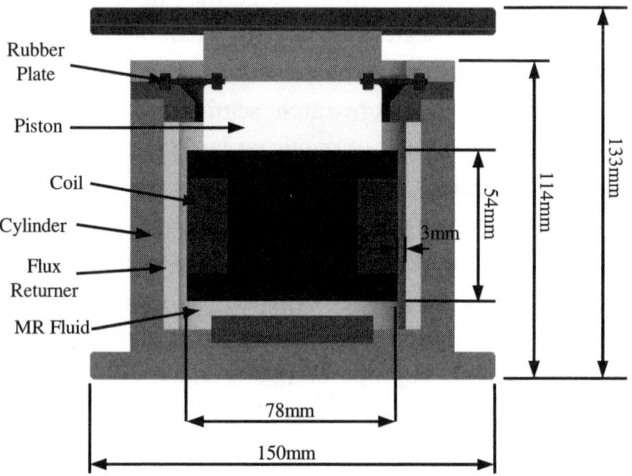

Figure 3. Design of the MR damper.

directly used as an orifice. The proposed MR damper's maximum needed stroke is ±5mm which is quite smaller than conventional MR damper's stroke. Therefore, rubber plate is adopted instead of sealing between piston rod and cylinder. Rubber plate is fixed on the piston and the cylinder each side and stretched according to the piston movement as shown in Fig. 2. So this MR damper doesn't make any dry friction. Also the rubber plate works as a gas chamber which is needed for the volume compensation. Therefore, the design of the proposed MR damper is much simpler than the conventional MR damper.

Figure 3 shows the design of the proposed MR damper. The gap (orifice) between piston head and cylinder is 3mm and rubber plate is adopted. MR fluid manufactured by RMS company in Korea is filled in the MR damper. Also

Figure 4. Photograph of the manufactured MR damper.

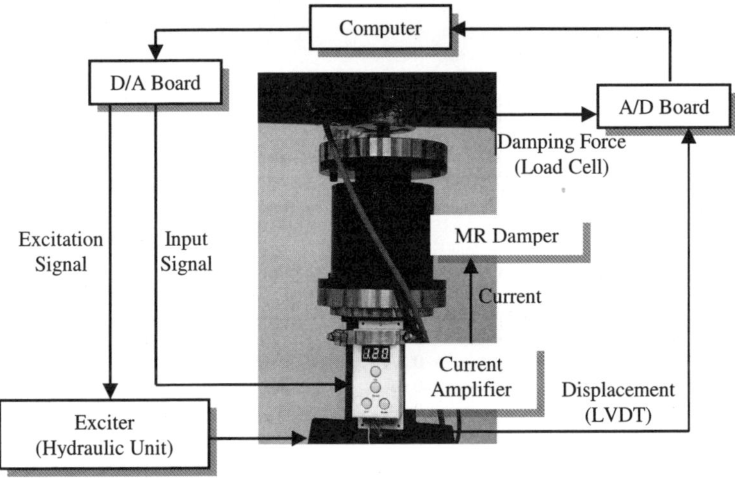

Figure 5. Experimental configuration for damping force measurement of the MR damper.

design variables are optimized for maximizing damping force. Based on the design variables, the MR damper is manufactured as shown in Fig. 4.

4. Experimental Evaluation

Figure 5 shows the experimental configuration for damping force measurement of the MR damper. The MR damper is excited by amplitude ±4.0mm and frequency 3.0Hz using a hydraulic exciter. This excitation condition is similar

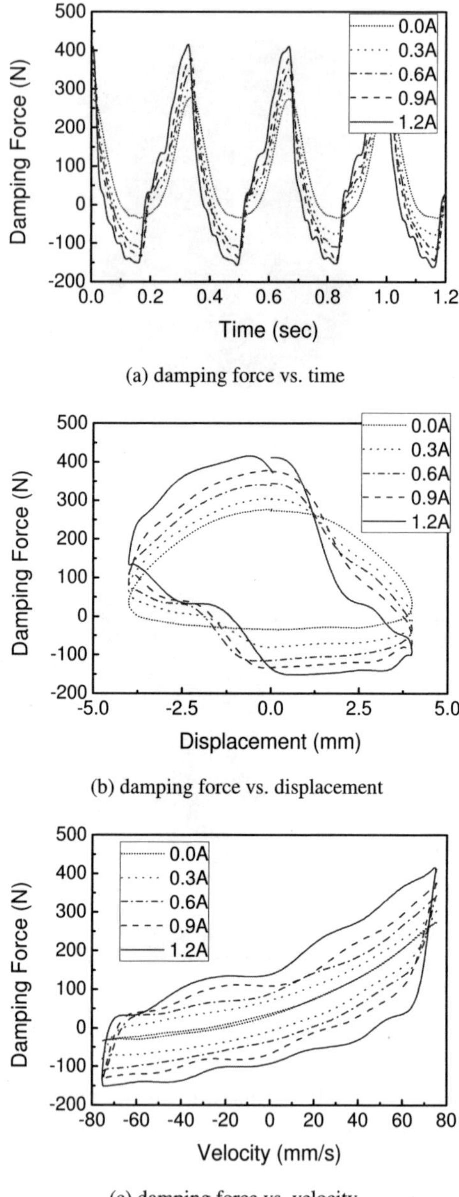

(a) damping force vs. time

(b) damping force vs. displacement

(c) damping force vs. velocity

Figure 6. Damping force characteristics.

with transient vibration from the stage moving. The damping force of the MR damper is measured by a loadcell and MR damper movement is measured by

LVDT (linear variable differential transformer). Input signal is generated from a computer DAQ system, and this signal is fed back to the current amp and applied to the MR damper.

Figure 6 presents the measured field-dependent damping force characteristics of the MR damper. As shown in Fig. 6 (a), as the current (magnetic field) increases the damping force increases. Figure 6 (b) presents damping force versus displacement graph, and Fig. 6 (c) presents damping force versus velocity graph. The large hysteretic behavior is seen in the Fig. 6 (c). This behavior needs to be considered for high accuracy vibration control.

When the current is not applied, the maximum damping force is 277.08N and the minimum damping force is -34.52N. And the maximum field (1.2A) is applied, the maximum damping force is 415.81N and the minimum damping force is -161.71N. It means the damping force range is almost doubled when we control the MR damper. This damping force range is suitable for control the vibration of ultra-precision manufacturing system.

5. Conclusion

In this work, a new design of MR damper for integrated isolation mount was proposed and its performance was verified. The dry-frictionless MR damper was originally designed for the integrated isolation mount. After manufacturing the MR damper, experimental characteristics were evaluated. From the measured damping force characteristics, we can conclude that the proposed MR damper is suitable for the integrated isolation mount system.

References

1. S. T. Smith and R. M. Seugling, *PRECIS ENG.* **30**, 245 (2006).
2. B. Bringmann and P. Maglie, *CIRP ANN-MANUF TECHN.* **58**, 343 (2009).
3. M. Rahman, A. B. M. A. Asad, T. Masaki, T. Saleh, Y. S. Wong and A. S. Kumar, *INT J MACH TOOL MANU.* **50**, 344 (2010).
4. T. Kato, K. Kawashima, T. Funaki, K. Tadano and T. Kagawa, *PRECIS ENG.* **34**, 43 (2010).
5. X. Huang, S. J. Elliott and M. J. Brennan, *J SOUND VIB.* **263**, 357 (2003).
6. S. R. Hong, S. B. Choi and M. S. Han, *INT J MECH SCI.* **44**, 2027 (2002).
7. X. Wang and F. Gordaninejad, *SMART MATER STRUCT.* **18**, 095045 (2009).
8. M. S. Seong, S. B. Choi and Y. M. Han, *SMART MATER STRUCT.* **18**, 074008 (2009).

MRF-BALL-CLUTCH

MARCO JACKEL

Fraunhofer Institute for Structual Durability and System Reliability LBF
Bartningstraße 47, 64289 Darmstadt, Germany

MICHAEL MATTHIAS

Fraunhofer Institute for Structual Durability and System Reliability LBF
Bartningstraße 47, 64289 Darmstadt, Germany

BJÖRN SEIPEL

Fraunhofer Institute for Structual Durability and System Reliability LBF
Bartningstraße 47, 64289 Darmstadt, Germany

Beside active dampers, rotating brake and clutch systems are the main field of application for magnetorheological fluids (MRF). Within this paper an idea is shown that uses the well known technology of MRF dampers and combines it with the also well known technology of common mechanical ball clutches. The result of this fusion is a new type of MRF-clutch design that has a lot of advantages compared to traditional MRF clutch designs. Additionally to show the potential of the idea for common applications, two demonstrator structures that had been investigated will be presented.

1. Traditional Principles of MRF-Clutches

Today MRF are technically used for adjustable dampers e.g. within car suspensions systems, switchable engine mounts, etc. and for rotating brake and clutch systems. Usually there are two main designs of these clutch systems, the "disk" and the "bell" (Fig. 1). Both consists of two rotatable parts with a small gap in between connected to the input respectively output side (light blue). This gap is filled up with MRF (red) that transmits the power by interaction of the surfaces and the fluid. An electromagnetic system (green) provides the necessary field stress to change the viscosity of the MRF.

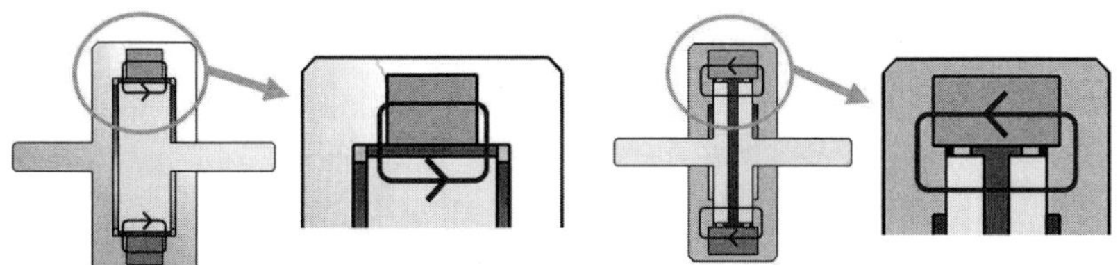

Figure 1: MRF-Clutch, "Bell"-Design (left) and "Disk"-Design (right)

Although the aforementioned designs are very popular, the performance of these clutches has room for improvement. Especially clutches that are designed to transmit higher torques, e.g. some hundred Nm in typical automotive applications, need to have mechanical parts with large surfaces to interact with the MRF. As the corresponding volume of the MRF is also very big, the electromagnetic system has to be really strong to provide a sufficient magnetic field. If a smaller electromagnetic system should be used, often a MRF with a higher basic viscosity is applied to transmit the torque. Here however the ratio between minimum and maximum transmittable torque is limited.

Another disadvantage of the traditional MRF clutch designs is caused by the rotation of the mechanical parts that are in contact with the MRF. All passes to the outside have to be sealed very carefully. This is very challenging as the components of the MRF are highly abrasive and thus bad for the durability of the seals.

2. Basic Principle of MRF-Ball-Clutch

Within this chapter the idea of a completely new type of MRF clutch will be described. It was developed at the Fraunhofer LBF [1] to avoid the disadvantages of traditional MRF clutch designs.

As it is a combination of a ball safety coupling mechanism and an axial operating MRF-actuator, it was called "MRF-Ball-Clutch" (principle see Fig. 2). In contrast to common clutch designs where the torque is transmitted by the MRF, here mechanical coupling devices act as power transmitter. Balls (4) which are circularly located with positive locking in countersinks (5) on the front of two opposite placed discs perform this task. By the geometrical interaction of these balls and the two disks an applied torque produces a force in the direction of the rotational axis which lets increase the distance between input and output disk. If the torque exceeds a certain level the balls are no longer able to rest in their countersinks and the power transmission is stopped immediately.

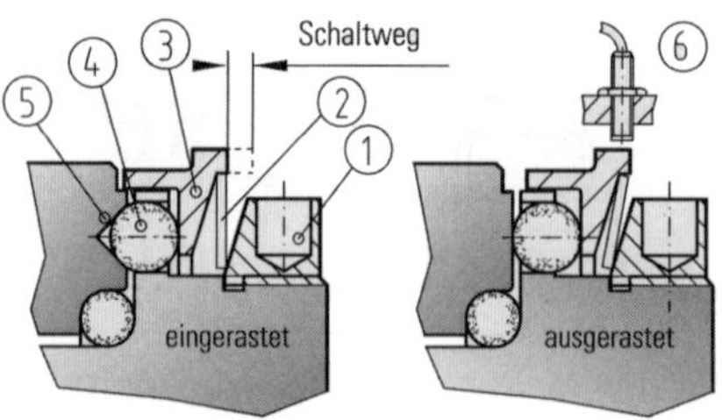

Figure 2: Principle of a ball clutch; locked (left) and unlocked (right) (Source: R+W Antriebselemente GmbH)

To adjust the maximum transmittable torque, a counter force is required that hinders the axial movement of the clutch disks. This force is provided by an axial operating MRF actuator, which is designed similar to a common single-tube damper (Fig. 3). Here a solenoid coil generates a magnetic field in the fluid gap and by this the MRF is more or less hindered to flow through. The piston of the damper as well as the clutch disks are locked in their current position.

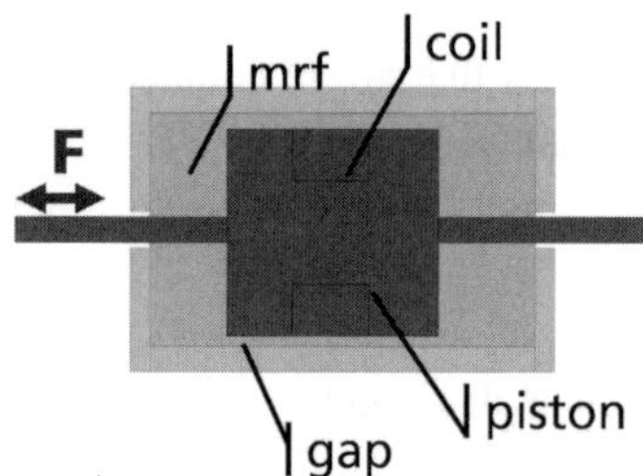

Figure 3: Principle of a MRF single tube damper

3. Example Application Torque Wrench

To show the potential of the developed clutch concept, a functional demonstrator was designed which will be presented within this chapter.

3.1. *Design of the mechanical and electric parts*

As example an electrically adjustable torque wrench was chosen to represent a typical application.

Figure 4 shows a cross-section of the main parts. On the left side the "ball clutch" is represented by two hardened steel disks with 90° countersinks and 6 steel balls of 8 mm. The balls are arranged on a circle with a diameter of 36 mm. On the right side the MRF actuator is realized by a low carbon steel piston with a coil of 0.5 mm copper wire and a tube housing also made of low carbon steel. The counterforce of the piston (typically 5.6 kN) is transmitted by a guided rod to the clutch side. To readjust the system, a "weak" retention spring is located at the right end of the rod.

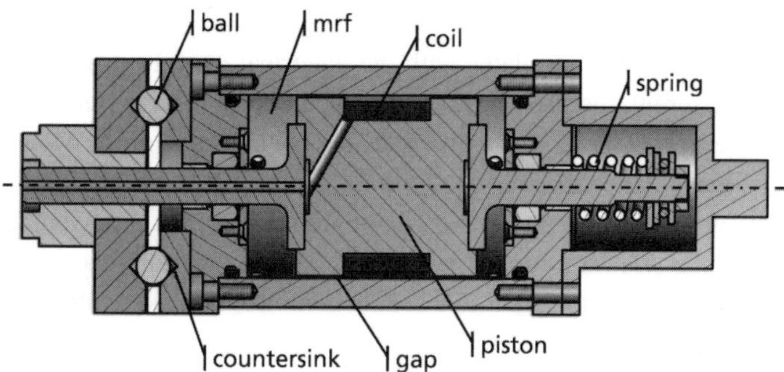

Figure 4: MRF torque wrench, cross section

The finally realized demonstrator structure (Fig. 5) consists of the MRF actuator, a handle bar with force sensor, a connector for a socket and an electronic unit. With the demonstrator it is possible to tighten a screw exactly with a predefined torque between 5 and 100 Nm.

Figure 5: MRF torque wrench, realized demonstrator structure

3.2. *Controller strategies*

There are two main strategies to control the maximum applied torque.

3.2.1. *Open Loop*

Here the release-torque correlates with the electrical current in the coil. Before tightening the screw, the user sets the electrical current to a certain level. If the axial force generated by the balls overrides the counterforce of the MRF actuator the clutch opens. The advantage of this strategy is that no additional torque sensor and only a low cost electronic circuit are needed. The disadvantage is that the setting of the release torque is not very reliable as the influence of changing system characteristics e.g. by particle sedimentation is not taken into account.

3.2.2. *Closed Loop*

For the closed loop control the torque wrench has to be equipped with a force sensor. The electrical current in the coil is always set to a maximum value. Before tightening the screw, the user sets a desired value for the fastening torque. While using the wrench, the current torque measured by the sensor is compared with the nominal value. If the two values are equal, the electrical current in the coil is switched of and the screw could not be tightened any more. Compared to the open loop control here the effort for the electronic parts (sensing and evaluation of the torque) is higher. A significant advantage of the closed loop control is certainly a much better reliability of the system, a faster response time and a higher durability of the mechanical parts (balls and countersinks).

Figure 6 shows a comparison of the simulated and the measured torque plotted over the electrical current in the coil. As the relation between the electrical current in the coil and the release torque is not really constant, the closed loop control strategy should be used for practical applications.

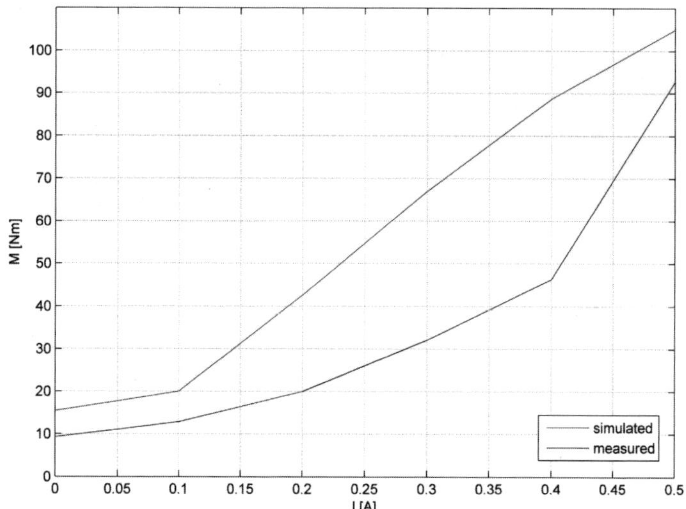

Figure 6: comparison of simulated and measured torque

3.3. *MRF-torque-wrench conclusion*

With the investigated demonstrator it was possible to approve the functional capability of the developed clutch concept. Compared to mechanical clutch systems, the demonstrator showed that it is possible to transmit a precise online adjustable torque of 5 to 100 Nm by a small MRF-clutch with very low electrical power consumption. The thermal stress of the MRF is much lower than in common MRF "disk" or "bell" clutch systems. With respect to this promising results there seems to be a large field of applications for the described new MRF clutch concept e.g. for machine tools, hand-operated power tools or in the automotive sector.

4. MRF-Clutch for E-Mobility / Automotive Application

To further investigate the applicability of the aforementioned MRF-ball-clutch technology for e-mobility and automotive tasks, currently a safety clutch for the Fraunhofer AutoTram (Fig. 7, l.) is being developed.

Figure 7: Fraunhofer Auto Tram (left) and power unit (right)

The target is to replace the passive clutch between a Diesel engine and an electric generator within the power unit (Fig. 7, r.) by a MRF clutch. This clutch should work as safety component and should be able to decouple the two main systems, even under maximum load conditions. As the power unit consists of a 180 kW V8 Diesel-engine with a maximum torque of typically 560 Nm (operating ratio = 1.2 → ~700Nm) and a maximum speed of 4000 rpm, the clutch design is very challenging.

Figure 8 shows a CAD model of the clutch. The design of the MRF actuator unit is similar to the design of the torque wrench actuator unit. The size, number and position of the balls are adapted to the different requirements. The clutch has a diameter of 260 mm, an over all length of 300 mm and a mass of about 12 to 15 kg. The power consumption of the electric parts (coil) is only at about 6 W! Compared to traditional active MRF and non MRF clutches these first values show a high potential of the new technology.

Figure 8: CAD Model of the MRF safety clutch

5. Conclusion

The newly developed MRF-Ball-Clutch design combines the technology of MRF dampers with the technology of common mechanical ball safety clutches. The Application of the design in a torque wrench is approved and currently there are ongoing works at Fraunhofer to develop a high torque safety clutch for an automotive application.

The main advantages of the MRF-Ball-Clutch in comparison with "classic" MRF clutches (disc or bell design) are the low power consumption and the low installation space. The advantages over the common ball safety clutches are the online adjustable release torque and the emergency stop.

Future applications are primarily fast switching safety clutches, for example for machine tools or automotive applications.

References

1. Patent pending DE102009034055 "Drehmomentbegrenztes Kupplungselement sowie Verfahren zum drehmomentbegrenzten Kuppeln" M. Jackel, M. Matthias, B. Seipel, Fraunhofer Institute for Structural Durability and System Reliability LBF (2009).

MAGNETORHEOLOGICAL VALVE IN SERVO APPLICATIONS

ESA KOSTAMO

Engineering Design and Production, Aalto University, Sähkömiehentie 4
Espoo, 02150, Finland

JARI KOSTAMO

Engineering Design and Production, Aalto University, Sähkömiehentie 4
Espoo, 02150, Finland

JYRKI KAJASTE

Engineering Design and Production, Aalto University, Sähkömiehentie 4
Espoo, 02150, Finland

MATTI PIETOLA

Engineering Design and Production, Aalto University, Sähkömiehentie 4
Espoo, 02150, Finland

In this paper the servo property of a high performance magnetorheological valve will be evaluated by closing the pressure feed back loop. The magnetorheological valve developed for this study has two separately controllable fluid flow channels and is especially designed for high frequency applications. A state space model for the magnetorheological valve from the control signal to the pressure output will be indentified. The indentified model is used for tuning a PID controller and in simulating the closed loop system. Finally the controller will be implemented to a control computer and the pressure output will be controlled in a real time control loop. By analyzing the dynamic and static results of the magnetorheological servo valve, it can be stated that the magnetorheological valve has a good potential for pressure and force control applications.

1. Introduction

The fact that the change of the magnetorheological (MR) fluids yield stress can be achieved in a fraction of a millisecond makes magnetorheological fluids very attractive choice for different kind of vibration control and actuation applications [1]. A unique property for the magnetorheological devices is that characteristics of the device can be varied in a wide range by controlling the electrical current inducing a magnetic field. This enables for instance a novel design of a magnetorheological hydraulic valve without moving and wearing parts.

The contribution of this study is to examine the servo property of a high performance magnetorheological valve by closing the pressure feedback loop. The combined model of the MR valve and power electronics is estimated and validated by measuring the frequency response of the system. Further on the model of the system is used for tuning of the controller. By implementing the controller to the control computer the static and dynamic performance of the magnetorheological servo valve can be measured and the operating bandwidth can be defined.

2. Materials

2.1. *Magnetorheological fluid*

The magnetorheological fluid used in this study was manufactured by Lord Corporation. MRF-132DG is a hydrocarbon-based fluid that has a high resistance to hard settling and is developed for energy dissipative applications. The magnetic field induced yield stress is given by the manufacturer as a graph defining the yield stress as a function of magnetic field strength. According to the manufacturer, the fluid reaches its maximum yield stress of 45 kPa at the field strength of approximately 280 kAmp/m. However a good manner is to design the fluid to reach the maximum of 35–40 kPa yield stress in the MR device because at the higher levels of yield stress the fluid starts to act considerably non-linearly and finally saturate [2].

2.2. *Experimental valve*

The prototype valve designed for this study is a square shaped and has two separately controllable fluid gaps where the fluid can be activated. The core parts inside the valve are mounted into aluminum body and manufactured of laminated steel to avoid eddy currents during fast magnetic field transients.

Figure 1 shows the MR valve with overlaid hydraulic diagram demonstrating the functional parts of the system. Three pressure sensors P_1, P_2 and P_3 are mounted to the valve to measure the pressure in the fluid supply line, actuator connection, and tank line. A small high pressure accumulator is connected to the valve supply line to ensure sufficient fluid supply during pressure transients. The function of the low pressure tank line accumulator is to prevent the pressure rise in the tank line when the fluid is bursted out of the valve.

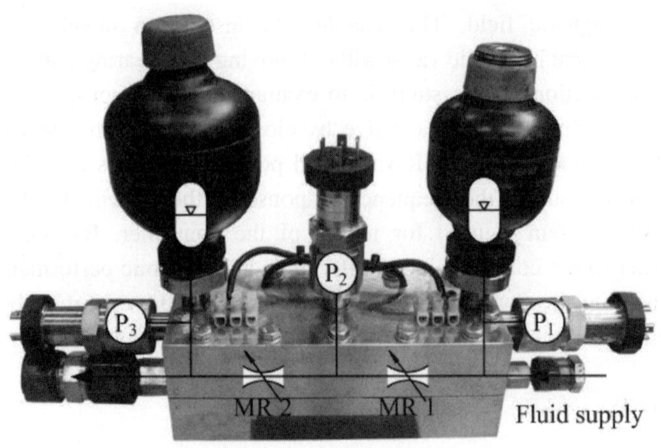

Figure 1. Experimental magnetorheological valve

The output pressure, P_2, can be varied by controlling the currents of the electromagnets that activate the MR fluid in the fluid gaps. As the valve is mounted to a magnetorheological test setup with a constant fluid pressure supply the maximum pressure in the actuator connection can be achieved when the fluid gap MR 2 is fully activated and the fluid gap MR 1 is deactivated.

The magnetic circuit of the valve was designed in a way that the fluid can be nearly saturated by the electric current of two amperes. The magnetic field calculations were carried out by using finite element method, where the MR fluid's non-linear property of permeability as a function of magnetic field strength can be taken into account. However, the more detail results of the magnetic analysis of the valve are not presented here for brevity [3].

The static hydraulic performance of the valve was defined to be able to generate pressure difference of 10 MPa in the actuator connection when the fluid gap MR 2 is fully activated. In the design phase the Bingham behavior of the fluid was assumed and the relevant fluid models are presented in references [4-6]. The final design parameters of the developed valve are listed in the table 1.

Table 1. Design parameters of the magnetorheological valve

Dimension	Value
Height of the MR fluid gap, g	0,4mm
Length of the MR fluid gap, L	30mm
Width of the MR fluid gap, w	40mm
Number of Ampere turns, N	80

3. Results and Discussion

3.1. *Model identification*

The modeling of the magnetorheological system, consisting of the valve and the current amplifiers, was carried out by black box modeling. For the model identification an open loop frequency response of the MR system was measured by using sine excitation. The bandwidth of the excitation signal was selected to limit between 25 Hz and 2000 Hz and it was divided evenly in 100 frequency steps. Frequencies below 25 Hz were estimated to belong to the static operating range of the valve and frequencies over 2000 Hz were assumed to be beyond the dynamics of the MR system. The pressure level used in the identification was set to 4 MPa.

After iterative model evaluations a sixth order state space model was chosen to estimate the response of the MR system. In general the state presentation of a system can be formulated as

$$\dot{x}(t) = Ax(t) + Bu(t)$$
$$y(t) = Cx(t) + Du(t) \tag{1}$$

where $x(t)$ is the state vector, $u(t)$ is the control vector, $y(t)$ is the output vector, A is the state matrix, B is the input matrix, C is the output matrix and D is the feed forward matrix. The identified model resulted 93,7 % correspondence between the open loop response and the theoretical results.

In figure 2 both the measured response of the open loop MR system and the output of the model are compared and an equivalency with minor error can be observed.

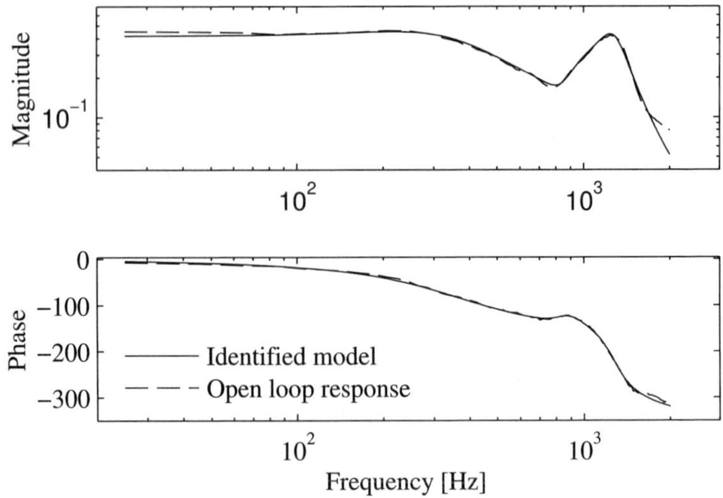

Figure 2. Open loop response of the MR system and the identified model

3.2. *Controller*

The identified model of the MR system was imported to Matlab Control System Toolbox to simulate and to tune the response of the controlled system. The controller was based on a standard PID structure and the controller gains were tuned by using an internal model control (IMC) based tuning algorithm. The IMC algorithm provides an easy framework for tuning a controller with only a single adjustable parameter that is related to the speed of the response. In this study the performance of the closed loop system was estimated by simulating the step response of the system and as tuning criteria the response time and the over shoot of the output signal was used [7].

To realize the measurements of the real time controlled closed loop system National Instruments LabVIEW environment was used. The tuned controller was imported to the control software to control the actuator pressure of the MR system with the control frequency of 10 kHz.

3.3. *Closed loop measurements*

3.3.1. *Frequency response*

The pressure frequency response of the closed loop MR system was measured by using the same sine excitation signal between 25 Hz and 2000 Hz. The measured results are presented in figure 3 where the open loop and the closed loop responses are compared in a Bode diagram. By taking the amplification coefficient of the pressure transducers into account the level of the closed loop magnitude graph in the low frequency range can be calculated to correspond with the output pressure level of 7.8 MPa

The first examination of the results proves that the open loop system has a very flat frequency response before the magnitude response starts to descent. This indicates that there exist no disturbing mechanical resonances in the valve construction, and the MR system is well suited for the control applications. Natural frequency that appears at the frequency of over 1 kHz is assumed to be dominated by the characteristics of the hydraulic circuit of the valve.

When the closed loop results are analyzed, it can be found that the controlled MR system has also very flat magnitude response up to 300 Hz. Further analysis of the results show that the -3 dB level of the closed loop system, where the magnitude of the response has dropped to the half of the low frequency level, can be found at the frequency of 445 Hz. From the phase diagram it can be seen that the controlled system is a little bit slower compared to the open loop system. On the other hand by closing the feedback loop the effect of the mechanical resonance at higher frequencies to output pressure can be diminished.

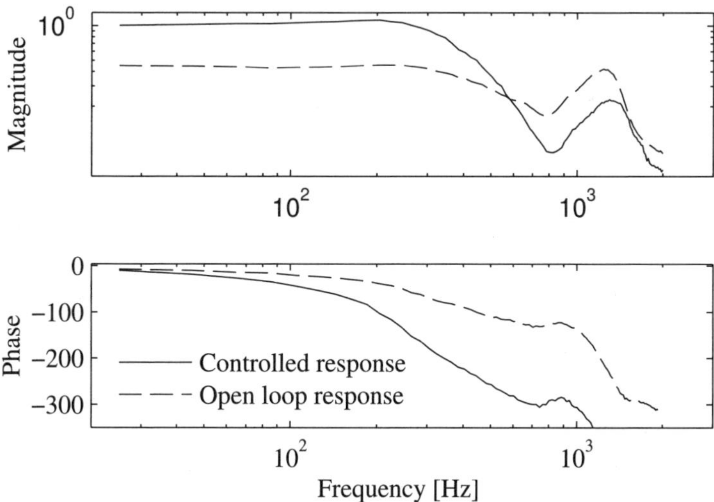

Figure 3. Frequency response of the open loop and closed loop systems

3.3.2. *Static response*

The static response of the open loop and controlled magnetorheological valve system was measured by using a 1 Hz sine excitation. The control voltage in open loop and closed loop measurements were scaled to correspond with the pressure sensor scale and thereby both measurements are comparable in the same figure scale. The pressure offset at the 0 V control signal was chosen to be approximately in the halfway, *i.e.* 4.9 MPa, of the maximum operating pressure of the valve. The measured results are compared in figure 4.

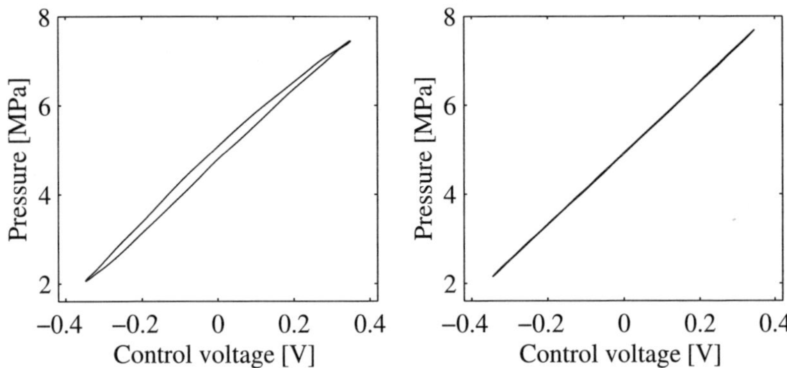

Figure 4a. Static response of the open loop measurements, 4b. Static response of the closed loop systems

From the static results it is interesting to notice that open loop pressure response of the system is fairly linear with only small hysteresis. The hysteresis in the pressure signal is assumed to be caused by the remanent magnetism of the core parts of the valve. However, a system with this small hysteresis could be used in some applications without closed loop pressure feedback information.

In the case of controlled results it can be found that the pressure response of the MR system is very linear with no hysteresis. The effect of the integrating factor in the controller can be noticed as disappearance of the slight nonlinearity in the open loop response and as the improvement of the accuracy of the output pressure in proportion to the control signal. It is also essential that by closing the pressure feedback loop the output pressure of the system becomes more robust against changes in process *i.e.* in the fluid itself. These changes may be caused, for instance, by contamination of the fluid, temperature variations or change in the air ratio of the fluid. This is in contrast to the open loop system, where these changes in the fluid properties may cause large changes to the output pressure.

4. Conclusions

In this paper the pressure control properties of a magnetorheological valve have been studied by closing the pressure feedback loop.

The open loop frequency response of the MR system was measured by using sinusoidal excitation and a sixth order state space model for the system was identified based on the frequency data. The identified model was used for tuning a standard PID controller and simulating the step response of the closed loop system. As a controller a standard PID controller was chosen. Finally the controller was implemented to a real time control computer.

The frequency response measurements proved that with magnetorheological technology the fluid flow and pressure can be controlled up to several hundred Hertz with the pressure difference of 7.8 MPa. The dynamic performance of the closed loop MR system, developed in this study, was documented to reach -3 dB magnitude level at the frequency of 445 Hz. The static measurements of the closed loop system demonstrated how the hysteresis in the pressure response can be removed and a very linear response can be reached by using a controller with an integrating factor.

By the frequency response and the static results of the closed loop MR system, documented in this study, it has been demonstrated that a magnetorheological technology has a strong potential for high frequency servo applications with good accuracy. In future work possible applications could be, for example, in the field of high frequency pressure or force control, active vibration control in smart structures, fast semiactive dampers and active shock control.

References

1. F. D. Goncalves, M. Ahmadian and J. D. Carlson, *SMS,* **15,** 75, 2006
2. Lord Co., *Product specification,* 2009
3. J. T. Kostamo and J. T. Kajaste, *FPNI,* 2006
4. G. Bossis, S. Lacis, A. Meunier and O. Volkova, *JMMM,* **252,** 224, 2002
5. J. Kostamo, E. Kostamo, J. Kajaste and M. Pietola, *PTMC,* 2007
6. F. D. Goncalves, J. H. Koo and M. Ahmadian, *SVD,* **38,** 203, 2006
7. D. E. Rivera, M. Morari and S. Skogestad, *Ind Eng Chem Process Des Dev,* **25,** 252, 1986

APPLICATION OF MAGNETO-RHEOLOGICAL FLUIDS FOR A MINIATURE HAPTIC BUTTON

TAE-HEON YANG

Mechanical Engineering, KAIST, Guseong-dong, Yuseong-gu
Daejeon, 305-701, Korea

JEONG-HOI KOO

Mechanical Engineering, Miami University, 501 East High Street
Oxford, Ohio 45056, USA

SANG-YOUN KIM

Computer Science & Engineering, Korea University of Technology and Education,
Gajeon-ri, Byeongcheon-myeon, Dongnam-gu
Cheonan-si, 330-708, Korea

DONG-SOO KWON

Mechanical Engineering, KAIST, Guseong-dong, Yuseong-gu
Daejeon, 305-701, Korea

Haptic units in small consumer electronic products (such as hand-held devices) require miniaturization of tactile and kinesthetic modules. However, it is quite challenging, particularly minimizing the size of kinesthetic actuators. Thus, this study investigates a miniature haptic button actuated by magnetorheological (MR) fluids with an aim to convey kinesthetic information or realistic button sensations to users in small electronic devices. To this end, a prototype haptic button was designed and constructed. The design focus was to maximize the resistive force generated by the fluids in a given size by using multiple operating modes of MR fluids. In order to evaluate the performance of the prototype button, a test setup consisting of a micro stage and a precision load cell was constructed. Using the setup, the resistive force of the button was measured by varying the indented depth and the input current. The results show that the force change (defined by the ratio of the difference between the maximum force and the minimum force to the maximum force) is over 72 % for all indented depths (up to 1.5 mm). This change is sufficient to create various button sensations. In other words, the proposed haptic button can offer a range of stiffness change that can be conveyed to human operators.

1. Introduction

For realistic haptic interaction, both tactile and kinesthetic information should be simultaneously conveyed to users. While kinesthetic information refers to sensory data that are obtained through receptors of joints, muscles, as well as ligaments, tactile information refers to sensory data acquired through receptors of skin. Generally, users rub (tactile) and press (kinesthetic) target objects when they try to perceive an object. Therefore, to convey a more realistic haptic sensation, both tactile and kinesthetic information should be simultaneously presented to users. However, most of haptic studies for hand-held devices have focused on stimulating tactile sensation because the tactile modules can be constructed in small-size.

Haptic devices for kinesthetic sensation mostly use AC/DC motors [1-4]. Since the motors tend to cause instability problems and consume "high" power, the motor-based devices can hardly be incorporated into the hand-held electronic products. Moreover, most smart materials are used only for tactile feedback applications, and little research has been performed towards kinesthetic display applications. Therefore, this study proposes a miniature haptic button based on MR fluids to convey kinesthetic sensation. This is because MR devices require low power, offer fast response time, and eliminate instability problems.

2. Design and Fabrication of an MR Haptic Button

2.1. *Working Principle of the Haptic Button*

Fig. 1 illustrates the working principle of the proposed haptic button. Upon pressing the button, MR fluids flow out of the gap between the plunger and the york, as shown in Fig. 2(b). After releasing the display, the deformed elastic spring returns back to the original configuration, bringing the plunger, which is attached to the spring, back to its original position. During the motion of the plunger, the MR fluids can create various resistive forces depending on the input current supplied to the solenoid coil.

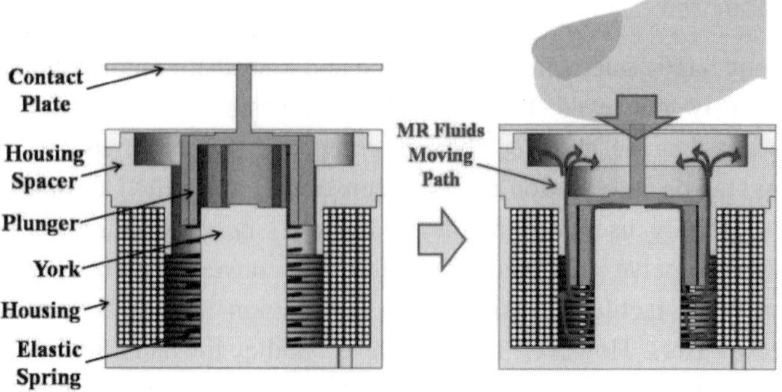

Figure 1. Schematic illustration of the working principle of the haptic button:
(a) before pressing and (b) after pressing.

2.2. *Design of the Haptic Button*

In order to miniaturize the haptic button, the design focused on (1) integrating multiple operating modes of MR fluids to maximize the resistive force produced by MR fluids and (2) selecting "optimal" design parameters to minimize the size of the button. To do this, a series parametric and numerical studies was performed using analytical equations along with an FEM tool. Fig. 2 shows a result obtained by the FEM analysis. The magnetic field induced by the solenoid coil passes from the yoke to the housing via the plunger. As shown in the zoomed figure, when the plunger moves to the downward direction, the MR fluids between the plunger and the yoke become squeezed (squeeze mode). The MR fluids flowing through the gaps between the plunger and the york as well as between the york and the housing produce yield stress (the mixed mode - flow and direct shear mode).

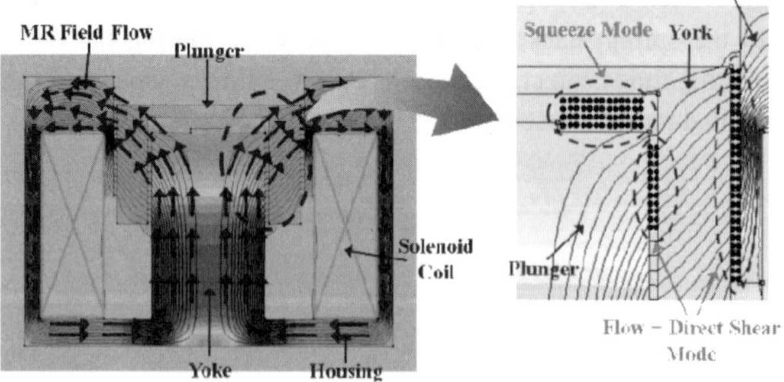

Figure 2. FEM simulation of the haptic button using multiple mode.

3. Fabrication of the Haptic Button

Fig. 3 shows the constructed, prototype haptic button along with its components. The button consists of the solenoid coil, the elastic spring, the plunger, the housing with the york, the housing spacer, the cover and the contact plate. The solenoid coil and the elastic spring are placed into the housing surrounding the york. The plunger vertically moves along the york. After filling MR fluids in the housing, the housing spacer and the cover seal the housing. The contact plate is fixed to the tip of the plunger. The size of the prototype haptic button (diameter-10mm × height-12mm) is considerably smaller than conventional devices based on AC/DC motors. The button stroke is 1.5 mm.

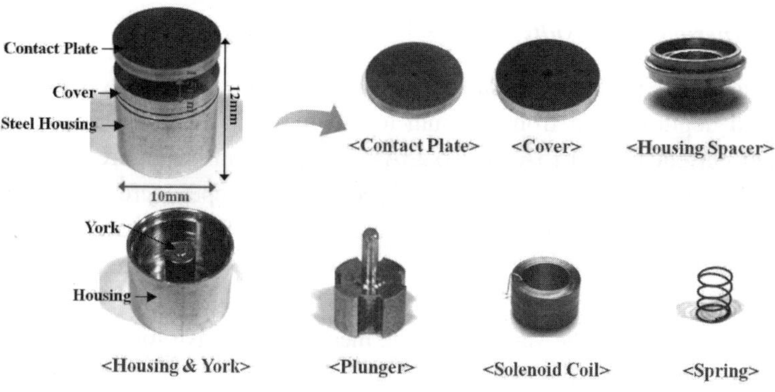

Figure 3. Fabricated haptic button.

4. Evaluation of the Haptic Button

4.1. *Experimental Setup*

To evaluate the performance of the prototype button, a test setup was constructed (see Fig. 4). It consists of a micro stage, a single point load cell (CAS BCL-1L), and an indicator (CAS CI-5010A). The haptic button is placed on the micro stage which moves along the vertical direction (Z-axis). The applied load and the intended depth of the haptic button were measured by varying the input current and the position of the micro stage.

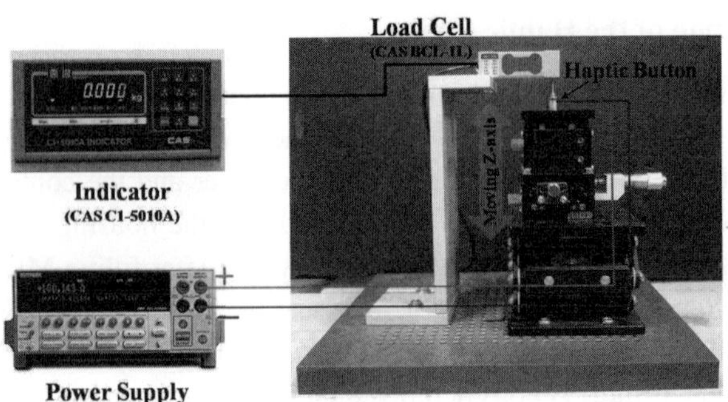

Figure 4. Experimental Setup.

4.2. *Experimental Results*

Fig. 5(a) shows the variation of the measured resistive force with respect to the indented depth (varied from 0.1 mm to 1.5 mm – the button stroke) and the applied current (0A, 0.15A, and 0.25A). As shown in the figure, the resistive force increases as the input current and the indented depth increases. To further analyze the results, the measured force rate (Qv) is defined, for a given indented depth, as the ratio of the difference between the maximum and minimum resistive force (Pv) to the maximum force (Lv).

Fig. 5 (b) shows the variation of the measured resistive force rate (Qv). The minimum measured force rate is about 72% with the input current of 0.15A and the indented depth of 0.1mm (the force varied from 0.5 N with 0A to 1.7N with 0.15A). For a force range between 0.5N and 200N, force JND (Just Noticeable Difference), which is equivalent to the measured force rate, is about 10% [5]. Therefore, the proposed haptic button with at least 72% force rate can generate various button sensations which can be conveyed to users.

Fig. 6 demonstrates that the prototype haptic button can create a force-displacement profile, which represents that of a typical spring-loaded button. When an actual button is pressed, the resistive force increases linearly with respect to the indented depth until the spring is fully compressed. The resistive force then suddenly drops to nearly zero when the button is released. By adjusting the input current, the haptic button can realize this button force profile, enabling users feel realistic button clicking sensation.

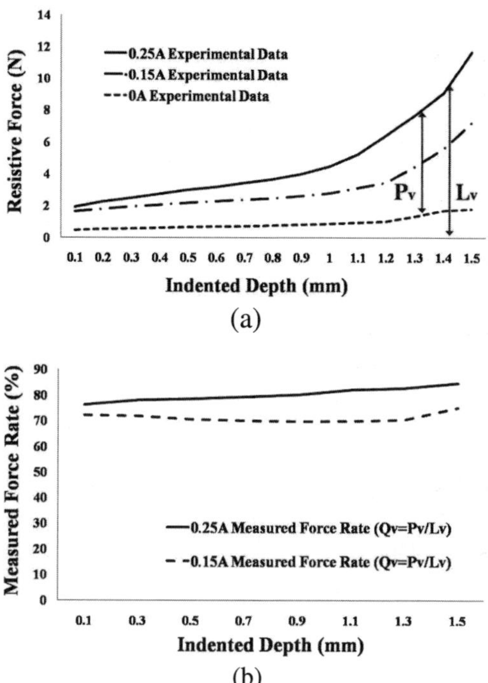

(a)

(b)

Figure 5. (a) Measured force versus indented depth.
(b)Measured force Rate versus indented depth.

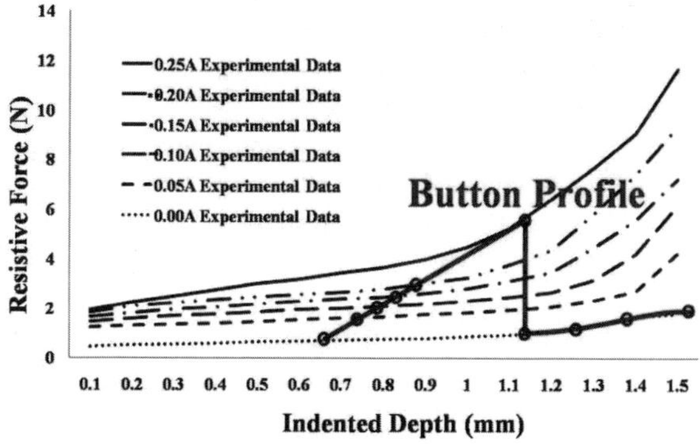

Figure 6. Button profile generation by varying the input current.

5. Conclusion

This paper has presented a new miniature haptic button based on MR fluids. In the proposed button, multiple operating modes of MR fluids are integrated into the design to maximize the resistive force. After constructing a prototype haptic button, a series of experiments was conducted using a micro stage setup to evaluate the performance of the prototype. The results indicate that the measured force increases as the indented depth and the input current increase. The results further indicate that the minimum measured force rate is around 72 %, which is sufficient to generate several steps of force or stiffness changes for realistic button sensations. In summary, this study demonstrated that the MR fluid-based haptic button can be used as a kinesthetic actuator for small electronic devices. Moreover, this work paved the way to developing haptic displays or panels based on MR buttons that can convey realistic haptic sensations over a continuous surface, such as a touch screen.

Acknowledgments

The work was partially supported by the Korea Research Foundation Grant Funded by the Korean Government (MOEHRD, Basic Research Promotion Fund) (KRF-2008-521- D00416).

References

1. K. Fujita, H. Ohmori, "A new softness display interface by dynamic fingertip contact area control", In Proceedings of the 5th World Multiconference on Systemics, Cybernetics and Informatics, Orlando, Florida, October 23, pp.78-82 (2000).
2. A. Song, D. Morris, J. E. Colgate, "Haptic Telemanipulation of Soft Environment without Direct Force Feedback", Proceedings of the 2005 IEEE International Conference on Information Acquisition June 27 - July 3, Hong Kong and Macau, China (2005).
3. H. Iwata, H. Yano, R. Kawamura, "Array Force Display for Hardness Distribution", Proceedings of the 10th Symp. On Haptic Interfaces For Virtual Environment & Teleoperator Systems.
4. M. Bianchi, E. P. Scilingo, A. Serio and A. Bicchi, "A new softness display based on bi-elastic fabric", Third Joint Eurohaptics Conference and Symposium on Haptic Interfaces for Virtual Environment and Teleoperator Systems, Salt Lake City, UT, USA, March 18-20, (2009).
5. L.A. Jones, "Stiffness Sensing", Human and Machine Haptics, MIT Press (2000).

APPLICATION OF A MAGNETORHEOLOGICAL ELASTOMER TO DEVELOP A TORSIONAL DYNAMIC ABSORBER FOR VIBRATION REDUCTION OF POWERTRAIN

NGA HOANG and NONG ZHANG

School of Electrical, Mechanical and Mechatronics Systems
University of Technology, NSW 2007

WEIHUA LI

School of Mechanical, Materials & Mechatronic Engineering
University of Wollongong, Wollongong, NSW 2522, Australia

HAIPING DU

School of Electrical, Computer and Telecommunications Engineering
University of Wollongong, NSW 2522, Australia

This study presents an application of magnetorheological elastomer (MRE) on the development of an Adaptive Tuned Vibration Absorber (ATVA) for torsional vibration reduction of vehicle powertrains. The MRE used for the ATVA development consists of a silicone polymer, silicone oil and magnetic particles with the weight fractions are 60%, 20%, 20%, respectively. The experimental testing was conducted to obtain MRE properties such as elastic modulus and damping ratio. Also, effective formulas for elastic modulus, damping ratio were derived to facilitate ATVA design. Numerical simulation show that the ATVA works effectively for powertrain vibration reduction.

1. Introduction

Magnetorheological elastomer (MRE) is a smart material whose mechanical property can be magnetically controlled. Thus, MRE is a potential material for developing ATVAs (Adaptive Tuned Vibration Absorbers) [1-3].

A vehicle powertrain is an essential component of automobiles for delivering power from engine to vehicle tires. It consists of an engine, a clutch, a transmission gearbox and driveline components, [4]. Obviously, there are vibrations occurred in powertrains which are part of noise, vibration, and harshness of vehicles. Especially, when the excitation fluctuation frequency of an engine coincides with or is close to a powertrain frequency. Thus, the vibrations are needed to be minimized.

The objective of this study is to develop an ATVA using a MRE for powertrain vibration minimization. By tuning the ATVA parameters such as stiffness and damping coefficients, the powertrain frequency can be shifted to avoid the resonance. This work consists of two main sections. The first section presents experiment set-up for measuring the MRE mechanical properties such as elastic modulus, damping ratio. The second section proposes an ATVA concept design and numerical simulations to validate the ATVA effectiveness.

2. A Magnetorheological Elastomer and its Vibration Characteristics

2.1. *MRE preparation and experiment set-up*

The MRE consists of a rubbery silicone polymer matrix, silicone oil which serves as a plasticizer and magnetic particles with the weight fractions is 60%, 20%, 20%, respectively. The size of magnetic particles is 6-7 μm. To enhance the MR effect, the mixed material was placed in a magnetic field before being cut into rectangular prisms samples 16×16×45mm for testing. This MRE material was fabricated at University of Wollongong, Australia.

The test was conducted at Dynamic and Solid Mechanics Laboratory, University of Technology, Sydney. The experimented set-up for measuring the young modulus is shown in figure 1.

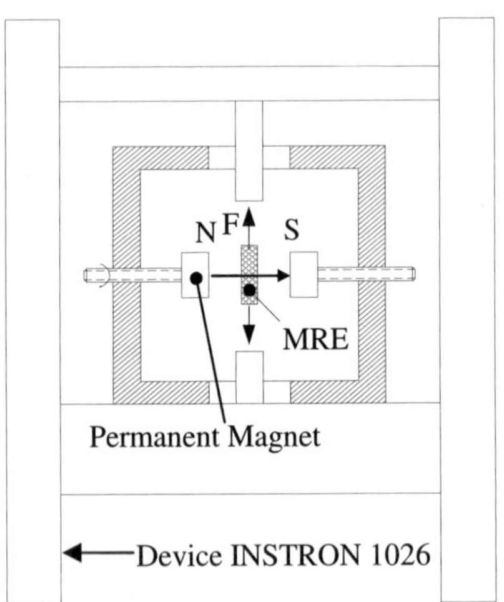

Figure 1. The schematic diagram for measuring MRE Young's modulus E

MRE samples were placed in the middle of two permanent cylindrical magnets D-D50H12.5-N45 (disc 50mm diameter× 12.5mm high). By tuning the distance between the two magnets, the magnetic field applied to the MRE can be varied. The MRE magnetic field density is measured by BELL 610 Gauss-meter.

The MRE Young's modulus is measured using the equation:

$$E = \frac{\sigma}{\varepsilon} = \frac{F / A_0}{\Delta l / l_0} = \frac{F l_0}{A_0 \Delta l} \tag{1}$$

Where E is the Young's modulus, F is the applied force; A_0 is the original cross-sectional area through which the force is applied; Δl the length change of the MR specimen; l_0 is the original specimen length.

To measure the MRE damping ratio, a weight is attached to the MRE specimen. By measuring the vibration attenuation, the damping ratio can be calculated as:

$$\varsigma = \frac{\delta}{\sqrt{(2\pi)^2 + \delta^2}} \tag{2}$$

The logarithmic decrement $\delta = \ln(x_1/x_2)$. x_1, x_2 are vibration amplitudes measured one cycle for the vibration of the weight.

2.2. *Experimented results and the proposed model*

The experiment data and the proposed model of MRE Young's modulus and damping ratio are shown in figures 2, 3.

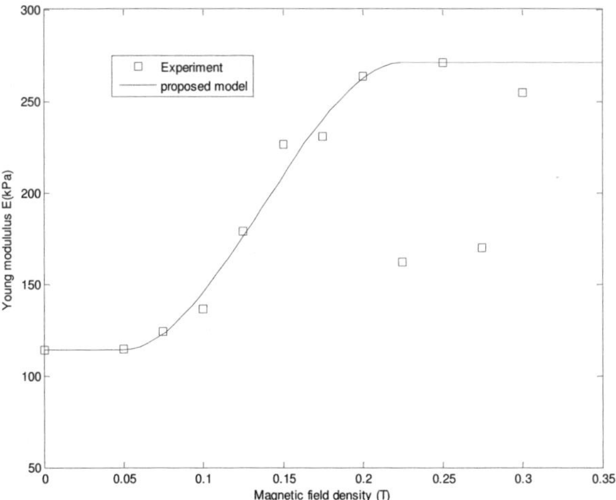

Figure 2. Young's modulus proposed model and experiment data

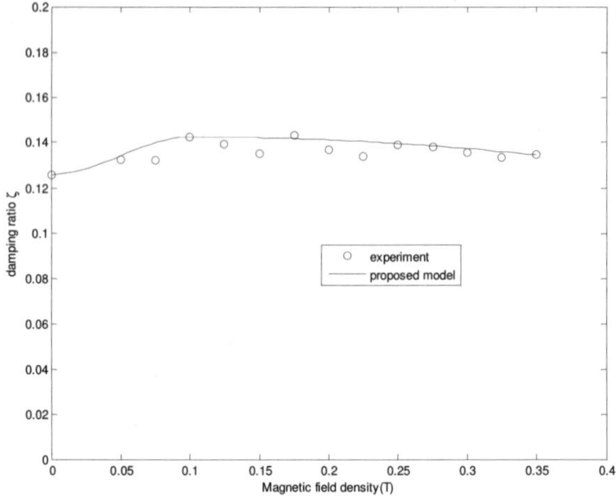

Figure 3. Damping ratio proposed model and experiment data

In which the Young's modulus is approximated by following equation:

$$E = E(B) = \begin{cases} E_0 & B \le B_0 \\ E_0 + (E_{max} - E_0)\dfrac{B^2}{B_S^2}(3 - 2\dfrac{B}{B_S}) & 0 < B < B_S \\ E_{max} & B \ge B_S \end{cases} \quad (3)$$

Here B_0=0.05T is the value of magnetic field density, from which the MRE material initially effected, B_S =0.225T is the saturated point, E_0=114.2kPa, E_{max}=270.9kPa.

The damping ratio is proposed as:

$$\varsigma = \begin{cases} 0.126 + 4.9794B^2 - 33.1958B^3 & 0 \le B < B_C \\ 0.1412 + 0.0256B - 0.1281B^2 & B_C \le B \le B_{max} \end{cases} \quad (4)$$

B_C=0.1T, B_{max}=0.35T. It can be seen that the experimented data and the proposed model are in good agreement. The model will be used for the ATVA design.

3. An ATVA Proposed Design

An ATVA proposed design is shown in figure 4, in which the inner cylinder is fixed on the rotating shaft and the MRE samples are put into the gaps between the inner and outer cylinders. The MRE samples cause elastic forces so that the cylinders can vibrate to each other. The electromagnetic coil can provide a magnetic field 0.35T for the MRE.

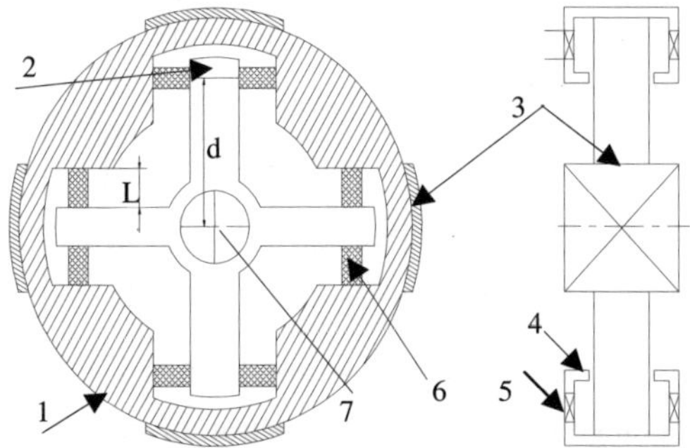

Figure 4. ATVA proposed design: 1, 2: inner and outer cylinder; 3 magnetic circuit; 4: steel core; 5: electromagnetic coil; 6: MRE; 7: shaft

ATVA natural and damped frequencies are expressed as:

$$f_n = \frac{1}{2\pi}\sqrt{\frac{k_A}{J_A}} \tag{5}$$

$$f_d = f_n\sqrt{1 - \varsigma_A^2} \tag{6}$$

In which $k_A = 8\dfrac{EA}{L}d$ and $c_A = 2\omega_n J_A \varsigma_A = 4\pi f_n J_A \varsigma_A$.

Here ς_A is MRE damping ratio calculated by equation (4). With J_A=0.15kgm^2, d=0.2m, L=0.035m, A=0.016×0.016m^2, the ATVA frequency in figure 5.

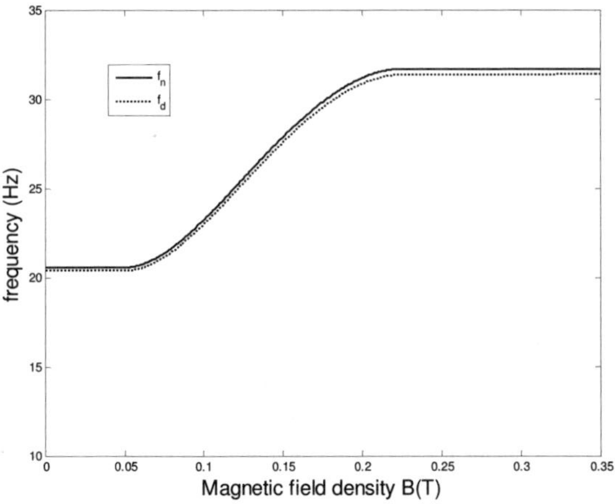

Figure 5. ATVA frequency

184

4. A Numerical Simulation

Hoang et al [5, 6] proposed a simplified model of powertrain model consisting of inertias, stiffnesses and dampings as shown in figure 6.

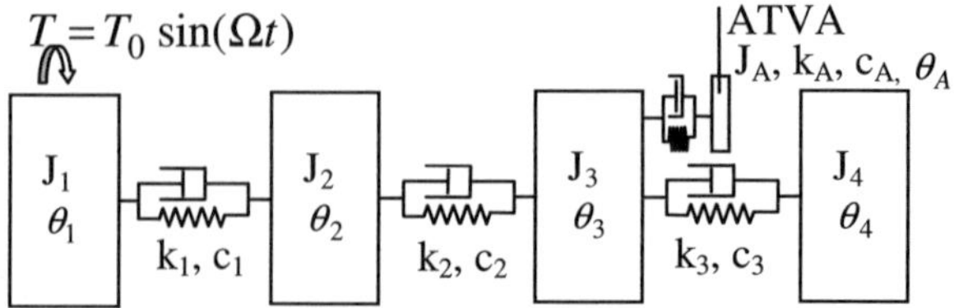

Figure 6. A powertrain model with an ATVA

By using Lagrange's equation, the equation of motion of the system before and after adding the ATVA can be expressed as the same equation as below:

$$\mathbf{J\ddot{\theta} + C\dot{\theta} + K\theta = T} \qquad (7)$$

where $\mathbf{\theta}$ and $\mathbf{T}$ are vectors of generalized coordinates and external torque. $\mathbf{J}$, $\mathbf{K}$ and $\mathbf{C}$ are inertial, stiffness and damping matrices, respectively. By solving equation (7), powertrain frequencies, steady response can be obtained.

To investigate the effectiveness of the ATVA, the powertrain vibration parameters are set as $J_1=0.82$, $J_2=0.2$, $J_3=0.4$, $J_4=40 \text{kgm}^2$; $c_1=20$ $c_2=10$, $c_3=10 \text{Nms/rad}$; $k_1=120000$, $k_2=75000$, $k_3=30350 \text{Nm/rad}$.

Let $T_0=5 \text{Nm}$, the powertrain steady responses for the fundamental frequency of powertrain $f_1=23 \text{Hz}$ before and after adding the ATVA are shown in figure 7.

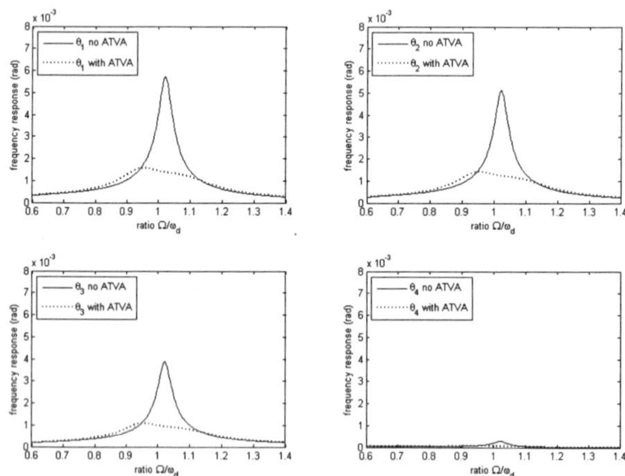

Figure 7. Powertrain vibration mode shapes

It can be seen that the powertrain vibration frequency response are reduced significantly. It confirms that the ATVA works as design.

5. Conclusions

A MRE was fabricated and tested for determining the stiffness and damping properties. It was developed a torsional ATVA for powertrain vibration control. To facilitate the ATVA design, valid formulas for elastic modulus and damping ratio are derived. With the derived formulas, stiffness and damping coefficients were calculated effectively from the magnetic field density. Thus, ATVA frequency can be tuned properly. Numerical simulations show that by using the ATVA, powertrain frequencies are shifted away from resonance frequency so that powertrain response is reduced significantly. The ATVA is a useful demonstrated device for powertrain vibration reduction.

Acknowledgments

The support from Tongfei Tian, University of Wollongong; Christopher Chapman and Michael Tran, University of Technology, Sydney for the experiment testing is gratefully acknowledged.

Reference

1. J.D. Carlson and M. R. Jolly 10 555–69 (2000).
2. G. Y. Zhou J. Smart Mate Struct 12 139-46 (2003).
3. A. A. Lerner and K.A. Cunefare J. Intelligent Mate Systems and Struct 19 551-563 (2008).
4. N. Zhang, A. Crowther, D.K. Liu and J. Jeyakumaran, Proc. of the Instn of Mech Engrs Part D: Journal of Automobile Engineering 217 461-73 (2003).
5. N. Hoang, N. Zhang N and H. Du, J. Smart Mate Struct 18(074009) (2009).
6. N. Hoang, N. Zhang N and H. Du, *Proc. SPIE 7643-36*, San Diego, March 7-11 (2010).

RESEARCH ON MAGNETORHEOLOGICAL ELASTOMER ABSORBER AND ITS IMPACT TEST[*]

XIAOMIN DONG[†]

State Key Laboratory of Mechanical Transmission, Chongqing University, Chongqing, 400044, China, xmdong@cqu.edu.cn

MIAO YU and LIXI ZHU

College of Opto-Electronic Engineering, Key Lab of Opto-Electronic Technology and System of Education Ministry, Chongqing University, Chongqing, 400044, China

In this study, a new magnetorehological elastomer (MRE) based absorber is proposed and its vibration isolation performance is investigated. The MRE absorber with a compact structure is firstly designed in order to accomplish the maximization of the variable stiffness range. The working characteristics of the MRE are then measured. On the basis of the experimental data, the control model of the MRE is also formulated. Finally, the drop test is carried out to check the actual impact isolation performance.

1. Introduction

Magnetorheological (MR) materials include Magnetorheological fluid (MRF), MR foams and magnetorheological elastomer (MRE), whose rheological properties can be controlled by the application of an external magnetic field [1]. The most common MR material is MRF. There are magnetically polarizable particles suspended in the viscous fluid. In recent years, the MRF technology has made significant advancements [2] and many applications such as the automotive suspension vibration control [3], the earthquake resistance [4], clutch [5] etc. However, MRF are prone to particle settling with time due to the density mismatch of particles and the carrier fluid, which may degrade the MR effect. Therefore, another smart material, MRE, has received much attention in recent years.

As solid analogs of MRF, magnetorheological elastomer (MRE), which includes a wide variety of composite materials, typically consist of magnetically

---

[*] **Foundation item:** Projects (60804018, 50830202) supported by the National Natural Science Foundation of China; Project (200902292) supported by the Special Post- doctoral Fund of People's Republic of China; Project (20090191110011) supported by Doctoral Fund of Ministry of Education of China.

[†] Corresponding author.

polarizable particles dispersed in a polymer medium with many advantages. Compared to MRF, MRE can avoid some disadvantages such as settling of particles normally associated with MRF. MRE do not require channels or seals to hold or prevent leakage and thus can avoid the particle sedimentation associated with MRF. The valuable characteristic of MRE is that the mechanical properties of MRE such as the storage and loss modulus can be altered reversibly by the application of an external magnetic field. The field dependence of the mechanical properties enables the construction of controllable elastomeric components. As a result, MRE possesses many potential engineering applications for vibration control in damping and vibration isolation systems such as engine mounts and suspension bushings to reduce noise and vibration.

Due to the little adjustment range of variable stiffness, there are only few application reports about MRE when comparing to the MRF. A tunable automotive mounts and bushings based on MRE was developed by Ginder et al [6]. They found that the suspension resonances excited by torque variation could be suppressed by shifting the resonance away from the excitation frequency. Another similar adaptive tuned vibration absorber based on MRE was studied by Deng et al. [7]. The results show that the natural frequency of the absorber based on MRE can be tuned from 55 to 82 Hz. In their later work [8], the shift-frequency capability of the adaptive absorber was theoretically and experimentally evaluated. The results demonstrate that the natural frequency of the proposed absorber can be tuned from 27.5 Hz to 40 Hz. Collette et al.[9] investigated numerically two systems based on MRE according to the measured characteristics from published studies. The numerical results show that the commandability of the elastomer improves the isolation performance of the MRE isolator under a frequency varying harmonic excitation, and decreases the stress level of the MRE dynamic vibration absorber under a random excitation.

Consequently, the main contribution of this study is to propose a new adaptive variable stiffness absorber based on MRE and evaluate its vibration control performance. To accomplish it, a new vibration absorber based on the fabricated MRE is developed and its control model is formulated on the basis of measured working characteristics. Then the drop test is carried out to validate the proposed the MRE absorber.

2. Development of MRE Absorber

In this study, an adaptive variable stiffness absorber based on MRE is designed and manufactured to realize variable stiffness characteristics of the absorber. It is schematically shown in Fig.1. The absorber consists of a coil, two smart MRE, the outer cylinder and the piston. Two MREs are clung to both the outer cylinder and the piston, respectively. The magnetic field is generated by the coils and can be tuned by tuning the coil current. By the motion of piston, the shear deformation of MRE will appear. The magnetic field exits in the MRE which is perpendicular to the motion of the piston after the current is applied to the coil. The shear modulus of the MRE is altered when subjected to the magnetic field. Thus, the elastic force of the absorber is controlled by the intensity of the magnetic field.

In absence of magnetic field, the absorber produces an elastic force only caused by the 704 silicone rubber' shear resistance. If a certain level of magnetic field is supplied to the absorber, the MRE absorber produces an additional and variable elastic force owing to the increased shear modulus of the MRE. The elastic force of the MRE absorber can be continuously tuned by controlling the intensity of the magnetic field. To simplify the analysis of the MRE absorber, a Kelvin-Voigt model is adopted in this study. The model consists of a constant stiffness spring, a variable stiffness spring and a damper with constant damping coefficient in parallel. In the shear direction, k is given by [7],

$$k = \frac{GA}{h} = \frac{(G_0 + \Delta G)A}{h} = \frac{G_0 A}{h} + \frac{A}{h} 36\phi\mu_r\mu_0\beta^2 H_0^2(\frac{R}{d})^3\varsigma = k_e + F_{MRE} \qquad (1)$$

where G denotes the shear modulus of MRE and consists of two terms, an initial shear modulus, G_0, without a magnetic field and a shear modulus increment, ΔG, while applying a magnetic field; A is the shear area, and h is the thickness of MRE; k_e is the equivalent stiffness coefficient, c_e denotes the equivalent damping coefficient and F_{MRE} is the controllable elastic force of the MRE stiffness part. For simplification, the F_{MRE} can be described as a quadratic function of the input current, $F_{MRE} = dI^2 + eI + f$, and d, e, f are constants obtained by fitting the measured experimental data.

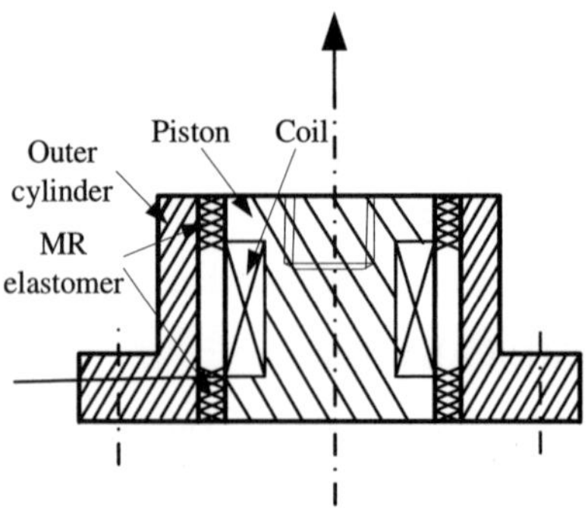

Fig.1 Schematic configuration of MRE absorber

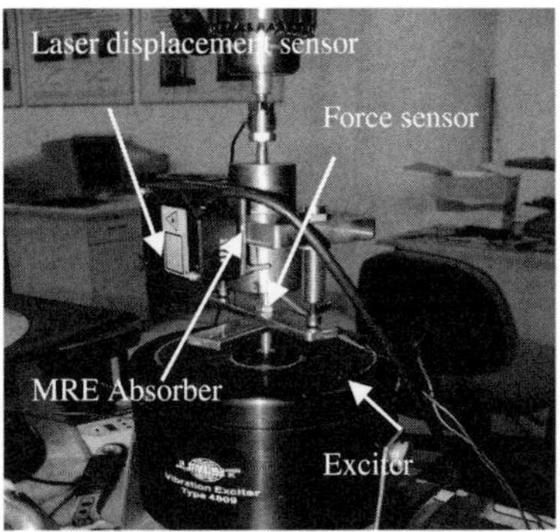

Fig.2 Test set up for MRE absorber

To measure working features of the MRE absorber, the experimental platform has been set up, which is shown in Fig.2. A vibrator produced by B&K Co. is used to drive the MRE absorber; the displacement and force are measured. The velocities are calculated using a central differences approximation. Sinusoidal command signals are used. In this study, the excitation frequencies are 25, 50 and 75 Hz and the displacement amplitudes have the same value of 0.5 mm. Various constant currents are applied to the MRE absorber to observe the characteristics of the absorber. The applied input current is from 0 to 2 A with the increment of 0.5 A. For brevity, only the force-displacement hysteresis under the excitation frequency of 25 Hz is given in Fig. 3. The equivalent stiffness of MRE absorber is calculated and shown in Fig. 4. From the figures 3-4, it can be seen that the elastic force will increase with the increase of input current. The enclosed area of the force-displacement remains invariable and is independent of input current, which represents the damping characteristics can not be controlled by the application of magnetic field.

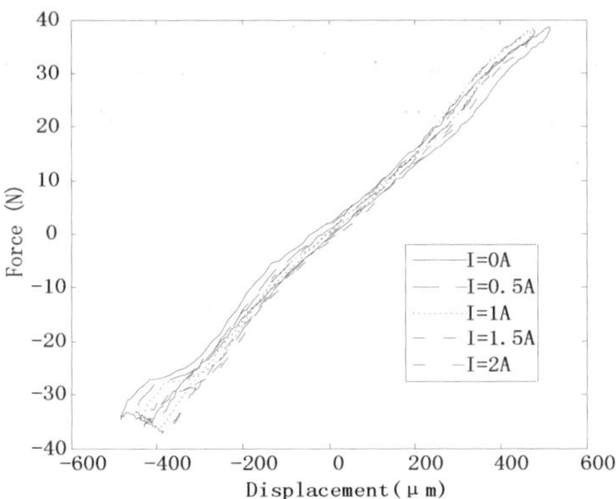

Fig. 3 The relationship between force vs. displacement

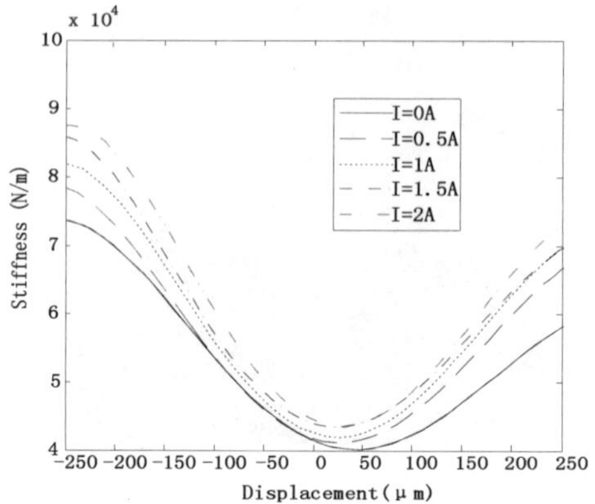

Fig.4 The relationship between stiffness vs. displacement

3. Drop Test and Result

To validate the proposed MRE absorber, a drop test is carried out, which is shown in Fig. 5. The spring mass and unsprung mass are connected by the MRE absorber. During the test, the absorbing system is sliding along the guide way vertically. The impact force between the unsprung mass and ground is measured by a force sensor. The acceleration of spring mass is acquired by the acceleration sensor. To reduce the impact acceleration of spring mass, human simulated intelligent control strategy is adopted [10]. As comparison, the response of the absorbing system without control is also measured. The height of drop test is 10 cm for each test. Figure 6 shows the comparison result of sprung mass under on-control and off-control. According to the figure, it can be seen that the maximum acceleration peak of sprung mass under control can be reduced by 12.1 m/s^2 compared to that of the off-control. That also suggests that the maximal impact energy can be effectively reduced. However, a much longer adjusting time is needed for the on-control.

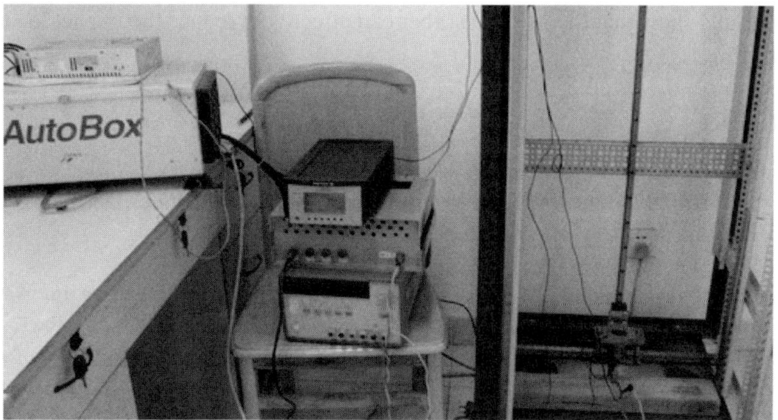

Fig.5 Setup for drop test

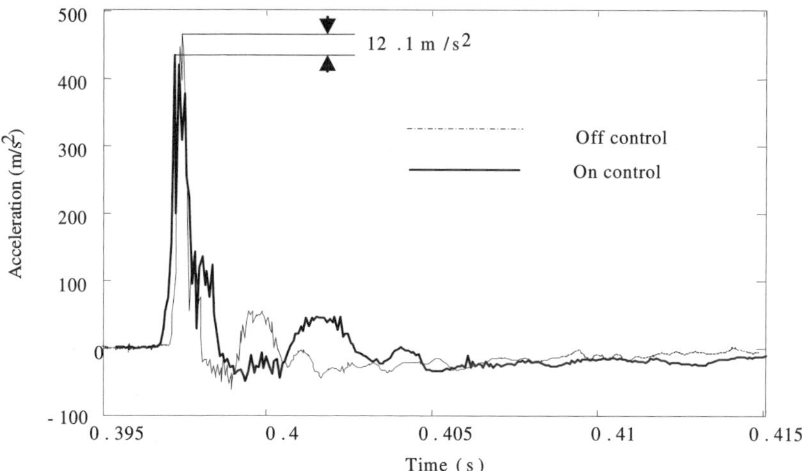

Fig.6 Acceleration time history of sprung mass

4. Conclusion

In this study, a new variable stiffness absorber based on MRE was developed and its vibration absorbing capability was evaluated. The MRE absorber was firstly developed. The working characteristics of the MRE absorber were measured and then its control model was formulated. Finally, a drop test was performed to evaluate the vibration absorbing ability of the MRE absorber. The test results show that the MRE absorber can effectively reduce the impact load of sprung mass.

Reference

1. J.D. Carlson and M.R. Jolly, *Mechatronics*, **4-5**, 10(2000).
2. J.H. Koo, F. Goncalves and M. Ahmadian, *Shock Vib. Dig.* **3**, 38(2006).
3. B. Ioan, *J. Magn. Magn. Mater.* 241(2002).
4. Z.D. Xu and Y.Q. Guo, *J. Intel. Mat, Syst. Str.* 17(2006).
5. F.K. Gordaninejad, M. Barkan and X.J. Wang, *Proceedings of the SPIE* (2007).
6. J.M. Ginder, M.E. Nichols, L.D. Elie and S.M. Clark, *Proceedings of the SPIE* (2007).
7. H.X. Deng, X.L. Gong and L.H. Wang, *Smart Mater. Struct* 15(2006).
8. H.X. Deng and X.L. Gong, *Commun. Nonlinear Sci* 13(2008).
9. C. Collete, K. Kroll, G. Saive, V. Guillemier, M. Avraam and A. Preumont, *J. Phys.* 149(2009).
10. L.X. Zhu, *Chongqing University dissertation*, (2009).

DYNAMIC TESTING AND MODELING OF AN MR SQUEEZE MOUNT

XIN-JIE ZHANG[1,2], ALIREZA FARJOUD[2], MEHDI AHMADIAN[*,2],
MICHAEL CRAFT[2] and KONG-HUI GUO[1]

[1]*State Key Laboratory of Automobile Dynamics Simulation, Jilin University, Changchun, China, 130022*

[2]*Center for Vehicle Systems and Safety, Virginia Tech, Blacksburg, VA 24060, USA*

Magneto-rheological fluid squeeze mode investigations at CVeSS have shown that MR fluids show large force capabilities in squeeze mode. Theoretical and experimental study of MR fluids in squeeze mode has previously been limited to quasi-static conditions. It was found that MR fluids in squeeze mode can be used in a wide range of applications such as engine mounts and impact dampers. In these applications, MR fluid is flowing in a dynamic environment due to the transient nature of inputs and system characteristics. The research presented in this paper undertakes the problem of dynamic testing and modeling of MR fluid squeeze mounts. Dynamic tests of the MR mount are studied for different densities of magnetic field, frequencies and excitation amplitudes. An effective mathematical model of MR mount for steady state was built which contains inertial effect. The results show the compression force and the area of the hysteresis loop increase when increasing the density of magnet field or excitation amplitude. Also, the inertia effect becomes more significant for higher displacement frequencies.

Keywords: Magneto-Rheological (MR) fluid, MR dynamic mount, squeeze flow, inertia effect, compression mode

1. Introduction

The essential characteristic of a magneto-rheological (MR) fluid, consisting of a suspension of micron-sized ferromagnetic particles in a non-magnetic carrier fluid, is the rapid and reversible transition from the state of a Newtonian-like fluid to the behavior of a stiff semi-solid by applying a magnetic field of 0.1–0.4T. Due to this rheological behavior, MR fluids are currently being used in a wide variety of applications such as: controllable dampers, rotary brakes [1],

---

[*] Corresponding author, E-mail: ahmadian@vt.edu, Phone: (540)-231-1408

seismic vibration mitigation devices [2], and prosthetic devices [3]. MR fluid offers three modes of operation: direct shear mode, valve mode, and squeeze mode. The latter is of particular interest due to its large force capabilities, which is still not fully understood and therefore expected to give rise to new industrial applications.

"Squeeze flows" are flows in which a material is compressed between two parallel plates and thus squeezed out radially. Stefan [4] was the first to investigate the squeeze flow of Newtonian fluids. The work was extended to power law fluids [5]. Since then, various aspects of squeeze flows have been studied by a range of investigators focusing on rheometry, parameter identification, constitutive equations, numerical simulation, etc. A good review of the squeeze flow rheometry and its applications was done by Engmann et al [6]. An ER squeeze mode was applied in a prototype automotive engine mount [7]. Behavior of MR fluids in squeeze mode under oscillatory conditions was investigated in [8]. It was found that damping performance of an MR damper can be obtained by increasing magnetic field density or displacement amplitude. Changing the frequency of the system also contributed to various behaviors of MR fluids. The peak compression force was found to increase as the number of cycles is increased.

A novel MR squeeze mount was designed and built at CVeSS [9]. This MR squeeze mount is studied for different densities of magnetic field, frequencies and excitation amplitudes. An effective mathematical model of MR mount for steady state conditions is built that contains inertial effect.

2. Experimental Testing

To obtain a thorough understanding of dynamic behavior of MR fluid in squeeze mode, an electromagnetic linear actuator (EMA) dynamometer is employed to test an MR squeeze mount. The Roehrig-EMA dynamometer is run by a desktop computer via Roehrig-Shock 6.0 software. The hardware within the EMA measures input displacement and velocity, while the load cell measures force. The linear actuator has a range of 0.0098-6.985 in. and is able to produce harmonic inputs up to 100 Hz. Additionally, an Interface brand loadcell is able to measure forces of up to 2000 lbf. The electromagnet power supply is an EXTECH 80w switching DC power supply. The experiment setup for testing the MR squeeze mount is shown in Figure 1. An MR squeeze mount containing Lord MRF-132DG is tested, as seen in Figure 2. Inputs are sine wave displacement signals incremented from 1Hz to 20 Hz in 1 Hz steps and test amplitudes are 0.02 and 0.04 in. Initial gap of the mount plates is 0.1

in. The electromagnet for magnetic field creation uses an EXTECH 80w switching DC power supply. The magnetic field is simulated by a finite element modeling software, FEMM. Each test frequency has ten cycles and the results shown in this paper are the ninth cycle from each test run.

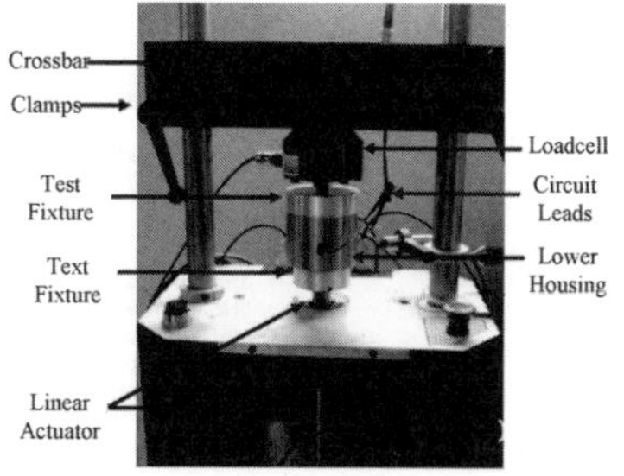

Figure 1: Test setup for MR squeeze mount and magnetic system in Roehring EMA Dynamometer [10]

Figure 2: MR squeeze mount

Figure 3 shows the influence of electric current on the relationship between vertical force and displacement. Input displacement amplitude of Figure 3a is 0.02 in. Negative displacement values show that the MR mount gap is smaller than the initial gap of 0.1 in. Figure 3b shows an amplitude of 0.04in. It shows that increasing the current always results in a higher compression force and the rebound force doesn't increase as much as the compression force. Decreasing the initial gap between two poles can also increase the compression force tremendously.

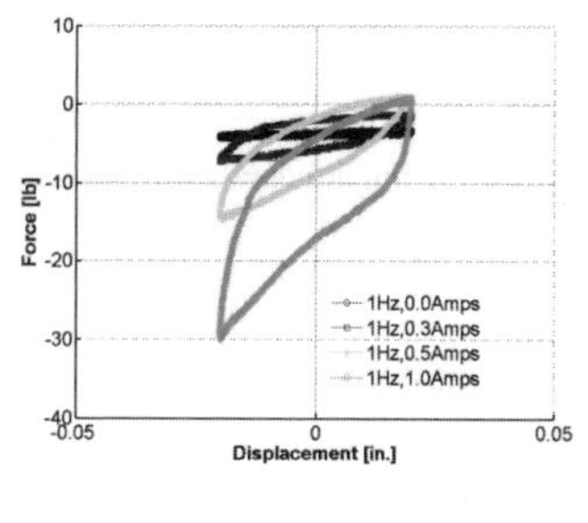

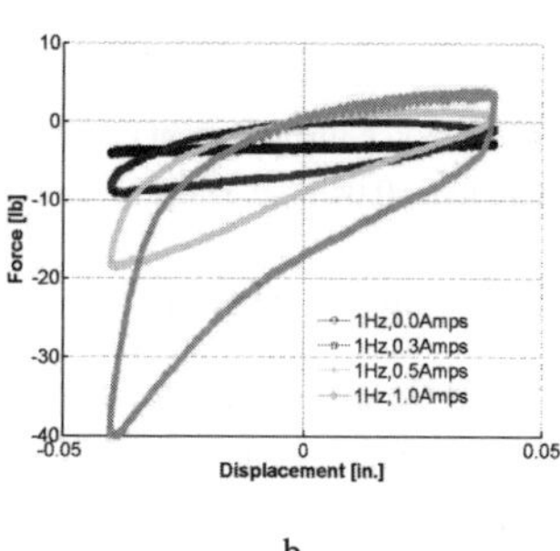

Figure 3: Force-displacement curves for different currents,
a: amplitude=0.02in., b amplitude=0.04in.

Figure 4 considers the effect of frequency on the behavior of loop curves of the MR mount. The amplitude in Figure 4a and 4c is 0.02 in and 0.04in. for Figure 4b and 4d. Figure 4c and Figure 4d show that increasing the frequency will slowly increase the tension and compression forces. There are still some transient properties in Figure 4a and Figure 4b even after eight cycles since the damping is much lower than Figure 4c and Figure 4d.   At higher frequencies or larger displacement amplitudes, it takes a longer time for the transient response of the MR squeeze mount to decay.

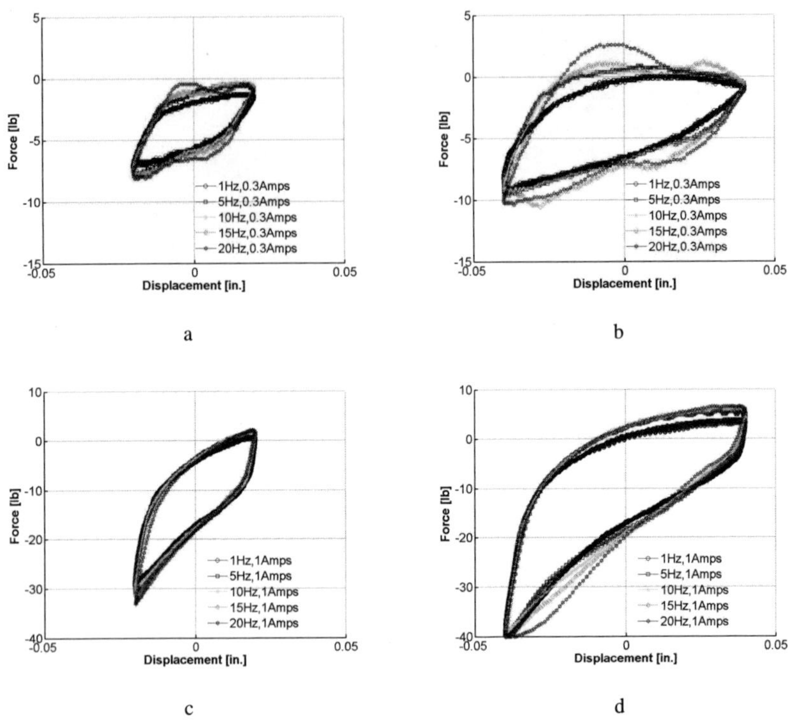

Figure 4: Force-displacement curve for different frequencies, a: amplitude=0.02in., 0.3amps, b: amplitude=0.04in., 0.3amps, c: amplitude=0.02in., 1.0amps, d: amplitude=0.04in., 1.0amps.

3. Mathematical Model

A standard theory of lubrication is applied to the equations governing fluid flow in which the aspect ratio h/R is small. h is the gap between two poles and R is the radius of the MR squeeze mount.   The governing equation is

$$\rho\frac{\partial u}{\partial t}+\frac{dp}{dr}=\frac{\partial \tau}{\partial z} \tag{1}$$

Where u is the radial velocity, dp/dr is the radial pressure gradient, and τ is shear stress in the fluid. Integrating equation (1) across the gap gives

$$\rho\int_{h_u}^{h_l}\frac{\partial u}{\partial t}dz+(h_l-h_u)\frac{dp}{dr}=\tau(h_l)-\tau(h_u) \tag{2}$$

h_l is the z-coordinate of the lower pole and h_u is the z-coordinate of the upper pole. Applying the condition of global conservation of mass and boundary conditions for u on the pole plates, equation (3) is obtained. Differentiating equation (3) with respect to time and replacing in equation (2) will results in equation (4).

$$2\pi r \int_{h_u}^{h_l} u\,dz = -\pi r^2 (\dot{h_l} - \dot{h_u}) \tag{3}$$

$$\frac{dp}{dr} = \frac{\tau(h_l) - \tau(h_u) + \frac{1}{2}\rho r(\ddot{h_l} - \ddot{h_u})}{h_l - h_u} \tag{4}$$

Using initial condition, $p(R) = 0$, and integrating equation (4), the total vertical force from MR mount is found.

$$F_{mount} = -\frac{\pi}{h_l - h_u}\int_0^R r^2[\tau(h_l) - \tau(h_u)]dr - \frac{\pi R^4 \rho(\ddot{h_l} - \ddot{h_u})}{8(h_l - h_u)} \tag{5}$$

The second term on the right side of equation (5) is the inertia due to the oscillatory flow. The first term is the same as [9], so the final equation is as following:

$$F_{mount} = \pi R^2 \tau[\frac{-1}{K_H} + \frac{1}{2}\frac{(h_l - h_u)^2 e^{\frac{2K_H R}{h_l - h_u}}}{K_H{}^3 R^2} + \frac{3}{2}\frac{R^2 \eta(\dot{h_l} - \dot{h_u})}{(h_l - h_u)^3 \tau}$$
$$-\frac{1}{2}\frac{(h_l - h_u)^2}{K_H{}^3 R^2} - \frac{(h_l - h_u)}{K_H{}^2 R}] - \frac{\pi R^4 \rho(\ddot{h_l} - \ddot{h_u})}{8(h_l - h_u)} \tag{6}$$

Using equation (6), the behavior of the MR squeeze mount force was compared against test data as shown in Figure 5. τ is found based on the magnitude of field density from FEMM. η is found from the manufacturer data and K_H is found with a heuristic process. The rebound tension force hasn't been considered in this model since it is much smaller. The simulation results can clearly capture the trend of the test data as shown in Figure 5. Figure 6 shows the model results for the behavior of MR mount under different frequencies. It shows that the loop curve of the MR mount with 50Hz displacement input has

198

slightly changed compared to 1 Hz. The same trend also can be seen in test results, as seen in Figure 4. In other words, the inertia effect becomes more significant at higher displacement frequencies, although more modeling and testing must be performed to more closely match this data.

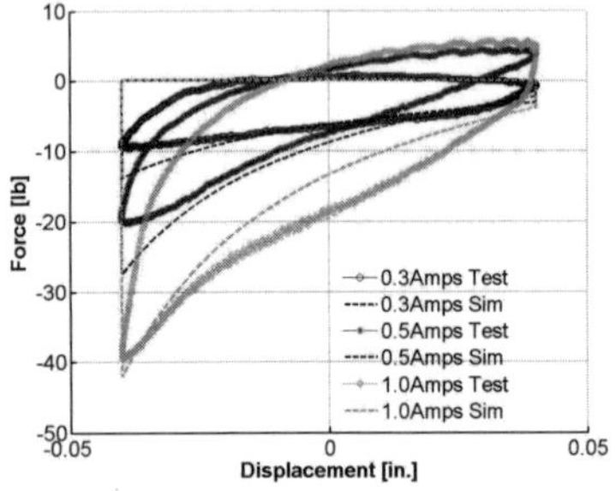

Figure 5:   Comparison between test data and simulation results at 5Hz

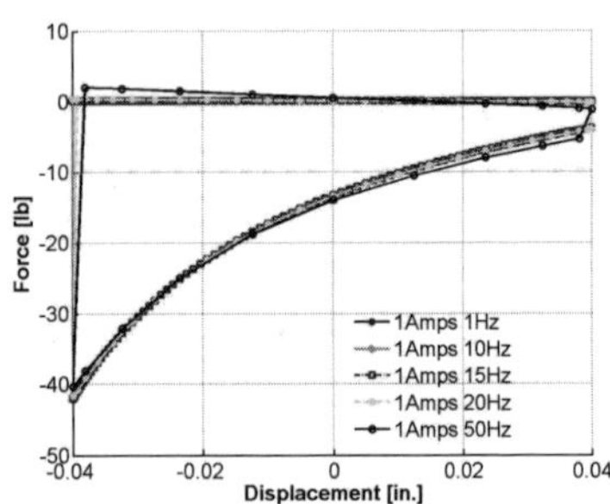

Figure 6:   Model results for force-displacement curve at different frequencies

4. Conclusion

A novel MR squeeze mount was designed and built at CVeSS. This MR squeeze mount was tested with different densities of magnetic fields, frequencies and displacement amplitudes. An increase in magnetic field or displacement amplitude results in increase of the compression force and the area of the hysteresis loop. Increasing the frequency will slowly increase the tension and compression forces. At higher frequencies or displacement amplitudes, it takes a longer time for the transient response of the MR squeeze mount to decay. A dynamic mathematical model was developed, which considered the inertial effect and was validated by the test data. The test data and simulation results show that inertia effect becomes more significant at higher displacement frequencies.

Acknowledgements

Xin-jie Zhang is funded by the China Scholarship Council (CSC).

Reference

1. W. Li and H. Du, "Design and experimental evaluation of a magnetorheological brake," The International Journal of Advanced Manufacturing Technology, vol. 21, 2003, p. 508-515.

2. S. Cho, H. Jung, and I. Lee, "Smart passive system based on magnetorheological damper," Smart Materials and Structures, vol. 14, 2005, p. 707-714.

3. J.Z. Chen and W.H. Liao, "Design, testing and control of a magnetorheological actuator for assistive knee braces," Smart Materials and Structures, vol. 19, 2010, p. 035029.

4. G.J. Dienes and H.F. Klemm, "Theory and application of the parallel plate plastometer," Journal of Applied Physics, vol. 17, 1946, p. 458.

5. Stefan, J. and K. Sitzgber, Versuche Uber Die Scheinbare Adhasion. Sitz. Kais. Akad. Wiss. Math. Natur. Wien., 1874. 69(2): p. 713-735.

6. J. Engmann, C. Servais, and A. Burbidge, "Squeeze flow theory and applications to rheometry: A review," Non-Newton. Fluid Mech., vol. 132, 2005, p. 1-27.

7. E.W. Williams, S.G. Rigby, J.L. Sproston and R. Stanway, "Electrorheological fluids applied to an automotive engine mount," Journal of Non-Newtonian Fluid Mechanices, 47 (1993):p. 221-238.

8. S. Vieira, C. Ciocanel, P. Kulkarni, a. Agrawal, and N. Naganathan, "Behaviour of MR fluids in squeeze mode," International Journal of Vehicle Design, vol. 33, 2003, p. 36.

9. Michael J. Craft , Mehdi Ahmadian, Alireza Farjoud, William C.T. Burke, Clement Nagode, "Force characteristics of a modular squeeze mode magneto-rheological element," Active and Passive Smart Structures and Integrated Systems 2010, Proceedings of the SPIE, Volume 7643, p. 764313-764313-11 (2010).

10. Brian Mitchell Southern, "Design and characterization of tunable magneto-rheological fluid-Elastic mounts," M.S. Thesis, Virginia Polytechnic Institute and State University, Blacksburg, VA, 2009.

EXPERIMENTAL INVESTIGATION OF MR SQUEEZE MOUNTS

ALIREZA FARJOUD[*], MICHAEL CRAFT, WILLIAM BURKE and
MEHDI AHMADIAN

*Center for Vehicle Systems and Safety (CVeSS), Virginia Tech, Blacksburg, VA 24060,
USA*

Based on results obtained from testing MR fluid in a squeeze mode rheometer, a novel compression-adjustable element has been fabricated and tested, which utilizes MR fluid in squeeze mode. While shear and valve modes have been used exclusively for MR fluid damping applications, recent modeling and testing with MR fluid has revealed that much larger adjustment ranges are achievable in squeeze mode. Utilizing squeeze mode, an MR squeeze mount was developed and tested. Test results show the device was capable of varying the compression force from less than 8lbs to greater than 800lbs when the pole plates were 0.050" apart. Test results showed that the mount tends to achieve higher forces after each repeat of test. This behavior, called "clumping", was studied and solutions to minimize this effect are discussed.

1. Introduction

The use of MR fluids in commercial devices has become increasingly common over the past decade. The vast majority of these devices operates with the MR fluid in valve or shear mode, typically in dampers [1-3]. While the use of MR fluid in these modes is well understood, the less documented squeeze mode can be applied to MR fluid which has shown promising behavior for device usage. Recent research at CVeSS demonstrated that a tested MR fluid in squeeze mode was capable of a large range of adjustment. The force required to yield the fluid at a gap size of 0.20" in squeeze mode was approximately seven times the force needed to yield the same fluid in direct shear mode [4].

For an MR device to operate in squeeze mode, MR fluid is compressed between two end plates that also serve as magnetic poles for a magnetic field to pass through. As the fluid is compressed, the available volume between these end plates continuously becomes smaller, requiring a means for the MR fluid in the gap to escape to avoid hydraulic lock-up.

Figure 1 shows an MR squeeze mount developed at CVeSS. Each pole plate is consisting of two parts that clamp to each other and provide modularity as well as self-containment.

---

[*] Corresponding author, email: farjoud@vt.edu

Figure 1 MR squeeze mount

2. MR Squeeze Mount Membrane Development

The MR fluids used for MR mount testing are oil-based MR fluids. Because the rubber is also made from oil-based material, it was observed that MR fluid and rubber were reacting with one another over time. This reaction changed the physical properties of the MR mount during testing. It was observed that the rubber became softer, lost some elasticity and slowly began releasing oil after reacting with the MR fluid. Figure 2 shows a side-by-side comparison of rubber before and after reacting with MR fluid. As can be seen, the shape of the rubber has completely changed, most likely due to the relaxation of internal stresses created during the manufacturing process of the rubber.

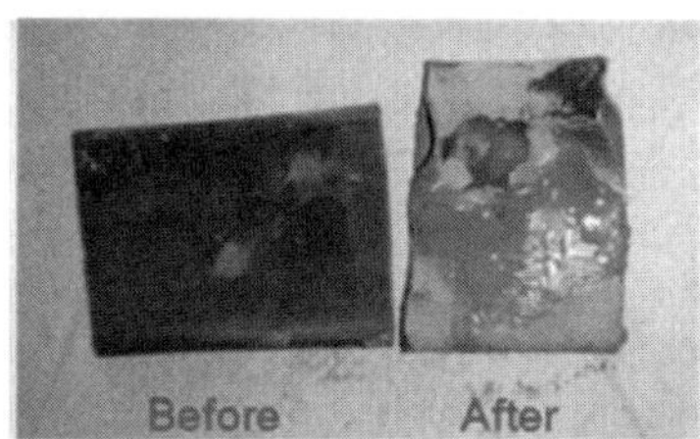

Figure 2 Mount rubber before and after reaction with MR fluid

This material incompatibility was studied and a new flexible membrane was fabricated with a commercially available two-part polyurethane. Polytek 74-30 RTV polyurethane was used for molds. A rotary mold was made to fabricate the thin-walled membrane. To make a membrane, a finite amount of the mixed, two-part polyurethane was added to a cylinder with sealing end caps. This mold was then spun at high speed until the rubber was set. After some time, the rotary mold was stopped and the membrane was left to cure. The membrane thickness can easily be adjusted with the rotary mold, although the thickness range of interest was 0.010-0.030".

3. Squeeze Mode Rheometer

MR fluid needs a magnetic field to activate and change properties. In order to provide a magnetic field in the fluid gap of the MR mount, it is placed in a squeeze mode rheometer designed and built previously by the authors [5] and shown in Figure 3.

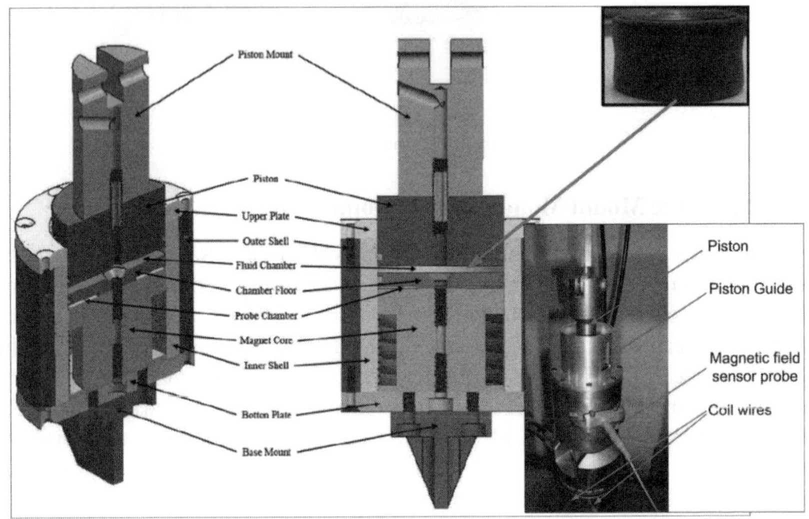

Figure 3 Squeeze mode rheometer used to test the MR squeeze mount

The rheometer is capable of measuring the magnetic field directly, during testing. This is made possible by implementing a probe chamber in the rheometer design. The Gauss-probe chamber is located directly below the MR chamber and separated by a disk sealed with an O-ring (Figure 3). The Gauss-probe is inserted from the side of the rheometer through a hole in the rheometer body. The probe can measure the magnetic field density along the radius of the rheometer.

4. Magnetic Field Study

The magnetic field density in the MR mount should be homogeneous to prevent any undesired effect on the behavior of the fluid in the mount. If the magnetic field density is not homogenous along the radius, iron particles in the fluid will tend to form agglomerations in the regions with higher magnetic field density. The magnetic field density distribution in the mount was studied using Finite Element Method. Figure 4 shows the magnetic field density distribution in the squeeze mode rheometer and MR mount. As shown, it can be seen that the magnetic field density is constant in the MR mount fluid gap between the mount pole plates.

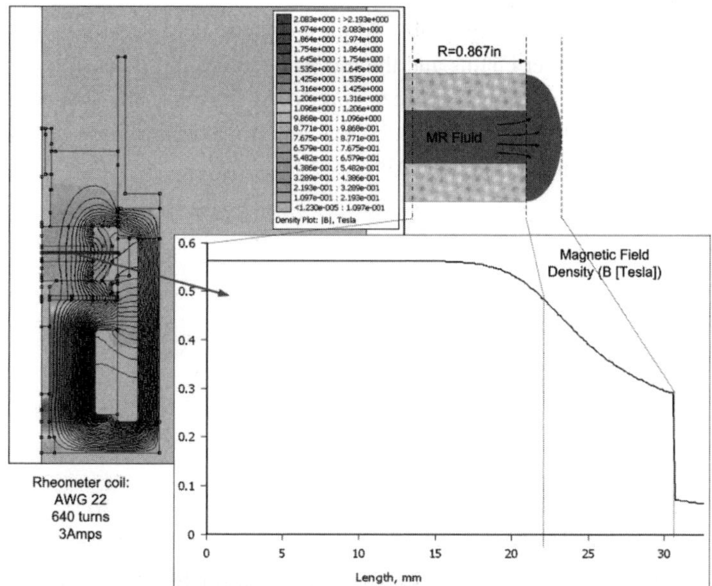

Figure 4 Magnetic field simulation in the squeeze mode rheometer and MR mount

In order to validate the magnetic field density simulations, measurements from the magnetic probe in the rheometer were compared to FEM simulations and good agreement was observed.

5. MR Mount Quasi-Static Testing

MR squeeze mount has been extensively tested under quasi-static and dynamics conditions. In this paper, only the quasi-static test results are presented and discussed. The test procedure is as follows: First, the mount is placed in the rheometer and the initial desired gap is adjusted. Then the electric power supply is turned on and the desired electric current is adjusted. Finally, the piston starts to move down and squeezes the mount.

There are two ways of actuating the load frame used for testing. One way is to control the displacement. This method is called the "displacement-controlled" method. Alternatively, the force can be controlled. The results that will be presented in this paper are based on the first approach, "displacement-controlled" mode. The MR mount gap size was decreased from an initial value of 0.25" to a final value of 0.05" under various magnetic field strengths and the force was measured for each case. The piston speed was chosen to be 0.005 in/sec to ensure quasi-static condition. Figure 5 shows a typical test result for squeeze testing of the MR mount. At 0 Amp, the magnetic field density is 0 Tesla and the only resistive force is friction and the mount rubber compliance.

This resistance is not more than 8lbs. At the magnetic field density is increased, MR fluid forms stronger chains and the mount becomes more resistive to piston movement resulting in higher squeeze forces. At 3 Amps, the magnetic field density in the mount is more than 0.5 Tesla and a maximum force of 800lbs is achieved at minimum gap size of 0.05".

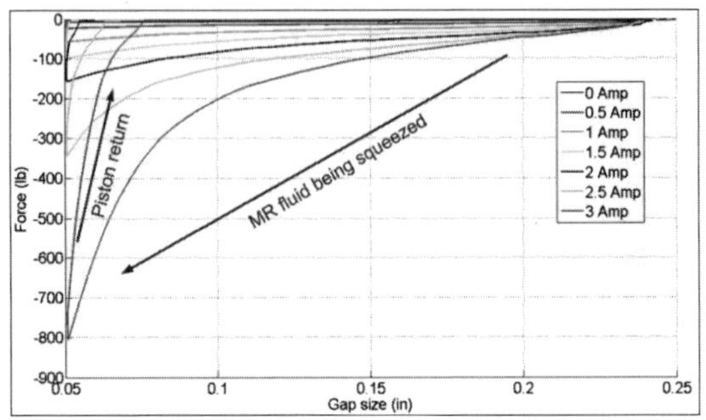

Figure 5 Displacement control test results of the MR squeeze mount

Test results show a very large range of controllable force (8lbs-800lbs) along a short stroke (0.2"). This large range of achievable force can be used in a wide variety of applications requiring large controllable force in small operational envelopes.

6. Clumping Behavior

While testing the MR squeeze mount, it was observed that if the same test is repeated without mixing the MR fluid in the mount, the achieved force will increase after each test run. Figure 6 shows a time history of the achieved squeeze force during 5 cycles of test repetition. It can easily be seen that the force is increased drastically after 5 cycles.

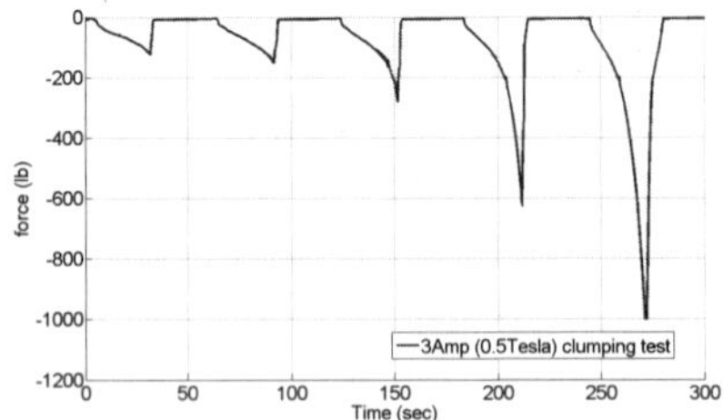

Figure 6 Force vs. time plot for MR squeeze mount during 5 cycle of test repetition

The behavior seen in Figure 6, called "clumping behavior" is suspected to be due to particle aggregation in MR fluid during squeeze process. As the fluid is squeezed, the magnetic iron particles are trapped in the magnetic field in the gap while the carrier oil is freely flowing radially outward. This caused iron particles to form aggregates in the mount gap and separate from the carrier oil. This effect is intensified after each test repeat, as shown schematically in Figure 7.

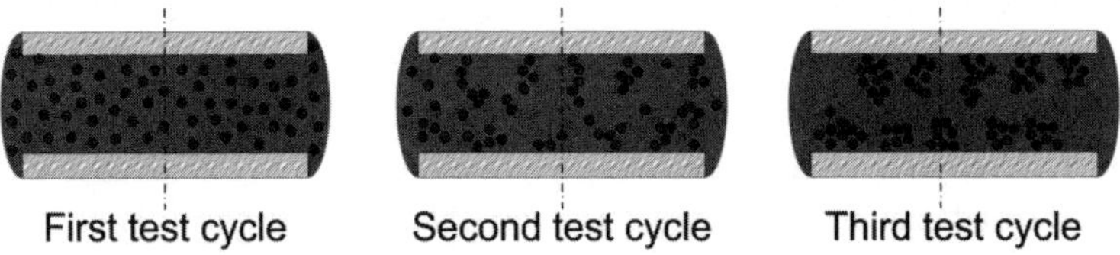

Figure 7 Clumping behavior as the MR squeeze mount is squeezed in cycles

Clumping behavior is quantified using the following definition,

$$\% \text{ Clumping}=\frac{\text{Avg. squeeze force per inch squeeze}|_{\text{Last Cycle}} - \text{Avg. squeeze force per inch squeeze}|_{\text{First Cycle}}}{\text{Avg. squeeze force per inch squeeze}|_{\text{First Cycle}} \times \text{Number of cycles}} \times 100 \qquad (1)$$

For test results shown in Figure 6, percentage clumping is found to be 172% showing that after each test cycle maximum force is increased by an average value of 172% of the maximum force in previous test.

A great deal of attention was given to reducing the clumping effect and increasing the test repeatability of the MR squeeze mount. It was found that a magnetic chirp during piston return can greatly reduce the amount of clumping seen in tests. As the piston is moving up returning to the initial position and before starting the next squeezing cycle, a magnetic chirp is applied to the MR mount. A sine wave was chosen to chirp the fluid. Figure 8 shows a comparison between a 5 cycle clumping test without magnetic chirp (shown in Figure 6) and the same test with a magnetic chirp during piston return. Chirp amplitude was +/-1 Amp and frequency was 1Hz. As shown in Figure 8, the current chirp greatly reduces the clumping. Clumping was reduced from 172% to 29.5% by using a sine current chirp during piston return. This reduction is because during piston return, the chirp can cause the iron particles to escape aggregates and form new chains. However, the chirp could not completely eliminate clumping. Also, it should be noted that no pattern was found showing a relation between chirp signal properties (amplitude and frequency) and reduction in clumping.

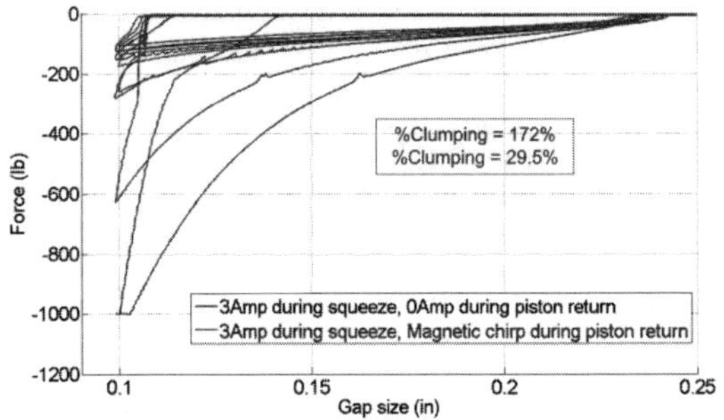

Figure 8 Effect of magnetic chirp during piston return on the clumping behavior

7. Summary and Conclusion

This paper presented quasi-static experimental test results from testing a novel MR squeeze mount. The MR mount was tested in a squeeze mode rheometer capable of testing the mount under various conditions. Magnetic field study of the mount and the rheometer was presented and validated. Design of the rheometer ensures a homogenous magnetic field density in the MR mount. Test results showed that the MR mount was capable of achieving a large range of controllable force along a short stroke. Also, a clumping behavior was observed during testing. This behavior is due to separation of iron particles from carrier oil in the MR fluid. Clumping behavior was studied extensively and it was found that a magnetic chirp can help the clumped particles re-disperse in the MR fluid. Although this chirp greatly reduces the clumping, it could not completely eliminate clumping. Research continues at CVeSS on clumping effect.

References

1. Carlson, J.D., D.M. Catanzarite, and K.A. St. Clair. *Commercial Magneto-Rheological Fluid Devices.* in *5th International Conference on Electro-Rheological Fluids, Magneto-Rheological Suspensions and Associated Technology, July 1995, International Journal of Modern Physics B.* 1996. Sheffield, UK: World Scientific.

2. Jolly, M.R., J.W. Bender, and J.D. Carlson, *Properties and Applications of Commercial Magnetorheological Fluids.* Journal of Intelligent Material Systems and Structures, 1999. **10**(1): p. 5–13-5–13.

3. Phulé, P.P. and J.M. Ginder, *Synthesis and Properties of Novel Magnetorheological Fluids Having Improved Stability and Redispersibility.* in 6th International Conference on ER Fluids and MR Suspensions and Their Applications, 1997.

4. Farjoud, A., M. Ahmadian, and R. Cavey. *Rheometer Characterization of Mr Fluids in Squeeze Mode.* in *Active and Passive Smart Structures and Integrated Systems 2009, 9 March 2009.* 2009. USA: SPIE - The International Society for Optical Engineering.

5. Farjoud, A., R. Cavey, M. Ahmadian, and M. Craft, *Magneto-Rheological Fluid Behavior in Squeeze Mode.* Smart Materials and Structures, 2009. **18**(9): p. 095001-095001.

PERFORMANCE EVALUATION OF AN MR DAMPER-BASED SEMIACTIVE CONTROL SYSTEM OPERATED BY AN ELECTROMAGNETIC INDUCTION DEVICE

DONG-DOO JANG

Department of Civil and Environmental Engineering, KAIST,
Daejeon 305-701, South Korea

HYUNG-JO JUNG

Department of Civil and Environmental Engineering, KAIST,
Daejeon 305-701, South Korea

JEONG-HOI KOO

Department of Mechanical and Manufacturing Engineering, Miami University,
Oxford, Ohio 45056, USA

This study experimentally investigates the dynamic performance of a semiactive control system consisting of an MR damper and an electromagnetic induction (EMI) device. The EMI system, consisting of permanent magnets and coils, is capable of converting reciprocal motions (kinetic energy) of the MR damper into useful electrical energy (electromotive force or emf). The emf signal is an alternating voltage signal, proportional to the velocity of the motion. Thus, the EMI can act as a relative velocity sensor for the implementation of a control algorithm for the MR damper system. However, this sensing capability of the EMI has not been evaluated in control experiments with a building structure. Thus, this study intends to investigate the performance of an MR damper system, solely operated by an EMI device (without a conventional velocity sensor) using a three storey shear building structure under scaled historic earthquake loadings. To this end, an MR damper and an EMI device are installed in the building structure, and a semiactive control algorithm, maximum energy dissipation algorithm, is implemented based on the EMI signal. The experimental results show that the MR-damper system with an EMI device outperformed a passive "optimal" control system, implying that the EMI can replace conventional velocity sensors in MR damper systems.

1. Introduction

Magnetorheological (MR) dampers are one of semiactive control devices, which provide controllable damping force. Since the mid-1990s, MR dampers have received considerable attention in structural control of civil engineering structures because of their mechanical simplicity, high dynamic force range, and low operating power [1-2]. Despite these benefits, the conventional MR damper based semiactive control systems still require sensors, controllers and external power supplies, which may cause maintenance and cost problems, especially for large-scale civil structures. To address these issues, MR damper systems with electromagnetic induction (EMI) devices have been studied [3]. Consisted with permanent magnets and solenoid coils, the EMI device is capable of converting their relative reciprocal motions into electrical energy according to the Faraday's law of induction. The converted electrical energy or electromotive force (emf) can be used as an alternative power source for an MR damper. Moreover, it can be used as a relative velocity sensor because the converted emf signal is proportional to the relative velocity between the permanent magnets and the coils. Jung et al. investigated the feasibility of the EMI system as a velocity sensor [4], and they suggested that the "EMI sensor" can be useful for control algorithms that require the sign change of the velocity signals.

The objective of this study is to implement a control algorithm using a emf signal from an EMI for an MR damper system and to evaluate the performance of the MR-EMI system in reducing vibrations of a three storey shear building structure under scaled historic earthquake loadings. For the control algorithm, the maximum energy dissipation algorithm (MEDA) is used because it is commonly used for MR damper systems. Moreover, it only requires velocity sign change information along with the damper force to implement.

2. Semiactive Control System based on MR Damper and EMI Device

2.1. *MR-EMI System*

The structure's vibration causes the relative motion between two parts of an EMI system as shown in Figure 1, and then the electric energy is induced according to the Faraday's law of electromagnetic induction. The induced electrical energy can be used as a power for an MR damper to change its damping characteristics depending on the response of the structure.

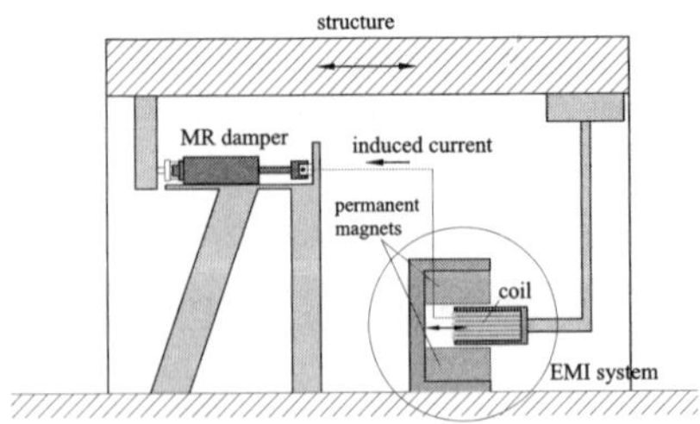

Figure 1 Conceptual diagram of the smart passive system

The EMI system can also be used as a relative velocity sensor because the converted emf signal is proportional to the relative velocity between the permanent magnets and solenoid coils in the EMI system. According to the Faraday's law of induction, the induced voltage can be estimated as

$$\varepsilon = -NB\frac{dA}{dt} = -NB\frac{d(wx)}{dt} = -NBw\frac{dx}{dt} = -NBw\dot{x} \tag{1}$$

where ε is the induced electromotive force (emf) in volts (V); N is the number of coil turns; A is the area of cross section; B is the magnetic field; w is the coil width; and x is the relative displacement between the permanent magnets and coils. Equation (1) indicates that the induced voltage is proportional to the velocity of the structure (see Figure 1). Therefore, it can be considered as a velocity signal for the MR damper system. In other words, the EMI system can act as a velocity sensor for semiactive control algorithms that require only the information of the relative velocity to implement them. In this study, the Maximum Energy Dissipation Algorithm (MEDA) is used to control the MR-EMI system.

2.2. *Maximum Energy Dissipation Algorithm (MEDA)*

A seismically excited structure controlled with MR dampers can be modeled as

$$M\ddot{x} + C\dot{x} + Kx = \Lambda f - M\Gamma \ddot{x}_g \tag{2}$$

where, M, C and K are mass, damping and stiffness matrices, respectively; x is vector of the relative displacements of the structure; $\ddot{x}_g$ is the ground acceleration; $f = [f_1, f_2, \cdots, f_n]^T$ is control forces generated by MR dampers; Γ is the column vector of ones; and Λ is matrix defining the MR dampers position. This equation can be written in state-space form as;

$$\dot{z} = Az + Bf + E\ddot{x}_g$$
$$y = Cz + Df + v \tag{3}$$

where z is the state vector; y is the measured output; and v is the measurement noise vector. MEDA can offer simplicity of control design by employing the Lyapunov's direct approach. Lyapunov's direct approach requires the use of a Lyapunov function, denoted $V(x)$, which must be a positive definite function of the states of the system x. Jansen and Dyke (2000) instead considered a Lyapunov function that represents the relative vibratory energy in the structure as [5];

$$V = \frac{1}{2}x^T K x + \frac{1}{2}\dot{x}^T M \dot{x} \tag{4}$$

According to Lyapunov stability theory, if the rate of change of the Lyapunov function $\dot{V}(x)$ is negative semidefinite, the origin is stable in the sense of Lyapunov. Thus, in developing the control law based on Lyapunov stability theory, the goal is to choose control inputs for each device that will result in making the rate of change of the Lyapunov function as negative as possible. Using Equation (4), the rate of change of the Lyapunov function is then,

$$\dot{V} = x^T K \dot{x} + \dot{x}^T M (-C\dot{x} - K x - M\Gamma \ddot{x}_g + \Lambda f) \tag{5}$$

In this expression, the only way to directly effect $\dot{V}$ is through the last term containing the force vector f. To control this term and make $\dot{V}$ as large and negative as possible, the following control law is obtained.

$$v_i = V_{max} H\left(-\dot{x}\Lambda_i f_i\right) \tag{6}$$

where, v_i is input voltage for i-th MR damper; and V_{max} is maximum voltage; and $H(\cdot)$ is Heaviside step function.

In conventional MR damper systems, a velocity sensor is used to determine the sign term of $-\dot{x}\Lambda_i$, or the relative velocity across damper. However, in this study, the velocity term is determined by the emf signal induced by the EMI device to evaluate if the EMI can replace conventional velocity sensors.

3. Experimental Verification

3.1. *Experimental Setup*

In order to verify the control performance of the proposed MR-EMI system, a scaled three-story shear building model including an MR damper attached between ground and the first floor is considered through the shaking table test' as shown in Figure 2. The test structure model is constructed of steel and has a height of 105 cm and a total of 48.27 kg, which is distributed evenly between the three floors. The first three natural frequencies of the structure are 2.05 Hz, 5.57 Hz and 8.4 Hz. The MR damper used in the experiment is a controllable friction damper (RD-1097-01) from Lord Corporation. Maximum force level is approximately ±100N, and maximum allowable voltage is 10 volts. A set of 5x5 cm^2 permanent magnets rated at 0.6 Tesla, and the solenoid coil having 2,630 turns and coil diameter of 0.5mm is used in the EMI system. In the experiment, the test structure is subjected to four scaled historic earthquakes (El Centro, Kobe, Hachinohe, and Northridge earthquakes).

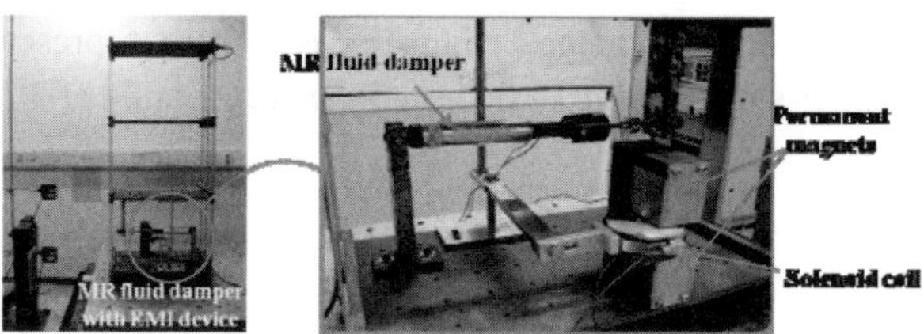

Figure 2 Experimental setup

3.2. *Performance Criteria Index*

To systematically evaluate the control performance, a set of 4 normalized evaluation criteria are defined as follows: peak inter-story drift at 1st floor (J1) , peak acceleration at 3rd floor (J2), RMS inter-story drift at 1st floor (J3) and RMS acceleration at 3rd floor (J4).

$$J1 = \frac{\max_t |d_1(t)|}{\max_t |\hat{d}_1(t)|} \qquad J2 = \frac{\max_t |a_3(t)|}{\max_t |\hat{a}_3(t)|} \qquad J3 = \frac{RMS(d_1(t))}{RMS(\hat{d}_1(t))} \qquad J4 = \frac{RMS(a_3(t))}{RMS(\hat{a}_3(t))} \qquad (7)$$

where d_1 is inter-story drift at 1st floor, a_3 is acceleration at 3rd floor and $\wedge$ represents the uncontrolled case.

3.3. *Experiment Results*

Before conducting the test using the proposed MR-EMI system, an optimal passive control case is investigated by operating the MR damper in pass mode. In this case, a constant input voltage is provided to the MR damper, from 0V to 10 V. Figure 3 shows the evaluation criteria variation with varying the input voltage with an increment of 1 V. As shown in Figure 3b, the overall minimum average of evaluation criteria is obtained at 4 V. Hence, a passive MR damper with 4V is considered as an "optimal" passive case, whose performance will be compared with that of the proposed MR-EMI system.

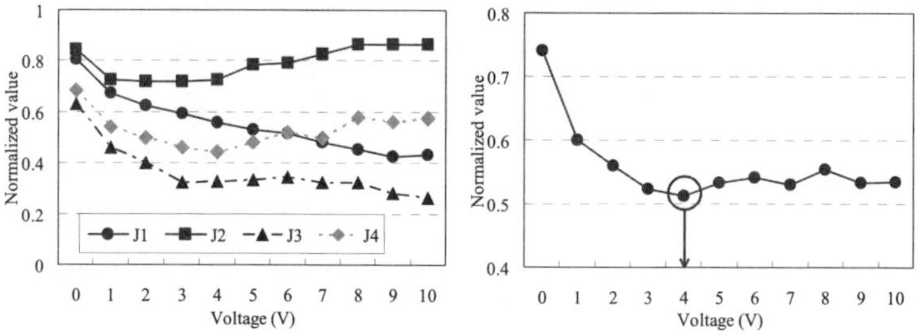

(a) Average values for all earthquake   (b) Mean value of all the evaluation criteria

Figure 3 Evaluation criteria variation with respect to the input voltage in the passive control system

As an example of performance comparisons, Figure 4 presents the time histories of the inter-story drift at 1st floor and the acceleration at 3rd floor under Kobe earthquake.

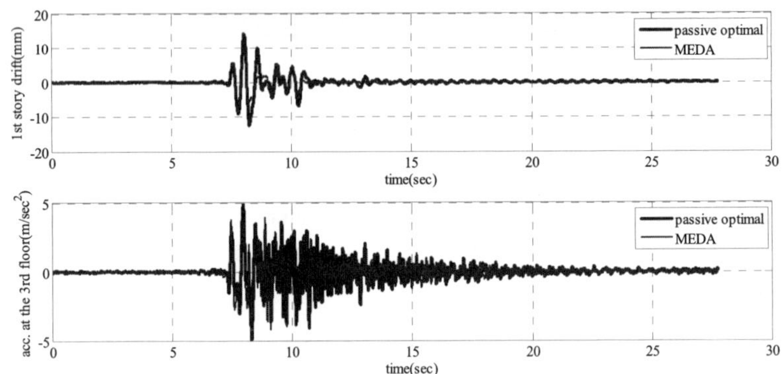

Figure 4 Time histories of inter-story drift at 1st floor and acceleration at 3rd floor under Kobe

Figure 5 presents evaluation criteria of proposed system compared to passive optimal case. The proposed MR-EMI system shows much better control performance related with inter-story drift at 1st floor (J1, J3), while shows slightly less control performance associated with acceleration at 3rd floor (J2, J4). However, the overall performance (see "Mean" in Figure 5) of the proposed system is better than the passive optimal case. The mean value is the average of the four evaluation criteria.

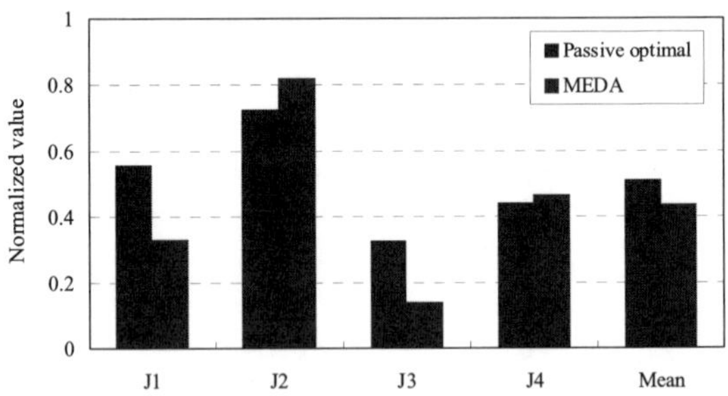

Figure 5 Evaluation criteria of proposed system compared to passive optimal case

4. Conclusion

In this paper, the control performance of an MR damper augmented with an EMI system was evaluated. The EMI system was used to measure the relative velocity across the MR damper, and MEDA was employed to regulate the input voltage of the MR damper. Using a shaking table setup with a three-story building structure, the control performance of the proposed system was compared with passive optimal MR damper system. The experimental results showed that the proposed system with the MEDA control, which was implemented by a velocity signal from the EMI has better control performance than passive optimal MR damper system. The results suggest that the EMI can replace conventional velocity sensors used in MR damper systems.

Acknowledgments

This research was supported by the Korea Science and Engineering Foundation (KOSEF) grant funded by the Korea government (MEST) (Grant No. R11-2003-101-03001-0) and a grant (07High Tech A01) from High tech Urban Development Program funded by Ministry of Land, Transportation and Maritime Affairs of Korean Government.

References

1. Dyke SJ, Spencer BF, Sain MK, Carlson JD. Modeling and control of magnetorheological dampers for seismic response reduction. Smart Mater Struct. 1996. 5:565–575.
2. Dyke SJ, Spencer BF, Sain MK, Carlson JD. An experimental study of MR dampers for seismic protection. Smart Mater Struct 1998. 7(5):693–703.
3. Cho SW, Jung HJ, Lee IW. Smart passive system based on magnetorheological damper. Smart Mater Struct. 2005. 14:707-714.
4. Jung HJ, Jang DD, Cho SW, Koo JH. Experimental verification of sensing capability of an electromagnetic induction system for an MR fluid damper-based control system. Proc. 11th Int. Conf. Electrorheological fluids and magnetorheological suspensions, Dresden, Germany. 2008.
5. Jansen LM, Dyke SJ. Semi-active control strategies for MR dampers: A comparative study. Jour. Eng. Mech., ASCE, 2000. 126: 795-803.

INVESTIGATION ON ITS VIBRATION-REDUCTION AND SHOCK-RESISTANT PROPERTIES OF A GUN RECOIL MECHANISM BASED ON MR DAMPER[*]

HONGSHENG HU[†]

Department of Mechanical and Electrical Engineering, Jiaxing University
Jiaxing, 3104001, China

JIONG WANG

School of Mechanical Engineering, Nanjing University of Science and Technology
Nanjing, 210094, China

YANCHENG LI

Faculty of Engineering and Information Technology, University of Technology Sydney
Sydney NSW 2007, Australia

MR damper has brought out new challenges for development of the recoil mechanisms and vibration stability control of weapons because of its good electromechanical coupling performances. At present, it has been an urgent task during automatic firing to ensure its dynamic performance and its reliability of gun recoil mechanism under continuous fastly impact. For recoil mechanisms applications, MR dampers are desired to provide optimal damping force to control the recoil dynamics, so that large peak of recoil forces can be avoided with a certain limited stroke, and the firing stillness and stability are ensured. According to its vibration and shock mechanics process of gun recoil mechanism, the measurement method of its vibration-reduction and shock-resistant properties of gun recoil mechanism based on MR damper is analyzed. The results show that a gun recoil mechanism based on MR damper is quite a good vibration-reduction and shock-resistant equipment when the vibration and shock energy dissipation by damp is considered.

1. Introduction

When gun and automatic weapon are launching, there are strong impact load produced by the high temperature, high pressure powder gas's instantaneous effect, which will have a great effect on its firing accuracy and stability of gun. With its rapidly development of modern weapon system toward larger power,

---

[*] This work is supported by National Defense Foundation of China, China Postdoctoral Science Foundation funded project (20080431099), etc.

[†] Work partially supported by grant 50675106 of the China National Science Foundation.

lightweight and high maneuverability, the traditional passive hydraulic damping buffer device has shown its deficiency and limitedness, and hasn't meet higher performance requirements. For recoil mechanisms applications, MR dampers are desired to provide optimal damping force to control the recoil dynamics, so that large peak of recoil forces can be avoided with a certain limited stroke, and the firing stillness and stability are ensured.

The Smart Materials and Structures Laboratory of NUST have been engaged in its design of gun recoil mechanism equipped with MR damper from the year of 2002 and developed a series of MR shock absorbers (Shen, 2007; Zhang, 2008). However, a systematic architecture for MR damper subjected to impact load has still not been formed for many years, including its structure design, dynamic response model and controlling method. In this paper, on the background of the recoil mechanisms applications, its vibration-reduction and shock-resistant properties of a gun recoil mechanism based on MR damper is investigated in depth.

2. Principle of MR Gun Recoil Damper and Test Rig

According to Figure 1, its movement equation of MR damper equipped with gun can be described as followers (Shen, 2007; Zhang, 2008):

$$m_h \frac{d^2 X}{dt^2} = m_h \frac{dv_0}{dt} = F_{pt} - F_R \tag{1}$$

Where, F_{pt} is the impact load produced by firing, m_h is the recoil mass, and F_R is the recoil force, $F_R = F_{\Phi h} + F_f + F_T$, F_f =200N.

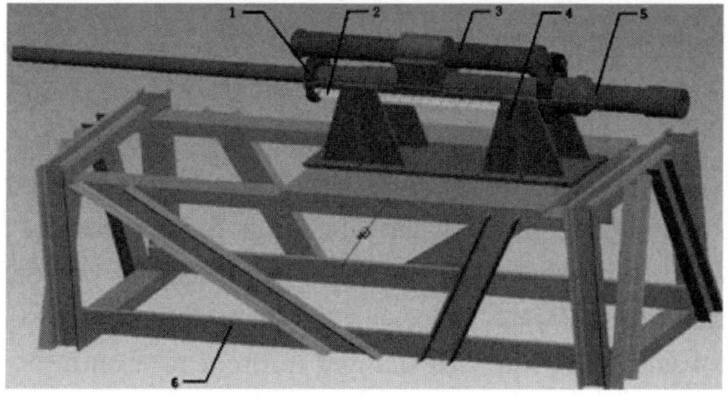

Figure 1. Physical model of MR gun recoil damper.

The simulation experimental platform for MR gun recoil damper is designed and developed in Figure 2. This developed test rig includes some test devices:

216

industry computer, dSPACE simulation system in real time, current controller and rig-testing.

According to Figure 2, In order to accurately investigate its vibration-reduction and shock-resistant properties performance of MR gun recoil damper, five kinds of signal were chosen to describe its behaviors, which are impact force, displacement, velocity, acceleration, and pressure in the MR gun recoil damper. Force sensor was installed at junction of piston rod and the ground supports. Pressure sensor installed in working chamber of damper was used to measure pressure of MR gun recoil damper. Displace and speed sensor under the bench of MTS was used to monitor displace and speed of MR shock absorber. Considering the impact duration is very short, a sample frequency of 20 KHz is used to ensure all the dynamics will be captured.

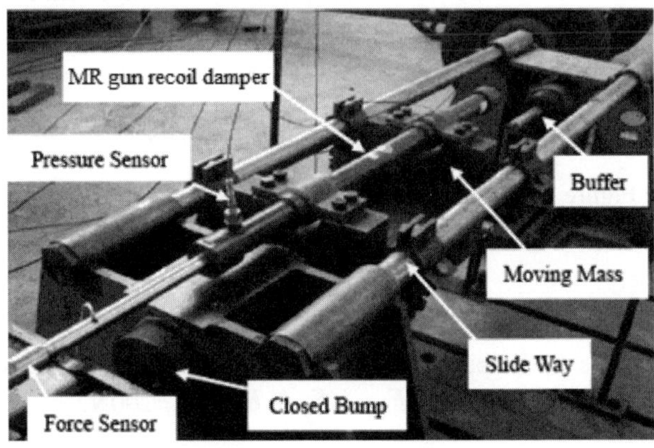

Figure 2. Test bench of MR gun recoil damper.

3. Shock Response Analysis

Figure 3~Figure 5 is respectively a series of shock response curves for MR damper subject to impact load, where damping force measured by sensor is compared with working chamber pressure. The amount of powder for this test was 2g. A series of tests were done with the applied current of 0A, 0.5A, 1A, 1.5A, 2A that was applied into MR damper in order to ensure the effectiveness of test data, each test were done three times. The test data are filtered by low-pass Butterworth IIR filter in Matlab environment. The trigger signal is its displacement response and its value is set 0.12m.

According to the above test response curves, the following conclusions can be concluded:

(1) Its peak will increase along with the current, and the stroke reduces gradually, and last time of recoil movement reduces. It means that in the whole dynamic process its recoil stroke of MR damper can be controlled by the changing current to achieve the purpose of the adjustable recoil stroke.

(2) Its changeable process of moving mass's velocity can be divided into two stages. The first stage is excited by the impact load; its velocities of moving mass rises suddenly in addition, correspond to two peak values caused by the electric igniter and the gunpowder. Velocity's peak value will drop along with current increase in the second stage after the excited disappeared, and, the acceleration increases.

(3) There are two stages for moving mass's movement acceleration. The former is the movement acceleration process which is caused by the excited impact load. The latter is the moving mass decelerates until to the static state under damping force. There exists a high frequency vibration in its changing curve of acceleration response. The reason is that 45 steel's counterweight mass increased the system attachment rigidity.

(4) According to its damping force response of piston rod and pressure response of MR gun recoil damper, may see that its effective reduce-vibration range of damping force exists an interval between its peak value and movement's end. Its changing trend of recoil force is similar to cavity pressure response, yet the latter is smoother. Based on its changing trend of recoil force, there is an apparent shake before movement of moving mass ends. The reason includes signal interference and the negligible additional stiffness.

(5) According to its changing trend curve, its adjustable range of recoil force is equal to 1152N and the adjustable rate is 49.34%. Its adjustable range of cavity pressure 17.19bar and the adjustable rate is 38.01%.

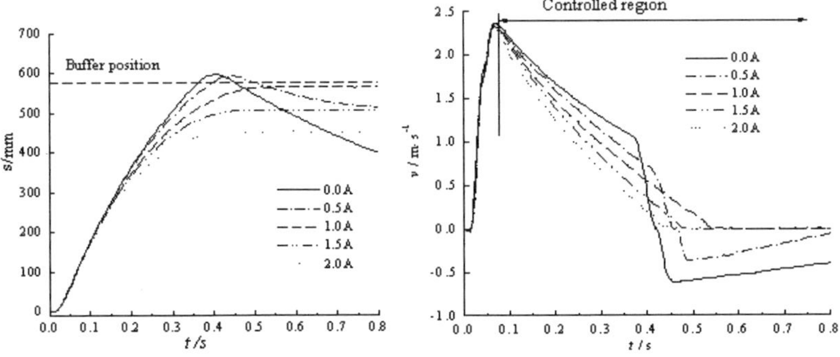

Figure 3. Stroke and velocity response of MR damper under impact load.

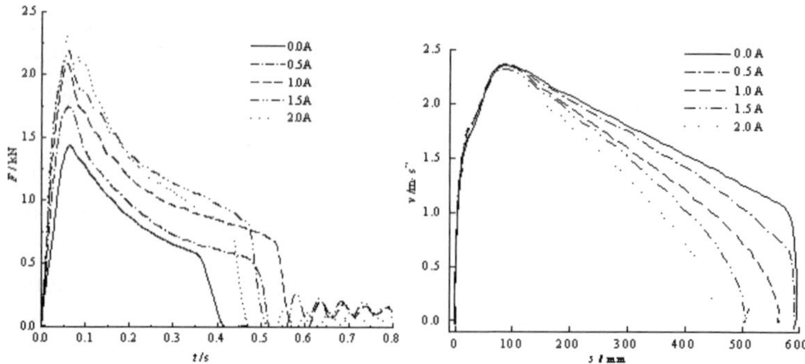

Figure 4. Damping force response and velocity vs stroke of MR damper under impact load.

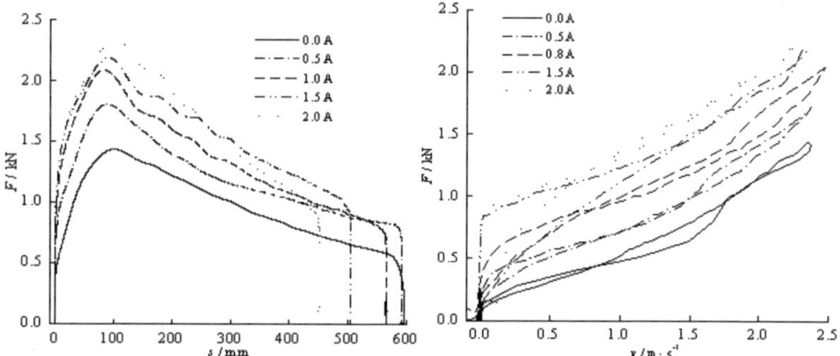

Figure 5. Damping force vs stroke and velocity of MR damper under impact load.

Because in the process of gunpowder's impulse, recoil force is only a weak impulse compared with impact. Figure 6 shows both of comparisons for test and simulation acceleration signal power spectrum. According to the power spectrum curve, parameter E and τ is respectively equal to 81.4 and 50.

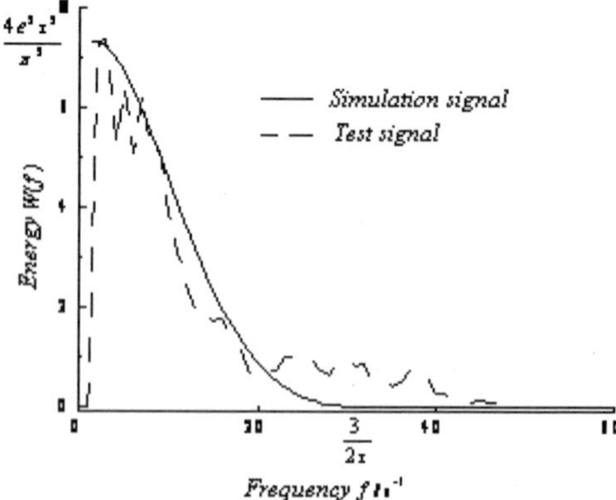

Figure 6. Energy spectrum comparison of test and simulation acceleration.

4. Controllability Analysis

Figure 7 show its recoil response of 30mm caliber gun recoil damper, including damping force vs time, stroke vs time and velocity vs time. Experimental results show that damping force, stroke and velocity respectively has a corresponding variation as the applied current into MR gun recoil damper. As its increment of the applied current, its peak of damping force become larger, and its peaks of d the stroke are decreasing.

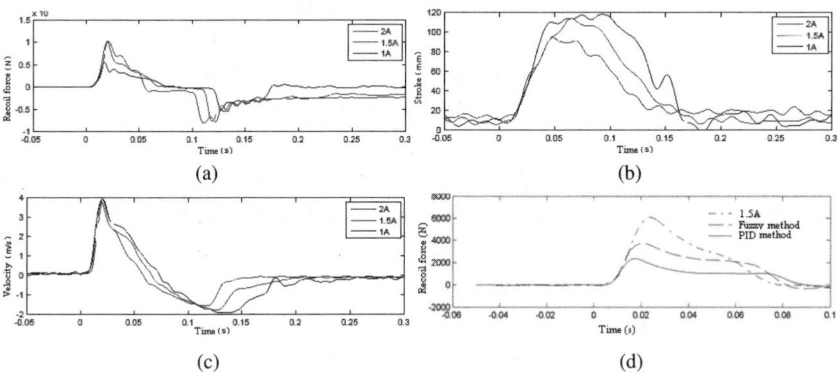

Figure 7. Dynamic response of 30mm caliber equipped with MR damper.

From Figure 7, its stroke of 30mm caliber gun decreases 30% when the applied current into MR gun recoil damper changes from 1A to 2A. While the applied current into MR gun recoil damper decreases, its peak value of recoil damping force wound gradually decrease and its stroke increase. It is seen that the MR damper against the recoil has smart adjustability to the recoil of the artillery.

In order to test its controability of MR gun recoil damper under impact load, their effects of different controlling policies are all proved by experimental results. Figure 7(d) compares its recoil damping force and stroke response of MR gun recoil damper under PID controlling method, fuzzy control method and fixed current 1.5A. Further, it can be seen that compared with 1.5A fixed current, both of recoil damping forces under PID controlling method and fuzzy controlling method decrease slowly after its peak value, and recoil damping force keeps steady and has a better degree of fullness.

According to the above analysis, in shooting test, no matter the recoil controlled by PID or the fuzzy, either of them is steadier than not controlled. Compare with the shooting test, impact bench test and simulation result, it can be seen that effect of impact test control is very close to a "platform effect". Effect of the shooting test is quite steady, and its peak value is also small. However, there is still an obvious peak value, and its "platform effect" also needs for further research.

5. Conclusions

At present, little research has been focused on its performances of MR gun recoil damper, especially its controllability. Based on the impact bench test results and the shooting results, the following conclusions are made:

(1) The designed MR damper has an obvious controllability under the condition of artillery recoil mechanism's strong impact, and its recoil stroke may reduce above 50%. MR gun recoil damper can be applied into different shooting environments and achieve a good controability, including different types of artillery, different firing angles and charging powder, and so on.

(2) Application of MR damper in the artillery recoil mechanism can effectively reduce its recoil damping force and adjust its recoil damping force curve smoother, which will reduce its recoil stroke and achieve a good feasibility.

(3) Its energy consumption of MR gun recoil damper increases along with the electric current increases. Suitable control method may cause its energy consumption more average distribution in the entire range. Application of a reasonable control method into MR gun recoil damper will obviously change its effect and has the good validity.

(5) Its successful application of MR gun recoil damper can be helpful to achieve the purpose of reducing stress of artillery bench, enhancing fire stability and reducing its vibration of barrel.

References

1. Lee, D.Y., Choi, Y.-T, Wereley, N.M, *Journal of Intelligent Material Systems and Structures*, **13**, 4525(2002).
2. M. Ahmadian and J. Gravatt., *Proceedings of SPIE – On Smart Structures and Materials*, 5386(2004).
3. Ou jinping, *Structural vibration active, semi-active and intelligent control*, (2003).
4. J. A. Norris, M Ahmadian, *Proceedings of ASME–On International Mechanical Engineering*, 1(2003).

GAP-SIZE EFFECT OF THE COMPACT MR FLUID BRAKE[*]

TAKEHITO KIKUCHI[†]

Graduate School of Science and Engineering, Yamagata University
4-3-16 Jonan, Yonezawa 992-8510 Japan

In the previous works, we suggested a compact MR fluid brake (CMRFB) in which multi-layered disks and narrow gaps of 50 μm are used. This device has a great advantage in its small size and high torque. However, there are unignorable errors between analytical results and experimental results in its torque. To clarify the reason of this phenomenon, and decide the practical limitation of the gap-size for the CMRFB, we developed new CMRFB whose gap-size is 100 μm. In this paper, we analytically and experimentally compared torque characteristics of these two types of CMRFB. The braking torques of these brakes show approximately same level. According to the comparison of analytical and experimental results for 50 micro-gaps brake shows a large error. Flow observation tests shows a possibility that the fluid is not sufficiently filled into the 50μm-gap.

1. Introduction

Magnetorheological fluids (MRF) are kinds of functional fluids, whose rheological properties change rapidly, stably and repeatedly with application of the magnetic field [1]. By using this material and an electromagnet, we can develop torque controllable brake with high performance (high speed and stable).

As a first generation of the rotary MRF brakes, many researchers [2-3] have developed single-discal (or single-cylindrical) brakes, in which shear stress of the MRF generated on the surface of a disc (or a cylinder) is transformed into the braking torque. However, these devices require large area of the MR Fluid layers to generate large torque, and it caused large size of the devices.

Then, as a second generation of the MRF brakes, Kavlicoglu B., et al. [4] have applied a multi-plate (multi-layered disk) structure for a high-torque MRF brakes. However, their device has millimeters-sized gap (0.5~1.0mm) of the MRF layers and it causes a high magnetic resistance. The high magnetic resistance, at the same time, requires a high power supply for a sufficient generation of the magnetic field to excite an effective viscosity change of the MRF.

---

[*] This work is partly supported by the JAPAN Grant-in-Aid for Scientific Research, No. 21760198.

[†] E-mail: t_kikuchi@yz.yamagata-u.ac.jp

In terms of the gap-size, Ossur Inc. [5] released a controllable prosthetic knee joint that contains an MRF brake with the multi-plate structure and several ten micro-sized gaps of the MRF layers. It is a good idea to utilize the micro-sized gaps for reduction of the magnetic resistance and power supply; however detailed characteristics of the MRF between micrometer-sized gaps have not been unveiled. Jonsdottir F., et al. [6] presented influence of parameter variations on the design of the Ossur's MRF brakes; however this paper includes only theoretical calculations without experimental results.

In our previous study [7], we developed a compact MRF brake (CMRFB) which has the multi-layered disk structure and narrow gaps for MRF layers. The basic structure and appearance of the CMRFB are shown in Fig.1. A coil is winded around the output shaft and it generates the magnetic flux shown by the dashed line in the drawing. Multi-layered disks are fixed on the casing and the output shaft alternately, and the MRF is filled between these disks. We set 50μm as the gap-size in the initial version of the CMRFB. By reducing the gap-size, we could reduce the magnetic resistance of the magnetic circuit and realized a compact size (52mm (D)*32mm (H)), light weight (237g) and high torque (5Nm). We compared analytical torques and experimental torques of this brake. We could estimate qualitative characteristics of the torque from analytical results. However, there are unignorable errors between analytic results and experimental results in its torque.

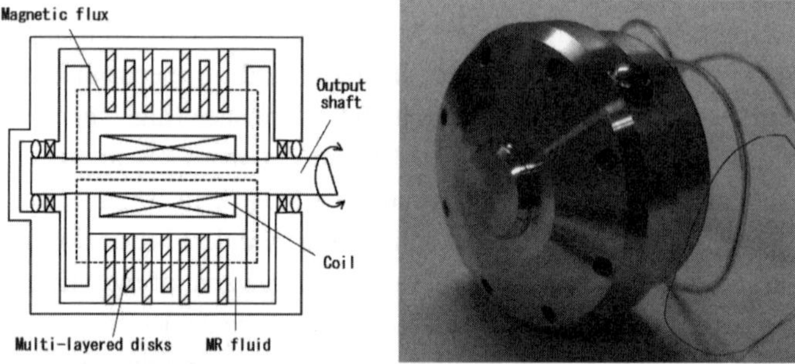

Figure 1. Compact MR Fluid Brake: Basic structure (left) and appearance (right).

The objective of this study is to clarify the reason of these errors, and decide a practical limitation of the gap-size for the CMRFB. We developed new CMRFB which have the same structure of the previous CMRFB but different gap-size of 100μm. In this paper, we analytically and experimentally compare these torque characteristics and discuss gap-size effect in the CMRFB.

2. Analytical Torques of the CMRFBs

The braking torque of the CMRFB was analytically estimated with magnetostatic analysis and material characteristics. Fig 2 shows size-parameters of rotational parts of the CMRFB. The multi-layered disks and MRF layers are located in the box framed by a dashed line. The common design parameters of the multi-layered disks are shown in the table 1. These parameters are based on the previously developed 5Nm-Class CMRFB reported in the reference [7]. Analytical braking torque was calculated for each gap size. Magnetostatic analysis was conducted with the finite element method (FEM) software (ANSYS, Ver.12). We used commercially available MRF (Lord Corp., 140CG). Therefore, magnetic characteristics provided by the material maker [8] were used in the analysis. The MRF is sandwiched between stator-disks and rotor-disks with overlapping width of 5mm from outer edge of the rotor disks.

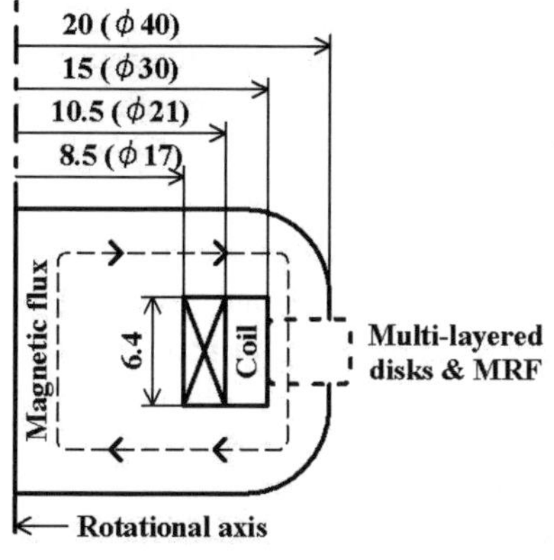

Figure 2. Size parameter of the CMRFB.

Table 1. Design parameters of multi-layered disks.

Diameter of disk [mm]	40
Overlapping width [mm]	5
Number of MR fluid layer	18
Turning number of coil	191

Figure 3 shows analytic results for each gap-size. The horizontal axis indicates electric current applied to the coil [A], and vertical axis indicates output torque of the devices [Nm]. As shown in this figure, the increasing gap-size causes the reduction of the maximum torque; however, due to the non-linear magnetic characteristics of the materials, the gap-effect becomes small in high current region. For example, torque ratio of 50μm-gap to 100μm-gap at 0.4A of the electric current is about 1.6; however, that at 1.0A is about 1.1.

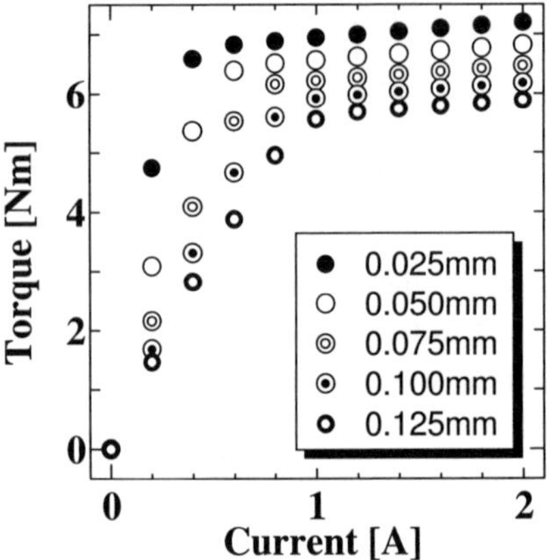

Figure 3. Analytic results on braking torque of the CMRFB for each gap-size.

3. Experimental Torques of the CMRFBs

We newly developed three same CMRFBs for 50µm-gaps and 100 µm-gaps each, and conducted experiments to examine static and dynamic performances of the braking torque. The casing of the brake was fixed on an immovable plate and the output shaft was driven by the servo-motor in the tests.

Figure 4 shows static toques of the CMRFB for each device. The straight line and dashed line show analytical results for 50µm-gaps and 100 µm-gaps, respectively. The black marks and white marks show experimental results for three units of 50µm-gaps and 100 µm-gaps, respectively. As shown in these

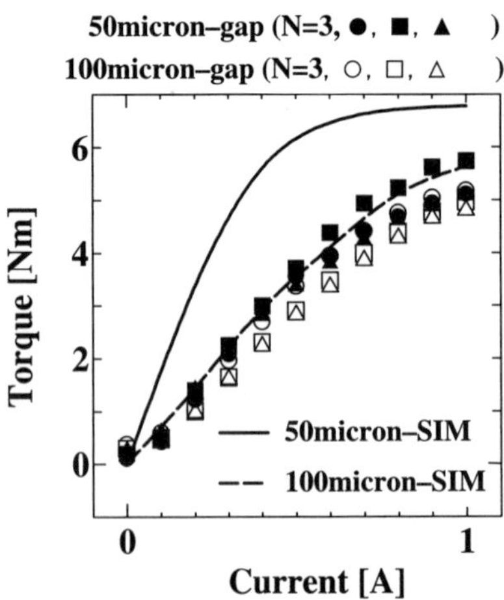

Figure 4. Comparison between analytical and experimental static torque of the CMRFB.

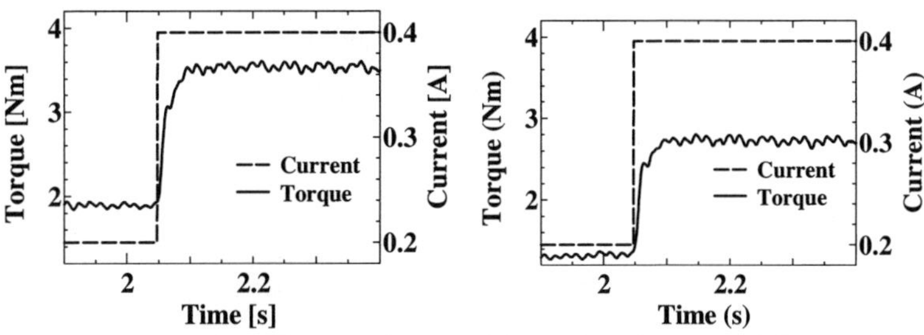

Figure 5. Step responses of the 50 μm-gap CMRFB (left) and the 100 μm-gap CMRFB (right).

graphs, the simulation and experiments on the 100 μm-gaps devices have good similarity. On the other hand, there are unignorable errors between a simulation and experiments on the 50 μm-gaps devices. Especially, the real braking torque of the 50μm-gaps is almost same level of the 100 μm-gaps.

Figure 5 shows dynamic torque characteristics of each device. In the dynamic tests, the output shafts of these brakes were driven at 3 rad/s by the servo-motor. Electric current was controlled from 0.2A to 0.4A as a stepwise function. The averages of these braking torque are different. However, they have same level of response times (about 10ms).

4. Discussion

In order to confirm filling conditions of the MRF between the narrow gaps, we conducted filling tests to single-disk flow test cells shown in the Fig.6. Basically same structure of the CMRFB was applied in this test cell, but the number of the disk is only one. The single disk was sandwiched with acrylic plates and spacers to make 50 and 100 μm-gaps. We prepared two testing cell for 50 and 100 μm-gaps, and connected with an injector containing the MRF via tubes. We put this setup in the vacuum furnace and set the temperature at 50 degrees C. The potential head of the injector was slowly risen up to the top of the cavity, and left it quietly for 2 hours. This method is same as the way we fill the MRF into the CMRFB. Lord MRF-140CG was used.

Fig.6 shows testing cells after the flow tests. We confirmed some voids and color heterogeneity in the 50μm-gap cell. In the case of the 100μm-gap, we could not confirm any void in the cell. Therefore, one of the reasons of the large error between the estimated torque and real torque of the 50μm-gap CMRFB is insufficient filling of the MRF into the narrow gap. According to the results, we can say that a practical limitation of the gap-size for the MRF in the CMRFB is about 100 μm, at least greater than 50 μm.

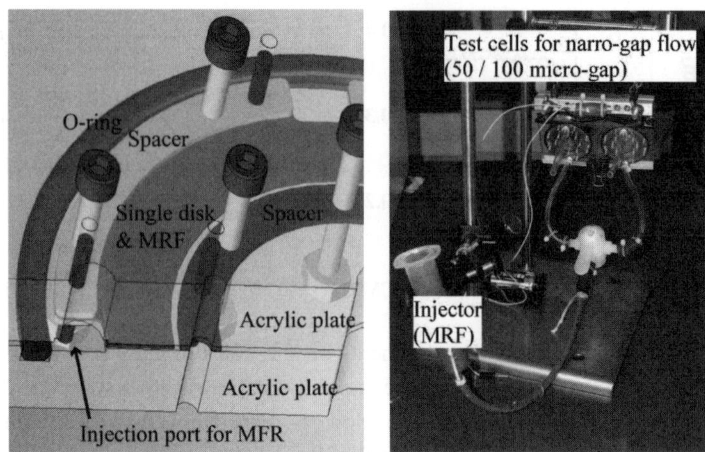

Figure 6. Structure of flow test cell (left) and experimental setup (right).

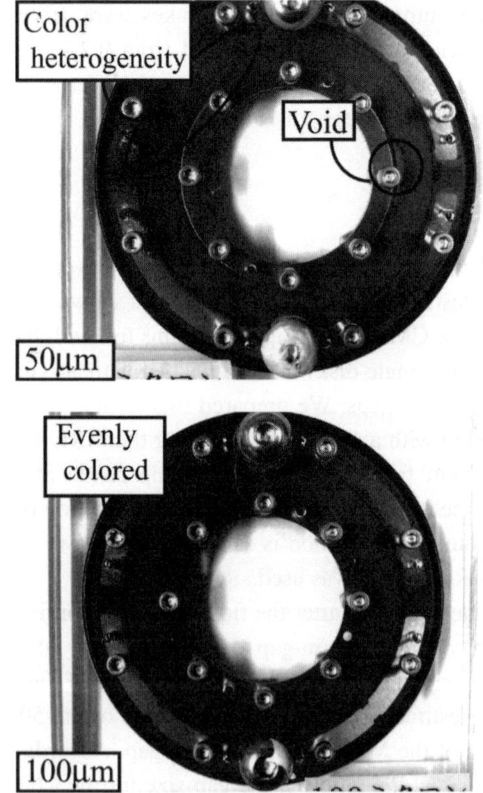

Figure 7. Results of filling tests for 50μm-gap (upper) and 100 μm-gap (lower).

5. Conclusion

We developed two types of the compact MR fluid brake (CMRFB) with 50 / 100 μm-gap for the MRF layers. We confirmed unignorable errors between a simulation and experiments on the 50 μm-gaps devices. According to the filling test with single-disk test cell, insufficient filling of the MRF was confirmed in the 50 μm-gap cell, but not in the 100 μm-gap cell.

References

1. J.D. Carlson, M.R. Jolly, Mechatronics, **10**, 555 (2000).
2. S. Choi, S. Hong, C. Cheong, Y. Park, Journal of Intelligent Material Systems and Structures, **10**, 615 (1999).
3. U. Lee, D. Kim, N. Hur, D. Jeon., Journal of Intelligent Material Systems and Structures, **10**, 701 (1999).
4. B. Kavlicoglu, F. Gordaniejad, C. A. Ecrensel, Y. Liu, N. Kavlicoglu, F. Fuchs, Journal of Intelligent Material Systems and Structures, **19**, 235 (2008).
5. Ossur Inc., http://www.ossur.com/
6. F. Jonsdottir, E. T. Thorarinsson, H. Palsson, K. H. Gudmundsson, Journal of Intelligent Material Systems and Structures, **20**, 659 (2009).
7. T. Kikuchi, K. Otsuki, J. Furusho, H. Abe, J. Noma, M. Naito and N. Lauzier, Advanced Robotics, **24**, 1489 (2010).
8. Lord Corp.: http://www.lord.com/

MR ELASTOMERS ISOLATOR FOR SEAT VIBRATION CONTROL

WEIHUA LI

School of Mechanical, Materials and Mechatronic Engineering, University of Wollongong, Wollongong, NSW 2522, Australia

XIANZHOU ZHANG

Defence & Aerospace, G H Varley Pty Limited, TOMAGO, NSW, 2322, Australia

HAIPING DU

School of Electrical, Computer and Telecommunications Engineering, University of Wollongong, NSW 2522, Australia

Driver fatigue is one of the leading factors contributing to road crashes. Environmental stresses such as unwanted seat vibration is a key contributor to fatigue. This paper presents the design and development of a MR elastomers isolator for seat suspension system. By altering the system's stiffness and damping, the vehicle's vibration energy input to seat is reduced, which then suppresses the seat's response. Results shown that the proposed isolator can reduce vibration further comparing with a passive isolation system, indicating the significant potential of its application in vehicle seat vibration control.

1. Introduction

Fatigue is one of the leading factors contributing to road crashes. CARRS-Q estimates that fatigue is the primary contributing factor in 6% in all crashes, 15% in all fatal crashes and 30% of fatal crashes on rural roads nationally [1]. Ride vibration has a significant influence on the driver's fatigue, health and safety [2]. For example, vibration transfers to seat and the human body are known to be a major source of discomfort for the driver. In commercial vehicles, driver exposure to low-frequency and high-amplitude vibration is a major factor in health disorders [3]. Seat vibration energy is concentrated at low frequency below 10 Hz, and to which the human-body is very sensitive [4]. Low frequencies can affect the body though and can be very unhealthy. They can stretch the nerves and lead to serious fatigue, such as dizziness, balance disorder and blindness. Seat suspension has been employed as a simple and effective method to attenuate unwanted vibrations, however, it is very hard to isolate low frequencies by using a conventional isolator [5].

For the last decade, a very attractive semi-active suspension system featuring electrorheological (ER) fluid and magnetorheological (MR) fluid has been developed by several groups [6-9]. For example, a semi-active seat suspension system featuring electrorheological (ER) fluids has been investigated for vibration isolation with various approaches [6,7]. This type of semi-active seat suspension has several advantages such as fast response time, continuously controllable damping force and low energy consumption compared with the active seat suspension system.

However, these systems are only effective for a narrow frequency range. This is because either ER fluid or MR fluid only has controllable-damping capability. Different from MR/ER fluids, MR elastomers have controllable modulus [10]. For a low-frequency and high-amplitude vibration, just like car seat vibration, controllable modulus should be much more effective than controllable damping. Because the damping force is depended on the speed and damping factor, and elastic force is depended on the amplitude and stiffness. Controllable modulus can provide large control force at low-frequency and high-amplitude vibration. By now, there are few researches about this.

The main contribution of this paper is to present a stiffness controllable isolator. The behavior of isolator was experimentally evaluated. The isolator was installed in a seat suspension system. The system was modeled and controlled by non-resonance control algorithms. Based on the displacement at seat frame and driver's head, vibration control effects of the isolator are evaluated under chirp, bump and random road conditions.

2. MRE Isolator

2.1. *Fabrication and characterization of MREs*

The Room Temperature Vulcanizing (RTV) silicon rubber (HB Fuller Company, Germany) and silicone oil were chosen as matrix. The average diameter of carbonyl iron particle is about 5μm (BASF Company, Germany). After being cured about 24 hour at the room temperature under a constant magnetic flux density 1 Tesla, MRE samples were prepared. The volume fraction of iron powder, silicone oil and silicone rubber is about 1:1:1.

The MR effect was evaluated by measuring the dynamic shear modulus at different magnetic field using Physica MRD 180 Magneto Rheolgical Device (Anton Paar Companies, Germany), equipped with an electromagnet kit. Both steady state and dynamic properties of the MREs were reported in our previous study [11]. The results demonstrate that MREs could be used as a novel variable stiffness device for tuned vibration control.

2.2. *Design the MRE isolator*

The schematic of the proposed MRE seat isolator is shown in Figure 1. The coils, core and base form the electromagnet, which is used to generate varying magnetic fields. The nonmagnetic rings restrict the magnetic line to cross the MR elastomers and come into being a magnetic circuit. The current intensity in the coils can be controlled by electrical power to adjust the magnetic field intensity. As such, the modulus of MREs can be controlled in real time, i.e. the stiffness of the spring can be adjusted to fit the needs.

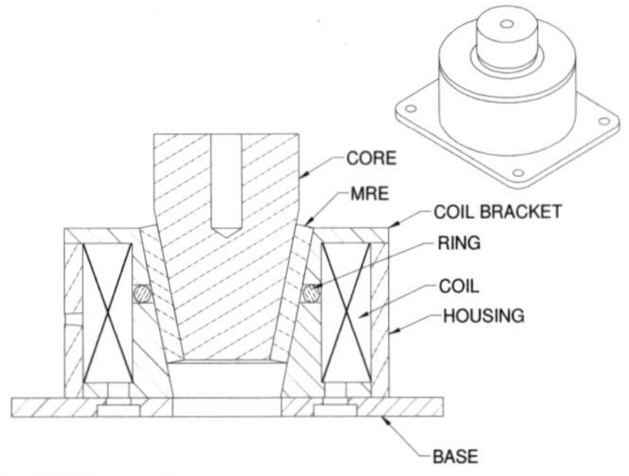

Figure 1. Schematic of the MRE seat isolator

3. Performance Testing and Mathematical Modeling

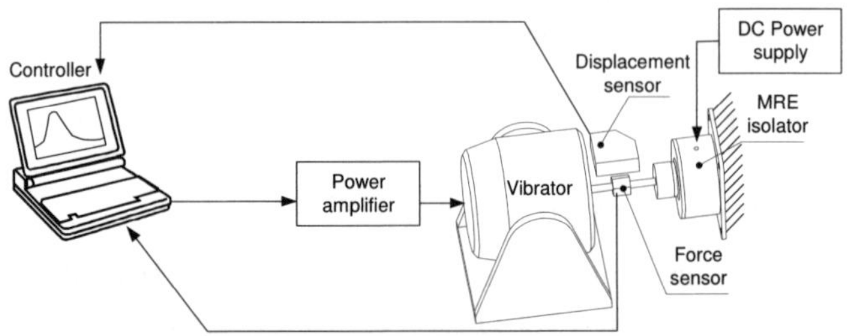

Figure 2. Dynamic testing the MRE isolator

A vibration testing system was constructed to characterize the MRE device, as shown in Figure 2. The device's base is fixed on a vibration exciter. The device is forced to vibrate by the exciter (Type: JZK-5, SINOCERA PIEZOTRONICS, INC. China), which is driven by a fluctuating source from a power amplifier (YE5871-100W) whose signal is provided by the Data Acquisition (DAQ) board (Type: LabVIEW PCI-6221, National Instruments Corporation. U.S.A) and computer. Force sensor (Type: CL-YD-302, SINOCERA PIEZOTRONICS,

INC. China) monitors the force generated by exciter. The displacement sensor (ILD1700-10) measures the isolator displacement. The signals from force sensor and displacement sensor are amplified by charge amplifiers (YE5851) and processed and transferred by DAQ to the computer. A GW laboratory DC power supply (Type: GPR-3030D, TECPEL CO., LTD. Taiwan) can adjust the input current on the MR valve in order to control the magnetic field intensity of the isolator and change its dynamic performance.

Prior to each testing, a preload 15N was applied on the MR isolator and set the zero displacement position. The isolator worked in a compression situation. By applying harmonic inputs (5Hz in this experiment) with varying coil currents from 0A to 3.0A, the response forces versus displacement loops at various magnetic fields were obtained and shown in Figure 3. The effective stiffness and damping were obtained by using the same method in [12]. Our calculation indicated that the equivalent stiffness of the MRE isolator at 3A control current is about three times as that in zero field.

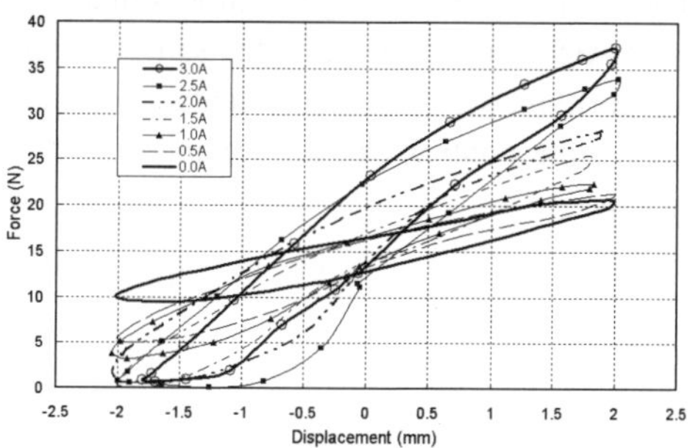

Figure 3. Force versus displacement loops at various magnetic fields

4. Model of MRE Seat Suspension with Driver

A seat-driver model consisting of the semi-active MRE isolator suspension and human-body is shown in Figure 4. The excitation input from the road is transmitted to the cabin floor. The proposed MRE seat suspension system consists of the MRE isolator, seat frame and cushion. The MRE isolator is modeled with the equivalent stiffness and damping. The cushion and body are modeled by a set of linear springs and dampers whose characteristics are constant. Therefore, the governing equation of the motion of the MRE seat suspension with human-body can be obtained as follow:

$$m_1\ddot{x}_1 = -c_{eff}(\dot{x}_1 - \dot{x}_0) - k_{eff}(x_1 - x_0) - c_2(\dot{x}_1 - \dot{x}_2) - k_2(x_1 - x_2)$$
$$m_2\ddot{x}_2 = -c_2(\dot{x}_2 - \dot{x}_1) - k_2(x_2 - x_1) - c_3(\dot{x}_2 - \dot{x}_3) - k_3(x_2 - x_3)$$
$$m_3\ddot{x}_3 = -c_3(\dot{x}_3 - \dot{x}_2) - k_3(x_3 - x_2)$$

$$(1)$$

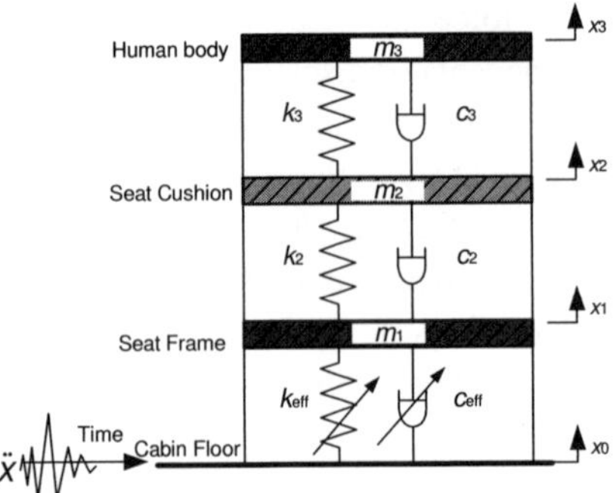

Figure 4. Vibration model of the MRE seat suspension system with human-body

5. Controller Formulation

The control strategy based on non-resonance theory was used to control the isolator [12, 13]. The main idea of the algorithm was that the stiffness of system should be increased if $x(t)$ and $\dot{x}(t)$ are in the same direction, vice verse. For this simple control algorithm, the long time calculation of feed back signals can be avoid and the real time control can be obtained easily. In this paper, the controller is designed as

$$i(t) = \begin{cases} 1 & x_2(t)\dot{x}_2(t) > 0 \\ 0 & x_2(t)\dot{x}_2(t) \leq 0 \end{cases} \quad \text{and} \quad k_{eff}(t) = \begin{cases} k_0 & x_2(t)\dot{x}_2(t) > 0 \\ 3k_0 & x_2(t)\dot{x}_2(t) \leq 0 \end{cases} \tag{2}$$

where $i(t)$ is the control signal.

6. Performance Evaluation and Discussion

In order to compare the control effect, parameters similar to reference [4] were selected. In this simulation, m_1, m_2 and m_3 are 15, 1 and 70 kg; k_2 and k are 18000 and 90000 N/m; c_2 and c_3 are 200 and 2064 Ns/m, respectively. c_{eff} can be adjusted to 830Ns/m and considered as a constant. k_{eff} is 31000N/m without control and 93000N/m with maximum control current.

6.1. *Single sine sweeping waves excitation*

Figure 5 presents the comparison of theoretical results of MRE isolator in time domain. The excitation signaL is a sine sweeping waves with amplitude of 1 cm from 1 to 20Hz during 20s. Figure 5(a) shows the displacement of the body with passive isolator. Figure 5(b) denotes the response of non-resonant controlled system. It is clear that the amplitude ratio of the non-resonant controlled system is lower than that of the passive system, i.e. the controlled MR isolator has a better isolation effect.

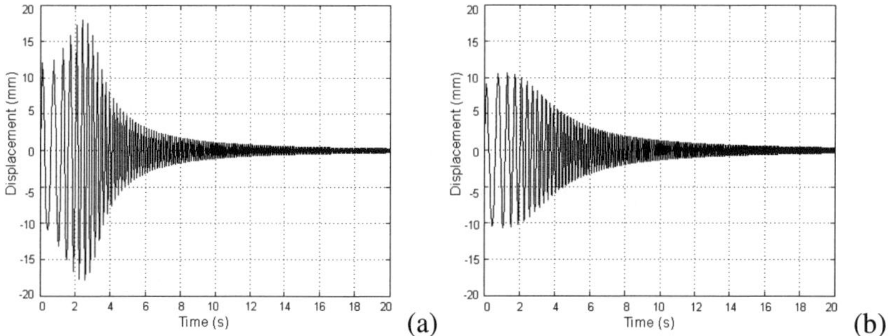

Figure 5. Theoretical results of the MRE isolator under chirp excitation in time domain

6.2. *White noise excitation*

A white noise excitation was induced to evaluate the isolation effect. Figure 6(a) and (b) show the displacement of the body with passive isolator and MRE isolator, respectively. The MRE isolator can reduce the response peak of the body.

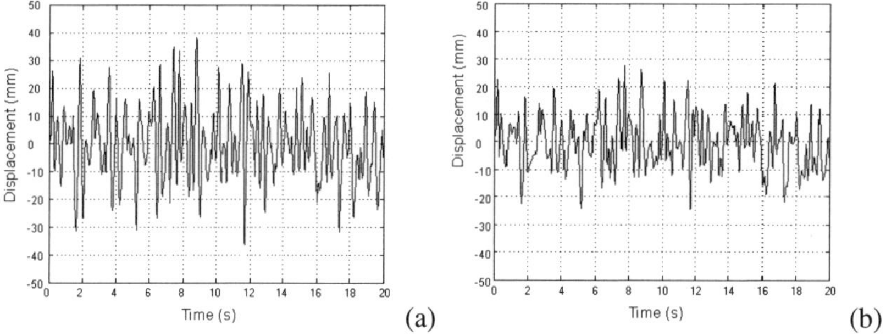

Figure 6. Theoretical results of the MRE isolator under white noise excitation in time domain

6.3. *Bump excitation*

According to reference [4], a bump profile is used to reveal the transient response characteristic (As shown in Figure 7 (a)). The response of passive and MRE isolator are shown in Figure 7 (b). It can be assured that the MRE isolator can reduce the response of the body while compare with passive system.

234

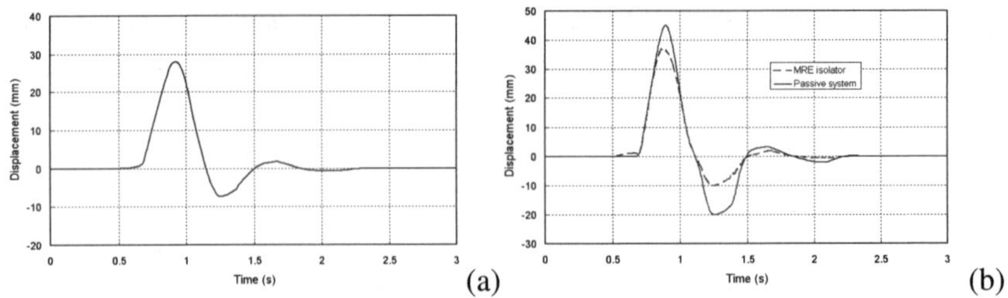

Figure 7 Theoretical results of the MRE isolator under bump excitation in time domain

7. Conclusion

This paper describes the development of a MRE isolator and verifies its application for seat vibration control with simulation analyses. The isolator's stiffness is actively controlled to establish a non-resonant state against base excitations, thus suppressing the seat's response. Results indicated that the MRE isolator can reduce vibration further than the passive isolation system, demonstrating the significant potential of its application in vehicle seat vibration control.

References

1. Legislative Assembly of Queensland: Parliamentary Travelsafe Committee. Driving on empty: Fatigue driving in Queensland. Report No. 43. Queensland Government: Brisbane (2005).
2. G.S. Paddan, M.J.Griffin, *J. Sound Vibrat.*, **253**, 215(2002).
3. F. Amirouche, L.Palkovics, J. Woodrooffe, *Transportation Systems, ASME*, 277(1994).
4. S.B. Choi, Y.M. Han, *J. Sound Vibrat.*, **303**, 391(2007).
5. R.T. Tong, F. Amirouche, *Veh. Syst. Dyn.*, **33**, 578 (1999).
6. S.B. Choi, Y.M. Han, *Int. J. Veh. Des.*, **31 (2)**, 201 (2003).
7. S.B. Choi, J.H. Choi, Y.S. Lee, M.S. Han, *J. Dyn. Syst. Meas. Contr.*, 125 (1), 60 (2002).
8. C.Y, Liao, W.H. Liao, J. Vib. Contr., 8 (4), 527 (2002).
9. Y.T. Choi, N.M. Wereley, J. Aircraft., 42 (5), 1288 (2005).
10. H.X. Deng, X.L. Gong, L.H. Wang, *Smart Mater. Struct.*, 15 (5), N111 (2006)
11. W.H. Li, Y. Zhou, T.F. Tian, *Rheol. Acta*, 49, 733 (2010).
12. T. Kobori, M. Takahashi, T. Nasu, N. Niwa, *Earthquake Eng. Struct. Dyn.*, **22**, 925(1993).
13. N.G. Pnevmatikos, L.F. Kallivokas, C.J. Gantes, *Eng. Struct.*, **26**, 471(2004).

MAGNETORHEOLOGICAL FLUIDS AND DAMPERS FOR VIBRATION CONTROL OF THE TIANXINGZHOU BRIDGE[*]

H. B. CHENG[1, 3], J. W. TU[2], W. L. QU[2], Q. J. ZHANG[1] and N. M. WERELEY[3]

[1]*State Key Laboratory of Advanced Technology for Materials Synthesis and Processing, Wuhan University of Technology, Wuhan, 430070, China*

[2]*Hubei Key Laboratory of Roadway Bridge and Structure Engineering, Wuhan University of Technology, Wuhan, 430070, China*

[3]*Smart Structures Laboratory, Dept. of Aerospace Engineering, University of Maryland, College Park, MD, 20742, USA. wereley@umd.edu*

For implementing intelligent control of the longitudinal vibration for the Tianxingzhou Bridge, which is a double-decker railway and highway cable-stayed bridge, twelve MR dampers were fabricated and identified, and installed between the deck and lower cross beam of the bridge. The results show that the MR fluid, manufactured in a 500 l quantity, exhibited high stability, high yield stress and suitable viscosity. The MR dampers had excellent mechanical characteristics; the maximum damping force was 537 kN for an applied current of 1.0 A, and the damping increased by a factor of 18 when the current was increased from 0 A to 1.0 A. Simulation results show that installing these twelve MR dampers maintains the longitudinal vibration below the maximum allowable value of 30 mm, even for a worst case scenario when two trains brake at the same time, in the same direction and in the middle of the bridge span.

1. Introduction

Magnetorheological (MR) fluids are smart fluids that bring potentially revolutionary capabilities to various fields, such as automobile industry [1], aerospace systems [2], precision optical polishing [3], vibration control of civil infrastructure systems [4]. In this study, we report on a new engineering application of MR fluids – utilization of MR dampers for intelligent control of longitudinal vibration, induced by braking trains, of the Tianxingzhou Bridge.

The Tianxingzhou Bridge is a floating deck-type cable-stayed bridge, with two levels of decks: the upper deck carries six lanes of automobile traffic and the lower deck carries four lines of railway (two freight and two passenger rail lines). The allowable maximum longitudinal displacement of the bridge deck is

---

[*] This work was partly supported by a grant from the Major State Basic Research Development Program of China (973 Program) (No. 2010CB227104) and by self-determined and innovative research funds of WUT (No.2010-II-013)

30 *mm*. However, this maximum allowable value will be exceeded when the trains apply emergency brakes (a rare event) in the absence of a vibration mitigation strategy. Therefore, suppressing longitudinal vibration of the bridge is urgently required to ensure the safety and serviceability of the bridge. In this study, we seek to suppress longitudinal vibration of the bridge using MR dampers. Accordingly, we synthesized and characterized 500 liters of MR fluid, fabricated and identified behavior of twelve MR dampers, and calculated the control effect of intelligent control with MR dampers. Finally, twelve MR dampers have been successfully installed between the decks and lower cross beams of two towers.

2. Experimental Procedures

MR fluid was synthesized using procedures outlined in [5]. The composition included silicone oil, composite magnetic particles [6, 7, 8], Fe_3O_4 nanoparticles and additives (including anti-wear additives, and lubricants).

The settling rate of the MR fluid was tested by conducting a stability test [9]. The test set-up is shown in figure 1 (left). The rheological properties of the MR fluid were characterized using a rheometer at room temperature, equipped with an electromagnet. The test set-up is shown in figure 1 (middle), which includes a rheometer (ARES, TA Co.), a set of magnetic field supplies and controller, and a pair of parallel-plate clamps. The mechanical properties of the MR dampers were evaluated for a range of applied current using a dynamic loading testing system using conventional methods [10]. The test-up is shown in figure 1 (right).

3. Results and Discussion

3.1. *Characterization of MR fluid*

3.1.1 Sedimentation stability. The sedimentation velocity of particles within the MR fluid was determined by exploiting the magnetic properties of MR fluid particles. The principle and detailed procedures are described in [9]. The sedimentation velocity of the MR fluid is defined as the rate at which the mudline descends due to particle settling, until the particles pack tightly at the bottom of the sample cell without further sedimentation. Because the magnetic permeability of MR fluid is highly dependent on the volume fraction of fluid, therefore, measuring the change of magnetic permeability of MR fluid by an inductance-based solenoid sensor can track the mudline location of the settling fluid [9]. The result is shown in figure 2 (left). The sedimentation velocity of the MR fluid was measured to be $0.00032 \mu m/s$.

The long-term stability of MR fluids was experimentally investigated through repeatedly measuring the yield stress and off-state viscosity of MR fluids before and after static storage over a two year period [11]. By comparing the yield stress and off-state viscosity of MR fluids before and after storage, any impact of storage can be characterized. The experimental results did not show noticeable changes in yield stress and off-state viscosity after storage for a two year period. Detailed results were reported in our previous work [11]. Therefore, from these test results of sedimentation velocity and long-term stability of the MR fluid, it can be concluded that the synthesized MR fluid exhibits high stability.

3.1.2 Yield stress. The field-induced yield stress of the MR fluids was tested with magnetic flux density ranging from 0 to 0.5 T at the shear rate of 1 s^{-1} and is shown in figure 3 (middle). It can be seen that the yield stress of MR fluid dramatically raise with increasing applied magnetic field strength from 0 to 0.5 T. The maximum yield stress was more than 70 kPa at 0.5 T.

3.1.3 Off-state viscosity. The stress-strain behavior of the Bingham viscoplastic model can be used to describe the behavior of MR fluids [12]. In this model, the plastic viscosity, η, is defined as the slope of the measured shear stress, τ, versus shear strain rate, $\dot{\gamma}$. Thus, the total stress is given by

$$\tau = \tau_y \, \mathrm{sgn}(\dot{\gamma}) + \eta \dot{\gamma} \tag{1}$$

Here, τ_y is the yield stress induced by the magnetic field. In the off-state (no magnetic field), equation (1) can be simplified:

$$\tau = \eta \dot{\gamma} \tag{2}$$

where τ is the shear stress, and η is the off-state viscosity of MR fluid. According equation (2), the off-state viscosity of the MR fluid can be calculated based on the slope of shear stress versus shear strain rate data. The off-state viscosity of the MR fluid was determined by measuring the relationship between shear stress and shear strain rate. The shear stress increases linearly with shear rate as shown in figure 2 (right), the slope is 0.89, i.e. the off-state viscosity of the MR fluid is 0.89 *Pa s*.

238

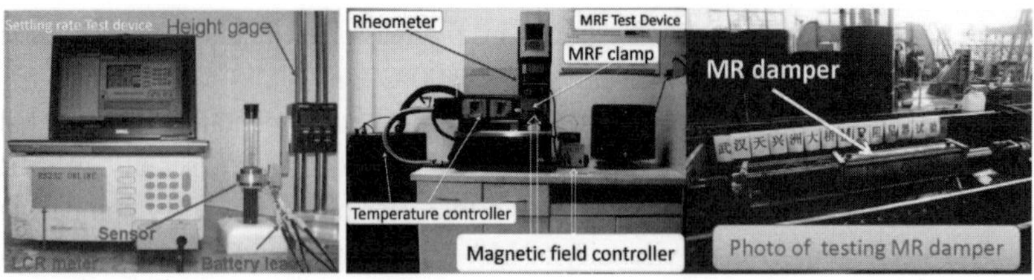

Figure 1. Photos of test set-up for sedimentation velocity (left) and rheological property (middle) of MR fluid (left) and for MR damper (right)

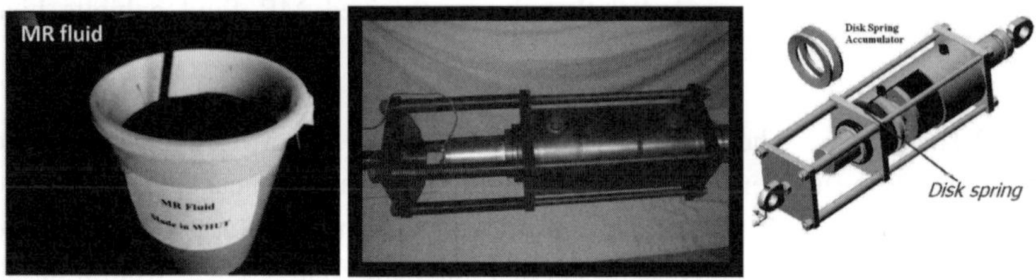

Figure 2. Photos of MR fluid (left) and MR damper (middle). MR damper design schematic (right)

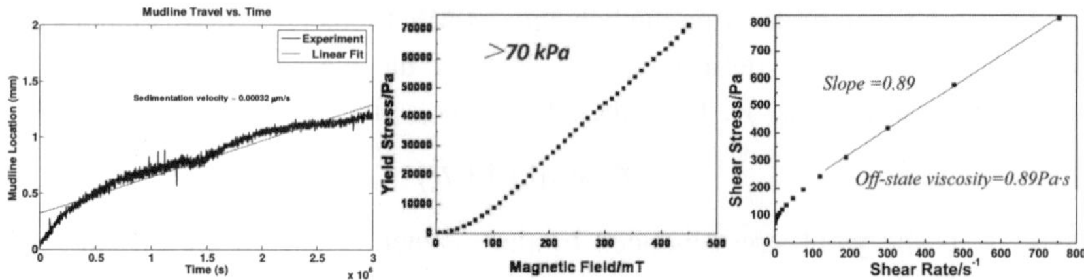

Figure 3. Settling velocity (left), yield stress (middle) and off-state viscosity (right) for MR fluid

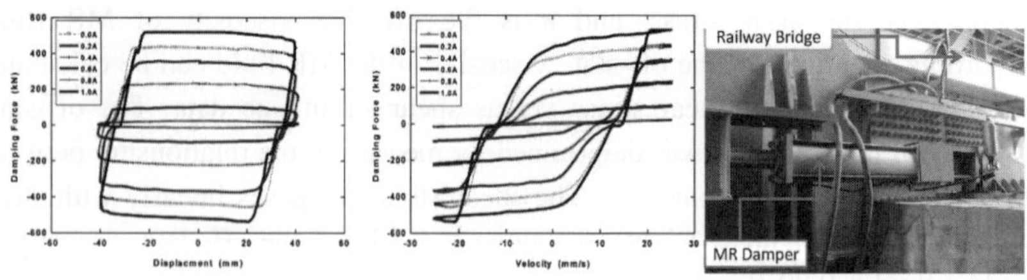

Figure 4. Force-displacement curve (left) and Force-velocity curve (middle) of MR damper. One of twelve MR dampers installed between the deck and lower cross beam at each tower (right)

3.2. Fabrication and identification of MR damper

3.2.1 Fabrication of MR damper. Twelve 500 *kN* full-scale MR dampers were fabricated, in which a built-in disc-spring accumulator was adopted, and stainless steel plates were installed at both ends of the piston to effectively prevent magnetic flux leakage. The design schematic is shown in figure 2 (right); the photo of MR damper is shown in figure 2 (middle). The MR damper size parameters are as follows: cylinder inner diameter D = *320 mm*; cylinder outer diameter D_0 = 360 *mm*; piston diameter d = *160 mm*; stroke = ±150 *mm*; effective length L = 120 *mm*; gap size h = 2 *mm*; coil turn numbers (turn) 3×1 coil = 800; MR fluid volume of 40 *l*. The detailed design was discussed previously [13].

3.2.2 Identification of MR damper. The mechanical properties of MR dampers over a range of current/magnetic field were evaluated using a dynamic loading testing system [10]; the test set-up is shown in figure 1(right). The sinusoidal displacement excitation, x = A *sin* $(2\pi ft)$, was adopted for this test, where x is the relative displacement of both ends of the MR damper piston rod, A and f are the amplitude and frequency of the displacement excitation, respectively. A group of data is selected randomly with the excitation amplitude A = 40 *mm* and the excitation frequency f = 0.1 *Hz*. The current increased from 0 to 1 A with an incremental step 0.2 A. The hysteretic curve between the damping force and displacement is shown in figure 4 (left), while the hysteretic curve between the damping force and velocity is shown in figure 4 (middle). It can be seen that the maximum damping force of the MR damper is 537 *kN* when the current reaches 1.0 A (designed value is ≥ 400 *kN*); damping ratio reaches 18 times from 0 A to 1.0 A. The durability of the MR dampers was assessed, including steady state and dynamic mechanical property stability, was assessed during an endurance tests lasting for 1.86 million cycles. The measurements of yield force and damping exhibited good repeatability over the duration of this test. These test results demonstrate that the fabricated MR dampers can meet the engineering requirements of this bridge application.

3.2.3 Installation of MR dampers. The purpose of installing MR dampers is to reduce the longitudinal vibration responses of the bridge deck to excitations of train braking. We installed twelve 400 *kN* MR dampers between the deck and lower cross beam of two towers.

Consider the scenario where two freight wagons, with an initial braking speed of 80 *km h^{-1}*, are traveling in the same direction along two tracks and at the same time, but different position, we calculated the longitudinal maximum displacement of deck and longitudinal maximum bending moments at the bottom of tower both without and with the twelve 400 kN MR dampers. The results are shown in table 1. For all nine cases, the longitudinal maximum displacements of the deck are less than the allowable value of 30 mm proposed in the design under intelligent control with MR dampers; but all are larger than 30 mm if no MR dampers are implemented. Mitigation for the most critical case 7, are especially striking when MR dampers are implemented: (1) the maximum longitudinal displacements of the bridge deck is reduced by 87.0%, from 149.2 mm to 19.4 mm; and (2) the maximum bending moment at the bottom of the towers under intelligent control is reduced by 44.5%, from $8.18 \times 10^5 kN\ m$ to $4.54 \times 10^5 kN\ m$. Therefore, from these results it can be concluded that the intelligent control system using MR dampers can effectively reduce the longitudinal displacement and bending moment responses of the bridge, thereby, ensuring serviceability and safety of the cable-stayed Tianxingzhou Bridge.

The twelve MR dampers have been successfully installed between the deck and lower cross beam of the two towers (see figure 4 right). Performance of the intelligent control systems will be assessed *in situ* during the operational lifetime of the system.

Table 1. Longitudinal maximum displacement of deck (D/mm) and bending moments at the bottom of tower ($B/10^5 kN\ m$) under nine cases of train braking

Case		1	2	3	4	5	6	7	8	9
Braking Stopped	Freight train line 1	Entry end	Middle	Exit end	Entry end	Entry end	Entry end	Middle	Middle	Exit end
position	Freight train line 2	None	None	None	Entry end	Middle	Exit end	Middle	Exit end	Exit end
D/mm	No	53.8	74.6	68.2	107.5	104.7	104.5	149.2	134.4	136.3
D/mm	Installing	6.2	7.6	7.0	13.1	14.3	11.8	19.4	14.4	15.1
$B/10^5 kN\ m$	No	2.4	4.10	2.86	5.86	5.36	4.42	8.18	5.14	5.70
$B/10^5 kN\ m$	Installing	1.64	2.17	2.01	3.20	3.42	3.00	4.54	3.84	3.95

4. Conclusion

MR fluids were synthesized and were validated by experimental tests to be highly stable, have high yield stress, and suitable off-state viscosity. MR dampers were designed and fabricated for engineering application to the Tianxingzhou Bridge for controlling its longitudinal vibration response. The test results show that the MR damper exhibited high damper force, 537 *kN* at 1.0A; the damping range reaches 18 times at 1.0 A current. The calculated results show that installing twelve MR dampers can effectively control the longitudinal response of the floating deck-type railway bridge under excitations of braking trains, and ensure the serviceability and safety of the Tianxingzhou cable-stayed bridge.

References

1. G.Z. Yao, F.F. Yap, G. Chen, W.H. Li and S.H. Yeo, *Mechatronics,* **12**(7), 963-973 (2002).
2. G. J. Hiemenz, W. Hu, and N. M. Wereley, *AIAA J. Aircraft.* **45**(3), 945–953 (2008).
3. W.I. Kordonski, D. Golini, P. Dumas, S.J. Hogan and S.D. Jacobs, *Proc. SPIE,* **3326**, 527 (1998).
4. J.D. Carlson and B.F. Jr. Spencer, *Proc. 3rd Internat. Conf. on Motion and Vibr. Control, Chiba, Japan, Sept.* **Vol. III 35**, 40 (1996).
5. H.B. Cheng, J. Zhang, W.X Gao, Q.J. Zhang, J. M. Wang and R. Z. Yuan, *CN Pat.No. ZL* 200610124728.0 (2006).
6. H.B. Cheng, W.X. Gao, Q.J. Zhang, S. Xu, W.Y. Zhao and R.Z. Yuan, *CN Pat. No. ZL* 200610124790.X (2006).
7. H.B. Cheng, J.M.Wang, Q.J. Zhang and N.M. Wereley, *Smart Mater. Struct,* **18** 085009 (6pp) (2009).
8. H.B. Cheng, L. Zuo, J.H. Song, Q.J. Zhang, and N.M. Wereley, *J. Appl. Phys.,* **107**, 09B507 (2010).
9. G.T. Ngatu and N.M. Wereley, *IEEE Trans Magn,* **43**(6), 2474-2476 (2007).
10. G.Q. Yang, *PhD Dissertation University of Notre Dame, Indiana, USA* (2001).
11. H.B. Cheng, P. Hou, Q.J. Zhang and N.M. Wereley. *Journal of Physics: Conference Series* **149**, 012043 (2009).
12. B.F. Spencer, S.J. Dyke, M.K. Sain and J.D. Carlson. ASCE *Journal of Engineering Mechanics,* **3**, 230-238 (1997).
13. W.L. Qu, S.Q. Qin, J.W. Tu, J. Liu, Q. Zhou, H.B. Cheng and Y.L. Pi, *Smart Mater. Struct.,* **18**(12), 125003 (20pp) (2009).

PERFORMANCE EVALUATION OF HAPTIC CUE ACCELERATOR UTILIZING MAGNETORHEOLOGIAL CLUTCH

H. G. LEE

Department of Automotive Engineering, Daeduk University, University Department, Daejeon 305-715, Korea

Y. M. HAN, M. S. SEONG and S. B. CHOI[*]

Department of Mechanical Engineering, Inha University Incheon 402-751, Korea

This paper proposes a driver supportive device with haptic cue function to achieve optimal gear shifting in manual transmission vehicles. This function is implemented on accelerator pedal by utilizing magnetorheological (MR) clutch mechanism. In order to achieve this goal, an MR fluid-based clutch is devised to be capable of rotary motion of accelerator pedal. The proposed MR clutch is then manufactured and its transmission torque is experimentally evaluated according to field intensity. Then the manufactured MR clutch is integrated with accelerator pedal to establish the haptic cue device. A virtual environment emulating 4-cylinder 4-stroke engine is constructed and communicated with the haptic cue device. Control performances such as torque tracking are experimentally evaluated via simple feed-forward controller.

1. Introduction

Recently, fuel economy gets much emphasis in automobile engineering to reduce energy cost and global emission. Several factors can affect on the fuel economy of vehicles such as vehicle weigh, aerodynamic drag, regenerative braking, traffic condition and personal driving style. Among them, a gear shifting timing caused by personal driving style is significantly related with driving power and fuel consumption. When driving automatic transmission vehicles, the vehicle starts with low gear, and then automatically goes up to top gear. On the other hand, manual transmission vehicles need driver's decision for proper gear shifting timing before gear shifting action. Many researchers have widely studied on optimal gear shifting of vehicle transmission to enhance fuel economy [1, 2]. Most of these works are conducted with automatic transmission such as

---

[*] Corresponding Author

continuously variable transmissions (CVT). However, in manual transmission, research works considering optimal gear shifting are still very limited because the gear shifting highly depends on personal driving style.

A haptic technology based on the sense of touch provides a new communication method. Many researchers have studied on various haptic devices for vehicles. Aoki and Murakami developed a haptic pedal to feedback the estimated road condition to a driver [3]. Kobayashi et al. proposed an accelerator pedal device which provides haptic feedback to avoid rear-end collision situation of a vehicle [4]. Recently, by adopting a haptic function utilizing a controllable magnetorheological (MR) brake into acceleration mechanism, the authors suggested a driver supportive system to achieve an optimal timing of gear shifting for manual transmission vehicles [5]. However, a passive brake mechanism can only provide torques during increasing pedal angle by pushing accelerator pedal. This imposes restriction on practical implementation of the proposed system. Therefore, the main contribution of this work is to enhance the haptic accelerator by adopting MR cultch mechanism instead of MR brake mechanism. It can provide haptic force in practical driving situations even though vehicles slowly accelerate after pushing the accelerator pedal at a certain angle.

2. Haptic Cue Accelerator

A haptic technology utilizing MR fluids can provides a new driver supportive function that a driver can recognize an optimal timing of gear shifting with minimum fuel consumption. This can be achieved by adopting a haptic function into acceleration mechanism because the most meaningful gear shifting occurs when accelerating the vehicles. Figure 1 shows the schematic configuration of the proposed haptic cue accelerator, which is devised by adopting the MR fluid and rotary type of clutch mechanism. The accelerator pedal is connected to a cylindrical electromagnetic disk which has a flux guide and an electromagnetic coil. The electromagnetic disk is assembled to housing with a specific gap fully filled with MR fluid. A geared AC motor rotates the housing. In this work, a commercial MR product has been employed as a controllable fluid. When a magnetic field is applied to the MR fluid, the shear stress (τ) can be expressed by

$$\tau = \tau_y(\cdot) + \eta\dot{\gamma} \qquad (1)$$

where η is the dynamic viscosity, γ is the shear rate, and $\tau_y(\cdot)$ is the dynamic yield stress of the MR fluid.

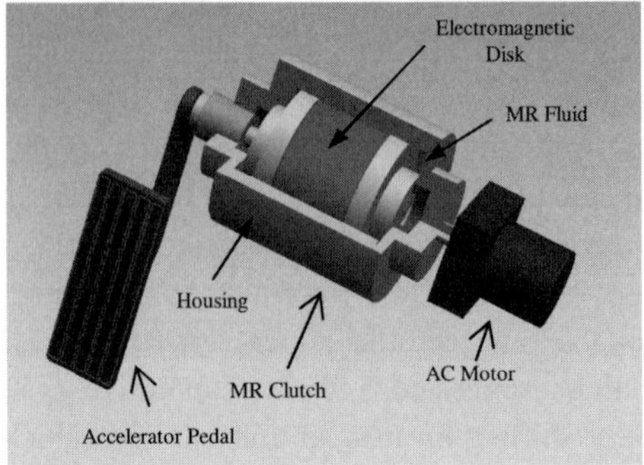

Figure 1. Schematic configuration MR haptic cue accelerator.

Among several operation modes of MR fluid-based device, the shear mode is significantly developed for the rotary motion. The produced torque by the MR brake is mathematically described as summation of the following controllable torque, viscous torque and friction torque:

$$T = 4\pi R^2 d\tau_y(\cdot) + \frac{4\pi\eta R^3 d\dot{\theta}}{h} + T_f \qquad (2)$$

where $\dot{\theta}$ is the rotational velocity of the haptic cue device in rotary motion. R and h are the outer radius and gap height of the electromagnetic disk, respectively. d is the length of the flux guide. T_f is the friction torque.

Among several torque terms in Eq. (2), the control torque has a large effect on the performance of the MR haptic cue device. It is desirable that the relative control torque to total torque takes a large value as much as possible to have large control torque. Therefore, the values of design parameters has been determined to maximize the relative control torque using finite element (FE) analysis. Figure 2(a) shows the manufactured MR haptic cue device. The disk height and radius are 0.006m and 0.04m, respectively, and the gap size is 0.001m. A mild steel (S45C) is adopted for the flux guide as ferromagnetic material. The other part was made of paramagnetic aluminum (Al6061). The optimized coil area was winded by copper wire with 0.75mm diameter. Its number of coil turns is 500. The electromagnetic disk was assembled to the housing with a specific gap fully filled with the MR fluid. Figure 2(b) presents the produced torque characteristics of the manufactured haptic cue device. A variable speed AC motor with controller is used to drive. A rotational torque transducer which has a capacity of 10Nm is employed to measure the produced

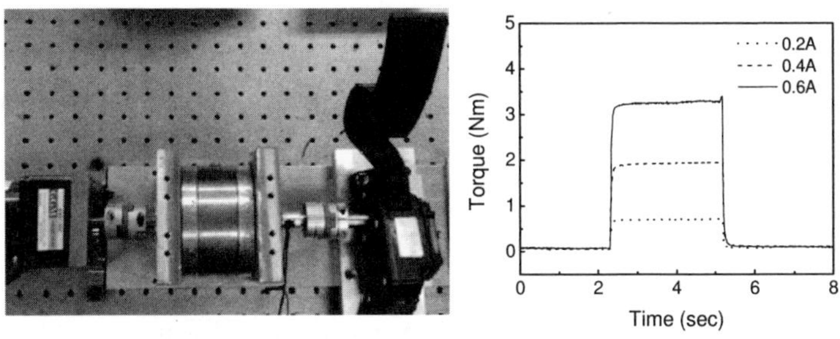

(a) manufactured device   (b) produced torque

Figure 2. The MR haptic cue accelerator.

torque. The measured responses (scatter symbol) are compared with the predicted responses (line). Under input current of 0.6A and rotational velocity of 120deg/s, the MR haptic cue device produces 3.28Nm.

3. Haptic Control Algorithm

The MR haptic cue accelerator interacts with the virtual vehicle through driver. When a driver pushes the accelerator pedal, its positional information is transmitted to the virtual vehicle, which generates cue signals for gear shifting. The input current is then determined with the torque control algorithm and cue signal. Finally the MR haptic cue accelerator reflects reaction torque to the driver in order to notify appropriate gear shifting timing. In the mean time, the driver can determine to follow the suggested timing or not. In this work, the cue signal for optimal gear shifting is generated at a specific engine speed. After generating a cue signal in the virtual environment, the desired torque determined from a torque map which has the relationship between the torque and the engine speed. Then, the haptic controller should follow the desired reaction torque.

A feed-forward torque control algorithm, that is simple but very effective, is formulated for the haptic cue in this work. The feed-forward controller can be formulated by inversion of the controllable torque model. From Eq. (2), the controllable torque is given by

$$T_c = 4\pi R^2 d\tau_y(H) \tag{3}$$

In the above, the dynamic yield stress of the MR fluid can be expressed by magnetic field strength (H) by αH^β. This has been obtained from experimental Bingham properties as follows : $0.024H^{1.398}$

In order to achieve the controllable desired torque trajectory, the controllable torque T_c should follow the desired torque as follows:

$$T_c = 4\pi R^2 d \cdot \alpha H^{\beta} = T_d \tag{4}$$

By inversion of the above model, the control input can be expressed in terms of magnetic field strength as follows:

$$H = \left(\frac{T_d}{\alpha \cdot 4\pi R^2 d} \right)^{\frac{1}{\beta}} \tag{5}$$

However, the torque model in Eq. (2) has the uncontrollable terms such as the viscous term and friction term. Therefore the desired torque trajectory is determined by eliminating the fluid viscous torque term, friction torque term and pedal torque term from the torque map output upon the measured position information as follows:

$$T_d = T_m - 4\pi \eta R^3 d \dot{\theta} / h - T_f \tag{6}$$

where T_m is the desired reaction torque from the torque map. Finally, once the magnetic field strength as control input is determined, the input current to be applied to the MR haptic cue device can be calculated by [15]

$$I = \frac{2h}{N} \left(\frac{T_m - 4\pi \eta R^3 d \dot{\theta} / h - T_f}{\alpha \cdot 4\pi R^2 d} \right)^{\frac{1}{\beta}} \tag{7}$$

where N is the number of coil turn.

4. Performance Evaluation and Concluding Remarks

Figure 3 shows the experimental apparatus to evaluate control performance of the proposed MR accelerator that interacts with the virtual vehicle realized by dSPACE and Matlab Simulink. The accelerator pedal connected to the electromagnetic disk of the MR device is pushed by a cam system driven by an AC motor which corresponds to a vehicle driver. When the accelerator pedal is pushed, the pedal position is obtained by an incremental encoder which has resolution 3600 pulse per 1 revolution. The feed-forward controller is then activated for the MR haptic cue device to reflect the desired torque to a driver.

Figure 4 presents torque tracking results for gear shifting. In the beginning, the vehicle cruised in a steady speed at which engine and vehicle speeds are about 1800rpm and 5m/s, respectively. It is assumed that it takes about 0.5s until the driver recognizes the haptic cue and tries to change the gear stage. The rotational speed of the vehicle engine increases by pushing the pedal to change the gear. When the engine speed meets the threshold of 2500rpm, a cue signal and the corresponding desired torque trajectory are generated. Then the MR haptic cue device was activated to follows them. As clearly observed from the results, the desired torque trajectory from the map is well followed by the feed-forward controller after cue signal. Figure 5 shows torque tracking results at fast pushing speed of the pedal. The engine speed slowly increases after pushing pedal. From the results, the torque for haptic cue was successfully reflected to the human driver, even though the pedal is kept at constant angle position. The response time is favorable and the final tracking error before releasing the pedal is below 0.11Nm in both cases. In the mean time, input current is shown in

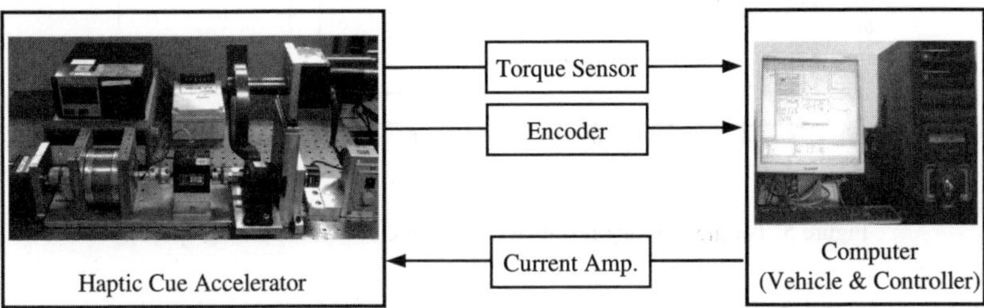

Figure 3. Experimental apparatus for the haptic cue control.

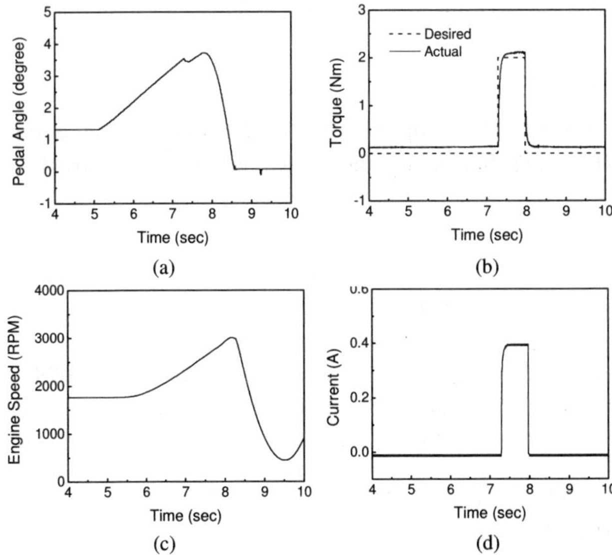

Figure 4. Torque tracking results for the haptic cue (slow pushing speed).

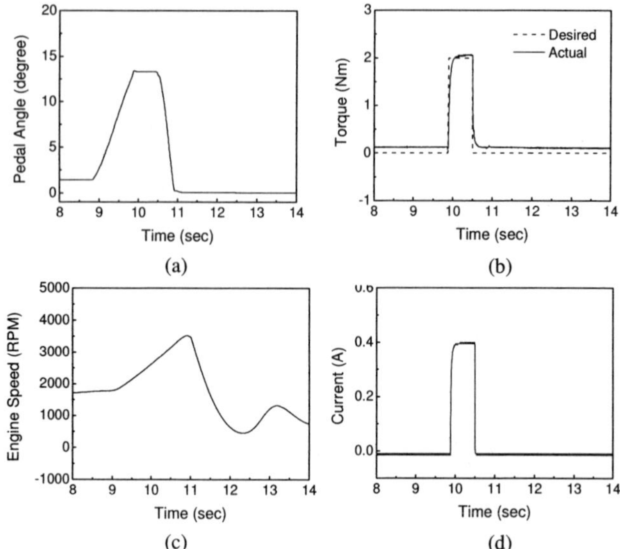

Figure 5. Torque tracking results for the haptic cue (fast pushing speed).

Figures 4(d) and 5(d) whose magnitude is about 0.39A. It is noted that the initial torque about 0.078N/m corresponds to the nominal transmission torque by the clutch at zero field. From these results, it can be assured that the proposed haptic cue accelerator successfully cues driver to change gears at optimal moment.

Acknowledgments

This work was partially supported by the National Research Foundation of Korea (NRF) grant funded by the Korea government (MEST) (No. 2010-0015090).

References

1. M. A. Haj-Fraj and F. Pfeiffer, *Journal of the Franklin Institute-Engineering and Applied Mathematics* **338**, 371-390 (2001).
2. S. Sakaguchi, E. Kimura and K. Yamamoto, *SAE* **1999-01-0754** (1999).
3. J. Aoki and T. Murakami, *AMC'08 The 10th international workshop on advanced motion control, Centro Santa Chiara, Trento, Italy*, 777-782 (2008).
4. Y. Kobayashi, T. Kimura, T. Yamamura, G. Naito, and Y. Nishida, *SAE* **2006-01-0572**, (2006).
5. Y. M. Han, K. W. Noh, Y. S. Lee and S. B. Choi, *Smart Materials and Structures* **19(7)**, 075016 (2010).

STUDY OF SENSING CAPABILITY OF MR ELASTOMERS

TONGFEI TIAN and WEIHUA LI

School of Mechanical, Materials and Mechatronic Engineering, University of Wollongong, Wollongong, NSW 2522, Australia

In this paper, the graphite based Magnetorheological Elastomer (Gr MREs) is presented to study the sensing capability of MREs. With the help of graphite, the giant resistance of MREs is reduced to kΩ level. When either the magnetic field or the external force is applied on to the Gr MREs, its resistance is changed. This finding opens up possibilities for the design of a force sensor combining Gr MREs. A representative unit model is addressed to express the phenomenon theoretically.

1. Introduction

A sensor is a device that measures a physical quantity and converts it into a signal which can be read by an observer or by an instrument. Sensors are used in everyday objects such as a Three-axis gyro, Accelerometer, Proximity sensor and Ambient light sensor which are all used in the iPhone 4. A force sensor is one kind of sensor used to convert force to another readable signal, which has been widely used in various industries. In force sensors there are many typical techniques of using a force collector such as diaphragm, piston, bourdon tube, piezoresistive strain gauge, capacitive, electromagnetic, piezoelectric sensors to measure strain or deflection due to applied force over an area.

Magnetorheological elastomers (MREs) are smart materials where polarised particles are suspended in a non-magnetic solid or gel-like matrix. There are two kinds of MREs which are anisotropic and isotropic. In anisotropic samples, polarised particles can be arranged as chains in polymer media such as silicon rubbers and natural rubbers. The Gr MREs show resistance changing when an external load is applied to MRE samples. By this feature of MREs, the external stress signal can be converted to a resistance signal, which can be used potentially in a force sensor.

Kchit and Bossis [1] found that the initial resistivity of metal powder at zero pressure is about 108 Ωcm for pure nickel powder and 106 Ωcm for silver coated nickel particles. Wang et al. [2] proposes a phenomenological model to understand the impedance response of MREs under mechanical loads and magnetic fields. Bica [3] found that MRE with graphite micro particles (~14%)

is electroconductive. The magnetoresistance has an electric resistance whose value diminishes with both the increase of the intensity of the magnetic field and with the compression force. Li et al. [4] introduced graphite into conventional MREs and found a MRE sample with 55% carbonyl iron, 20% silicon rubber and 25% graphite powder has the best performance.

Based on the microstructure of MREs, Wang et al. [2] proposed an equivalent circuit model to interpret the impedance measurement results. Li et al. [4] presented the development of a new force sensor with MR elastomer as a sensing element. MREs with different composition were manufactured and measured with a modified Rheometer. Bica [5] reported that the electrical conductivity of MR elastomer can be used for achieving magnetoresistors, magnetic field sensors, transductors of mechanical distortions, strains, among others. The electrical conductivity of MRE can be modified in a magnetic field.

2. MRE Fabrication

The materials used for the Graphite Magnetorheological Elastomers (Gr MREs) are: silicone rubber (Selleys Pty. LTD), silicone oil, type 378364 (Sigma-Aldrich Pty. LTD), carbonyl iron particles, type C3518 (Sigma-Aldrich Pty. LTD) and graphite powder, type 282863 (Sigma-Aldrich Pty. LTD). The particle sizes of graphite powder are less than 20 μm, the iron particles' diameter is between 3 μm and 5 μm.

The MREs with various graphite weight fractions (Gr %) were fabricated to compare the effect of graphite on to MREs. Table 1 shows the compositions of all Gr MRE samples. For each composition, there are two samples namely isotropic and anisotropic samples.

Table 1 Components of Gr MRE samples

	Graphite based MREs							
Sample No.	1	2	3	4	5	6	7	8
Carbonyl iron	10g	10g	10g	10g	10g	10g	10g	10g
Silicone oil	3g	3g	3g	3g	3g	3g	3g	3g
Silicone rubber	3g	3g	3g	3g	3g	3g	3g	3g
Graphite	0g	1g	2g	3g	3.5g	4g	4.5g	5g
Graphite weight fraction (Gr %)	0	5.88	11.11	15.79	17.95	20	21.95	23.81

3. Experimental Setup and Results

Fig. 1 shows a schematic of the experimental device used. In this setup a long plastic plate is used to hold the weight to apply the load to the Gr MRE samples. A Gaussmeter (HT201, Hengtong magnetoelectricity CO., LTD) is used to test the intensity of the magnetic field. A multimeter (Finest 183, Fine Instruments Corporation) measures the resistance in the Gr MRE samples.

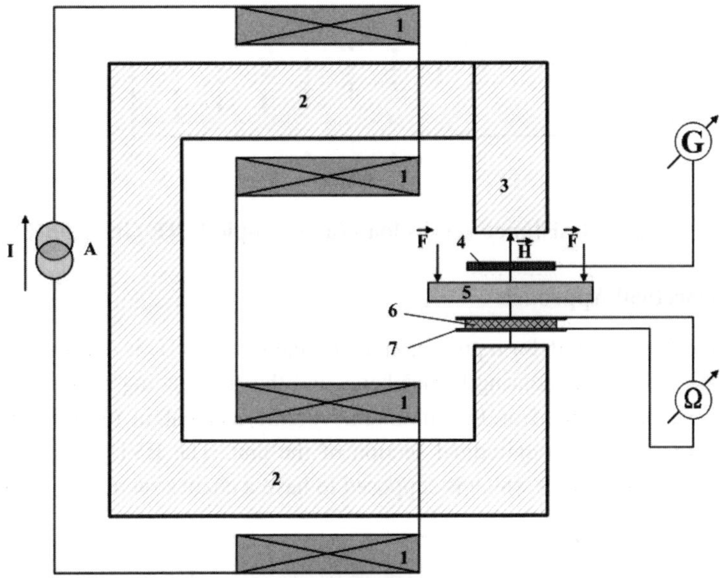

Fig. 1 Sketch of the experimental device
1- coils; 2- electrical magnetic; 3- moving magnetic pole; 4- Gaussmeter probe; 5- plastic plate; 6- Gr MRE sample; 7- measuring copper plates; G- Gaussmeter; Ω- Multimeter; F- external force applied on Gr MRE samples

After testing the resistance of all the anisotropic MRE samples only the samples whose graphite weight fraction are no less than 15% are detectable by Finest 183 multimeter. Thus in all the anisotropic MRE samples tested there are only three of them, whose graphite weight fractions are 20%, 21.95% and 23.81%, respectively where resistance can be detected. The data of anisotropic MRE with graphite weight fraction 21.95% is shown in Fig. 2. As can be seen in Fig. 2, in a fixed magnetic field when the external load increases from 0 to 10 N, the resistance reduces. With small load the resistance changes significantly but it decreases slowly when the load is more than 5 N.

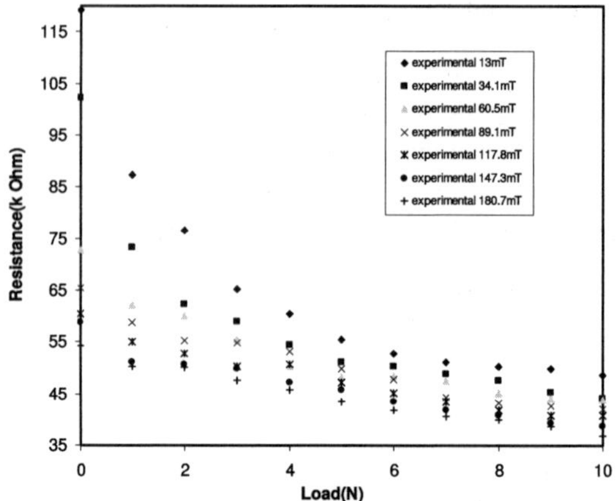

Fig. 2 Resistance versus load (anisotropic MRE Gr 21.95%)

4. Theoretical Approach

From the Dipole Model a representative volume unit (RVU) is derived. A RVU consists of two neighbouring hemispheres and the surrounding polymer matrix, which can be regarded as the minimum volume element in the conventional MRE. Fig. 3 shows longitudinal section of the unit. The RVU is a model for ideal anisotropic MREs which is supposed to have a chain structure.

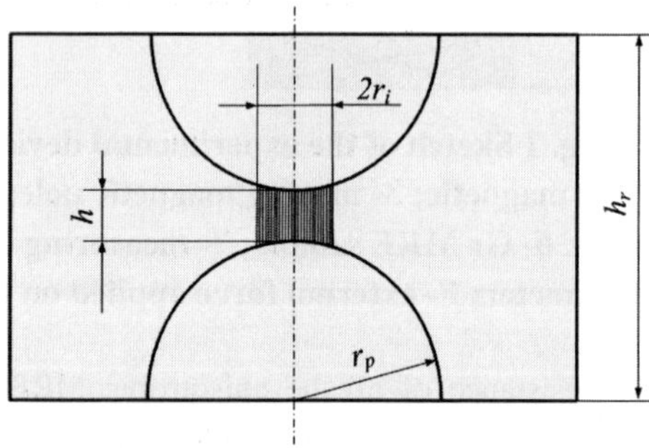

Fig. 3 The longitudinal section of RVU

The insulating matrix film between two neighboring iron particles is very thin, across which the electrical field assisted tunnel current, j_t, exists

$$j_t = \alpha E_{loc}^2 \exp\left(-\frac{\beta}{E_{loc}}\right) \text{ and } E_{loc} = \frac{h_r E}{h} = \frac{(2r_p + h)E}{h} \approx 2r_p \frac{E}{h}$$

The total current density is the sum of the tunnel density and the conduction density $j = j_t + j_c = \alpha E_{loc}^2 \exp\left(-\dfrac{\beta}{E_{loc}}\right) + \sigma_f E_{loc}$.

For the two particles in each RVU, both the magnetic attraction and external loading compress the thin film between the two adjacent particles. Thus, the radius r_i is updated as $r_i = r_{i0} + r_p\left((\sigma_0 + \sigma)^{1/3} - \sigma_0^{1/3}\right)\cdot\left(\dfrac{3\pi\left(1-v^2\right)}{2E_p}\right)^{1/3}$ where

$\sigma1$ and $\sigma2$ are compressive stress and magnetic attraction, respectively.

The conductivity of MREs is given by

$$\sigma_m = 3\phi \cdot \left[\dfrac{2\alpha}{h^2}E\exp\left(-\dfrac{h\beta}{2r_p E}\right) + \dfrac{\sigma_f}{r_p h}\right]\cdot\left[r_{i0} + r_p\left((\sigma_0 + \sigma_1 + \sigma_2)^{1/3} - \sigma_0^{1/3}\right)\cdot\left(\dfrac{3\pi\left(1-v^2\right)}{2E_p}\right)\right]^2$$

From the graphite MRE, two parameters, λg and λi, were introduced in the model. These two parameters were used to represent the effects of the graphite weight fraction and iron particle weight fraction. The final MRE resistance is given by

$$R_g = \dfrac{\lambda_g \lambda_i l}{3A\phi \cdot \left[\dfrac{2\alpha}{h^2}E\exp\left(-\dfrac{h\beta}{2r_p E}\right) + \dfrac{\sigma_f}{r_p h}\right]\cdot\left[r_{i0} + r_p\left((\sigma_0 + \sigma_1 + \sigma_2)^{1/3} - \sigma_0^{1/3}\right)\cdot\left(\dfrac{3\pi\left(1-v^2\right)}{2E_p}\right)\right]}$$

By substituting the constant parameters into the formula, Rg can be simplified to

$$R_g = \dfrac{4.78 * 10^{-13} \times \lambda_g \lambda_i}{10^{-8} + 0.000002\left((2000 + \sigma_1 + \sigma_2)^{1/3} - 2000^{1/3}\right)\times 0.001166}$$

5. Comparation and Analysis

To show the comparison between the experimental result and the theoretical result (anisotropic MRE with graphite weight fraction 21.95%) clearly, each four series at both the experimental result and the theoretical result are picked out to be compared in Fig. 4. It shows that the experimental result and theoretical result couldn't match each other perfectly, however, the trends of both the experimental result and the theoretical result are the same. Similarly some other comparison can be made from the data. At the fixed external load, when the magnetic field intensity increases the resistance of the sample decreases. The data under three external loads such as 1 N, 5 N and 10 N were chosen to compare the experimental and theoretical result in Fig. 5.

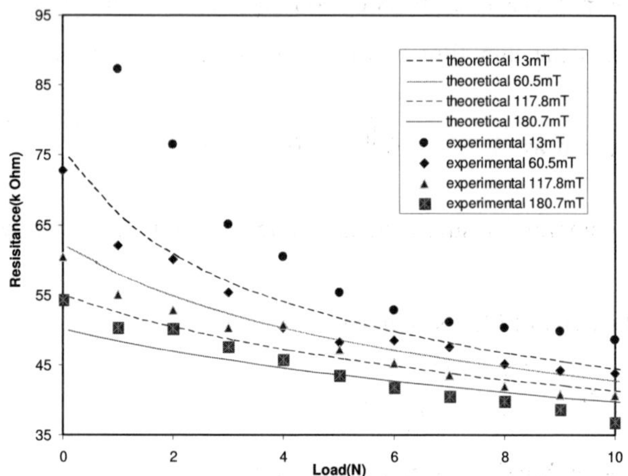

Fig. 4 Comparison between experimental result and theoretical result (anisotropic MRE Gr 21.95%)

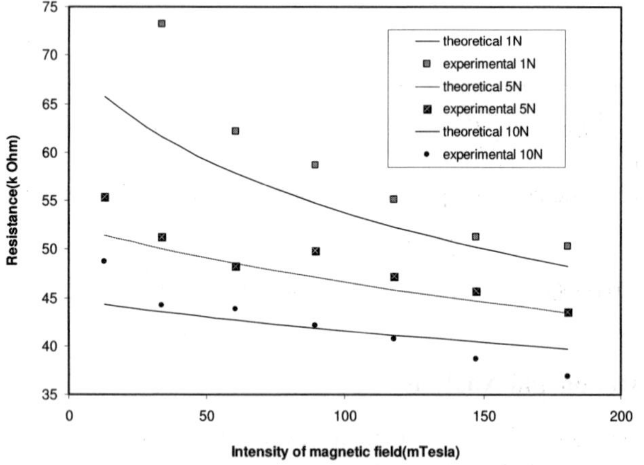

Fig. 5 Resistance changing at a fixed load (anisotropic MRE Gr 21.95%)

The Fig. 5 shows the changing of the resistance when the external force is fixed. Along with the raising of magnetic field intensity, the sample's resistance decreases. The higher external load applied leads to lower resistance of Gr MREs.

6. Conclusion

In this study, a representative volume unit was presented to show the resistance of ideal anisotropic MREs. Experimental results proved the feature of MREs, namely the resistance of MREs, is dependent on the external force and applied magnetic field. At a certain magnetic field the force signal can be converted to a resistance signal. This shows the potential for using MREs as the key components in a force sensor.

References

1. Kchit, N. and Bossis, G., 2009. Journal of Physics D-Applied Physics, 42(10): p. 5505-5505.
2. Wang, X.J., Gordaninejad, F., Calgar, M., Liu, Y.M., Sutrisno, J. and Fuchs, A., 2009. Journal of Mechanical Design, 131(9): p. 6.
3. Bica, I., 2009. Materials Letters, 63(26): p. 2230-2232.
4. Li, W.H., Kostidis, K., Zhang, X.Z., Zhou, Y. and Ieee. 2009. in 2009 Ieee/Asme International Conference on Advanced Intelligent Mechatronics, Vols 1-3. New York: Ieee.
5. Bica, I., 2010. Journal of Industrial and Engineering Chemistry, 16(3): p. 359-363.

II. MATERIALS TECHNOLOGY

THE COMPOSITION ANALYSIS OF THE POLAR-MOLECULE TYPE ELECTRORHEOLOGICAL FLUIDS

S. H. YANG, X. GAO, C. X. LI, Q. WANG*, X. J. NIU, G. SUN and K. Q. LU

*Beijing National Laboratory for Condensed Matter Physics
and Key Laboratory of Soft Matter Physics, Institute of Physics,
Chinese Academy of Sciences, Beijing 100190, China
E-mail: qwang@iphy.ac.cn

The yield stress of the electrorheological fluid based on the powder of the precursor of $CaTiO_3$, the main component of which is calcium oxalate monohydrate, has being measured for different treating temperature. It is found that the yield stress decreases dramatically as the treating temperature increases to 160 °C. The components of the vaporized materials during heating process are analyzed by thermogravimetry-mass spectrum method and also thermogravimetry-infrared spectrum method. Both of the methods show that the vaporized material is almost water before the temperature rising to 200 °C. These results suggest that the water adsorbed in the dispersed particles plays a decisive role in this type of electrorheological fluids.

Keywords: Electrorheological effect; polar molecule; water.

1. Introduction

In 2003, a new electrorheological (ER) fluid with extremely high yield stress, which is known as giant electrorheological (GER) fluid, has been reported [1]. The yield stress of GER fluid can reach about 200 kPa, which value is about two orders higher than the upper limit predicted by pure dielectric theory [2]. Since then, a series of GER fluids made by the powders of the precursor of titanate, e.g., $CaTiO_3$, $SrTiO_3$, $LaTiO_3$ or even TiO_2, have been synthesized [3-7]. Among them, the ER fluid based on the powders of the precursor of $CaTiO_3$ (referred as CTO) shows the best comprehensive properties (e.g., high yield stress, low current density, high production etc.).

Unlike the dielectric ER fluids, the GER effect is not only determined by the dielectric properties of the dispersed and dispersing phase, but also strongly concerned with the additives adsorbed in the dispersed particles.

260

However, up to now, neither the key component of the additives nor the mechanism for the GER effect is known clearly. In this paper, the composition analysis was carried out for the CTO powders to determine the key component of the additives that is responsible to the GER effect.

2. Yield Stress with Different Drying Temperature

The CTO powder is synthesized by co-precipitation method from tetrabutyl titanate, anhydrous calcium chloride and oxalic acid [8]. It is a well-known method for producing $CaTiO_3$. However, for producing CTO, the obtained white precipitation is dried at much lower temperature. For producing $CaTiO_3$, the precipitation is needed to be dried at the temperature higher than 800 °C, while for producing CTO, the standard drying process is that heating the powders firstly to 50 °C for 24 hours and then to 120 °C for 2 hours. It is found that the final products seriously depend on the

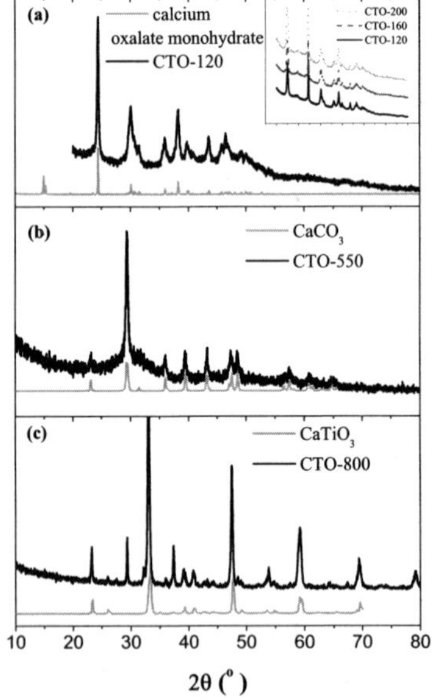

Fig. 1. X-ray diffraction pattern for the CTO powders finally dried at (a) 120 °C, (b) 550 °C and (c) 800 °C. The standard X-ray diffraction pattern for $CaC_2O_4 \bullet H_2O$, $CaCO_3$ and $CaTiO_3$ are also plotted for comparison. The inset show X-ray diffraction pattern for the CTO powders finally dried at 120 °C, 160 °C and 200 °C, respectively.

drying process. Figure 1 is the X-ray diffraction pattern for the final products by use of 120 °C, 550 °C and 800 °C second drying process (referred as CTO-120 or simply CTO, CTO-550 and CTO-800, respectively). It is found that the observable crystal structure for CTO-120 is that of calcium oxalate monohydrate $(CaC_2O_4 \bullet H_2O)$ together with some amorphous compounds [see Fig. 1(a)]. For CTO-550 and CTO-800, the observable crystal structures change to that of $CaCO_3$ [Fig. 1(b)] and $CaTiO_3$ [Fig. 1(c)], respectively. The CTO based GER fluid is made by mixing the CTO-120 powders with 10# silicon oil, which shows strong GER effect [see Fig. 2(a)].

For CTO-550 and CTO-800, the material and its structure change a lot from that of CTO-120, so the dielectric properties may also change. It is also found that the ER fluids made by them do not show any GER effect. To analyze the CTO based GER fluids, we restrict in changing the final drying temperature in the range between 120–200 °C, in which the material and its structure observed by X-ray diffraction is almost unchanged [see the inset of Fig. 1(a)]. However, the yield stress and the current density for the ER fluids made by them change seriously (see Fig. 2). As illustrated in Fig. 2(a), the yield stress of the ER fluids made by CTO-120 is 77 kPa at 5 kV/mm, while it decreases to 29 kPa for that made by CTO-160 and

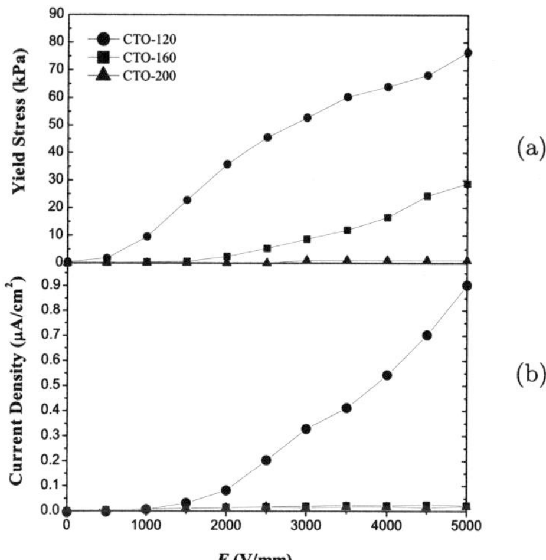

Fig. 2. Yield stress (a) and current density (b) as a function of electric field for the ER fluids made by the CTO powders dried at different temperatures.

further decreases to only about 1 kPa for that made by CTO-200. The current density for the ER fluids made by CTO-160 and CTO-200 is only about 0.02 μA/cm^2 at 5 kV/mm, much lower than that of the ER fluids made by CTO-120 (0.90 μA/cm^2 at 5 kV/mm) [see Fig. 2(b)].

Above experiments imply that at least one necessary factor for the GER effect is lost when we change the final drying temperature from 120 °C to 200 °C. The variation of the dielectric constant of CTO can be excluded, since the X-ray diffraction shows that they contain similar materials and have similar structure. The only possible factor should be focused to the changes in the additives adsorbed in the CTO particles.

3. Components Lost in Heating Process

The CTO powders are synthesized at relatively lower temperature, so many additives will remain. In general, these additives are adsorbed in the CTO particles and are sensitive to the environment. These are verified by the variation of the yield stress with the final drying temperature in the last section. In this section, we heat the CTO-120 powders continuously and try to find the component of the vaporized materials.

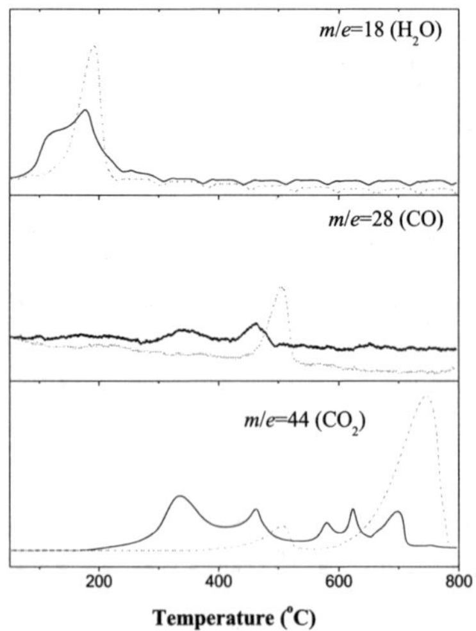

Fig. 3.   Mass spectrum of CTO-120 (solid line) and high purity CaC$_2$O$_4$•H$_2$O (dash line).

Firstly, thermogravimetry-mass spectrum (TG-MS) method is used to measure the CTO-120 powders with heating rate 5 °C/min. For comparison, high purity calcium oxalate monohydrate, which is the main material of the CTO-120 powders observed by X-ray diffraction, is also measured. The synchronous mass spectra are shown in Fig. 3. From Fig. 3(a), we can see that the vaporized materials are almost water vapor in low temperature range (below 200 °C). From the shape of the mass spectrum of CTO-120, we can imagining that the spectrum is composed of two peaks in between 100 °C and 200 °C, one starts at about 100 °C and the other starts at 160 °C. The peak starts at 100 °C should correspond to the vaporization of free water, and the peak starts at 160 °C correspond to the vaporization of crystallized water in calcium oxalate monohydrate. This can be verified by that the position of the second peak of the mass spectrum of water for CTO-120 is almost coincided with that of the single peak of calcium oxalate monohydrate. It is known that for calcium oxalate monohydrate, the crystallized water start to vaporize at about 160 °C and then calcium oxalate monohydrate becomes calcium oxalate after most of the crystallized water is lost. Above 200 °C, CO and CO_2 molecules can be found, but there is almost no similarity of the mass spectrum between CTO-120 and calcium oxalate monohydrate because the basic materials contained in CTO-120 is changed from calcium oxalate monohydrate. It should be stressed that the vaporized materials above 200 °C are not related to the GER effect, so the only material that can be attributed to the GER effect is water.

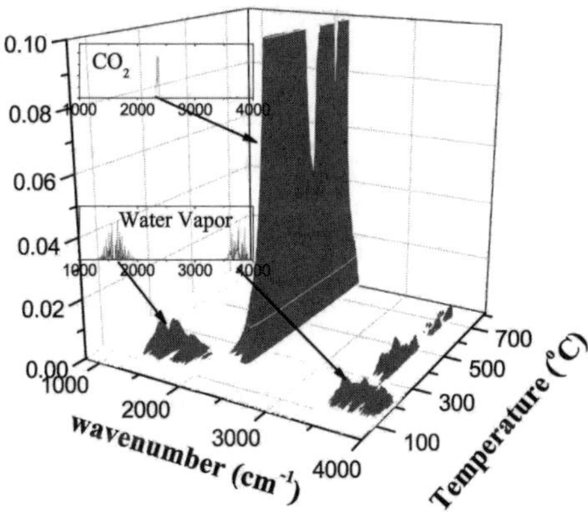

Fig. 4.   TG-IR curves for CTO-120 as function of both temperature and wave number. The insets show standard IR spectrum for water vapor and CO_2, respectively.

Thermogravimetry-Infrared spectrum (TG-IR) method is also used to measure CTO-120 powders with the same heating rate used in TG-MS method. The synchronous IR spectra are shown in Fig. 4. Figure 4 shows more clearly that the vaporized materials before CTO-120 being heated to 200 °C are almost water vapor, and after 200 °C, a large amount of CO_2 is vaporized.

Now, we have found that the GER effect decreases prominently with the vaporization of water adsorbed in the CTO particles, especially with the vaporization of the crystallized water. Since the crystallized water in calcium oxalate monohydrate is stable below 160 °C, it is expected that the CTO based ER fluid can retain its high yield stress up to 160 °C. For the temperature higher than 160 °C, the high yield stress property disappears quickly with the vaporization of the crystallized water.

4. Conclusions

The CTO based GER fluid is synthesized by different treating temperature. It is show that the GER effect decreases seriously as the treating temperature increases and vanishes completely when the treating temperature increases to 200 °C. To analyze the components of the additives adsorbed in CTO, TG-MS and TG-IR methods have been used. The results show that the vaporized materials are almost water vapor when the heating temperature is below 200 °C. This suggests that water molecules, especially the crystallized water in calcium oxalate monohydrate, are responsible to the GER effect.

Acknowledgments

This work is supported by the Key Item of Knowledge Innovation Project of Chinese Academy of Sciences (Grant No. KJCX2-YW-M07), the National Basic Research Program of China (Grant No. 2009CB930800) and National Natural Science Foundation of China (Grant Nos. 10674157 and 10875166).

References

1. W. J. Wen, X. X. Huang, S. H. Yang, K. Q. Lu and P. Sheng, *Nature Materials* **2**, 727 (2003).
2. H. R. Ma, W. J. Wen, W. Y. Tam and P. Sheng, *Adv. Phys.* **52**, 343 (2003).
3. K. Q. Lu, R. Shen, X. Z. Wang, G. Sun, W. J. Wen and J. X. Liu, *Chin. Phys.* **15**, 2476 (2006).
4. R. Shen, X. Z. Wang, Y. Lu, D. Wang, G. Sun, Z. X. Cao and K. Q. Lu, *Adv. Mater.* **21**, 4631 (2009).

5. J. G. Cao, M. Shen and L. W. Zhou, *J. Solid State Chem.* **179**, 1565 (2006).
6. Y. C. Cheng, J. J. Guo, G. J. Xu, P. Cui, X. H. Liu, F. H. Liu and J. H. Wu, *Colloid. Polym. Sci.* **286**, 1493 (2008).
7. Y. C. Cheng, X. H. Liu, J. J. Guo, F. H. Liu, Z. X. Li, G. J. Xu and P. Cui, *Nanotechnology* **20**, 055604 (2009).
8. K. Q. Lu, R. Shen and X. Z. Wang, *WO patent*, application no. 2007147347.
9. X. Z. Wang, R. Shen, D. Wang, Y. Lu and K. Q. Lu, *Mater. Des.* **30**, 4521 (2009).
10. K. Q. Lu, R. Shen and X. Z. Wang, *WO patent*, application no. 2007147348.

LANTHANUM TITANATE NANOPARTICLES ER FLUIDS WITH HIGH PERFORMANCE[*]

DE WANG[1,2], RONG SHEN[1†], SHIQIANG WEI[2] and KUNQUAN LU[1]

[1]*Beijing National Laboratory for Condensed Matter Physics and Key Laboratory of Soft Matter Physics, Institute of Physics, Chinese Academy of Sciences, Beijing 100190, China*

[2]*National Synchrotron Radiation Laboratory, University of Science and Technology of China, Hefei 230029, China*

A new type of electrorheological (ER) fluid consisting of lanthanum titanate (LTO) nanoparticles is developed. The ER Fluids were prepared by suspending LTO powder in silicon oil and the particles are fabricated by wet chemical method. This ER fluid shows excellent ER properties: The static yield stress reaches over 150kPa under 5kV/mm with linear dependence on the applied DC electric field, and the current density is below 10μA/cm2. In order to investigate the affect factor on the ER behavior the LTO powder were heated under different temperatures. The ER performances of both two particles treated under different temperatures were compared and the properties of those particles were analyzed with TG-FTIR technique. It was found that the static yield stress of the suspensions fell from over 150kPa to about 40kPa and the current densities decreased prominently as the rise of the heating temperature. TG-FTIR analysis indicated that organic groups remained in the particles such as alkyl group, hydroxyl group and carbonyl group and etc, may contribute to the ER effect significantly. The experimental results are helpful to understand the mechanism of the high ER effect and synthesize better ER material.

1. Introduction

Electrorheological fluid is a kind of novel intelligence materials consisting of micron or nanometer scale solid particles with high dielectric property and insulated oil. Under an external DC electric field, the apparent viscosity of ER fluids will rise rapidly and reversibly [1-4], and transforms form liquid-like state to solid-like due to chain structure of polarized particles. The ER fluids will

---

[*]This work is supported by the National Natural Science Foundation of China Grant 10674156, the National Basic Research Program of China Grant 2004CB619005 and 2009CB930800, the Knowledge Innovation Project of Chinese Academy of Sciences Grant KJCX2-YW-M07, and the Instrument Developing Project of the Chinese Academy of Sciences，Grant No. YZ200758.
[†]Rong Shen, E-mail: rshen@aphy.iphy.ac.cn, De Wang, E-mail: manks@mail.ustc.edu.cn

become back the low viscosity state when the electric field is removed. Because of the continuous controllable and reversible ER properties, the ER fluids have attracted a lot of attentions [5-7].

To satisfy the requirement of the application for an ER fluid the shear stress should be high enough and the leaking current density should be low under an electric field, while the viscosity must be low at zero field. However, the yield stress of most ER fluids was only about several kilopascals in the past half century from 1940s to 1990s, which could not meet the demands for the applications. A mass of ER materials including organic and inorganic particles [6-11] were created in past two decades. Especially after the giant ER effect was discovered. A large number of polar molecule dominated ER materials were developed [12, 13]. The yield stress of which has far broken the limit of the traditional ER fluids, and some of which reached over 200kPa [7, 12, 13]. In a polar molecule ER fluid, the interaction of the polar molecule and polarized particle is greatly higher than that between the neighboring particles [13, 14]. So adding polar molecules or polar group could highly enhance the ER performance.

In this article, we report a new type of ER fluids with suspending La-Ti-O (LTO) based powder in silicon oil, and it illustrates marvelous properties: high shear stress, low current density and long-time stability.

2. Experiment

Nanometer scale LTO powders were prepared by a simple co-precipitation procedure [15]. Lanthanum chloride was dissolved in deionized water, and oxalic acid and tetrabutyl titanate were suspended in absolute alcohol. Then the water solution was added into alcohol solution drop by drop with strong stirring under a constant temperature. White precipitation appeared immediately in the mixed solution when the lanthanum chloride solution was dropped in. The suspension was aged in a water bath for 4 hours, and then the precipitation was filtrated, washed, and dried in a drying oven at 50°C for 2 days and at 80°C or 120°C for 2 hours. The morphology and size of the particles were detected by a field emission scanning electron microscope. The suspension was prepared by dispersing dried powders in silicon oil. The static yield stress and electric current density were measured by using a rheometer fabricated by our group. The TG-FTIR analysis was carried out to observe the influence of the drying temperature of the LTO nanoparticles.

3. Results and Discussion

Figure 1 illustrates the SEM photograph of LTO particles, from which we can see that most particles are approximate spherical grain, and their grain size is less than 100 nanometers.

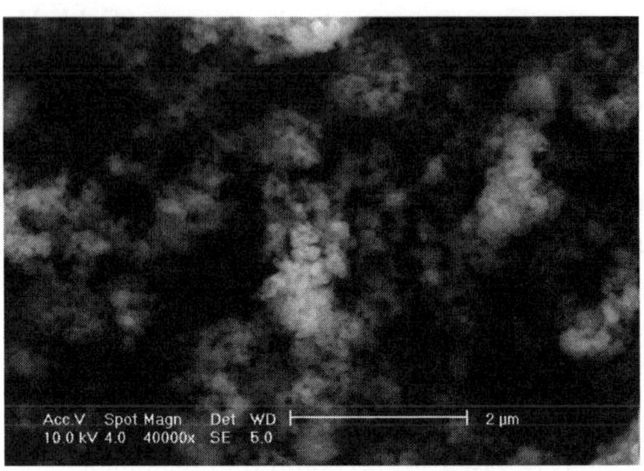

Figure 1. SEM photographs of LTO nanoparicles.

All the electrorheological data was collected with a at room temperature. Figure 2 shows the static yield stress of the LTO/silicon oil ER fluids, of which the particles were dried at 80°C and 120°C respectively. The yield stress of both sample rises lineally with the increase of the applied DC electric field. The sample using the powders dried at 80°C exhibits higher ER performance than another sample, of which the yield stress can reach about 170kPa at 5000V/mm electric field. The ER effect is much lower, about 40kPa at 5000V/mm, for the sample dried at 120°C. This means that the drying temperature rises from 80°C to 120°C, the ER effect decreases with the temperature increasing. However the electric current density of these two samples shown in the inset of Figure 2 reveals an opposite behavior, which increases with the drying temperature increasing. For instance the current densities are 0.4μA/cm^2 and 6μA/cm^2 for the samples dried at 80°C and 120°C at 5000V/mm respectively. There is nearly one order difference on the current density for the ER fluids consisted of the particles treated at different temperatures, although all current densities follow an exponential dependence on the electric field.

The high yield stress and its linear dependence on the electric field, as well as the non Ohmic behavior of the current density indicate that LTO based ER fluid does not belong the conventional dielectric ER fluid [16] but the polar molecule dominated ER (PM-ER) fluid. The big difference of the ER performance arising from the treating temperature of the particles indicates the

LTO powder's components and structure may appear a remarkable vibration with the temperature.

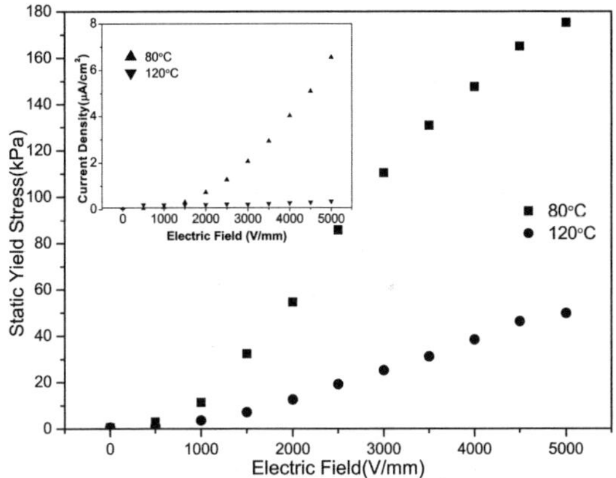

Figure 2. Yield stress of ER fluid for the LTO powder dried at 80°C and 120^{0}C with the volume fraction about 60%. The current densities of the samples are shown in the inset.

To study the influence of the drying temperature, TG-FTIR co-analysis was conducted [17]. The sample was put in thermo gravimetric analyzer, and the gases vaporized during the TG analysis process were transformed through a heated pipe to a FTIR spectrometer to get FTIR spectra of the decomposition products synchronously. Coupling the TG curve and IR we can identify the components of decomposition products at a specified temperature. Both particles mentioned above were sent for FG-FTIR test, and the results are shown in Figure 3 and Figure 4. As seen the TG curves illustrated in Fig. 3 when the temperature climbs over 100°C, the TG curve of 80°C dried powder declines rapider than that of 120°C dried one. Obviously, in between 100°C to 150°C a decomposition procedure occurs for the powder dried at 80°C and is also demonstrated by the simultaneous FTIR analysis showing a fast increase of the decomposition products among this temperature region. The FTIR spectra of decomposition products vaporized from the TG measurement at 120°C of the particles dried at 80°C was illustrated in Fig. 4 (a). Compared with the standard FTIR spectra of CO_2 and water which were shown in Fig. 4 (b) and (c), all the decomposition products between 100°C to 150°C are water and CO_2. Combining the TG and FTIR results, we think that lots of components with organic polar group such as alkyl group, hydroxyl group and carbonyl group were decomposed to form water and CO_2 during the heating process. Both the ER behavior and the TG-FTIR analysis prove that the high performance of the LTO based ER fluid comes from the organic polar groups in the particles. The lower yield stress for the ER fluid consisted of the particles dried at 120°C possibly is due to that some polar groups are decomposed during the drying procedure at higher temperature. With the decrease of polar groups, the ER effect falls down quickly, but the current density declines beneficially. Maybe a proper heating temperature may agree a balance in both properties.

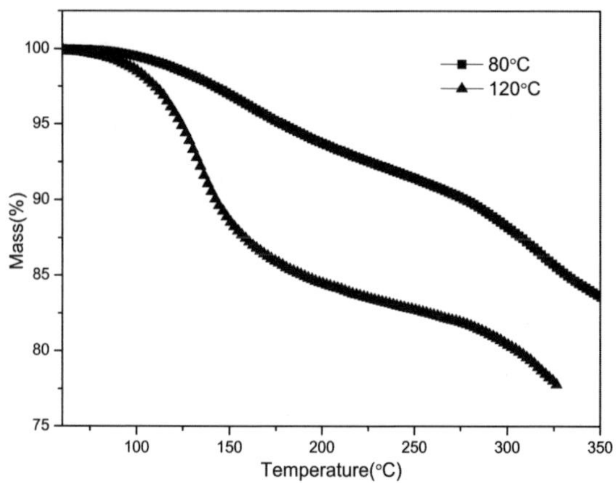

Figure 3. TG curve of 80°C and 120°C dried particles.

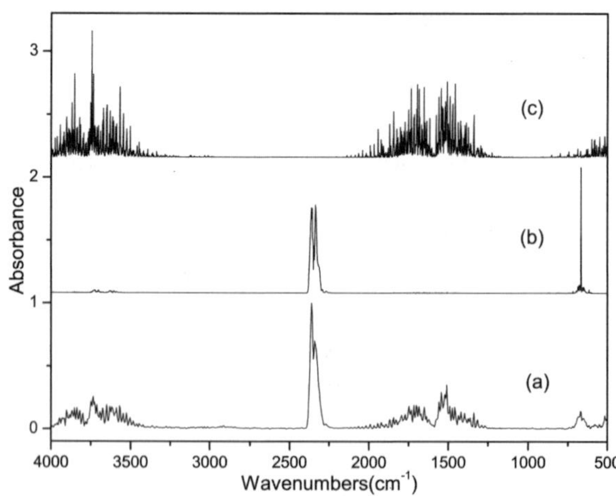

Figure 4. FTIR spectra: (a) Gases vaporized at 120°C of 80°C dried powder; (b) CO_2; (c) Water.

Also the hydraulic oil was used to replace the silicon oil, and the LTO/hydraulic oil suspension exhibited better zero viscosity and fluidity. A further experiment is needed to investigate the influence of the oil.

4. Conclusions

A novel LTO based ER fluids with high performance is developed through a simple convenient and cheap wet chemical method, and its ER properties could reach the demand of potential applications.

References

1. W. M. Winslow, *Journal of Applied Physics,* **20,** 1137 (1949)
2. T. C. Halsey and W. Toor, *Physical Review Letters,* **65,** 2820 (1990)
3. T. C. Halsey, *Science,* **258,** 761 (1992)
4. X. Q. Gong, J. B. Wu, X. X. Huang, W. J. Wen and P. Sheng, *Nanotechnology,* **19,** 165602 (2008)
5. D. R. Gamota and F. E. Filisko, *Journal of Rheology,* **35,** 1411 (1991)
6. X. P. Zhao, Q. Zhao and X. M. Gao, *Journal of Applied Physics,* **93,** 4309 (2003)
7. W. Wen, X. Huang and P. Sheng, *Applied Physics Letters,* **85,** 299 (2004)
8. F. E. Filisko and L. H. Radzilowski, *Journal of Rheology,* **34,** 539 (1990)
9. W. J. Wen, W. Y. Tam and P. Sheng, *Journal of Materials Science Letters,* **17,** 419 (1998)
10. C. Wei, Y. H. Zhu, Y. Jin, X. L. Yang and C. Z. Li, *Materials Research Bulletin,* **43,** 3263 (2008)
11. H. J. Choi and M. S. Jhon, *Soft Matter,* **5,** 1562 (2009)
12. W. J. Wen, X. X. Huang, S. H. Yang, K. Q. Lu and P. Sheng, *Nature Materials,* **2,** 727 (2003)
13. K. Q. Lu, R. Shen, X. Z. Wang, G. Sun, W. J. Wen and J. X. Liu, *Chinese Physics,* **15,** 2476 (2006)
14. R. Shen, X. Z. Wang, Y. Lu, D. Wang, G. Sun, Z. X. Cao and K. Q. Lu, *Advanced Materials,* **21,** 4631 (2009)
15. X. Z. Wang, S. Rong, W. J. Wen and K. Q. Lu, *International Journal of Modern Physics,* **B19,** 1110 (2005)
16. H. Ma, W. Wen, W. Y. Tam and P. Sheng, *Physical Review Letters,* **77,** 2499 (1996)
17. S. Materazzi, *Applied Spectroscopy Reviews,* **32,** 385 (1997)

FLEXIBLE FERROMAGNETIC FILAMENTS AS ARTIFICIAL CILIA

A. CĒBERS* and R. LIVANOVIČS

University of Latvia, Zeļļu-8, Rīga, LV-1002, Latvia
E-mail: aceb@tesla.sal.lv

The model of an artificial cilia as a flexible ferromagnetic filament in a rotating magnetic field is proposed. Numerical algorithm for the simulation of its behavior is developed and the characteristic shapes of the filament with one fixed end under the action of a rotating field are found. It is concluded that ferromagnetic filaments may be used as mixers in microfluidics.

Keywords: Ferromagnetic filament; Rotating field; Mixer; Artificial cilia.

1. Introduction

Recently synthesized ferromagnetic filaments[1] enable the creation of different microdevices driven by magnetic field. It is possible to create self-propelling devices which develop thrust due to periodic formation of loops in the AC magnetic field.[2] Magnetic filaments created by linking superparamagnetic beads and driven by rotating field may be used as mixers in microfluidics.[3,4] Another possibility is the creation of arrays of magnetic filaments which mimic the surface of ciliated microorganisms for the enhancement of mass transport. Nanorods arrays of PDMS-ferrofluid composite materials are synthesized and their behavior under the action of rotating field investigated in.[5] Fixed arrays of chains of superparamagnetic particles in the rotating field are experimentally investigated in[6] and behavioral similarity with ciliated microorganisms is demonstrated. Here basing on developed theoretical models[7,8] and numerical algorithms we investigate the motion of the ferromagnetic filament in rotating field attached by one end to solid wall.

2. Model and Numerical Method

The theoretical model for the dynamics of the ferromagnetic filament is developed in[7,8] and is based on the expressions for the force $\vec{F}$ and torque

$\vec{T}$ in the cross-section of the filament. It is assumed that magnetization is along the tangent vector of the filament. In the Frenet frame the relations for $\vec{F}$ and $\vec{T}$ read

$$\vec{F} = A\frac{d}{dl}\left(\frac{1}{R}\right)\vec{n} + A\frac{1}{R}\tau\vec{b} + \left(\Lambda - \frac{1}{2}A\frac{1}{R^2}\right)\vec{t} - M(\vec{H} - \vec{t}(\vec{t}\cdot\vec{H})) \; ; \quad (1)$$

$$\vec{T} = -A\frac{1}{R}\vec{b} \, , \quad (2)$$

where $\vec{t}$ is the tangent vector to the center line of the filament, $\vec{n}, \vec{b}$ are the normal and binormal of the center line respectively, l is the arclength of the center line, R is the radius of the curvature of the center line and τ its torsion. Λ characterizes tension arising due to the inextensibility of the filament, A is the bending modulus, $M\vec{t}$ is the magnetization per unit length of the filament, $\vec{H}$ is the magnetic field strength. The contribution of the twist is neglected in (1) and (2).

Using the Frenet equations $d\vec{t}/dl = -\vec{n}/R$ and $d\vec{n}/dl = \vec{t}/R + \tau\vec{b}$ the relation (1) may be rewritten as follows

$$\vec{F} = -A\frac{d^3\vec{r}}{dl^3} - M\vec{H} + \tilde{\Lambda}\vec{t}, \quad (3)$$

where $\tilde{\Lambda} = \Lambda + M\vec{t}\cdot\vec{H} - 3A/2R^2$. Tildes further are omitted.

The equation of motion in the Rouse approximation reads

$$\zeta\vec{v} = \frac{d\vec{F}}{dl} \, . \quad (4)$$

Since the magnetic field is homogeneous then representing the filament with one fixed and clamped end by $p+3$ marker points at the distance h from each other the equation (4) approximating the tangent vector by finite differences in discretized version reads

$$\zeta h\vec{v}_i = \vec{f}_i + \Lambda_i(\vec{r}_{i+1} - \vec{r}_i)/h - \Lambda_{i-1}(\vec{r}_i - \vec{r}_{i-1})/h \; (i = 1,...,p) \quad (5)$$

and for the free end

$$\zeta h\vec{v}_{p+1} = \vec{f}_{p+1} - \Lambda_p(\vec{r}_{p+1} - \vec{r}_p)/h + \vec{f}^m_{p+1} \quad (6)$$

Here

$$\vec{f}_i = -Ah\frac{d^4\vec{r}}{dl^4}\,|_i; \; (i = 1,...,p); \; \vec{f}_{p+1} = A\frac{d^3\vec{r}}{dl^3}\,|_{p+1} + \vec{f}^m_{p+1} \quad (7)$$

and the first two markers are fixed ($\vec{r}_{-1} = 0$; $\vec{r}_0 = h\vec{e}_z$). Due to inextensibility of the filament $p+1$ constraints are satisfied

$$g_k = (\vec{r}_{k+1} - \vec{r}_k)^2 = h^2, \; (k = 0,...,p) \quad (8)$$

As a result introducing $p + 1 \times 3(p + 1)$ matrix $J = \partial g_k / \partial \vec{r}_i$ equations (5) and (6) are rewritten in vectorial forms as follows (the upper index ^{T} denotes transposed matrix)

$$\zeta h \vec{v} = \vec{f} - J^T \cdot \Lambda (2h)^{-1} . \tag{9}$$

Since $J \cdot \vec{v} = 0$ we have

$$\Lambda (2h)^{-1} = (J \cdot J^T)^{-1} \cdot J \cdot \vec{f}$$

and equation of motion reads (I is the unit matrix)

$$\zeta h \vec{v} = (I - J^T \cdot (J \cdot J^T)^{-1} \cdot J) \cdot \vec{f} . \tag{10}$$

Further notation of the projection operator is introduced $P = I - J^T \cdot (J \cdot J^T)^{-1} \cdot J$. Approximating force $\vec{f}$ by central differences and taking into account that $\vec{r}_{-1} = 0$; $\vec{r}_0 = h \vec{e}_z$ for the force $\vec{f}$ we have $\vec{f} = \vec{f}^e + \vec{f}^c$, where $\vec{f}^e$ is elastic force due to the bending of the filament and $\vec{f}^c$ is a reaction force due to the fixed end. The expressions for $\vec{f}^e$ and $\vec{f}^c$ are as follows

$$\vec{f}_i^e = -(\vec{r}_{i+2} - 4\vec{r}_{i+1} - 4\vec{r}_{i-1} + \vec{r}_{i-2} + 6\vec{r}_i)/h^3; \quad \vec{f}_i^c = 0; \quad (i = 3, ..., p) \tag{11}$$

$$\vec{f}_1^e = -(\vec{r}_3 - 4\vec{r}_2 + 6\vec{r}_1)/h^3 ;$$

$$\vec{f}_2^e = -(\vec{r}_4 - 4\vec{r}_3 + 6\vec{r}_2 - 4\vec{r}_1)/h^3 ;$$

$$\vec{f}_1^c = 4\vec{e}_z/h^2 ;$$

$$\vec{f}_2^c = -\vec{e}_z/h^2 ;$$

$$\vec{f}_{p+1}^e = -(\vec{r}_{p-1} + \vec{r}_{p+1} - 2\vec{r}_p)/h^2 ;$$

$$\vec{f}_{p+1}^c = 0 .$$

Introducing $3(p + 1) \times 3(p + 1)$ stiffness matrix C we have $\vec{f}^e = C \cdot \vec{r}$.

As a result implicit scheme for time evolution of the coordinates of the marker points reads

$$\vec{r}_{t+\tau} - \vec{r}_t = \tau P \cdot C \cdot (\vec{r}_{t+\tau} - \vec{r}_t) + \tau P \cdot (C \cdot \vec{r}_t + \vec{f}^c + \vec{f}^m) \tag{12}$$

and the coordinates of the marker points at the next time step are obtained by inversion of the matrix $I - \tau P \cdot C$ as follows

$$\vec{r}_{t+\tau} = \vec{r}_t + \tau (I - \tau P \cdot C)^{-1} \cdot P(C \cdot \vec{r}_t + \vec{f}^c + \vec{f}^m) \tag{13}$$

Equations are put in dimensionless form introducing the following scales,- for length the length of the filament L, for time the characteristic elastic relaxation time $\tau_e = \zeta L^4 / A$, and elastic force A/L^3. The dynamics of the filament in the rotating field is controlled by two parameters: the magnetoelastic number $Cm = MHL^2/A$ and $\omega \tau_e$.

If the fixed end of the filament is clamped in z axis direction for small perturbation of the filament in x, y directions we have equations

$$\frac{\partial x}{\partial t} = -\frac{\partial^4 x}{\partial l^4}; \quad \frac{\partial y}{\partial t} = -\frac{\partial^4 y}{\partial l^4} \tag{14}$$

at following boundary conditions $x(0,t) = y(0,t) = 0$; $\partial x/\partial l(0,t) = \partial y/\partial l(0,t) = 0$; $\partial^2 x/\partial l^2(1,t) = \partial^2 y/\partial l^2(1,t) = 0$ and $\partial^3 x/\partial l^3(1,t) = -Cm\cos(\omega t)$; $\partial^3 y/\partial l^3(1,t) = -Cm\sin(\omega t)$, where the dimensionless frequency of rotating in x, y plane field $\tilde{\omega}$ is determined according to $\tilde{\omega} = \omega \tau_e$ (tildas further are omitted).

Solution of the boundary problem in a steady case reads

$$\xi = x + iy = \sum_{k=0}^{3} A_k \exp(\alpha_k l) \tag{15}$$

where $\alpha_k = \omega^{1/4} \exp(-i\pi/4 + k\pi i/2)$, $(k = 0, ..., 3)$ and A_k is found from the solution of the set of linear algebraic equations.

3. Numerical Simulation Results

Magnetic field rotating in x, y plane induces the bending of the filament which by adjusting the viscous and magnetic torques rotates with the frequency of the field. If Cm values are small enough the shapes found numerically may be compared with the analytical solution (15). In Fig. 1 for $Cm = 10$ and $\omega = 400$ it is shown that the agreement of the shapes found numerically with those calculated analytically in the framework of the linear theory approach is quite good. Sequence of the three dimensional configurations of the filament for one period of the rotating field is shown in Fig. 2 (projection on x, y plane) and Fig. 3 (projection on x, z plane).

It should be remarked that approach to steady regime occurs by transient oscillatory regime. This is illustrated by Fig. 4 where the maximal curvature of the filament calculated by formula $| \, d\vec{r}/dl \times d^2\vec{r}/dl^2 \, |$ at $Cm = 288$ and $\omega = 4800$ is shown. Curvature is calculated by interpolating the coordinates of the marker points by cubic splines. Configurations of the filament close to steady state for those values of the parameters in projection on x, y plane for one period of rotating field are shown in Fig. 5. By comparison Fig. 2 and Fig. 5 we see that by increasing the value of magnetoelastic number the length of the bent part of the filament increases. As an interesting feature of these simulations we should remark that filament adjust the length of its bent part to achieve the synchronous with rotating field regime.

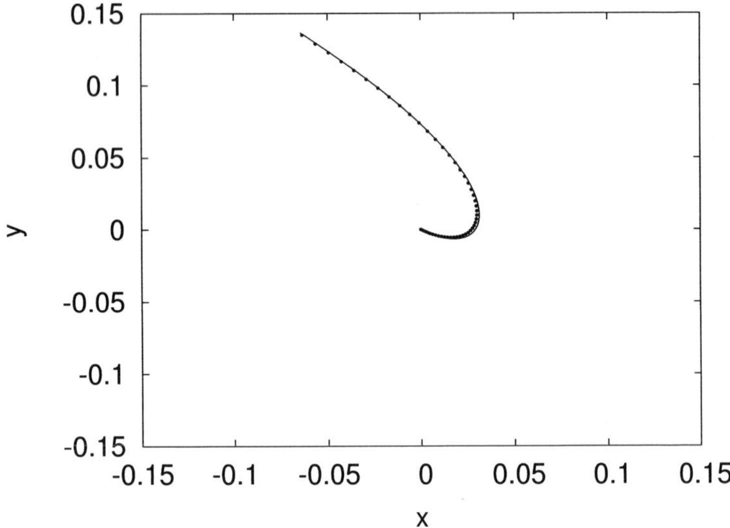

Fig. 1.   Comparison of numerical calculation (solid circles) and analytical solution (solid line). $Cm = 10$ and $\omega = 400$. Numerical configuration is obtained for $p = 300$ from initial configuration with time step $\tau = 10^{-4}$.

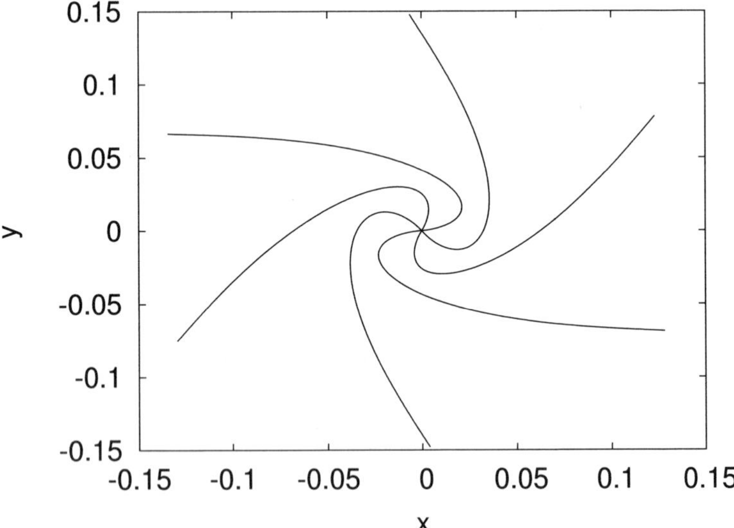

Fig. 2.   Configurations of filament for one period of rotating field in projection on x, y plane. Time interval between subsequent configurations is 0.0026 in dimensionless units. $Cm = 10$ and $\omega = 400$.

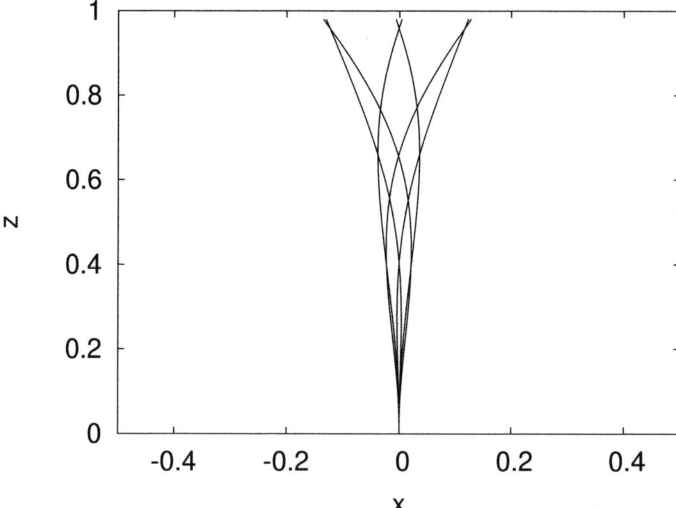

Fig. 3. Configurations of filament for one period of rotating field in projection on x, z plane. Time interval between subsequent configurations is 0.0026 in dimensionless units. $Cm = 10$ and $\omega = 400$.

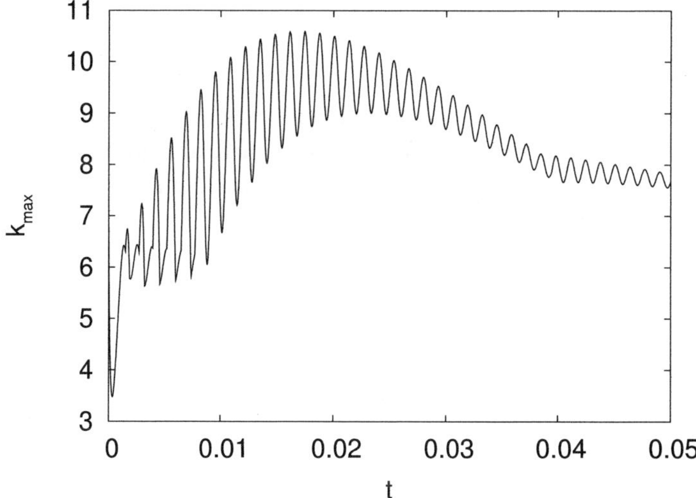

Fig. 4. Maximal curvature of filament in transient oscillatory regime. $Cm = 288$ and $\omega = 4800$.

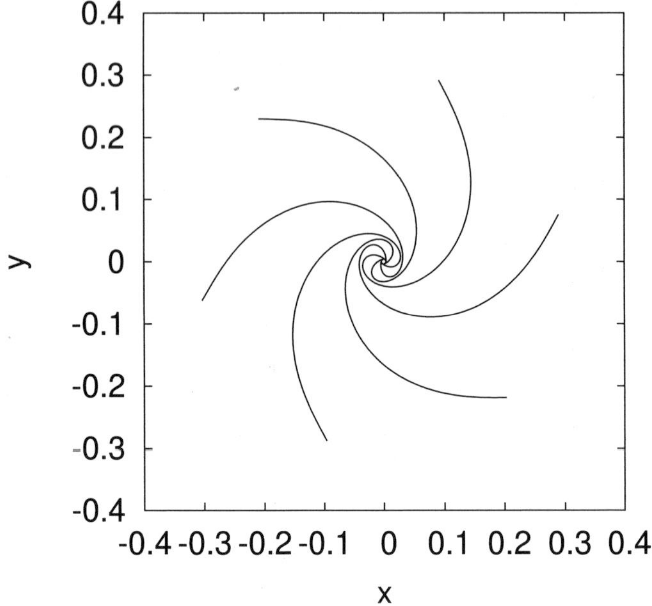

Fig. 5. Shapes of filament in projection on x, y plane for one period of rotating field. Subsequent configurations are at time interval $2.2 \cdot 10^{-4}$ in dimensionless units. $Cm = 288$ and $\omega = 4800$.

4. Conclusions

It is shown that flexible ferromagnetic filaments may be used as the mixers for microfluidics. Proposed model and numerical algorithm allow us to calculate the shape of the filament in dependence on the magnetoelastic number and the frequency of the rotating field. For large frequency of the rotating field transient oscillatory regime of the filament motion is found.

Acknowledgments

The work is supported by the grant of University of Latvia ESS2009/86.

References

1. K. Ērglis, M. Belovs, and A. Cēbers, *JMMM* **321** 650 (2009).
2. M. Belovs, and A. Cēbers, *Phys. Rev. E* **79** 051503 (2009) (*Erratum: Phys. Rev. E* **79** 069906(E) (2009)).
3. S. L. Biswal, and A. P. Gast, *Anal. Chem.* **76** 6448 (2004).
4. T. G. Kang, M. A. Hulsen, P. D. Anderson, J. M. J. den Toonder, and H. E. H. Meijer, *Phys. Rev. E* **76** 066303 (2007).

5. B. A. Evans, A. R. Shields, R. Lloyd Caroll, S. Washburn, M. R. Falvo, and R. Superfine, *Nano Letters* **7** 1428 (2007).
6. M. Vilfan, A. Potočnik, B. Kavčič, N. Osterman, I. Poberaj, A. Vilfan, and D. Babič, *PNAS* **107** 1844 (2010).
7. M. Belovs, and A. Cēbers, *Phys. Rev. E* **73** 051503 (2006).
8. A. Cēbers, *J. Phys.: Condens. Matter* **15** S1335 (2003).

SYNTHESIS OF TITANATE/POLYPYRROLE COMPOSITE ROD-LIKE PARTICLES AND THE ROLE OF THE CONDUCTING POLYMER ON ELECTRORHEOLOGICAL EFFICIENCY

MIROSLAV MRLIK, VLADIMIR PAVLINEK, QILIN CHENG and PETR SAHA

*Polymer Centre, Tomas Bata University in Zlin, TGM 275, 76272
Zlin, Czech Republic*

This study was aimed on the preparation of titanate/polypyrrole core-shell rod-like composite particles. The mere titanate rod-like particles were prepared as a core material and PPy was polymerized on their surface in different amounts. Rheological measurements showed that under an applied external electric field, shear stress of these materials significantly increased with amount of PPy in the shell layer. The yield stresses obtained from the Cho-Choi-Jhon model were correlated with dielectric properties of suspensions. Polarizability as a measure of particle polarization obtained from Havriliak-Negami model of dielectric spectra increases with the content of PPy in the samples. Furthermore, role of particle concentration and silicone oil viscosity was also investigated.

Keywords: Electrorheology, rods, conducting polymer

1. Introduction

Since an electrorheological (ER) phenomenon was discovered by Winslow in 1947[1], ER materials and their suspensions showed significant progress. ER suspensions are typically two-phase systems consisting of semiconducting particles randomly suspended in an insulating medium e.g. silicone oil. Such suspensions belong to the group called "smart materials" which includes materials whose properties can be influenced by certain external stimuli[2, 3]. In this case, stimulus is application of an external electric field. The response of the suspended particles in oil to the external electric field is in several milliseconds and appears as a reversible change between solid and fluid-like states. This transition is caused by dramatic change in suspension structure due to formation of chains or columns of polarized particles in the direction of intensity of electric field and, consequently, rheological properties (viscosity, yield stress) change as well.

The ER properties of the conducting polymers themselves and combination of the inorganic core and polymer shell have been investigated[4-7]. On the other hand the influence of amount of conducting material polymerized on the surface of the rod-like particles has not been studied in detail.

In the preceding study[8], the positive influence of the conducting polymer polymerized on the surface of titanate core on the ER properties in general has been confirmed. The presented investigation was aimed at the relationship between amount of conducting polypyrrole synthesized on the surface of titanate rods and electrorheological and dielectric properties at various viscosity of suspension medium.

2. Experimental

2.1. *Materials*

Ammonium persulfate (APS, $(NH_4)_2S_2O_8$, 98 %) and cetyltrimethyl ammonium bromide (CTAB) were purchased from Aldrich Chemicals company. TiO_2 powder (Degussa P25, Germany) and NaOH were used as received without further treatment. Silicone oils Lukosiol M200, viscosity η_c = 194 mPa.s, density d_c = 0.970 g cm^{-3}, relative permittivity ε' = 2.89, loss factor tg δ = 0.0001 and Lukosiol M15, viscosity η_c = 14.5 mPa.s, density d_c = 0.970 g cm^{-3}, relative permittivity ε' = 2.85, loss factor tg δ = 0.0001 were purchased from Chemical Works Kolin, Czech Republic. Pyrrole (Py, 98 %, Aldrich Chemicals, USA) was distilled twice under reduced pressure before the use.

2.2. *Synthesis of TNt and TNt/PPy composite particles*

TNt rods were synthesized via hydrothermal chemical process[9]. To create TNt/PPy composite particles, CTAB (1.845 g) was dispersed in 100 mL of distilled water. Solution was stirred for 30 minutes and after addition of 2 g of TNt nanorods suspension was sonicated for additional 20 minutes at room temperature. Then the mixture was transferred into the three-neck flask and cooled down to 0-5 °C with intensive stirring. Later, Py was added into the flask at different amounts. After few minutes, initiator APS was added into the mixture dropwise. Composition of sample particles is shown in Table 1. Mixture was kept at temperature between 0-5 °C for 8 hours and another 12 hours under room temperature. Then, black precipitate was filtered by distilled water. Conductivity of TNt/PPy particles was reduced by immersion in a fivefold molar excess of 1 M ammonium hydroxide for 5 hours. After this procedure, particles were filtered and dried at 80 °C until constant weight[10].

Table 1: Composition of the samples

Sample	TNt nanorods [g]	Pyrrole [ml]	APS [g]
S1	2	0	0
S2	2	1	3.32
S3	2	2	6.64
S4	2	4	13.28

2.3. *Preparation of suspensions*

Dried particles of all samples were dispersed mechanically with a glass rod in silicone oil in concentrations of 5, 10, 15 % (w/w). Before each ER measurement the suspensions were sonicated for 30 seconds.

2.4. *Determination of electrorheological properties*

Rheological measurements were performed under controlled shear rate mode (CSR) using a concentric cylinder viscometer (Bohlin Gemini, Malvern Instruments, UK). Suspensions were placed into the Couette cell with a rotating inner cylinder of 14 mm diameter and stationary outer cylinder separated by 0.7 mm gap. The instrument modified for ER experiments was connected with DC high-voltage source TREK (TREK 668B, USA) with electric field strength 0.5-3.0 kV mm^{-1}. Before each measurement at different electric field strengths, the built-up structures of particles were destroyed by shearing of the sample at shear rate 20 s^{-1} for 80 s. All experiments were performed at 25 °C.

2.5. *Determination of dielectric properties*

Frequency dependences (in the range of $10 - 10^5$ Hz) of relative permittivity and dielectric loss factor of 5 % (w/w) suspensions were carried out with Hioki (3522 RLC, Hi Tester, Japan) at room temperature.

3. Result and Discussion

3.1. *Formation of suspension structures in electric field*

The synthesis provided rod-like particles with lengths of tens of micrometers and hundreds of nanometers in diameter. At rest, in the absence of the external electric field, the dispersed rods were randomly oriented in silicone oil (Fig. 1a).

In the presence of electric field the polarized particles orient in direction of electric field strength and a chain-like structure was formed (Fig. 1b).

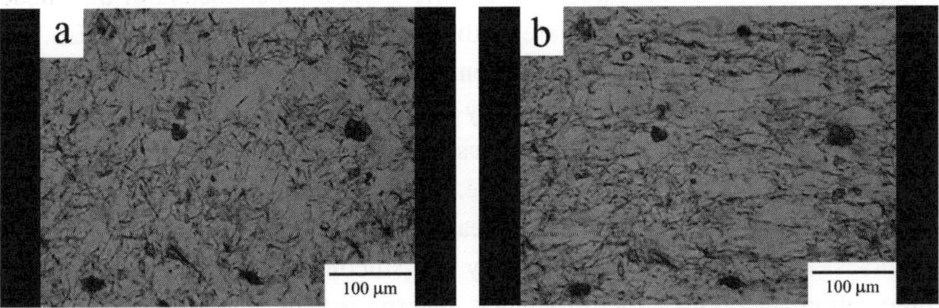

Fig. 1: The suspension of the TNt rod-like particles in the absence (a) and in the presence of electric field (b).

3.2. *Rheological properties*

In the absence of electric field suspensions of all particle samples (S1-S4) exhibit nearly Newtonian behavior, only slight deviation at higher PPy coating might be observed at lower shear rates (Fig. 2a). On the other hand, under applied external electric field, strong pseudoplastic behavior appeared, while the yield stress increased with amount of the PPy polymerized in the shell layer (Fig. 2b).

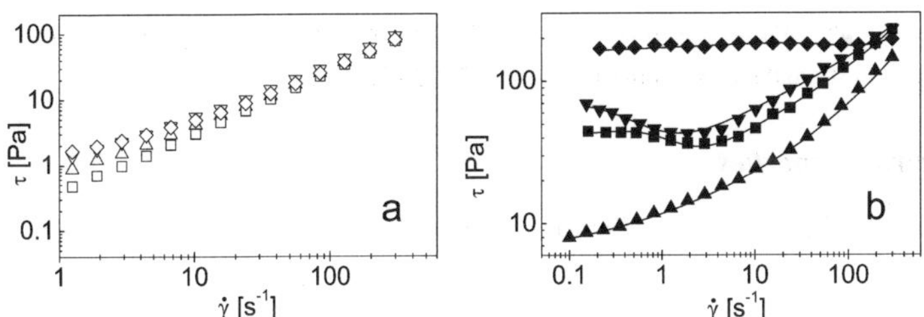

Fig. 2: The dependence of shear stress, τ, on shear rate g at 0 kV/mm (a) and 3 kV/mm (b) for 5 % (w/w) suspensions of samples S1 ($\triangle$, $\blacktriangle$), S2 ($\square$, $\blacksquare$), S3 ($\triangledown$, $\blacktriangledown$) and S4 ($\diamondsuit$, $\blacklozenge$) in silicone oil M200.

Rheological data in the Fig. 2b were fit with Cho-Choi-Jhon model[11] (eq. 1).

$$\tau = \frac{\tau_y}{1+(t_1\dot{\gamma})^\alpha} + \eta_\infty\left(1+\frac{1}{(t_2\dot{\gamma})^\beta}\right)\dot{\gamma} \qquad (1)$$

284

Where τ_y is yield stress, α is related to decrease in the shear stress, β is exponent and falls $0<\beta<1$, t_1 and t_2 are time constants, and η_∞ is the shear viscosity at a high shear rate in the absence of an electric field. Six parameters of this model were determined using least square method in Solver option of MS EXCEL. The most effective sample S4 was chosen for elucidation the effect of particle concentration and silicone oil viscosity on the ER intensity. It has been found that increasing concentration of particles increased viscosity of suspensions both in the absence, η_0, and in the presence, η_E, of electric field. Thus, the relative viscosity, η_E/η_0, of suspensions decreases significantly with concentration of particles (Fig. 3a). As the viscosity of silicone oil used for suspension preparation increases, ER efficiency determined by the relative viscosity decreases (Fig. 3b).

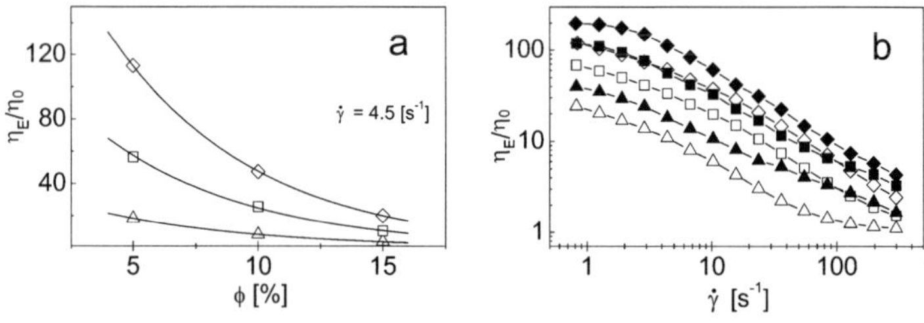

Fig. 3: The dependence of the relative viscosity, η_E/η_0, on the concentration of the sample S4 suspended in silicone oil M15 (a) and dependence of the relative viscosity, η_E/η_0, on the shear rate, g, of suspensions of the sample S4 in silicone oil M200 *(open symbols)* and M15 *(solid symbols)* at various electric field strengths E (kV/mm): ($\triangle$) 1, ($\square$) 2, ($\diamondsuit$) 3.

3.3. *Dielectric properties*

Dielectric spectra in Fig. 4 were fit with Havriliak-Negami (HN) model[12] (eq. 2),

$$\varepsilon_{HN}^*(\omega) = \varepsilon_\infty' + \frac{\Delta\varepsilon'}{\left(1 + (i\omega \cdot t_{rel})^a\right)^b} \tag{2}$$

Where, ε'_∞ is the high frequency dielectric constant, ε_s' is the static dielectric constant, $\Delta\varepsilon' = (\varepsilon_s' - \varepsilon'_\infty)$ is polarizability, ω the angular frequency ($\omega = 2\pi f$), t_{rel} is the relaxation time, a and b are shape parameters which describe the symmetric and asymmetric broadening of the dielectric function, $0<a$, $a.b<1$. The parameters a and b are related to the limiting behavior of the dielectric function at low and high frequencies. The parameters of this model were determined using least square method in Solver option of MS EXCEL and are shown in the Table 2.

Table 2: Parameters of Havriliak-Negami model

Parameters/sample suspensions	S1	S2	S3	S4
ε'_{∞}	2.58	2.68	2.80	2.74
$\Delta\varepsilon'$	0.79	1.45	1.49	1.77
t_{rel}	0.389	0.18	0.07	0.003
a	0.99	0.94	0.92	0.54
b	0.11	0.16	0.24	0.78
$\varepsilon'_{\mathrm{s}}$	3.37	4.14	4.29	4.52

It is clear that polarizability, $\Delta\varepsilon'$, increased with PPy content deposited on the surface of TNt rod-like particles (Table 2, Fig. 4a). Dielectric loss factor, ε'', exhibited local maxima and the position of the maxima was shifted to the higher frequencies due to increasing content of the PPy polymerized in the shell layer (Fig. 4b). Thus, higher polarizability, $\Delta\varepsilon'$, and shorter relaxation time, t_{rel}, were reflected in electrorheological performance of suspensions.

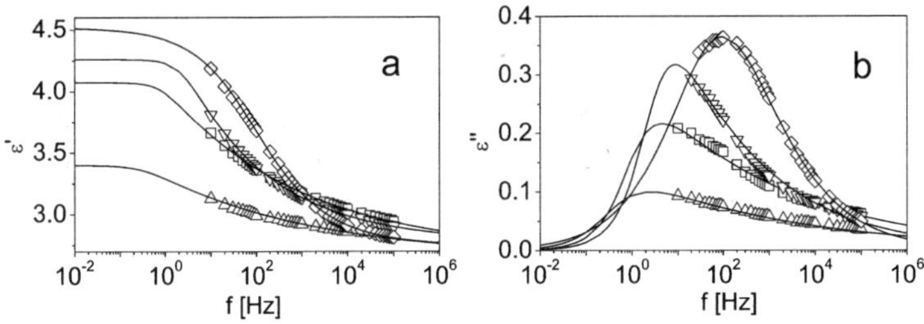

Fig. 4: The frequency spectra of (a) relative permittivity, ε', and (b) dielectric loss factor, ε'', of 5 % (w/w) suspensions of particles S1 ($\triangle$), S2 ($\square$), S3 (∇) and S4 ($\lozenge$) in silicone oil M200. Solid lines represent Havriliak-Negami model fit (eq. 2).

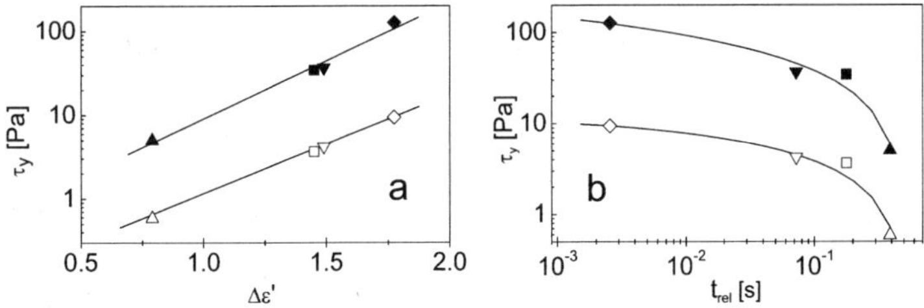

Fig. 5: The dependence between (a) yield stress, τ_y, and polarizability, $\Delta\varepsilon'$, (b) yield stress, τ_y, and relaxation time, t_{rel}, at two voltages 0.5 kV mm^{-1} *(open symbols)* and 3 kV mm^{-1} *(solid symbols)* for suspensions of samples S1 ($\triangle$), S2 ($\square$), S3 (∇) and S4 ($\lozenge$).

3.4. *Correlation between rheological and dielectric properties*

In order to correlate electrorheological and dielectric properties, yield stress, τ_y, obtained from eq. 2 as a function of $\Delta\varepsilon'$ and t_{rel} obtained from HN model were plotted (Fig. 5). Yield stress, τ_y, as a measure of rigidity of suspension structure under applied electric field, increases with increasing $\Delta\varepsilon'$ (Fig. 5a) and decreasing t_{rel} (Fig. 5b) which indicates that enhanced polarizability and shorter relaxation time caused by PPy coating positively influence ER performance.

4. Conclusion

The results showed that the amount of the PPy polymerized on the surface of the TNt rod-like particles significantly enhances particle polarizability and decreases relaxation time which is reflected in higher ER performance. The ER efficiency decreases with increasing particle concentration and is higher for suspensions in silicone oil with lower viscosity.

Acknowledgments

The authors acknowledge the financial support of the Ministry of Education, Youth and Sports of the Czech Republic (MSM 7088352101) and the Grant Agency of the Czech Republic (202/09/1626).

References

1. W. M. Winslow, *J. Appl. Phys.* **20**, 1137 (1949).
2. H. Block, and J. P. Kelly, *J. Phys. D-Appl. Phys.* **21**, 1661 (1988).
3. T. C. Jordan, and M. T. Shaw, *IEEE Trns. Electr. Insul.* **24**, 849 (1989).
4. F. F. Fang, B. M. Lee, and H. J. Choi, *Macromol. Res.* **18**, 99 (2010).
5. H. J. Choi, and M. S. Jhon, *Soft Matter* **5**, 1562 (2009).
6. A. Lengalova, V. Pavlinek, Q. L. Cheng, and P. Saha, *Int. J. Mod. Phys. B* **21**, 4883 (2007).
7. M. Stenicka, V. Pavlinek, P. Saha, N. V. Blinova, J. Stejskal, and O. Quadrat, *Colloid Polym. Sci.* **287**, 403 (2009).
8. Q. L. Cheng, V. Pavlinek, Y. He, C. Z. Li, and P. Saha, *Colloid Polym. Sci.* **287**, 435 (2009).
9. Y. Lan, X. P. Gao, H. Y. Zhu, Z. F. Zheng, T. Y. Yan, F. Wu, S. P. Ringer, and D. Y. Song, *Adv. Funct. Mater.* **15**, 1310 (2005).
10. X. T. Zhang, J. Zhang, W. H. Song, and Z. F. Liu, *J. Phys. Chem. B* **110**, 1158 (2006).
11. M. S. Cho, H. J. Choi, and M. S. Jhon, *Polymer* **46**, 11484 (2005).
12. S. Havriliak, and S. Negami, *Polymer* **8**, 161 (1967).

INFLUENCE OF ALKYL CHAIN LENGTH ON CARBOXYL-GROUP-IMMOBILIZED HOLLOW POLYANILINE SPHERES DISPERSED SUSPENSION UNDER AN ELECTRIC FIELD

YOUNG GUN KO

Energy Mechanics Center, Korea Institute of Science and Technology, 39-1 Hawolgok-dong, Wolsong-gil 5, Seoul 136-791, Korea

BO HYUN SUNG

Energy Mechanics Center, Korea Institute of Science and Technology, 39-1 Hawolgok-dong, Wolsong-gil 5, Seoul 136-791, Korea

UNG SU CHOI[*]

Energy Mechanics Center, Korea Institute of Science and Technology, 39-1 Hawolgok-dong, Wolsong-gil 5, Seoul 136-791, Korea, E-mail: uschoi@kist.re.kr

Here, we have fabricated carboxyl-group-immobilized hollow polyaniline (PANI) spheres with various alkyl chain lengths. The hollow PANI sphere was fabricated by the extraction of polystyrene (PS) spheres after polymerization of aniline on PS sphere. And then, carboxyl-group was immobilized on the surface of hollow PANI sphere. After the application of an electric field, the shear stress of carboxyl-group-immobilized hollow PANI spheres was higher than hollow PANI spheres in the following order: PANI adipate > PANI glutarate > PANI succinate > PANI malonate > PANI.

1. Introduction

The elctrorheological (ER) material is one kind of intelligent (smart) materials which can respond to an external environmental stimulus in a timely manner, producing a useful effect [1]. The rheological properties of ER fluids are controllable through the application of an electric field, showing useful and special function with the effect of reversibility [2].

Positive ER materials have rheological properties that dramatically increase with the applied electric field [3]. Among various positive ER materials, polyaniline (PANI) has been used due to the semi-conducting properties [4]. However, it has demerit of the fast sedimentation and the low viscosity after applying the electric field. In the previous study, hollow PANI sphere and the introduction of carboxyl group onto its surface were suggested to solve the problem of the sedimentation and the low viscosity, respectively [5,6].

---

[*]Corresponding author.

However, the influence of alkyl chain length was not well studied. Here, we synthesized the carboxyl-group-immobilized hollow PANI spheres with various alkyl chain lengths (the number of CH_2: 1-4) to investigate their effects on ER property.

2. Experimental

2.1. *Synthesis of Carboxyl-group-immobilized Hollow PANI Spheres*

The mono-dispersed 100 nm polystyrene (PS) was synthesized by emulsion polymerization using a free radical initiator according to Dufour's method [7]. Styrene (b.p. 145-146 °C, 99%, Aldrich) was distilled at reduced vacuum to remove traces of the inhibitor 4-*tert*-butylcatechol (b.p. 285-286 °C). After distillation, the monomer was put into nitrogen-filled amber bottles, and stored in the refrigerator, until required. Colloidal polystyrene (PS) spheres were synthesized by an emulsion polymerization using a free radical initiator (potassium persulfate, $K_2S_2O_8$, 98%, Aldrich). Styrene, deionized (DI) water, $K_2S_2O_8$, and sodium dodecyl sulfate (SDS, 99%, Aldrich) were added into the reaction flask, and then the polymerization was carried out in aqueous solution at 70 °C for 7 h under a nitrogen atmosphere. The reaction mixture was agitated using a twin-paddled overhead stirrer at 350 rpm. The products were finally put into a freezing dryer for drying.

Ammonium persulfate ($(NH_4)_2S_2O_8$, APS, 98%, Aldrich) was dissolved in the polyvinyl alcohol (PVA, Mw = 85,000 – 146,000, 98% hydrolyzed, Aldrich) stabilized PS particles in a screw-cap bottle with stirring. The reaction mixture was acidified to pH 0.7 for APS, and the initiator/monomer molar ratios were fixed 1.25. Aniline (99.9%, Aldrich) was added via syringe, and the polymerization was allowed to proceed for 24 h at 0 °C. The pH value was maintained at 0.7 using a pH stat with 1N HCl aqueous solution during the polymerization. And the HCl-doped PS-PANI composites were converted to the emerladine base form by treating it with NH_4OH (28% NH_3 in water, >99.99%, Aldrich) aqueous solution for 12 h. The synthesized particles was then washed three times using DI water to remove the initiator, the unreacted monomer, and the oligomer. The products were finally put into a freezing dryer for approximately 2 days for drying.

The extraction of PS from PANI-coated PS sphere with tetrahydrofuran (THF, 99.5%, Aldrich) was allowed under stirring at room temperature for 7 days. After 7 days, the particles were centrifuged at 21,000 rpm for 10 min and resuspended in THF. The step was repeated many times, and then washed with DI water. The resulting black residues were dried at a freezing dryer.

The carboxyl-group-immobilized hollow PANI spheres were synthesized according to Mitsunobu et al [8]. Dried hollow PANI spheres, reagents having the dicarboxyl group (malonic acid, succinic acid, glutaric acid, and adipic acid; All reactants were purchased from Aldrich), triphenyl phosphine (TPP, 99%,

Aldrich), diethyl azo dicarboxylate (DEAD, 99%, Aldrich), and DI water were put into a round flask with nitrogen purging and reacted at 60 °C oil bath for 5 h with stirring. After a reaction, carboxyl-group-immobilized hollow PANI sphere was washed with DI water and dried at a freezing dryer.

2.2. *Particles Characterizations*

The morphology study of hollow PANI spheres was observed by field-emission scanning electron microscopy (FE-SEM, Hitachi S-4200) and transmission electron microscopy (TEM, ISI DS-130) instrument operating at 10 kV. The particle size distribution was examined by dynamic light scattering (DLS, BI9000AT, Brook Heaven Co. Ltd.).

2.3. *Suspension Preparation and Electro-rheological Measurements*

The ER fluids were prepared by dispersing the modified hollow PANI spheres into silicon oil, whose viscosity was 30 cS at 25 °C. The silicon oil was dried using molecular sieves before use, and the particle concentration was fixed at 30 vol%. The rheological properties of the suspension were investigated in a static DC field using a Physica Couette-type rheometer (Physica US200) with a high-voltage generator. The measuring unit was of a concentric cylindrical type, with 1 mm gap between the bob and the cup. The shear stress for the suspensions was measured under a shear rate of 1 - 1000 s^{-1} and the electric fields of 0-3 kV/mm.

3. Results and Discussion

Various carboxyl-group-immobilized hollow PANI spheres were synthesized to confirm the effect of alkyl chain length as shown in Figure 1. Firstly, mono-dispersed PS sphere was synthesized as the core material with the emulsion polymerization. And then, the semi-conducting polymer PANI was coated onto the PS core. The PS core was removed by using THF. Finally, the synthesized hollow PANI sphere was reacted with dicarboxylic acids with various numbers of CH_2. The numbers of CH_2 in malonic acid, succinic acid, glutaric acid, and adipic acid are 1, 2, 3, and 4, respectively.

The morphology and particle size of PS sphere, PS-PANI composite sphere, and hollow PANI sphere were investigated by FE-SEM, TEM and DLS as shown in Figure 2. The diameters of the PS sphere and the PANI-coated PS sphere were ca. 100 nm and ca. 200 nm, respectively. DLS curve of the hollow PANI sphere showed the highest peak at the diameter of 200 nm as shown in Figure 2(c). Its diameter is same with the diameter of the PANI-coated PS sphere. TEM image of hollow PANI sphere is shown in inset of Figure 2 (c). It was found that the PS core could be successfully removed by dissolution in THF. After the immobilization of carboxyl groups on the surface of hollow PANI spheres, the diameter of modified hollow PANI spheres is also ca. 200 nm. The carboxyl-group immobilized hollow PANI sphere showed the similar size of the diameter with the bare hollow PANI sphere (data not shown) due to the short length of coupled molecules having carboxyl group.

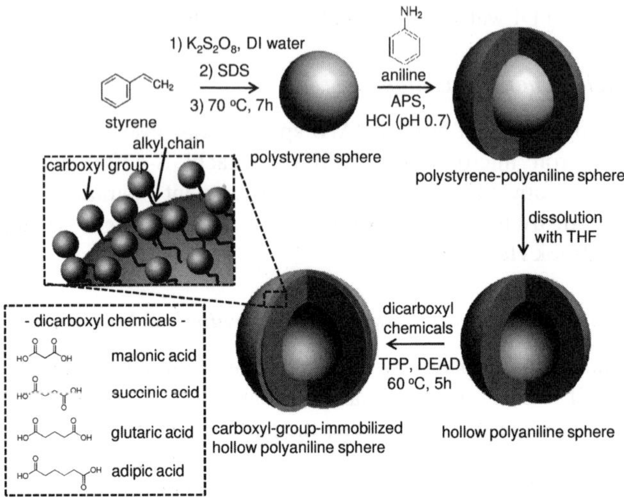

Figure 1. Schematic illustration of the synthesis of carboxyl-group-immobilized hollow PANI spheres.

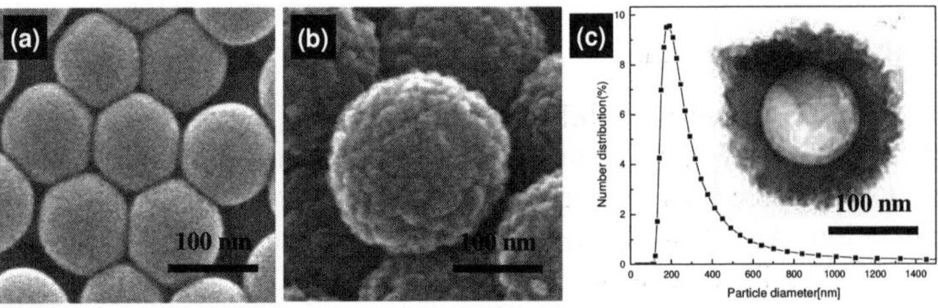

Figure 2. FE-SEM images of (a) PS sphere and (b) PANI-coated PS sphere; and (c) particle size distribution of hollow PANI sphere (inset; TEM image of hollow PANI sphere).

Shear stress curves as a function of shear rate for hollow PANI sphere and carboxyl-group-immobilized hollow PANI spheres under an electric field of 3 kV/mm were shown in Figure 3. All materials showed the typical Bingham plastic behavior although they showed some deviations. In the region of the low shear rate, the deviations of the value of shear stress between hollow PANI sphere and carboxyl-group-immobilized hollow PANI spheres were small. However, their deviations increased with the increase of the shear rate. The carboxyl-group-immobilized hollow PANI spheres showed higher shear stress than hollow PANI sphere. In principle, the conducting mechanism of PANI type semi-conductors is based on the flow of electrons or charge carriers along the conjugated π system PANI backbone [9]. The polar group of carboxyl group on hollow PANI sphere may enhance the ER behavior by playing the role of electronic donor. Especially, the shear stress increased with the increase of the number of CH_2 in carboxyl-group-immobilized hollow PANI spheres increase. It may be seems that a synergetic effect of van der Waals force between CH_2 and high mobility of CH_2 may make the bonds between polarized ER particles endure shear stress at high shear rate under an electric filed.

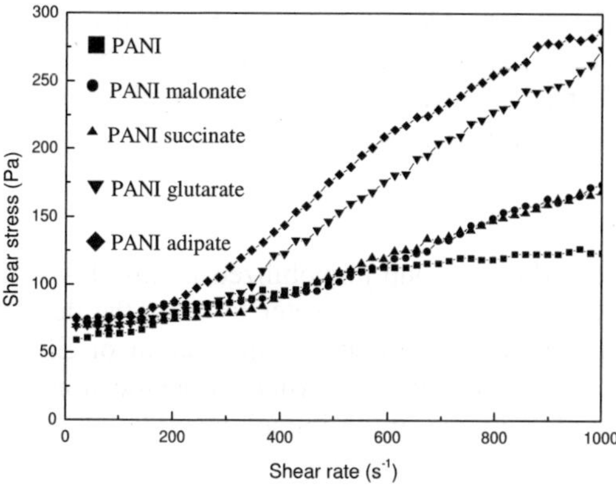

Figure 3. Shear stress vs. shear rate for hollow PANI sphere and carboxyl-group-immobilized hollow PANI spheres.

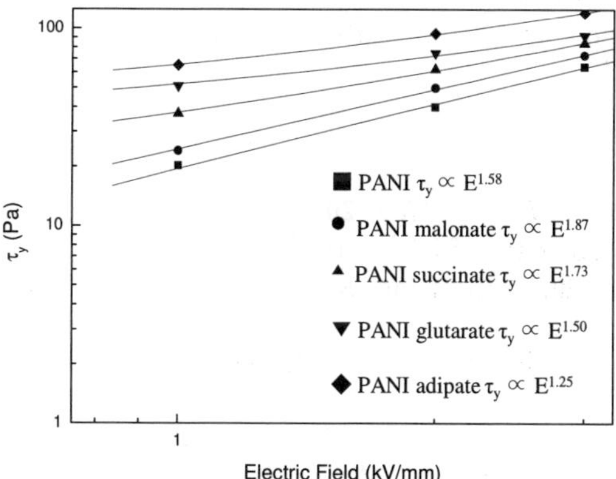

Figure 4. The yield stress for hollow PANI sphere and carboxyl-group-immobilized hollow PANI spheres under various electric field strengths.

The yield stresses (τ_y) obtained from various logarithmic shear stress versus shear rate curves also plotted as a function of electric strength (E) for five kinds of ER fluids as shown in Figure 4. All carboxyl-group-immobilized hollow

PANI spheres showed higher yield stress than hollow PANI sphere. The yield stress increased as the number of CH_2 in carboxyl-group-immobilized hollow PANI spheres increase under all electric fields. The change of yield stress with different electric filed can be represented by the following equation $\tau_y = E^a$. The a values of hollow PANI sphere, hollow PANI malonate sphere (CH_2: 1), hollow PANI succinate sphere (CH_2: 2), hollow PANI glutarate sphere (CH_2: 3), and hollow PANI adipate sphere (CH_2: 4) are 1.58, 1.87, 1.73, 1.50, and 1.25, respectively. This result differ from the theoretical prediction that τ_y is proportional to the electric filed strength E^2. Especially, the a value decreased with the increase of the number of CH_2 in the carboxyl-group-immobilized hollow PANI sphere. The difference is due to several factors, such as the degree of the polarization, the polarization rate, etc.

4. Conclusions

In this study, the carboxyl-group-immobilized hollow PANI spheres dispersed suspensions were fabricated without water to solve the device erosion and the particle sedimentation which inhibit the application of ER fluids to practical utilizations in active control devices. After the application of an electric field, the shear stress of carboxyl-group-immobilized hollow PANI spheres was higher than hollow PANI spheres in the following order: PANI adipate > PANI glutarate > PANI succinate > PANI malonate > PANI. The yield stress of all carboxyl-group-immobilized hollow PANI spheres showed better than hollow PANI sphere. The yield stress increased as the number of CH_2 in carboxyl-group-immobilized hollow PANI spheres increase under all electric fields. By this result, we can know that the carboxyl group and the CH_2 chain enhance the ER effect.

References

1. W. M. Winslow, *J. Appl. Phys.* **20**, 1137 (1949).
2. T. C. Halsey, *Science* **258**, 761 (1992).
3. W. Wen, X. Huang, S. Yang, K. Lu and P. Sheng, *Nature Mater.* **2**, 727 (2003).
4. P. Hiamtup, A. Sirivat and A. M. Jamieson, *J. Colloid Interface Sci.* **325**, 122 (2008).
5. B. H. Sung, U. S. Choi, H. G. Jang and Y. S. Park, *Coll. Surf. A: Physicochem. Eng. Aspects* **274**, 37 (2006).
6. B. H. Sung, Y. G. Ko and U. S. Choi, *Coll. Surf. A: Physicochem. Eng. Aspects* **292**, 217 (2007).
7. M. G. Dufour and A. Guyot, *Coll. Polym. Sci.* **281**, 105 (2003).
8. O. Mitsunobu, *Synthesis* **1**, 1 (1981).
9. X. Li and K. Su, *Mater. Rev.* **5**, 70 (2000).

ER-ACTIVITY OF SUSPENSIONS BASED ON HYDRATED METAL OXIDES

KOROBKO E.V. and BEDIK N.A.

A.V. Luikov Heat and Mass Transfer Institute, National Academy of Sciences of the Republic of Belarus
15 Brovka Str., Minsk 220072 Belarus

ESHENKO L.S.

Belarusian State Technological University
13ᵃ Sverdlov Str.t, Minsk 220006 Belarus

The thermal stability of ERF based on oxyhydroxides of metals has been established. It increases in the row of metals Al<Fe<Cr<Ni. It is shown that hydrates of oxyhydroxides of polyvalent metals that are characterized by a laminated structure and by the presence of a potential charge carriers as a result of the formation of the hydrogen bonding of the OH groups of unstructured water molecules are efficient thermally stable fillers of ERFs for various technological facilities.

1. Introduction

A dispersed phase in an electrorheological fluid (ERF) plays a basic role in ensuring the rheological response to electric action. The dispersed phase of an ERF may consist of particles of different materials: inorganic, organic, polymer, segnetoelectrics, liquid crystals, ets. An urgent problem here is the creation of ERFs capable of operating at elevated temperatures that they can be used in various technological facilities subjected to heating. Of particular interest are hydrated oxides of polyvalent metals meeting a wide range of requirements imposed upon electrically sensitive fillers of dispersions in which water is an activator and favors the manifestation of the ER effect. The importance of the presence of a certain amount of water as an activator was noted in a number of works [1-7] in which only hydrated oxides of metals were objects of investigation, but also other compounds such as aluminosilicates of neolithic type, ion-exchange resins, ets.

Hydrated oxides and oxyhydroxides of metals contain different forms of water, in particular, a structural one in the form of OH groups presenting an element of the crystal lattice of hydroxide and bound with cations by ionic (partially covalent) bond, as well as a nonstructural one in the form of interlayer,

"

adsorbed H_2O molecules that are united by the H bond in the compound into endless chains and lattices. It is believed that their role in the processes of proton transfer and, consequently, in proton conductivity responsible for the electrorheological effect must be very appreciable. It was shown in our works [8-10] that the ERF based on hydrated aluminum oxide in the form of pseudo boehmite containing a unstructured water in the form of interlayer molecules with strong hydrogen bonds possess a high ER activity and thermal stability.

The aim of the present work was the study of the ER activity of suspensions which as a dispersed phase contain hydrated oxides of polyvalent metals, in particular, of aluminum, iron, chromium, and nickel of specified composition.

2. Technique and Materials

Experimental investigation of the rheological properties of ERF was carried out using the electrorheological cell of the «Physica MCR 301» rheometer of Anton Paar Company (Austria) with continuous strain in the shear rate range from 0.01 to 1000 s^{-1} and electric field strengths from 0 to 6 kV/mm at a temperature of 20 °C, as well as at 20 s^{-1} in the temperature range from 20 to 100 °C and electric field strength $0 - 4$ kV/mm. The measuring cell of the device represents a system of coaxial cylinders consisting of the outer fixed cylinder with a cylindrical rotor immersed into it. The investigated medium was placed in the annular gap between the cylinders and it was thermostated with the aid of the Peltier system. The device was controlled with the aid of the Rheoplus/32 V3.40 program. The results of the automatic processing of data were put out onto the computer monitor and were recorded in a file.

The ERF fillers based on oxihydroxides of aluminum, iron, chromium, and nickel were obtained by the method of chemical deposition with subsequent stages of the ageing of sediments, their washing, drying, and thermal treatment. In hydrolysis of monomer compounds, in particular, of the salts of indicated metals in an alkali medium polymerized products are formed that ultimately determine the composition and structure of the solid phase being formed. A specimen of hydrated nickel oxide was produced in the presence of the NaClO oxidizer at pH $7.2 - 7.5$ till the formation of black slightly solvable sediment nickel (III) NiOOH, on ageing of which the $Ni_3O_2(OH)_4$ compound is formed.

The content of H_2O in specimens was determined thermogravimetrically. The phase composition was determined radiographically on a 08 Bruker AXS Advance firm diffractometer. The obtained oxyhydroxides of metals were used for preparing suspensions containing 10 wt. % of a filler in a mineral oil. Surfactants were not added for the purity of experiments.

3. Results

The characterization of the oxyhydroxides of metals as the dispersed phase of an ERF are presented in Table 1.

Table 1. Characterization of Specimens of the Dispersed Phase for an ERF

No	Dispersed phase	Phase composition	Molar ratio with $Me_2O_3:H_2O$
1	aluminum oxyhydroxide	pseudo boehmite $AlOOH$	1:1.50
2	iron oxyhydroxide	getit $FeOOH$	1:1.27
3	hydrated chromium oxide	slightly crystallized phase of chromium oxyhydroxide $CrOOH$	1:2.40
4	hydrated nickel oxide	mixture of nickel hydroxide $NiOOH$ and nickel hydroxide $Ni(OH)_2$	1:1.46*

* – in terms of $Ni_2O_3 \cdot NiO$

At follows from Table 1 that the dispersed phase for preparing an ERF was oxyhydroxides of polyvalent metals containing a structural H_2O (mole or H_2O / mole of metal oxide) and unstructured H_2O, the amount of which in spite of the identical conditions of production depends on the metal nature. The least hydrated state for the compounds investigated is typical of iron oxyhydroxide, the maximum one is for chromium which can be attributable to both its imperfect structure, low degree of crystallization, involvement of the roentgenoamorphous phase, and to the specific features of cation properties.

Attention is particularly drawn to the filler based on specimen No 4 which was a mixture of hydrated nickel (II) and nickel (III) oxides. As it follows from Table 1, the content of unstructured water in this compound is about 0.5 mole and it comparable with the content of unstructured water in pseudo boehmite.

This group of substances refer to hydrated of variable composition, the content of water in which, in the first place of unstructured one, is the function of temperature and pressure. Oxyhydroxides of metal can be considered as aquacomlexes, in which water molecules participate in the formation of H bonds and are coordinated by a metal. Proceeding from the chemical composition of the compounds investigated, their properties and structure, with allowance for the geometric, dynamic and energy characteristics, it is possible to assume that the unstructured water in their composition has different forms of H_2O molecules and energies of binding between them by the H-bonding mechanism. The above-described hydrated oxides of metals may have several types of H bonds with participation of water protons, both localized in the laminated structure of the oxyhydroxide and adsorbed on the active centers of the surfaces of particles, which is responsible for the proton conductivity of a solid body and favors creation of efficient fillers for ERFs.

Figure 1 presents the flow curves of ERF based on the obtained fillers in the absence of an electric field at E=6 kV/mm.

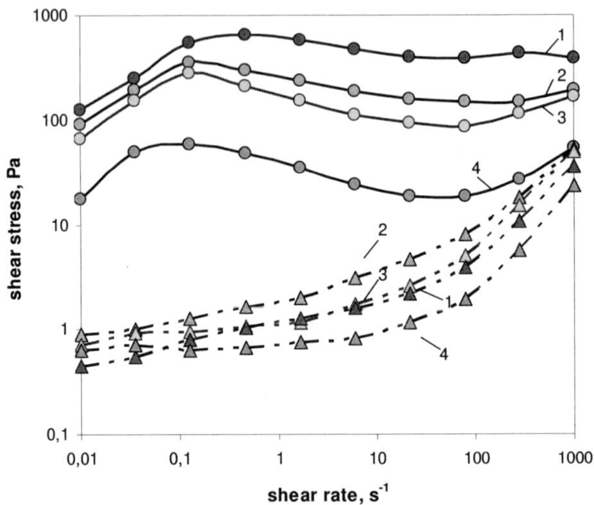

Figure 1. Shear stress vs. shear rate for an ERF based on metal oxyhydroxides at T=20°C: dash-dotted lines, E=0 kV/mm; solid lines, E=6 kV/mm; 1, 2, 3, 4, ER fluids based on specimens Nos. 1-4, respectively.

It is seen that the existence of the electric field substantially influence the shear stress of the ERFs based on metal oxyhydroxides. Thus, the shear stress of an ERF at 6 kV/mm increase in the row of oxyhydroxides Ni<Cr<Fe<Al at all shear rates (Fig. 1). The difference in the ER activity is in all likelihood due to the state of the molecules of unstructured water characterized by the electrostatic influence of the cation field on the proton of water molecules.

Figures 2 and 3 present the temperature dependence of the ER activity and current density of ERF based on the oxyhydroxides of the indicated metals.

According to the experimental data, for all the investigated ERFs based in specimens of oxyhydroxides of metals the characteristic feature is the increase in the ER activity with temperature. For each composition of a filler of ERF there are threshold values of temperature and electric field strength above which the current density exceeds the value 50 $\mu A/cm^2$ that leads to automatic switching off of a high-voltage source. For example, the ERF based on specimen No 1 endures heating up to 50 °C at 4 kV/mm (Fig. 2), but at 2 kV/mm the same suspension endures heating up to 100 °C, with its ER activity increasing in the entire range of temperatures from 20 °C to 100 °C.

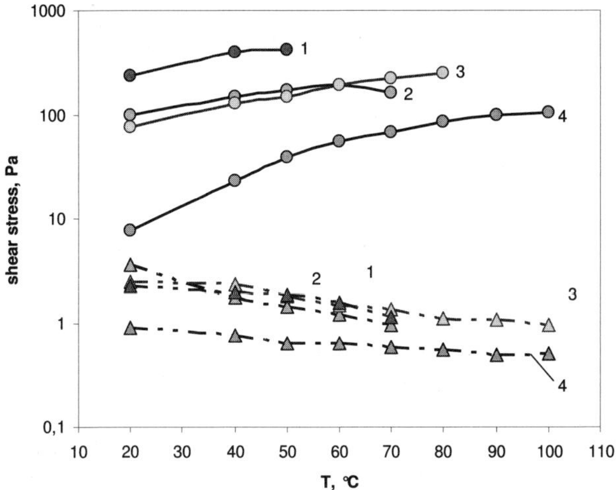

Figure 2. Shear stress vs. temperature of ERF based on metal oxyhydroxides at a shear rate of 20 s^{-1}: dash-dotted lines, E=0 kV/mm; solid lines, E=4 kV/mm; 1, 2, 3, 4, ER fluids based on specimens Nos. 1-4, respectively.

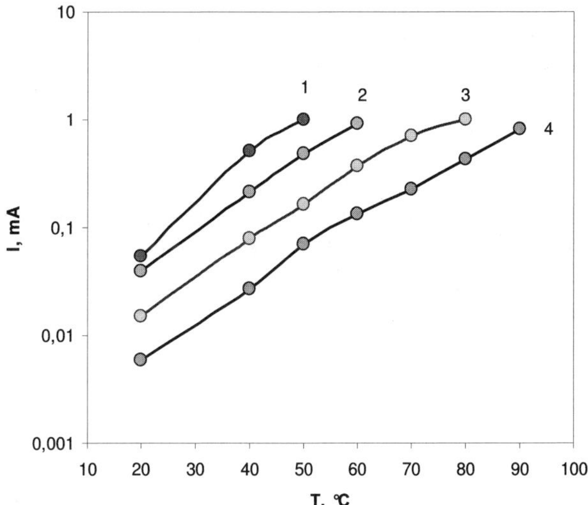

Figure 3. Current density vs. temperature of ERF based on metal oxyhydroxides at a shear rate of 20 s^{-1} at E=4 kV/mm: 1, 2, 3, 4, ER fluids based on specimens Nos. 1-4, respectively.

Of interest is the behavior of the ERF based on specimen No 4 in the process of heating. At 20 °C and E=4 kV/mm the ER activity of the specimen is the lowest ($\tau_E/\tau_{E=0}$=10) in the investigated group of fillers (Fig. 2), the current density is close to zero (Fig. 3). Heating leads to substantial activation of the filler and to an increase in the ER activity of the ERF at 100 °C $\tau_E/\tau_{E=0}$=125. At such a strength of the electric field none of the investigated fluids based on the oxyhydroxides of metals withstood such heating because the current density in the cell exceeded 50 $\mu A/cm^2$.

4. Discussion

The increase in the ER activity of fluids with temperature is attributable to the increase in the concentration of the charge carriers H^+, the source of which are the unstructed water molecules that were protonated in the electric field and that posses the thermal stability.

A comparison of the results of studying the ER activity of the investigated ERFs based on the oxyhydroxides of metals shows that their proton conductivity differs greatly under the action of the electric field. This means that the character of binding of unstructured molecular water depends on a number of specific features, such as the electrostatic field of the cation and its electron structure, the protodonating capabilities of the surface hydroxyl groups, the hydrate structure, the degree of hydration and others. To explain the obtained dependences of the ER activity on the composition of the dispersed phase of ERF it is possible to assume that aluminum oxyhydroxides with a structure of pseudo boehmite possesses a higher concentration of potential charge carriers in view of the polarizing action of the cation field and formation of the hydrogen bonding of OH groups of the unstructured water molecules that may represent the centers of binding of subsequent H_2O molecules. In this case, the transfer of H^+ protons in polymolecular layers of H_2O bound by the H bonding with the coordinationally bound H_2O molecules is responsible for the surface proton conductivity.

According to [11], in oxyhydroxides of transient metals the hydrogen bonding of hydrate water with coordinationally bound OH groups appears to be stronger because of the electrostatic influence of the field of cations on the protons of water molecules. This leads to an increase in the donating capability of the OH groups of water molecules of forming the hydrogen bond.

It follows from the analysis of experimental data that a decrease in the thermal stability of oxyhydroxides in the row Ni<Cr<Fe<Al in heating is directly connected with the force of interaction of water molecules with the indicated cations. The most stable is the nickel-containing hydrate complex. The different ER activity of suspensions based on the indicated oxyhydroxides both at 20 °C and in heating can be due to a number of factors, in particular, to the number of water molecules bound with the cation by the donating capability of the OH groups of water molecules, the coordination number, etc.

5. Conclusions

The thermal stability of ERF based on oxyhydroxides of metals has been established. It increases in the row of metals Al<Fe<Cr<Ni. It is shown that hydrates of oxyhydroxides of polyvalent metals that are characterized by a laminated structure and by the presence of a potential charge carriers as a result of the formation of the hydrogen bonding of the OH groups of unstructured water molecules are efficient thermally stable fillers of ERFs for various technological facilities.

References

1. Tian Hao, Electrorheological suspensions, *Adv. Colloid Interface Sci.*, **97**, 1–35 (2002).
2. D. L. Klass and T. W. Martinek, Electroviscous fluids. II. Electrical properties, *J. Appl. Phys.*, **38**, No. 1, 75–80 (1967).
3. E. V. Korobko, V. A. Marshak, A. A. Makhanyok, T. A. Pokhodina, L. Zhou, and Chang-Ho Kim, Structural mechanics of ERF's within a wide temperature range, in: *Proc. MIF-96* [in Russian], Vol. 6, Minsk (1996), pp. 221–226.
4. Zh. Shengbin., W. T. Winter, and A. J. Stipanovic, Water-activated cellulose-based electrorheological fluids, *Cellulose*, **12**, 135–144 (2005).
5. Z. P. Shul'man, R. G. Gorodkin, M. M. Ragotner, Z. A. Novikova, and T. A. Demidenko, A means of obtaining diatomite electrorheological suspensions, Author's Certificate 1404515, MPK C 09 K 3/00, *Byull.* No. 23 (1988).
6. A. A. Baran, A. D. Matsepuro, V. M. Nosov, Z. P. Shul'man, and V. P. Yashcheritsyn, Working medium for electrostatic engines, Author's Certificate 898574, MPK H 02 N 1/10. *Byull.* No. 2 (1982).
7. E. V. Korobko, V. I. Dubkova, and M. M. Ragotner, Electrorheological properties of suspensions dispersely reinforced with a fibrous and powdery component, in: *Proc. 6th Nat. Conf. on Mechanics and Technology of Composite Materials*, Sofia (1991), pp. 9–12.
8. E. V. Korobko, L. S. Eshchenko, N. A. Bedik, and G. M. Zhuk, Investigation of the electrorheological sensitivity of suspensions based on hydrated aluminas, *Kolloidn. Zh.*, **69**, No. 2, 201–205 (2007).
9. E. V. Korobko, L. S. Eshchenko, N. A. Bedik, Influence of Temperature on the Electrorheological Response of Dielectric Suspensions Based on a Water-Containing Filler, *J. of Engineering Physics and Thermophysics,* Vol. 82, No. 4, 743-748 (2009).
10. E. V. Korobko, L. S. Eshchenko, N. A. Bedik, G. M. Zhuk, and V. K. Gleb, Flowing composition with electrorheological properties, Patent of the Republic of Belarus No. 10460 (2008).
11. G. Tsundel. Hydration and intermolecular interaction, Moscow, (1972).

MOLECULE-BASED ELECTRORHEOLOGICAL MATERIAL, ELECTRORHEOLOGICAL PERFORMANCE AND MOLECULE STRUCTURE

LI HUO[a,b], FU-HUI LIAO[b], JUN-RAN LI[b*] and SHAO-HUA ZHANG[c]

[a]*Department of Biochemistry, Baoding University, Baoding, Hebei 071000*

[b]*Beijing National Laboratory for Molecular Science, (State Key Laboratory of Rare Earth Materials Chemistry and Applications), College of Chemistry and Molecular Engineering, Peking University, Beijing 100871, China*

[c]*School of Vehicle and Transmission Engineering, Beijing Institute of Technology, Beijing, 100081, China*

Molecule-based electrorheological (ER) materials as a novel type of ER materials, a series of compounds using melamine ($C_3N_6H_6$) as the substrate, have been synthesized using orthophosphoric acid (H_3PO_4), oxalic acid ($H_2C_2O_4$), 4-toluenesulfonic acid (PTA, $C_7H_8O_3S$) and 5-sulfosalicylic acid (SSA, $C_7H_6O_6S$), respectively, as starting materials. The ER performance and dielectric property of materials have been studied. The results show that these materials have ER activity. The unusual relationship between dielectric property and ER property of these materials was found and discussed. The composition and structure of molecule are the dominant factors, the function group plays an important role, in influencing ER performance of the molecule-based ER material.

1. Introduction

Electrorheological (ER) materials have been widely recognized as a promising class of materials because of their potential applications.[1,2] The origin of the ER effect has been intensively discussed in many articles. The results from a series of researches[3-6] have shown that particle polarization is potentially responsible for the interaction force that leads to the rheological change of ER fluids. T. Hao[7] indicated in his review that a generally accepted concept for the positive ER effect is that the particles form fibrillated chains, which cause an abrupt increase in the rheological parameters. He also pointed out that polarization processes are extremely important in an ER system.[8] The conductive, dielectric,

---

*Corresponding author. Tel.: +(86)-10-62753517, fax: +(86)-10-62751708.
E-mail: lijunran@pku.edu.cn

and surface properties of several water-free polymer or inorganic material based ER fluids were investigated by T. Hao, A. Kawai and F. Ikazaki.[9] They found that interfacial polarization, rather than other types of polarization, would contribute to the ER effect; and indicated that large interfacial polarization can produce a large amount of surface charges, which could result in the particles turning to form a fibrillation structure. They proved that a large dielectric loss is required for a good ER material, because only a material having a large dielectric loss could give a large interfacial polarization once it is dispersed into a liquid.[9] Therefore, interfacial polarization is believed to play an important role, and the particle dielectric property should be dominant in the ER effect.

In order to obtain a material with a strong ER effect, a wide spectrum of ER materials have been synthesized and studied.[8] Unlike regular materials, which are constructed with covalent bonds, ionic bonds or metal bonds between atoms or ions, molecule-based solid materials are assembled through weak interactions such as hydrogen bonds, π-π interactions, van der waals forces and/or coordination bonds. Therefore, molecule-based solid materials with any specific functionality can be synthesized by selecting suitable molecules and an appropriate assembly method. The research progress on molecule-based solid function materials with optical, electrical or magnetic properties has been reported.[10,11] Molecule-based magnetic materials have been studied extensively in the fields of chemistry, physics and material science.[12-14] Molecule-based ER materials have not been reported previously except for research conducted in our lab.[15-18]

It is well known that molecule structure and composition in the particle, which would influence the physical and chemical natures of a material, are essential to the dielectric and polarization properties of the material. Therefore, it is possible to modify a material's dielectric and polarization properties by adjusting the synthetic procedure such that an ER material having a desired microstructure and composition is produced. It is necessary to research further the influence of the composition and structure on the ER performance of a particle material. In order to better understand the mechanism of the observed ER effect, and to find an ideal ER material with an enhanced ER performance and is economical and facile, we selected melamine as the substrate, and synthesized a series of molecule-based ER materials. The effects of the composition and structure of the molecules on the ER performance of these materials have been investigated, and the relationship between the dielectric property and ER effect is discussed in this paper.

2. Experimental

All reagents were provided by Beijing Chemistry Reagent Co. (China) and used without further purification.

The $C_3N_6H_6 \cdot HA$ (HA=H_3PO_4, $H_2C_2O_4$, PTA and SSA) materials were synthesized by using following process: first, $C_3N_6H_6$ (10mmol) was dissolved in hot water, the aqueous solution of HA (15mmol for $H_2C_2O_4$, PTA and SSA, 25mmol for H_3PO_4) was added drop-wise to the hot solution of $C_3N_6H_6$, the clear solution was then stirred under heating for 30 min. The white sediment was obtained by adding adequate acetone in the solution. After filtering and washing with deionized water and acetone, the white solid product was comminuted and roasted at 60°C for 48 hours and dried in vacuum for 72 hours at 50°C. The $C_3N_6H_6 \cdot HA$ materials were thus obtained.

The dried particle materials were mixed quickly with dimethyl silicone oil (density $\rho=0.98 g \cdot cm^{-3}$ and viscosity $\eta=98 mPa \cdot s$ at $25°C$) under stirring, and ultrasonically dispersed for five minutes, yielding the ER fluid (25wt%) samples. The suspensions were then put in the gap between the cylinders of the apparatus as soon as possible for ER measurements. The ER experiments were carried out using a German Rotary Viscometer (Type HAAKE CV20). In this study, the sample shear stresses (τ) and viscosities have been determined under different electric field strengths (E, dc field) at a given temperature ($20°C$) and a shear rate (γ) range of $0\text{-}300s^{-1}$.

The suspensions (25wt%) of the materials were used to investigate their dielectric properties. The capacitance C and dielectric loss tangent (tanδ) at room temperature under various frequencies (f) were obtained on a HP4274A Multi-frequency LCR Meter. The dielectric constant (ε) was derived from the measured C according to the conventional relation, $\varepsilon=C \cdot d/(\varepsilon_0 \cdot S)$, where ε_0 is the dielectric constant of vacuum i.e. $8.85 \times 10^{-12}\, F \cdot m^{-1}$, and d is the thickness of the gap between the electrodes and S is the contact area of the electrodes. The element analyses of materials were performed on an Elementar Vario EL instrument.

3. Result and Discussion

3.1. *Composition of material*

Fig. 1 shows the XRD patterns of materials. Comparing the XRD patterns of $C_3N_6H_6 \cdot HA$ materials to that of $C_3N_6H_6$ material, the characteristic peaks of $C_3N_6H_6$ material are not present in the XRD patterns of $C_3N_6H_6 \cdot HA$ materials, indicating that $C_3N_6H_6 \cdot HA$ is not an aggregate composed of $C_3N_6H_6$ and HA, but a compound resulted from the reaction between $C_3N_6H_6$ and HA, thus not a mixture. Therefore, the XRD patterns of the materials can serve as evidence for the formation of the $C_3N_6H_6 \cdot HA$ compounds. The results from elemental analyses (See Table 1) can prove that the compositions of $C_3N_6H_6 \cdot HA$ compounds are $C_3N_6H_6 \cdot H_3PO_4$, $C_3N_6H_6 \cdot H_2C_2O_4$, $C_3N_6H_6 \cdot C_7H_8O_3S$ ($C_3N_6H_6 \cdot PTA$) and $C_3N_6H_6 \cdot C_7H_6O_6S$ ($C_3N_6H_6 \cdot SSA$), respectively.

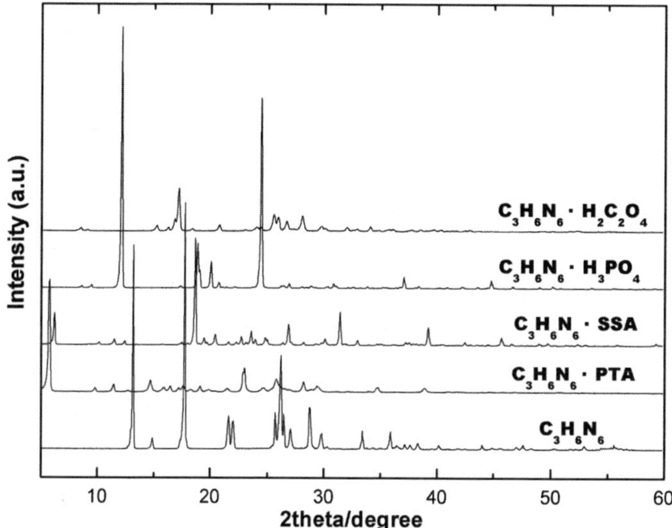

Fig. 1 XRD patterns of materials.

Table 1 The results from element analyses

Composition of $C_3N_6H_6 \cdot HA$	Content of element[*]		
	C(wt%)	N(wt%)	H(wt%)
$C_3N_6H_6 \cdot H_3PO_4$	15.87(16.06)	37.65(37.48)	4.11(4.05)
$C_3N_6H_6 \cdot H_2C_2O_4$	27.74(27.78)	38.94(38.88)	3.73(3.73)
$C_3N_6H_6 \cdot C_7H_8O_3S$ ($C_3N_6H_6 \cdot PTA$)	39.81(40.25)	28.75(28.18)	4.31(4.73)
$C_3N_6H_6 \cdot C_7H_6O_6S$ ($C_3N_6H_6 \cdot SSA$)	34.85(34.89)	24.43(24.41)	3.51(3.51)

[*]Calculated values were put in parentheses

3.2. *ER property of material*

Fig. 2 illustrates the shear stresses as a function of the electric field strength (E, dc field) at γ=300s^{-1} for the suspensions of materials. In order to obtain a clear comparison between the ER performances of these materials, three types of shear stresses were employed: the shear stresses of the suspension with and without applied electric field (τ_E and τ_0), and relative shear stress (τ_r), which is defined in our work as the ratio of the shear stress at an electric field to zero-field shear stress ($\tau_r=\tau_E/\tau_0$). Relative shear stress was also used to represent the magnitude of the ER activity. The corresponding data at E=4.2kv/mm and γ =300s^{-1} are listed in Table 2.

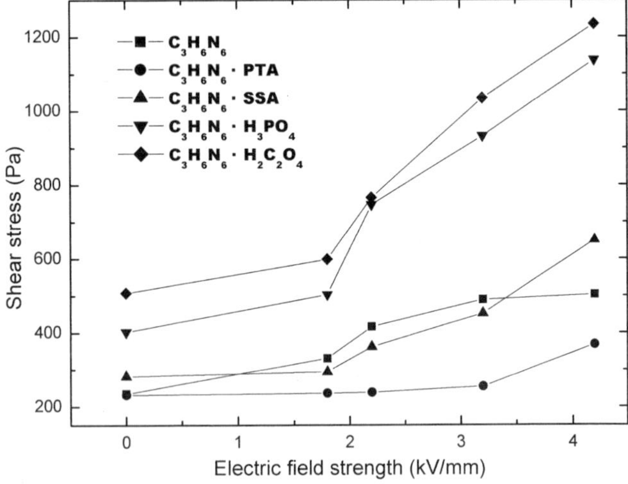

Fig. 2 Shear stresses as a function of the electric field strength at γ=300s^{-1} for the suspensions of materials.

Table 2 ε, tanδ, τ_0, τ_E and τ_r of material suspensions

Sample	ε	tanδ	τ_E	τ_0	τ_r
$C_3N_6H_6$	2.90	0.11	502.4	235.4	2.13
$C_3N_6H_6 \cdot H_3PO_4$	2.85	0.10	1137	402.7	2.82
$C_3N_6H_6 \cdot H_2C_2O_4$	1.92	0.059	1234	507.8	2.43
$C_3N_6H_6 \cdot PTA$	1.35	0.0098	368.2	232.2	1.59
$C_3N_6H_6 \cdot SSA$	1.83	0.015	649.6	282.1	2.30

In the comparison between the zero-field shear stresses (τ_0) of the suspensions (See Fig. 2 and Table 2), those of $C_3N_6H_6 \cdot HA$ materials are higher than that of $C_3N_6H_6$ material except for that of $C_3N_6H_6 \cdot PTA$ material, which may be related to presence of hydrogen bond interaction between the compound molecules. The particles can be aggregated because there is the hydrogen bond interaction between the molecules on surfaces of particles, which would induce an increase of τ_0 of the suspension. Fig. 3 illustrates the structural formulas of $C_3N_6H_6$ and HA molecules. The reaction between compounds $C_3N_6H_6$ and HA belongs to a neutralization reaction between weak acid and weak alkali. In the reactions, an amido group (-NH$_2$) in $C_3N_6H_6$ molecule combines with the sulfonic group (-SO$_3$H) in PTA molecule or SSA molecule, a carboxyl group (-COOH) in $H_2C_2O_4$ molecule, a hydroxy group (-OH) in H_3PO_4 molecule, respectively, the compounds $C_3N_6H_6$·HA thus were obtained. Therefore, in the studied materials, the presence of hydrogen bond interaction should be: between amido groups (-NH$_2$) in $C_3N_6H_6$ molecules or in $C_3N_6H_6$·HA (HA=PTA, SSA, H_3PO_4 or $H_2C_2O_4$) molecules, between carboxyl groups (-COOH) in $C_3N_6H_6$·$H_2C_2O_4$ molecules or in $C_3N_6H_6$·SSA molecules, between hydroxy groups (-OH) in $C_3N_6H_6$·H_3PO_4 molecules or in $C_3N_6H_6$·SSA molecules, respectively. In other words, there can be only hydrogen bond interaction between amido groups (-NH$_2$) on the surfaces of particles in the suspensions of $C_3N_6H_6$ and $C_3N_6H_6$·PTA materials, which results in a smaller degree of particle aggregation in the suspension due to weaker intermolecular hydrogen bond in the comparison with other $C_3N_6H_6$·HA materials. As a result, the τ_0 values of the suspensions of $C_3N_6H_6$ and $C_3N_6H_6$·PTA particle materials are close and the least.

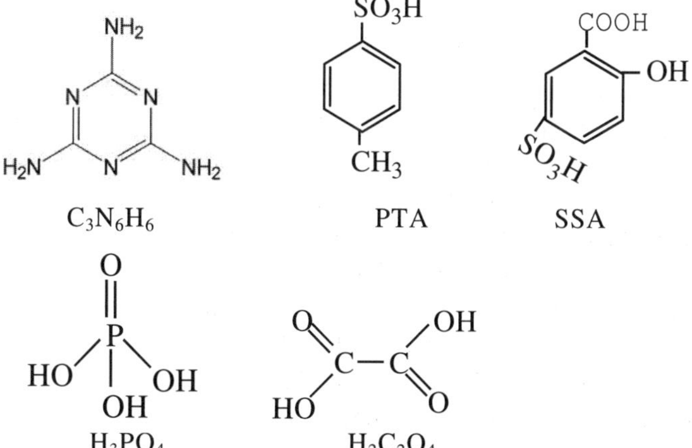

Fig. 3 Structural formulas of melamine ($C_3N_6H_6$), 4-toluenesulfonic acid (PTA, $C_7H_8O_3S$), 5-sulfosalicylic acid (SSA, $C_7H_6O_6S$), orthophosphoric acid (H_3PO_4) and oxalic acid ($H_2C_2O_4$) molecules.

The difference on the zero-field shear stresses for the suspensions of $C_3N_6H_6 \cdot H_2C_2O_4$, $C_3N_6H_6 \cdot H_3PO_4$ and $C_3N_6H_6 \cdot SSA$ materials may be explained by the fact that: there is stronger intramolecular hydrogen bond between -COOH and -OH in $C_3N_6H_6 \cdot SSA$ molecule than between -OH and -OH in $C_3N_6H_6 \cdot H_3PO_4$ molecule, there is not intramolecular hydrogen bond in $C_3N_6H_6 \cdot H_2C_2O_4$ molecule, thereby intermolecular hydrogen bond is the strongest for $C_3N_6H_6 \cdot H_2C_2O_4$ molecules and it is weakest for $C_3N_6H_6 \cdot SSA$ molecules among $C_3N_6H_6 \cdot H_2C_2O_4$, $C_3N_6H_6 \cdot H_3PO_4$ and $C_3N_6H_6 \cdot SSA$ molecules. Therefore, the hydrogen bond interaction between the molecules on surfaces of particles is the strongest for $C_3N_6H_6 \cdot H_2C_2O_4$ particle material, hence the degree of particle aggregation is the largest in the suspension of $C_3N_6H_6 \cdot H_2C_2O_4$ particle material, so the τ_0 value of the suspension is the largest. Moreover, the suspension of $C_3N_6H_6 \cdot H_3PO_4$ material has larger τ_0 value than that of $C_3N_6H_6 \cdot SSA$ material.

From Fig. 2 and Table 2, we can see that the shear stresses of the suspensions of $C_3N_6H_6$ and $C_3N_6H_6 \cdot HA$ materials increase with increasing electric field strength, showing that these materials have ER effect. However, τ_r value of the suspension of $C_3N_6H_6 \cdot PTA$ material is very small (1.59), it is smaller than that (2.13) of $C_3N_6H_6$ material, indicating that the formation of $C_3N_6H_6 \cdot PTA$ has decreased ER activity of $C_3N_6H_6$ material. The reason may be the decrease of $-NH_2$ as polar function group and the presence of methyl ($-CH_3$) as non-polar function group after the integration of $C_3N_6H_6$ and PTA, which makes a fall of interface polarization in the suspension. The shear stresses under electric field strengths of E<3.5kV/mm are smaller, and the τ_r value at E=4.2kV/mm is slightly larger, for the suspension of $C_3N_6H_6 \cdot SSA$ material than for the suspension of $C_3N_6H_6$ material. The presence of strong intramolecular hydrogen bond between $-COOH$ and $-OH$ as polar function groups in $C_3N_6H_6 \cdot SSA$ molecule should be responsible for lower ER effect of $C_3N_6H_6 \cdot SSA$ material. The strong intramolecular hydrogen bond between the polar function groups can influence the polarization of the molecule, sequentially result in the descent of interface polarization in the suspension, in an applied electric field. The results from Fig. 2 and Table 2 show that the formations of compounds $C_3N_6H_6 \cdot H_3PO_4$ and $C_3N_6H_6 \cdot H_2C_2O_4$ can clearly enhance ER performance of $C_3N_6H_6$ material. The $-OH$ in $C_3N_6H_6 \cdot H_3PO_4$ molecule and $-COOH$ in $C_3N_6H_6 \cdot H_2C_2O_4$ molecule should be dominant in heightening the ER activity of the material. These polar function groups in the compounds can effectively improve interface polarization in the suspensions. The suspension of $C_3N_6H_6 \cdot H_3PO_4$ material has larger τ_r value than that of $C_3N_6H_6 \cdot H_2C_2O_4$ material, which can be owing to larger τ_0 value of the suspension of $C_3N_6H_6 \cdot H_2C_2O_4$ material than that of $C_3N_6H_6 \cdot H_3PO_4$ material.

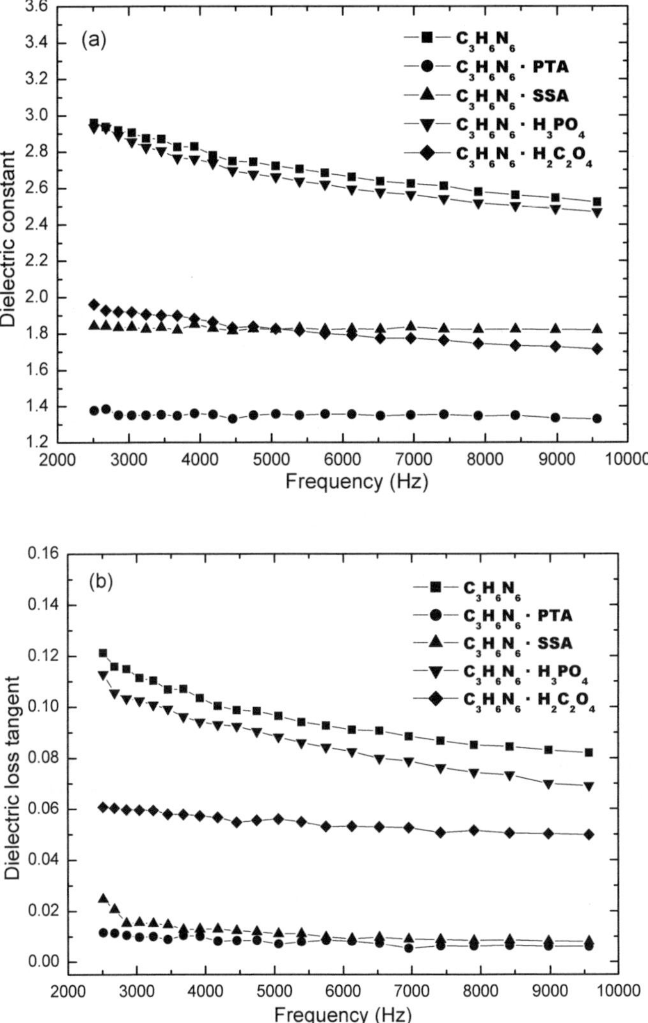

Fig. 4 The frequency dependence of dielectric constant (a) and dielectric loss tangent (b) of material suspensions.

3.3. *Dielectric properties of material*

The frequency dependence of dielectric constant (ε) and dielectric loss tangent ($\tan\delta$) of the material suspensions are illustrated in Fig. 4. Table 2 lists the ε and tanδ values at f=3.0kHz. The results from Fig. 4 and Table 2 show that the ε and tanδ values of $C_3N_6H_6 \cdot PTA$ suspension having the lowest ER performance are the smallest. The ε and tanδ values of $C_3N_6H_6$ suspension are the largest although its ER performance is not the best. The $C_3N_6H_6 \cdot H_3PO_4$ suspension, the ER activity (τ_r) of which is the highest, has the largest ε and tanδ values among the $C_3N_6H_6 \cdot HA$ suspensions. The tanδ value of $C_3N_6H_6 \cdot SSA$ suspension is slightly larger than that of $C_3N_6H_6 \cdot PTA$ suspension. The tanδ value of $C_3N_6H_6 \cdot H_2C_2O_4$ suspension is larger than that of $C_3N_6H_6 \cdot SSA$ suspension and its ε value is in close approximation with that of $C_3N_6H_6 \cdot SSA$ suspension, the τ_r values of both, 2.43 for $C_3N_6H_6 \cdot H_2C_2O_4$ suspension and 2.30 for $C_3N_6H_6 \cdot SSA$ suspension, are also close.

Hao et al.[9] indicated that a large dielectric loss is required for a good ER material, because only a material having a large dielectric loss could give a large interfacial polarization once it is dispersed into a liquid, which can impel the particles to turn and form fibrillation chains along the electric field direction. Moreover, Hao and his co-workers[9] proposed an empirical criterion for selecting ER material: the particle dielectric loss tangent should be approximately 0.10 at 1000Hz, and the larger the particle dielectric constant, the stronger the ER effect. However, the experimental results in regard to the dielectric property of the material and its ER effect in the comparison between the studied materials do not accord entirely with Hao's conclusions mentioned above. The unusual phenomenon can be related to molecule structure of compound in the material. The electronic displacement polarization can be dominant on influencing the dielectric property for a molecule-based material. There is a large nonlocalized π bond π_9^{12} in $C_3H_6N_6$ molecule. The large nonlocalized π bond would be decreased or destroyed with forming $C_3H_6N_6 \cdot HA$ molecules. Therefore, the electric polarizability of $C_3H_6N_6$ molecule is the largest due to the presence of a large nonlocalized π bond π_9^{12}, consequently ε and tanδ values of the suspension of $C_3H_6N_6$ material are the largest among these molecule-based materials[19]. The relative magnitude of the ε and tanδ values is consistent with the relative magnitude of the τ_r values for different $C_3H_6N_6 \cdot HA$ suspensions, which accords with Hao's conclusions mentioned above. Here it should be noticed that the ε values of both $C_3N_6H_6 \cdot H_2C_2O_4$ suspension and $C_3N_6H_6 \cdot SSA$ suspension are very close and the τ_r values of both are also close.

It is worthy to study further the relationship between the dielectric property and ER effect and molecule structure for molecule-based ER materials.

4. Conclusions

The ER performance and dielectric property of a series of molecule-based ER materials, $C_3N_6H_6$ ($C_3N_6H_6$=melamine), $C_3N_6H_6 \cdot$PTA (PTA=4-toluenesulfonic acid, $C_7H_8O_3S$), $C_3N_6H_6 \cdot$SSA (SSA=5-sulfosalicylic acid, $C_7H_6O_6S$), $C_3N_6H_6 \cdot H_3PO_4$ (H_3PO_4=orthophosphoric acid) and $C_3N_6H_6 \cdot H_2C_2O_4$ ($H_2C_2O_4$=oxalic acid) materials, were studied. The results show that the composition and structure of molecule are the dominant factors, the function group plays an important role, in influencing ER performance of the molecule-based ER material.

Acknowledgements

This Project is supported by the scientific research foundation of Baoding University (2009003), the National Natural Science Foundation of China (20023005, 29831010) and the National Key Project for Fundamental Research (G1998061305).

References

1. T. C. Halsey, *Science* **258**, 761 (1992).
2. H. Tada, Y. Saito, M. Hirata, M. Hyodo and H. Kawahara, J. Appl. Phys. 73, 489 (1993).
3. H. Block and J. P. Kelly, *J. of Phys. D: Appl. Phys.* **21**, 1663 (1998).
4. C. W. Wu and H. Conrad, *Phys. Rev. E* **56**, 5789 (1997).
5. M. Whitte, W.A. Bullough, D. J. Peel and R. Froozian, *Phys. Rev. E* **49**, 5249 (1994).
6. T. Hao, *Appl. Phys. Lett.* **70**, 1956 (1997).
7. T. Hao, *Advanced Materials* **13**, 1847 (2001).
8. T. Hao, *Adv. in Colloid and Interface Science* **97**, 1(2002).
9. T. Hao, A. Kawai and F. Ikazaki, *Langmuir* **14**, 1256 (1998).
10. J. Becher. and K. Schaumburg, *Molecular Engineering for Advanced Materials* Kluwer Academic Publishers, Dordrecht, (1995).
11. D. Braga, F. Grepioni and A. G. Orpen, *Crystal Engineering: From Molecules and Crystals to Materials* Dordrecht, Kluwer Academic Publishers, (1999).
12. O. Kahn, *Molecular Magnetism* VCH, (1993).
13. E. Coronado, P. Delhaes, D. Gatteschi and J. S. Miller, *Molecular Magnetism: From Molecular Assemblies to the Device* NATO ASI Series, (1995).
14. J. S. Miller, M. Drillon, *Magnetism: Molecules to Materials* Vol. I -V, Wiley-VCH, (2002-2005).

15. Y. L. Jia, Y. L. Shang, Y. Ma, J. R. Li, J. Wang, S. H. Zhang and M. X. Li, *J. Rare Earths* **25** Suppl., 9 (2007).
16. Y. L. Jia, Y. L. Shang, Y. Ma, J. R. Li, J. Wang, S. H. Zhang and M. X. Li, *J. Rare Earths* **25** Suppl., 24 (2007).
17. Y. L. Shang, Y. L. Jia, Y. Ma, J. R. Li, S. H. Zhang and M.X. Li, *Korea-Australia Rheology Journal* **22,** 43 (2010).
18. Y. L. Jia, L. Huo, Y. Ma, J. R. Li, S. H. Zhang, M.X. Li, *J. Alloys and Compounds* **478**, 538 (2009).
19. J. H. Lin, X. P. Jing, Y. Li, Y. X. Wang, *Inorganic Material Chemistry* Peking University Press, (2006).M. Barranco and J. R. Buchler, *Phys. Rev.* **Cf22**, 1729 (1980).

IMPACT OF ELECTRORHEOLOGICAL BEHAVIOUR OF PVB SOLUTIONS ON THE PROCESS OF ELECTROSPINNING

P. SVRCINOVA[*] and P. FILIP

*Institute of Hydrodynamics, Acad. Sci. Czech Rep., Pod Patankou 5,
Prague, 166 12, Czech Republic*

D. LUBASOVA and L. MARTINOVA

*Tech. Univ. of Liberec, Fac. of Textile Engng., Dept. of Nonwovens, Studentska 2,
Liberec, 461 17, Czech Republic*

The aim of this contribution is to find a correlation between shear rheological characteristics of polymer solutions and spinnability. Specifically it describes the suitability of various polyvinylbutyral (PVB) solutions for the process of electrospinning in which, under a strong electrostatic field, fibres are generated and deposited on a template as a non-woven sheet.

1. Introduction

In electrospinning, polymer solution or melt is ejected from a tip by a strong electric field and deposited as a fibrous mat on a grounded collector. As the charged jet travels in air, its diameter decreases due to high extension rates and simultaneous effect of stretching of the jet and evaporation of the solvent. A typical electrospinning process is described extensively in the literature [1-4]. With respect to their small diameters, the electrospun fibres have a large specific surface and the potential applications of fibres are in various areas such as filtration, nanocomposite material [3,5,6]. Viscosity, concentration, surface tension of polymer solution, molecular weight, intensity of electric field strength, temperature range among the principal parameters affecting fibre formation.

The viscosity of polymer solution has an impact on diameter of fibres, morphology and path of jet. The effect of viscosity, conductivity and surface tension has been investigated by Fong et al. [7]. Wang et al. [8] performed elongation rheology measurements with polymer solutions and they found that a slower rate of capillary thinning is expected to correlate with better spinnability of the polymer solution.

---

[*] Corresponding author: e-mail: svrcinova@ih.cas.cz, tel.: +420 233 109 029

Winslow [9, 10] was the first who started research of rheological behaviour of liquids under a presence of electric field. Using quasielastic light scattering (QELS) Price et al. [11] analyzed the effect of external electric field on the dynamics of polymer chains. They found that in presence of an electric field there is a polarisation effect which is dependent on a difference in permittivity between the polymer segments and solvent.

This contribution describes the suitability of various polyvinylbutyral (PVB) solutions for the process of electrospinning. As a criterion shear rheological behaviour of these materials is taken into account, specifically a course of a curve viscosity ratio η / η_0 vs. electric field strength E. The symbols η and η_0 represent shear viscosities of a solution in the presence and absence of an electric field, respectively. In spite of high elongation rates that are imposed on jets during electrospinning, the present work focuses on a correlation between shear rheology of a polymer solution and its spinnability.

2. Experiment

The polyvinylbutyral (PVB, M_w=60,000 g/mol; Mowital, Kuraray Specialities Europe (KSE)) was dissolved in methanol, ethanol, isopropanol and butanol as 6, 10 and 14 wt% solution (basic characteristics in Table 1) at 25°C. For viscosity measurements there was used a rotational rheometer Anton Paar MCR 501 equipped with an electrorheological cell. The electrospun fibres were obtained from the solution at 30kV with a tip-to-collector distance of 10 cm (=300 V/mm). The surface characteristics of the prepared nanofibre sheets were observed with a scanning electron microscope Vega TS 5130.

Table 1. Basic characteristics of the solvents used and polyvinylbutyral (PVB).

	Relative permittivity [-]	Specific conductivity [S/m]	Surface tension [mN/m]	Density [g/cm^3]	Hansen solubility parameters		
					δ_D [MPa$^{1/2}$]	δ_P [MPa$^{1/2}$]	δ_H [MPa$^{1/2}$]
Methanol	32.7	$1.5*10^{-7}$	22.12	0.7899	15.1	12.3	22.3
Ethanol	24.5	$1.35*10^{-7}$	21.9	0.785	15.8	8.8	19.3
Isopropanol	19.9	$58*10^{-7}$	21.38	0.7813	16	6.8	17.4
Butanol	17.5	$9.12*10^{-7}$	24.5	0.806	16	5.7	15.8
PVB	3.60	$1*10^{-9}$	-	1.09	18.6	4.4	13.0

2.1. *Hansen solubility parameters*

The rheological measurement has a tight connection with solubility determined through Hansen solubility parameters [12] δ_D (representing energy from dispersion bonds between molecules), δ_P (representing energy from polar bonds between molecules), and δ_H (representing energy from hydrogen bonds between molecules) representing each molecule, see Table 1. These three parameters can be taken as orthogonal coordinates for a point in three dimensional so-called Hansen space. The nearer two molecules are in this space, the more likely they are dissolved into each other. This is exactly the situation for PVB solved in isopropanol or butanol. Both materials exhibit shorter distances in the Hansen space in comparison with ethanol and methanol, see Figure 1.

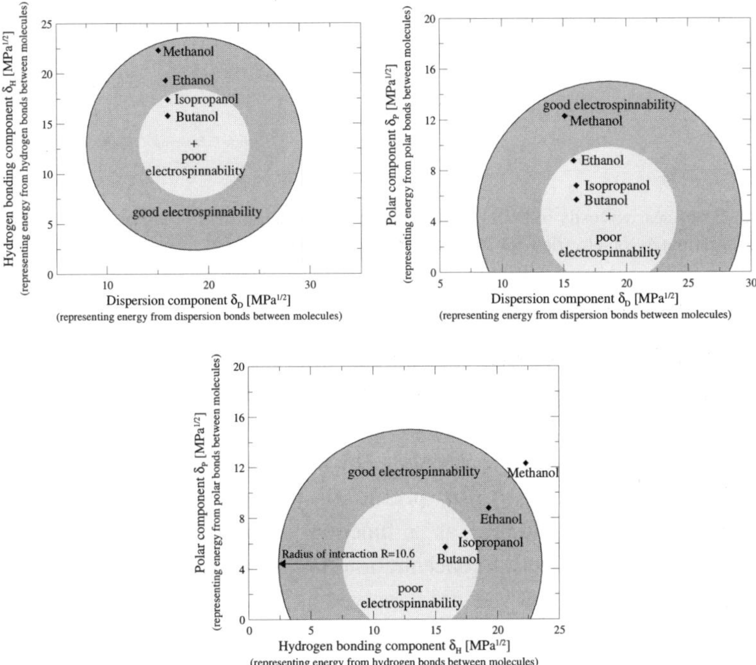

Figure 1. Hansen solubility parameters for PVB and solvents.

2.2. *Shear viscosity measurements*

The shear viscosities of all solutions were measured in the ramp mode using bob and cup arrangement, diameters of 27 and 17 mm in absence and presence of electric field, respectively. Figure 2a depicts shear viscosity of PVB dissolved in alcohols as 10 wt% solutions. PVB dissolved in good solvents shows shear thinning behaviour at high shear rates $> 100s^{-1}$ and its zero shear viscosity is higher. On the contrary, for poor solvents the viscosities are constant, zero shear viscosity is lower. For each solution there was determined the curve relating the viscosity ratio η/η_0 to the electric field strength E (Figure 2b). The good solvents of PVB exhibit an almost constant curve of η/η_0 vs. E, whereas viscosity curves for the poor solvents exhibit enhancement.

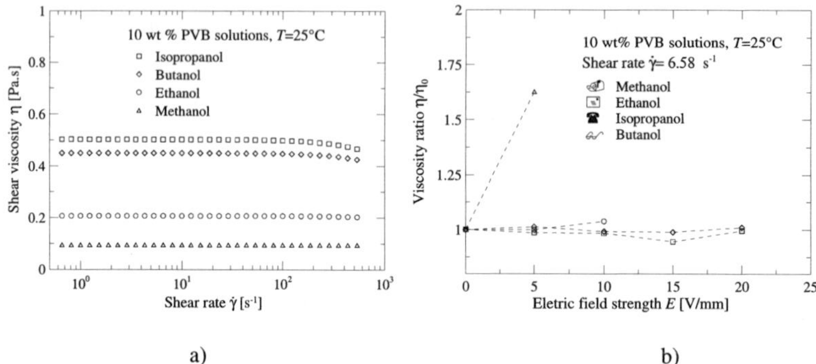

a)   b)

Figure 2. a) Shear viscosity of PVB solutions (absence of electric field strength), b) Dependence of viscosity ratio η/η_0 on electric field strength for PVB solutions.

2.3. *Effect of concentration on viscosity enhancement*

Gupta et al. [13] showed that uniform and bead-free fibres were obtained with increasing concentration ($c/c^* > 6$) of polymer solution.

In this case the critical chain overlap concentration, c^*, was calculated from $c^* \sim 1/[\eta]$, where the intrinsic viscosity $[\eta]$ was determined from a slope in the plot of reduced viscosity η_{red} (a specific viscosity-to-concentration ratio, also denoted as viscosity number) as a function of concentration c in the low concentration limit. Effect of concentration on viscosity of polymer solutions is depicted in Figure 3. PVB was not dissolved as 14 wt% solution in butanol and isopropanol due to high viscosity, formation of gel.

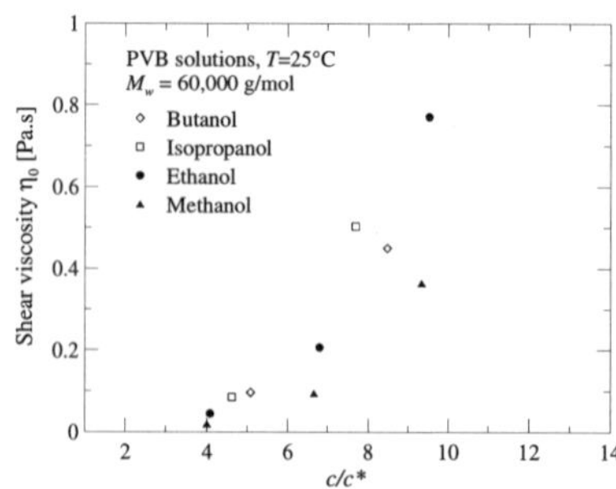

Figure 3. Zero shear viscosity vs. c/c^* for PVB in different solvents.

Figure 4 shows slightly increasing viscosity curves for PVB dissolved in methanol (poor solvent) whereas the viscosity curves of PVB in butanol (good solvent) are constant. The enhancement of viscosity is dependent on time and this effect is not reversible.

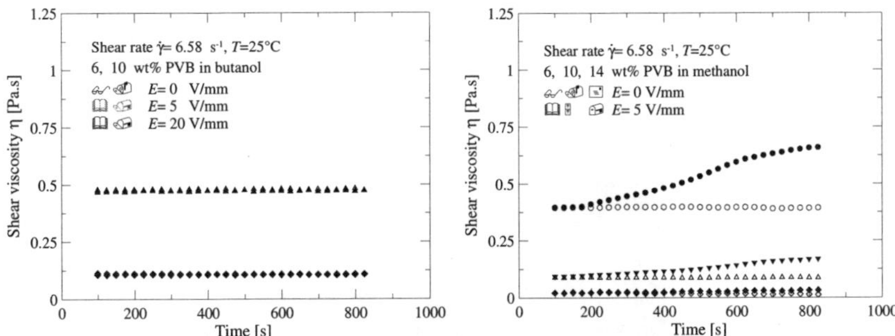

Figure 4. Enhancement of shear viscosity of PVB dissolved in methanol (poor solvent), no change of shear viscosity values with electric field strength for butanol (good solvent).

2.4. *Electrospun fibres of PVB*

Quality of spinnability of these materials was evaluated using the SEM analysis. Effect of concentration of polymer solution in electrospinning process is depicted in Figures 5-7. The fibres were obtained from 6 wt% PVB solutions of isopropanol, ethanol and methanol, in contrast to PVB in butanol where no fibres were obtained. Fibres with higher diameters were obtained from PVB dissolved in ethanol and methanol (poor solvents) as 14 wt% solutions.

This shows that with increasing curve η/η_0 vs. E the spinnability of the respective solution improves.

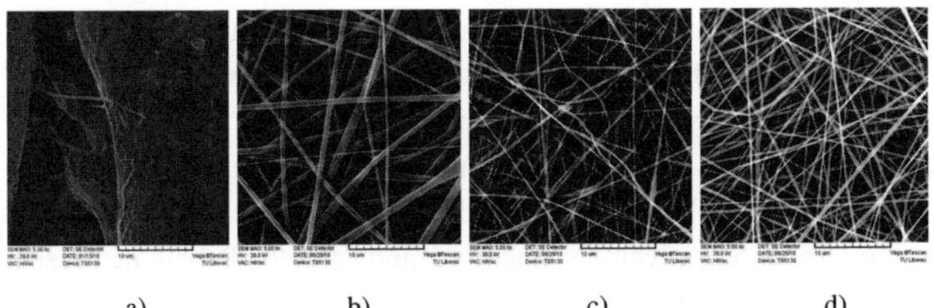

a)　　　　　　b)　　　　　　c)　　　　　　d)

Figure 5. SEM micrographs of PVB electrospun from 6 wt% solution:
a) PVB in butanol,　b) PVB in isopropanol,　c) PVB in ethanol,　d) PVB in methanol.

316

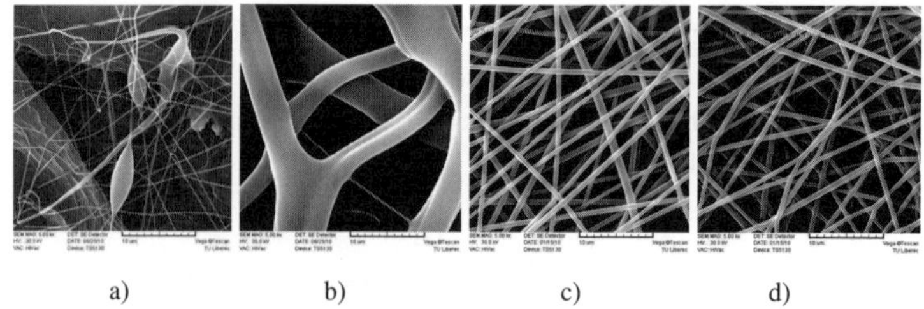

a)   b)   c)   d)

Figure 6. SEM micrographs of PVB electrospun from 10 wt% solution:
a) PVB in butanol, b) PVB in isopropanol, c) PVB in ethanol, d) PVB in methanol.

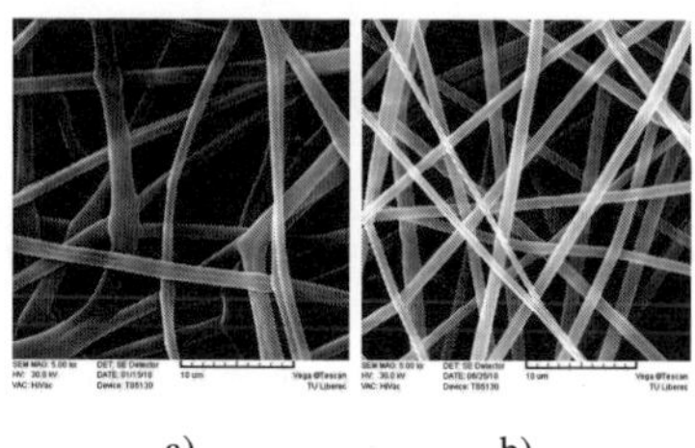

a)   b)

Figure 7. SEM micrographs of PVB electrospun from 14 wt% solution:
a) PVB in ethanol, b) PVB in methanol.

3. Results and Discussion

Impact of the solvents on spinnability is validated through the SEM images (Figures 5-7). From PVB dissolved in butanol and isopropanol (good solvents) there were obtained no fibres, whereas for PVB dissolved in ethanol and methanol (poor solvents) there were developed fibre sheets. The tight connection to spinnability and solubility determined through the Hansen solubility parameters was validated, see Figure 1. The shape of the polymer chain depends on solubility of polymer in a solvent. If the individual polymer chains in good solvent are free to uncoil and stretch, viscosity of a solution is high. On the contrary when polymer chains stay coiled and grouped together into microscopic clusters then this topology reflects in lower viscosity. If the external electric field is applied, the polymer chains uncoil and stretch, therefore the viscosity of solution increases and the process of electrospinning is much easier. If the polymer chain in good solvent is uncoiled and stretched then in presence of electric field polymer chain will not react.

A series of the papers documented an influence of concentration of the solutions on the quality of electrospun fibres. In this case the uniform bead-free fibres from PVB dissolved in poor solvent were obtained at the concentration range $4 < c/c^* < 10$. PVB dissolved in poor solvents as 6, 10, 14 wt% solutions are suitable for the process of electrospinning. It was found that viscosity enhancement of a polymer solution in the presence of an electric field increases with concentration of a polymer solution. The behaviour of two good solvents was different when compared with two poor solvents, which increased viscosity in presence of an electric field. The uniform fibres were obtained from PVB dissolved in isopropanol as 6 wt% solution ($c/c^* \sim 4$), whereas polymer droplets were obtained from PVB dissolved in butanol at the same concentration.

4. Conclusion

The electrorheological properties of the individual solvents play a substantial role in the process of electrospinning. It was found the correlation between shear viscosity and spinnability. Shear viscosity enhancement in a presence of an electric field corresponds to good spinability. Structure of nanofibres was documented by the SEM images.

Acknowledgement

The authors wish to acknowledge GA AS CR for the financial support of Grant No. A200600703.

References

1. S.Ramakrishna, K.Fujihara, W.E.Teo, T.C.Lim, Z.Ma, *An Introduction to Electrospinning and Nanofibres, World Sci. Pub. Co.*, Singapore 2005.
2. A.L.Andrady, *Science and Technology of Polymer Nanofibers, John Wiley & Sons,* New Jersey 2008.
3. Z.-M.Huang, Y.-Z.Zhang, M.Kotaki, S.Ramakrishna, *Comp.Sci.Tech.* **63** (2003), 2223-2253.
4. D.H.Reneker, A.L.Yarin, *Polymer* **49** (2008), 2387-2425.
5. J.Doshi, D.H.Reneker, *J.Electrostat.* **35** (1995), 151-160.
6. P.Gibson, H.Schreuder-Gibson, D.Rivin, *Colloids and Surfaces A: Physicochemical and Engineering Aspects*, **187-188** (2001), 469-481.
7. H.Fong, I.Chun, D.H.Reneker, *Polymer* **40** (1999), 4585-4592.
8. M.Wang, A.J.Hsieh, G.C.Rutledge, *Polymer* **46** (2005), 3407-3418.
9. W.M.Winslow, 1947, *U.S. Patent* 2,417,850.
10. W.M.Winslow, *J.Appl.Phys.* **20** (1949), 1137-1140.
11. C.Price, N.Deng, F.R.Llyond, H.LI, C.Booth, *J.Chem.Soc., Faraday Trans.* **91** (1995), 1357-1362.
12. C.M.Hansen, *Hansen Solubility Parameters: A User's Handbook - 2nd ed., CRC Press, Taylor & Francis Group*, Boca Raton 2007.
13. P.Gupta, C.Elkins, T.E.Long, G.L.Wilkes, *Polymer* **46** (2005), 4799-4810.

EFFECT OF LUBRICANT ADDITIVES ON THE OXIDATION STABILITY OF LORD® MR FLUIDS

DANIEL E. BARBER

Chemical Research and Development, LORD Corporation,
110 LORD Drive, Cary, NC 27511

A series of experiments were conducted to determine the impact of various lubricant additives on the oxidative stability of LORD MR fluids. Oxidative stability was assessed by measuring the oxidation induction time (OIT) by pressure differential scanning calorimetry, and experiments were designed using DOE methodology to determine the effects of and interactions between various fluid components. A variety of anti-oxidants, anti-friction agents, and metal passivators were screened in several designed experiments. Metal passivators and anti-friction agents had little effect on OIT, but a combination of particular primary and secondary anti-oxidants was observed to extend the OIT of a test MR fluid threefold. MR fluids with OIT of greater than two hours were prepared, as compared to the test fluid with an OIT of about 50 minutes.

1. Introduction

Magnetorheological (MR) fluids are suspensions of magnetic particles, usually micron-sized iron particles, in a carrier liquid that undergo a reversible change from free-flowing, nearly Newtonian liquids to semi-solid, Bingham plastic materials upon exposure to a magnetic field [1-2]. Transition of the fluid between the on-state and off-state rheologies occurs reversibly in less than one millisecond [3], and the controllability of the fluid has enabled the commercial development of various devices, particularly semi-active dampers and rotary brakes [4-5].

Relatively few published articles have considered the failure modes of MR fluids during use. Carlson has described the phenomenon of in-use thickening (IUT), which is an increase in the off-state viscosity of the MR fluid due to mechanical degradation and formation of nanometer-sized iron oxide particles during use [6]. Ulicny et al. described MR fluid degradation that resulted in lower on-state forces in a fan clutch application, attributed to oxidation of the iron particle surface and formation of a less-magnetic oxide layer of about 100-200 nm thickness [7]. Several patents from LORD Corporation [8] describe the use of various additives to improve the durability of LORD MR fluids, including anti-oxidants, anti-wear agents, and friction reducers similar to those used in the lubricant industry.

The thermal stability of the chemical components of MR fluids may become more important in applications with high energy dissipation (e.g., clutches or high-energy dampers) or as MR fluids are used in higher-temperature environments [9]. The purpose of the current research was to determine the effect on MR fluid oxidation stability of some of the lubricant additive types described in the patent literature [8].

2. Experimental Details

All additives used in the experiments below were gifts from R. T. Vanderbilt Co., Inc. Experiments were designed using design-of-experiments (DOE) methodology and analyzed using Minitab® 15 software.

2.1. *OIT Measurement*

Oxidative stability of MR fluid was assessed by measuring its oxidation induction time (OIT) according to a procedure modified from ASTM D6186-98. OIT was measured using a TA Instruments Q-10 differential scanning calorimeter fitted with a high-pressure cell (PDSC). A sample of MR fluid (10.0 ± 0.5 mg) was placed in the pressure cell of the PDSC. The cell was pressurized with oxygen gas (99%) to 525 ± 25 psi, and the outflow rate was adjusted to 100 cm^3/min. The cell temperature was increased from 50°C to 180°C at 30°C/min, and then maintained at 180°C until the oxidation peak was observed. Software analysis of the onset point of the oxidation curve yielded the OIT.

2.2. *MR Fluid Sample Preparation*

A stock MR fluid was prepared from proprietary ingredients by mixing iron powder, base oils, and suspension aids such that the iron concentration was about 26% v/v. To 100-g portions of the stock MR fluid were added the necessary amounts of the various additives required according to the experimental design, and the samples were mixed thoroughly. Additive amounts are weight percentages unless otherwise indicated.

3. Results and Discussion

Table 1 lists the additives used in the current study and their typical roles in lubricants [10]. The first three entries in the table are hindered phenol, diphenylamine (two different types), and dithiocarbamate anti-oxidants, respectively, and are representative of the anti-oxidant types used in lubricating oils. The tolutriazole compound, Vanlube 887, is described as a synergist with phenol and/or dithiocarbamate anti-oxidants [R. T. Vanderbilt product literature]. Entry 4 is described as an anti-friction compound and was expected to have no effect on OIT. The last three additives are metal passivators and were chosen for their potential to passivate the high iron powder surface area, which has the potential to catalyze oxidation reactions.

Table 1. Additive types used in the current study.

	Commercial Name	Additive Type	Function
1	Vanlube PCX	2,6-Di-*t*-butyl-*p*-cresol	Primary antioxidant (radical scavenger)
2	Vanlube NA, Vanlube 961	Alkylated diphenylamines	Primary antioxidant (radical scavenger)
3	Vanlube 7723	Methylene bis(dibutyl-dithiocarbamate)	Secondary antioxidant (peroxide decomposer), extreme pressure agent
4	Vanlube 7611M	Ashless phosphorodithioate	Antiwear agent
5	Vanlube 887	Tolutriazole compound in oil	Antioxidant, synergist with phenols and/or dithiocarbamates
6	Vanlube 601	Heterocyclic sulfur-nitrogen compound	Non-ferrous metal passivator, corrosion inhibitor
7	Vanlube 871	Dimercapto-thiadiazole	Film-forming metal passivator, antioxidant/antiwear
8	Cuvan 303	Amino-alkylated tolutriazole	Non-ferrous metal passivator, corrosion inhibitor

3.1. *Anti-Oxidant Screening Experiment*

Additives 1-5 in Table 1 were used in a screening experiment to find the most effective anti-oxidant(s); Vanlube NA was chosen as the diphenylamine in this first experiment. Using DOE methodology, each additive was added to every sample at either a low (0.1%) or a high (0.5%) level, giving a total of 32 ($= 2^5$) fluid samples in all possible additive combinations. The OIT was initially measured only for those eight samples that constituted a ¼ factorial experiment (five variables in eight experiments) to screen for main effects. Data analysis indicated that the dibutyldithiocarbamate, diphenylamine, and tolutriazole synergist had the largest effects. Therefore, the OIT was measured for an additional seven samples, for a total of 15 samples that would represent a full factorial experiment in these three variables with nearly full replication. Because the other two additives also varied in these samples, any significant effects from them would be manifest in the final data analysis.

Figure 1 shows the main effects plot from the data analysis. The values plotted are the average OIT for the given levels of a particular additive, regardless of the concentration of all other additives, as compared to the global average for all 15 samples (the horizontal line at about 52 minutes in all graphs). For example, the methylene bis(dibutyldithiocarbamate) additive had the greatest effect overall, with a low value of about 39 minutes and a high value of about 67 minutes. The statistical significance of an additive's effect is given by a parameter called the P-value, which measures the likelihood that the difference in averages for the low and high additive levels is a result of experimental variation. In practice, only variables with a P-value less than 0.05 are considered to be statistically significant. For the current experiment, the effects of the phosphorodithioate and the 2,6-di-t-butyl-p-cresol were not statistically significant. The tolutriazole was also found to have a small but significant effect at low levels of the dibutyldithiocarbamate.

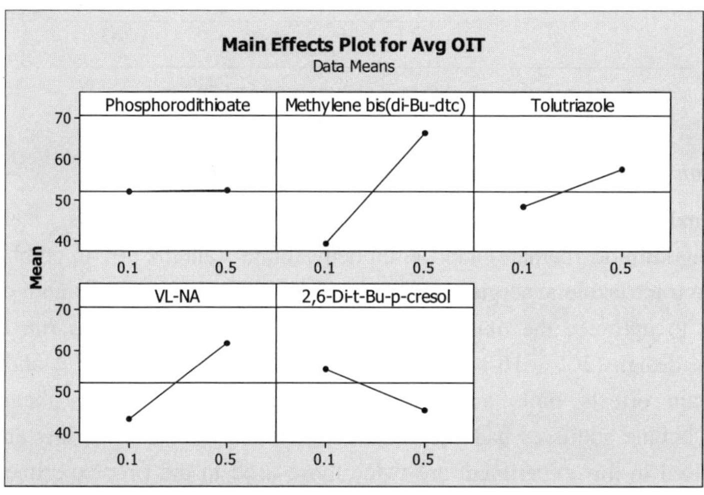

Figure 1. Main effects plot for the anti-oxidant screening experiment.

3.2. *Metal Passivator Effects*

In a second experiment, the three metal passivators (additives 6-8, Table 1) were tested at high (1%) and low (0.2%) levels in combination with an anti-oxidant package containing 0.5% of the dithiocarbamate, 0.1% of the tolutriazole synergist, and 0.5% of either Vanlube NA or Vanlube 961. The experiment was conducted as a full-factorial experiment ($2^4 = 16$ samples). All variables were found to have statistically significant effects, with no significant interactions. Figure 2 shows the main effects of the additives. The thiadiazole and amino-alkylated tolutriazole additives were found to have positive effects, and the diphenylamine anti-oxidant Vanlube 961 was found to have a stronger effect than Vanlube NA, possibly due to solubility differences.

322

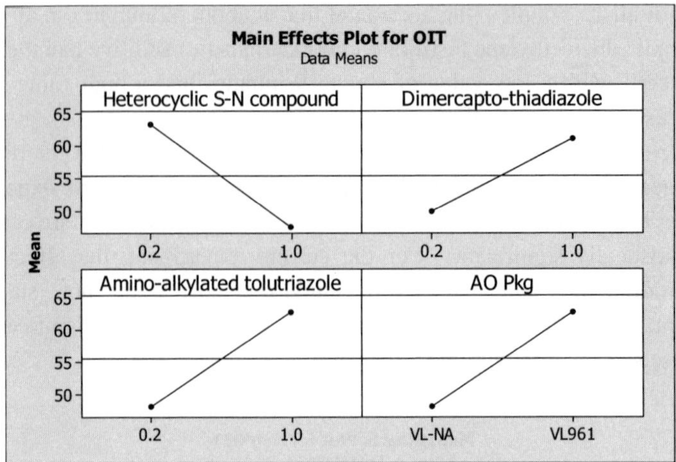

Figure 2. Main effects for metal passivator and diphenylamine additives.

3.3. *Combined Anti-Oxidant/ Metal Passivator Evaluation*

The final experiment combined the three best anti-oxidants, methylene bis(dibutyldithiocarbamate) and the diphenylamine Vanlube 961 in combination with the tolutriazole synergist, with the two best metal passivators above in an attempt to optimize the oxidative stability. The experiment was run as a ½ factorial design (2^{5-1} = 16 samples) at low and high levels of 0.2% and 1.0%. The main effects plots are shown in Figure 3. Only the phenol and diphenylamine additives had statistically significant effects. The low and high levels used in this experiment are twice those used in the first experiment (see Figure 1), and the OIT averages for these two anti-oxidants are also approximately doubled, indicating that the response is linear for these additives in this concentration range regardless of the other additives present. The sample with high levels of both the dithiocarbamate and the diphenylamine and low levels of the other three additives had an OIT of 164 minutes, showing that MR fluids with high oxidative stability can be prepared.

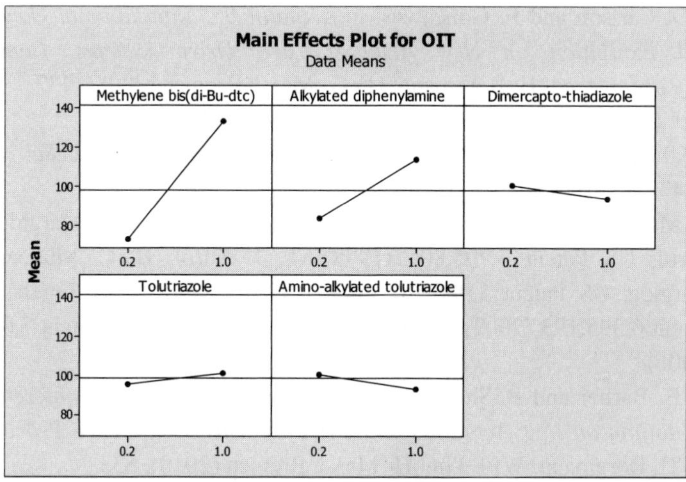

Figure 3. Main effects for combined anti-oxidant/ metal passivator experiment.

4. Conclusions

In the final model, a combination of primary and secondary anti-oxidants was most effective at improving the oxidation stability of the MR fluid. The specific type of diphenylamine additive was also important, likely indicating an influence of additive solubility in the MR fluid. Using this combination of additives, MR fluids with high oxidative stability were successfully prepared.

Acknowledgments

The author gratefully acknowledges student intern Delina Medhin for assistance in preparing samples and measuring OIT and Marilyn Yorke-Smith of LORD Library Services.

References

1. T. Black, J. D. Carlson, in *Synthetics, Mineral Oils, and Bio-Based Lubricants*, ed. L. R. Rudnik, CRC Press (2005), 565.
2. F. D. Goncalves, J.-H. Koo, M. Ahmadian, *Shock & Vibration Digest* **38**, 203 (2006).
3. F. D. Goncalves, J. D. Carlson, *Intl. J. Mod. Phys. B* **21**, 4832 (2007).
4. J. D. Carlson, in *Proceedings of the 10th International Conference on Electrorheological Fluids and Magnetorheological Suspensions*, ed. F. Gordaninejad, O. A. Graeve, A. Fuchs, D. York, World Scientific (2006), 389.

5. J. D. Carlson and F. Goncalves, in *Actuator 08: International Conference and Exhibition on New Actuators and Drive Systems, Conference Proceedings*, ed. H. Borgmann, HVG- Messe Bremen (2008), 477.

6. J. D. Carlson, *J. Intel. Matl. Syst. Struct.* **13**, 431 (2002).

7. J. Ulicny, M. P. Balogh, N. M. Potter, R. A. Waldo, *Matl. Sci. & Eng.* **A443**, 16 (2007).

8. B. Munoz, US Patent 5,683,615 (1997); B. Munoz, A. J. Margida, T. J. Karol, US Patent 5,705,805 (1998); T. J. Karol, B. C. Munoz, A. J. Margida, US Patent 5,906,767 (1999); K. A. Kintz, T. L. Forehand, US Patent 6,395,193 (2002); K. A. Kintz, T. L. Forehand, US Patent 7,087,184 (2006).

9. D. E. Barber and P. Sheng, in *Actuator 10: International Conference and Exhibition on New Actuators and Drive Systems, Conference Proceedings*, ed. H. Borgmann, WFB GmbH- Messe Bremen (2010), 533.

10. J. Braun, in *Lubricants and Lubrication, 2nd Edition*, ed. T. Mang and W. Dresel, Wiley-VCH GmbH (2007), 88.

MAGNETORHEOLOGICAL FLUIDS WITH CARBONYL AND WATER ATOMIZED IRON POWDERS

ANTONIO J F BOMBARD[*] and JOÃO VICTOR R TEODORO

Instituto de Ciências Exatas, Universidade Federal de Itajubá, Av. BPS, 1303 Itajubá/MG, 37.500-903, Brazil

Our aim in this work was propose the use of a ternary blend of two carbonyl iron powder CIP, mixed with water atomized iron powder (WAIP), to reduce the off-state viscosity, without prejudice of MRF performance in terms of yield stress and torque output. The idea of mix water atomized iron powder with carbonyl iron powder is not new. The US Pat. # 5,900,184 by Weiss et al (1999) describes that a binary blend, half-to-half, can reduces the viscosity of MRF in the absence of magnetic field, and increase the torque output under field.

1. Introduction

One challenge in formulating good magnetorheological fluids (MRF) is to reach high volume fractions, keeping the "off-state" viscosity as low as possible. Several approaches are used to reduce the viscosity of MRF in the absence of magnetic field. Another challenge is to find a magnetic material as good as carbonyl iron powder (CIP), but cheaper than it for MRF.

On one hand, to maximize the yield stress and/or torque output of a MRF under applied magnetic field, one must to increase the volume fraction of magnetic dispersed phase as much as possible. On the other hand, it is well known that viscosity of any dispersion increases exponentially with the volume fraction of dispersed phase. This is not different for MRF, in the absence of magnetic field.

Lee describe that dispersions, based on ternary mixtures of spheres with uniform size, will reduces their viscosity, but only above size ratios above 1:5:25 (fine:medium:coarse), and the key to minimize the viscosity is not only the size ratio, but also the amount ratio, maximizing the packing and minimizing the viscosity. However, CIP do not have uniform size, but follows a log-normal size distribution.

---

[*] Corresponding author: antonio.bombard@gmail.com

"

326

2. Experimental

Ternary blends were prepared with 3 types of iron powders: A) water atomized powder Atomet 1001 (Quebec Metal Powders, Canada), that was sieved and only the fraction $53 < \varnothing < 63$ μm was used. B) carbonyl iron powder with D50 = 9 μm and C) carbonyl iron powder with D50 = 1.1 μm, according the manufacturer (BASF SE).

All MRF formulations were prepared with a total of 90% w/w of iron powder. The base liquid was a poly-α-olefine oil. Commercial dispersing agent and an organoclay (same amount in all formulations) were also used. MRF with only one each CIP were highly viscous, but measurable in the rheometer, at low shear at least. MRF with only Medium CIP, above 10 s-1, was not measurable. It created artifacts. On the other hand, MRF with only WAIP was impossible to measure. It seemed like wet sand.

Redispersibility tests with and without magnetic field were made at 25°C. Sedimentation stability was evaluated allowing MRF formulations to settle in test tubes during 1 week, at rest.

3. Results and Discussion

Figure 1 shows that the combination of sizes reduces substantially the viscosity, with the blend viscosity 10 fold lower than that of MRF based on only each powder alone.

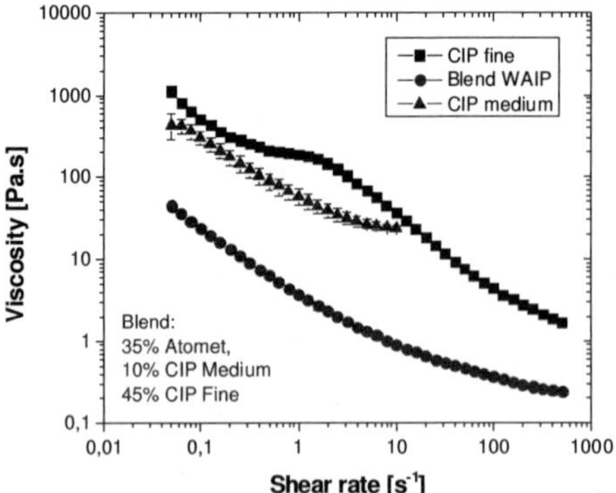

Figure 1. Viscosity curves of 3 MRF prepared with only fine or only medium CIP, and a blend with fine, medium CIP and WAIP.

Figure 2 shows magnetosweep curves for MRF. The relevance of the different amounts of WAIP is evident on the shear stress.

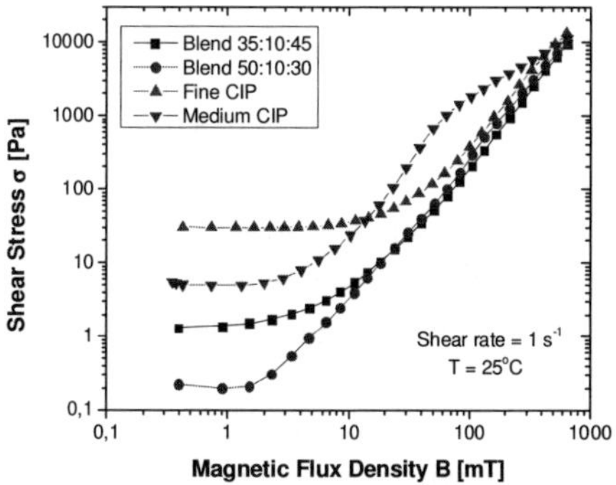

Figure 2: Magnetosweep curves for MRF prepared with only fine or only medium CIP, and two blends with fine, medium CIP and WAIP. Steady shear rate = 1 s-1. Blend ratio is: Atomet/Medium/Fine

One can see above some advantages of ternary blend. The MRF prepared with 35% WAIP, 10% CIP medium and 45% CIP fine, is more efficient, regarding magnetosweep at low, constant shear rate 1 s-1.

The figures 3 and 4 shows side by side the redispersibility of some selected blends, after 1 week settling at rest.

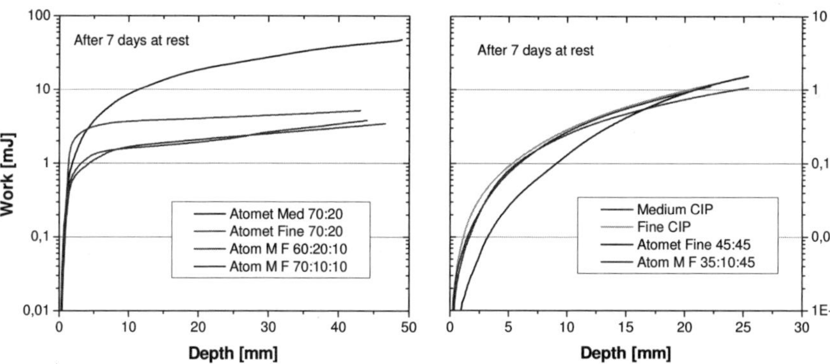

Figure 3. Redispersibility test: Work versus Depth after 1 week at rest. Blends with 60 / 70% WAIP.

Figure 4. Redispersibility test: Work versus Depth after 1 week at rest. Blends with 35 / 45% WAIP and CIP's only.

328

One can see above that the redispersibility of blends with WAIP > 45% is poor (Fig. 3). Redispersibility of blends with ~ 40% WAIP is comparable of MRF with CIP only (Fig. 4).

4. Conclusions

The redispersibility of MRF is dependent of WAIP's amount and combination of iron powders particles sizes.

The type of ternary blend with 35% WAIP, 10% CIP medium and 45% CIP fine is more efficient and shows to us an evidence in the economic scope is great significant. Actually, the WAIP kilogram's price is ~ US$ 1.15. While the CIP kilogram's price is ~ € 20.

If one will propose the use of blend ternary in different technologies and researches applications, there will be a save of 32%, according to current prices. At side some example applications of MRF.

Drawbacks? Wear and abrasion must be investigated in the future.

Acknowledgments

Mr. Teodoro wishes to thanks FAPEMIG for his undergraduate fellowship and METALÚRGICA MOCOCA S.A which kindly provide the air tickets to attend ERMR-2010. Bombard acknowledges FAPEMIG for the Grants APQ-2676-5.02/07, APQ-00531/2008 and financial support to attend ERMR-2010. We also thank Mr. Claude Gelinas (Québec Metal Powders) and Dr. Kieburg (BASF) who kindly gave us the *Atomet 1001* WAIP and CIP samples.

References

[1] Weiss, KD; Carlson, JD and Nixon JD; 1999; *US Patent 5,900,184.*
[2] Lee DI; 1970; J. *Paint Technology* **42**: *579 – 587.*
[3] Kieburg C et al 2007 *Proceedings 10th Int. Conf. On ER fluids and MR suspensions* - Lake Tahoe, USA. World Sci. Pub.; pp. 101-107.

FeNbVB ALLOY PARTICLES SUSPENDED IN LIQUID GALLIUM: INVESTIGATING THE MAGNETIC PROPERTIES THE MR SUSPENSION

GJERGJ DODBIBA, KENJI ONO, HYUN SEO PARK, SEIJI MATSUO and
TOYOHISA FUJITA

*Department of Systems Innovation, Graduate School of Engineering,
The University of Tokyo, Japan*

A MR suspension was prepared by dispersing silica-coated iron alloy particles into a liquid gallium. In other words, the iron alloy particles of 30 to 50 nm in diameter were first prepared and then coated with silica. Next, the particles were then suspended in a liquid Ga (assay: 99.9999%). In addition, the magnetic properties of the synthesized particles and suspension under the influence of the magnetic field were investigated. One of the main findings of this study is that the prepared powder showed a temperature sensitive of magnetization within the testing temperature range of 293 - 353 K. The saturation magnetization of silica-coated FeNbVB particles was about 0.55 T, whereas the saturation magnetization (297 K) of the synthesized MR suspension was 0.019 T.

1. Introduction

Generally speaking, magneto-rheological (MR) fluids are suspensions that contain micron-sized particles (ca. 10 μm) of ferromagnetic materials, as opposed to the nano-sized particles (ca. 10 nm) that are usually used in preparing magnetic fluid (MF). The most commonly used magnetic particles are made of ferrite [1-3], however, they exhibit a relatively low saturation and magnetization. Nevertheless, magnetization of ferrites is temperature sensitive, which make them useful for certain applications. One of the main methods for producing ferrites is wet-grinding in a ball-mill in the presence of a suitable surfactant until the ferrite is in a colloidal state. Centrifugation is then employed to remove larger particles which could lead to agglomeration and sedimentation. In addition, the co-precipitation method, which has been the subject of many patents and publications [4], is also widely used for the preparation of particles of magnetite, maghemite, etc.

Several studies have been carried out on how to prepare iron alloy particles [5-6]. Usually, iron alloy particles are prepared by mechanical alloying [7-8], since it is relatively simple and cost-effective. On the other hand, little is known about the chemical method on how to prepare the iron alloy particles.

Nevertheless, there are few reports which indicate that the particles prepared by chemical synthesis may have better magnetic properties due to their smaller particle size and uniform size distribution when compared with mechanical alloying [5, 6]. In this work, we chose to prepare the FeNbVB alloy particles by chemical synthesis.

Carrier liquid is another very important component of a smart fluid. Generally speaking, carrier liquids are organic or polar solvents. Magnetic particles suspended in liquid are then stabilized by using surfactant. Typical organic solvents are toluene and cyclohexane, which are good dispersant. However, they exhibit relatively high toxicity, and high volatility. Water, on the other hand, is a typical polar solvent, which unfortunately is not very useful for certain types of applications, where the temperature of the surrounding is relatively high. However, water-based smart fluids are useful for biomedical applications because they are non-toxic. A third alternative could be a liquid metal. Using a liquid metal as carried liquid could be of benefit, since a liquid metal has a relatively high thermal, electrical conductivity, and boiling point.

Some researchers have prepared magnetic fluid by dispersing iron particles into mercury, [9]. However, mercury (Hg) is toxic and therefore, when possible, its use should be avoided. Hence, liquid gallium (Ga) could be an environment-friendly alternative. In addition, gallium has a wide temperature range in the liquid state, especially when compared with mercury. Its melting point is lower than 30 oC, whereas its boiling point is higher than 2200 oC. Moreover, both the thermal and electrical conductivity of gallium are also higher than those of mercury, or any conventional organic or aqueous carrier liquids. Thus, these are the reasons, why liquid gallium could be a useful carrier liquid for preparation of a new type MR suspension. Consequently, the aim of this research is: (1) to prepare temperature sensitive FeNbVB alloy particles coated with silica and suspend them in liquid gallium; and (2) to investigate the magnetic properties of the suspension.

2. Experimental

Figure 1 depicts the steps we followed in order to prepare the liquid gallium MR suspension. First, the synthesis of iron alloy particles was carried out, and then the particles were coated with silica. Finally, we suspended the silica-coated iron alloy particles in liquid gallium (Ga assay: 99.9999 %). The concentration of particles in liquid was kept constant at 3.0 mass%.

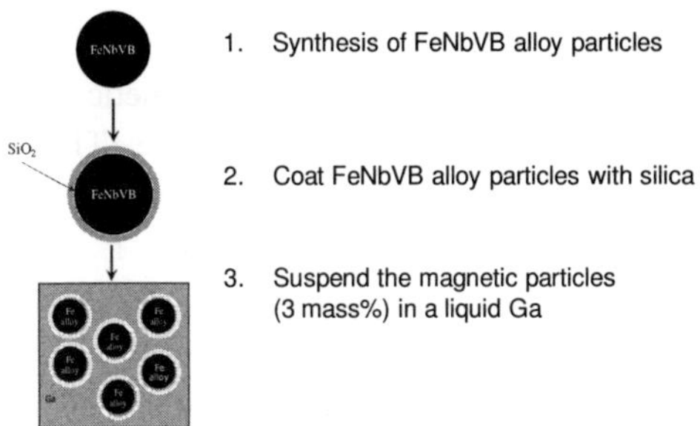

Figure 1. Schematic of the experimental procedure for the MR suspension.

2.1. *Preparation of FeNbVB alloy particles*

Figure 2 shows the experimental procedure for preparation of iron alloy magnetic particles by chemical synthesis. Reagents of niobium fluoride (NbF_5), iron (II) chloride tetrahydrate, ($FeCl_2 \cdot 4H_2O$) and sodium borohydride ($NaBH_4$) were dissolved separately in distilled water, whereas ammonium vanadate (NH_4VO_3) was dissolved in a mixture of water and 0.5 M solution of sulfuric acid. Next, niobium fluoride (NbF_5), ammonium vanadate (NH_4VO_3) and iron (II) chloride tetrahydrate ($FeCl_2 \cdot 4H_2O$) were mixed in a flask and a 336 mM solution of sodium borohydride ($NaBH_4$) was then added. The FeNbVB particles were therefore formed due to a reduction process. Finally, the synthesized particles were collected by means of filtration, washed several times with ethanol, and then dried in a desiccators at room temperature.

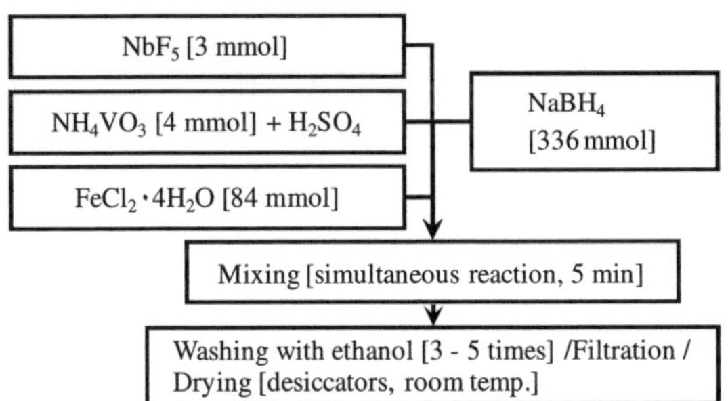

Figure 2. Experimental procedure for preparing FeNbVB alloy particles.

2.2. *Preparation of Silica-Coated FeNbVB alloy particles*

Generally speaking, "naked" metal powders are not easily dispersed in a liquid metal. Previous studies have already suggested that magnetic nano-particles are

332

easier dispersed in a liquid metal if they are coated with silica. Sol-gel process is widely adopted for preparation of silica-coated magnetic particles, [10-12]. In a typical reaction, a mixture of tetraethyl orthosilicate (TEOS) that is the silica precursor and water is used. The process is typically carried out at room temperature and at relatively high pH (ca. pH 10), Eq. 1.

$$Si(OC_2H_5)_4 + 2H_2O \leftrightarrow SiO_2 + 4C_2H_5OH \tag{1}$$

The reaction (Eq. 1) between the precursor and water occurs in two steps. The first step (Eq. 2) is a hydrolysis reaction, followed by a condensation (or polymerization) reaction, (Eq. 3). The hydrolysis results in the formation of an M-OH bond while the condensation reaction occurs once the OH groups are created, [13].

$$Si(OC_2H_5)_4 + 4H_2O \leftrightarrow Si(OH)_4 + 4C_2H_5OH \tag{2}$$

$$Si(OH)_4 \leftrightarrow SiO_2 + 2H_2O \tag{3}$$

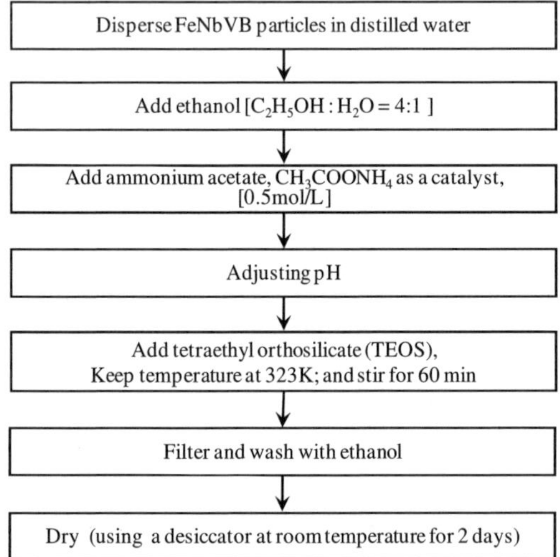

Figure 3. Experimental procedure for preparing silica-coated FeNbVB alloy particles.

Figure 3 shows the procedure for preparing silica-coated FeNbVB alloy particles. The coating of FeNbVB nanoparticles with silica was carried out in alcohol/water (4:1 v/v %) mixture containing FeNbVB particles as nucleation centers. The 0.5 M of ammonium acetate (CH_3COONH_4) was added in the sample as a catalyst. Then, tetraethyl orthosilicate, TEOS, ($Si(OC_2H_5)_4$) was added to this mixture. After stirring for 60 min., particles were filtered, washed with ethanol and dried in a desiccator for 2 days. During the process the temperature was kept constant at 323 K.

2.3. *Measurements*

In this work, the crystalline structure of iron alloy particles was analyzed by X-ray diffraction (XRD, *Rigaku-2400*), whereas the morphology and the size of the particles was analyzed by Transmission Electron Microscopy, equipped with energy dispersive X-ray spectroscopy (TEM-EDS, *JEM-2000FX*). The chemical composition of the synthesized iron alloy particles, on the other hand, was analyzed by Inductive Coupled Plasma Optical Emission Spectrometry (ICP-OES, Optima 5300 DV, PerkinElmer). Next, a vibrating sample magnetometer (VSM) was employed to measure the magnetization of the samples as a function of magnetic flux density in order to determine the saturation magnetization. In this work, the calibration of VSM was carried out by using a Ni sample as reference, whereas the maximum magnetic field applied to samples was 0.9 T.

3. Results and Discussions

This figure shows the XRD pattern of FeNbVB and silica-coated FeNbVB alloy particle. Only Fe was observed by XRD, indicating the amorphous structure of other elements.

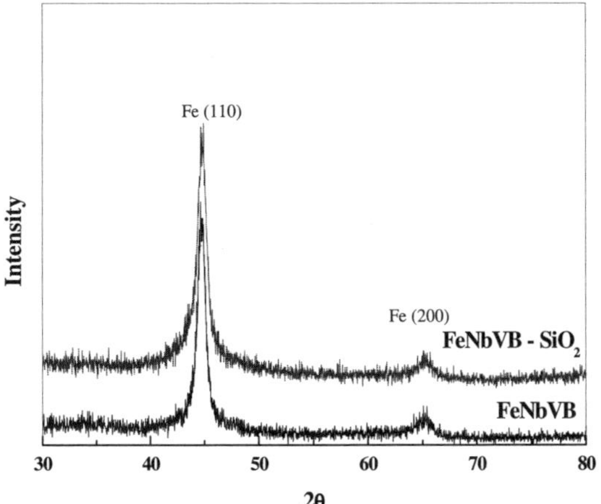

Figure 4. XRD pattern of FeNbVB particles before and after silica coating.

TEM images and EDS spectra for FeNbVB alloy particles, before and after the silica-coating process are shown in Figure 5. The size of FeNbVB particles is between 30 - 50 nm. In addition, an amorphous type of silica shell with a thickness of about 10 nm was also observed on the silica-coated FeNbVB particles (b).

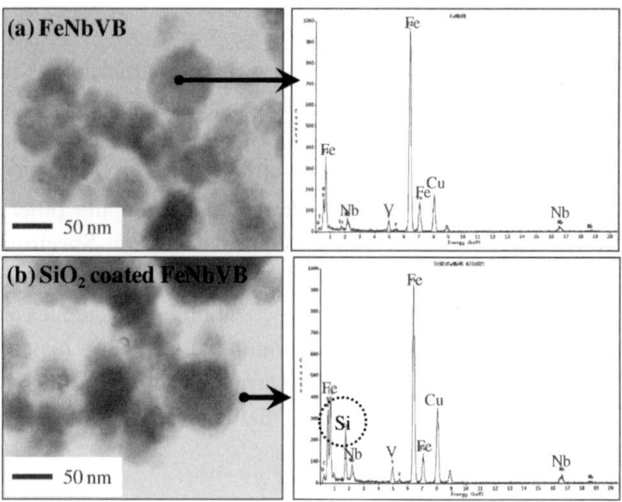

Figure 5. TEM images and EDX spectra of FeNbVB particles before (a) and after (b) silica coating.

Table 1 shows the results of the chemical analysis, measured by ICP-OES after dissolving the sample using an acidic solution. The content of niobium and vanadium in sample was 2.8 and 3.7 mol% respectively. The specific surface area of the powder was about 10.6 m^2/g, and the density was about 7500 kg/m^3.

Table 1. Elemental composition of the synthesized FeNbVB alloy particles, measured by ICP-OES

	Fe	80.28
	Nb	2.84
Elements, (mol%)	V	3.75
	B	13.13
Total, (mol%)		100

Figure 6 shows the magnetization curves for both iron alloy (Fig. 6.a.) and silica coated iron alloy particles, (Fig. 6.b.) as a function of magnetic field and at magnetic flux density. The measurements were carried out at various temperatures, which ranged between 293 and 353 K. It can be seen that the magnetization increased with increasing of magnetic flux density, though the magnetization values for silica coated particles are smaller due to the existence/presence of the non-magnetic layer of silica.

Figure 7, on the other hand, shows the saturation magnetization for the FeNbVB and silica coated FeNbVB nano-particles as a function of temperature, when the magnetic flux density was kept constant at 0.9 T. These results suggest that there is a temperature dependency of the magnetization for both FeNbVB and silica-coated FeNbVB particles; and this tendency is similar, since the rate of magnetization for both cases was 0.001 T/K.

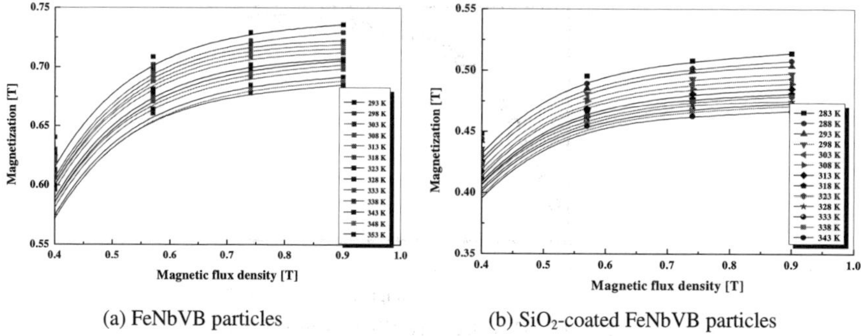

(a) FeNbVB particles   (b) SiO₂-coated FeNbVB particles

Figure 6. Magnetization curves of (a) FeNbVB and (b) silica-coated FeNbVB nanoparticles, measured between 293 K and 353 K.

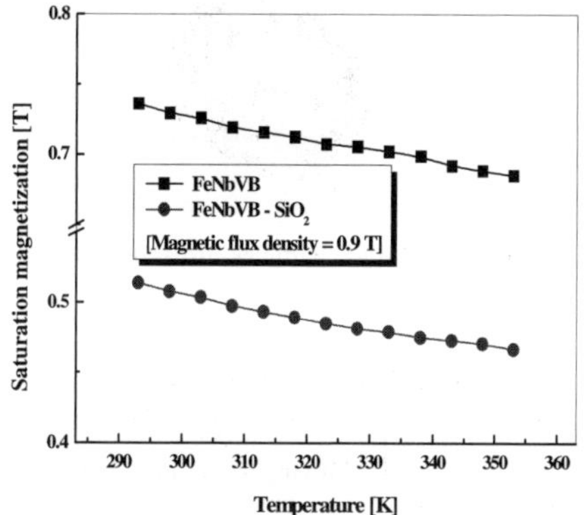

Figure 7. Saturation magnetization as a function of temperature, measured at 0.9 T.

Figure 8 shows the magnetization curve of SiO2 coated FeNbVB magnetic particles suspended in liquid gallium. In other words, this is the magnetization curve of the MR suspension. We found that the saturation magnetization of the suspension was 0.019 T.

Finally, we observed the vertical movement of the synthesized MR suspension (Fig. 9) under the influence of the magnetic field. There were two distinctive modes of operation. The first one is the movement of the suspension upward from the lowest position, and the second one is the movement of the suspension downward from the highest position. Table 2 tabulates various parameters, which were adjusted at the given values in order to ensure the vertical movement of the suspension.

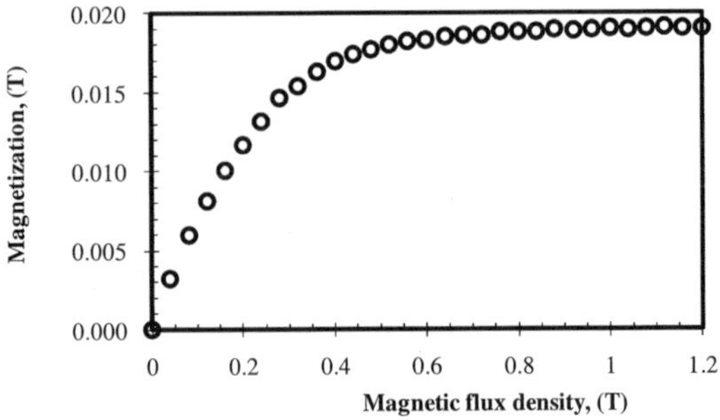

Figure 8. Magnetization curve of SiO2 coated FeNbVB alloy magnetic particles suspended in liquid gallium.

Figure 9. Photo of silica-coated FeNbVB alloy particles suspended in liquid gallium.

Table 2. Value of various parameters

Parameters	Mode 1: Moving upward from the lowest position	Mode 2: Moving downward from the highest position
Magnetic flux density at starting point:	0.210 T	0.142 T
Magnetic field gradient:	4,304 kA/m^2	2,201 kA/m^2
Magnetization of MR suspension:	0.0122 T	0.0092 T
Temperature:	298 K	298 K
Applied current:	15 A	4 A
Applied voltage:	24 V	6 V

4. Conclusions

In this work we prepared temperature sensitive FeNbVB alloy particles coated with silica and suspend them in liquid gallium. The size of FeNbVB alloy particles, prepared by chemical synthesis was 30 - 50 nm, whereas the saturation magnetization was about 0.75 T. The prepared FeNbVB alloy particles were then coated with a silica layer, which thickness was about 10 nm. We found that the prepared powder showed a temperature sensitive of magnetization within the testing temperature range of 293 - 353 K. The saturation magnetization of silica-coated FeNbVB particles was about 0.55 T; whereas the saturation magnetization (297 K) of the synthesized MR suspension was 0.019 T. Finally, we observed that the MR suspension easily moved under the influence of the magnetic field and therefore it can be employed at various application were a relatively high electrical or thermal conductivity of the smart material is required.

Acknowledgments

The research was carried out with the support of a FY2010 Grant-in-Aid for Scientific Research (S), provided by Japan Society for the Promotion of Science (JSPS).

References

1. W. J. Schuele, and V. D. Deetscreek, Ultra-fine particles, *Kuhn W E (ed.) Wiley*, New York, (1963)
2. J. Smit, and H. P. J. Wijn, Ferrites, *Wiley*, New York., (1959)
3. J. L. Dormann, and M. Nogues, *J. Phys: Condens. Mat.*, **2**, 1223-1237, (1990)
4. S. E. Khalafalla, and G. W. Reimers, *U. S. Patent* 3,764,540, (1973)
5. D. Eckert, K. H. Muller, A. Handstein, J. Schneider, R. Grössinger and R.Krewenka, *IEEE Trans. Magn.*, **26**, 1834-1836, (1990)
6. S. Sun, C. B. Murray, D. Weller, L. Folks and A. Moser, *Science*, **287**, 1989-1992, (2000)
7. J. J. Kim, Y. Choi, S. Suresh and A. S. Argon, *Science*, **295**, 654-657, (2002)
8. H. W. Kwon and I. C. Jeong, *Physica status solidi (a)*, **201(8)**, 1921-1925, (2004)
9. E. Dubois, J. Chevalet, and R. Massart, *J. Molecular Liquids*, **83**, 243-254, (1999)
10. I. Hap, and E. Matijevic, *J. Coll. Int. Science*, **192(1)**, 104-113, (1997)
11. J. Y. Ying and T. Sun, *J. Electroceramics*, **1**, 219-238, (1997)
12. S. Castillo, M. Morán-Pineda, V. Molina, R. Gómez, and T. López, *Applied Catalysis B: Environ*, **15**, 203-209, (1998)
13. M. Gopal, W. J. Moberly Chan and L. C. De Jonghe, *J. Mat. Scie.*, **32**, 6001-6008, (1997)

NOVEL ELECTRORHEOLOGICAL FLUIDS WITH POLYMER PARTICLES CONTAINING ORGANIC DOPANTS

YANINA REICHERT and HOLGER BÖSE[*]

*Fraunhofer-Institut für Silicatforschung ISC, Neunerplatz 2,
D-97082 Würzburg, Germany*
[*]*E-mail: boese@isc.fraunhofer.de*

In a novel approach, electrorheological (ER) fluids based on doped polyurethane (PUR) particles were prepared and described. The dopants were electrically polar and polarizable organic substances either as guest molecules or as modifying agents of the PUR network. The ER properties of these ER fluids were investigated and compared with those of salt-doped and non-doped reference samples. It was found that organic compounds, in particular those with a highly conjugated π-electron system, as guest molecules of PUR delivered remarkable improvements of the ER activity. In some ER fluids, the PUR particles were modified by covalent bonding of polar organic molecules to the polymer network. These modifying compounds are equipped with one or several reactive hydroxyl or amine groups. Furthermore, the influence of the number of chemical bonds of the modifying agents to the PUR network on the ER properties was also studied. Upon the application of modifying agents, which serve as plasticizer in the PUR particles, ER fluids with significantly improved electrorheological properties were obtained. The most promising ER fluid exhibited an extraordinary high shear stress of more than 10 kPa and simultaneously a very low current density below 4 μA/cm^2 at 40 °C, which strongly improves the two basic features of ER activity in comparison to the known salt-doped ER fluids.

1. Introduction

Most electrorheological (ER) fluids are suspensions of electrically polarizable particles in a carrier liquid. In the presence of an electric field, ER fluids exhibit a strong increase of their shear stress due to the formation of mechanically resilient chain-like structures of the polarized particles along the electric field. The particle polarization is generally caused by differences in the conductivity and/or dielectric properties between the particle and the liquid medium [1].

ER fluids have attracted much attention in the past because they represent a promising class of materials for a large number of applications like automotive shock absorbers, industrial dampers, hydraulic drives, fitness devices and haptic operation tools [2, 3]. Some products based on ER fluids have already been commercialized [4].

For most applications, a lot of requirements have to be met by the ER fluid in order to be exploitable in commercial products. The most relevant features of a good ER fluid are high shear stresses and low current densities in the electric field as well as a low base viscosity in the absence of the field. These properties give rise to high damping forces, low power consumption and a large variability of the shear force. Other important characteristics concern the sedimentation stability, the redispersibility, the response time and the accessible temperature range. However, the major challenge in the development of ER fluids is, to combine all these requirements in one single material.

A well known type of ER fluids is a suspension of doped polyurethane particles in silicone oil. The PUR particles are doped with an inorganic salt, which is soluble in the polymer due to the high polarity of polyether chains of the used polyol [5]. The arising ions are the source of the electric polarizability of the particles, which causes the ER activity. Preferred salts are LiCl and $ZnCl_2$, which generate different ion conductivities in the particles [6].

These PUR based ER fluids have a well balanced combination of properties in terms of shear stress and current density in the electric field, base viscosity without field and redispersibility. However, higher shear stresses and simultaneously lower current densities in the electric field could enhance the benefit for many applications, since the dimensions of ER device and the required electric power may be reduced.

The introduction of organic dopants instead of inorganic salts into the polymer network represents a novel approach and is herein described. Preferred organic species are those with a high dipole moment which can be aligned in the electric field and/or with high polarizability. Both mechanisms can lead to strong polarization of the particles and enhance the ER activity. The dopants are distinguished between compounds which are solved in the PUR particles due to intermolecular interactions (guest molecules) and those which are covalently bonded to the polymer (modifying agents). The purpose of this work was to study the influence of various organic dopants in the PUR particles on the electrorheological properties of the corresponding ER fluids and to correlate the effect with their chemical structures.

2. Experimental

The PUR particles with a size range between ca. 1 µm and 15 µm were prepared by suspension polymerization. In this synthesis, the particles were directly formed in the carrier liquid. The concentration of the particles was 50% by weight. As the continuous phase of all ER fluids, silicone oil with a viscosity of 5 mm^2/s at 25 °C was selected. The key educts of the PUR particles were polyethertriol and toluene diisocyanate. As organic dopants of PUR, polar aromatic compounds bearing donor and acceptor groups were chosen due to their comparably high permanent dipole moments between 3 D and 13 D as well as their high polarizabilities.

In addition to the donor-π-acceptor system, the modifying agents possessed free OH- and NH-groups in order to react with the isocyanate functions of the toluene diisocyanate molecules, leading to covalent bonding to the polyurethane system. The used dopants with their chemical structures and numbering are depicted in Fig. 1.

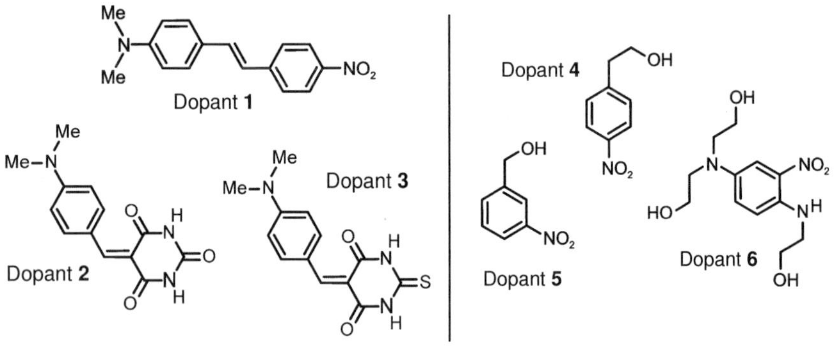

Figure 1. Applied dopants as guest molecules: **1** 4-dimethylamino-4'-nitrostilbene, **2** 5-[[4-(dimethylamino)phenyl]methylene]barbituric acid, **3** 5-[[4-(dimethylamino)phenyl]methylene]-2-thio-barbituric acid (left) and as modifying agent with covalent bonding to the polymer: **4** 4-nitrophenetyl alcohol, **5** 3-nitrobenzyl alcohol, **6** 2,2'-[4-(2-hydroxyethylamino)-3-nitrophenylamino]diethanol (right)

The rheological properties of the prepared ER fluids were investigated using a rheometer MC 100 from Anton Paar Physica equipped with a concentric cylinder measuring system and a high voltage source. The standard measuring conditions were DC electric field strengths between 0 kV/mm and 5 kV/mm, 40 °C temperature and 10 s^{-1} shear rate. Furthermore, the viscoelastic properties of the solid PUR samples were measured at 25 °C using a rheometer MC 300 from Anton Paar Physica equipped with a plate-plate measuring system. The amplitude sweep was carried out between 0.0001 and 1 at 1 Hz frequency.

3. Results and Discussion

3.1. *Reference ER fluids*

In the first step, the already known ER fluids with LiCl and $ZnCl_2$ as inorganic dopants as well as a sample with undoped PUR particles were studied as references. Fig. 2 shows the dependence of the shear stress and the current density, respectively, on the electric field strength. The highest shear stress was obtained for the ER fluid with the LiCl dopant, but the current density (70 $\mu A/cm^2$ at 5 kV/mm) and the corresponding required electric power were relatively high. In contrast to this, the ER fluid with the $ZnCl_2$ dopant had strongly reduced shear stresses and current densities. Finally, the undoped sample had an intermediate shear stress, which is referred to impurities in the PUR polymer, and the lowest current density of all reference ER fluids (below 3 $\mu A/cm^2$). The ER properties vary with the shear rate applied, but the general conclusion from the three reference ER fluids is the same.

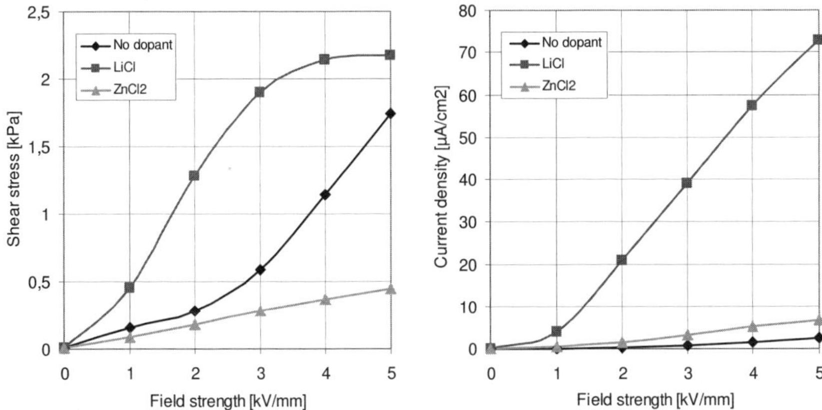

Figure 2. ER fluids with salt doped and undoped PUR particles: Shear stress vs. field strength (left) and current density vs. field strength (right), all measurements at shear rate 10 s^{-1} and temperature 40 °C

3.2. *ER fluids with guest molecules as organic dopants*

In the next step, the ER fluids, whose PUR particles contain organic compounds without covalent bonding to the PUR network, were investigated and their properties were compared with those of the reference ER fluids. The results are compiled in Fig. 3. Obviously, the ER fluid containing the stilbene derivate **1** exhibits remarkable electrorheological properties, i. e. a high shear stress of about 5 kPa and simultaneously a very low current density of below 4 $\mu A/cm^2$ at 5 kV/mm. Dopant **1** differs from the other compounds in the lack of free hydrogen atoms, which could be a source of proton conductivity in the presence of an electric field.

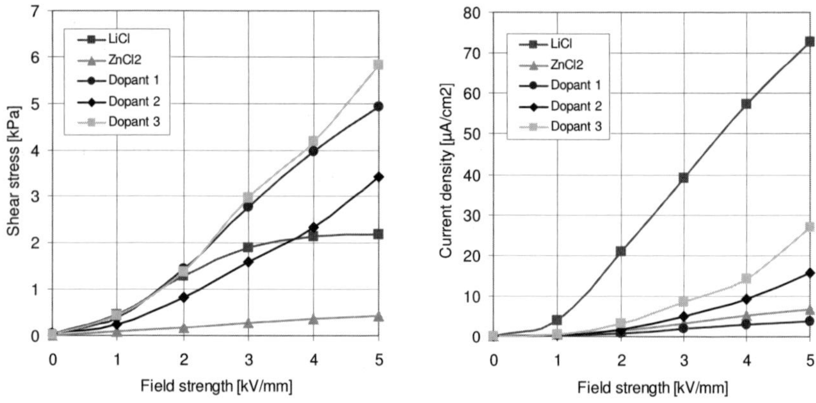

Figure 3. ER fluids with PUR particles containing guest molecules as organic dopants: Shear stress vs. field strength (left) and current density vs. field strength (right), all measurements at shear rate 10 s^{-1} and temperature 40 °C

Furthermore, it is noticeable from Fig. 3, that the thiobarbituric acid derivative **3** dispersed in PUR particles provides stronger shear stresses and current densities of the respective ER fluid in the electric field than the barbituric acid derivative **2**. This effect is attributed to the higher polarizability of the thiocarbonyl group in comparison to the carbonyl group. Consequently, it could be shown that the chemical structure of the doping agent, namely its polarizability, affects the attainable ER activity.

In addition, it was found that the PUR system may become softer by the use of organic dopants. Applying differential scanning calorimetry (DSC), the glass transition temperatures (T_g) of polyurethane plates were determined. A lowering of T_g for PUR containing the barbituric acid derivative **2** compared to the respective pure PUR was detected. The majority of the investigated doping agents exhibited a so-called plasticizer effect, thus improving the flexibility and ductility of the respective PUR system, which leads in most cases to advanced ER performance. It is assumed that due to the higher flexibility of the organically doped PUR particles, an orientation of the permanent dipoles along the electric flux lines is supported. The result of this dipole alignment is a higher polarization of the doped PUR particles as well as an enhancement of the ER activity.

3.3. *ER fluids with modifying agents as organic dopants*

In the final step, ER fluids with PUR particles containing covalently bonded organic compounds were the subject of investigation. The selected organic compounds had either one or several hydroxyl or amine groups, which are capable to react with the diisocyanate under formation of a covalent bond to the PUR network.

In the case of the integration of the modifying agents 4-nitrophenethyl alcohol (**4**) and 3-nitrobenzyl alcohol (**5**), which both have only one free OH-function for the addition to the PUR structure, the polymer chain propagation stopped. As a result, the formed PUR had a lower degree of crosslinking. This plasticizer effect was confirmed by oscillatory shear tests (see Fig. 4) and glass transition temperature measurements.

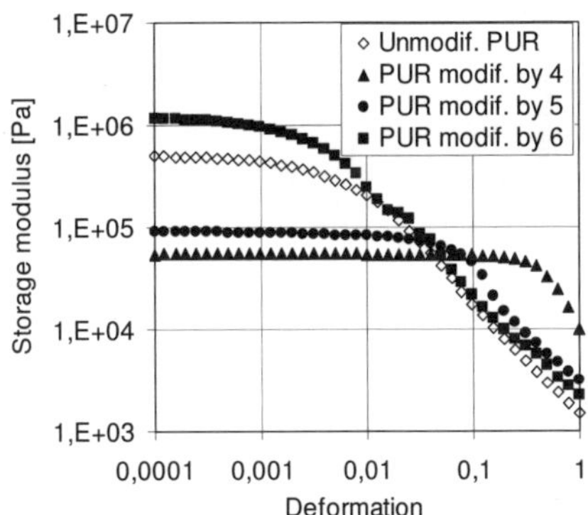

Figure 4. Solid PUR samples containing modifying agents as organic dopants: Storage modulus vs. deformation (left) and photo of a sample in the rheometer (right)

In contrast, the application of the modifying agent **6** with more than two functional group resulted in an increase of the crosslinking density of the PUR framework, i.e. the PUR became stiffer. This effect caused a rise of the glass transition temperature in comparison to unmodified PUR. As another evidence of the enhancement of the PUR crosslinking density, an increase of the storage modulus in the oscillatory measurements was observed. Fig. 4 displays the storage modulus of solid PUR samples modified by organic dopants **4**, **5**, and **6** as well as of unmodified PUR versus the deformation. In contrast to the modifier **6**, the compounds **4** and **5** caused a decrease of storage modulus compared to unmodified PUR.

The influence of the modifying agents on the ER properties of the corresponding ER fluids is shown in Fig. 5. The particles modified by plasticizers (compounds **4** and **5**) exhibited a remarkable ER activity, namely shear stresses of about 4.4 kPa and 5.4 kPa, and current densities of only 4.2 $\mu A/cm^2$ and 1.3 $\mu A/cm^2$, respectively. One reason for the enhancement of the ER effect of these ER fluids may be the softness of the polymer, resulting in an less hindered orientation of the electric dipoles in the presence of the electric field.

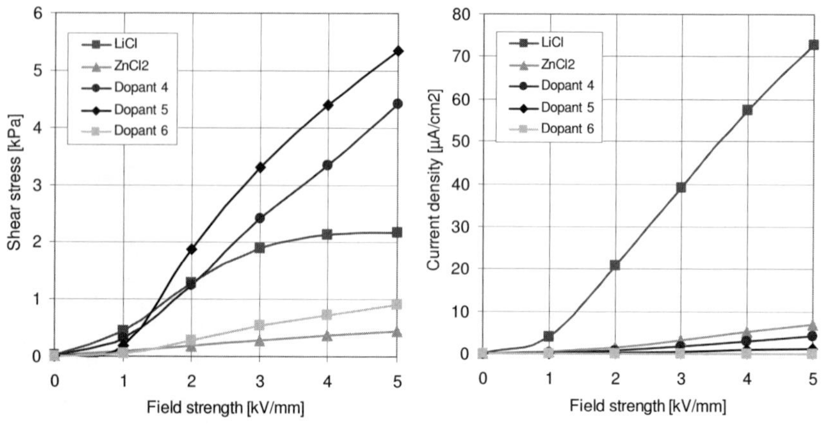

Figure 5. ER fluids with PUR particles containing modifying agents as organic dopants: Shear stress vs. field strength (left) and current density vs. field strength (right), all measurements at shear rate 10 s^{-1} and temperature 40 °C

In contrast to the plasticizers **4** and **5**, the modifying agent **6** at its highest concentration caused a dramatic decline of the shear stress (see Fig. 5).

Simultaneously, from the hardening of the PUR particles a very low current density of the ER fluid was found.

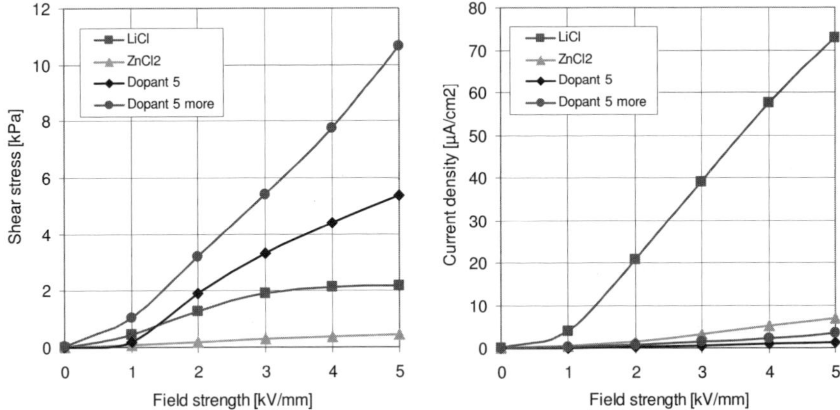

Figure 6. ER fluids with PUR particles containing reactive dopant 5 in two concentration: Shear stress vs. field strength (left) and current density vs. field strength (right), all measurements at shear rate 10 s^{-1} and temperature 40 °C

The most promising ER effect was observed in case of an ER fluid, whose PUR particles were modified with 3-nitrobenzyl alcohol (**5**) in a higher concentration than before (see Fig. 5). The shear stress of this ER fluid was over 10 kPa, which is about a doubling with respect to the best other ER fluids, and its current density was below 4 μA/cm^2 (see Fig. 6). This result exhibits a major improvement of the ER performance compared to the known salt-doped reference ER fluids.

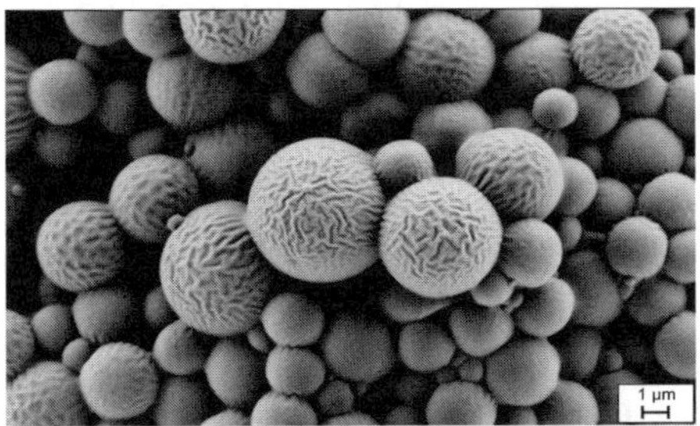

Figure 7. Surface of PUR particles containing a softening organic modifying agent

It was also found, that with increasing concentration of the modifying agents like **5** and **4**, the surface of the formed spherical PUR particles became rough (see Fig. 7). A possible consequence of this morphology could be, that the mechanical slipping resistance between the particles against shear is enhanced and hence the shear stress of the ER fluid.

Another important result of the investigations was the very low base viscosity of the ER fluids without the application of an electric field. The base viscosity of the organically modified ER fluids is about 40 mPas, which is comparable to that of the salt-doped ER fluids. In combination with the figure of the shear stress in the field, a factor of increase of the shear stress by the electric field is obtained, which is extremely high and could surmount the corresponding figures of all other known ER fluids.

4. Conclusions

Novel ER fluids with PUR particles containing organic dopants with high dipole moments and high polarizabilities were synthesized. Their electrorheological properties were studied and compared with those of salt-doped and undoped ER fluids. Strongly enhanced shear stresses and reduced current densities in the electric field could be achieved with organically doped ER fluids.

The ER activity depends strongly on the type and concentration of the organic dopant. A significant improvement on the electrorheological performance was found for organic modifying agents, which are covalently bonded to the PUR network with only one bond. This improvement is referred to a plasticizer effect, which softens the polymer network. Furthermore, the formation of a rough particle surface morphology seems to enhance the ER activity.

The most powerful ER fluid reached a shear stress of more than 10 kPa and a current density of less than 4 μA/cm at an electric field strength of 5 kV/mm. Together with the low base viscosity, the ER fluid exhibits an excellent combination of properties, which offer a huge potential for a large number of applications.

Acknowledgments

Financial support of this work by the German Ministry for Education and Research is gratefully acknowledged.

References

[1] M. Parthasarathy, D. J. Klingenberg, *Mater. Sci. Eng.* **R17**, 57 (1996)

[2] N. Markis, S. Burton, D. P. Taylor, *Smart Mater. Struct.* **5**, 551 (1996)

[3] H. Böse, H.-J. Berkemeier, *J. Intelligent Material Systems and Structures* 10, 714 (2000)

[4] www.fludicon.com

[5] E. Wendt, R. Bloodworth, *EP 0824128 B1 / 1997.*

[6] M. Gurka, D. Adams, L. Johnston and R. Petricevic, *J. of Physics: Conference Series* **149**, 012008 (2009).

MAGNETIC AND MAGNETORHEOLOGICAL PROPERTIES OF NANOFIBER SUSPENSIONS

A. GÓMEZ-RAMÍREZ and M. T. LÓPEZ-LÓPEZ

*Departamento de Física Aplicada, Universidad de Granada, Campus de Fuentenueva
s/n Granada, 18071, Spain*

P. KUZHIR

*Laboratoire de Physique de la Matière Condensée, Université de Nice, Parc Valrose,
Nice 06108, France*

J. D. G. DURÁN and F. GONZÁLEZ-CABALLERO

*Departamento de Física Aplicada, Universidad de Granada, Campus de Fuentenueva
s/n Granada, 18071, Spain*

In this work the preparation and characterization of magnetorheological (MR) fluids constituted by CoNi nanofibers (56 nm length, 6.6 nm width) are reported. The properties of these new fluids were characterized by usual techniques (including magnetometry and magnetorheology). The results where compared with those obtained for conventional magnetic suspensions constituted by CoNi nanospheres. We found a remarkable effect of particle shape on the magnetic properties of the compressed powders: suspensions of nanofibers showed a weaker magnetic response than suspensions of nanospheres at low and medium applied field. This difference is probably due to the strong influence that nanofiber orientation could have on the demagnetizing field and, thus, on the internal field that governs the magnetization of the particles. This result was corroborated by means of finite element method simulations. Steady-state and oscillatory experiments were carried out in the presence of applied fields in order to characterize the MR response of the fluids. We found that the MR response of suspensions of nanofibers was considerably enhanced as compared to suspensions of nanospheres, in agreement with the results reported previously for microfibers.

1. Introduction

Due to their large potentiality for many technological [1] and biomedical [2] applications, there is an increasing scientific interest in the preparation of new magnetorheological (MR) fluids with novel properties. The synthesis of non-spherical magnetic particles and their use in the preparation of new MR fluids is one of the most recent approaches proposed in this direction [3-5]. For example, several works have been devoted in the last years to the preparation of MR fluids composed of magnetic microfibers [6-10]. It has been reported that these new fiber-based MR fluids present an enhanced stability against particle settling and a higher yield stress as compared to conventional (based on spherical particles) MR fluids. In the present work magnetic nanofibers and nanospheres are used for the preparation of MR fluids. The effect of particle shape on both the magnetic and the MR response of MR fluids is investigated. An interesting effect of nanofiber orientation on the magnetic properties of the suspensions is found. Furthermore, an enhancement of the MR properties takes place when nanofibers are used instead of nanospheres as solid phase in MR suspensions. Such enhancement is considerably higher than that reported previously for the case of microparticles [6].

2. Experimental Methods

The CoNi particles used in this work, both nanofibers and nanospheres, were synthesized as described in Ref. [3] by means of the polyol process [11, 12]. Essentially, the synthesis process consisted of the reduction of cobalt and nickel ions in a polyol medium. The size of the particles was controlled by using an appropriate nucleating agent; in the case of nanofibers ruthenium was used, while platinum was employed for nanospheres. On the other hand, the shape of the particles was controlled by the concentration of NaOH in the medium. Figure 1 shows a high resolution transmission electron microscopy (HREM) picture of the particles synthesized.

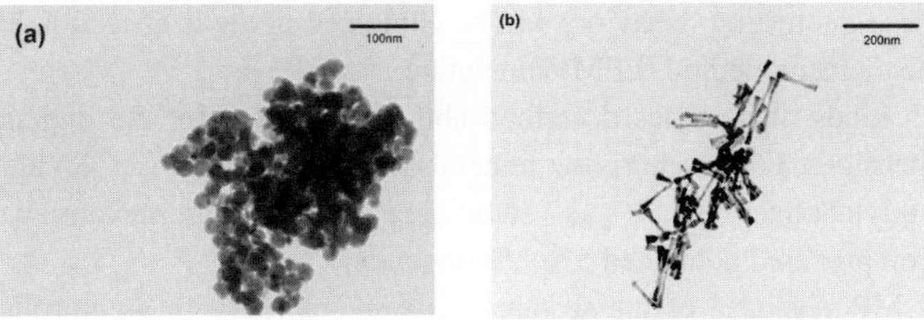

Figure 1. HREM picture of the synthesized nanoparticles. (a) nanospheres, (b) nanofibers.

The size distribution of the particles was obtained from HREM pictures. For nanospheres, the mean diameter was 24 nm, while for nanofibers the mean width and length were 6.6 nm and 56 nm, respectively. Note the high aspect ratio of fibers, which is higher than 8. Energy dispersive X-ray spectroscopy and X-ray diffraction were used to characterize the chemical composition and crystal structure of the particles. As a result of these experiments it could be concluded that the chemical composition was $Co_{81}Ni_{19}$ for nanospheres and $Co_{71}Ni_{29}$ for nanofibers, and that in both cases cobalt presented a hexagonal structure, while nickel structure was a face-centered cubic one [3].

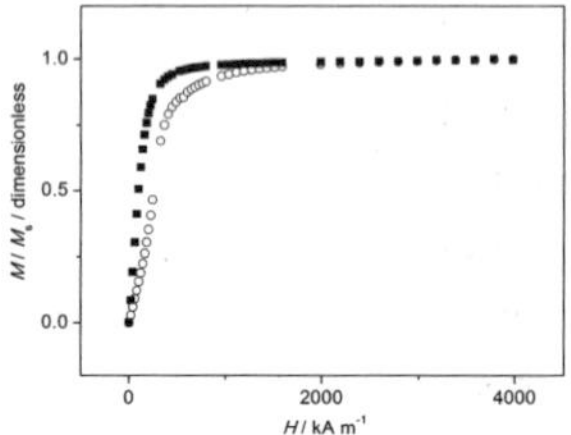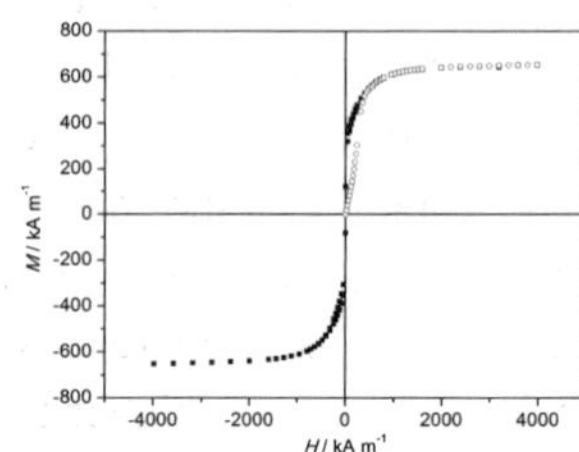

Figure 2. (a) First magnetization curve normalized by M_s ■, nanospheres; o nanofibres, (b) hysteresis loop (■) and first magnetization curve (o) for nanofibers. Taken from Ref. [3].

The characterization of the properties of the particles was completed with their magnetic behavior, measured with a Squid Quantum Design Magnetometer. The first magnetization curve and the hysteresis loop were obtained (see Figure 2). From these curves the saturation magnetization (M_s) and the initial susceptibility (χ_i) were calculated. For nanofibers: $M_s = 626$ kA/m, $\chi_i = 2.7$; for nanospheres: $M_s = 743$ kA/m, $\chi_i = 9.2$. Besides, in is interesting to note (see Figure 2a) that the values of M/M_s are considerably lower for nanofibers than for nanospheres at low and medium field. In addition, it is seen in Figure 2b that for nanofibers, the first magnetization curve does not fall inside of the hysteresis loop. These interesting behaviors will be explained in the next section by means of a finite element method (FEM) simulation.

Both kinds of particles described above were used for the preparation of MR suspensions. For this purpose, mineral oil was used as carrier liquid and 1-α-phosphatidylcholine (PDC) as stabilizing additive (surfactant). All the suspension prepared contained 5 vol.% of solids.

The MR response of the suspensions was studied with a controlled-stress magnetorheometer (MCR300, Physica-Anton Paar). Steady-state and dynamic measurements were carried out. The system geometry was a plate-plate with a gap of 0.35 mm, and the temperature was set at 25 °C in all cases. All the suspensions were mechanically stirred before placing them on the magnetorheometer. A pre-shear of 30 s at a shear rate of 100 s^{-1}, and a waiting time of 30 s with no rate applied were allowed right before starting the measurements. During the waiting time the magnetic field was applied. In the case of steady state measurements a ramp of shear rates from from 0 to 500 s^{-1} was applied. For oscillatory measures the amplitude of the stress was varied in the range 0.1-2000 Pa with a frequency of 1Hz. From these experiments the viscosity, yield stress and viscoelastic muduli were obtained as a function of the applied magnetic field.

3. FEM Simulation of the Magnetic Properties

In the previous section it was shown that the values of M/M_s are smaller for nanofibers than for nanospheres at low and medium field. In addition, for nanofibers, the first magnetization curve does not fall inside of the hysteresis loop. In this section we justify these interesting behaviors by FEM simulation performed by using FEMM software package [13]. With this aim, we considered a fibers-in-air suspension of fibers aligned in the vertical direction (see Figure 3) containing a solid volume concentration of 13 %, similar to that estimated for the suspensions used in the magnetometry measurements. For each value of the intensity of the magnetic field, we considered two directions of application (of the field): perpendicular and parallel to the fiber axis. The magnetic induction in the fiber suspension was calculated (by FEM simulation) for each field intensity and direction of application. As an example, Figure 3 shows the results obtained for an applied magnetic field $B = 75$ mT. As observed, when the magnetic field is parallel to the axis of the fibers, the magnetic induction in the fiber volume is higher than when it is applied perpendicularly. Similar results were obtained for the different magnetic field considered in the range 0 to 100 mT. Consequently, we can conclude that fiber orientation plays an important role in the magnetization of concentrated suspensions of fibers. This is a consequence of the fact that the demagnetizing field is higher when the external magnetic field is applied in the perpendicular direction to the fiber axis than when it is applied in the parallel one. Such simulation explains the results obtained experimentally (Figure 2). For the experimental measurements, the powders were poured in capsules and compressed with a plug. In the case of spheres the compression was very good, but for nanofibers it was rather poor due to their anisotropic shape. When the field was initially applied most of the fibers were misaligned with it, and consequently their demagnetizing field was considerably high, thus their magnetization considerably reduced. As the field was progressively increased, some of the fibers could align with it, but not all of them, justifying a lower magnetization in the fibers volume than for spheres at low and medium field (as observed in Figure 2a). This mechanism also explains that the first magnetization curve of nanofibers is not inside its hysteresis loop (Figure 2b): experimental measures of the magnetization loop began at the highest field and, consequently, most of fibers were immediately oriented with it. During the rest of the measure the demagnetizing field was lower than that induced in the measurements of the first magnetization curve, which started at zero field, thus with most of the fibers misaligned with the field.

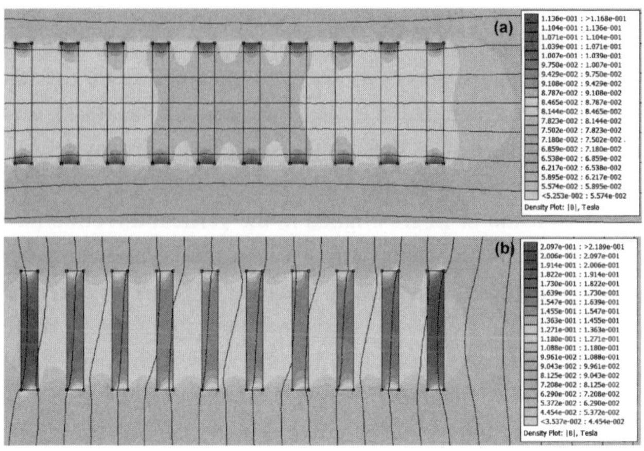

Figure 3. FEM simulation. Distribution of the magnetic field B in the fiber suspension, when an external field of 75 mT is applied perpendicular (a) and parallel (b) to the axis of fibers. Fibers are depicted by the vertical rectangles.

4. Experimental Results of the MR Tests

From shear stress vs. shear rate curves (not shown here for brevity), it could be concluded that these kinds of suspensions (both suspensions of nanofibers and suspensions of nanospheres) presented a typical MR behavior: increase of the shear stress with the magnetic field and appearance of field-dependent yield stress. The curves of the storage (G') and loss (G'') moduli vs. the amplitude of the shear stress, obtained by oscillatory measurements, were also typical of MR fluids: a pseudoplateau, which could be associated to the viscoelastic linear region (VLR), was observed at low shear amplitude, followed by a sharp drop at higher amplitudes. The values of both G' and G'' increased as the magnetic field was increased.

In order to focus on the differences in the MR behavior due to particle shape, it is convenient to normalize the characteristic MR magnitudes (yield stress, and G' and G'' corresponding to the VLR) by the square of the magnetization of the suspensions, since nanofibers and nanospheres present different magnetic properties (see Figure 2). Figure 4a shows the normalized yield stress as a function of the external field. From this figure it could be concluded that the yield stress of the nanofiber suspension is much higher than that of the nanosphere suspension, probably due to the existence of solid friction between fibers, as it is the case for suspensions of microparticles [14]. Nevertheless, the enhancement of the yield stress attributed to the fiber-like shape in the case of nanoparticles is more important than that reported for microparticles [15].

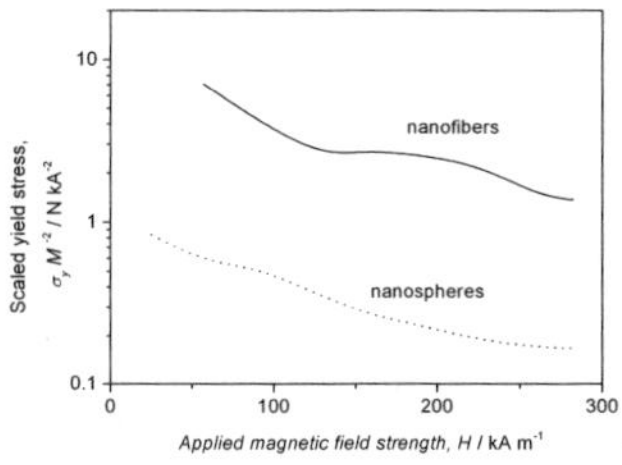
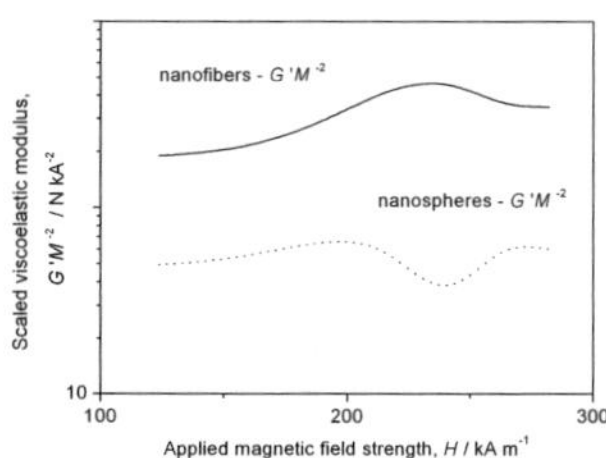

Figure 4. Normalized yield stress (a) and storage modulus (b) as function of the applied magnetic field strength, for nanofiber and nanosphere suspensions containing 5 vol.% of solids.

In Figure 4b the normalized storage modulus corresponding to the VLR is plotted against the intensity of the applied field. A similar behavior is obtained for the loss modulus, not shown here. As observed, as it was the case for the yield stress (Figure 4a), there is an important enhancement of the storage modulus when nanofibers are used instead of nanospheres.

5. Conclusions

It can be concluded that an interesting magnetic behavior arises in suspensions of nanofibers, based on the effect of fiber orientation on the demagnetizing field. In addition, an enhancement of the yield stress and the viscoelastic moduli is found when nanofibers are used instead of nanospheres, as solid phase in MR fluids.

Acknowledgments

This work has been supported by projects P08-FQM-3993 and P09-FQM-4787 (Junta de Andalucía, Spain) and FIS2009-07321 (MICINN, Spain). A.G.-R. and M.T.L.-L. also acknowledge financial support by Secretaría de Estado de Universidades e Investigacion (MEC, Spain) and by the University of Granada (Spain), respectively.

References

1. C.Ross, *Annu. Rev. Mater. Res.*, **31**, 203 (2001).
2. J.D.G. Durán, J.L. Arias, V. Gallardo and A.V. Delgado, *J Pharm Sci.*, **97**, 2948 (2008).
3. A. Gómez-Ramírez, M.T. López-López, J.D.G. Durán and F. Gónzalez-Caballero, *Soft Matter*, **5**, 3888 (2009).
4. L. Chen, C. Zhao, Y. Zhou, H. Peng, Y. Zheng, *J. Alloys and Comp.*, **504**, L46 (2010).
5. H. Shahnazian, D Graf, D. Y. Borin and S. Odenbach, *J. Phys. D: Appl. Phys.*, **42**, 205004 (2009).
6. M. T. López-López, G. Vertelov, P. Kuzhir, G. Bossis and J. D. G. Durán, *J. Mater. Chem.*, **17**, 3839 (2007).
7. R. C. Bell, J. O. Karli, A. N. Vavreck, D. T. Zimmerman, G. T. Ngatu and N. M. Wereley, *Smart Mater. Struct.*, **17**, 015028 (2008).
8. J. de Vicente, J.P. Segovia-Gutierrez, E. Andablo-Reyes, F. Vereda and R. Hidalgo-Álvarez, *J. Chem. Phys.*, **131**, 194902 (2009).
9. D. T. Zimmerman, R. C. Bell, J. A. Filer, J. O. Karli and N.M. Wereley, *Appl. Phys. Lett.*, **95**, 014102 (2009).
10. O. Padalka, H. J. Song, N. M. Wereley, J.A. Filer and R. C. Bell, *IEEE T. Magn.*, **45**, 2275 (2010).
11. D. Ung, G. Viau, C. Ricolleau, F. Warmont, P. Gredin and F. Fiévet, *Adv. Mater.*, **17**, 338 (2005).
12. P. Toneguzzo, G. Viau, O. Acher, F. Fiévet-Vincent, and F. Fiévet, *Adv. Mater.*, **10**, 1032 (1998).
13. http://femm.foster-miller.net/.
14. P. Kuzhir, M. T. López-López and G. Bossis, *J. Rheol.*, **53**, 127 (2009).
15. M. T. López-López, P. Kuzhir and G. Bossis, *J. Rheol.*, **53**, 115 (2009).

RHEOLOGY OF MAGNETIC FIBER SUSPENSIONS

PAVEL KUZHIR, GEORGES BOSSIS and ALAIN MEUNIER

Laboratory of Condensed Matter Physics, University of Nice-Sophia Antipolis, Parc Valrose,
Nice, 06108, France

MODESTO T. LOPEZ-LOPEZ and ANA GOMEZ-RAMIREZ

Departamento de Fisica Aplicada, Universidad de Granada, Campus de Fuentenueva s/n Granada, 18071, Spain

This work reports the experimental and theoretical results on the steady-state shear flow of the suspension of magnetic fibers in the presence of a homogeneous magnetic field perpendicular to the flow. In experiments, we did not observe a significant static yield stress. However, the experimental flow curves show a steep initial section corresponding to gap-spanning aggregates formed in the fiber suspension under a magnetic field. At higher Mason numbers, aggregates are not more confined by the walls and the flow curves become linear manifesting a Bingham behavior with the dynamic yield stress growing with the magnetic field intensity. This yield stress for the fiber suspension appears to be about three times the yield stress for the suspension of spherical particles. Such difference is explained in terms of the enhanced magnetization of the aggregates composed of fibers compared to the aggregates composed of spherical particles.

1. Introduction

Magnetorheological (MR) fluids belong to a class of smart materials, which mechanical properties can be tuned by means of external magnetic fields. A liquid-to-solid transition and appearance of a significant yield stress in these fluids upon application of a magnetic field is referred to as a magnetorheological effect. This effect has found industrial applications in active damping systems and in finishing of optical surfaces [1,2]. Nowadays, sedimentation of MR particles and reduced lifetime of the existing MR fluids are the major problems affecting the performance of the MR smart devices. Ngatu et al. [3] has proposed to use elongated magnetic particles together with spherical ones in order to improve the sedimentation stability of the MR fluids. Besides the sedimentation

356

stability, the MR fluids based on rod-like magnetic particles experience a few times higher MR effect as compared to conventional MR fluids composed of spherical particles. Methods of synthesis of magnetic rod-like particles, their characterization and MR behavior of the suspensions based on these particles are reported in papers by Bell et. al [4,5], Vereda et al. [6], de Vicente et al. [7], Lopez-Lopez et al. [8,9], Gomez-Ramirez et al. [10]. An enhanced MR effect observed in magnetic fiber suspensions has been recently explained by either solid friction between fibers [11] or by enhanced dipolar interactions between them [7]. In this paper, we present new experimental results on the steady shear flows of fiber suspensions under a magnetic field and develop a new micromechanical model of these suspensions taking into account both magnetic and hydrodynamic interactions.

2. Experiments

The MR fluids used are composed of cobalt micro-fibers dispersed in silicon oil of a viscosity $\eta_0=0.479$ Pa·s at a volume fraction of $\Phi=5\%$. The fibers were synthesized by a reduction of the cobalt ions in polyol liquid in the presence of an external magnetic field [8]. The fiber morphology is shown on optical microscopy pictures in Fig. 1a. The fibers are rather polydisperse and are aggregated into small flocks while dispersed in a silicon oil. Their mean length and diameter are $2l=37$ and $2a=4.9$ µm, respectively. So the mean fiber aspect ratio is $l/a=7.6$. The magnetization curve was fitted to a Fröhlich-Kennelly law: $M=\chi_i M_S H/(M_S+\chi_i H)$, with $\chi_i=70$ – initial magnetic susceptibility of fibers and $M_S=1366\pm8$ kA/m – the saturation magnetization.

The flow curves of the fiber suspensions were measured using a Haake RS150 control stress rheometer in a plate-plate geometry, with the plate diameter being 35 mm and the gap between plates 0.2 mm. A homogeneous magnetic field of intensity H_0, ranging from 0 to 30.6 kA/m, was generated by using a specially designed solenoid. A stress cycle was applied to the suspensions: first, a growing stepwise stress ramp was applied from 0.1 Pa to some maximum value depending on the magnetic field intensity; then, the same procedure was repeated but decreasing the stress. The measurements at increasing and decreasing shear stress were carried out in order to check the possible hysteresis of the flow curves, which could be a manifestation of the solid friction between fibers.

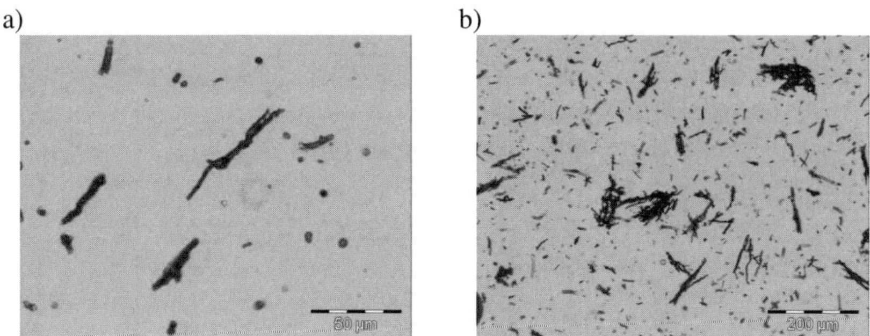

Figure 1. Optical microscopy images of the magnetic fibers in a silicon oil. Figure (a) shows fiber's morphology and figure (b) - fiber's polydispersity and formation of flocks under colloidal forces.

The experimental flow curves of the fiber suspension are shown in Fig. 2 for different magnetic field intensities. We see that the flow curves have two straight sections with different slopes, the left one with a steep slope and the right one with a less steep slope. The steepness of the initial part of the flow curve is likely due to the presence of the gap-spanning aggregates, which offer a high hydraulic resistance to the flow. The small-slope section of the flow curve corresponds to the shear rates $\dot{\gamma} > 50$ s^{-1} and is attributed to the regime of free aggregates not touching the walls and oriented close to the direction of streamlines. At this regime, flow curves are linear and almost parallel to each other, which corresponds to the Bingham rheological law, $\tau = \tau_Y + \eta\dot{\gamma}$, with τ_Y being the dynamic yield stress and η the plastic viscosity. The rounded part of the flow curves corresponds to the transition between the regime of confined aggregates to that of free aggregates.

In experiments, we observe a relatively small static yield stress – the real threshold stress at zero shear rate, corresponding to the failure of the suspension structure and to the flow onset. It is, at least, one order of magnitude lower than the dynamic yield stress. The smallness of the static yield stress could be understood by a poor adhesion of the aggregates to the rheometer plates, made of titanium with a roughness around 1 micron. So, when the suspension is sheared, the aggregates are supposed to slide on the plates with only a low solid friction, which probably gives a negligible contribution to the shear stress.

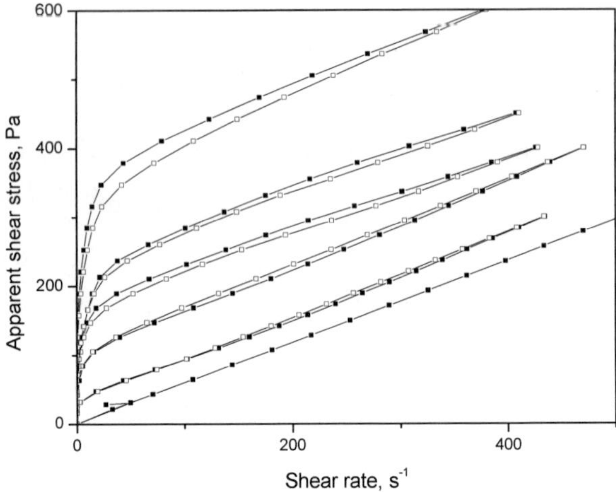

Figure 2. Flow curves of the suspension of magnetic fibers under magnetic fields of different intensities: from bottom – to top curve: H_0=0, 6.11, 12.2; 18.3; 24.4 and 30.6 kA/m. Full and open symbols stand respectively for increasing and decreasing magnetic fields.

Notice finally, that the experimental flow curves have a small hysteresis, which could come from the solid friction between fibers. Nevertheless, the smallness of this hysteresis could mean that the hydrodynamic forces dominate over the forces of solid friction.

3. Theory

To predict the dynamic yield stress developed in the fiber suspension under a magnetic field, we adapt the model of drop-like aggregates developed by Shulman et al. [12] for a conventional MR fluid to the suspension of rod-like particles. The principal hypotheses of this model are as follows. Under the action of an external magnetic field, the fibers are supposed to form drop-like aggregates with high aspect ratio. These aggregates are displaced with the shear flow; they do not rotate but are inclined by the hydrodynamic torque. The angle θ of their inclination with respect to the field direction is defined by the balance between magnetic and hydrodynamic torques. The aggregates are also subjected to a tensile hydrodynamic force trying to break them but the magnetic attractive forces between particles consolidate the aggregate such that its size (or rather aspect ratio, r) is defined by the balance between these forces. We use Ginder's model [13] for calculation of magnetic interactions between closely spaced particles with high magnetic permeability. Once the microstructural parameters, θ and r, are obtained, we calculate the shear stress generated in the suspension by using Batchelor's slender body theory [14] adapted to the case of particulate system with external torques [15]. The shear stress is the sum of the solvent stress, $\eta_0\dot\gamma$ and of the particle stress. The latter is roughly proportional to $\eta_0\dot\gamma r^2$, and since $r^2 \propto \dot\gamma^{-1}$, the particle stress appears to be independent of the shear rate. So, the particle stress is associated to the dynamic yield stress, τ_Y. In experiments this stress is defined by an extrapolation of the linear part of the flow curve onto the zero shear rate. The final result for the yield stress reads:

$$\tau_Y = \Phi\mu_0 H^2 \left\{ \left(\frac{2M_S}{3H} \frac{\chi_f^2(1-\Phi)}{2+\chi_f(1-\Phi)} \right)^{1/2} \sin^2\theta\cos^2\theta + \frac{\chi_f^2(1-\Phi)}{2+\chi_f(1-\Phi)}\sin\theta\cos^3\theta \right\}, \qquad (1)$$

where μ_0 is the magnetic permeability of free space, H is the internal magnetic field, χ_f is the magnetic susceptibility of fibers.

In order to compare the MR effect of fiber suspensions with the one of suspensions composed of spherical particles, we adapt our model to the suspension of spheres and obtain the same expression (1), in which Φ must be replaced by Φ/Φ_a and χ_f by χ_a, with $\Phi_a\approx0.64$ – the internal volume fraction of aggregates of spherical particles and χ_a is the magnetic susceptibility of these aggregates.

The field dependence of the yield stress of both kinds of MR fluids is presented in Fig. 3. Both in experiments and in theory, we observe a few times enhancement of the yield stress in fiber suspensions as compared to the suspension of spheres at the same volume fraction. Such difference is explained in terms of higher magnetization of aggregates composed of fibers compared to aggregates composed of spheres. In fact, the demagnetization field is lower in aggregates composed of aligned fibers, so the magnetization of these aggregates is higher.

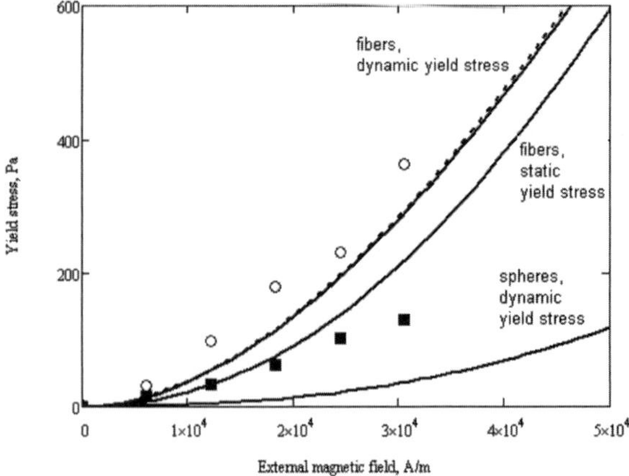

Figure 3. Yield stress of the suspensions of magnetic fibers and of magnetic spheres as function of the intensity of an external magnetic field. Circles and squares stand for experimental results on fibers and spherical particles, respectively; lines correspond to the theory.

In Fig. 3, we see that the theory predicts reasonably well the yield stress in fiber suspensions but underestimates it significantly in suspensions of spherical particles. This is likely because our model underestimates magnetic interactions between spherical particles.

We have also plotted in Fig. 3, the theoretical curve corresponding to the static yield stress in fiber suspensions. This quantity was calculated using our previous model [11], in which we considered a quasi-static deformation of the fiber suspension and took into account a solid friction between fibers characterized by a friction coefficient ξ, which is of the order of unity. Interestingly, this model gives a rather close result on the yield stress compared to the model presented in this paper, even though both theories are based on completely different grounds.

4. Conclusions

In this study, we have confirmed the enhanced MR effect generated in suspensions of rod-like particles, compared to those of spherical magnetic particles, which is attributed to a stronger magnetization of aggregates composed of rod-like particles. The flow curves of the fiber suspension have two linear sections with different slopes. The initial steep-slope section corresponds to the gap-spanning aggregates and the final small-slope section does to smaller aggregates almost aligned with the flow. We have developed a theoretical model predicting a dynamic yield stress in fiber suspensions taking into account hydrodynamic interactions. This model gives a reasonably good correspondence with experimental results.

Acknowledgments

This work has been supported by Biomag (PACA), Eureka E! 3733 Hydrosmart project, FIS2009-07321 (MICINN, Spain), P08-FQM-3993, P09-FQM-4787 (Junta de Andalucía, Spain) and Cooperation Program between CNRS and BRFFR (N° 23178, France-Belarus). One of the authors (M.T.L.-L.) also acknowledges financial support by the University of Granada (Spain).

References

1. J.D. Carlson., D.M. Catanzarite and K.A. St. Clair, Int. J. Mod. Phys. B, **10**, 2857-2865(1996).
2. W.I. Kordonski and S.D. Jacobs, Int. J. Mod. Phys. B, **10**, 2837-2848 (1996).
3. Ngatu G., N.M. Wereley, J. Karli, and R.C. Bell. Smart Materials and Structures. **17**, 045022 (8 pp). DOI: 10.1088/0964-1726/17/4/045022 (2008).
4. R.C. Bell, E.D. Miller, J.O. Karli, A.N. Vavreck and D.T. Zimmerman, Int. J. Mod. Phys. B **21**, 5018-5025 (2007).
5. Bell, R.C., J.O. Karli, A.N. Vavreck, D.T. Zimmerman, G.T. Ngatu and N.M. Wereley, Smart Mater. Struct. **17**, 015028 (2008).
6. F. Vereda, J. de Vicente, and R. Hidalgo-Álvarez, Langmuir **23**, 3581-3589 (2007).
7. J. de Vicente, J.P. Segovia-Guitérrez, E. Anablo-Reyes, F. Vereda and R. Hidalgo-Alvarez, J. Chem. Phys. **131**, 194902 (2009).
8. M.T. López-López, G. Vertelov, G. Bossis P. Kuzhir and J.D.G. Durán, J. Mater. Chem., **17**, 3839-3844 (2007).
9. M.T. López-López, P. Kuzhir and G. Bossis, J. Rheol., **53**, 115-126 (2009).
10. A. Gómez-Ramírez, M.T. López-López, J.D.G. Durán and F. González-Caballero, Soft Matter, **5**, 3888-3895 (2009).

11. P. Kuzhir, M.T. López-Lópcz and G. Bossis, J. Rheol., **53,** 127-151 (2009).

12. Z.P. Shulman, V.I. Kordonski, E.A. Zaltsgendler, I.V. Prokhorov, B.M. Khusid and S.A. Demchuk. Int. J. Multiphase Flow, **12**, 935-955 (1986).

13. J.M. Ginder, L.C. Davis and L.D. Elie, Int. J. Mod. Phys. B **10,** 3293-3303 (1996).

14. G.K. Batchelor, J. Fluid. Mech. **46**, 813-829 (1971).

15. V.N. Pokrovskiy, "Statistical mechanics of diluted suspensions", Nauka, Moscow (1978).

THE ROLE OF PARTICLES ANNEALING TEMPERATURE ON MAGNETORHEOLOGICAL EFFECT

M. SEDLACIK*, V. PAVLINEK and P. SAHA

Polymer Centre, Tomas Bata University in Zlin, TGM Sq. 275, 762 72 Zlin, Czech Republic

P. SVRCINOVA and P. FILIP

Institute of Hydrodynamics, Academy of Sciences of the Czech Republic, Pod Patankou 5, 166 12 Prague 6, Czech Republic

The spinel cobalt ferrite ($CoFe_2O_4$) nanoparticles were prepared via a sol-gel method followed by the annealing process. Their structural, magnetic and magnetorheological (MR) properties depending upon the annealing temperature were investigated. The X-ray diffraction analysis revealed that the higher annealing temperature, the larger grain size of $CoFe_2O_4$ particles resulting in larger magnetic domains in particles. The saturation magnetization, determined via a vibrating sample magnetometry, increased with annealing temperature and, in contrast, the coercivity decreased. The rheological behavior of $CoFe_2O_4$ particles based MR suspensions determined under the small-strain oscillatory shear flow in magnetic field showed that higher annealing temperature reflects in larger changes of rheological properties.

1. Introduction

An important field of current research is represented by magnetorheological (MR) suspensions. These smart fluids enable an abrupt change of rheological properties such as yield stress, viscosity or viscoelastic moduli by transforming from liquid to solid-like structure under an applied magnetic field within milliseconds [1].

Cobalt ferrite ($CoFe_2O_4$) as a typical example of ferromagnetic materials attracts a serious attention of researches in many areas due to its favorable properties like moderate saturation magnetization, high coercivity, as well as a good hardness and an excellent chemical stability. Especially, moderate saturation magnetization favors cobalt ferrite for successful application as a dispersed phase of intelligent magnetorheological suspensions. On the other hand, the high coercivity is induced by the large anisotropy of Co^{2+} ion due to

---

* Corresponding author:
  e-mail: msedlacik@ft.utb.cz, tel.: +420 576 031 452, fax: +420 576 031 444

the spin–orbit coupling. Coercivity is an important feature in magnetic hyperthermia, the efficient and harmlessness method in medicine in cancer treatment causing the necroses of tumour cells due to heat generation by AC magnetic field on magnetic nanoparticles introduced into the diseased tissue [2]. Thus, from economic and time points of view it is very interesting to change the magnetic properties of one material dependent upon resulting application by small variation during the preparation.

To increase the variability of the use of $CoFe_2O_4$ particles, this work deals with the effect of temperature of the annealing process followed after the sol–gel synthesis of $CoFe_2O_4$ particles on their magnetic properties and rheological behavior of MR suspensions based on such particles.

2. Experimental

2.1. *Preparation of CoFe₂O₄ particles*

All the chemicals, except distilled water, used for preparation of ferrite particles were of reagent grade and purchased from Sigma–Aldrich (USA).

As can be seen in Fig. 1, cobalt ferrite particles were obtained via sol–gel method with subsequent annealing [3]. Briefly, the synthesis process starts with the preparation of solution consisting of cobalt acetate tetrahydrate (1/200 mol) and iron(III) nitrate nonahydrate (1/100 mol) dissolved in 2–methoxyethanol (100 ml) and water (15 ml). The solution was sonicated for 30 min and then refluxed at 70 °C for 12 h. Afterwards the formed gel was dried at 100 °C for 24 h and ground in a ball mill (Retsch MM 301, Retsch GmbH, Germany). Final step, and crucial for resulting magnetic properties of $CoFe_2O_4$ powders, was annealing in muffle furnace (LAC LMH 07/12, LAC, USA) with heating rate of 1 °C/min up to 400 °C or 850 °C and holding for 3 h. Different annealing conditions were also studied but above mentioned appeared as the best variants from $CoFe_2O_4$ properties and economical point of view.

Structure of prepared particles was determined with an X–ray diffractometer (X'Pert PRO, Philips, The Netherlands) with CuK_α radiation ($\lambda = 0.154$ nm) in reflection mode and 2θ ranged from 25° to 70°. Measurement of magnetic properties of $CoFe_2O_4$ was carried out at 25 °C using a vibrating sample magnetometer (VSM, Lakeshore, USA) at high magnetic field of 12 kOe.

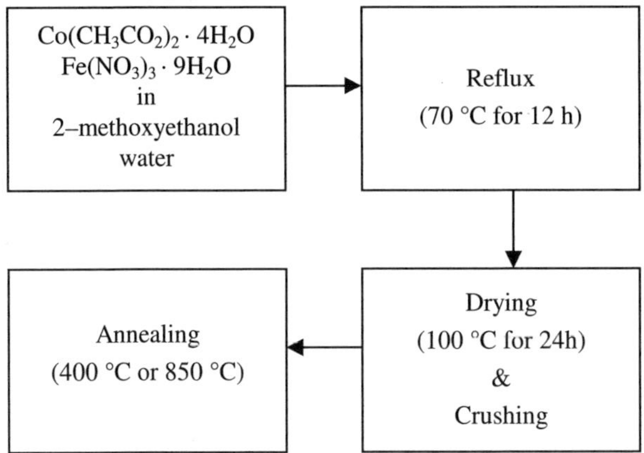

Figure 1. Preparation procedure of CoFe$_2$O$_4$ particles.

2.2. *Preparation of MR suspensions*

CoFe$_2$O$_4$ particles were dispersed by ultrasound in silicone oil (Lukosiol M15, Chemical Works Kolin, Czech Republic; viscosity $\eta = 14.5$ mPas, density $d = 0.965$ g/cm^3) to form stable suspensions with a particle weight fraction of 40% for materials annealed at 400 °C or 850 °C, respectively.

Measurements of rheological properties of the prepared MR suspensions were carried out using a rotational rheometer Physica MCR501 (Anton Paar GmbH, Austria) with Physica MRD 180/1T magneto–cell at 25 °C. True magnetic flux density (0–300 mT) was measured using a Hall probe. A parallel–plate measuring system with a diameter of 20 mm and gap of 1 mm was used. The dynamic oscillation tests were performed by dynamic strain sweeps and frequency sweeps. The strain sweep was carried out with applied strains from 10^{-5} to 0.1 at angular frequency of 62.8 rad/s under a magnetic field to determine the linear viscoelastic region. Then, the rheological parameters were obtained from frequency sweep tests (0.628–62.8 rad/s) at fixed strain amplitude in the linear viscoelastic region.

3. Results and Discussion

3.1. *Characterization of CoFe$_2$O$_4$ particles*

The XRD patterns indicating the structure of CoFe$_2$O$_4$ particles annealed at 400 °C or 850 °C are shown in Fig. 2. All the diffraction peaks can be according to JCPDS card No. 22–1086 and JCPDS card No. 01–1121 indexed to the face centre structure of CoFe$_2$O$_4$. Moreover, no impurity peaks are detected. From

Fig. 2 can be further seen the increase in intensity of the major peaks implying the growth of the grain size of $CoFe_2O_4$ particles with increasing annealing temperature. Based on the Scherrer formula [4], the $CoFe_2O_4$ particle size was calculated as 16 nm and 35 nm for particles annealed at 400 °C or 850 °C, respectively.

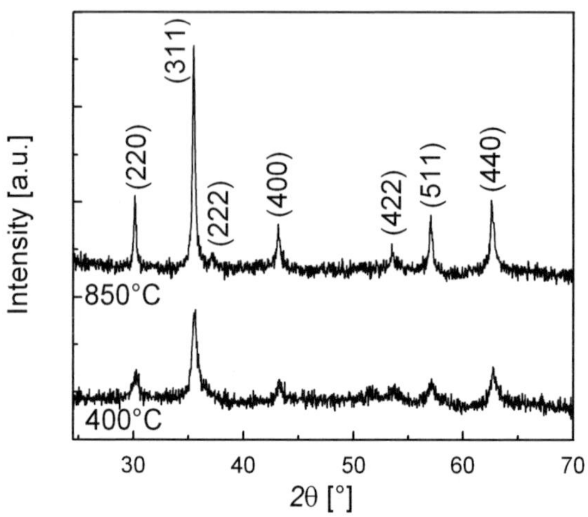

Figure 2. X–ray diffraction patterns of the $CoFe_2O_4$ particles annealed at 400 °C or 850 °C.

The magnetic properties of prepared $CoFe_2O_4$ particles are depicted in Fig. 3. The magnetization saturations of annealed particles, 36 or 63 emu/g (annealed at 400 °C or 850 °C), are lower in comparison with bulk $CoFe_2O_4$ (72 emu/g) [5] confirming the nanocrystalline nature of prepared particles. In contrast to magnetization saturation values, the coercivity is higher for particles annealed at lower temperature. The existence of coercivity in the samples can be attributed to the large residual strain and defects in $CoFe_2O_4$ particles caused by milling during their preparation [6]. The differences in the magnetic properties of prepared $CoFe_2O_4$ can be due to different particle sizes (magnetic domains) depending upon annealing temperature.

3.2. *Rheological properties of MR suspensions*

An abrupt change of a MR suspension internal structure under an external magnetic field causes a transition from nearly Newtonian liquid to solid-like state in several milliseconds. A possible method for the investigation of such transition is a dynamic oscillation test, in which storage, G', and loss, G'', moduli describe the change of the suspension internal structure. First of all, the range of the independency of G' and G'' on strain amplitude, so called linear viscoelastic region, must be determined and that was until strain amplitude $\gamma = 2\times10^{-4}$ in our experiment.

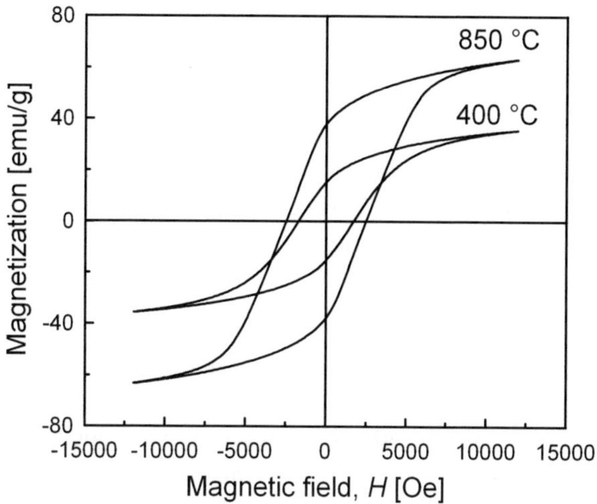

Figure 3. Magnetization curves of CoFe$_2$O$_4$ particles annealed at 400 °C or 850 °C.

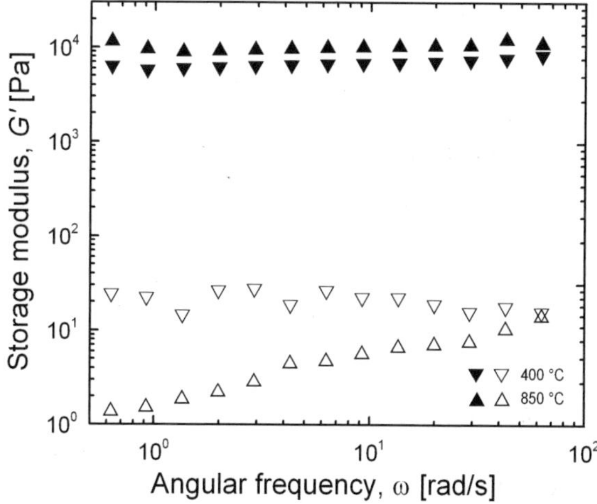

Figure 4. Frequency dependence of storage modulus, G', for 40 wt.% suspension of CoFe$_2$O$_4$ annealed at 400 °C or 850 °C at different magnetic flux densities: 0 mT (*open symbols*); 260 mT (*solid symbols*).

Figure 4 shows storage modulus, G', for MR suspensions of two kinds of $CoFe_2O_4$ particles differing in the annealing temperature as a function of angular frequency at a small strain amplitude of 2×10^{-4} in the linear viscoelastic region. In the absence of a field, G' is dependent on the frequency, especially in case of particles annealed at 850 °C. However, it increases significantly and becomes independent of the frequency upon a magnetic field. This is a typical feature of MR suspensions and can be explained by the formation of rigid internal structure of magnetized particles imposed to magnetic field. Suspension of $CoFe_2O_4$ particles annealed at 850 °C exhibits higher G'. Moreover, the higher annealing temperature of particles results in lower G' of MR suspensions in the absence of magnetic field and thus in higher MR efficiency.

4. Conclusions

The synthesis of cobalt ferrite via sol–gel method with subsequent annealing enable to tailor properties of such particles in dependence on required application, which can also be effective from economical point of view. It means that $CoFe_2O_4$ particles annealed at higher temperature are good candidates for classical magnetorheology while those annealed at lower temperature can be used in magnetic hyperthermia.

Acknowledgments

This work was financially supported by the Ministry of Education, Youth and Sports of the Czech Republic (MSM 7088352101) and the Grant Agency of the Czech Republic (202/09/1626).

References

1. D. Susan–Resiga, *J. Intell. Mater. Syst. Struct.* **20**, 1001 (2009).
2. D. H. Kim, D. E. Nikles, D. T. Johnson and C. S. Brazel, *J. Magn. Magn. Mater.* **320**, 2390 (2008).
3. J. G. Lee, J. Y. Park and C. S. Kim, *J. Mater. Sci.* **33**, 1965 (1998).
4. A. Patterson, *Phys. Rev.* **56**, 978 (1939).
5. A. E. Berkowitz and E. Kneller, *Magnetism and Metallurgy.* New York: Academic Press (1969).
6. M. V. Limaye, S. B. Singh, S. K. Kothari, et. al., *J. Phys. Chem. B* **113**, 9070 (2009).

EFFECT OF FE₃O₄ NANOPARTICLES ON THE PROPERTIES OF BIDISPERSE MAGNETORHEOLOGICAL FLUIDS[*]

H. B. CHENG[1,2], C. ZHOU[1], Y. GAO[1], Q. J. ZHANG[1] and N. M. WERELEY[2]

[1]*State Key Laboratory of Advanced Technology for Materials Synthesis and Processing, Wuhan University of Technology, 430070, Wuhan, China*

[2]*Smart Structures Laboratory, Dept. of Aerospace Engineering, University of Maryland, College Park, MD, 20742, USA. wereley@umd.edu*

Fe_3O_4 nanoparticles (NPs) were synthesized via chemical co-precipitation in an aqueous solution, and were characterized by scanning electron microscope (SEM). Bidisperse magnetorheological (MR) fluids were prepared with Fe_3O_4 NPs, silicone oil and micron scale carbonyl iron particles (CIPs). Their rheological properties were experimentally investigated via a rheometer equipped with an electromagnet. Results show that the sedimentation stability and yield stress of the bidisperse MR fluids were increased when adding small concentrations of Fe_3O_4 NPs.

1. Introduction

Magnetorheological (MR) fluid is a smart fluid that changes its rheological characteristics upon application of magnetic field [1-4]. MR fluids can be used in adaptive damping systems for aircraft [5], civil engineering applications [6] and vehicle applications [7]. Generally, a highly stable fluid is desired for applications, such as crashworthiness, where the fluids are seldom used. Therefore, a key issue is the need for an MR fluid that resists sedimentation. MR fluids typically utilize spherical micron-sized magnetic particles (e.g. iron microspheres) suspended in a carrier fluid such as silicone oil or water [8]. Unfortunately, such ferromagnetic particles are prone to sedimentation in MR fluids in the absence of constant or frequent remixing, because of the density difference between magnetic particles and carrier liquids. Furthermore, the separated particles may agglomerate, resulting in undesired, tightly bound particle clusters, which make re-dispersion very difficult, possibly rendering the

---

[*] This work was supported by a grant from the Major State Basic Research Development Program of China (973 Program) (No. 2010CB227104) and by self-determined and innovative research funds of WUT (No. 2010-II-013)

MR fluids ineffective. To overcome the sedimentation problem, several strategies have been proposed such as using composite magnetic particles or polymeric core/shell-structured magnetic particles, as well as adding surfactants and nanoparticles, such as carbon nanotubes [9-12]. The stability of MR fluids greatly improved using these methods; however, these strategies can produce unfavorable tradeoffs. Generally, eliminating the sedimentation velocity and ratio of MR fluids by replacing iron particles with polymeric core/shell-structured magnetic particles, would bring about a decrease in yield stress. This results from the introduction of nonmagnetic polymer, such as polystyrene and polymethacrylate, in either the core or shell of the composite particles, so that, although the particle density is reduced, the saturation magnetism, M_s also decreases. The maximum yield stress of MR fluids is proportional to M_s^2. Similarly, substituting carbonyl iron particles by iron nanoparticles typically decreases yield stress [1, 8]. In this study, we synthesized Fe_3O_4 nanoparticles (NPs) via chemical co-precipitation and prepared a series of bidisperse MR fluids using the synthesized Fe_3O_4 NPs, as well as carbonyl iron particles (CIPs), suspended in a silicone oil. By comparing their stability, off-state viscosity and MR characteristics to a MR fluid synthesized using the same volume fraction of neat CIPs suspended in a silicone oil, the effects of the Fe_3O_4 NPs on the properties of the MR fluids are investigated.

2. Experimental Procedures

Fe_3O_4 NPs were synthesized via a chemical coprecipitation technique [13] with $FeCl_3·6H_2O$ and $FeCl_2·4H_2O$, and NaOH (analytic grade, Shanghai Chemical Reagent Co.). The $FeCl_3$ and $FeCl_2$ reaction liquids were prepared by adding $FeCl_3·6H_2O$ and $FeCl_2·4H_2O$ into de-ionized water and stirring to complete dissolution. The NaOH solution was prepared by dissolving NaOH into de-ionized water. The sodium hydroxide was added by dropwise into the $FeCl_3$ and $FeCl_2$ reaction liquid (the mole ratio of $FeCl_3/FeCl_2=2$), under vigorous continuous stirring and inert nitrogen atmosphere, at 60~70℃, until the pH value of reaction liquid reached about 9. The color of the reaction solution shifted from orange to black and a black precipitate started to appear, indicating magnetite formation. The reaction mixture was heated to 80℃ under stirring and kept it for 2h for ageing. The chemical reaction of Fe_3O_4 precipitate was:

$$2FeCl_3+FeCl_2+8NaOH=Fe_3O_4 +4H_2O+8NaCl \qquad (1)$$

The unreacted solutions, NaCl and solvent were separated from the black precipitates of this reaction by centrifuging and washing with de-ionized water and ethanol several times. Finally, the black precipitate was dried in a vacuum oven at 50℃ for 24h; and the black Fe_3O_4 NPs were obtained.

The crystal structure of the synthesized Fe_3O_4 NPs was investigated by powder X-ray diffraction (XRD) and proved to have the inverse cubic spinel structure of Fe_3O_4. Mean diameter of NPs was nominally 15 nm. The particle size and morphology were imaged by a field emission scanning electron microscope (FESEM) (Hitachi S4800 FESEM, Japan). The particle size distribution of dilute Fe_3O_4 NPs in an ethanol solution was measured via laser scattering method (ZetaPLAS, Brookhaven Instruments Corporation, America).

Bidisperse MR fluids were prepared by mixing 70 vol% silicone oil, 30 vol% carbonyl iron (CI) particles (Tiayi Co., diameter: 1~6 μm) and the Fe_3O_4 NPs 0.0, 0.5, 1.0, 1.5, 2.0 wt% according to the total weight of silicone oil and CI particles, respectively. The mixtures were blended with a high-speed emulsifier at a rate of 10000 rpm for 15 min. MR fluid samples were stored in a plastic bottle and labeled as MR fluid sample #a, #b, #c, #d and #e, corresponding to 0.0, 0.5, 1.0, 1.5, 2.0 wt% of Fe_3O_4 NPs, respectively.

The magnetorheological properties of MR fluids were characterized using a strain-controlled rheometer (ARES model by TA Instruments) equipped with an electromagnet. In this measurement, parallel-plate fixtures with diameters of 20 mm at a gap of 0.3 mm were used, and the magnetic field direction was set to be perpendicular to the flow direction.

3. Results and Discussion

3.1. *Particle size distribution of Fe_3O_4 NPs*

The morphology of the dried Fe_3O_4 NPs imaged via a field emission scanning electron micrograph (SEM) is shown in Figure 1. From the SEM image, single Fe_3O_4 particles are nominally 15 nm or less, agreeing with the result calculated using the Scherrer equation based on XRD pattern. These NPs have a strong tendency to aggregate; large granular aggregates are composed of many smaller particles, and larger clusters also formed via larger particles sticking together. Even the smaller particles are composed of many tiny (a few nanometers) nanoparticles. The size distribution of the dried Fe_3O_4 NPs ranges from 15 nm and 60 nm. The particle size distribution was dominated by particles ranging from 15-20 nm, with a small number ranging from 20-40 nm, although there were still several 50 nm or more agglomerates of smaller particles.

372

Size distribution was also measured via a laser scattering method, which presented a normal distribution (shown in figure 2). The particles sizes of dilute Fe_3O_4 NPs/alcohol solution are mainly between 20~60 nm, and the mean size is about 38 nm. Comparing with the SEM image, both size distribution measurements are well correlated, and further reveals that the dried Fe_3O_4 NPs consisted of differently sized agglomerates, which were difficult to separate into distinct particles even after extended treatment with ultrasonic waves prior to particle size measurements.

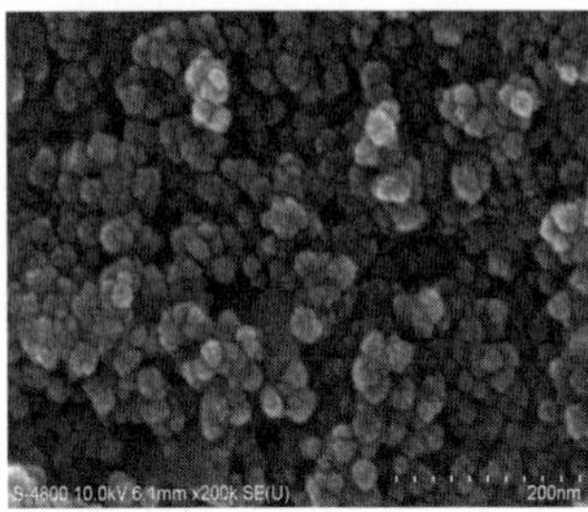

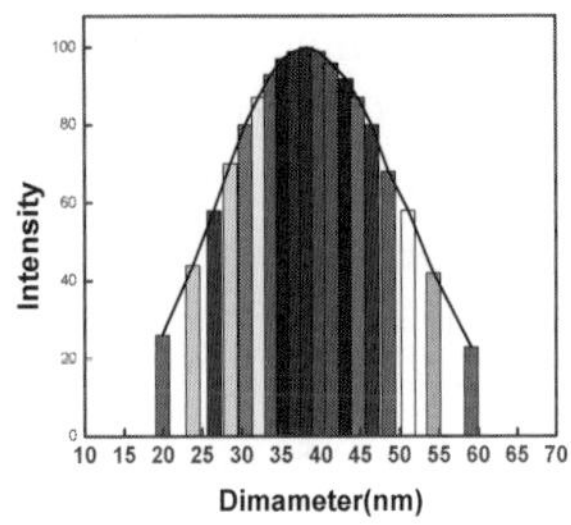

Figure 1. SEM images of Fe_3O_4 NPs.   Figure 2. Particle size distribution of Fe_3O_4 NPs.

3.2. *Effect of Fe_3O_4 NPs on properties of MR fluids*

3.2.1. *Sedimentation stability*

The effect of Fe_3O_4 NPs on the sedimentation stability of MR fluids was investigated by a traditional observational method [14]. Each MRF sample was transferred to 10 ml graduated test tubes and placed vertically in a holder. The volume of the supernatant liquid can be obtained by observing the phase boundary between the supernatant liquid and the concentrated suspension at given times, until reaching an asymptotic value. The ratio of sedimentation is defined as [14]:

$$\text{Sedimentation ratio } (\%) = \frac{\text{volume of the supernatant liquid}}{\text{volume of the entire suspension}} \times 100\%$$

(2)

The sedimentation ratios of MRFs experimentally measured via this stability test are represented graphically in Figure 3 (left). Figure 3 (left) shows that settling rates of all MRF samples were relatively fast during the first 10 hours, and then slowed down, so that after 24 hours, the sedimentation rate tended to zero. Both settling rate at the initial time, and final sedimentation ratio, decreased as the NP concentration increased. Of key interest was that a minimum final sedimentation ratio was presented for an NP concentration of 1.5 wt%, but decreased for a higher NP concentration of 2.0 wt%. This implies that there is an optimal NP concentration to effect reductions in settling rate. Under these experimental conditions, the optimal NP concentration was found to be around 1.5 wt%. When increasing NP concentration up to this optimal value of 1.5 wt%, both the settling rate and sedimentation ratio decrease as NP concentration increases component. In contrast, as for the sample with an NP concentration of 2.0 wt%, both settling rate and sedimentation ratio increased. The final sedimentation ratio of sample #d (1.5 wt% NP concentration) was 16.5%, which was 58.8 % lower than that of sample #a (no NPs). Analogous results were also observed in our previous work [1, 8].

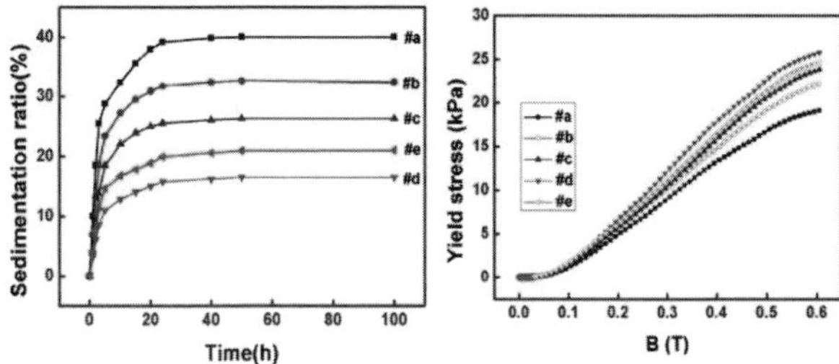

Figure 3. Sedimentation ratio versus time (left) and yield stress versus magnetic flux B (right) for MRFs. ($\varnothing$=0.3, Fe_3O_4 NPs (#a ■0%), (#b●0.5%), (#c ▲1.0%), (#d ▼1.5%), (#e ◀2.0%).

3.2.2. *Yield stress*

The field-induced yield stress for MRFs with varying Fe_3O_4 NP concentration is presented in Figure 3 (right), which shows that the yield stresses of all MRFs increase with the magnetic flux from 0T to 0.6T, which is typical of MR fluids. However, the yield stress as a function of magnetic field varies for the five MRFs, even though the volume fraction, 30 vol%, of microparticles is the same for all MRF samples before adding NPs. The yield stress curves suggest that adding small concentrations of Fe_3O_4 NPs into a microparticle-based MRF can significantly enhance its yield stress. The enhancement effect is increases in proportion to NP concentrations up to 1.5 wt%. The yield stress (25.78 kPa) of sample #d (1.5 wt%) is 34.8% greater than the yield stress (19.13 kPa) of sample #a (no NPs).

In our prior work [15], it was shown that nanoparticles appear to fill in the interparticle spaces that appear in the network structures manifested in the monodisperse fluids. These distinct chains suggest the possibility that a bidisperse fluid can be synthesized with a yield stress that is substantially greater than an MRF with microparticles alone. In addition, the nanoparticles in the MRFs will tend to redisperse themselves by thermal convection. When the applied magnetic field is small, the particles in MRFs will disperse loosely in the carrier fluid so that in inter-microparticle space is not populated by the naoparticles so that the yield stress enhancement effect is not strongly manifested. As the magnetic flux is increased, microparticles form chains, and the nanoparticles migrate to the microparticles thereby forming a closely packed structure, Inter-microparticle spaces will be subject to intrusions by NPs thereby increasing the localized permeability and enhancing the yield stress. For concentrations of NPs beyond a critical vol% concentration, excess NPs will block direct contact of microparticles so that the nanoparticles tend to behave as a lubricant, and not as a filler. Therefore, there is a critical concentration of nanoparticles beyond which the yield stress enhancement effect is reduced.

3.2.3. *Off-state viscosity*

Off-state viscosity of MRF samples decreased steadily with shear rate and presented shearing thinning behavior (Figure 4). The apparent viscosity increased as NP concentration increased, without exception. The increase in off-state viscosity at low shear rate was relatively large (nominally 90 Pa s between sample #a and #e at 0.1 s^{-1}). This augmentation of off-state viscosity reduced substantially as shear rate increased, demonstrating that the NPs tend to enhance the off-state viscosity. The demonstrated improvements in the off-state viscosity are due to this enhanced thixotropy.

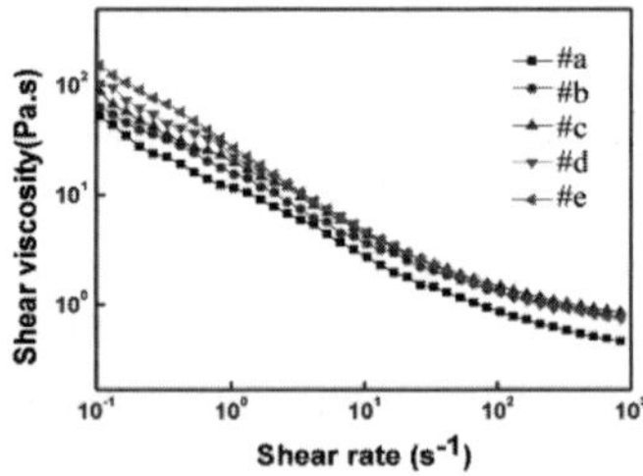

Figure 4. Off-state shear viscosity versus shear rate for MRFs ($\varnothing$=0.3, Fe$_3$O$_4$ NPs (#a ■0%), (#b●0.5%), (#c ▲1.0%), (#d ▼1.5%), (#e ◀2.0%).

4. Conclusion

Fe$_3$O$_4$ nanoparticles (NPs) were synthesized via chemical co-precipitation and used to prepare bidisperse MR fluids with carbonyl iron particles (30 vol%)

suspended in silicone oil (70 vol%). Magnetorheology of the bidisperse MRFs were shown to be highly sensitive to the added concentration of NPs. First, there was a substantial yield stress enhancement effect, which peaked for the MRF sample containing 1.5 wt% of NPs. The sedimentation stability was also increased, and also peaked for the MRF sample containing 1.5 wt% of NPs. In contrast, the off-state viscosity increased monotonically as NP concentration increased.

References

1. Chaudhuri, A., Wang, G., and Wereley, N.M. *Int. J. Mod. Phys. B*, 19 (7, 8 & 9), 1374-1380 (2005).
2. Chen, M.C., Kim, Y.N., Li, C.C., and Cho, S.O. *J. Phys. Chem. C*, 112, 6710-6716 (2008).
3. Lim, S.T., Cho, M.S., Jang, I.B., Choi, H.J. and Jhon, M.S. *IEEE Trans. Magn*, 40, 3033-3035 (2004).
4. Guerrero-Sanchez, C., Lara-Ceniceros, T., and Jimenez-Regalado, E. *Adv. Mater.,* 19, 1740-1747 (2007).
5. Kamath, G.M., Wereley, N.M. and Jolly, M.R. *J. American Helicopter Soc.,* 44(3) 234-248 (1999).
6. Spencer, B.F., Nagarajaiah S. 129(7) 845-856 (2003).
7. Choi, S.B., Nam, M.H. and Lee, B.K. *J. Intell. Mater. Syst. Struct.* 11(12) 936-944 (2000).
8. Ngatu, G.T. and Wereley, N.M. *IEEE Trans Magn*, 43(6) 2474-2476.
9. Gorodkin, S.R. 2000. *Rev. Sci. Instrum*, 71, 2476-2480 (2007).
10. Fang, F.F., Choi, H.-J. and Jhon, M.S. *Colloids and Surfaces A: Physicochem* 351:46-51 (2009).
11. Min S. Cho, Sung T., Lim, In B. Jang, Choi, H.J. and Jhon, M.S. *IEEE Trans. Magn*, 40(4) 3036-3038(2004).
12. Ko, S.W., Lim, J.Y., Park, B.J., Yang, M.S. and Choi, H.J. *J. Appl. Phys,* 105: 07E703 (2009).
13. Kim, D.K., Zhang, Y., Voit, W., Rao, K.V. and Muhammed, M. *J. Magn. Magn. Mater.*, 225, 30-36 (2001).
14. Cheng, H.B., Wang, J.M., Zhang, Q.J., and Wereley, N.M. *Smart Mater. Struct,* 18 085009 (6pp) (2009).
15. Wereley, N.M. Chaudhuri, A., Yoo, J.-H., John, S., Kotha, S., Suggs, A., Radhakrishnan, R., Love, B. and Sudarshan, T.S., *J. Intell. Mater. Systems and Structures*, **17**(5):393-401 (2006).

MAGNETORHEOLOGICAL FLUIDS EMPLOYING SUBSTITUTION OF NONMAGNETIC FOR MAGNETIC PARTICLES TO INCREASE YIELD STRESS

L.A. AHURE[1], S. LUSTIG[1], N.M. WERELEY[1] and J. ULICNY[2]

[1]*Dept. of Aerospace Engineering, University of Maryland, College Park, MD 20742 USA*

[2]*GM Global R&D, Manufacturing Systems Research Lab, Warren, Michigan 48092, USA*

Magnetorheological fluids (MRFs) were characterized in this study to investigate the effect of replacing magnetic particles with nonmagnetic micro-scale glass beads to increase yield stress, while also reducing density and settling rate. Two MRF samples were prepared. MRF-40 had a Fe concentration of 40 vol% while the second fluid had a total Fe concentration of 37 vol% and a glass bead concentration of 3 vol%. A comparative study of MRF characteristics was conducted to determine the impact of the nonmagnetic glass beads on yield force, as well as viscosity, and settling rate. Both MRFs were characterized as follows: 1) magnetorheology as a function of magnetic field, 2) sedimentation analysis conducted with an inductance-based sensor, and 3) cycling of a small-scale damper undergoing sinusoidal excitations at frequencies of 1 Hz for characterization and 4 Hz for fatigue tests. Further investigation of the glass beads was conducted to examine their durability post-cycle. The goal of this study is to analyze the effectiveness and feasibility of replacing magnetic particles with micro-scale glass beads in MR fluids as a means to reduce the specific gravity and increase the yield stress.

1. Introduction

Magnetorheological fluids (MRFs) are suspensions of magnetic particles in a carrier fluid that have the ability to change properties when a magnetic field is applied. MRFs are used in an increasing range of applications such as, semi-active vibration absorbing systems and primary vehicle suspensions[1,2]. Carbonyl iron (CI) powder has been the primary magnetic dispersant used to prepare MRFs because such powders lead to MRFs with high yield stress. Unfortunately, particles in such MRFs do settle, and in the absence of remixing, the particles descend to the bottom the container. What is desired is an MRF that essentially does not settle, but instead provides the full MR effect the first time it is used even in the absence of remixing. Consequently, methods have been proposed to reduce settling, such as chemical modification of particles, or substitution of micro-scale magnetic particles for nano-scale particles[3-5]. The goal of our study was to determine the impact of substituting nonmagnetic glass particles for magnetic CI particles in MRFs. An experimental evaluation of the impact of the glass beads on yield stress, specific gravity, and settling rate was performed.

2. Experimental Procedures

2.1. *Rheology Testing*

MRFs characterized in this study used the same carrier fluid and Fe microparticles. The first fluid, MRF-40, had a density of 3.53 g/cm³, a Fe concentration of 40 vol%, and no glass beads. The second fluid, MRF-37, had a density of 3.26 g/cm³, a Fe concentration of 37 vol%, and a glass bead concentration of 3 vol%. The viscosity and yield stress of both MRF samples were determined using a Paar Physica MCR 300 parallel disk rheometer with an MR cell. In the absence of magnetic field, a test was run to measure off-state viscosity. A 0.15 ml MRF sample was placed in the rheometer test volume, which had a 0.5 mm gap separating the rotating disk from the platen. The flow curves (shear stress vs. strain rate) were then measured. To measure flow curves in the presence of magnetic field, a 0.3 ml MRF sample was placed between the platen and rotating disk separated by a 1 mm gap. The current was increased from 0.2A to 4A, and the magnetic field lines were oriented normal to the parallel plates. For this study, the fluid flow curves were characterized using the Bingham Plastic (BP) model[6], whose constitutive law is:

$$\tau = \tau_y + \mu\dot{\gamma} \tag{1}$$

In eq. 1, τ_y is the yield stress, μ is the post-yield viscosity, and $\dot{\gamma}$ is the shear rate. The post-yield viscosity is the slope of the high shear rate asymptote of the BP model, and the yield stress is the intercept of the high shear rate asymptote to the shear stress axis. In the BP model, the MR fluid exhibits a rigid behavior in the pre-yield region (i.e. below the yield stress, τ_y), and then exhibits viscous flow behavior in the post-yield region[6].

2.2. *Sedimentation Testing*

Sedimentation tests were performed in order to measure sedimentation rates. Sedimentation tests were conducted with an inductance-based sensor fabricated in-house[6], a GW INSTEK LCR-816 inductance meter, and a height gage to mount and move the sensor (fig.1). A uniformly dispersed MR fluid was contained in a ¼-inch test tube and placed in a vertical position. As settling progresses, there is a distinct boundary, or "mudline," between the clarified carrier fluid above and the MRF below. The permeability of a volume of fluid in the test tube was measured using a sensing coil, and data were logged via computer. As the mudline travels

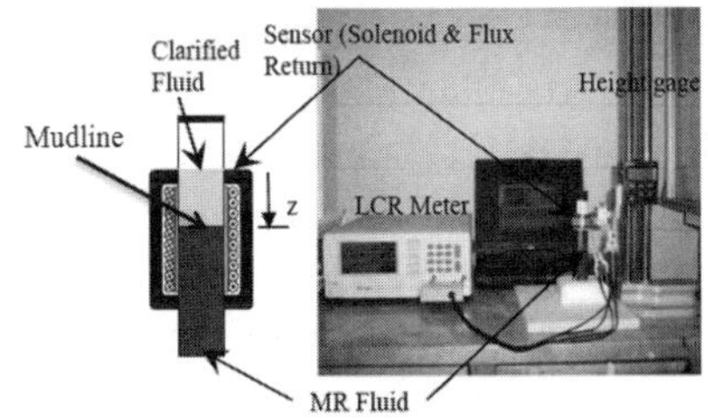

Figure 1: Sedimentation setup[6].

downwards through the sensing coil, the permeability of the fluid volume contained therein decreases. The slope of this curve yields the sedimentation rate.

2.3. *Damper Testing*

The yield force and post-yield viscosity of an MR damper filled with MRF-37

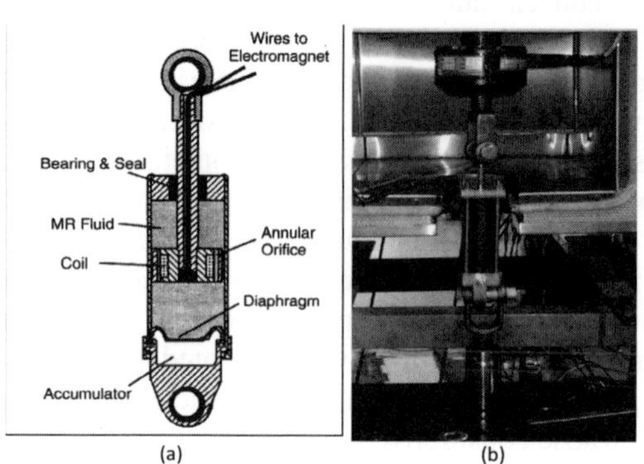

Figure 2: (a) Cross section of MR damper (b) Setup of MR damper on MTS machine.

were determined by analyzing the behavior of a modified Rheonetic SD-1000-2 MR damper from Lord Corporation, which was tested under sinusoidal loading at frequencies of 1 Hz and 4 Hz. The MR damper and test setup on a servo-hydraulic Testing machine (MTS-810) are depicted in fig. 2a and 2b, respectively. Testing was done while varying an electric current (from 0A to 3A) to generate magnetic field in the electromagnet, and activating the MRF. Force vs. displacement data were analyzed using the Nonlinear BiViscous (NBV) model[7]. The model is piecewise continuous in velocity and assumes that the MR fluid is plastic in both the preyield, C_{pr}, and the post-yield, C_{po}, conditions with the pre-yield damping being much greater than the post-yield damping. In fig. 3, the yield force, F_y, is represented by the high velocity asymptote intercept with the force axis. The piecewise continuous equations describing the model are shown in eq.2.

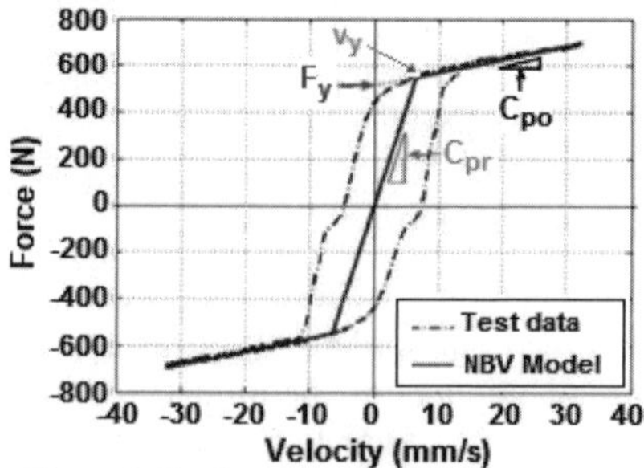

Figure 3: Nonlinear Biviscous (NBV) Model.

$$f(t) = \begin{cases} C_{po}v + F_y & v \geq v_y \\ C_{pr}v & -v \geq v \leq v_y \\ C_{po}v - F_y & v \leq v_y \end{cases} \quad (2)$$

The pre-yield velocity is:

$$v_y = \frac{F_y}{C_{pr} - C_{po}} \quad (3)$$

3. Results

3.1. *Magnetorheology*

The flow curves are shown in fig. 4. The BP model was used to characterize the flow curves for the viscosity and yield stress of both fluids. From fig. 5a, the MRF-37 (with glass beads) had a 22% higher off-state viscosity, and from fig. 5b, both fluids exhibited a variation in maximum yield stress of only 10%.

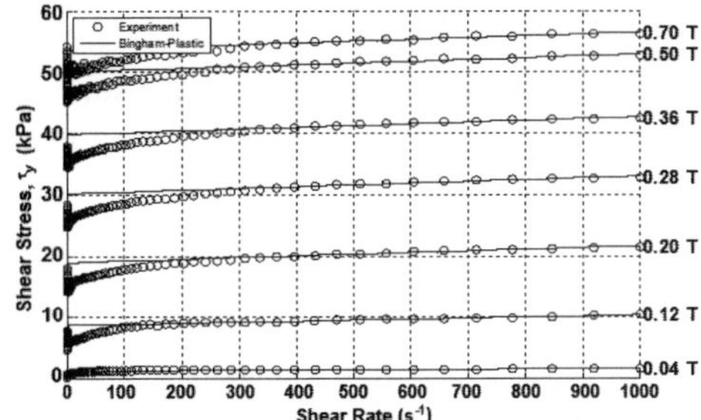

Figure 4: Shear stress as a function of shear rate.

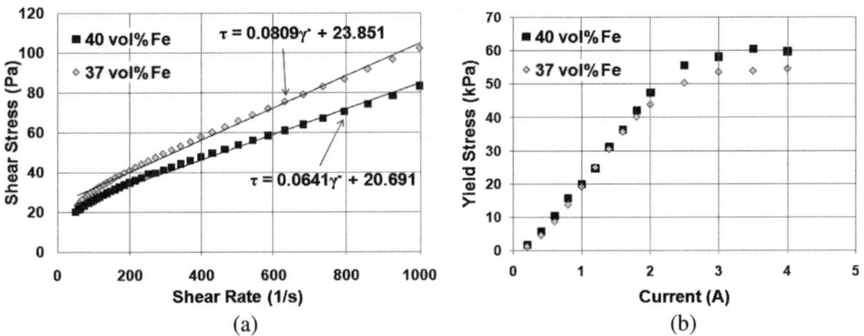

(a)   (b)

Figure 5: Shear stress (Pa) as a function of shear rate (s^{-1}) (b) Yield stress (in kPa) as a function of current values (in A).

3.2. *Sedimentation*

The settling rates of both MR fluids were measured with an inductance based sensor. The data was measured over the period of time the mudline took to descend by 6.35 mm. The mudline of the MRF-37 (with glass beads) had a slower sedimentation rate (0.041 µm/s) than the fluid with no glass beads (0.043 µm/s), as expected, due to its lower specific gravity. The mudline descent of the two MRFs, MRF-37 and MRF-40, are compared in fig. 6.

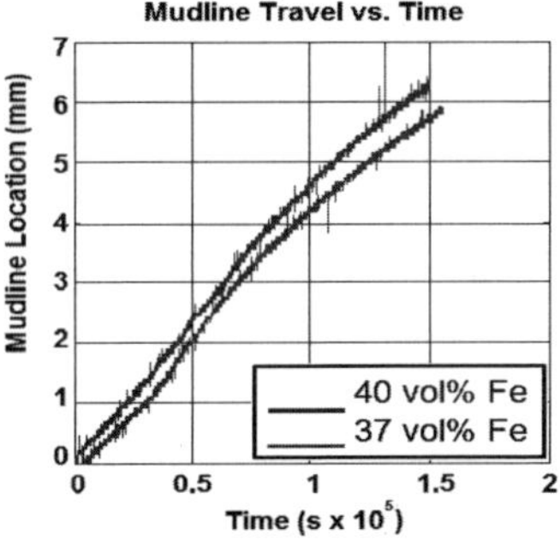

Figure 6: Mudline descent as a function of time for MRF-37 (glass beads) and MRF-40 (no glass beads).

3.3. *Damping Behavior*

The MRF-37 (with glass beads) was studied using a linear stroke MR damper[8]. Data shown in fig. 7a and 7b are the force vs. displacement and the force vs. velocity curves, respectively, reconstructed using the NBV model. As the

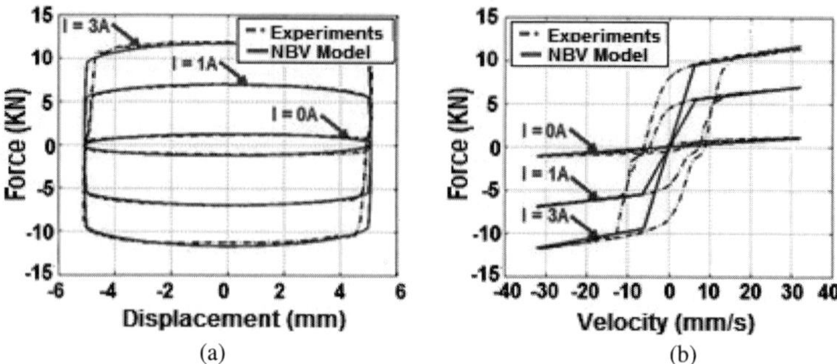

Figure 7: Sinusoidal loading at 1 Hz. (a) Force vs. displacement (b) Force vs. velocity.

applied current increased the damping capacity also increased, which is depicted by the area enclosed by the force vs. displacement sinusoidal response in fig. 7a. The NBV model was used to characterize yield stress and post-yield damping using force vs. velocity data at 1 Hz as in fig. 7b.

An endurance test was also carried out where the damper was cycled at 4 Hz for a total of 518,400 cycles. The yield force and post-yield damping were characterized by fitting the NBV model to the force vs. velocity data taken at 1 Hz during a number of breaks in the endurance tests indicated by the symbols in fig. 8. For 518,400 cycles, the peak value of the yield force followed similar

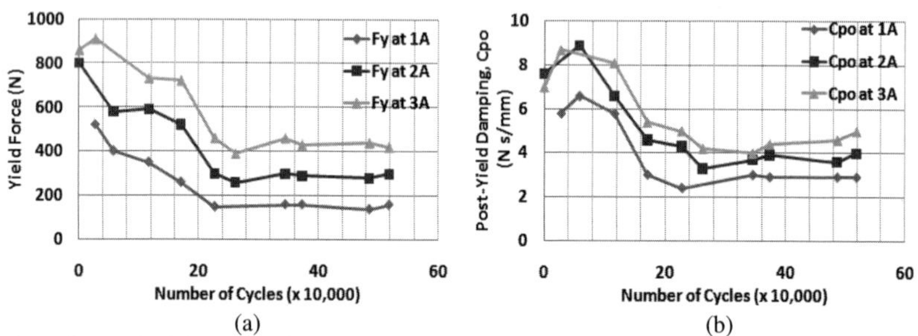

Figure 8: (a) Yield force as a function of number of cycles (b) Post-yield damping as a function of the number of cycles.

trends for each current value, that is, the force decreased markedly until nominally 250,000 cycles at 4 Hz had been completed, and leveled off and remained relatively constant (fig. 8a). The post-yield damping (fig. 8b) also decreased until about 250,000 cycles at 4 Hz, and then leveled off as well. The ratio of maximum to minimum yield force at a particular value of current value is plotted as a function of number of cycles (fig 9). It was observed that adding glass beads to the MR fluid improved the yield force by factors ranging from 2 to 3.2 for current values between 1A to 3A.

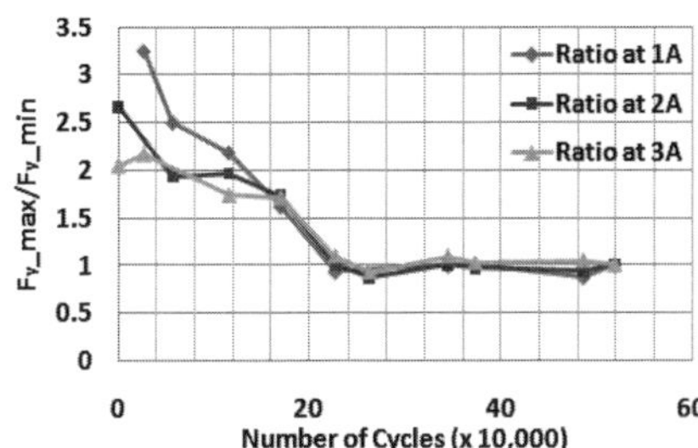

Figure 9: Ratio of maximum to minimum yield force vs. number of cycles.

The durability of the glass beads during the endurance test was examined under a microscope with a magnification of x25 before and after 518,400 of cyclic loading at 4 Hz. Photomicro-graphs of the glass beads, both before and after the endurance test are shown in fig. 10a and fig.10b, respectively. The glass beads were mechanically separated from the MRF using a solvent. Before cycling, the micrometer-scale glass beads were spherical and undamaged. After the endurance test, the glass spheres were completely crushed and no longer visible in the MRF. Therefore, the damping motion caused the glass beads to break into ultra-fine pieces, which did not contribute to an increase in the MR fluid off-state viscosity.

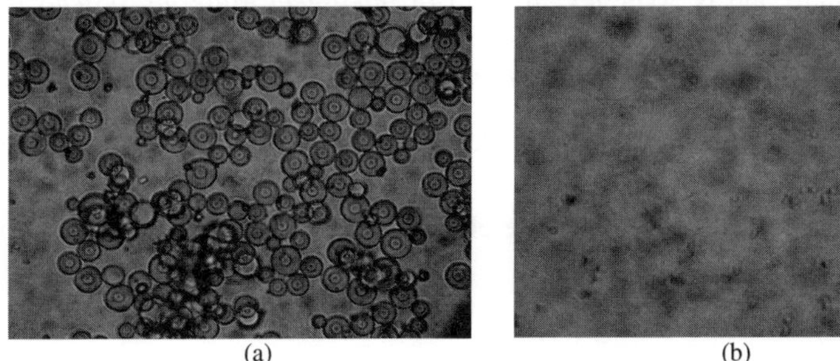

(a)   (b)

Figure 10: (a) Glass beads pre-cycle magnified x25 (b) Glass beads post-cycle magnified x25.

4. Conclusions

The behavior of two MRFs was investigated in this study. The first fluid, MRF-37, had 37 vol% of iron powder, 3 vol% glass beads, and 60 vol% carrier fluid. The second fluid, MRF-40, had 40 vol% of iron powder, and 60 vol% carrier fluid. Based on this study the following conclusions are made:

1. MRF-37 (with glass beads) presented a substantial enhancement (increase) in yield stress in the as-mixed fluid damper cycling tests. The glass beads more than doubled the yield stress at high field strengths (here 2-3 A in the electromagnet) in the damper tests suggesting that nonmagnetic fillers can greatly increase damper yield force.

2. MRF-37 (with glass beads) had an off-state viscosity 22% greater than the MRF-40 (no glass beads).

3. After subjecting MRF-37 to an endurance test (i.e. 518,400 cycles of sinusoidal loading at 4 Hz), the yield stress enhancement effect was eliminated because the glass beads were eroded or crushed to very small sizes. This further supports the conclusion that the glass beads are the strong contributing factor to the yield stress enhancement.

4. MRF-37 would provide a lower specific gravity fluid with a much higher yield stress at full field, thereby providing performance improvements for applications where the MR device is intended for single or infrequent use. For cases where extensive cycling would be required, a passive filler that is not subject to erosion, as glass beads are, would be more appropriate.

References

1. J. Carlson and M. Jolly, *Mechatronics*, **Vol. 10**, pp. 555 (2000).
2. J. Rabinow, *U.S. Patent 2*, **Vol. 575**, pp. 360 (1951).
3. B. Park, K. Song, and H. Choi, *Material Letters*, **Vol. 63**, No. 15, pp. 1350-1352 (2009).
4. Z. Cao, W. Jiang, X. Ye, and X. Gong, *Journal of Magnetism and Magnetic Materials*, **Vol. 320**, No. 8, pp. 1499-1505 (2008).
5. A. Chaudhuri and N. Wereley, *Journal of Intelligent Material Systems and Structures*, **Vol. 17**, No. 5, pp. 393-401 (2006).
6. G. Ngatu and N. Wereley, *IEEE Transactions on Magnetics*, **Vol. 43**, No. 6, pp. 2474-2476 (2007).
7. R. Stanway, *Smart Materials & Structures*, **Vol. 5**, No. 4, pp. 464-482 (1996).
8. R. Snyder, M. Kamath, and N. Wereley, *AIAA Journal*, **Vol. 39**, No. 7, pp. 1240-1253 (2001).

IMPACT OF MORPHOLOGY AND PRE-STRAIN ON DYNAMIC STIFFNESS AND DAMPING OF CO-BASED MAGNETORHEOLOGICAL ELASTOMERIC COMPOSITES

O. PADALKA, H. J. SONG and N. M. WERELEY

Department of Aerospace Engineering, University of Maryland,
College Park, MD 20742, USA

J. A. FILER II and R. C. BELL

Department of Chemistry, Pennsylvania State University,
Altoona, PA 16601, USA

Magnetorheological elastomeric (MRE) composites composed of a silicon rubber matrix with dispersed Co particles of different morphology and weight fraction: (a) spherical microparticles of 10, 30, and 50 wt% and (b) nanowires of 10 wt% were subjected to compressive pre-strain with normalized amplitudes of 1, 2, or 3 % while held in the magnetic cell. The deformation frequency ranged from 0-20 Hz, while the magnetic flux density was fixed at discrete values of 0, 0.1, and 0.2 T. Our investigation of the spherical microparticle-based composites show that the dynamic stiffness and equivalent damping increase with the particle weight fraction for all strain amplitudes. The most significant magnetorheological (MR) effect on dynamic stiffness is observed for 10 wt% samples at strain amplitude of 1 %. This effect highly decreases with both weight fraction and strain amplitude. The MR effect on equivalent damping is much higher than that on dynamic stiffness and it only slightly decreases with particle weight fraction and strain amplitude. To assess the dependence of MR properties on Co particle morphology, the 10 wt% spherical microparticle- and nanowire-based composites were compared. The dynamic stiffness and equivalent damping coefficient values are much higher for the nanowire-based MRE compared to the spherical microparticle-based MRE for all strain amplitudes. However, the MR effect on dynamic stiffness and equivalent damping coefficient is slightly smaller for the nanowire-based composite.

1. Introduction

Magnetorheological elastomeric (MRE) composites is a class of so-called smart materials whose mechanical properties can be controlled by an external magnetic field [1]. They consist of magnetically-polarizable particles dispersed in a non-magnetic matrix. The composite magnetorheological (MR) effect depends on particle magnetic properties, such as permeability and magnetic saturation, as well as on particle morphology, volume fraction, and distribution. These dependencies were widely investigated mostly for the Fe and its oxide

384

microparticle-based composites [2, 3, 4, 5]. Moreover, the MR effect depends on the strain amplitude, which is due to the dependence of magnetic force on the distance between the dipoles [2]. Recently, we have investigated the stiffness and damping properties of the aligned MREs filled with 10 wt% Fe, Co, and Ni nanowires under normalized strain amplitude of 1, 2, and 3% [6]. The MR effect on the dynamic stiffness was found to be the most significant for 1% strain amplitude for all investigated composites. Another of our studies on MREs utilizing Fe and Co microparticles and nanowires of the same weight fraction showed that the nanowire-based composites have much higher dynamic stiffness and equivalent damping coefficient values compared to the microparticle-based composites [7]. We have also found that the 10 wt% Co nanowire-based MRE have better damping capacity than the Fe nanowire-based composite of the same weight fraction.

In the present study we investigate the field dependent characteristics (dynamic stiffness and damping) of the aligned MREs with two different Co particle morphologies (spheres and nanowires) as a function of strain amplitude and magnetic flux density.

2. Experimental Samples and Setup

The investigated MRE composites were composed of a silicon rubber matrix with dispersed Co particles of different morphology and weight fraction: (a) spherical microparticles (1.6 μm in diameter) of 10, 30, and 50 wt% and (b) nanowires (12 ± 3 μm long, 321 ± 44 nm in diameter) of 10 wt%. Light microscope micrographs of the manufactured samples showed more densely packed microstructure of the nanowire-based composites than the spherical based composites of the same weight fractions [7]. This is due to the reduced interparticle spacing in the nanowire-based composites, resulting from a larger number of nanoparticles compared to microparticles. Cylindrical samples (25.4 mm long, 5 mm in diameter) were prepared by mixing particles with the uncured silicone rubber matrix, and then injected into a tubular mold. An external magnetic field was applied along the sample longitudinal axis at the time of curing to polarize the particles resulting in chain formation, thereby maximizing MR effects.

The samples were tested in an Instron test machine (type 8841) modified with a magnetic cell that produces a variable magnetic field along the sample longitudinal axis. Details of the experimental setup can be found in references 3 and 7. The MRE samples were subjected to compressive pre-strain with normalized strain amplitudes of 1, 2, or 3 % while held in the magnetic cell. The deformation frequency ranged from 0-20 Hz, while the magnetic flux density was fixed at discrete values of 0, 0.1, or 0.2 T. The experimental stress-strain curves were used to calculate the values of the dynamic stiffness and equivalent damping coefficient for combinations of these conditions. As examples of experimental data, Fig. 1 shows force-displacement curves of the Co spherical microparticle-based composite under 1% strain amplitude. The dynamic stiffness is calculated as the slope of the major axis of the stress-strain curve. The equivalent damping coefficient quantifies the energy dissipation in the samples and it is calculated as $C_{eq}=E/(2\pi^2 f X_0^2)$ [8], where E is the energy dissipated per cycle, f is the deformation frequency, and X_0 is the strain amplitude.

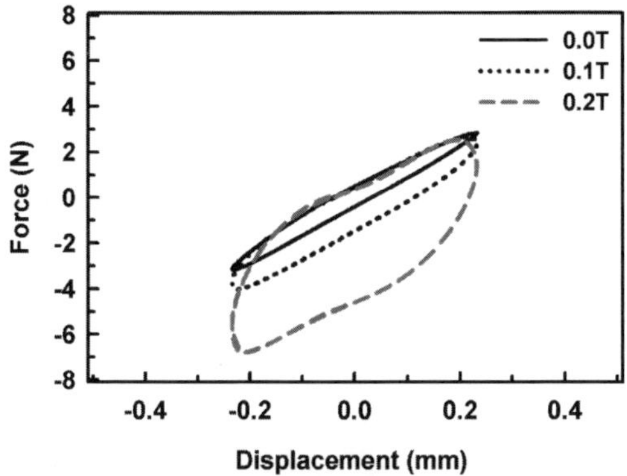

Figure 1. Force vs. displacement curves of the Co spherical microparticle-based composite (30 wt%) under 1% strain amplitude, magnetic flux density of 0, 0.1, and 0.2 T, and 1 Hz cyclic deformation.

3. Experimental Results and Discussion

3.1. *Stiffness and damping properties versus weight fraction of the dispersed Co particles*

Our investigation of the properties of the spherical microparticle-based MREs shows that the dynamic stiffness (Fig. 2) and equivalent damping (Fig. 3) in the absence of applied magnetic field increase with the particle weight fraction for all strain amplitudes. In contrast, these parameters significantly decrease with increase of strain amplitude from 1 to 2 % and then show softening behavior at higher strains.

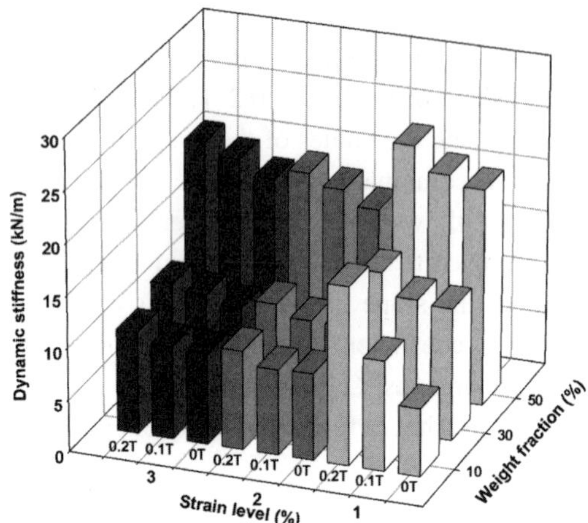

Figure 2. Dependence of the dynamic stiffness of Co spherical microparticle-based composites on filler weight fraction, strain amplitude, and magnetic flux density under 1 Hz cyclic deformation.

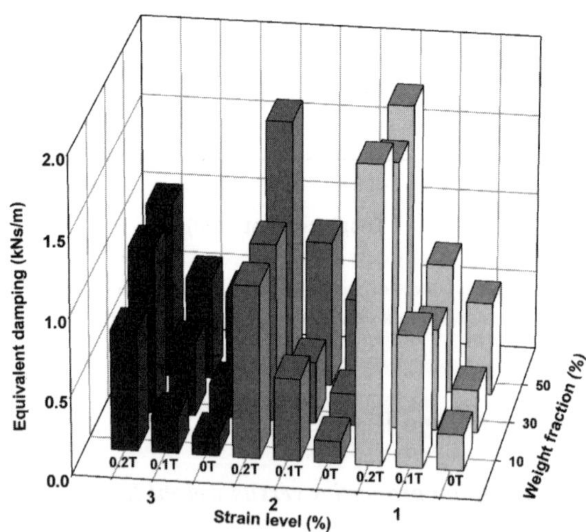

Figure 3. Dependence of the equivalent damping of Co spherical microparticle-based composites on filler weight fraction, strain amplitude, and magnetic flux density under 1 Hz cyclic deformation.

The MR effect on the dynamic stiffness can be defined as the ratio of the dynamic stiffness under the maximum magnetic flux density, which is 0.2 T, to the dynamic stiffness in the absence of applied magnetic field. The most significant MR effect on dynamic stiffness is observed for 10 wt% samples at strain amplitude of 1 % (as seen from Fig. 2). This effect highly decreases with both weight fraction and strain amplitude and only slightly depends on deformation frequency. The MR effect on equivalent damping, defined in the same manner, (as seen from Fig. 3) is much higher than that on dynamic stiffness. The effect only slightly decreases with particle weight fraction and strain amplitude and remains significant even for the highest values of weight fraction and strain amplitude.

3.2. *Stiffness and damping properties versus Co particle morphology*

To assess the dependence of MR properties on Co particle morphology, the 10 wt% spherical microparticle- and nanowire-based composites were compared. The dynamic stiffness (Fig. 4) and equivalent damping coefficient (Fig. 5) values in the absence of an applied magnetic field are much higher for the nanowire-based MREs as compared to the microparticle-based MREs for all strain amplitudes, which is attributed to the larger contact area between the matrix and nanowires and the lattice structure of the nanowires that results in more rigid composites of the same weight fraction as the microspheres. However, the MR effect on dynamic stiffness and equivalent damping coefficient is slightly smaller for the nanowire-based composite. Past studies using iron particle displayed a slight increase in the MR effect for nanowire-based MREs over spherical microparticle-based MREs [7]. However, the cobalt particles used in these studies were not quite spherical. It is this shape anisotropy that is the likely cause for this difference in behavior as previously observed and warrants further study. In addition, a similar dependence of the MR effect on strain amplitude was observed for the nanowire-based MREs as was observed for the spherical-based MREs.

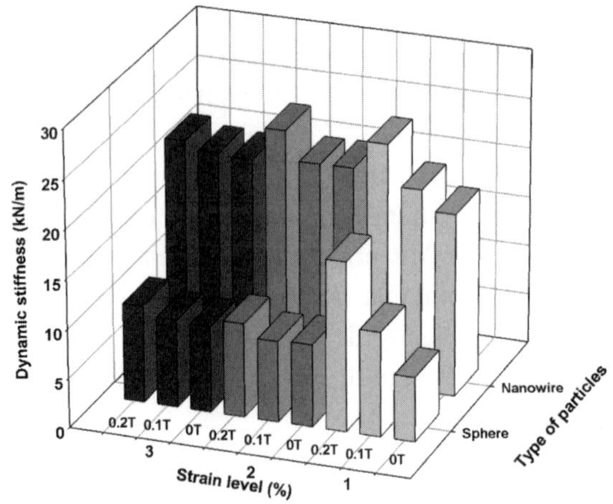

Figure 4. Dependence of the dynamic stiffness of 10 wt% Co spherical microparticle- and nanowire-based composites on strain amplitude and magnetic flux density under 1 Hz cyclic deformation.

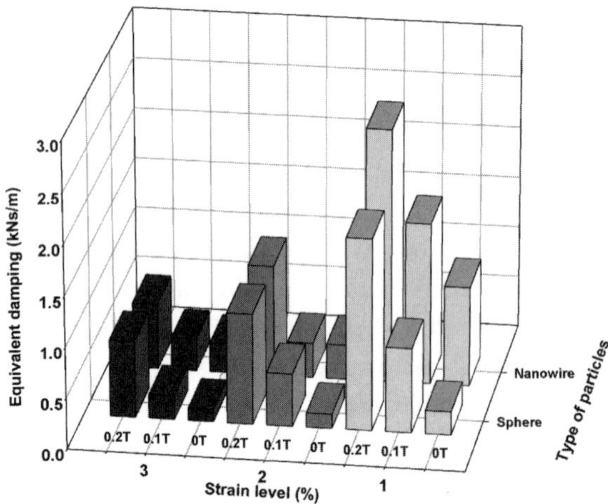

Figure 5. Dependence of the equivalent damping of 10 wt% Co spherical microparticle- and nanowire-based composites on strain amplitude and magnetic flux density under 1 Hz cyclic deformation.

4. Conclusions

In this study, the influence of particle morphology and pre-strain on dynamic stiffness and equivalent damping properties of MRE composites was investigated. From the presented results the following can be concluded:

1. Dynamic stiffness and equivalent damping increase with the particle weight fraction for all strain amplitudes.
2. The dynamic stiffness and equivalent damping significantly decrease with increase of strain amplitude from 1 to 2 %, which is followed by softening behavior of these parameters at higher strains.
3. The most significant MR effect on dynamic stiffness is observed for 10 wt% samples at strain amplitude of 1 %. This effect highly decreases with both weight fraction and strain amplitude and only slightly depends on deformation frequency.
4. The MR effect on equivalent damping is much higher than that on dynamic stiffness. The effect only slightly decreases with particle weight fraction and strain amplitude.
5. The dynamic stiffness and equivalent damping coefficient values are much higher for the nanowire-based MRE compared to the spherical microparticle-based MRE.
6. The MR effect on dynamic stiffness and equivalent damping coefficient is reduced for the nanowire-based composite.

Acknowledgments

The authors acknowledge funding support from the National Science Foundation (CBET-0755696).

References

1. J. D. Carlson, M. R. Jolly, *Mechatronics,* **10**, 555 (2000).
2. M. Kallio, T. Lindroos, S. Aalto, E. Järvinen, T. Kärnä, T. Meinander, *Smart Mater. Struct.* **16**, 506 (2007).
3. H. J. Song, N. M. Wereley, R. C. Bell, J. L. Planinsek, and J. A. Filer II, *J. of Phys., Conference Series* 11, (2009).
4. B. K. Woods, N. M. Wereley, R. Hoffmaster, and N. Nerssesian, *Internat. J. Modern Phys. B* 21, 5010 (2007).
5. X. Guan, X. Dong, J. Ou, *J. Magn. Magn. Mater.* **320**, 158 (2008).
6. O. Padalka, H. J. Song, N. M. Wereley, J. A. Filer II, and R. C. Bell, *IEEE Transactions on Magnetics* 46 (b), 1 (2010).
7. H. J. Song, O. Padalka, N. M. Wereley, R. C. Bell, *AIAA Adaptive Structures Conference,* (2009).
8. W. J. Li, G. Z. Yao, G. Chen, S. H. Yeo, F. F. Yap, *Smart Mater. Struct.* **9**, 95 (2000).

PREPARATION AND ELECTRORHEOLOGICAL EFFECT OF ACETAMIDE-MODIFIED TITANATE NANOTUBE SUSPENSIONS INSTRUCTIONS

YUCHUAN CHENG[*], JIANJUN GUO, XUEHUI LIU, GAOJIE XU and PING CUI

Ningbo Institute of Material Technology and Engineering, Chinese Academy of Sciences, Ningbo, 315201, P. R. China
[*]*yccheng@nimte.ac.cn*

Titanate nanotube (TN) synthesized by hydrothermal reaction of titania nanoparticles in alkali solution, then the TNs were modified by acetamide. X-ray diffraction analyses, scanning electron microscopy and Fourier transform infrared spectrometry are used to determine the structure of the TNs. Under DC electric field, the yield stress of these acetamide-modified TN suspensions shows large yield stress. It is also found that the ER effect is sensitive to the wettability of the TNs.

1. Introduction

Electrorheological (ER) fluid is a smart suspension composed of polarizable solid particles dispersed non-polar liquid medium, of which rheological properties are drastically changed by an applied electric field[1-4]. Because of its rapid response (< 10 ms) to electric field and simple control, ER fluid has attracted much interest for applying in various devices such as clutches, dampers and actuators[5,6]. Nevertheless, the broad applications of ER devices have been hampered by the poor ER activity of most available ER fluids. Thus, improving the ER capability of ER fluids is of urgent significance for the further development of electromechanical industry.

Many recent studies show that the shape effect of particles could greatly affect the ER activity of materials[7-12]. Since the discovery of carbon nanotubes in 1991[13], one-dimensional (1D) nanostructures (rods, wires, and tubes) have attracted extensive attention due to their fundamental significance in investigating the dependence of various physicochemical properties on dimensionality and size reduction, as well as potential applications in numerous fields. Titania is widely used in a great variety of applications such as photocatalysis, solar energy conversion, sensors, and photovoltaic cells. As a kind of environmentally friendly semiconductor with relatively high dielectric constant and polarization ability, titania has been considered as the potential high-performance ER materials. In this paper, we prepared uniform acetamide-modified titania nanotubes (TN), and found these materials had a high ER effect. X-ray diffraction analyses (XRD), scanning electron microscopy (SEM), transmission electron microscope (TEM), Fourier transform infrared (FTIR) spectrometry, and interface tension/contact angle (CA) measurement were utilized to characterize the component, structure, and morphology of the TN materials. Furthermore, the influence and e of wettability on the ER performance between the TN and silicone oil was examined.

2. Experimental

2.1. *Chemicals*

TiO_2 nanoparticles (P25, average particle size$\approx$ 20 nm) were purchased from EVONIK-degussa Company (Germany). Acetamide (AR), and chloroacetamide (AR), NaOH (AR), HCl (AR), methanol (AR), ethanol (AR) were purchased from Sinopharm Chemical Reagent Company (Shanghai, China). (3-amino-propyl) trimethoxysilane (APTES, >97%) was obtained from Alfa Aesar China Company (Tianjin, China).Silicone oils (η=50 mPa·s) were obtained from Hangping Company (Beijing, China). All the chemicals cited above were used as received without further purification. The water used in this work was deionized (DI) water form a Millipor-Q purification system (Millipore, USA) of resistivity 18.2 MΩ·cm.

2.2. *Preparation of TN*

In a typical synthesis, 1.5 g of P25 and 70mL of aqueous NaOH (10 mol/L) were placed into a Teflon-lined autoclave. The mixture was stirred to form a milk-like suspension, sealed and hydrothermally treated at 150 $^{\circ}$C for 24 h. The precipitate was separated by filtration and washed with DI water until a pH value near 8 was reached. The precipitate was then ground in alcohol followed by ultrasonic-assisted dispersion. After a second filtration and alcohol washing step, the sample was keeping under vacuum at 60 $^{\circ}$C.

2.3. *Preparation of acetamide-modified TN*

The TNs were dispersed in a dilute solution of APTES in methanol (5 vol%), and stirred for 4 h at room temperature. The amino-coated TNs were separated by filtration and washed with methanol. Then the precipitates were dispersed in the dilute solution of chloroacetamide in methanol (5 vol%). After stirred for

4 h, the white precipitate was obtained by filtration, followed by washing with ethanol and DI water several times and keeping under vacuum at 60 $^{\circ}$C.

2.4. *Preparation of ER fluids*

All the silicone oils were dried at 120 $^{\circ}$C for 2h before the experiment to avoid the influence of moisture. The ER fluids were formed by mixed TNs or P25 with silicone oils. Subsequently, the mixed suspension was dry at 60 $^{\circ}$C for 1 h. Concentration of the ER fluids is denoted as the ratio of the solid particles volume to the total volume of the ER fluid.

2.5. *Characterization*

The morphology and grain size of the samples were examined by a Hitachi S4800 field emission scanning election microscope (SEM) and FEI Tecnai G2 F20 transmission electron microscope (TEM).X-ray diffraction (XRD) patterns were obtained with a Bruker D8 Advance/Discover diffractometer, using CuKα radiation. All measurements were taken using a generator voltage of 40 kV and a current of 40 mA. Fourier transform infrared (FTIR) spectra were recorded on Nicolet 6700 spectrometer using 32 averaged scans at 4 cm^{-1} resolution. The wetting properties of the samples were determined by measuring GBX digidrop, interface tension/contact angle measure equipment. The yield stress was measured by a circular-plate type rheometer (Haake RS6000, 15 mm in diameter), which was determined by the maximum shear stress value at which a kink occurs in the strain-stress curve. Experimental data were collected with the help of the software package Rheowin. The suspensions were placed in the gap between two plates separated by a distance of 1.00±0.001 mm. The maximum voltage of the DC high-voltage generator (SPELLMAN SL300) was 10 kV, and the current limitation was 3.0 mA. All measurements were carried out at ambient temperature.

3. Results and Discussions

3.1. *Characteristic*

The morphology of the samples was investigated by SEM and TEM. Fig. 1b is the low magnification SEM image of as-prepared TNs. From the image, we can observe a large quantity of tubular materials with narrow size distribution. The diameters of the tubular materials are around 8–10 nm and the lengths range from several tens to several hundreds of nanometers. High-resolution TEM in

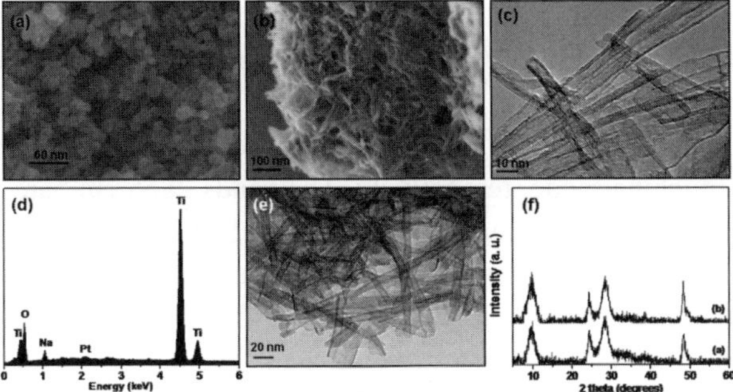

Figure 1. SEM images of (a) P25 nanoparticles, (b) TNs; and TEM images of (c) TNs, (e) acetamide-modified TNs; (d) EDS analysis of TN; (f) XRD patterns of TNs and acetamide-modified TNs.

Fig. 1c shows that the nanotube is multilayered structure with a hollow inner diameter of 4 nm and an interlayer distance of 0.9 nm. The energy-dispersive X-ray spectroscopy (EDS) analysis (Fig. 1d) of individual TN exhibits the existence of Na, Ti and O elements. The morphology of acetamide-modified TNs is not observed obvious distinction. (Fig. 1e) The powder XRD patterns (Fig. 1d) of unmodified TNs and acetamide-modified TNs show that there are four peaks at $9.6°$, $24°$, $28°$, and $48°$, respectively, which is the characteristic peak for $H_4Ti_4O_{10}$ (JCPDS No. 38-0699). According the EDS analysis, we consider that the TN is described as $Na_xH_{4-x}Ti_4O_{10}$. The interlayer distance of these TNs is about 0.93 nm according to the low angle diffraction peak, which is consistent with the result of TEM. The XRD patterns of TNs after acetamide modifying is similar to that of unmodified TNs, which indicates that the acetamide molecules coating treatment dose not induce the interlayer of TNs.

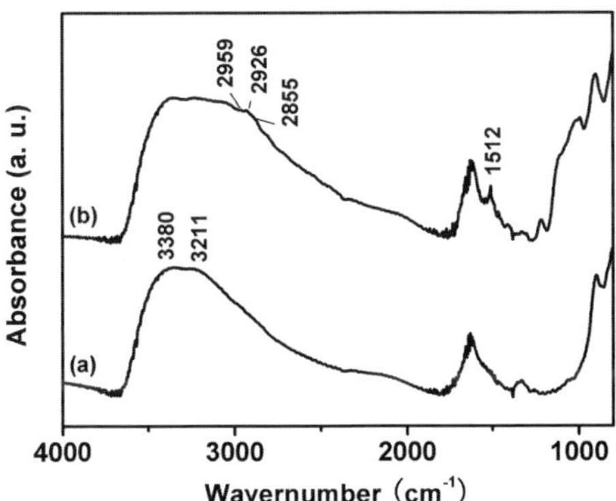

Figure 2. IR spectra of (a) unmodified TNs and (b) acetamide-modified TNs.

Fig. 2 gives FT-IR spectra of the unmodified TNs (a) and acetamide-modified TNs (b), respectively. The broad band around 3380 cm^{-1} is assigned to O-H stretching vibration mode. Compared to Fig. 2a, some new bands are found in Fig. 2b. The bands at 2959 and 2926 cm^{-1} are assigned to $-CH_3$ and $-CH_2$ asymmetric stretching mode, respectively. The bands 2855 cm^{-1} are attributed to $-CH_2$ symmetric stretching mode. The bands at 1512 cm^{-1} correspond to amide II mode. The emergence of these bands proves that the TNs are really capped by acetamide molecules. It is well know that the frequencies of $-CH_2$ streching modes are sensitive to the conformation of an alkyl chain. The bands appear near 2918 and 2848 cm^{-1}, if the chain is highly ordered, while the disorder increase, the bands shift upward. From Fig. 2, we suggest that alkyl chains on the TNs surface are disorder.

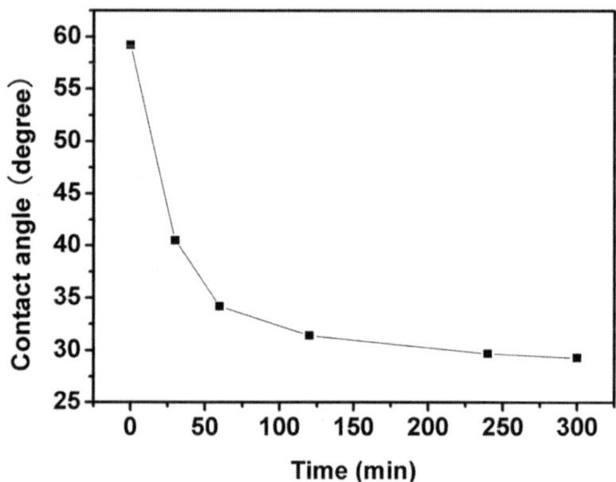

Figure 3. The immersion time as a function of contact angle of acetamide-modified TN films.

The CA is usually referred as the wettability of the sample surface. The CA plotted as a function of time of the TNs immersed in a chloroacetamide methanol solution is presented in Fig. 3. It is found that the decrease in the value of CA is very rapid for the first few minutes and stabilizes with the maximum value at 240 min after the immersion for the chloroacetamide solution. Consequently, the optimized immersion time in experiment is 4 h. The above results clearly show the kinetics of the adsorption. The acetamide molecules and the silicone oil molecules form hydrogen bonding network, which give rise to the favorable effect of the wettability of TNs that is closely related with anti-electric breakdown, stability and strong polarizatio[14]. As the addition to immersion time, the molecule amount of chloroacetamide adsorbed on the TNs increase. Therefore, the increase of the wettability is rapid in first range. When the adsorption of chloroacetamide molecule on particle surface is saturated, the best wettability of TNs presents. After this, further increasing immersion time does not affect the wettability of TNs.

3.2. *Electrorheological property*

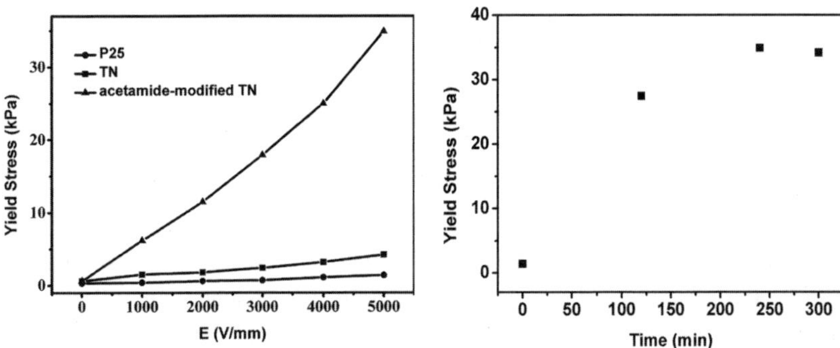

Figure 4. (a) Yield stress of ER fluids containing P25 (circle), TN (square) and acetamide-modified TN (triangle) as a function of the electric field strength, (b) yield stress of acetamide-modified TN ER fluid (35 wt%) as a function of the immersion time at 5 kV/mm.

Fig. 4a shows the yield stress of P25 nanoparticles, unmodified TNs, and acetamide-modified TNs based ER fluids under different DC electric field at γ =0.1 s^{-1}. The concentration of particles in ER fluid is 35 wt%. The ER effects of ER fluids containing the P25 nanoparticles and unmodified TNs are very low. Meanwhile, the acetamide-modified TNs ER fluids display notable ER activity, their yield stress is up to 30 kPa (5 kV/mm). It indicates that acetamide coating can enhance evidently the ER performance of TNs materials.

The yield stress of acetamide-modified TNs ER fluids is affected by the time of TNs immersed in a chloroacetamide methanol solution. Fig. 4b shows the dependence of the yield stress on the immersion time. The yield stress increases with increasing time first, then stabilizes with the maximum value. The result is in good agreement with the change of the wettability of TNs. It indicates the wettability between TN and silicone oil is an important factor for enhancing the ER effect, good wettability to silicone oil maybe increases the contact area and interaction among TNs in the silicone oil.

4. Conclusion

We fabricated acetamide-modified TN composites by hydrothermal reaction and self-assembled process. The morphology, structure, and ER characteristric of the composites were also investigated. It is found that the ER effect of suspensions shows a strong dependence on the wettability of acetamide-modified TNs.

Acknowledgments

This research was supported by the National Basic Research Program of China (2009CB930801), the National Natural Science Foundation of China (10904155 and 21003145), the Knowledge Innovation Project of Chinese Academy of Sciences (KJCX2.YW.M07), the Zhejiang Provincial Natural Science Foundation of China (D4080489 and Y4090044), and the Ningbo Natural Science Foundation (2009A610031 and 2010A610170), the CAS/SAFEA International Partnership Program for Creative Research Teams. We also express our gratitude to the aided program for Science and Technology Innovative Research Team of Zhejiang Province and Ningbo Municipality (2009B21005).

References

1. W. M. Winslow, *J. Appl. Phys.* **20,** 1137 (1949).
2. D. J. Klingenberg and C. F. Zukoki, *Langmuir* **6,** 15 (1990).
3. T. Tao, *Adv. Mater.* **13,** 1847(2001).
4. T. C. Halsey, *Science* **258,** 761 (1992).
5. N. M. Wereley, *J. Intell. Mater. Syst. Struct.* **19,** 257 (2008).
6. C. S. Zhu, *Int. J. Mod. Phys. B* **19,** 1577 (2005).
7. R. C. Kanu and M. T. Shaw, *J. Rheol.* **42,**657 (1998).
8. J. T. Hu, T. W. Odom and C. M. Lieber, *Acc. Chem. Res.* **32**, 435(1999).
9.   (10)G. M. Whitesides and B. Grzybowski, *Science* **295**, 2418 (2002).
10. J. B. Yin and X. P. Zhao, *Colloids Surf. A* **329**, 153 (2008).
11. A. J. Mieszawska, G. W. Slawinski and F. P. Zamborini, *J. Am. Chem. Soc.* **128**, 5622 (2006).
12. K. Tsuda, Y. Takeda, H. Ogura and Y. Otsubo, *Colloids Surf. A* **299**, 262 (2007).
13. S. Iijima, *Nature* **354**, 56 (1991).
14. B. X. Wang and X. P. Zhao, *Adv. Funct. Mater.* **15**, 1815 (2005).

EFFECT OF HEAT TREATMENT ON POLAR MOLECULE DOMINATED-TIO₂ ELECTRORHEOLOGICAL FLUIDS

XUEHUI LIU, JIANJUN GUO[*], YUCHUAN CHENG, GAOJIE XU[†] and PING CUI

Division of Functional Materials and Nano Devices, Ningbo Institute of Material Technology & Engineering, Chinese Academy of Sciences, Ningbo 315201, China
[*]jjguo@nimte.ac.cn and [†]xugj@nimte.ac.cn

In this paper, we studied the effects of heat treatment on the ER performance, electrical and dielectric properties of a type of polar molecule dominated-TiO₂ ER fluids. It was found that the permittivity of unheated TiO₂ ER fluids was large and there was a clear dielectric loss peak between 10^3 to 10^4 Hz. However, for heat-treated ER fluids, the permittivity moves up to a relative low value from 40 to 10^4Hz. No dielectric relaxation process was observed at low frequency range. The distinct changes of polar molecule dominated-TiO₂ electrorheological fluids before and after heat treatment are due to the presence of water absorbed on the surface of particles, which play an important role in improving the ER effects.

1. Introduction

Electrorheological (ER) fluids are known as smart suspensions composed of polarizable solid particles in a non-conducting liquid medium that exhibit drastic changes in their rheological properties, including a large increase in apparent viscosity [1–4]. Because of their controllable viscosity and short response time, ER materials have huge potential for applications in dampers, clutches, brakes, mechanical sensors, valves and actuators [5, 6]. But some of the technical problems such as low shear stress, poor temperature stability and particle sedimentation, have greatly limited their industrial applications [7-8].

In recent publications, considerable efforts have gone into the preparation of different suspended particles with the aim of improving the ER yield stress [9–12]. Among them, TiO₂ is a good ER material for its high dielectric constant and has been paid much attention to its ER suspension. However, the dried TiO₂ based suspensions have quite low ER responses, usually of only several kPa. When these particles absorb a small amount of water, their ER response can be substantially improved. It is obvious that water absorbed onto the particulate phase influences the bulk particle permittivity and conductivity, which govern the ER performance essentially.

In this paper, we prepared a kind of polar molecule dominated-TiO_2 particles via simple sol-gel method in our previous study [13]. The aim of our paper is to investigate the effects of heat treatment on the polar molecule dominated-TiO_2 ER fluids at low temperature (80°C). It is interesting to found that although the ER activity decreased, it is still showing high yield stress. What is more, the leaking current density was below 1 $\mu A/cm^2$ under electric field of 5 kV/mm, which helps us to design high-performance ER fluids with low current density that are important to develop an ideal ER fluid for industrial applications.

2. Experiments

2.1. *Preparation of polar molecule dominated-TiO₂ ER fluids*

The TiO_2 nanoparticles were prepared by means of a sol-gel method [13]. In brief, titanium butoxide [$Ti(C_4H_9O)_4$] was dissolved in ethanol (C_2H_5OH) at a volume ratio of [$Ti(C_4H_9O)_4$] / C_2H_5OH = 1: 2. A small amount of acetic acid (HAc) was added to avoid precipitation. The deionized water was dropped into the TBT solution with vigorous stirring and the volume ratio of H_2O and C_2H_5OH was kept at 1:5. After the dropping, the suspension was stirred for 20h. The white precipitates were filtrated and washed with water and ethanol. The white precipitates were dried in vacuum at 60°C for 16h and 120°C for another 4h in order to remove physical adsorbed water.

The ER suspensions of TiO_2 in silicone oil were prepared by the following steps. The silicone oil (50 mPa at 25°C) was heated at 120°C for 2h. The dehydrated TiO_2 powder and silicone oil were immediately mixed before the rheological measurements. The mixed suspension was heated at 80°C for 4h for purposes of comparison.

2.2. *Characterization*

The microstructures of the products were obtained by transmission scanning electron microscopy (TEM, Tecnai G2 F20, FEI Ltd., American). Fourier transform infrared (FT-IR) spectra were recorded on a Nicolet 6700 transform infrared spectrometer using 32 averaged scans at $4cm^{-1}$ resolution. The static yield stress was measured by a circular plate type rheometer (Haake RS6000, 15 mm in diameter). The suspensions were placed in the gap between two plates separated by a distance of 1.00±0.001 mm. The shear stress was measured in the controlled shear stress (CS) mode, in which the deformation was monitored while the stress was continuously increased. The static yield stress was determined by the maximum shear stress value at which a kink occurs in the strain–stress curve. The dielectric relaxation spectra of all prepared ER Fluids were examined using a NOVOCONTROL GmbH BDS40 dielectric analyzer. Frequency of AC electric fields ranged from 40Hz to 1 MHz. A bias electrical potential of 1 V was applied to the ER fluids during the measurement.

3. Experimental Results

3.1. *Material characteristics*

The size and morphology of TiO_2 particles were examined by TEM, and the results were shown in Fig. 1. It is observed that the TiO_2 nanoparticles show nearly spherical shape and some irregular agglomerates. The particle diameter ranges from 150 to 250nm. The selected area electron diffraction (SAED) pattern of TiO_2 nanoparticles is shown in the inset of Fig. 1. It is clearly observed that the TiO_2 power behaves as an amorphous pattern giving rise to no diffraction spots that can be indexed.

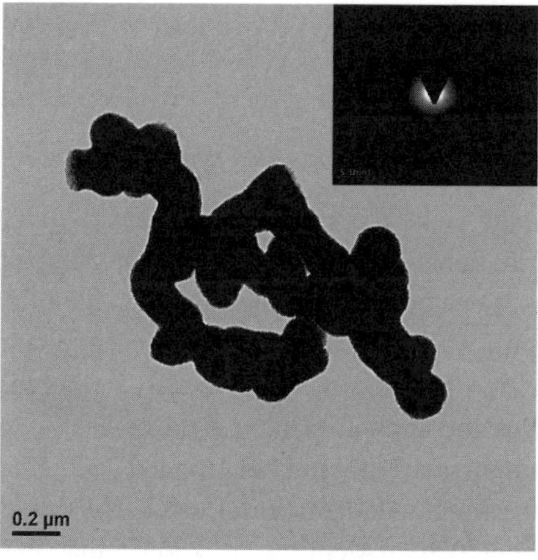

Figure 1. TEM image of TiO_2 nanoparticles; inset: a selected-area electron diffraction pattern.

Fig. 2 gives FT-IR spectra of as-prepared TiO_2 nanoparticles. The broad band around 3400 cm^{-1} is assigned to O–H stretching vibration mode. The band at 1630 cm^{-1} is attributed to the O–H bending vibration, which confirmed the

400

presence of molecular H_2O [14]. The peaks in the region of 1300–1500cm^{-1} is assigned to the bending vibrations of C–H bonds of CH_3 and CH_2 groups of BtOH and TBT. At 1100 and 1033cm^{-1} appear the characteristic Ti–O–C stretching vibrations of TBT [15]. The band at 500–800cm^{-1} is assigned to the Ti–O stretching vibrations. The FT-IR spectroscopy indicates it is difficulty to obtain a completely hydrolysed $[Ti(C_4H_9O)_4]$ molecule.

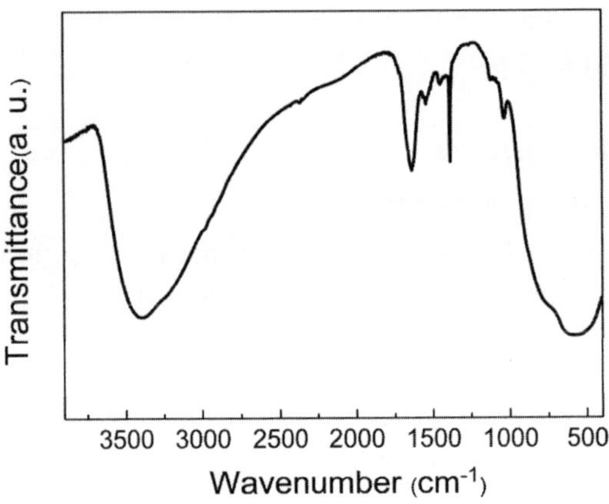

Figure 2. FT-IR patterns of as-prepared TiO2 nanoparticles.

3.2. *Rheological and electrical properties of the prepared ER fluids*

Fig. 3a shows the static yield stress and current density of TiO_2 ER fluids under different DC electric fields without heat treatment. The concentration of the sample is 26 vol. %. It can be seen that the sample exhibits strong yield stress up to 70 kPa (5 kV/mm), while the corresponding leakage current density is very high, about 142 $\mu A/cm^2$. In contrast, we have also measured the yield stress and current density versus the applied electric field of heat-treated TiO_2 ER fluids (Fig. 3b). The maximum yield stress of heat-treated TiO_2 ER fluids is close to 45 kPa (5 kV/mm) at the same volume fraction, while the leaking current density is very slow, only about 0.80 $\mu A/cm^2$. It is known that the ER effect is very sensitive to the interaction of polar molecules and polar groups absorbed on TiO_2 nanoparticles. Therefore water absorbed onto the particulate phase has been observed to dramatically enhance the performance of fluids compared to 'dry' ER fluids at comparable electric fields. To confirm the existence of water absorbed on the surface of particles, the mixed ER fluids without heat treatment in a small test tube were immediately placed in big glass tube. The mouth of the glass tube were sealed and then stored at a temperature of 80°C for 4h. After cooling to room temperature, liquid droplets were found on the tube wall. A small mount of anhydrous cupric sulfate power was added on the liquid droplets and it turned to blue instantly. From experimental results, it is evident that the liquid droplets on the glass tube wall were mainly polar molecular H_2O. Although our sample was dried at 120°C for 4h, there was still free water absorbed on the surface of the particles due to its large surface area and porosity. There is water inside the porous core almost inevitable from the fabrication process [16]. The absorbed water and lattice water can cleared away at higher heating temperature (500 °C), but all absorbed polar molecule or polar groups are desorbed off [10].

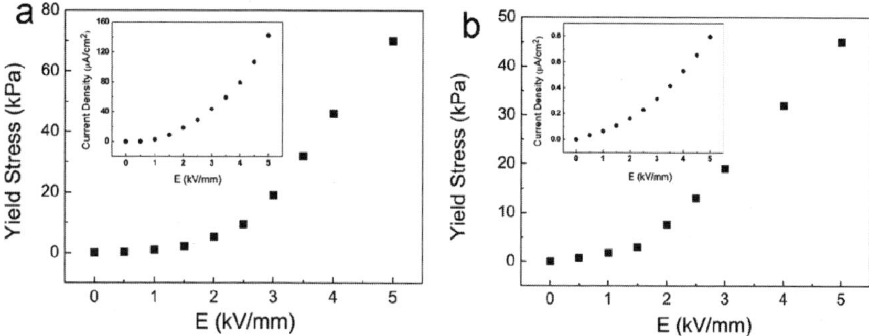

Figure 3. Yield stress and current density of unheated TiO_2 (a) and heat-treated TiO_2 (b) ER fluids as a function of applied electric field (26 vol %).

3.3. *Dielectric properties of the prepared ER fluids*

The particle polarization is considered to responsible for the ER effect in many investigations about ER mechanisms. It was well known that a good ER effect requires two important factors, including a dielectric relaxation peak in the ε'' within an adequate frequency range of 10^2 –10^5 and a large dielectric difference, $\Delta\varepsilon'$ ($\Delta\varepsilon = \varepsilon'100Hz – \varepsilon'100kHz$). The dielectric relaxation peak is related to a suitable polarization response and $\Delta\varepsilon'$ is related to achievable polarizability. Because the relaxation frequency of an ER fluid is proportional to the rate of polarization of the ER fluid, the polarization rate is too high or too slow, a good ER effect can not be obtained [17].

Fig. 4 shows the dielectric spectra for ER fluids made with unheated TiO_2/silicone and heat-treated TiO_2/silicone, respectively. It is found that no relaxation peak is observed in the heat-treated TiO_2 ER suspension in the measured frequency range of 40 Hz – 1 MHz and the dielectric constant of the heat-treated mixtures are very low. However, unheated TiO_2 ER fluid increases

$\Delta\varepsilon'$ and induces a dielectric loss peak at about 2 kHz; this has been attributed to the enhancement of charge carriers due to the existence of water. Our results suggest that the presence of water essentially modifies the dielectric constants. The function of the adsorbed water is supposed to create mobile charge carriers on the surface of the particles. The migration of these charge carriers causes an interfacial polarization to induce ER effect under electric field. Therefore, the large improvement in interfacial polarization is responsible for the enhancement of the ER activity of unheated TiO_2 ER fluids. Unlike water- containing materials, water-free ER materials have intrinsic charge carriers in either the bulk particles or their surfaces that can move locally to induce bulk or interfacial polarization under an applied electric field [14].

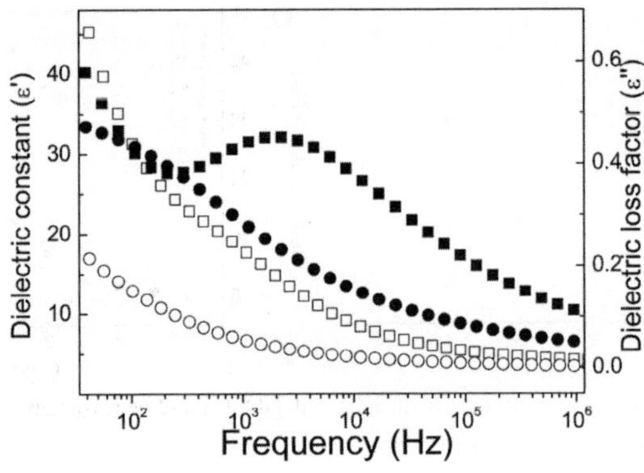

Figure 4. Dielectric spectra of unheated TiO2 ER fluids (squares) and heat-treated TiO2 ER fluids (circles) (the open symbols represent the dielectric constant ε', and the solid symbols represent the dielectric loss factor ε''; the volume fraction of the particles is 26%.

4. Conclusions

In this work, we have chosen polar molecule dominated-TiO_2 ER fluids and study the effects of heat treatment on their rheological, electrical and dielectric properties. The absence of polar molecule H_2O exhibits a good explanation for decreased yield stress and current density after heat treatment. The unheated TiO_2 ER fluids have high dielectric constant and high relaxation frequency which show a strong electrorheological effect due to large interfacial polarization. But the shortcomings of water make it unsuitable for many applications, so our study provides a better way to dispose moisture of TiO_2 ER fluids and still obtain high electrorheological properties simultaneously.

Acknowledgements

This research is funded by the National Basic Research Program of China (2009CB930801), the National Natural Science Foundation of China (10904155, 21003145), the Knowledge Innovation Project of Chinese Academy of Sciences (KJCX2.YW.M07), Zhejiang Provincial Natural Science Foundation of China (D4080489, Y4090044) and Ningbo Natural Science Foundation (2009A610031, 2010A610170). We also express our gratitude to the aided program for Science and Technology Innovative Research Team of Zhejiang Province and Ningbo Municipality (2009B21005).

References

1. W.M. Winslow, J. Appl. Phys. 20, 1137 (1949).
2. T.C. Halsey, Science 258, 761 (1992).
3. T. Hao, Adv. Mater. 13, 1847 (2001).
4. D.J. Klingenberg and C.F. Zukoki, Langmuir 6, 15 (1990).
5. T. Hao, Adv. Colloid Interface Sci. 97, 1 (2002).
6. N.M. Wereley, J. Intell. Mater. Syst. Struct. 19, 257 (2008).
7. Y. Cheng, X. Liu, J. Guo, F.H Liu, Z. Li, G. Xu and P. Cui, Nanotechnology 20, 055604 (2009).
8. Y. Cheng, J. Guo, G. Xu, P. Cui, X. Liu, F. Liu and J. Wu, Colloid. Polym. Sci. 286, 1493 (2008).
9. L. Xu, W.J. Tian, X.F. Wu, J.G. Cao, L.W. Zhou, J.P. Huang and G.Q. Gu, J. Mater. Res. 23, 409 (2008).
10. K.Q. Lu, R. Shen, W.Z. Wang, G. Sun, W.J. Wen and J.X. Liu, Int. J. Mod. Phys. B 21, 4798 (2007).
11. F. Liu, G. Xu, J. Wu, Y. Cheng, J. Guo and P. Cui, Smart. Mater. Struct. 18, 125015 (2009).
12. X.P. Zhao and J.B. Yin, Chem. Mater. 14, 2258 (2002).
13. X. Liu, J. Guo, Y. Cheng, G. Xu, Y. Li and Cui P, Rheol. Acta 49, 837 (2010).
14. M.J. Velasco, F. Rubio, J. Rubio and J.L. Oteo, Spectrosc. Lett. 32, 289 (1999).
15. S. Doeuff, M. Henry, S. Sanchez and J. Livage, J. Non-Cryst. Solids. 89, 296 (1987).
16. X. Huang, W. Wen, S. Yang and P. sheng, Solid State Commun. 139, 581 (2006).
17. A. Kawai, Y. Ide, A. Inoue and F. Ikazaki, J. Chem. Phys. 109, 4587 (1998).

III. PHYSICS MECHANISM

MICROSCOPIC MECHANISM OF THE GIANT ELECTRORHEOLOGICAL EFFECT

SHUYU CHEN, XIANXIANG HUANG, WEIJIA WEN and PING SHENG[*]

Department of Physics and William Mong Institute of Nano Science and Technology, Hong Kong University of Science and Technology, Clear Water Bay, Kowloon, Hong Kong, China

NICO F. A. VAN DER VEGT

Center of Smart Interfaces, Technical University of Darmstadt, Petersenstrasse 32, 64287 Darmstadt, Germany

Electrorheological (ER) fluids are a type of colloids whose rheological characteristics can be varied reversibly with the application of an electric field. The traditional ER mechanism is based on induced polarization arising from the dielectric constant contrast between the suspended solid particles and the fluid. However, for induced polarization there is a maximum value for the dimensionless electric susceptibility $\chi \sim 0.24$. That implies an upper bound for the traditional ER effect. For permanent molecular dipoles, χ can be as high as 4-50. Thus, one to two orders of magnitude higher electric energy can be gained by harnessing the molecular dipoles. Indeed, the recent discovery of the giant electrorheological (GER) effect, in nanoparticles of urea-coated barium titanate oxalate suspended in silicone oil, has shown that the theoretical upper bound of the traditional ER effect is no longer applicable to this new type of electrorheological fluid. A phenomenological model of the GER mechanism, based on aligned molecular dipoles of urea molecules in the contact regions of the nanoparticles, yields an adequate account of the observed effect but without a microscopic picture on how this can occur. More recently, by using molecular dynamics (MD) to simulate the urea-silicone oil mixture confined between two bounding surfaces of a nanocontact, we find that the urea molecules, which has a molecular dipole moment of 4.6 Debye, can form aligned dipolar filaments that penetrate the oil film to bridge the two substrates, with the attendant lowering of the aligning field for the urea dipoles. This phenomenon is explainable on the basis of a 3D to 1D crossover in urea molecules' microgeometry, realized through the confinement effect provided by the oil chains. The resulting electrical energy density is shown to give an excellent account of the observed yield stress variation as a function of the electric field.

---

[*] Corresponding author: sheng@ust.hk

408

1. Introduction

1.1. *Induced dipole vs. molecular dipole*

Electrorheological (ER) fluid comprises of solid particles dispersed in an insulating liquid. The traditional ER effect is based on induced polarization under an applied electric field $\vec{E}$, caused by the dielectric constant contrast between the suspended solid particles and the liquid. There is an extensive literature on the traditional ER effect, see, e.g., R. Tao [1], D. J. Klingenberg [2], H. Conrad [3], G. Bossis [4], H. R. Ma et al. [5], and J. E. Martin et al. [6]. Reviews can be found in [7,8].

Let ε_s denote the complex dielectric constant of the solid particles and ε_ℓ that of the liquid; then for spherically shaped particles, the induced dipole moment may be expressed as:

$$\vec{p} = \frac{\varepsilon_s - \varepsilon_\ell}{\varepsilon_s + 2\varepsilon_\ell} a^3 \vec{E} = \beta a^3 \vec{E}, \tag{1}$$

where a is the radius of the particles and β is the Clausius-Mossotti factor. The product $\beta a^3 = \alpha$ is denoted the polarizability. The (induced) dipole-dipole interaction between the particles means that the particles would tend to aggregate and form chains/columns along the applied field direction, leading to increased viscosity under shear. With increasing electric field the ER fluid may even turn into a weak solid characterizable by the existence of a yield stress that is proportional to the energy density $W = |-\vec{P} \cdot \vec{E}|$, where $\vec{P} = N\vec{p}$ is the polarization density, with N being the number density of the solid particles. The dependence of the yield stress on the electric field is necessarily quadratic. That is because for induced polarization $\vec{P} = \chi \vec{E}$, where $\chi = N\alpha$ is the dimensionless electric susceptibility. Hence $|-\vec{P} \cdot \vec{E}| = \chi E^2$. Also, since there is a maximum value of polarizability $\alpha = a^3$, reached at the limit of $\varepsilon_s / \varepsilon_\ell \to \infty$, it follows that there is a maximum value for the dimensionless electric susceptibility $\chi = N\alpha \simeq a^3 / (4\pi a^3 / 3) \sim 0.24$. A more rigorous theory [7], based on the effective dielectric constant formulation of the ER problem, can give an upper bound for the yield stress, expressible as $1.38\sqrt{R / \delta} \left(\varepsilon_l E^2 / 8\pi \right)$, with R the radius of the particles and δ the separation. Therefore, limited by the breakdown electric field, the yield stress achievable by most of the traditional ER effect is usually on the order of a few kPa.

Motivated by a desire to break the upper bound imposed by the induced polarization mechanism, the search naturally turns to permanent molecular dipoles, whose polarizability α_m differs from that expressed by Eq. (1) and is the result of competition between the alignment energy $-\vec{p}_o \cdot \vec{E}$ and thermal Brownian motion, where p_o denotes the molecular dipole moment. Since the dipole-field interaction energy is given by $-p_o E \cos\theta$, where θ denotes the angle between the field and the dipole moment, the Boltzmann factor is given by $\exp(p_0 E \cos\theta / k_B T)$. In 3D, the average dipole moment is therefore given by the Langevin function:

$$\langle p \rangle = \frac{\int_0^\pi p_0 \cos\theta \exp(p_o E \cos\theta / k_B T) \frac{1}{2} \sin\theta \, d\theta}{\int_0^\pi \exp(p_o E \cos\theta / k_B T) \frac{1}{2} \sin\theta \, d\theta} = p_o \left(\coth(x) - \frac{1}{x} \right), \quad (2)$$

where $x = p_0 E / k_B T$. For $p_0 = 4.6$ D and $T=300$ K, $x=1$ only when $E= 2.5 \times 10^6$ V/cm. For electric field much smaller than that, $\langle p \rangle \rightarrow (p_o^2 / 3k_B T)\vec{E}$ so that $\alpha_m = p_o^2 / 3k_B T$. That implies $\chi_m \sim$ 4-50, depending on the number density we choose. Thus there can be at least one to two orders of magnitude to be gained if the molecular dipoles can indeed be harnessed.

1.2. *Giant electrorheological effect and its phenomenological model*

Persistent experimental efforts have led to the discovery of urea-coated nanoparticles of barium titanate oxalate ($NH_2CONH_2 \oplus BaTiO(C_2O_4)_2$) which, when dispersed in silicone oil, exhibits ER effect orders magnitude larger than those based on the induced polarization mechanism, exceeding the upper bound value by a large factor [9]. The GER effect also displays a *linear electric field dependence* of the yield stress, implying the GER mechanism must involve a polarization saturation phenomenon. It has also been observed that the GER effect scales as $1/R$, implying a surface-area dependence [10]. But perhaps the most intriguing aspect is that the GER effect is highly sensitive to whether the dispersing oil can wet the solid particles [11,12]. From such experimental observations it can be deduced that the GER effect involves some form of synergistic interaction between the urea molecules and the oil.

We have developed a phenomenological GER model [9,13] that is based on the following four elements. (1) There is an electric field enhancement effect at the contact region, with an enhancement factor on the order of $\sim 10^2$ (estimated numerically by using the finite element method). (2) The molecular dipoles of urea can form aligned dipolar layers in the *contact region* between two coated nanoparticles, under a moderate microscopic electric field of 10^6-10^7 V/cm. (3) The equilibrium contact state is represented by the balance of the (attractive) electrostatic force with the (repulsive) elastic force; with the elastic deformation of two coated spheres in contact given by the Hertzian solution. (4) The shear stress is defined as the derivative of the total energy with respect to strain; and as the area of the contact region decreases under shear, the yield stress is given by the stress value at the point of separation (zero contact area).

In this model, there is only one adjustable parameter, given by the deformation modulus of the coating. It turns out that the value obtained from fitting the experimental data is ~0.1 GPa [13], similar to that for a liquid and agrees with the TEM observation that the coatings seem to be soft [9].

While the phenomenological model predicts both the linear dependence on the applied electric field as well as the surface area dependence on the magnitude of the effect (hence smaller particles would enhance the GER effect) [10], it fails to account *(1) how does the alignment of the molecular dipoles come about, and (2) what is the synergistic role of the silicone oil in the GER mechanism.* We have turned to the use of molecular dynamics (MD) simulation to seek a physical picture of the microscopic GER mechanism. In this paper, we show that in the contact region between two nanoparticles, urea molecules, which normally reside at the oil-particle interface, can form aligned dipolar filaments that penetrate the oil film to bridge the two substrates. The resulting electrical energy density is shown to give an excellent account of the observed yield stress variation as a function of the electric field.

2. Microscopic Mechanism of the GER Effect

We focus on the contact area of two spherical nanoparticles and model it with a 2.9 nm gap in which the (methyl-group terminated) silicone oil chains with 10 monomers per chain and urea molecules are sandwiched between two bounding surfaces populated by water molecules. Owing to the nonideal thermodynamic nature of the urea-water solution, it is known that urea molecules tend to predominantly aggregate at the surface of hydrophilic solid particles, forming a (nonuniform) liquid-like coating. MD simulations were performed with the package GROMACS. Details of the model and the methodology involved can be found in Ref. [14].

2.1. *Observation of molecular filament formation*

When the electric field was turned on, the urea molecules were seen to diffuse into the silicone oil layer from both sides, with their molecular dipoles generally aligned along the field direction, forming filaments that bridge the two sides of the gap. Figure 1 shows a snapshot taken from the simulation to illustrate the configuration of the system under a 0.2 V/nm electric field. The filament structure may be clearly discerned in Fig. 1(a). In Fig. 1(b) we explicitly delineate the hydrogen bonds between the urea molecules in the filaments. A typical dipole moment in the filament is >3 D along the direction of the electric field. In contrast, with a homogenous dispersion of liquid-like urea molecule, it was found that it requires at least 0.6 V/nm of electric field in order to align the urea molecules. Thus there is clearly a lowering of the required electric field for dipole alignment by a factor of 2 to 3. The formation of the filaments always

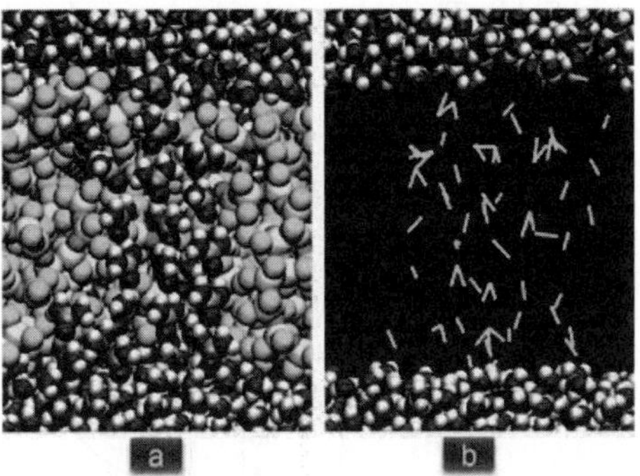

Figure 1. (a) Urea molecular filaments across silicone oil layer under 0.2 V/nm electric field. The methyl group is represented by a single green sphere. Oxygen, hydrogen, nitrogen, carbon, and silicon atoms are denoted by red, white, blue, navy blue, and yellow, respectively. Only part of the water boundary layer (red-white) is shown. (b) The hydrogen bondings (highlighted by bright yellow lines) between the urea molecules in the filament structure.

starts at the substrates, and no formation of filaments was ever observed in our simulations for gaps >10 nm. Thus the formation of filaments (and therefore the enhanced alignment) is inherently an interfacial effect with a decay length <4–5 nm. Figure 2(a) shows the results of varying the gap size and finding the field at which the filaments are no longer observed. It is seen that for large fields, the gap size saturates at 8–9 nm. This essentially sets a length scale for the GER effect. Owing to the small magnitude of this length scale, the GER effect may be considered an interfacial phenomenon, with its magnitude scaling [10] with the interfacial area.

2.2. *Confinement effect and the 3D to 1D crossover*

The formation of filaments, with the attendant lowering of the aligning field and the finite penetration length, may be attributed to the confinement effect exerted

412

by the oil chains, because of their hydrophobic nature. The confinement, which tends to decrease the entropic phase space of the urea dipoles, works in concert with the hydrogen bonding interactions (Fig. 1(b)) to significantly increase urea dipoles' sensitivity to the field. This effect may be mathematically formulated by noting that for 3D the thermally-averaged dipole moment along the field direction is given by Eq. (2), where for urea molecules $p_0 = 4.6$ Debye. But for 1D, it is given by $\langle p \rangle_{1D} / p_0 = \tanh\left(p_0 E / k_B T \right)$. At any given E, $\Delta p = \langle p \rangle_{1D} - \langle p \rangle_{3D}$ is *always positive*, as shown in Fig. 2(b); therefore $-\Delta p \cdot E$ provides a driving energy/force for the urea molecules to develop a more diffuse interface with the oil film, in the form of 1D filaments' penetration into the oil film (under an electric field). The presence of hydrogen bonds between the urea molecules in the filaments means that their energetic contributions (per urea

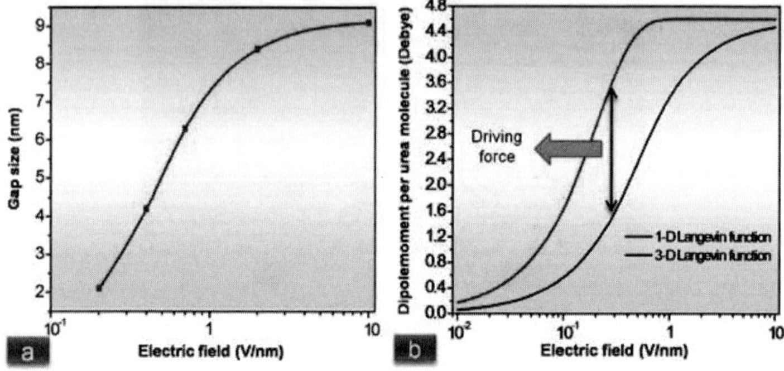

Figure 2. (a) Maximum thickness of the oil layer in the gap (beyond which no filament formation can be observed) plotted as a function of applied electric fields. (b) Dipolemoment per urea molecule along the field direction as a function of field in 3D (black curve) and 1D (red curve). The difference provides a driving force for forming the molecular filaments.

molecule) can help to counterbalance those in the (3D) urea molecular dispersion. The electrical alignment energy thereby emerges as the dominant factor. Moreover, as $E \rightarrow \infty$, we have $\Delta p \rightarrow k_B T / E$, which implies $-\Delta p \cdot E$ approaches a constant value, $-k_B T$, independent of E. This is consistent with the saturation behavior as $E \rightarrow \infty$, seen in Fig 2(a). This behavior also confirms the surface scaling aspect of the GER effect.

2.3. *Electrical energy density and comparison with GER data*

To obtain the relation between the yield stress as a function of the electric field, we calculate the total electrical energy density of the simulation box at different fields, obtained by calculating the total energy at different fields and subtracting off the zero-field total energy. The resulting difference ΔE is negative, indicating a large attraction between the two bounding surfaces. Plotted in Fig. 3 is ΔW, given by ΔE divided by the volume of our simulation sample, as a function of the electric field (red curve). Since the yield stress is directly proportional to ΔW as seen below, all the experimental yield stress vs. electric field data should be able to be scaled onto this relation, if the simulated effect is indeed the GER mechanism.

The electric field at the contact region of two nanoparticles is enhanced by a factor of $\alpha \sim 100\text{-}300$ (compared to the experimentally applied electric field), owing to the large dielectric constant of the nanoparticles and hence the field concentration effect. Thus 0.5 V/nm field in the gap region would correspond roughly to ~5000 V/mm (or less) of experimentally applied field. From Fig. 3, we obtain ~25 MPa to be the energy density at 0.5 V/nm. This energy density must be scaled by a volume dilution factor β for the energy density of the GER fluid, since the nanoscale gap considered here constitutes the region of closest approach between two nanoparticles. Hence the gap electrical energy should be

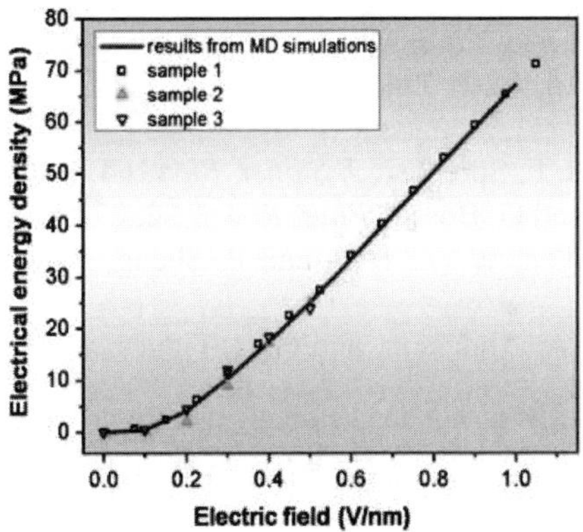

Figure 3. Electrical energy density plotted as a function of applied electric field. Scaled data from three sets of experiments are superposed on the curve. The electric field E is obtained from the measured field E_m by $E = \alpha E_m$, where α is the field enhancement factor (see text for details). The measured yield stress Y is related to the electrical energy density ΔW by $\Delta W = \beta Y / 10$, where β is the volume dilution factor. The values of α and β for samples 1 (Ref.9), 2 (Ref.10) and 3 (Ref.5) are (281.2, 7590), (100, 974), (100, 1010), respectively.

averaged over a volume on the order of d^3, d being the nanoparticle diameter. For $d \sim 50\text{-}100$ nm and our sample volume of 190 nm^3, $\beta \sim 1000\text{-}10{,}000$.

The stress may be expressed as $\tau = (1 / \beta)[\partial(\Delta W) / \partial\varepsilon]$, where ε denotes the strain. For a linear stress-strain relation $\tau = a\varepsilon$, it follows that $\Delta W = (\beta / 2)a\varepsilon^2$. Here the factor ½ may be replaced by a larger factor if the stress-strain relation deviates from linearity at large strain values, but such deviations can be absorbed into the β factor. Since yield stress $Y = \tau_y = a\varepsilon_o$, we have $Y = \tau_y = 2\Delta W / \beta\varepsilon_o$, where $\varepsilon_o = 0.2$ (radian) is the strain at the yield point [13]. By suitably scaling the experimental data by α and β values as indicated in the caption to Fig. 3, it is seen that all three data sets fall on the curve obtained from the MD simulations. As the values of α and β fall within the physically reasonable range, our simulation results thus offer a microscopic account of the GER effect.

Acknowledgments

This work is supported by HK Research Grants No. RGC602207 and No. RGC621006. Authors are grateful to the Max Planck Institute for Polymer Research where part of the work was performed.

References

1. R. Tao and J. M. Sun, *Phys. Rev. Lett.* **67**, 398 (1991).
2. D. J. Klingenberg, C. Z. Zukoski, and J. C. Hill, *J. Appl. Phys.* **73**, 4644 (1993).
3. H. Conrad and A. F. Sprecher, *J. Statist. Phys.* **64**, 1073 (1994).
4. H. J. H. Clercx and G. Bossis, *Phys. Rev. E* **52**, 2727 (1995).
5. H. R. Ma, W. Wen, W. Y. Tam, and P. Sheng, *Phys. Rev. Lett.* **77**, 2499 (1996).
6. J. E. Martin, R. A. Anderson, and C. P. Tigges, *J. Chem. Phys.* **110**, 4854 (1999).
7. H. Ma, W. Wen, W. Y. Tam, and P. Sheng, *Adv. in Phys.* **52**, 343 (2003).
8. P. Sheng and W. Wen, *Sol. State Comm.* **150**, 1023 (2010).
9. W. Wen, X. Huang, S. Yang, K. Lu, and P. Sheng, *Nature Materials* **2**, 727 (2003).
10. W. Wen, X. Huang, and P. Sheng, *Appl. Phys. Lett.* **85**, 299 (2004).
11. C. Shen, W. Wen, S. Yang, and P. Sheng, *J. Appl. Phys.* **99**, 106104 (2006).
12. X. Gong et al., *Nanotechnology* **19**, 165602 (2008).
13. X. Huang, W. Wen, S. Yang, and P. Sheng, *Solid State Commun.* **139**, 581 (2006).
14. S. Chen, X. Huang, Nico F. A. van der Vegt, W. Wen and P. Sheng, *Phys. Rev. Lett.* **105**, 046001 (2010).

MICRODYNAMICS OF MAGNETIC PARTICLES DISPERSED IN COMPLEX MEDIA

G. BOSSIS

LPMC (UMR6622), University of Nice-Sophia-Antipolis/CNRS ParcValrose, 06108 Nice, France

MODESTO T. LÓPEZ-LÓPEZ

Departamento de Física Aplicada, Universidad de Granada, Campus de Fuentenueva s/n Granada, 18071, Spain

A. ZUBAREV

Dpt of Physics, Ural State University, Ekaterinburg, Russia

The aggregation of magnetic particles in the presence of a magnetic field is the basic phenomenon which underlies all the physics of magnetorheological (MR) fluids. Although these interactions are well understood when the suspending fluid is a simple liquid, new MR fluids based on dispersion of magnetic microparticles in a ferrofluid or MR elastomers based on dispersion of magnetic particles in a rubber matrix, present some unusual properties which are not well described by conventional theories. We analyze in this work the motion of magnetic particles dispersed in a ferrofluid and submitted to a magnetic field and discuss the possible applications.

1. Introduction

It is well known that in the presence of a magnetic field, two ferromagnetic particles which are approximately aligned in the direction of the field, experience attractive forces. If they are free to move, they will come into contact, aligned on the field direction. This is the basic phenomenon that causes a quick and important change in the rheological properties of a magnetorheological (MR) suspension upon application of a magnetic field [1]. On the contrary, ferrofluids, which are stable dispersions of magnetic nanoparticles, show almost negligible rheological changes because magnetic attraction is dominated by Brownian motion [2]. Some attempts of increasing the strength of magnetic forces between microparticles have been done by dispersing them in a ferrofluid,

416

and several authors have actually demonstrated that the field-induced yield stress could be substantially increased in this way [3-5]. In addition, it has been shown that suspensions of ferromagnetic microparticles in ferrofluids also exhibit a better colloidal stability against irreversible aggregation and sedimentation [6-9].

We have tried to understand the cause of these unexpected properties by looking at the dynamics of two ferromagnetic microparticles in a ferrofluid in the presence of a magnetic field. Very surprisingly we have observed that, instead of coming into contact, they stop at a surface-to-surface distance of the order of the particle diameter [10]. We shall demonstrate the existence of a new repulsive force, which is due to a phase separation of the ferrofluid in the zone of high magnetic fields between the microparticles.

2. Interaction of Microparticles Inside a Ferrofluid

2.1. *Experimental*

Ferrofluids used in the experiments consisted of stable suspension of oleate-covered magnetite nanoparticles dispersed in kerosene. Details on its preparation are given elsewhere [4]. Magnetic microparticles used in this work were spherical nickel particles of average diameter 10 μm, supplied by Merck KgaA (Germany). The suspensions of microparticles dispersed in ferrofluids were prepared as described in Ref. [4]. The volume fraction of microparticles in the final suspensions was approx. 0.01 % in all cases. A drop of the suspension was placed between two glass slides and microscopic observations were performed. A magnetic field was applied to the samples with the help of a home-made electromagnet.

2.2. *Results*

The first aim of this work was to study experimentally the field-induced aggregation process between magnetic microparticles dispersed in a ferrofluid. With this goal, we tried to find in our suspensions, whenever possible, groups of a few particles that upon magnetic field application would interact without being subjected to relevant influences of other particles not belonging to the group of interest. Then, the time evolution of the particles was monitored. As an example, Figure 1 shows the aggregation process of two nickel particles upon application of a magnetic field of intensity 22 kA/m. The snapshot shown in Figure 1a was taken before the application of the field. As observed, an isolated particle is placed in a homogeneous ferrofluid carrier. Figure 1b shows a snapshot taken 1.0 s after the field was applied. Apart from a little increase of the diffuse aspect of the particle no other change is observed. On the other hand, a layer of nanoparticles attached to the microparticle is clearly deformed in snapshots shown in Figures 1c-d, where a clear motion of the microparticle is also observed. This deformation takes place as a remarkable thickening of the nanoparticle layer from the poles of the microparticles, along the field direction. As the particles approach each other (Figure 1e) a dense cloud of nanoparticles between the two microparticles is observed. As the equilibrium distance is reached this cloud becomes denser and smaller in size. In the stationary state (Figure 1f) this cloud of nanoparticles placed between the two microparticles has a diameter approx. twice the diameter of the latter ones and a spherical shape, centered in the imaginary line that links the centers of the microparticles.

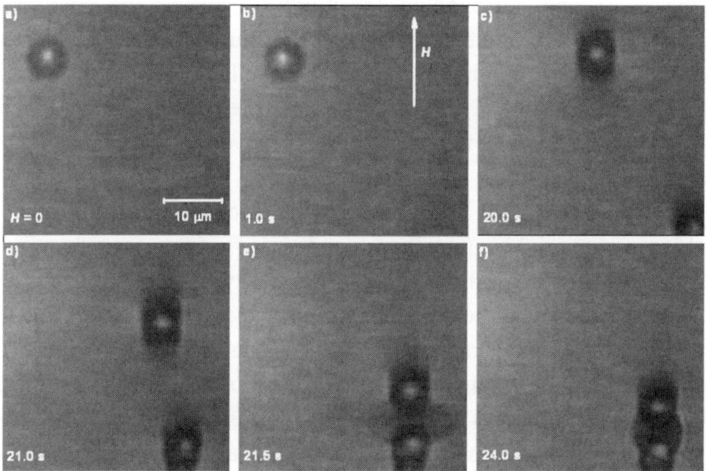

Figure 1. Snapshots of the aggregation of two nickel particles dispersed in a ferrofluid, upon application of a magnetic field of 22 kA/m. Direction of the field is indicated. Ferrofluid solid concentration was 5 % vol. Snapshot a) was taken prior to magnetic field application. Snapshots b) to f) were taken after field application. Time elapsed since field application is indicated in each snapshot.

At this point, we have already arrived to the surprising finding of our work: upon magnetic field application, the equilibrium position of two magnetic microparticles dispersed in a ferrofluid is not the surface-to-surface contact position, but a position where there exists a distance, about one diameter of the particles, between the particles surfaces. This interesting result could only be explained if there is a repulsive force between the microparticles. Otherwise, as a consequence of the dipolar magnetic attraction, the particles would be in contact.

2.3. *Theoretical model*

Theoretical description of the micron magnetizable particle dynamics in a carrier ferrofluid, taking into account the dense ferrofluid clouds surrounding the particles, presents very complicated mathematical problem. Here we focus on the explanation of the observed stop-effect. The interaction between two micron-sized particles is the sum of the magnetic interaction between them and the osmotic forces, which are provided by the ferrofluid nanoparticles surrounding the magnetizable microparticles. Because of the dense ferrofluid coats near the microparticles, magnetic interaction between them strongly differs from that in a homogeneous medium. Thus in order to determine the force of magnetic and osmotic interactions between the particles, first, one needs to determine shape and density of these coats. Figure 1 demonstrates quite sharp boundaries between the coats and surrounding ferrofluid. Therefore the coat can be considered as a domain of dense phase of ferrofluid which appears due to condensation of the ferrofluid nanoparticles. It is well known (see, for example, refs. [11-17]) that magnetic field stimulates the phase condensation in ferrofluids, although only the biggest particles of polydisperse ferrofluid interact strongly enough to undergo the phase condensation.

Since the local magnetic field is especially strong near the poles of the magnetizable microparticles, ferrofluid condenses mainly near these poles. Determination of the equilibrium shape and density of the ferrofluid dense domains, as well as magnetic field and the nanoparticles distribution inside them, represents very intricate mathematical problem, which can not be solved strictly. In order to get reasonable estimates, we use the following approximation. First we consider the ferrofluid as monodisperse, consisting only of the biggest particles, which are able to condense into dense phases under the action of high enough magnetic field.

Next we use the following iteration procedure. At the first step of this iteration we ignore the heterophase coats near the microparticles and consider them as immersed in the homogeneous monodisperse ferrofluid. In the frameworks of this approximation, by using classical results of magnetostatics [18], one can estimate the magnetic moment M of each microparticle using the following relation: $\mathbf{M}=\beta V\mathbf{H_0}$, where β is the magnetic contrast factor, V is the microparticle volume, and $\mathbf{H_0}$ is the magnetic field in the ferrofluid far from the microparticle. Having estimated $\mathbf{M}$, by using standard methods of magnetostatics, one can determine the local magnetic field $\mathbf{H(r)}$ distribution near the particle. This allows us, using condition of equality of the ferrofluid particle chemical potentials near the magnetizable microparticle and far from it, to evaluate the shape of the dense coat and concentration of the ferrofluid particles inside the coat. To this end we take advantage of formulas of Ref. [11] for the chemical potential χ of the ferrofluid particles as follows:

$$\chi = kT\left[\ln\left(\varphi\frac{\kappa}{sh\kappa}\right) + \varphi\frac{8 - 9\varphi + 3\varphi^2}{(1-\varphi)^3} - 8\left(L^2(\kappa)\lambda + \frac{1}{3}\lambda^2\right)\varphi\right] \tag{1}$$

$$\kappa(\mathbf{r}) = \mu_0\frac{mH(\mathbf{r})}{kT}, \qquad \lambda = \frac{\mu_0}{4\pi}\frac{m^2}{(d+2s)^3 kT}$$

Here, φ is the hydrodynamical (i.e. determined with allowance for surfactant layers) volume concentration of nanoparticles, $L(x) = \coth x - 1/x$ is the Langevin function, m is magnetic moment of the nanoparticle, d the diameter of its magnetic core, s is the thickness of the surfactant layer on its surface. Parameter κ is the dimensionless, reduced to thermal energy kT, energy of interaction between the nanoparticle and the local magnetic field H, and λ is the dimensionless parameter that characterizes the energy of magnetodipole interaction of the nanoparticles. Since in our experiments the volume concentration of microparticles was very low, we assumed that the ferrofluid particle concentration far from the microparticles was equal to the initial concentration φ_0 determined by the ferrofluid composition. The equilibrium concentration $\varphi(\mathbf{r})$ inside the dense domain has been determined from the condition $\chi(\varphi(\mathbf{r}), H(\mathbf{r})) = \chi(\varphi_0, H_0)$.

Let us now apply these calculations to a situation like that shown in Figure 1f, where the distance between the centers of the particles is approximately equal to $3a$ (a-radius of the microparticle). The magnitude of the magnetic field required to provoke phase separation, for the ferrofluid used in experiments, is estimated as $H_c=22$kA/m and we took for the far field $H_0=20$kA/m. With this value we actually obtained a dense zone around the particle whose extension is close to the experimental one. The calculated hydrodynamical volume concentration φ of the ferrofluid particles inside this shell is about 30-35%. This estimate looks quite reasonable.

In the second step of iteration we take into account that the magnetic permeability μ in the dense ferrofluid coats is higher than the permeability of the surrounding dilute ferrofluid. For both phases the ferrofluid relative permeability μ_f can be estimated as $\mu_f = 1 + 24\varphi\lambda L(\kappa)[1 + 8\lambda\varphi\partial L/\partial\kappa]\kappa^{-1}$ [11]. By using the program package FEMM we have recalculated the field distribution on the surface of the magnetizable particles. For simplicity we have supposed that the coat was homogeneous with permeability equal to the mean permeability $\bar{\mu}$ over this region. The calculated magnitude of $\bar{\mu}$ was 3.75. Let us introduce a spherical coordinate system with the origin in the center of the upper particle (Figure 1f), polar axis directed straight up, radius vector $\mathbf{r}$ and the angle θ between $\mathbf{r}$ and this polar axis. The force F acting on the upper microparticle of Figure 1f, can be calculated as an integral on the surface of the particle:

$$F = 2\pi a^2 \int_0^\pi \left[-p + \sigma_{rr} \cos\theta - \sigma_{r\theta} \sin\theta \right] \sin\theta\, d\theta \qquad (2)$$

Here p is the osmotic pressure of the condensed gas of ferrofluid particles, σ_{rr} and $\sigma_{r\theta}$ are components of the Maxwell stress tensor. The expression for the osmotic pressure is the following:

$$P_{os}(H,\phi) = \phi \frac{1+\phi+\phi^2-\phi^3}{(1-\phi)^3} - \left(\frac{4}{3}\lambda^2 + 4.\lambda.L\left(\frac{mH(r,\theta)}{kT}\right)\right)\phi^2$$

With L the Langevin function. Our calculations shows that for the case corresponding to Figure 1f the force, acting on the upper particles, is $F \approx 1.4 \cdot 10^{-8} N$. Positive sign of this force means that the microparticle repels from the second (down) microparticle. The physical reason of this repulsion is the oblate shape of the dense ferrofluid domain between the particles, and, consequently, the high demagnetizing factor of this zone which will tend to elongate like a ferrofluid drop in a magnetic field. In other words ,due to the dense ferrofluid zone inside the gap, the magnetic field inside the gap is much lower than in the absence of phase separation and can become lower than the average field on the upper side of the particle,giving rise to this repulsive magnetic force because the magnetizable particle is attracted towards the region with the highest magnetic field. On the other hand, when the particles are far apart their dense coats of ferrofluid do not interpenetrate (Figures 1c to 1e), so the magnetic field near the surfaces facing to the other particle is higher than near the opposite surfaces, and the particles attract. For a more exact calculation of this repulsive force and of the shape of the condensed domain of ferrofluid we plan to solve the magnetostatic equation with a permeability of the ferrofluid given by the local volume fraction $\phi(r,\theta)$ in order to get the local field $H(r,\theta)$,then calculate a new volume fraction from the equality of the chemical potential and iterate this process until it converges.

Acknowledgments

This work has been supported by Biomag project (Conseil Régional PACA, France), projects P08-FQM-3993 and P09-FQM-4787 (Junta de Andalucía, Spain), project FIS2009-07321 (MICINN, Spain), grants of Russian Agency of Education 2.1.1/2571 and 2.1.1/1535, grants of Russian Federal Target Program 02.740.11.0202, 02-740-11-5172 and NK-43P(4), and grants of Russian Fund of Fundamental Investigations 10-01-96002-Ural, 10-02-96001, 10-02-00034 and 08-02-00647. M.T.L.-L. also acknowledges financial support by the University of Granada (Spain).

References

1. G. Bossis, O. Volkova, S. Lacis and A. Meunier, *Lecture Notes in Physics* **594**, 186 (2002).
2. H. Shahnazian and S. Odenbach, *Int. J. Mod. Phys. B* **21**, 4806 (2007).
3. J. M. Ginder, L. D. Elie and L. C. Davis - USA patent US5549837 (1996).
4. M. T. López-López, J. de Vicente, G. Bossis, F. González-Caballero and J. D. G. Durán, *J. Mater. Res.* **20**, 874 (2005).
5. N. M. Wereley A. Chaudhuri, J. H. Yoo, S. John, S. Kotha, A. Suggs, R. Radhakrishnan, B. J. Love and T. S. Sudarshan, *Intell. Mater. Syst. Struct.* **17**, 393 (2006).
6. E. F. Burguera, B. J. Love, R. Sahul, G. T. Ngatu and N. M. Wereley, *J. Intell. Mater. Syst. Struct.* **19**, 1361 (2008).
7. M. T. López-López, P. Kuzhir, S. Lacis, G. Bossis, F. González-Caballero and J. D. G. Durán, *J. Phys.-Condes. Matter* **18**, S2803 (2006).
8. G. T. Ngatu and N. M. Wereley, *IEEE Trans. Magn.* **43**, 2474 (2007).
9. R. Patel and B. Chudasama, *Phys. Rev. E* **80**, 012401 (2009).
10. M. T. López-López, A. Yu. Zubarev, G. Bossis, *Soft Matter* (2010) DOI: 10.1039/C0SM00261E.
11. A. Yu. Buyevich and A. O. Ivanov, *Physica A* **193**, 221 (1993).
12. C. F. Hayers, J. Colloid Interface Sci. **52**, 239 (1975).
13. E. A. Peterson and A. A. Krueger, *J. Colloid Interface Sci.* **62**, 24 (1977).
14. J. C. Bacri and D. Salin, *J. Magn. Magn. Mater.* **9**, 48 (1983).
15. A. F. Pshenichnikov, *J. Magn. Magn. Mater.* **145**, 139 (1995).
16. M. F Islam, K. H Lin, D. Lacoste, T. C. Lubenski, and A. G. Yodh, *Phys. Rev. E* **67**, 021402 (2003).
17. L. Yu. Iskakova, G. A. Smelchakova and A. Zubarev, *Phys. Rev. E.* **79**, 011401 (2009).
18. L. D. Landau and E. M. Lifshits, *Electrodynamics of Continuous Media* (Pergamon Press, Oxford, 1984).

ENHANCING MAGNETORHEOLOGY

D. J. KLINGENBERG

Chemical and Biological Engineering and Rheology Research Center,
University of Wisconsin,
Madison, WI, 537056, USA
E-mail: klingen@engr.wisc.edu
www.engr.wisc.edu/che/faculty/klingenberg_daniel.html

J. C. ULICNY

General Motors Global R&D,
Warren, MI 48090-9055, USA
E-mail: john.ulicny@gm.com

Experimental and simulation results are presented that illustrate how the addition of non-magnetizable particles to a magnetorheological (MR) fluid can enhance the field-induced yield stress. Efforts to understand the underlying mechanisms are described.

Keywords: Magnetorheology; nonmagnetizable particle; yield stress.

1. Introduction

While many formulations of magnetizable particles in liquids or gels exhibit magnetorheological (MR) effects, it is often desirable to engineer or optimize various properties for specific applications. The field-induced yield stress of MR fluids can be enhanced by a variety of schemes or phenomena. These include using roughened surfaces,[1] precompression,[2] exploiting lamella formation,[3,4] using bidisperse suspensions,[5,6] and using fibrous particles.[7–9]

We recently described a new approach in which the field-induced yield stress can be enhanced by adding nonmagnetizable particles.[10] Here we briefly review the experimental observations reported in Ref. 10 that illustrate this phenomenon, and then show how this effect is observed for a variety of nonmagnetizable materials. Particle-level simulations in three dimensions show qualitatively similar behavior. We show how the addition of nonmagnetizable particles alters the suspension microstructure, which is likely related to the observed rheological changes.

The ability to increase the yield stress by adding nonmagnetizable particles is valuable because larger yield stresses allow for more powerful devices or smaller devices with the same power. Furthermore, replacing iron particles with nonmagnetizable particles can result in less expensive and less dense MR fluids.

2. Methods

The magnetizable particles used included small and large carbonyl iron particles (2 and 8 μm average diameter; BASF Corp., Mount Olive, NJ). MR fluids were formulated using a 50:50 mixture by mass of the small and large iron spheres (bidisperse fluids) The nonmagnetizable particles employed in most of the experiments were hollow glass beads (12 μm average diameter; Accumet Corp., Ossining, NY). Other nonmagnetizable materials were also employed, as described in the following section. The carrier fluid was a low molecular weight poly(α-olefin) (Spectrasyn 2; ExxonMobil, Fairfax, VA). For experiments without nonmagnetizable particles, small amounts of fumed silica (M5; Cabot Corp., Billerica, MA; 0.06 wt%) were added to help keep the iron particles in suspension.[11] The MR fluids were prepared by mixing the components until the mixtures were homogeneous. Yield stresses were determined at various magnetic flux densities by fitting shear stress–shear rate data to the Bingham model.[12] Results reported here are for yield stresses in the limit of magnetic saturation.

We use a point-dipole model and simulation method employed previously,[13,14] modified to consider suspensions composed of mixtures of magnetizable and non-magnetizable spheres. The suspension is treated as monodisperse, non-Brownian, neutrally-buoyant spheres (diameter σ) in a nonmagnetizable, Newtonian continuous phase (viscosity η_c). The magnetizable spheres have relative permeability μ_p and saturation magnetization M_s. The equation of motion is solved numerically, neglecting inertia. to determine the particle motion. Forces considered include the field-induced magnetostatic pair interaction forces (between magnetizable spheres), short-range repulsive forces, and hydrodynamic drag. For magnetizable spheres i and j, the magnetostatic force on sphere i due to sphere j at the relative position (r_{ij}, θ_{ij}) (spherical coordinates, with θ_{ij} measured relative to the ambient magnetic field) is

$$\boldsymbol{F}_{ij}^{\text{mag.}} = F_0 \left(\frac{\sigma}{r_{ij}}\right)^4 \left[\left(3\cos^2\theta_{ij} - 1\right)\boldsymbol{e}_r + \sin 2\theta_{ij}\boldsymbol{e}_\theta\right], \tag{1}$$

where $F_0 = (3/16)\mu_0\sigma^2\beta_M^2 H_0^2$ for linear magnetization, and $(\pi/48)\mu_0\sigma^2 M_s^2$ for saturated magnetization. Here $\mu_0 = 4\pi \times 10^{-7}$ N/A^2, and $\beta_M = (\mu_p - 1)/(\mu_p + 2)$. Short-range repulsive forces acting between all sphere pairs are modeled with the short-ranged function

$$\boldsymbol{F}_{ij}^{\text{rep.}} = -F_0 \exp\left[-\kappa(r_{ij}/\sigma - 1)\right]\boldsymbol{e}_r, \tag{2}$$

424

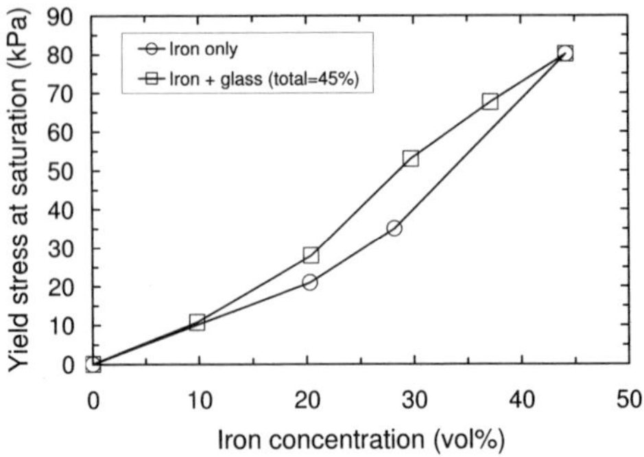

Fig. 1. Yield stress as a function of volume fraction of iron spheres (bidisperse) with and without added glass spheres. For the case of added glass spheres, the total volume fraction is constant at $\phi_T = \phi_M + \phi_N = 0.45$.

with $\kappa = 10^2$. Analogous expressions are employed for interactions between particles and bounding surfaces. Hydrodynamic forces are treated in the free-draining limit as $\boldsymbol{F}_i^{\text{hyd.}} = 3\pi\eta_c\sigma\left[\boldsymbol{u}^\infty(\boldsymbol{x}_i) - d\boldsymbol{x}_i/dt\right]$. The ratio of hydrodynamic to field-induced forces defines a dimensionless shear rate, $\dot{\gamma}^* = 3\pi\eta_c\dot{\gamma}/F_0$, which is 10^{-4} for results presented here.

Simulations were performed by randomly placing N_M magnetizable spheres and N_N nonmagnetizable spheres in a periodic cell of dimensions $(L_x, L_y, L_z) = (10\sigma, 5\sigma, 5\sigma)$. The volume fractions of the magnetizable and nonmagnetizable spheres are ϕ_M and ϕ_N, respectively; the total volume fraction of spheres is $\phi_T = \phi_M + \phi_N$. Shear flow is simulated to a strain of 5. The shear stress, calculated in a conventional manner,[14] becomes steady after a strain of 1. Results are averaged over the strain interval $\gamma = 1$–5, and over 10 different initial configurations for each set of parameters. The shear stress obtained at a dimensionless shear rate of $\dot{\gamma}^* = 10^{-4}$ is close to the yield stress;[15] here we use this shear stress as an estimate of the yield stress.

3. Results and Discussion

Experimentally-measured yield stresses at magnetic saturation are plotted as a function of ϕ_M in Fig. 1, for suspensions with only iron spheres, as well as for suspensions containing iron and glass spheres with the total volume fraction held fixed at $\phi_T = 0.45$. The yield stress of suspensions containing glass spheres exceeds that of suspensions containing only iron spheres at all ϕ_M.

Yield stresses at saturation for suspensions with iron particles ($\phi_M = 0.30$) and various nonmagnetizable particles ($\phi_N = 0.15$) are compared in Fig. 2. In all cases, the nonmagnetizable spheres cause a significant increase in the yield stress,

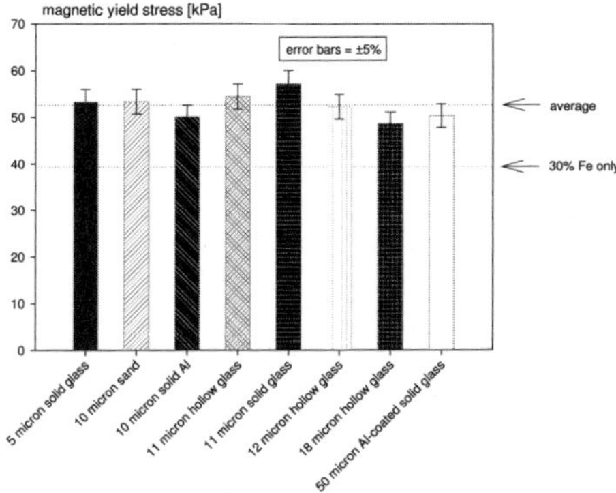

Fig. 2.   Yield stress at saturation for various nonmagnetizable materials.

with little difference among the results for different particle types. These results suggest that the enhancement arises from the nonmagnetizability of the particles, rather than some other material-specific nonmagnetic particle interaction.

The dimensionless yield stress $\langle \tau_0^* \rangle \equiv \langle \tau_0 \sigma^2 / F_0 \rangle$ obtained from simulations is plotted as a function of ϕ_M in Fig. 3, for suspensions composed of only magnetizable spheres, as well as for suspensions of magnetizable and nonmagnetizable spheres with the total volume fraction held fixed at $\phi_T = \phi_M + \phi_N = 0.45$. Similar to the experimental results, the nonmagnetizable spheres cause an increase in the yield stress. The simulation results for systems with nonmangetizable spheres exhibit a maximum in the yield stress, occurring at $0.35 < \phi_M < 0.45$ ($0 < \phi_N < 0.10$). This maximum is not observed with the experimental systems, although it is possible that a maximum may be observed if more data were obtained at concentrations approaching $\phi_M = 0.45$.

Insight into the mechanism of the enhancement can be obtained by considering the various contribution to the shear stress, or *partial stresses*.[16] The shear stress, expressed a sum over particles, can be separated into sums over each particle type,

$$\tau_{xz} = -\frac{1}{V} \sum_{i=1}^{N} z_i F_{x,i} = \left[-\frac{1}{V} \sum_{\substack{i=1 \\ \text{mag.}}}^{N_M} z_i F_{x,i} \right] + \left[-\frac{1}{V} \sum_{\substack{i=1 \\ \text{nonmag.}}}^{N_N} z_i F_{x,i} \right] \tag{3}$$

The first sum over only magnetizable spheres can describes the stress arising directly from the magnetizable spheres, and is labeled τ_{xz}^{MS}. The second sum over only nonmagnetizable spheres is labeled τ_{xz}^{MS}. By replacing the net force on a sphere $F_{x,i}$ by its sum over pair interactions, the stress can also be separated in

426

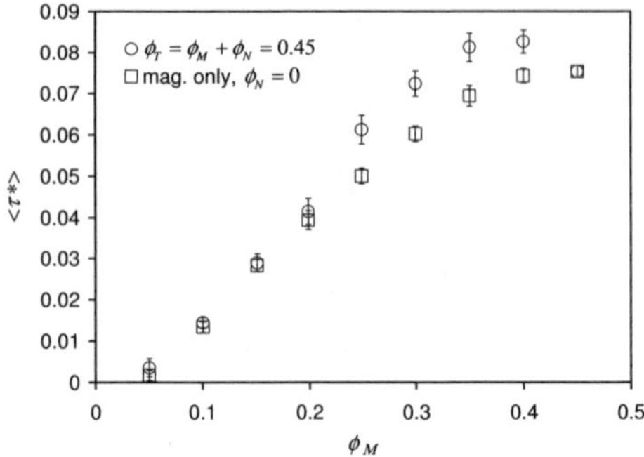

Fig. 3.   Yield stress as a function of ϕ_M with and without added nonmagnetizable spheres.

magnetic and repulsive force contributions,

$$\tau_{xz} = \left[-\frac{1}{V} \sum_{i=1}^{N} \sum_{j\neq i}^{N'} z_i F_{x,ij}^{\text{mag.}} \right] + \left[-\frac{1}{V} \sum_{i=1}^{N} \sum_{j\neq i}^{N''} z_i F_{x,ij}^{\text{rep.}} \right] \tag{4}$$

where the first term in square brackets, describing the magnetic force contribution to the stress, is labeled $\tau_{xz}^{\text{mag.}}$, and the second term in square brackets is labeled $\tau_{xz}^{\text{rep.}}$. The stress can be broken down in a variety of other ways, for example, by considering the magnetic and repulsive force contributions to τ_{xz}^{MS}.

The magnetic force contribution to the shear stress for suspensions with and without magnetizable spheres are compared in Fig. 4. For $\phi_M \gtrsim 0.20$, the magnetic force contribution is larger for the systems containing nonmagnetizable spheres. Because only the magnetizable spheres can contribute to the magnetic force contribution to the shear stress, these results suggest that the nonmagnetizable spheres cause a change in the microstructure of the magnetizable spheres.

To investigate changes in the microstructure, we examined measures of the distribution of clusters of spheres. One measure is the mean cluster size, defined

$$S_m \equiv \frac{1}{N} \sum_{k=1}^{N_c} m_k^2, \tag{5}$$

where N_c is the total number of clusters, and m_k is the number of spheres in cluster k. Two spheres are considered to be in contact if their center-to-center separation is $r_{ij}/\sigma \leq 1.05$. Clusters are identified using the algorithm described by Sevick et al.[17] Here we examine only clusters of magnetizable spheres, ignoring nonmagnetizable spheres in mixtures.

The mean cluster size of magnetizable spheres is plotted as a function of ϕ_M in Fig. 5 for suspensions composed of only magnetizable spheres, as well as for sus-

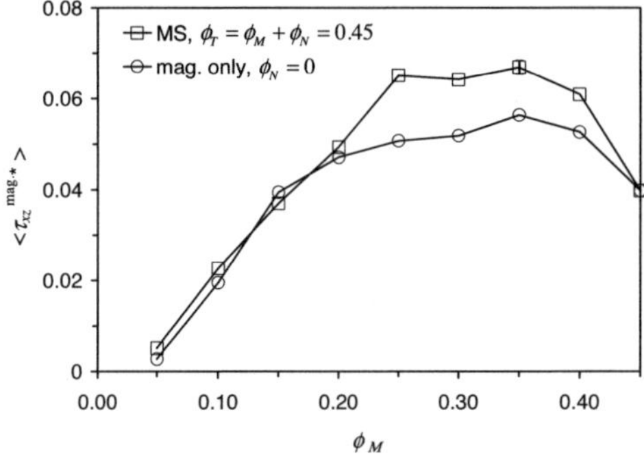

Fig. 4.   Magnetic force contribution to the yield stress as a function of ϕ_M with and without added nonmagnetizable spheres.

pensions of magnetizable and nonmagnetizable spheres with $\phi_T = \phi_M + \phi_N = 0.45$. The clusters of magnetizable spheres are significantly larger for the systems containing nonmagnetizable spheres, at least for intermediate ϕ_M.

The behavior of conventional MR fluids is commonly explained by a competition of magnetostatic and hydrodynamic forces, often expressed in terms of a Mason number (Mn $\propto \dot{\gamma}/H^2$). For small Mn, magnetostatic forces cause the formation of percolating, fibrous aggregates that transmit stresses from one bounding surface to another. At large Mn, hydrodynamic forces destroy the fibrous structures and the apparent suspension viscosity approaches that of nonmagnetized systems. The fact that added nonmagnetizable spheres increase the size of clusters of magnetizable spheres therefore suggests that the nonmagnetizable spheres create an additional mean force of attraction between magnetizable spheres, which may in part explain the enhanced yield stress for the mixed systems. Such a mechanism can potentially be tested by measuring forces between magnetizable spheres.

4. Conclusion

Experiments and simulations show that the yield stress of MR fluids can be enhanced by the addition of nonmagnetizable spheres. This phenomenon may be explained in part by an additional mean force of attraction between magnetizable spheres caused be the added nonmagnetizable spheres. The enhancement is independent of the type of nonmagnetizable spheres employed. This phenomenon can be exploited to produce improved, less expensive MR fluids with lower densities.

428

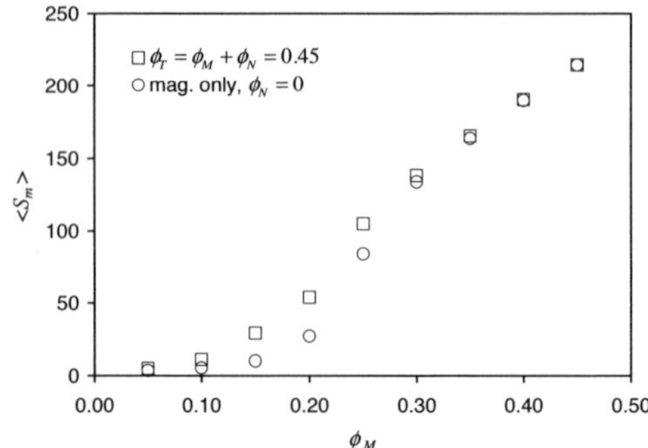

Fig. 5.   Mean size of clusters of magnetizable spheres as a function of ϕ_M with and without added nonmagnetizable spheres.

Acknowledgements

This work was supported in part by General Motors, and by the National Science Foundation (CBET-0932680).

References

1. A. L. Smith, J. C. Ulicny, L. C. Kennedy and T. Deuter, SAE 2010-01-0324, April 12, 2010.
2. X. Tang, X. Zhang and R. Tao, *J. Appl. Phys.* **87**, 2634 (2000).
3. S. Henley and F. E. Filisko, *J. Rheol.* **43**, 1323 (1999).
4. D. Kittipoomwong, D. J. Klingenberg, Y. M. Shkel, J. F. Morris and J. C. Ulicny, *J. Rheol.* **52**, 225 (2008).
5. R. T. Foister, U. S. Patent No. 5,667,715 September 16, 1997.
6. D. Kittipoomwong, D. J. Klingenberg and J. C. Ulicny, *J. Rheol.* **49**, 1521 (2005).
7. R. Kanu and M. T. Shaw, *J. Rheol.* **42** 657 (1998).
8. Y. Otsubo, *Coll. Surf. A* **153**, 459 (1999).
9. M. T. Lopez-Lopez, P. Kuzhir and G. Bossis, *J. Rheol.* **53** 115 (2009).
10. J. C. Ulicny, K. S. Snavely, M. A. Golden and D. J. Klingenberg, *Appl. Phys. Lett.* **96**, 231903 (2010).
11. J. C. Ulicny, A. L. Smith, M. A. Golden and C. A. Hayden, U. S. Patent 6,881,353, April 19, 2005.
12. J. C. Ulicny and M. A. Golden, *Int. J. Mod. Phys. B* **21**, 4898 (2007).
13. M. C. Heine, J. de Vicente and D. J. Klingenberg, *Phys. Fluids* **18**, 023301 (2006).
14. M. Parthasarathy and D. J. Klingenberg, *Rheol. Acta* **34**, 430 (1995).
15. D. J. Klingenberg, F. van Swol and C. F. Zukoski, *J. Chem. Phys.* **94**, 6160 (1991).
16. K. H. Ahn and D. J. Klingenberg, *J. Rheol.* **38**, 713 (1994).
17. E. M. Sevick, P. A. Monson and J. M. Ottino, *J. Chem. Phys.* **88**, 1198 (1988).

THEORETICAL ANALYSIS ON ELECTRO ADHESIVE EFFECT OF ER GEL

YOSUKE NAITO[*]

*School of Integrated Design Engineering, Keio University,
3-14-1 Hiyoshi, Kohoku-ku, Yokohama 223-8522, Japan*

TOJIRO AOYAMA and YASUHIRO KAKINUMA

*Department of System Design Engineering, Keio University,
3-14-1 Hiyoshi, Kohoku-ku, Yokohama 223-8522, Japan*

An electrorheological gel (ERG) is a functional material that changes its surface adhesive property according to the intensity of the applied electric field. It is a composite material consisting of ER particles and silicone gel. Under no electric field, the surface adhesion is low because the slippery ER particles protrude from the gel surface. When an electric field is applied to an ERG, silicone gel around ER particles rises and covers the ERG surface such that the surface adhesion becomes high. The surface adhesion of ERGs can be changed quickly and reversibly by adjusting the electric field. This unique property is called the electro adhesive (EA) effect. However, although this electro adhesive phenomenon has been experimentally confirmed, the physical theory has not been sufficiently developed. In this study, the theory of the EA effect based on electromagnetics is developed and validated using a numerical analysis. Simulation results show that ER particles sink into silicone gel, and silicone gel around ER particles heaves due to the Maxwell stress and gradient force of the applied electric field.

1. Introduction

An electrorheological fluid (ERF) is a functional fluid whose viscoelastic properties change swiftly and reversibly with the applied electric field [1]. This property is called the ER effect. Previous studies have proposed mechanical applications of ERFs, such as brake systems [2] and clutch systems [3]. However, ERFs have two disadvantages. First, the ER effect in dispersion-type ERFs [4] reduces due to the separation and sedimentation of ER particles over long time periods. Second, the necessity of a mechanical seal to prevent the leakage of ERFs makes the structure of the application device complex.

Electrorheological gels (ERGs) have been developed to overcome the abovementioned disadvantages of ERFs. An ERG is a functional material that changes its surface adhesive property rapidly according to the intensity of the

---

[*] Corresponding author.

applied electric field [5]. This unique property is called the electro adhesive (EA) effect. Fig.1 shows an ERG sandwiched between a sliding electrode and a fixed electrode. When no voltage is applied between the electrodes, the sliding electrode is supported by the ER particles protruding from the gel surface (Fig.2). Since the friction of ER particles is low, the sliding electrode moves freely on the surface of the ERG, as illustrated in Fig.1 (a). On the other hand, when a high voltage is applied, ER particles settle out and silicone gel heaves toward the sliding electrode. Consequently, the adhesive gel sticks to the sliding electrode, and the electrode is fixed to the ERG as shown in Fig.1 (b). Though the EA effect has been experimentally confirmed, the physical theory has not been sufficiently developed. In this study, the theory of the EA effect based on electromagnetics is developed and validated using a numerical analysis.

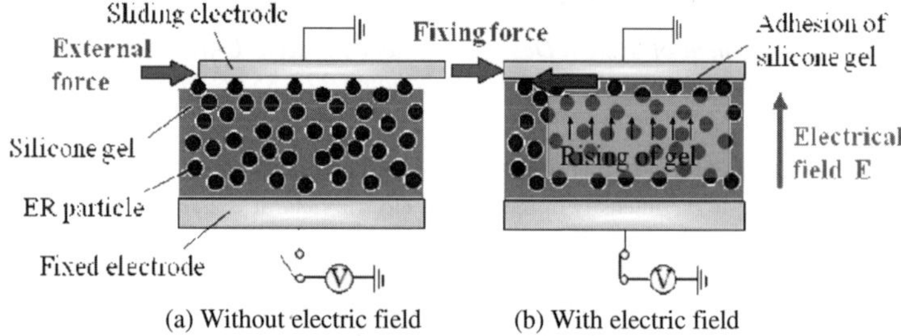

Figure 1. Structure of ERG.

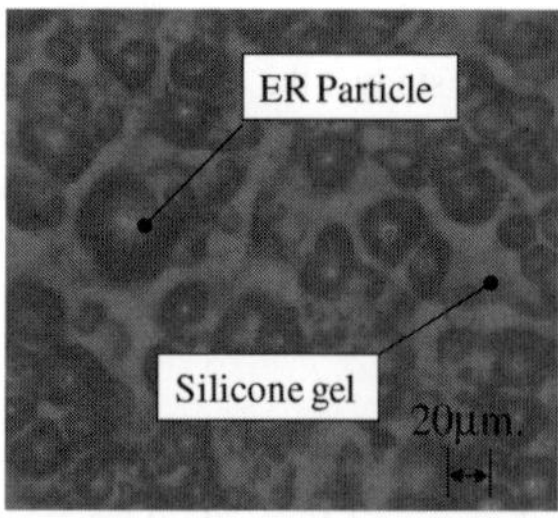

Figure 2. Surface of ERG under no electric field.

2. Three Forces Contributing to EA Effect

From the viewpoint of electromagnetics, it is suggested that three forces, i.e., the Maxwell stress, polarization force, and gradient force, act on ER particles and silicone gel when an electric field is applied to the ERG. These forces act ER particles and silicone gel as shown in Fig.3.

The Maxwell stress (F_m) acts on the interfaces between ER particles, silicone gel, and air, since this force is generated by the difference in the permittivity of dielectric substances. The polarization force (F_p) is an attractive force between the ER particles polarized under an electric field. The gradient force (F_g) is generated due to the difference in the electric field intensity. At the air gap with the lowest permittivity, the electric field is high and its intensity drastically increases around the contact point between the ER particles and the sliding electrode. From this distribution of electric field intensity at the air gap, it is suggested that both F_m and F_g are generated such that silicone gel is attracted toward the sliding electrode along the surface of ER particles.

Considering these three forces as dominant forces contributing to the EA effect, we estimate the mechanism of this effect. Though the ER particles at the surface are subjected to F_p and F_m, F_p is sufficiently high to sink the particles into silicone gel, as shown in Fig.3. On the other hand, silicone gel around the ER particles rises and sticks to the sliding electrode because of the resultant of F_g and F_m.

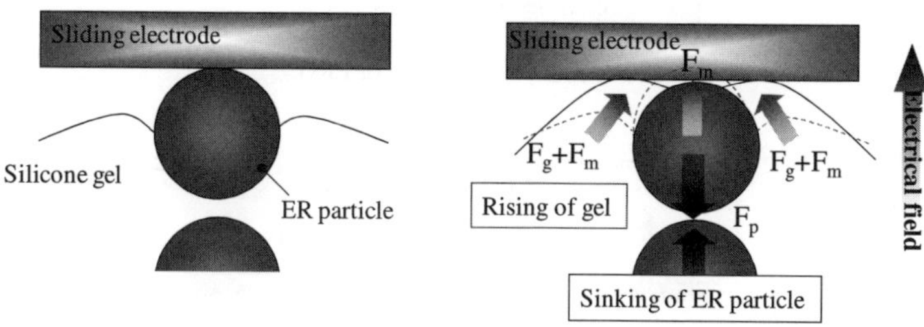

Figure 3. Estimated behavior of ER gel at interface under electric field.

3. Analysis of Static Electric Field

In order to verify the estimated mechanism of the EA effect based on the three forces, a numerical analysis is carried out using the finite element method (FEM). First, the distribution of the electric field at the interface between the ERG and the sliding electrode is investigated. The model for the numerical analysis is illustrated in Fig.4. The analysis conditions are as follows; diameter of ER particles = 20 μm, ERG thickness = 122 μm, air gap = 2 μm, applied electric field = 1000 kV/mm, relative permittivity of ER particles = 8, relative permittivity of silicone gel = 2.2, and relative permittivity of air = 1.

The finite element mesh consisted of 3-node trigonal elements (total element count: 20108). The mesh density is high at the air gap between the sliding electrode and the ER particles. The distribution of the electric field is shown in Fig.5 as a vector diagram. It is clear that the electric field intensity vectors are concentrated, and the gradient of electric field intensity becomes high at the air gap around the ER particles.

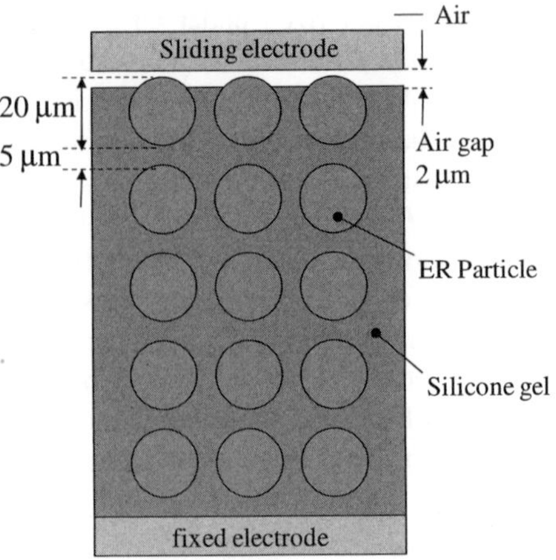

Figure 4. Analysis modeling of ERG.

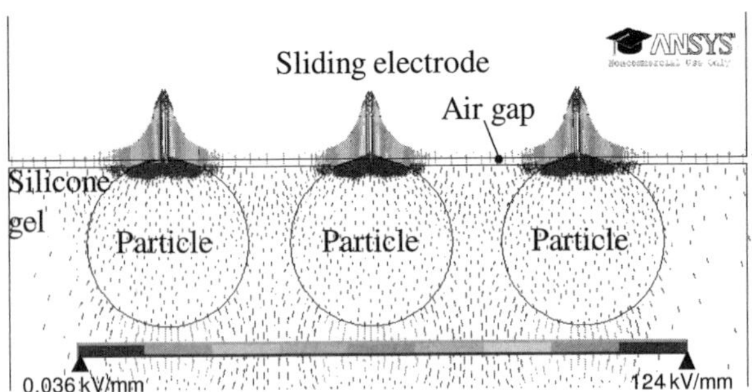

Figure 5. Distribution of electric field intensity at contact interface.

4. Theoretical Modeling of ERG

In order to construct the theoretical model of the ERG, equations of the three electric forces described in section 2 are derived. When two different dielectric substances exist under an electric field, the Maxwell stress acts from the high-permittivity substance to the low-permittivity one at the boundary of both the substances. When the dielectric substances are placed vertically against an electric field, the Maxwell stress F_n acting along the normal direction to the interface between the two substances is given by

$$F_n = \frac{1}{2}(\frac{1}{\varepsilon_2} - \frac{1}{\varepsilon_1})D_n^{\,2},$$

(1)

where

$\quad D_n \qquad$ Electric flux density in the normal direction to the interface.

When the dielectric substances are placed horizontally against an electric field, the Maxwell stress F_t is expressed as

$$F_t = \frac{1}{2}(\varepsilon_1 - \varepsilon_2)E_t^{\,2}, \tag{2}$$

where

$\quad E_t \qquad$ Electric field in the tangential direction to the interface.

In this analysis, in order to calculate the Maxwell stress F_m exerted on ER particles and silicone gel, it has to be considered that the electrical field and flux density passing through one dielectric substance enter the other one at a certain incident angle. The electrical field and flux density are refracted at the interface of the two different dielectric substances. Therefore, the Maxwell stress has to be calculated separately by dividing it into F_n and F_t. The two components of the Maxwell stress (F_n and F_t) at the interface are shown in Fig.6, and the Maxwell stress F_m is given as

$$F_m = F_n + F_t. \tag{3}$$

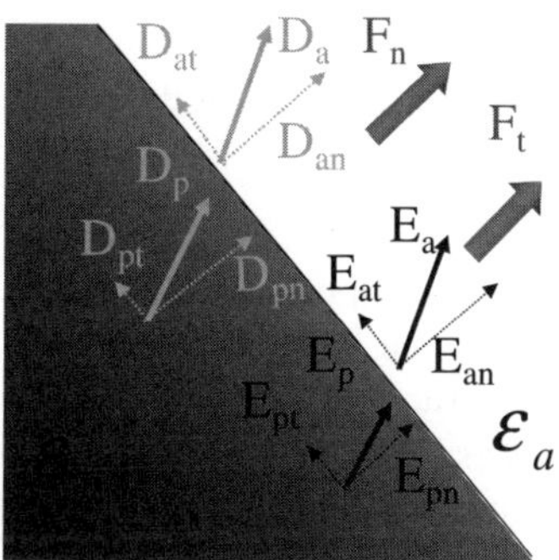

Figure 6. Maxwell stress acting on interface between ER particle and air.

On the other hand, the polarization force [6] between two ER particles is expressed as

$$F_p = \frac{3}{2}\pi\varepsilon_{gel}\beta^2 l^2 E^2, \quad \beta = \frac{\varepsilon_{particle} - \varepsilon_{gel}}{\varepsilon_{particle} + 2\varepsilon_{gel}}, \tag{4}$$

where

$\quad l \qquad$ Distance between the centers of the ER particles

$\quad E \qquad$ Electric field intensity.

This force is also exerted on ERGs as in the case of ERFs. The gradient force [7] acting on silicone gel at the air gap is given by

$$F_g = \frac{2\pi a^3}{A} \left(\frac{\varepsilon_{gel} - \varepsilon_{air}}{\varepsilon_{gel} + 2\varepsilon_{air}} \right) gradE^2 , \qquad (5)$$

where

A Projected area of ER particles
E Electric field intensity
a Radius of ER particles.

The gradient force, which is generated by the non-uniform electric field between the sliding electrode and the ERG surface, acts on silicone gel. The gel is easily deformed by the force because of its viscoelastic property.

The following stress-strain relationship is used to determine the displacement of ER particles and silicone gel due to these forces:

$$F/S = Y \cdot \Delta L/L , \qquad (6)$$

where

L Length of the material
Y Young's modulus of the material
S Cross-sectional area of the material.

5. Behavior Analysis

Electric field and flux density, which can be obtained by the static electric field analysis, are given in terms of the three forces by using Eqs. (1)–(5). On the basis of the three forces, the structure analysis is continuously carried out using Eq. (6). The numerical analysis conditions are listed in Table 1. As the boundary condition, the sliding surface of the sliding electrode is fixed in the X Y plane. The analysis result is shown in Fig.7. This result does not include the influence of gravity force. It can be confirmed that ER particles attract each other. Moreover, as shown in Fig.8, silicone gel rises around ER particles. These results show that the three forces mainly contribute to the EA effect.

Table 1: Numerical analysis conditions.

	air	gel	particle	electrode
Young's modulus [Pa]	-	1000	3.23e9	7.03e10
Poisson ratio	-	0.499	0.35	0.345
Element count			20108	

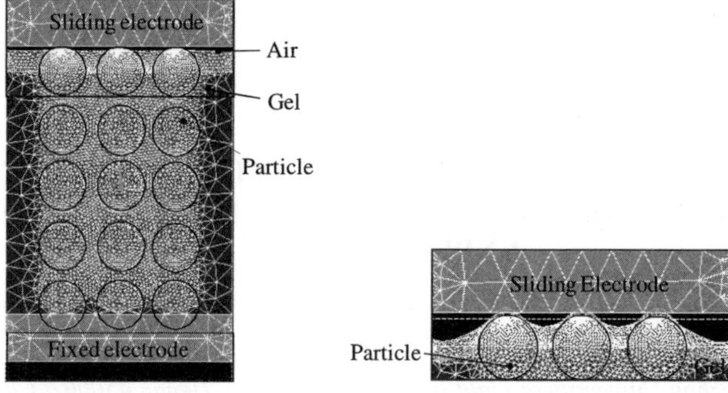

Figure 7. Result of behavior analysis.   Figure 8. Enlarged view of interface.

6. Conclusion

In this study, the theory of the EA effect based on electromagnetics is developed and verified using a numerical analysis. The results obtained are given below.

1. Electric field intensity vectors are concentrated, and the gradient of electric field intensity becomes high at the air gap around ER particles.
2. The theoretical model is developed on the basis of the three forces—Maxwell stress, polarization force, and gradient force—contributing to the EA effect.
3. The behavior analysis result shows that ER particles attract each other, and silicone gel rises around ER particles at the surface due to the three forces.

References

1. W. M. Winslow, *Journal of Applied Physics* **20**, 1137 (1946).
2. S. B. Choi et al., *J. Intelligent Mater. Syst. Struct.* **18** 1191 (2007).
3. T. Nakamura et al., *IEEE/ASME Trans. Mechatron* **10**, 154 (2005).
4. Y. Kakinuma et al., *Precision Engineering* **30**, 280 (2006).
5. Y. Kakinuma et al., *International Journal of Modern Physics B* **19** 1339 (2005).
6. T. Hanaoka, *T.IEE Japan*, **121-A**, 136 (2001).
7. A. Johnsen, K. Rahbeck, *I.I.E. Journal*, **61**, 713 (1923).

UNDERSTANDING ELECTRIC INTERACTIONS IN SUSPENSIONS IN GRADIENT AC ELECTRIC FIELDS I: EXPERIMENTAL[*]

YAN SHEN

Zhengzhou University of Light Industry, Zhengzhou, Henan 450002, China

ZHIYONG QIU

New York Center for Biomedical Engineering and the City College of the City University of New York, 140th Street and Convent Avenue, New York, NY 10031, USA

SHIGERU TADA

The National Defense Academy, Kanagawa 239-8686, Japan

When neutrally buoyant poly alpha olefin particles in corn oil were exposed to a gradient ac electric field generated by a spatially periodic electrode array, these particles experienced the negative dielectrophoresis and instability in all the suspensions of concentration range from 0.01% to 5% (v/v). One critical particle concentration was experimentally determined as 1% (v/v) below which the particles in corn oil were segregated to form island-like structures in the lower electric field regions; and above which, particles only formed straight stripes. The island-like structure was suspended in the lowest electric field area. Specially designed experiments with a suspension of 1.126% (v/v) confirmed that there exists particle instability. Anisotropic properties of electric interactions are responsible for particle instability in all the suspensions of different concentrations and island-like structures were formed only in the dilute suspensions in which the particle instability has enough space to be developed.

1. Introduction

Two previous papers of ours reported dielectrophoresis (DEP)-induced transport of particles in dilute suspensions in gradient electric fields [1-2]. The single particle model works well because interactions among particles are so weak that they are negligible. This paper will report our experimental studies of a suspension of neutrally buoyant poly alpha olefin particles in corn oil when the suspension is exposed to a gradient ac electric field.

---

[*] Corresponding Author: ZHIYONG QIU (LTSCLTSC@HOTMAIL.COM)

2. Experimental

2.1. *Setup and Suspension*

Shown in Figure 1 is one DEP chamber that consists of a transparent cover with one side ITO coated and grounded, a 3mm thick and 5mm wide close Teflon gasket as its four-side wall, and an insulating bottom plate with a spatially periodic electrode array. Its cavity is 150×70×3(mm). Eight pairs of 1.6×1.6 mm square brass bars act as the ground and high voltage electrodes that were inlaid into the insulating plate in parallel with a uniform gap of 2 mm. The top surface of each electrode was flattened and polished with fine sand paper and metal polishing agents. The chamber cover has two layers of a clear acrylic plate and ITO-coated glass. The ITO-coated glass and all the ground electrodes are electrically connected together and grounded. The electrode array was energized by a high voltage ac power supply. Mono-sized poly alpha olefin particles of mean diameter of 86.7μm were used for experimental studies. The particles have the exactly same density of 0.92g/cm^3 as corn oil only at 19°C. The dielectric mismatch factor (β) between particles and corn oil was measured with 4-volt excitation voltage as -0.143 in the frequency range from 50 to 10^5Hz. The viscosity of corn oil (η) is 0.060 Pas at 19°C. A CCD camera was used to record particle distribution with time when an ac voltage was applied to a suspension.

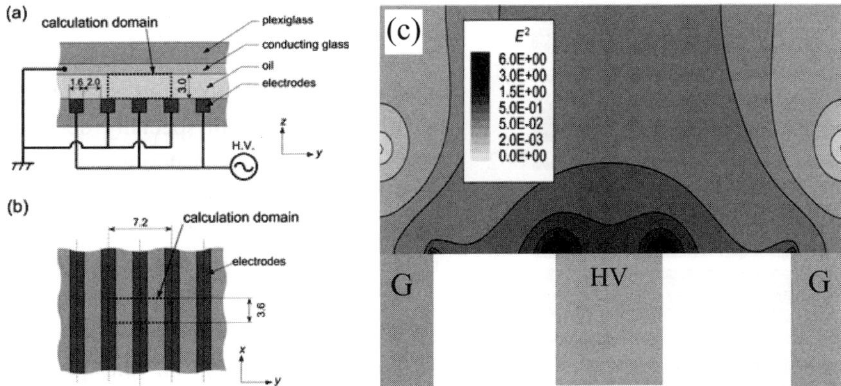

Figure 1. (a): Cross section view of the spatially periodical electrode array in the DEP chamber. (b): Top view of the electrode array. (c): Distribution of the square of electric field strength (E^2) in the calculation domain of 7.2x3.6x3.0 mm. HV and G denote the high voltage electrode and ground electrodes respectively. There is the lowest electric field at (X, 0, 1.245 mm) or (X, 7.2, 1.245 mm).

438

2.2. *Experimental Results*

Experiments of particle transport, aggregation and segregation induced by gradient ac electric fields were carried out with the DEP chamber and the suspension described above. A set of experimental photos is shown in Figure 2 for a suspension of 0.1% (v/v) to which a voltage of 5kV/100Hz was applied. The particles were transported to the lower electric field region because the dielectric mismatch factor of the suspension is negative, - 0.143. In the mean time, particles were segregated to form island-like structures that became remarkable on the last four photos. Although the single particle model works fine for dynamics of particle transportation induced by the DEP force, it is unable to describe dynamics of particle aggregation and segregation shown in Figure 2.

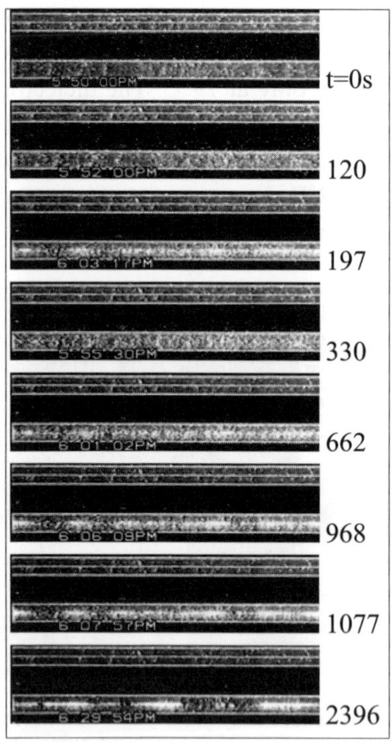

Figure 2. Particle transport, aggregation and segregation induced by an ac voltage of 5kV/100Hz in the DEP chamber. The suspension consists of poly alpha olefin particles and corn oil at 0.1% (v/v). All the images were taken when the camera was set to be normal to the plane of the electrode array. The voltage was applied to the suspension at 5:50PM. Time on each image was real time. One net exposure time is in the right side next to each photo.

Effects of particle concentrations of suspensions on particle transportation, instability, aggregation and segregation were scrutinized with specially designed experiments. The experiments were performed with suspensions of particle concentration of 0.01% to 5% (v/v). Experimental results are shown in Figure 3 for comparison. When the particle concentration is below 1% (v/v), for example, 0.1% (v/v), all the particles were collected around the calculated lower field area. However, these particles didn't form one uniform stripe, instead segregated into island-like structures. When the particle concentration was above 1% (v/v), for example, 5% (v/v), shown in Figure 3C, all the particles were collected around each calculated low field area and one uniform particle stripe was formed. So, one critical particle concentration for formation of the island-like structure was experimentally determined as 1% (v/v).

Further experiments were designed to confirm existence of the critical particle concentration of 1% (v/v). As shown in Figure 3B1, particles formed one uniform stripe in the suspension of 1.126% (v/v) at the 611th second after a voltage of 5 kV / 100 Hz was applied. Width of the particle stripe was measured as 0.618 of the electrode width. Being contrary to what is shown in Figure 3B1, Figure 3B2 shows that wavy structures appear on initially uniform particle stripes when some particles were removed from one end of the ground electrode with help of the edge effect of the electrode array. Width of particle stripes was not uniformly reduced from the central part to one end of the ground electrode; instead, saw tooth waveforms were developed on initially uniform particle stripes. The largest width of saw tooth stripes was measured as 0.561 of the electrode width. In Paper II [3], we will use here experimental data to establish quantitative relations among the particle instability, formation of island-like structures, particle concentration and electric field distribution, and confirm existence of one critical particle concentration of 1% (v/v).

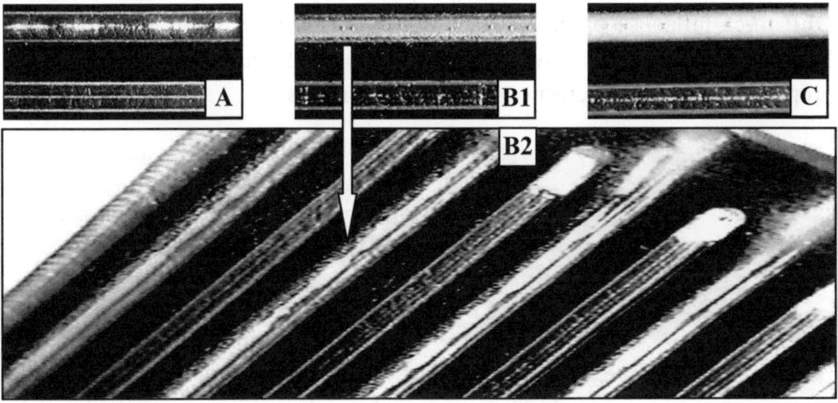

Figure 3. Particle transport, aggregation and segregation versus particle concentration of suspension. The excitation voltage is 5kV/100Hz for the sample suspensions in this figure. **A**: 0.1% (v/v) and exposure time of 2400s. **B1**: One uniform particle stripe was formed in a suspension of 1.126% (v/v) for an exposure time of 611s. **B2**: Wavy structures were developed on the uniform stripes when some particles were removed from one end of the ground electrode with help of the edge effect of the electrode array. The exposure time is 4200s. **C**: 5% (v/v) and exposure time of 1197s.

It should be pointed out that all the experiments except for the one shown in in Figures 3B1 and 3B2 were carried out with a close PTFE gasket and that the boundary condition of zero particle flux is valid.

Experimentally we determined locations of the island-like structures. We set the CCD camera at an angle of 30 ~ 45° between the camera axis and the plane of the electrode array. Figure 4 shows that these island-like structures were suspended somewhere above the ground electrode but didn't touch the ground electrode surface, which is in qualitative agreement with the calculation shown in Figure 1C.

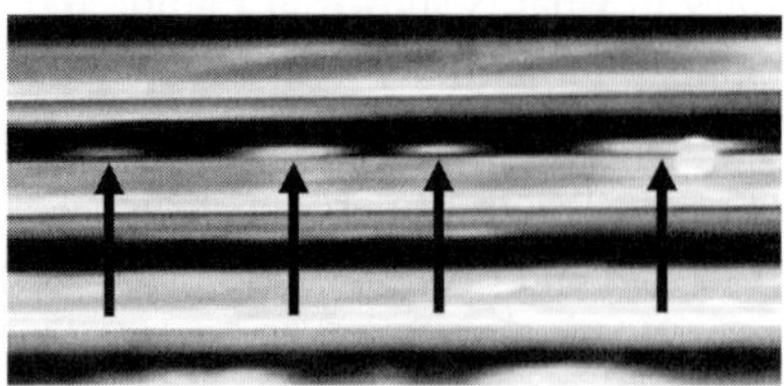

Figure 4. The island-like structures were located somewhere above the ground electrode, as arrows point. The continuous brighter stripes are the electrode surfaces. The CCD camera was set to be normal to the electrode direction and at an angle of 30 ~ 45° between the camera axis and the plane of the electrode array.

3. Anisotropic Properties of Electric Interactions

What force drives particles to segregate? Why are island-like structures formed only in dilute suspensions of 1% (v/v) or less? How does the transverse (that means the DEP force was in a family of planes normal to the electrode direction) DEP force influence particle instability or formation of island-like structures? We will qualitatively answer these questions via analyses of anisotropic properties of electric interactions among the particles.

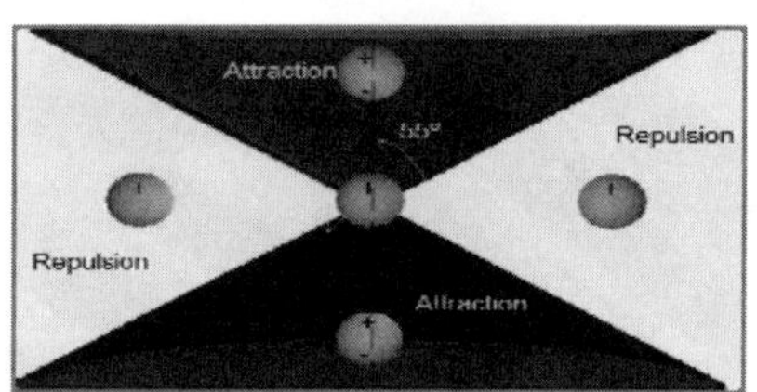

Figure 5. Critical interfaces that separate the attractive zone (dark) from the repulsive zone (light gray) for one two-particle system in a uniform external electric field.

As shown in Figure 5, given one two particle system in a uniform electric field, for convenience, one particle is called the host particle that is located on the common tip of two 110°cones that share the symmetric axis along the electric field and the other particle the guest particle that is located somewhere else. When the guest particle is within each cone, there exists an attractive interaction between them. The two particles will approach to each other and align along the electric field; otherwise, there exists a repulsive interaction between them and the two particles will be driven away from each other. So, the two cone surfaces act as critical interfaces that separate the repulsive zone from the attractive zone for the two particle system in one uniform electric field. Because the guest particle can not stably stay on the critical interfaces, the particle instability occurs and the critical interfaces are responsible for the particle instability.

Given one multi-particle system in a uniform external field, critical interfaces will no longer be perfect cone surface but they do exist and so do multi-attractive and repulsive zones. So the system is not stable. If one guest particle is in one attractive zone, it will be attracted to its host particle and align with it along the electric field; otherwise, the guest particle will be driven away from the host particle till the guest particle finds another attractive zone and aligns with another host particle along the electric field. As a consequence, many chains are formed in parallel along the electric field [4-7].

Even given one two-particle system in a non-uniform electric field, critical interfaces are no longer perfect cone surfaces either because a particle dipole moment becomes location dependent. However, there still exist critical interfaces. Given an instant, there are the critical interfaces that separate the attractive zone from the repulsive zone for the two particle system. Because no particle can stably stay on the critical interface, the particles must experience instability. Compared with a particle in a uniform electric field, a particle in a gradient electric field will feel *one* external force, a DEP force, from the gradient electric field. So, it depends combined effects of the dipole-dipole interaction and the DEP force where one particle goes.

Given one multi-particle system, when it is exposed to a gradient electric field, particles are involved in collective motion. On the one hand, some particles form short chains due to dipole-dipole interactions; the other hand, individual particles or short chains are transported by the DEP force. Sometimes, the DEP force do a favor for particle chaining, other time it takes some particles away from a short chain. So, it depends combined effects of the multi-body anisotropic dipole-dipole interactions and the DEP force where one particle or a short chain goes. In short, anisotropic properties of electric interactions are responsible for the particle instability.

We estimate relative strength of a driving force, if it exists, responsible for the particle segregation to the transverse DEP force. This driving force can be estimated as $F_{seg} = 6\pi\eta aV = 6\pi\eta aL / \tau_{seg}$ where τ_{seg} is the segregation time, a particle radius and L characteristic distance for particles to move during the segregation. The transverse DEP force F_{TDEP} can be written as $\langle F_{TDEP} \rangle = \langle 2\pi\varepsilon_0\varepsilon_l\beta a^3 \nabla E^2 \rangle$ where $\langle F_{TDEP} \rangle$ is to take average over the involved region and ε_l is the dielectric constant of corn oil. Therefore, the ratio is given as $\langle F_{seg} \rangle / \langle F_{TDEP} \rangle = (3\eta L)/(\varepsilon_0\varepsilon_l\beta a^2 \tau_{seg} < \nabla E^2 >)$. For the suspension of 1% (v/v) under 5kV/100Hz, $\langle F_{seg} \rangle / \langle F_{TDEP} \rangle \sim 0.01$. Compared with the average transverse DEP force, the possible driving force responsible for the particle segregation is so weak! Fortunately, we are still able to observe these tantalizing phenomena due to so weak a driving force in very dilute suspensions!

4. Conclusions and Further Work

The particle instability was recognized in all the suspensions of neutrally buoyant poly alpha olefin particles in corn oil at different particle concentrations in a gradient ac electric field generated by a spatially periodical electrode array. Island-like structures were observed in the suspensions of 1% (v/v) or less in a gradient ac electric field. The island-like structures were located in the lowest electric field region above each ground electrode. Experimentally, one critical particle concentration of suspensions was determined as 1% (v/v). When the particle concentration is 1% (v/v) or less, the particle instability is developed to form island-like structures; when the particle concentration is 1% (v/v) or higher, the particle instability was suppressed completely by the strong transverse DEP force and particles form a uniform straight stripe above each ground electrode.

In Paper II [3], we will present our understandings of these experimental findings via numerical simulations based on the continuous model [8] and the molecular dynamics (MD) model [9].

Acknowledgments

The authors gratefully acknowledged that this work was in part supported by NASA under grant No NAG3-2698 and PSC-CUNY Professional Development Fund three times. Some experiments were conducted in the Levich Institute of the City College of the City University of New York.

References

1. Z.Y. Qiu, N. Markarian, B. Khusid and A. Acrivos, *J. Appl. Phys.* **92**, 2829 (2002).
2. A. D. Dussaud, B. Khusid and A. Acrivos, *J. Appl. Phys.* **88**, 5463 (2000).
3. Y. Shen, S. Tada, D. Jacqmin, B.M. Fu and Z.Y. Qiu, *in this proceeding of ERMR2010*, (2010).
4. G. L. Gulley and R. Tao, *Phys. Rev. E,* **56**, 4328 (1997).
5. T. C. Halsey, *Science,* **258**, 761 (1992).
6. T. C. Halsey and J. E. Martin, *Scientific American,* **269**, 58 (1993).
7. M. Whittle and W. A. Bullough, *Nature,* **358**, 373 (1992).
8. A. Kumar, Z.Y. Qiu, A. Acrivos, B. Khusid and D. Jacqmin, *Phys. Rev. E,* **69**, 021402 (2004).
9. R. Tao and J. M. Sun, *Phys. Rev. Lett.* **67**, 398 (1991).

UNDERSTANDING ELECTRIC INTERACTIONS IN SUSPENSIONS IN GRADIENT AC ELECTRIC FIELDS II: SIMULATIONS AND APPLICATION EXPLORATION*

SHIGERU TADA[**]

The National Defense Academy, Kanagawa 239-8686, Japan

YAN SHEN[**]

Zhengzhou University of Light Industry, Zhengzhou, Henan 450002, China

DAVID JACQMIN

NASA Glenn Research Center, Cleveland, OH 44135, USA

BINGMEI FU and ZHIYONG QIU

New York Center for Biomedical Engineering and the City College of the City University of New York, 140th Street and Convent Avenue, New York, NY 10031, USA

We used numerical simulations of a continuous model and the molecular dynamics model to understand the particle instability, formation of island-like structures and existence of one critical particle concentration of 1% (v/v) for formation of island-like structures in the suspension in a gradient ac electric field reported in Paper I. The simulations of the continuous model show that the critical concentration of 1% (v/v) is the concentration of which the particles of a suspension are just fully filling the lower field region finally. According to the MD simulations, the particles instability does exist in the corn oil in a gradient ac electric field, anisotropic polarization interactions among the particles are responsible for the particle instability and have memory, and the memory is still kept even when the particles are transported by a dielectrophoresis force. The island-like structures can be regarded as signature of the memory. We explored possibilities to apply our findings in biomedical fields.

1. Introduction

In Paper I [1], we presented our experimental findings. Some questions remain to be answered. How does the particle instability happen? What force drives the particles to segregate? Why are the island-like structures formed *only* in dilute suspensions of 1% or less? How does the transverse DEP force that is normal to

---

* Corresponding Author: ZHIYONG QIU (LTSCLTSC@HOTMAIL.COM)

** Both the authors equally contributed to this work.

444

the electrode direction influence the particle instability in suspensions of different particle concentrations? These questions will be answered via numerical simulations of the continuous model [2] and the molecular dynamics (MD) model [3].

2. Calculation Domain

Shown in Figure 1 are the top view and cross section view of the electrode array as well as field distribution in the calculation domain of 7.2x3.6x3.0 mm. In the simulations based on the continuous model, we will only use half of the calculation domain due to symmetry. In the simulations based on the MD model, we will use the whole domain. The Laplace equation was solved in the calculation domain with the finite difference method and the boundary conditions for the Laplace equation were discussed in Ref. [4]. The computed electric field is plotted in Figure 1C that will be directly used for calculations of trajectories of particles based on the MD model because the particle concentration is low. All the parameters for the simulations are the same as those in Ref. [1].

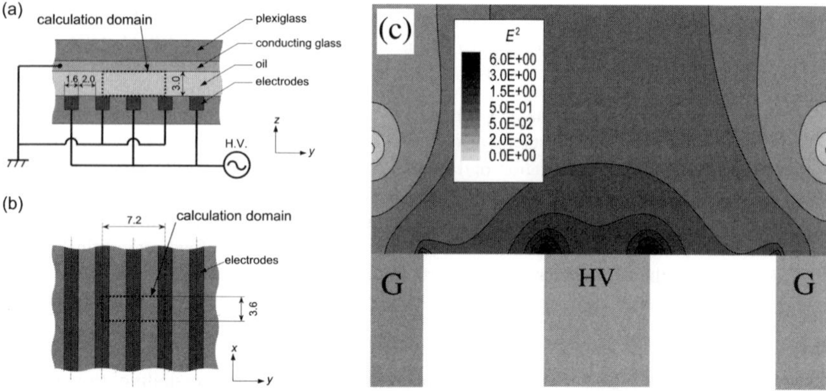

Figure 1. (a): Cross section view of the spatially periodical electrode array in the DEP chamber. (b): Top view of the electrode array in the chamber. (c): Distribution of the square of electric field strength (E^2) in the calculation domain. HV and G denote the high voltage electrode and ground electrodes respectively. There is the lowest electric field at (X, 0, 1.245mm) or (X, 7.2, 1.245mm).

3. Continuous Model

The continuous model was well developed and described in Ref. [2]. Here, we only discuss two sets of data for the two suspensions of concentrations of 0.1% (v/v) and 1% (v/v). Given a suspension, we calculate $\Phi(Y,Z,t)$ a spatial distribution of particle concentration with time in a gradient electric field. Mean final particle concentrations of a suspension in a gradient ac electric field are defined as $\langle\Phi(Y)\rangle = \left(\int_0^3 \Phi(Y,Z,t)dZ\right)/3$ and $\langle\Phi(Z)\rangle = \left(\int_0^{3.6} \Phi(Y,Z,t)dY\right)/3.6$ when t goes to infinite. $<\Phi(Y)>$ versus Y and $<\Phi(Z)>$ versus Z are plotted in Figure 2A and 2B for the suspensions of 0.1% (v/v) and 1% (v/v). Solid squares denote data for the suspension of 0.1% under 5kV/100Hz and all the particles are finally collected within a circle (in fact, one cylinder along the electrode) with a diameter being 0.5mm and its center being at (X, 0, 1.245 mm). Empty circles denote data for the suspension of 1% (/v/v) under 5kV/100Hz and all the particles are finally collected in an ellipse (in fact, one ellipse column along the electrode) with the long and short axes being 1.0mm and 0.9 mm and its center being at (X, 0, 1.245 mm).

Let's look at the distribution of E^2 shown in Figure 2C. There is a family of deep valleys in the curves of E^2 versus (Y, Z) and their centers are located at

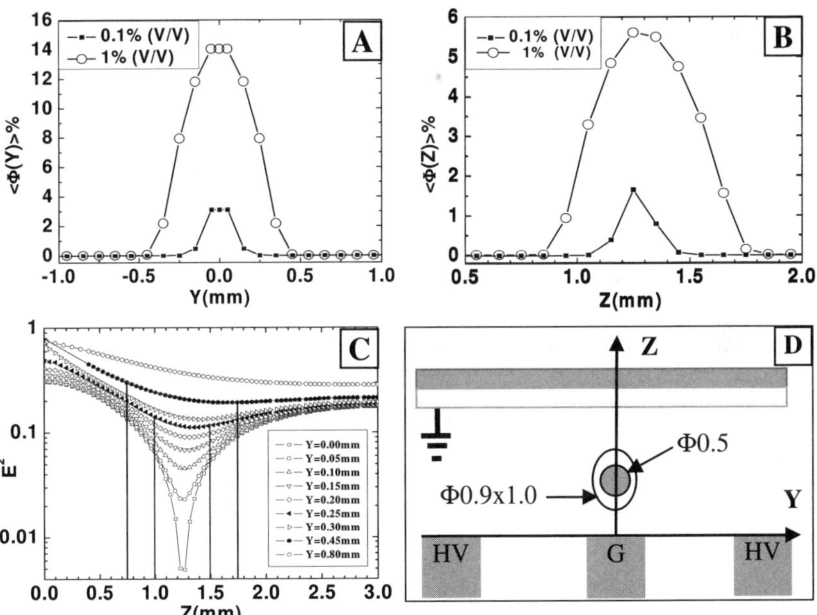

Figure 2. **A&B**: Calculated final distribution of mean particle concentration: $<\Phi(Y)>$ versus Y and $<\Phi(Z)>$ versus Z. The excitation voltage of 5kV/100Hz and dielectric mismatch factor $\beta = -0.143$ were used in the above calculations. Solid squares refer to the suspension of 0.1% (v/v) and empty circles refer to the suspension of 1% (v/v). **C**: Distribution of electric field strength (E^2) in Y-Z plane. **D**: Particles were finally collected in one cylinder of a diameter of 0.5 mm for the suspension of 0.1 % (v/v) and one ellipse column of long and short axes of 1.0 mm and 0.9 mm for the suspension of 1% (v/v). The cross sections of the two columns were shown here only.

446

(X, 0, 1.245 mm). The valleys rapidly become shallow with increase in Y and disappear at Y = 0.45mm. The area of valleys of E^2 is an approximate ellipse (column) with long and short axes being 1.00 mm and 0.90 mm, respectively and the center being at (X, 0, 1.245 mm). It will shed light on development of particle instability and formation of island-like structures to compare the distribution of electric field to the distribution of final mean particle concentration of the suspensions of 0.1% and 1% (v/v) under 5kV/100Hz.

For the suspension of 0.1% (v/v) under 5kV/100Hz, particles are finally distributed within the deep valley where there is very weak field, as shown in Figure 2C and 2D. For the suspension of 1% (v/v), the distribution of particles finally reaches the rim of the valley. If the particle concentration is further increased, particles not only finally take up the valley but the space beyond it also. Because there exists a very huge difference of the square of electric field intensity (E^2) between inside and outside the valley, the particles outside the valley are subjected to the much stronger transverse DEP force and so the particles are forced to enter the valley as many as possible. Even though there is the particle instability along the electrode direction in a suspension of 1% (v/v) or higher, the particle instability cannot be developed or no island-like structure forms, instead, the particles only form uniform straight stripes, because the transverse DEP force is strong enough to suppress the particle instability completely.

Now, we explain a transit from uniform stripes to saw tooth stripes for the suspensions of 1.126% (v/v) under 5kV/100Hz reported in Paper I [1]. When uniform straight stripes were formed at the 611th second, its width was 0.618 of the electrode width shown in Figure 3B1 of Ref. [1] that is larger than the critical width of 0.90 mm / 1.60 mm = 0.5625 shown in Figure 2. So, the DEP force dominated and the particle instability was suppressed completely. When some particles were removed and saw tooth waveforms appeared, the largest width was 0.561 of the electrode width shown in Figure 3B2 of Ref. [1] that is slightly less than the critical width of 0.5625 of the electrode width shown in Figure 2, all the particles were in the weak field areas and the DEP force was unable to suppress the particle instability completely. Next, we will reveal a driving force and how it initializes particle instability via the MD simulation [3].

4. Molecular Dynamics (MD) Model

Shown in Figure 3 is one two particle system in one electric field. Because the Brownian force and the effective gravitational force on Particle i are negligible, only the electric forces (F_i) and the Stokes' drag force govern dynamics of the suspension of neutrally buoyant particles and corn oil subjected to a gradient ac electric field. So the governing equation can be written as

$$m\frac{d^2 r_i}{dt^2} = F_i - 3\pi\eta d\frac{dr_i}{dt}$$

(1)

In Eq. (1), m is particle mass, d particle diameter and η viscosity of corn oil. F_i denotes all the electric forces on the i-th particle of

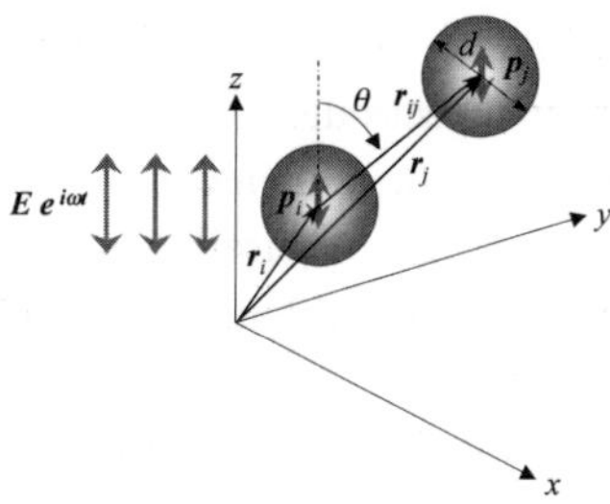

Figure 3. Two dielectric particles in an ac electric field.

$$F_i = \sum_{i \neq j} f_{ij} + f_{idep},$$

where

$$f_{ij} = -\frac{\partial \phi_{ij}}{\partial r}e_r - \frac{1}{r}\frac{\partial \phi_{ij}}{\partial \theta}e_\theta$$

$$\phi_{ij} = \frac{1}{4\pi\varepsilon_0\varepsilon_f}\left(\frac{p_i \cdot p_j}{r_{ij}^3} - \frac{3(p_i \cdot r_{ij})(p_j \cdot r_{ij})}{r_{ij}^5}\right), \quad r_{ij} = r_i - r_j$$

and

$$p_i = 4\pi\varepsilon_0\varepsilon_f\left(\frac{d}{2}\right)^3 \mathrm{Re}(\beta)E_{rms}$$

f_{ij} is the non-zero time-average dipole-dipole interaction force on the i-th particle from the j-th particle, and $f_{idep} = 2\pi\varepsilon_0\varepsilon_f\left(\frac{d}{2}\right)^3 \mathrm{Re}(\beta)\nabla E_{rms}^2(r_i)$ is the non-zero time-average DEP force from the gradient ac electric field. In the above equations, Re (β) is the real part of the dielectric mismatch factor of $\beta=(\varepsilon_p-\varepsilon_f)/(\varepsilon_p+2\varepsilon_f)$ between the particles and corn oil, ε_p and ε_f the dielectric constants of the particles and corn oil, respectively, E the ac electric field amplitude, and ε_0 the vacuum permittivity.

448

Simulations of the suspensions of 0.01% ~ 0.5% (v/v) have quantitatively reproduced our experimental observations of island-like structures in the lower electric field regions.

Shown in Figure 4 is a sequence of particle configuration with time in a gradient ac electric field. The upper sequence in Figure 4 is the cross section view of the particle configuration, and shows that the particles are leaving the high field area around the HV electrode for the lower field area shown in Figure 1C. A clear particle front formed after the electric field was applied for 10 ~ 20s. At 40th second, the particle front was broken down into two parts and each part was further becoming shorter with time and its particles were approaching to the lowest field area above each ground electrode.

The lower sequence in Figure 4 is the top view of the particle configuration, and shows that there are short chains in parallel and perpendicular to the electrode, above each ground electrode. Although these chains are becoming shorter with time, the distance among the chains was kept hardly changed till all the particles were collected into the lowest field area. Undoubtedly, particle segregation is regulated by the anisotropic dipole-dipole interactions among the particles. After 100s of particles in the gradient ac electric field, the evolution sequences reveal how island-like structures are being formed. In short, the

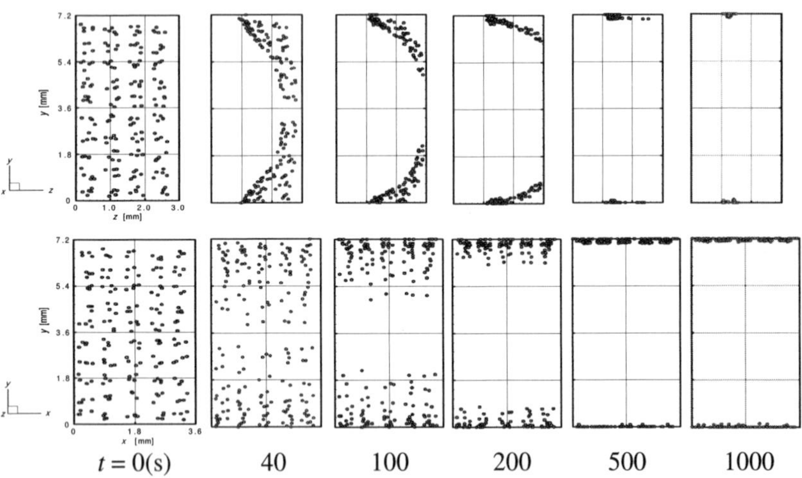

Figure 4. Evolution of particle configuration in a gradient ac electric field with time. In this simulation, the suspension concentration is 0.1% (v/v), the applied voltage is 5kV/100Hz, relative dielectric constants of particles and corn oil are 1.820 and 2.881, respectively, and Re (β) is -0.143. The viscosity of corn oil is 0.060 Pa s, the mean particle diameter is 86.7 μm and the density of particles or corn oil is 0.92 g/cm^3 at 19 °C.

anisotropic interactions among the polarized particles have memory. The memory is kept even when individual particles or short particle chains are transported by a strong, weak or extremely weak DEP force.

5. Conclusions and Further Work

In physics, we uncovered that anisotropic interactions among polarized particles are responsible for particle instabilities and formation of the island-like structures. The anisotropic interactions have memories and the memories are kept even when the particles are transported by the DEP force. The island-like structures can be regarded as signature of the memories. According to our widespread survey of references, the memory of anisotropic interactions among the polarized particles has profound implications to many dynamics processes in nature.

In application, we have verified that there is detectable difference between dielectric properties of cells, live and dead, in media. One application of our findings is to sort out dead human breast cancer cells from live ones before perfusion into a single microvessel to study their adhesion *in vivo* [5]. The other application is to online monitor cell live state after they are electroporated for developments of therapies.

Acknowledgments

The authors gratefully acknowledged that this work was in part supported by NASA under grant No NAG3-2698 and PSC-CUNY Professional Development Fund. The authors thank Dr. Junjun Mao for providing them with computing resources of the Levich Institute of the City College of the City University of New York for the MD simulations. The continuous model simulations were conducted in NASA Glenn Research Center.

References

1. Y. Shen, Z.Y. Qiu and S. Tada, *in this proceeding of ERMR2010*, (2010).
2. A. Kumar, Z.Y. Qiu, A. Acrivos, B. Khusid and D. Jacqmin, *Phys. Rev. E*, **69**, 021402 (2004).
3. R. Tao and J. M. Sun, *Phys. Rev. Lett.* **67**, 398 (1991).
4. Z.Y. Qiu, N. Markarian, B. Khusid and A. Acrivos, *J. Appl. Phys.* **92**, 2829 (2002).
5. S. Shen, J. Fan, B. Cai, Y. Lv, M. Zeng, Y. Hao, F. Giancotti and B.M. Fu, *J. of Exp. Physiology*, **95**, 369 (2010).

MULTIPLE SCATTERING APPROACHES
ON THE ELECTRORHEOLOGICAL FLUIDS

GANG SUN*, MINGCHUN JIAO, QIANG WANG and KUNQUAN LU

*Beijing National Laboratory for Condensed Matter Physics
and Key Laboratory of Soft Matter Physics, Institute of Physics,
Chinese Academy of Sciences, Beijing 100190, China
E-mail: gsun@iphy.ac.cn

The multiple scattering method is used in calculating the electric field distribution for the systems composed of spheres. According to the calculation result, we propose a new mechanism of the electrorheological effect and show that the yield stress induced by this mechanism exhibits the typical characteristics of that observed in experiments.

Keywords: Electrorheological effect; multiple scattering method; polar molecule.

1. Introduction

The electrorheological (ER) fluids are a kind of substances that exhibit an ability to change its rheological characteristics under the influence of the applied electric field. Since the inventions of ER fluids, the strength of the materials (as measured by the static yield stress) has been improved for generations, and some of the products have been successfully applied to commercial devices. The basic principle of ER fluids can result from the field-induced interaction force between the dispersed particles, which makes the particles align into chains or clusters so that the fluid is changed from the state of Newton-like fluid into that of a stiff semisolid exhibiting yield stresses. However, unlike the magnetorheological fluids, the high performance ER effect seems not attribute to the electromagnetic force, which is determined only by the dielectric properties of the dispersed and dispersing phase. For a high yield stress ER fluids [1], the experimental results show that the additives adsorbed in the dispersed particles plays very important roles. These kinds of ER fluids, including the hydrous ER fluids, generally show linear variation in yield stress when the electric field exceed certain threshold, which is controversial to the quadratic variation predicted by the dielectric theory. To explain the high yield stress occurring in these ER fluids, some phenomenological models are established, in which the water bridge model [2] and the polar-molecular model [3] are representative. The basic idea of these models is that the additives in the dispersed particles will redistribute when electric field is applied and play positive roles to increase the yield stress. The redistribution of the additives and its properties are determined by the local electric field, especially in the gap between particles, so it is seriously important in theoretical research to give an accurate calculation for corresponding system.

The calculation of the electric field for ER fluids system, which is usually approximated by a system composed of dielectric spheres, is not a trivial problem. The field induced force between two dielectric spheres align along the applied field has been studied both experimentally [4] and theoretically [5, 6]. The experimental work shows that the measured force is not only much higher than that predicted by the dipole approximation when the two spheres are closed up, but also prominently higher than that calculated by the improved dipole model (E corrections dipole model) and the finite-element approach (FEA). To explain the abnormal strong interaction force when the spheres are closed up, Cox et al developed a so-called re-expasion method by including higher order of polarization. Their results show a perfect agreement with that of the experiment [6]. This work implies that it is absolutely needed to include the higher order of polarization in the calculation.

The multiple scattering method [7] is a superior algorithm that can calculate the electric field accurately for the system composed of spheres. The key point of the method is to include the higher order of polarization up to an upper limit L_{max} in the calculation. The method can perform numerically and is able to use in much broader cases, such as, in the system of two spheres aligning obliquely to the applied field, or in the system of a chain of spheres, or even in the system composed of multi-scale spheres with arbitrary distribution etc., whereas the re-expasion method is restricted to the case of two spheres align along the field. In this paper, we use the multiple scattering method to calculate the electric field distribution for the ER fluids systems, and according to the calculation results we propose a new mechanism of the electrorheological effect.

2. Field Distribution Calculated by the Multiple Scattering Method

To introduce the multiple scattering method, we use it in the calculation of the field distribution for a system of two spheres with radius $R = 3.15$ mm and dielectric constant $\epsilon = 294.0$ separated by a small gap δ (see the inset of Fig. 1), of which the interaction force between the two spheres has been studied both experimentally [4] and theoretically [6]. Figure 1 plots the local electric field at the centre of the gap E_c normalized by the external applied field E_0 as a function of the highest order of polarization used in calculation, i.e., L_{max}. In Fig. 1, we can clearly see that the results seriously depend on L_{max} at first, and then tends to a stable value as L_{max} increases. For a smaller gap, larger L_{max} is needed to obtain stable results. In our computing environment, the largest L_{max} we can treated is about 30. Under the condition of $L_{max} = 30$, our multiple scattering method can give reasonable results if the gap is larger than about one thousandth of the diameter of the spheres, which is the only limitation of the method. Figure 1 also shows that the local electric field is much larger than external applied field if the gap is small and increases quickly as the gap decreases.

The multiple scattering method can used in almost all kinds of systems composed of spheres. Here we use it to calculate the electric field distribution for the typical structure of the ER fluids, i.e., a long chain (include

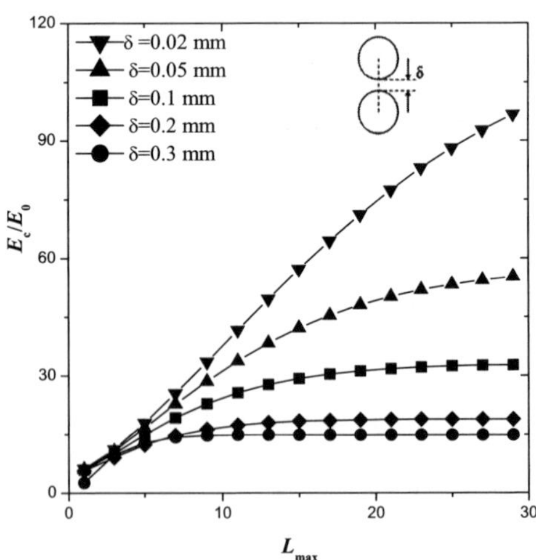

Fig. 1.   Local electric field at the centre of the gap as a function of L_{max} for a series of gap. Inset illustrates the system we treated in Figs. 1 and 2.

24 small spheres) between two electrodes (see the inset of Fig. 3). Since the multiple scattering method cannot treat the system including parallel plates, we use two big spheres with radius 100 times larger than that of the small spheres to replace the electrodes at the end of chain. The dielectric constant of the spheres (include the two larger spheres) is set to $\epsilon = 1920.0$ (correspond to that of $BaTiO_3$), and the gap between the spheres is set to $\delta = 0.005r$, where r is the radius of the small ball.

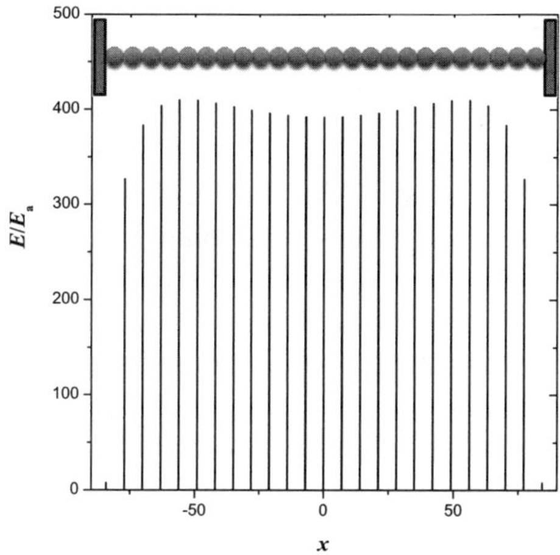

Fig. 2.   Normalized electric field distribution along the central axes of the chain. Inset illustrates the system we treated in Fig. 2.

The electric field distribution along the central axes of the chain is plotted in Fig. 2. In Fig. 2, the field is normalized by the average applied field E_a between two electrodes, which is obtained by the electric potential difference between two electrodes divided by the distance between them. From Fig. 2, we can see that the local field in the gap between small spheres is much higher than the average applied field. In general, the electric field in gap is about 300–400 times stronger than the average applied field. Since the value of the gap between the dispersed particles in the real ER fluids is unknown, we are satisfied with demonstrating that the field strength in the gap can be several hundred times larger than that of the average applied field. The effect of increment of the field will be more prominent if the dispersed particles close up further.

3. Yeld Stress Induced by Superpolarized State

In ER fluids experiments, the standard average applied field is about several kilovolt per millimeter, i.e., in the order of 10^6 V/m. This means that the field strength in the gap will reach about 10^8–10^9 V/m, which is an abnormal strong electric field. In such strong field, the polar molecules doped in the dispersed particles will not only go out of the particles and form a liquid bridge in the gap between the particles, but further the liquid bridge may change its liquidity and translate to a superpolarized state. A famous example that exhibit this transition is water. Many atomic force microscope (AFM) experiments show that a water bridge is often formed between a surface and an AFM tip [8]. Gómez-Moñivas et al use a continuum model and AFM data to demonstrated that water bridge formation is induced electrostatically when the electric field strength exceeds to a critical value in the range about 0.7–1.9×10^9 V/m [9]. This results is recently supported by molecular dynamics calculations, it is pointed out that the formation of a water bridge is marked by a first order discontinuity in the total energy of the system [10]. The superpolarized state can be considered as a kind of ER fluids with permanent dipole aligned along the electric field. It is known that this structure results in a yield stress by the similar mechanism of the classical ER fluids, though unlike the classical ER fluids, the yield stress for the superpolarized state doesn't vary with electric field because of the interaction being between the permanent dipoles (we neglect the reduction of the dipole moment of water molecular as the electric field increases). The strength of the yield stress for the superpolarized state is much stronger than that of the classical ER fluids because the interactions are in molecular level. According to Ref. [3], the yield stress for the superpolarized state can even be higher than 1 MPa.

From the previous discussion, we can image following physical process to explain the high yield stress attributed to the additives of polar moleculars. As the applied field increases to certain threshold value, the local field at the central line in the gap between the dispersed particles will firstly reach the critical value for superpolarized state, and hence, a thin pillar made of the polar molecules in the superpolarized state is formed between the dispersed particles. The pillar fixes the dispersed particles firmly because of the high yield stress of the superpolarized state. As the applied field increases further, the region of the electric field higher than the critical value for superpolarized state expands, and the pillar becomes thicker, so the dispersed particles are fixed more firmly.

According to above picture, the shear stress between two dispersed par-

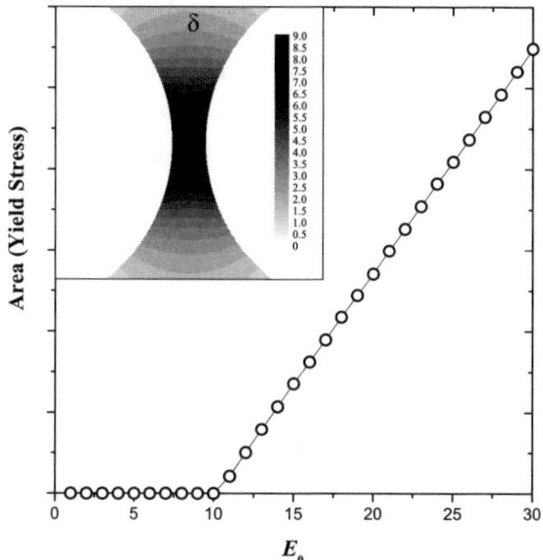

Fig. 3. Area of the cross section at the thinnest place of the pillar as a function of applied electric field. This value is proportional to the yield stress of the ER fuilds. Inset shows the electric field distribution in between the two shperes under unitary applied field strength.

ticles is proportional to the thickness (cross section) of the thinnest place of the pillar, which can be calculated from the electric field distribution under unitary applied field strength (inset in Fig. 3) because of the linearity between the local field and the applied field. The area of the cross section at the thinnest place of the pillar is plotted in Fig. 3 as a function of applied field. Figure. 3 exhibits the most typical characteristics of the yield stress observed in experiments, i.e., the yield stress is very weak when the applied field below a threshold and increase nearly linear when the applied field exceeds the threshold. We are satisfied with only giving a qualitative description for the yield stress induced by this mechanism. Since the superpolarized state occurs in extremely high field strength, there is no accurate data for its yield stress and also for its critical field at present. For this reason, the slope of the linear part of the yield stress and the threshold of the applied field in Fig. 3 can vary in a relatively broad range. However, a rough evaluation shows that the yield stress with several MPa and a critical field about 10^9 V/m for superpolarized state can reasonably explain the high yield stress occurring in the recent ER fluids.

4. Conclusions

We have used the multiple scattering method to calculate the electric field distribution for the ER fluids systems, and show that it is absolutely needed to include the higher order of polarization in the calculation for these systems. According to our calculation result, we propose a new mechanism of the electrorheological effect and show that the yield stress induced by this mechanism exhibits the typical characteristics of that observed in experiments.

Acknowledgments

This work is supported by the Key Item of Knowledge Innovation Project of Chinese Academy of Sciences (Grant No. KJCX2-YW-M07), the National Basic Research Program of China (Grant No. 2009CB930800) and National Natural Science Foundation of China (Grant Nos. 10674157 and 10875166).

References

1. W. J. Wen, X. X. Huang, S. H. Yang, K. Q. Lu and P. Sheng, *Nature Materials* **2**, 727 (2003).
2. J. E. Stangroom, *J. Stat. Phys.* **64**, 1059 (1991); *Phys. Technol.* **14**, 290 (1983).
3. K. Q. Lu, R. Shen, X. Z. Wang, G. Sun, W. J. Wen and J. X. Liu, *Chin. Phys.* **15**, 2476 (2006).
4. Z. Wang, Z. Peng, K. Lu and W. Wen, *Appl. Phys. Lett.* **82**, 1796 (2003).
5. L. C. Davis, *J. Appl. Phys.* **72**, 1334 (1992); *Appl. Phys. Lett.* **60**, 319 (1992).
6. B. J. Cox, N. Thamwattana and J. M. Hill, *Appl. Phys. Lett.* **88**, 152903 (2006).
7. F. Borghese, P. Denti, R. Saija, G. Toscano and O. I. Sindni, *Aerosol Sci. Technol.* **4**, 227 (1984).
8. J. Colchero, A. Storch, M, Luna, J. Gomez-Herrero and A. M. Baro, *Langmuir* **14**, 2230 (1998); H. Choe, M. H. Hong, Y. Seo, K. Lee, G. Kim, Y. Cho, J. Ihm and W. Jhe, *Phys. Rev. Lett.* **95**, 187801 (2005); T. Stifter, O. Marti and B. Bhushan, *Phys. Rev.* **B62**, 13667 (2000).
9. S. Gómez-Moñivas, J. J. Sáenz, M. Calleja and R. García, *Phys. Rev. Lett.* **91**, 056101 (2003).
10. T. Cramer, F. Zerbetto and R. García, *Langmuir* **24**, 6116 (2008).

MODELING APPROACH FOR THE PARTICLE BEHAVIOR IN MR FLUIDS BETWEEN MOVING SURFACES

D. GÜTH*, A. WIEHE and J. MAAS

Ostwestfalen-Lippe University of Applied Sciences,
Control Engineering and Mechatronic Systems,
Liebigstraße 87, 32657 Lemgo, Germany
**E-mail: dirk.gueth@hs-owl.de*
www.motion-ctrl.de

The influence of centrifugal accelerations on the particles in magnetorheological fluids due to high rotational speeds in brakes and clutches is often described with an undesirable increase of the off-state torque or an irreversible decomposition of the MRF when no magnetic field is applied. In this contribution this effect is studied by analyzing the flow fields caused by the rotary motion of the enclosed boundaries in radial and axial shear gaps from the fluid dynamics point of view. By an analytical approach for describing the particle behavior, changes in the particle distribution can be calculated. The approach is based on the simulation of several tracks of individual particles for different rotational speeds. The results of the simulation are compared with measurements performed on a test actuator for identifying a move of the particle concentration in radial shear gaps by measuring changes in magnetic flux density distribution.

Keywords: MRF; particle centrifugation; particle tracking; CFD.

1. Introduction

The use of magnetorheological fluids (MRF) in rotatory actuators is currently increasing more and more due to their advantages e.g. the smooth adjustable torque, the fast response time, the noiseless operation and the reduced design space of actuators compared to conventional dry friction based systems. Especially when running a brake or a clutch at high rotational speeds in idle mode without a magnetic field, the magnetorheological fluid is exposed to different forces caused by accelerating fields such as gravity, centrifugal forces and inertia force. Due to high differences in the density of the two MRF components carrier fluid and carbonyl iron powder (CIP) particles, these forces can cause a separation of the two-phase suspension, whereby the particles will concentrate in the direction of the resulting accelerations. When the centrifugal acceleration has the highest impact on the particles, these are carried to the outside of the shear gap. The effects of this particle centrifugation is often described in the literature[1] with an undesirable increase of the off-state torque or an irreversible decomposition of the MRF in the worst case. However, other investigations[3] mention that a conglomeration of particles at the outside area of shear gaps can be avoided due to a developing inward flow in the shear gaps. In this contribution the influence of high rotational speeds on the CIP in the MRF is modeled and analyzed based on the numerical solution for the fluid flow. A simulation of different tracks of the CIP is performed and superimposed for investigating and predicting a move in the concentration distribution. Finally, experiments on a special designed actuator are carried out for validating the results of the simulation.

2. Flow Field in Shear Gaps

Shear gaps in common MRF based actuators such as brakes or clutches are mainly distinguished between a radial (disk-shaped actuators) and an axial (cylinder-shaped actuators) oriented shear gap design. The decision for the design choice is often attributed to the particular application and the required design space. The basis for the subsequent modeling and simulation of the particle tracks are the flow profiles in the shear gaps for different rotational speeds respectively Reynolds numbers. These are now described from the fluid dynamics point of view for both cases of shear gaps.

2.1. *Disk-shaped actuators*

For analyzing the particle concentration, the flow profile in the shear gap has to be considered. In disk-shaped actuators these gaps are designed in radial orientation. The laminar steady state flow is based on the plane Couette flow between two rotating plates with a relative velocity ω and closed inner and outer surfaces. At higher Reynolds numbers Re the laminar azimuthal flow is superimposed by a radial flow due to centrifugal accelerations acting on the MR fluid. The fluid at the rotating boundary surface will be carried to the outside area developing a circulation flow because of continuity conditions in the bounded shear gap, see Fig. 1. The developing flow has already been studied for a rotating disk flow with the boundary conditions of a narrow gap size and an outside opened gap.[2] In section 3 it is investigated, how this circulation flow has an influence on the particle behavior concerning centrifugal effects by analyzing several particle tracks.

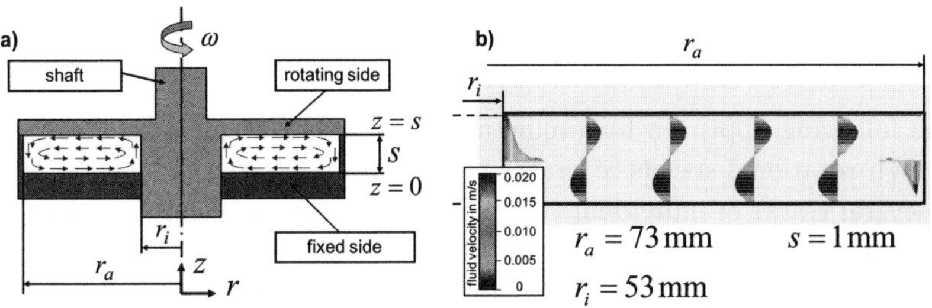

Fig. 1. Circulation flow in a radial shear gap shown: a) schematically for a typical actuator and b) as a numerical calculated flow field for MRF with given design parameters.

2.2. *Cylinder actuators*

Within the shear gap of cylinder-shaped actuators having a narrow gap size, a different flow profile compared to disk-shaped actuators is developed, called Taylor-Couette flow. For low angular speeds or low Reynolds numbers Re, the developed flow is a steady and purely azimuthal, circular Couette flow.[4] If the differential speed between inner and outer rotating cylinder is above a certain threshold, the Couette flow becomes unstable and a secondary steady state characterized by axisymmetric toroidal vortices emerges as depicted in Fig. 2. This behavior is also known as the Taylor vortex flow. Due to this kind of mixing effect, the developed flow profile can also cause an inherent effect for the homogeneity of the MR fluid, which is under the influence of centrifugal accelerations. To use this effect, it has to be considered that the faster rotating boundary is always placed in the inside.

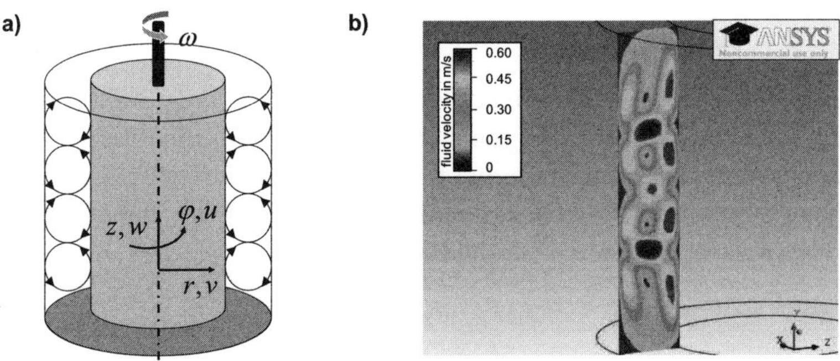

Fig. 2. Taylor vortex flow in an axial shear gap as a) a schematic representation and b) visualized for an exemplary numerical calculation.

3. Particle Tracking

3.1. *Fundamental approach*

The following approach for predicting the centrifugal effect in MR fluids at high rotational speeds of brakes or clutches is based on the calculation of several tracks of individual CIP particles. This way is chosen, because a calculation of the whole two phase MR fluid suspension, consisting of the carrier fluid and the CIP particles, is too complex, time-consuming or even impossible due to the high number of particles in the MR fluid. As a basis for the particle tracking calculation, the fluid flow profile $\vec{w}(u, v, w)$ depending on the flow velocity in azimuthal u, radial v and axial w direction is calculated with ANSYS CFX for a given actuator geometry (see Fig. 1b). The fluid should therefore be considered as a homogeneous fluid with properties of the suspension, while the containing single particles should been moved only by the interaction forces between fluid and particles. Due to the very high viscosity of the suspension of about $\eta = 0.5 Pa \cdot s$ and the very small particle size of about $d \approx 10 \mu m$ it is further assumed that other force actions like the particle-particle as well as the particle-fluid interaction can be neglected. Based on the modeling presented in the following subsection a numerically tractable model can be defined.

3.2. *Modeling*

For the calculation of a particle track all action of force on the particle must be considered. The forces responsible for the motion can be described by the balance of forces for the accelerated motion of a particle through a fluid with a constant velocity:

$$\vec{F}_G + \vec{F}_B + \vec{F}_I + \vec{F}_D = 0. \tag{1}$$

Therein $\vec{F}_G$ describes the weight, $\vec{F}_B$ the buoyancy force due to gravitational or centrifugal fields, $\vec{F}_I$ the inertia force and $\vec{F}_D$ the resistance force applied to the particle surface due to a relative motion between fluid and particle. The weight $\vec{F}_G$ considering the mass m_P of the particle and the gravitational acceleration $\vec{g}$ can be calculated by

$$\vec{F}_G = m_P \cdot \vec{g}. \tag{2}$$

The inertia force $\vec{F}_I$ describes the force on the particle mass m_P that acts in the opposite direction to the acceleration $d\vec{w}_P/dt$ in a straight line of a particle and the force due to a circular motion of a particle with the angular velocity ω_P and the radius position of the particle $\vec{r}$. It is expressed by:

$$\vec{F}_I = m_P \cdot \left(-\frac{d\vec{w}_P}{dt} + \vec{r} \cdot \omega_P^2\right). \tag{3}$$

For taking the pressure gradient into account, which is acting on a fluid and which is caused by acceleration fields, the static buoyancy force $\vec{F}_B$ with $grad\,(p) = \rho_f \cdot \vec{g}$ for a gravity field $\vec{g}$ and $grad\,(p) = \rho_f \cdot \vec{r} \cdot \omega_P^2$ for a centrifugal field can be described by

$$\vec{F}_B = -V_P \cdot grad\,(p), \tag{4}$$

where V_P is the volume of a particle and ρ_f the density of the fluid. The drag force $\vec{F}_D$ that considers pressure and surface forces caused by a relative velocity $\vec{v}_{rel} = \vec{w} - \vec{w}_P$ between fluid velocity $\vec{w}$ and particle velocity $\vec{w}_P$, as it is shown in Fig. 3a), can be calculated by

$$\vec{F}_D = c_d\,(\mathrm{Re}_d) \cdot A \cdot \frac{\rho_f}{2} \cdot |\vec{v}_{rel}| \cdot \vec{v}_{rel}, \tag{5}$$

with the drag coefficient c_d that is depending on the particle Reynolds number Re_d. To determine Re_d for a CIP particle the geometrical shape will be assumed as spherical. The introduced forces in (1)-(5) are considered in the three directions azimuthal, radial and axial, as it is depicted in Fig. 3b). For simulating the particle tracks, MATLAB/Simulink has been used. Three different characteristic rotational speeds at $\omega_1 = 50 rad^{-1}(\approx 500 rpm)$, $\omega_2 = 100 rad^{-1}(\approx 1000 rpm)$ and $\omega_3 = 300 rad^{-1}(\approx 3000 rpm)$ are analyzed and compared with the experimental results in section 4. The calculated particle track for 7 particles are shown as superimposed trajectories in the r-z-plane in Fig. 3c for ω_1. Beside the azimuthal movement of the particle it is shown that the circulating fluid flow is capable to move the particles in spite of the influence of centrifugal accelerations on a rotary track in the r-z-plane of the shear gap.

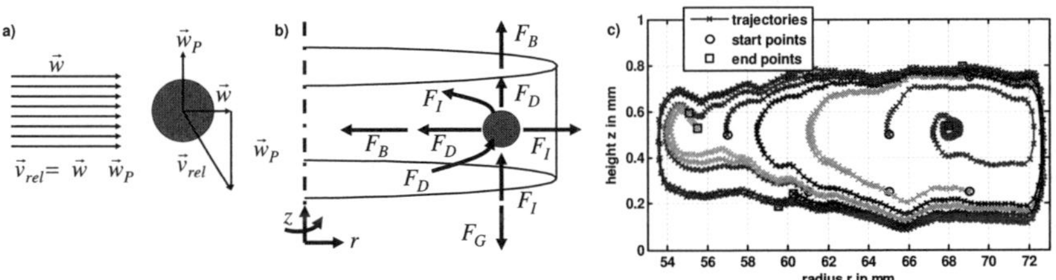

Fig. 3. a) Definition of the relative velocity $\vec{v}_{rel}$, b) forces acting on a single particle and c) exemplary trajectories of different calculated particle tracks for $\omega_1 = 500 rad/s$.

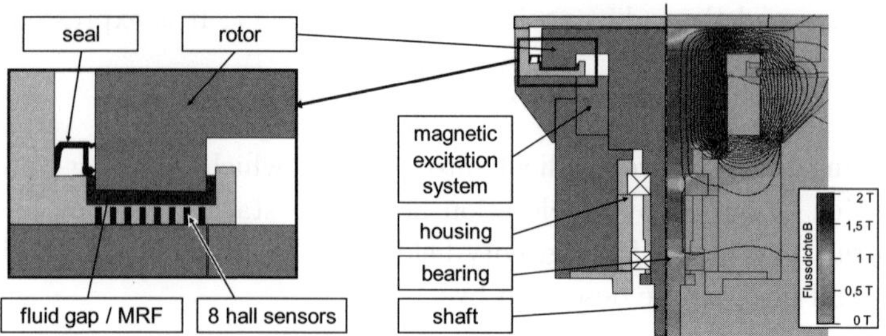

Fig. 4.   Test actuator with integrated sensors for measuring the magnetic flux density.

4. Experimental Setup

The achieved simulation results will be compared with measurements from a specially designed actuator for identifying shifts in the particle concentration in radial direction. The experimental method for the analysis is based on the measured changes of the magnetic flux density distribution. Therefore eight hall sensors are placed in radial direction as depicted in Fig. 4. A move of the CIP particles in radial direction will cause a change in the distribution of the magnetic flux density compared to a homogeneous distribution of the particles. The actuator is designed with the dimensions shown in Fig. 1.

5. Results

Based on the presented modeling approach, 32 trajectories have been calculated and superimposed. Therefore the dwell times of all particles has been analyzed as a function of the radius for a simulation time of $\Delta t = 10s$ shown in Fig. 5a). By the distribution for the three investigated rotational speeds ω_1, ω_2 and ω_3 a centrifugation can only be identified for ω_1 and with an decreasing trend for ω_2, resp. ω_3. Since the results for ω_3 show an equal distribution for the concentration, the circulation flow is sufficiently developed for carrying particles form the outside area back to the inside.

A comparable behavior can be identified by the measurement results. Therefore the magnetic flux density has been measured recently before and after a rotating cycle of $\Delta t = 10s$ at the considered rotational speeds ω_1, ω_2 and ω_3. The deviation of the flux density between these both measurements indicates a degree for the particle migration in radial direction. The used MRF had a concentration of about 15 Vol-%. With an increasing rotational speed the effect of centrifugation can be reduced. For the highest rotational

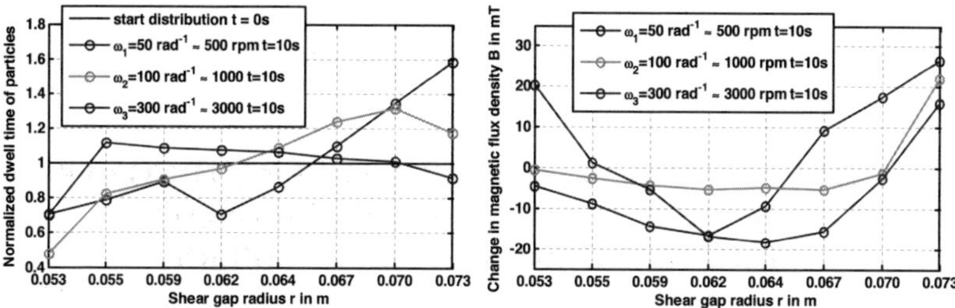

Fig. 5. a) Simulation results presented by the normalized dwell time of particles and b) measurement results with the deviation of the magnetic flux density after 10s for $\omega_1 = 50rad^{-1}$, $\omega_2 = 100rad^{-1}$ and $\omega_3 = 300rad^{-1}$.

speed ω_3, a migration of particles to the inside area was measured caused by the acting circulating fluid flow. The measurements, shown in Fig. 5b), confirm the occurrence of a partial homogenization in the particle concentration that validates the results of the particle track simulation.

6. Conclusion and Outlook

In this contribution the effect of particle centrifugation for radial shear gaps has been studied. By a modeling approach and performed measurements the centrifugation effect has been observed for lower rotational speeds. From a certain level an inherent fluid flow in radial direction is capable to reduce a particle centrifugation. In addition, the centrifugal effect should also be analyzed for axial shear gaps, too.

Acknowledgments

This contribution is accomplished within the project "HLD Fluidbrake", funded by the Federal Ministry of Education and Research (BMBF) of Germany under grant number 1713X09.

References

1. V. A. Neelakantan and G. N. Washington, *Modeling and Reduction of Centrifuging in Magnetorheological (MR) Transmission Clutches for Automotive Applications.*, J. of Int. Mat. Sys. Struct., Vol. 16 No. 9, pp. 703-711, 2005.
2. F. Schultz-Grunow, *Der Reibungswiderstand rotierender Scheiben in Gehäusen.*, ZAMM, Vol. 15, Issue 4, pp. 191, 1935.
3. D. Lampe,A. Thess and C. Dotzauer, *MRF- Clutch - Design Considerations and Performance.*, International Conference on New Actuators, P91, 1998.
4. V.N. Constantinescu, *Laminar Viscous Flow.*, Springer New York, 1995.

ADAPTIVE SLIDING MODE FAULT TOLERANT CONTROL FOR SEMI-ACTIVE SUSPENSION USING MAGNETORHEOLOGICAL DAMPERS[*]

XIAOMIN DONG[†] and ZHI GUAN

State Key Laboratory of Mechanical Transmission, Chongqing University, Chongqing, 400044, China, xmdong@cqu.edu.cn

MIAO YU

College of Opto-Electronic Engineering, Key Lab of Opto-Electronic Technology and System of Education Ministry, Chongqing University, Chongqing, 400044, China

This study presents adaptive sliding mode fault tolerant control for magneto-rheological (MR) suspension system considering the partial fault of MR dampers. After formulating the full car dynamic model featuring four MR dampers, the fault model of the MR damper due to the varying working temperature was derived. An adaptive sliding model fault tolerant control strategy was then proposed after the occurrence of a fault. Its performance was evaluated and compared under the bump road condition and presented in time domain. The results show that significant gains are made in the presence of partial fault of the MR damper. The control scheme could reduce the effect of the partial fault of the MR damper on the system performance.

1. Introduction

Compared to other actuator, the magnetorheological (MR) damper usually possesses higher reliability and even fail-safe ability [1]. In some sever cases such as outage, the MR damper can also provide basic damping force like a passive damper. However, some reasons will result in the occurrence of the partial fault of the MR damper during its long life period. These reasons include long term separation stability due to significant density mismatch between iron particle density and carrier fluid and temperature changes due to the heating associated with dissipation[2] [3]. Therefore, it is necessary to consider the effect of the partial fault in MR damper while formulating control strategy for a MR suspension system.

---

[*] **Foundation item:** Projects (60804018, 50830202) supported by the National Natural Science Foundation of China; Project (200902292) supported by the Special Post- doctoral Fund of People's Republic of China; Project (20090191110011) supported by Doctoral Fund of Ministry of Education of China.

[†] Corresponding author.

Consequently, the goal of this investigation is to insure, in the presence of a fault, the highest possible performance of the controlled system such as ride comfort and stability. To address these problems, a full vehicle suspension model equipped with four MR dampers considering partial failures of MR dampers is firstly constructed. The partial failure model of MR damper is then formulated. After estimating fault parameter of the partial failure model of MR damper, an adaptive sliding mode fault tolerant control algorithm is proposed to achieve satisfactory vibration suppression in the presence of the faults. Finally, its performance is evaluated and compared through the numerical simulation.

2. MR Suspension Model

The full vehicle dynamic model consists of vehicle body (spring mass) connected by the MR suspension system to four wheels (unsprung mass) [5]. The MR semi-active suspension system is 14 states nonlinear model; it can be written as a state space representation as follows:

$$\dot{x} = Ax + Bu + L\dot{w} \tag{1}$$

where $x \in R^{14}$ is the vector of state variables; $u=[-c_{fl}(\dot{z}_{fl}-\dot{z}_{ufl})+F_{dfl} -c_{fr}(\dot{z}_{fr}-\dot{z}_{ufr})+ F_{dfr} -c_{rl}(\dot{z}_{rl}-\dot{z}_{url})+F_{drl} -c_{rr}(\dot{z}_{rr}-\dot{z}_{urr})+F_{drr}]^{T}$ is the control damping force vector of four MR dampers and determined by the below discussed control strategies; A , B , L are constant matrices with appropriate dimensions, respectively.

3. Modeling of MR Actuator Partial Faults

Compared to other actuator faults caused by gravitational stability and agglomerative stability, those partial MR damper faults are mainly caused by the temperature variability for the vehicle suspension during working. This effect of temperature on MR damper performance could be particularly significant in continuously excited system such as a vehicle suspension [6].

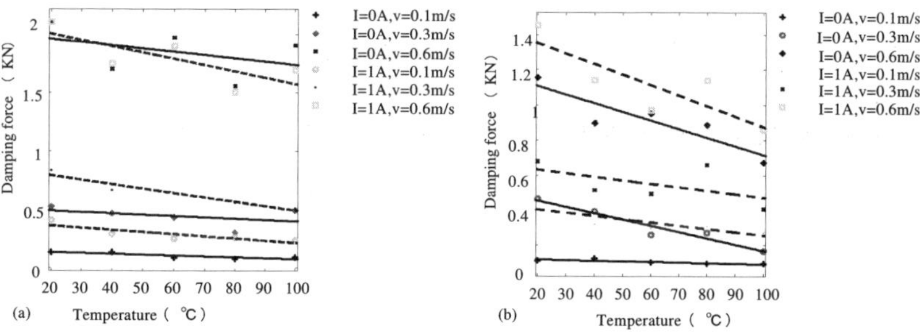

Figure 1 MR damper temperature test results[6]

466

In our prior work, the experiment of the temperature effect on a MR damper was investigated on MTS; the experiment was performed in an environmental temperature of 20℃. The tests are conducted under five temperature points from 20℃ to 100℃ with an increment interval of 20℃. The experimental results are shown in Figure 1, where Figure 1(a) is the status of compression stroke, and Figure 1(b) is that of rebound stroke.

From this figure, it can be found that the mean damping force will reduce with the increment of temperature; the effect of temperature on compression stroke is bigger than that on rebound stroke. Naturally, it is necessary to consider the temperature compensation while formulating a control scheme. However, it is difficult to apply the method of temperature compensation in the practical case. It is not easy to acquire the real time temperature in the inner space of a MR damper. Moreover, the temperature inside the MR damper is uneven and continuously changing while vehicle running. Therefore, it is feasible to look the damping force loss of MR damper as actuator partial fault and to formulate a fault tolerant control scheme.

To model the MR damper fault, four control effective factors γ^i ($i = fl, fr, rl, rr$, here fl, fr, rl and rr represent fore left suspension, fore right suspension, real left suspension and rear right suspension, respectively) are adopted to characterize the MR damper partial fault. $\gamma^i = 0$ denotes the ith healthy MR damper whereas $\gamma^i = 1$ corresponding to the total failure of the ith MR damper. Obviously, $0 < \gamma^i < 1$ represents partial loss in control effectiveness.

Considering the partial fault of MR dampers, the state space equation (1) can be written as

$$\dot{x} = Ax + Bu^f + L\dot{w} \tag{2}$$

where u^f denotes the control damping force vector with partial fault of four MR dampers and can be described as

$$u^f = \left[(1-\gamma^{fl})u_{fl},(1-\gamma^{fr})u_{fr},(1-\gamma^{rl})u_{rl},(1-\gamma^{rr})u_{rr}\right]^T \tag{3}$$

in which u_m ($m = fl, fr, rl, rr$) denotes the damping output of the healthy MR damper.

Considering the equation (4), the equation (3) can be written as follows

$$\dot{x} = Ax + B^f u + L\dot{w} \tag{4}$$

where $B^f = B(I - \Gamma_4)$, $\Gamma_4 = \begin{bmatrix} \gamma_{fl} & 0 & 0 & 0 \\ 0 & \gamma_{fr} & 0 & 0 \\ 0 & 0 & \gamma_{rl} & 0 \\ 0 & 0 & 0 & \gamma_{rr} \end{bmatrix}$.

To estimate the control effectiveness factors of MR damper, the equation (2) can also be written as follows

$$\dot{x} = Ax + Bu - \begin{bmatrix} b_1\gamma^{fl} & b_2\gamma^{fr} & b_3\gamma^{rl} & b_4\gamma^{rr} \end{bmatrix} \begin{bmatrix} u_{fl} \\ u_{fr} \\ u_{rl} \\ u_{rr} \end{bmatrix} + L\dot{w} \tag{5}$$

Equation (6) can further be written as

$$\dot{x} = Ax + Bu + D(u)\gamma + L\dot{w} \tag{6}$$

in which $D(u)$ is defined by $D(u) = BU$ and

$$U = \begin{bmatrix} -u_{fl} & 0 & 0 & 0 \\ 0 & -u_{fr} & 0 & 0 \\ 0 & 0 & -u_{rl} & 0 \\ 0 & 0 & 0 & -u_{rrl} \end{bmatrix}, \gamma = \begin{bmatrix} \gamma_{fl} \\ \gamma_{fr} \\ \gamma_{rl} \\ \gamma_{rr} \end{bmatrix}.$$

Due to difficulty in estimation of the real loss of control effectiveness γ for the MR damper, the control effective factors can be modeled as a random bias vector with a zero-mean white Gaussian noise sequence.

4. Fault Diagnosis

To derive controller reconfiguration in the presence of MR damper failures, it is necessary to estimate the control input matrix B^f in Equation (4). The matrix can be calculated through estimating the control effectiveness factors $\hat{\gamma} = \begin{bmatrix} \hat{\gamma}_{fl} & \hat{\gamma}_{fr} & \hat{\gamma}_{rl} & \hat{\gamma}_{rr} \end{bmatrix}^T$ by

$$\hat{B}^f = B(I - \hat{\Gamma}_4) \tag{7}$$

in which $\hat{\Gamma}_4 = \begin{bmatrix} \bar{\hat{\gamma}}_{fl} & 0 & 0 & 0 \\ 0 & \bar{\hat{\gamma}}_{fr} & 0 & 0 \\ 0 & 0 & \bar{\hat{\gamma}}_{rl} & 0 \\ 0 & 0 & 0 & \bar{\hat{\gamma}}_{rr} \end{bmatrix}$, $\bar{\hat{\gamma}}_m$ can be calculated through a two-stage adaptive kalman filtering algorithm [7].

5. Design of Reconfigurable Controller

Once the control effectiveness factors of MR dampers are estimated, an updating sliding mode control action is proposed in the following. The sliding mode control is adopted to control vehicle's heave, pitch and roll motions. The main advantage of the sliding mode control is its inherent insensitivity to variation in system parameters, external disturbances and modeling errors. In order to make all states of the suspension system in (2) to track the given desired trajectories at the same time, the sliding mode hyperplane is defined as follows,

$$s = e(t) + \Lambda \int e(t)dt \tag{8}$$

in which s is the sliding surface vector, Λ denotes a diagonal matrix which defines the slopes of sliding surfaces, $e(t)$ represents the state error vector and defined as $e(t) = x - x_d$.

   To guarantee the stability of system, a candidate Lyapunov function is given in terms of sliding hyperplane,

$$V(s) = \frac{1}{2}\frac{d}{dt}s^2 \le 0 \, , s \dot{s} \le 0 \tag{9}$$

First derivation of sliding surface function becomes,

$$\dot{s} = \dot{e} + \Lambda e = 0 \tag{10}$$

Considering $e(t) = x - x_d$, then

$$\dot{s} = Ax + B^f u + Lw - \dot{x}_d + \Lambda e \tag{11}$$

hence, the equivalent control term is written as

$$u_{eq} = B^{f^{-1}}[-Ax + \dot{x}_d - \Lambda e] \tag{12}$$

A corrective control term is adopted to meet the sliding condition for the controller. As a result, the overall controller with corrective control term will be derived as,

$$u = u_{eq} - B^{f^{-1}}[Ksat(\frac{s}{\Phi})] \tag{13}$$

where K denotes the corrective gain vector used for nominal case to guarantee a sliding regime on the switching surface vector s. $sat(\frac{s}{\Phi})$ is a saturation function.

6. Numerical Results

The effectiveness of the proposed scheme is demonstrated in this section through a vehicle with four MR dampers traveling along a bump road using MATLAB/SIMULINK. The system parameters of full vehicle can be referred in

[5]. The bump road is shown in Figure 2. The vehicle velocity is 20km/h. To simulate the partial fault of MR dampers due to variable temperature, the simulated scenario is the 50% reduction of the control effectiveness in the front right MR damper at 0.55s. The control performances of the MR suspension system under the bump input with/without reconfiguration control scheme are investigated. The results are shown in Figure 3-5.

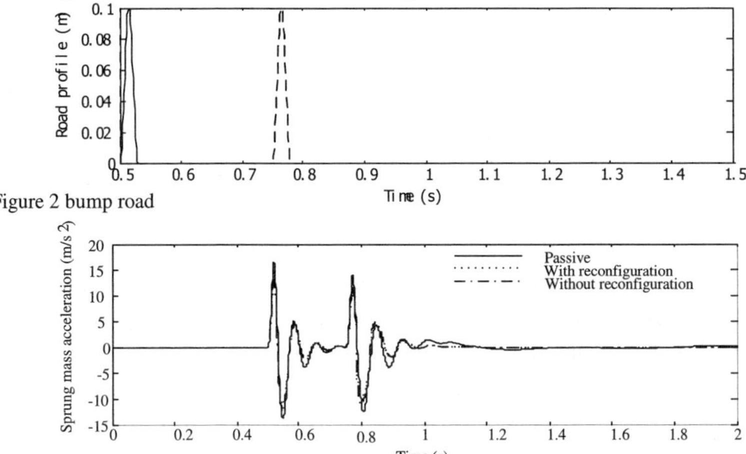

Figure 2 bump road

Figure 3 Response of sprung mass acceleration with partial fault of front right MR damper

From the control response in Figure 3-4, it can be found that the semi-active suspension can reduce the peak-peak values of acceleration, improve ride comfort and compensate the effect of temperature on the MR damper. The MR semi-active suspension system with reconfiguration can achieve better ride comfort improvement than that without reconfiguration. Moreover, the MR semi-active suspension system with reconfiguration has shorter adjusting time. It is also required to notice that dissimilar to active suspension, the partial fault in the MR damper can not cause the loss of system stability. The output damping force of front left MR damper has become smaller after the partial fault emerging.

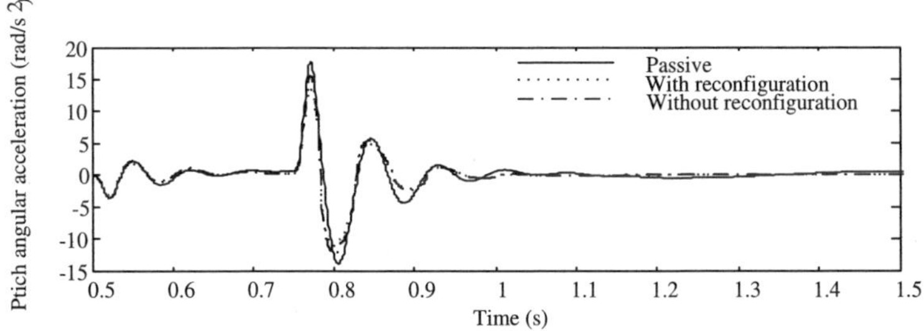

Figure 4 The response of pitch angular acceleration with partial fault of front right MR damper

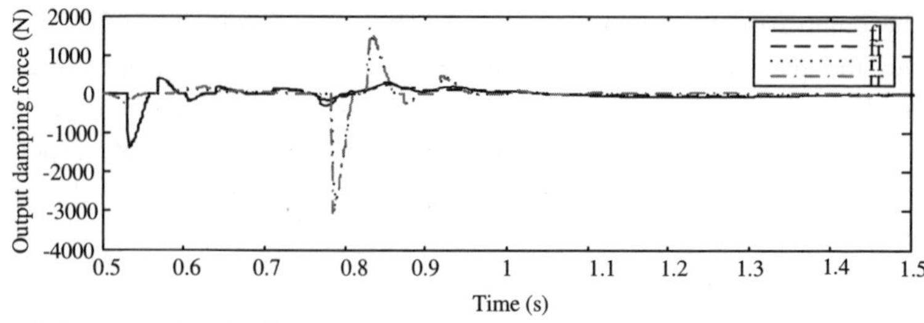

Figure 5 The output damping forces of four MR dampers

7. Conclusion

In this study, the integrated fault detection, diagnosis and reconfiguration control system design method has been applied to a MR semi-active suspension system. The partial fault of MR damper due to the variable temperature is investigated. A two-stage adaptive Kalman filter is adopted to estimate the control effectiveness factor of the MR damper. A sliding model fault tolerant control strategy is proposed after the occurrence of a fault. Simulation results show that the MR semi-active suspension system can guarantee the control performance during the partial fault of the MR damper emerging. Further work will consider more general MR damper faults in MR semi-active suspension systems including gravitational stability and agglomerative stability of fault.

References

1. G. D. Breese and F. Gordaninejad, *Int. J. Vehicle De.* **33**, 1-3(2003).
2. F. Gordaninejad and G. D. Breese, *J. Intel. Mat, Syst. Str.* **10**, 8(1999).
3. D. Batterbee, N.D. Sims, *J. Intel. Mat, Syst. Str.* **20**, 3(2009).
4. M. Yu, C.R. Liao, W.M. Chen and S.L. Huang, *J. Intel. Mat, Syst. Str.*, **17**, 8-9(2006).
5. X.M. Dong, M. Yu, Z.S. Li, C.R. Liao and W.M. Chen, *Smart Mater Struct*, 19(2009).
6. H.H. Zhang, C.R. Liao, M. Yu and W.M. Chen, *ANCRiSST Short Course on Smart Wireless Sensor Technology & Applications* (2002).
7. Y.M. Zhang and J. Jiang, *IEE Proceedings, Part D*, 149(2002).

IRREVERSIBLE EFFECTS IN MAGNETORHEOLOGICAL FLUIDS

S GORODKIN, R JAMES and W KORDONSKI

QED Technologies International, 1040 University Avenue, Rochester, NY 14607 USA

The effect of residual structure on properties of magnetorheological (MR) fluids, based on carbonyl iron (CI) particles, was investigated using a magneto-sweep technique and initial magnetic susceptibility measurements. The shear rate, particle size and the CI surface properties were among the experimental variables. A mechanical hysteresis, associated with the irreversible residual structure formation, was observed. The hysteresis was most pronounced at the creeping flow as compared to high shear conditions. Regardless of experimental settings, the level of the hysteresis was the same at high shear rate, while the fluids containing larger particles and higher CI concentrations demonstrated stronger residual effect at slow flow. The hysteresis loop was slightly wider for the silica treated particles suggesting that chemical (non-magnetic) interaction might contribute to the fluid irreversible behavior. Existence of some residual structure has been also shown in measurements of initial magnetic susceptibility of MR fluids. After exposure to a sequence of DC magnetic fields, fluids exhibited higher susceptibility presumably caused by the formation of residual structures of larger size than those in the initial state.

1. Introduction

An MR fluid is considered as a reversible structure material controlled by means of magnetic field [1-3]. However, in some instances, the reversibility of fluid properties, or restoration of the original state after relaxation or removal of the force field, might be incomplete due to existence of "residual" structure. The term "residual" here means a difference in MR fluid structure status between pre- and post-control conditions. This difference is assumed to originate from persistent particle agglomeration affected by particle properties including surface treatment. Similarly, irreversible structural effects were observed in experiments on permeability measurements with iron powder by J. de Vicente *et al.* [4].

In this paper some residual effects are demonstrated through appropriate rheological and magnetic measurements. Knowledge of fluid irreversible features can be important in prediction of the MR device performance especially for applications with cyclic control.

2. Problem Configuration

In an MR fluid under shear flow, the applied magnetic field causes individual iron particles to form extended structures. When brought into close contact by magnetic forces, the particles may also interact through non-magnetic attraction by direct adhesion [5] or chemically via adsorbed layers. After the magnetic field is removed, the shear forces pull particles apart and tend to destroy structure; however, non-magnetic interactions (or some magnetic interaction due to residual magnetization) may remain in effect, and these may prevent particles from complete dispersion. Thus, the residual structure of particle agglomerates persists in the absence of a magnetic field, and the agglomerate size depends on balance between shearing and attractive forces.

The objective of current study was to investigate experimentally the presence of residual structure and evaluate its influence on MR fluid behavior. Since rheological and magnetic properties of MR fluids are highly sensitive to the morphology of the structure [1, 6], we used both rheological and magnetic property responses to analyze residual structure effects for the fluid samples under cyclical field control conditions.

3. Experimental

3.1. *Fluid sample compositions*

The experimental MR fluids were composed using magnetically-soft carbonyl iron (CI) particles of different types and mean sizes obtained from BASF: namely HQ (1.1 um), EQ and EW (3.5 um), and CL (6 um). The EW particles were treated with silica (or "silicated") at the amount of 0.7 % w/w according to manufacturers' product specification, while the others did not have added silica. Silicone oil (the DMS-T05 oil from Gelest Inc., with addition of oleic acid from Aldrich) or water with appropriate stabilizing additives [7] were used as a carrier media.

3.2. *Magnetorheological measurements*

A commercial Anton Paar magnetorheometer, Physica MCR 301 with the plate-plate PP20/MRD measuring system and 1 mm working gap, was used in the study. A special anti-slip coating was applied to working surfaces of both plates to provide non-slip boundary conditions. An additional custom-made magnetic ring was placed to lift the removable magnetic bridge and account for coating thickness. The magnetic flux density in the working gap was measured with a Hall probe (1X probe with a model 9640 gaussmeter made by F.W. Bell) placed into the cavity below the stationary plate. Only oil based fluid samples were used for rheological experiments.

A magneto-sweep loop technique was employed to observe residual structure effects. The fluid sample was sheared at fixed shear rate, and the magnetizing current was varied gradually from zero to a maximum (5 A) and back to zero. The magnetic flux density changed accordingly from a residual 0.036 T to a maximum value of 0.475 T as shown in the Figure 1 below; and the corresponding shear stress was recorded. Note that the "instrument demagnetization procedure" was not applied in order to avoid the influence of hysteresis associated with the instrument's magnetic circuit on rheological measurements. Initially in the magneto-sweep loop method the fluid was pre-sheared for 5 sec before the current was applied. The current was then increased up to the maximum in 30 intervals over 60 sec, kept for 5 sec at the maximum value and then reduced back to zero with the same regime.

The value of mechanical hysteresis is defined as

$$HY_B = \left| \int_{B_{min}}^{B_{max}} \tau(B)dB - \int_{B_{max}}^{B_{min}} \tau(B)dB \right|_{\dot\gamma=const} \tag{1}$$

HY_B (in Pa*T units) is the space contoured by the upward and downward legs of the $\tau = f(B)$ function, and this was used as a response factor. Here B is magnetic flux density, τ is measured shear stress, and $\dot\gamma$ is shear rate. The hysteresis loop was assumed to originate from the presence of residual structure in the fluid. In practice, the raw shear stress data, gathered by magnetorheometer software, were transferred to a spreadsheet program for computation of the magnitude of the hysteresis, HY_B.

The shear rate, CI particle size and the CI surface treatment (i.e. "silicated" vs. "non-silicated") were among the experimental variables.

3.3. *Magnetic measurements*

The initial magnetic susceptibility was chosen as a parameter to characterize the magnetic properties of the fluid samples. Both aqueous and oil based fluid samples were used in these experiments. The susceptibility measurements for these fluids were performed using the Bartington MS2B system [6] with no applied field and after application of a series of low strength DC magnetic fields. A small volume cell was made for the instrument to enable the measurements. The cell was in the form of a coil of copper magnet wire wound onto a Delrin polymer "bobbin-like" core, with a small, central (0.27 cc) cylindrical cavity to be filled with a sample of MR fluid. The whole cell with coil fitted neatly into the sample chamber of the MS2B apparatus. This coil could be energized with up to ~ 0.9 A current from a power supply. A current of 0.8 A corresponded to a field ~ 0.035 T just outside the sample in the coil.

The susceptibility data were taken with the field "OFF", starting with zero field and after ~ 5 sec duration of applied magnetic field. The incrementally

474

increasing field was generated by application of increasing dc current through the solenoid coil.

In order to report the data as initial specific magnetic susceptibility (χ, SI units), we use the procedure presented in [6] and measured the mass of the sample, the percent moisture to estimate the specific gravity of the fluid sample and obtain the CI volume fraction and fluid sample volume.

4. Results and Discussion

4.1. *Mechanical hysteresis*

The magnetic field flux density range used in the experiments with oil based fluids is plotted in the Figure 1 as a function of applied magnetizing current. The measurements were taken in separate "calibrating" test without fluids in the gap by switching on a fixed current and reading the corresponding flux density from the gaussmeter. The flux density varies with a nearly linear dependence on the applied current.

The typical result as measured with the magnetorheometer for three different shear rates is shown in the Figure 2. The fluid developed lower shear stresses on the upward portion of magneto-sweep cycle compared to downward; thus, the hysteresis loop is formed. The area enclosed by the loop was significantly greater for the creeping flow of $\dot\gamma = 10^{-3}$ s^{-1}, but this area shrank at higher shear rates. The computed HY$_B$ values given in the chart emphasize that there may be an order of magnitude difference depending on the shear rate level.

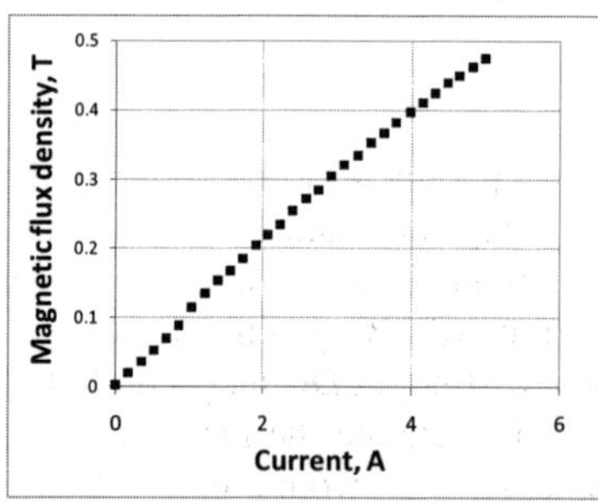

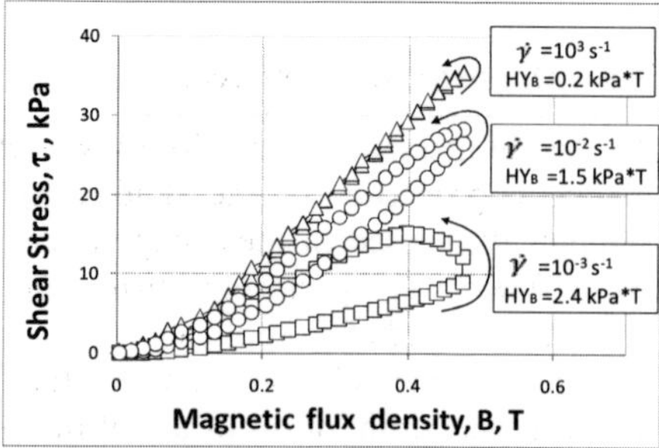

Figure 1. Magnetic flux density acting on the fluid samples as a function of applied current.

Figure 2. Typical hysteresis loops at different shear rates (EW CI 40 % v/v sample): squares-$\dot\gamma=10^{-3}$ s^{-1}; circles - $\dot\gamma=10^{-2}$ s^{-1}; triangles - $\dot\gamma =10^{3}$ s^{-1}. Arrows point to the direction of data acquisition

We relate the existence of mechanical hysteresis with difference in CI particle structures before and after the application of a stronger magnetic field. Stronger magnetic fields drive particles closer to each other and may facilitate non-magnetic interactions. It is reasonable to assume that at the same applied conditions, flux density and shear rate, the elements of the structure are more rigid on the downward leg producing higher stress. The adhesion of CI particles, reported in [4], and chemical coupling through adsorbed layers are hypothesized as a source of such an interaction presuming that the magnetically soft CI particles, used in our experiments, do not exhibit remnant magnetization [4,8]. When the field is released, the non-magnetic attraction remains in effect allowing residual structure with denser or bigger elements to persist.

The hysteresis loop area is assumed to reflect the total cumulative effect of residual structure on the fluid flow. Obviously, the higher shear rate tended to destroy the structure making its elements smaller and resulting in tighter loop and lower HY_B values. However, even when the shear rate was as high as $10^3\,s^{-1}$ some residual structure remained, and the hysteresis did not completely vanish as shown in the Figures 3-5 below. Regardless of experimental fluid conditions, the level of the hysteresis was nearly the same at high shear rate, while the fluids containing larger particles (Figure 3) and higher CI concentrations (Figure 4) demonstrated stronger residual effect at slow flow.

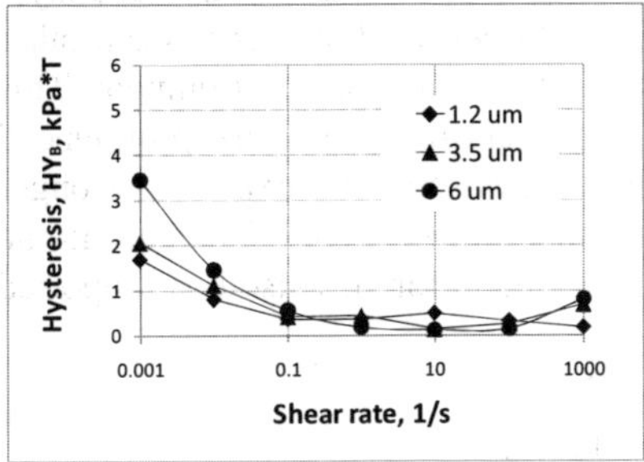

Figure 3. Hysteresis values vs. shear rate, impact of CI particle size (40 % v/v): diamonds - HQ, 1.2 um; triangles - EQ, 3.5 um; circles - CL, 6 um.

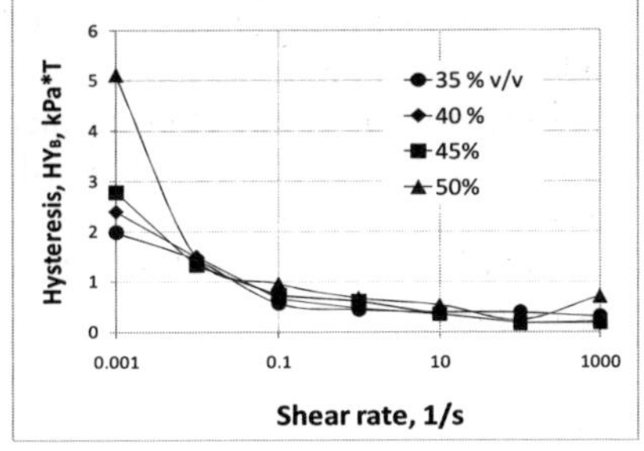

Figure 4. Hysteresis values vs. shear rate, impact of particles concentration (EW CI, 3.5 um size): circles - 35 % v/v; diamonds - 40 %; squares - 45 %; triangles - 50 %.

The area of the hysteresis loop was slightly more for the "silicated" particles compared to the non-treated particles of the same size and concentration as shown in the Figure 5 suggesting that chemical (non-magnetic) interaction might contribute to the fluid irreversible behavior.

The hysteresis was not seen if the measurement procedure was performed repeatedly for the same fluid sample that had been already subjected to magneto-sweep cycle; and the τ vs. B curve was coincident with downward curve of previous

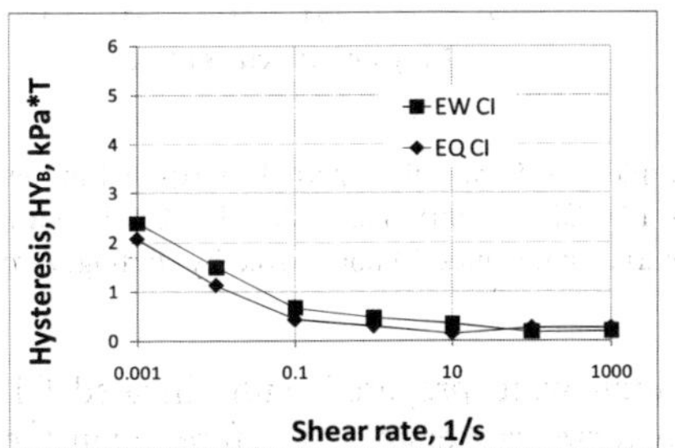

Figure 5. Hysteresis value vs. shear rate, impact of particles coating (40 % v/v, 3.5 um size): squares – silica coated EW CI; diamonds non-coated EQ CI.

hysteresis loop. After experiencing the magnetic field, persistent particle clusters seemed to be formed that then determined the subsequent rheological response.

4.2. *Magnetic properties*

The initial specific magnetic susceptibility χ for the four CI samples under investigation are shown by the intercept in Figure 6a and the values are correlated with particle size as reported by Gorodkin *et al* [5]. For all the MR fluids, prepared with these powders, the value of χ increases from the freshly mixed samples by about 10% for the aqueous fluids and by about 20% for the oil based fluids (see Figure 6b) as each 5 sec pulse in the magnetic flux is applied up to about 0.035 T. These results for both types of fluids are related to structure formation due to particle interactions that do not fully relax after the field pulse is "off". The oil, as it is suggested, develops a thinner adsorbed layer on the solid surfaces allowing the CI particles to approach closer to each other and form more dense agglomerates compared to the aqueous environment. This may explain the result that the increase in susceptibility is more pronounced with applied field for the oil based samples. Thus, one can expect a stronger irreversible effect with oil based fluids as compared to aqueous. The gain in the susceptibility due to particles' structure or agglomeration has also been reported earlier [4, 5].

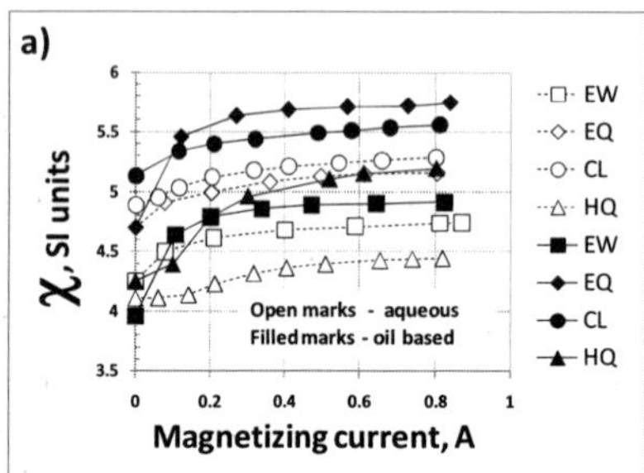

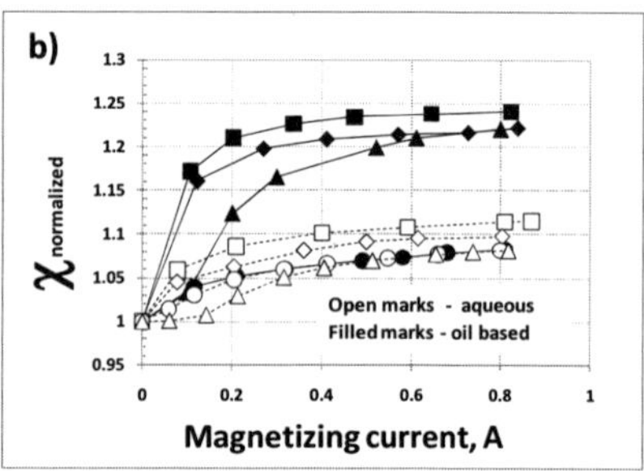

Figure 6. Initial magnetic susceptibility of fluid samples (40 % v/v of CI particles) exposed to low magnetic field of different flux densities: squares - EW CI (3.5 um); diamonds – EQ CI (3.5 um); circles – CL CI (6 um); triangles – HQ CI (1.2 um); a) actual values, b) normalized by starting, zero field, data points. Lines drawn to track data change.

The CI EW based fluid samples, which were prepared with silicated CI, appeared to exhibit a higher tendency to form irreversible structures than the non-silicated powders. Since EW and EQ products are based on the same iron particle except for the silica adhered to the CI core, the difference in χ between these samples must be due to the role of the silica incorporated in the EW powder. Perhaps hydrophobic silica, used by manufacturer for EW particle treatment, reduces hydration of particles in the magnetically induced structures, leading to somewhat "stickier" interactions and more MR hysteresis as demonstrated in the previous section.

5. Summary

In these experiments the rheological and magnetic properties of MR fluids under cyclic magnetic field excitation were studied. Some residual structure of iron particles was assumed to form under these conditions and to cause an irreversible flow effect, namely mechanical hysteresis, as well as a drift in the initial magnetic susceptibility. We found that:

- The hysteresis was most pronounced at the creeping flow of $\dot{\gamma} = 10^{-3}$ s^{-1} as compared to high shear conditions. At the same time, even when the shear rate was as high as 10^{3} s^{-1} the residual structure was not completely destroyed, and the hysteresis did not completely vanish.
- Regardless of experimental settings, the level of the hysteresis was the same at high shear rate, while the fluids containing larger particles and higher CI concentrations demonstrated stronger residual effect at slow flow.
- The hysteresis loop was slightly wider for the "silicated" particles suggesting that chemical (non-magnetic) interaction might contribute to the fluid irreversible behavior.

Also the existence of some residual structure has been shown in measurements of fluids initial magnetic susceptibility, which increases for both oil-based and aqueous MR fluids after exposure to the magnetic field.

Acknowledgments

We wish to acknowledge Mr. A. Sekeres of QED Technologies for his skillful support in making the solenoid test cell for measurement the effect of magnetic field pulses on the susceptibility of MR fluid samples and for modification of the Anton Paar magnetorheometer measuring system elements.

References

1. W. Kordonski *et al.*, *JMMM* **85**, 114 (1990).
2. X. Tang, X. Zhang and R. Tao, *J. Appl. Phys.* **87**, N5, 2634 (2000).
3. G. Bossis *et al.*, *JMMM* **252**, 224 (2002).
4. J. de Vicente *et al.*, *JMMM* **251**, 100 (2002).
5. L. Heim *et al.*, *J. Adhesion Sci. Technol.* **19**, 199 (2005).
6. S. Gorodkin, R. James and W. Kordonski, *J. Phys.: Conf. Series* **149**, 12051 (2009).
7. D. Golini *et al.*, *SPIE Conf. Optical Manuf. and Testing* **3782**, 80 (1999).
8. J. Arias *et al.*, *J. Colloid and Interface Sci.* **299**, 599 (2006).

NONLINEAR MODEL OF SQUEEZE FLOW OF MR FLUIDS USING PERTURBATION TECHNIQUES

ALIREZA FARJOUD[1,*], NIMA MAHMOODI[2] and MEHDI AHMADIAN[1]

[1]*Center for Vehicle Systems and Safety (CVeSS), Virginia Tech, Blacksburg, VA 24061, USA*
[2]*Department of Mechanical Engineering, the University of Alabama, Box 870276, AL 35487, USA*

This research is focused on mathematical modeling of MR fluids in squeeze mode. There is no universally accepted mathematical model of squeeze flow of MR fluids to date. In this research, the squeeze flow problem of MR fluids is solved using perturbation techniques and squeeze force, flow field, and plug flow region are determined.

1. Introduction

MR fluids are used in three distinct flow modes shown in Figure 1. In valve mode, MR fluid is flowing due to a pressure difference perpendicular to the magnetic field lines and the magnetic poles are stationary. In direct shear mode, one of the poles is moving with respect to other pole and MR fluid is being sheared. The least understood mode is squeeze mode, shown in Figure 1C. In squeeze mode, MR fluid is placed between two approaching plates and compressed while the magnetic field lines pass through the fluid. Recent research has shown that, in squeeze mode, MR fluids can provide a much larger range of controllable force in small operational envelopes compared to shear and valve modes [1, 2]. Devices using squeeze mode may take advantage of the very large range of adjustment that squeeze mode offers. Hence, MR devices utilizing squeeze mode can be stronger, smaller, and cheaper than currently available MR devices. Large structural vibration absorption systems, impact dampers, and engine mounts are examples of devices that can benefit from MR squeeze mode control.

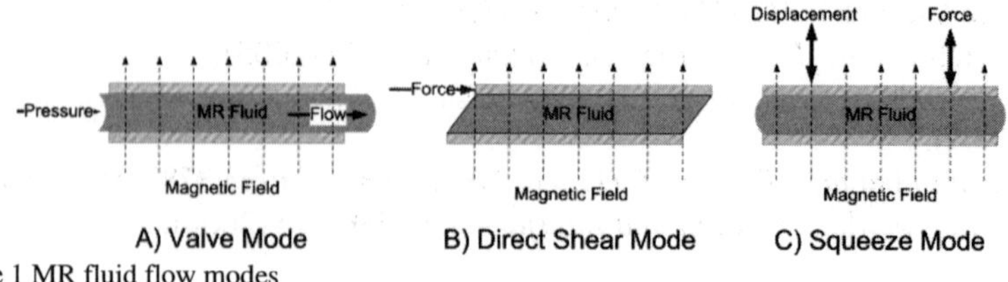

Figure 1 MR fluid flow modes

---

* Corresponding author, email: farjoud@vt.edu

In this paper, a novel and detailed mathematical model of the squeeze flow of MR fluids is presented using perturbation techniques, never done before. The model is capable of predicting flow field, shear rate distribution, pressure distribution, and squeeze force. The main advantage of this mathematical model is that the simulation results are obtained as semi-analytical equations which are suitable for design processes of MR squeeze devices.

2. Problem Statement and Assumptions

Consider an MR fluid contained between two parallel plates that are initially separated by a gap of height $2h_0$ (as shown in Figure 2). A normal force F is applied to the plates to cause the fluid in the gap to be squeezed out in the radial direction.

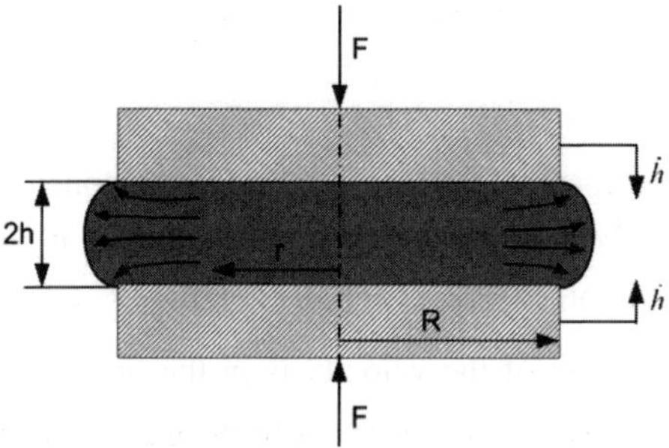

Figure 2 MR fluid contained between two parallel approaching circular plates

It is assumed that the quasi-static assumption can be used and the flow is axisymmetric. As a result, velocity and pressure are only functions of r- and z-coordinates, the θ-component of velocity is zero, and the plates' velocity, $\dot{h}$, is small. Additionally, it is assumed that inertial and gravitational effects are small and can be neglected. The velocity field (v_r, v_z), shear rate ($\dot{\Gamma}$) distribution, pressure (p) distribution, and squeeze force, F, are desired as functions of gap size.

The system of equations will be solved in non-dimensional forms in subsequent sections. The following parameters are defined to non-dimensionalize the equations.

$$V = \frac{-\dot{h}R}{h}, \quad \dot{\underline{\underline{\Gamma}}} = \frac{\dot{\underline{\underline{\gamma}}}h}{V}, \quad \underline{\underline{T}} = \frac{\underline{\underline{\tau}}}{\eta_0 \frac{V}{h}}, \quad \varepsilon = \frac{h}{R} \tag{1}$$

V is the characteristic velocity, $\underline{\underline{\tau}}$ is the stress tensor, $\dot{\underline{\underline{\gamma}}}$ is the rate of strain tensor, and $\dot{\underline{\underline{\Gamma}}}$ and $\underline{\underline{T}}$ are non-dimensional rate of strain and stress tensors, respectively. The aspect ratio of the gap is denoted by ε.

MR fluid is assumed to behave according to Papanastasiou's constitutive equation [3].

480

$$\underline{\underline{\tau}} = -\left[\eta_0 + \frac{\tau_y\left(1-e^{-n\dot{\gamma}}\right)}{\dot{\gamma}}\right]\underline{\underline{\dot{\gamma}}} \tag{2}$$

where, the sign convention of [4] is used. Off-state viscosity is shown as η_0, τ_y is yield stress, $\dot{\gamma}$ is shear rate, and n is the regularization parameter used in Papanastasiou's model. Using the parameters introduced in (1), Papanastasiou model can be written in non-dimensional form as

$$\underline{\underline{T}} = -\left[1 + \frac{Bn\left(1-e^{-N\dot{\Gamma}}\right)}{\dot{\Gamma}}\right]\underline{\underline{\dot{\Gamma}}} \tag{3}$$

where, $Bn = \tau_y h/(V\eta_0)$ is the Bingham number and $N = nV/h$ is the non-dimensional regularization parameter for Papanastasiou model.

3. System of Equations

Since the problem is axisymmetric, only r- and z-components of the equation of motion need to be solved. Because the plates' speed is $\dot{h}$, the z-component of velocity is in the order of $O(-\dot{h})$. Using the continuity equation, it can be easily seen that the r-component of the velocity is in the order of $O(-R\dot{h}/h)$. Hence, velocity components are expanded in terms of the gap aspect ratio (ε) as

$$v_r = V\left[\tilde{v}_{r0} + \varepsilon\tilde{v}_{r1} + \varepsilon^2\tilde{v}_{r2} + ...\right] \tag{4}$$

$$v_z = V\left[\varepsilon\tilde{v}_{z1} + \varepsilon^2\tilde{v}_{z2} + ...\right] \tag{5}$$

where, V is the characteristic velocity defined in (1) and the tilde sign denotes dimensionless quantities. Introducing the non-dimensional coordinates $\zeta = z/h$ and $\xi = r/R$, and using definitions in (1), the equations of motion and continuity can be written in non-dimensional form as

$$r-component: \quad 0 = -\varepsilon\frac{\partial P}{\partial \xi} - \left(\varepsilon\frac{1}{\xi}\frac{\partial}{\partial \xi}(\xi T_{rr}) - \varepsilon\frac{T_{rr}}{\xi} + \frac{T_{rz}}{\partial \zeta}\right) \tag{6}$$

$$z-component: \quad 0 = -\frac{\partial P}{\partial \zeta} - \left(\varepsilon\frac{1}{\xi}\frac{\partial}{\partial \xi}(\xi T_{rz}) + \frac{\partial T_{zz}}{\partial \zeta}\right) \tag{7}$$

$$continuity: \quad 0 = \frac{1}{\xi}\frac{\partial}{\partial \xi}\left(\xi\frac{v_r}{V}\right) + \frac{\partial}{\partial \zeta}\left(\frac{v_z}{V}\right) \tag{8}$$

$P = ph/(\eta_0 V)$ is the non-dimensional pressure. The relationship between stress components in equations (6) and (7), and velocity components is obtained from the constitutive equation (3). If gap ratio (ε) is small, shear rate can be expanded in terms of gap ratio,

$$\dot{\Gamma} = \frac{-\partial \tilde{v}_{r0}}{\partial \zeta} \left[1 + \frac{\partial \tilde{v}_{r1}/\partial \zeta}{\partial \tilde{v}_{r0}/\partial \zeta} \varepsilon + \frac{\left(\tilde{v}_{r0}/\xi\right)^2 + \left(\partial \tilde{v}_{r0}/\partial \xi\right)^2}{\left(\partial \tilde{v}_{r0}/\partial \zeta\right)^2} \varepsilon^2 + O\left(\varepsilon^3\right) \right] \tag{9}$$

Due to the quasi-static assumption, the shear rates encountered in the problem are small. This allows for expanding the viscosity in equation (3) in terms of non-dimensional shear rate, $\dot{\Gamma}$, which after combining with equation (9) results

$$\tilde{\eta}\left(\dot{\Gamma}\right) = A + B\varepsilon + C\varepsilon^2 \tag{10}$$

where,

$$A = 1 + BnN + \frac{1}{2}BnN^2 \frac{\partial \tilde{v}_{r0}}{\partial \zeta}, \quad B = \frac{1}{2}BnN^2 \frac{\partial \tilde{v}_{r1}}{\partial \zeta}, \quad C = \frac{1}{2}BnN^2 \frac{\left(\tilde{v}_{r0}/\xi\right)^2 + \left(\partial \tilde{v}_{r0}/\partial \xi\right)}{\partial \tilde{v}_{r0}/\partial \zeta} \tag{11}$$

The last variable that needs to be expanded is pressure. Performing a scaling analysis on the r-component of equation of motion reveals that the pressure should be expanded as

$$P = P_{-1}\varepsilon^{-1} + P_0 + P_1\varepsilon^1 + O\left(\varepsilon^2\right) \tag{12}$$

Inserting the expanded forms of velocities and pressure in equations (6) through (8) and equating the equal powers of ε, the following system of differential equations is obtained.

$$O\left(\varepsilon^{-1}\right): -\frac{\partial P_{-1}}{\partial \zeta} = 0 \tag{13}$$

$$O\left(\varepsilon^0\right): \begin{cases} -\dfrac{\partial P_0}{\partial \zeta} = 0 & (a) \\[2ex] -\dfrac{\partial P_{-1}}{\partial \xi} + \dfrac{\partial}{\partial \zeta}\left(A\dfrac{\partial \tilde{v}_{r0}}{\partial \zeta}\right) = 0 & (b) \\[2ex] \dfrac{1}{\xi}\dfrac{\partial}{\partial \xi}\left(\xi\tilde{v}_{r0}\right) + \dfrac{\partial}{\partial \zeta}\tilde{v}_{z1} = 0 & (c) \end{cases} \tag{14}$$

$$O\left(\varepsilon^1\right): \begin{cases} -\dfrac{\partial P_1}{\partial \zeta} + \dfrac{1}{\xi}\dfrac{\partial}{\partial \xi}\left(\xi A\dfrac{\partial \tilde{v}_{r0}}{\partial \zeta}\right) + 2\dfrac{\partial}{\partial \zeta}\left(A\dfrac{\partial \tilde{v}_{z1}}{\partial \zeta}\right) = 0 & (a) \\[2ex] -\dfrac{\partial P_0}{\partial \xi} + \dfrac{\partial}{\partial \zeta}\left(A\dfrac{\partial \tilde{v}_{r1}}{\partial \zeta} + B\dfrac{\partial \tilde{v}_{r0}}{\partial \zeta}\right) = 0 & (b) \\[2ex] \dfrac{1}{\xi}\dfrac{\partial}{\partial \xi}\left(\xi\tilde{v}_{r1}\right) + \dfrac{\partial \tilde{v}_{z2}}{\partial \zeta} = 0 & (c) \end{cases} \tag{15}$$

4. Boundary Conditions

Considering the assumptions mentioned earlier and assuming no slip at walls, the boundary conditions in terms of non-dimensional expanded variables are

$$\text{At } \zeta = 1: \tilde{v}_{z1} = -1, \ \tilde{v}_{zn} = 0 \ n > 1, \ \tilde{v}_{rn} = 0 \ n \geq 0$$

$$\text{At } \zeta = 0: \tilde{v}_{zn} = 0, \ \frac{\partial \tilde{v}_{rn}}{\partial \zeta} = 0 \ n \geq 0 \qquad (16)$$

$$\text{At } \zeta = 1 \ \& \ \xi = 1: P_{-1} = 0, \ P_0 = P_a = \left. P_a h \middle/ \eta_0 V \right., \ P_n + T_{rr,n} = 0 \ n \geq 1$$

$T_{rr,n}$ is the nth order term in the expansion of T_{rr}. The perturbation equations in (13) through (15) associated with boundary conditions in (16) now form a well posed system of differential equations and can be solved in a sequential manner.

5. Perturbation Solution

The system of differential equations in (13)-(15) are solved in a sequential manner to obtain the solutions for expanded velocity and pressure variables. These obtained solutions can then be combined together to result in velocity field, shear rate distribution, pressure distribution, and squeeze force. The r- and z-component of velocity are obtained as

$$\hat{v}_r = \frac{v_r}{V} = \tilde{v}_{r0} + \varepsilon \tilde{v}_{r1} = \frac{-3}{4} \xi \left(\zeta^2 - 1 \right), \quad \hat{v}_z = \frac{v_z}{V} = \varepsilon \tilde{v}_{z1} + \varepsilon^2 \tilde{v}_{z2} = \frac{1}{2} \left(\zeta^3 - 3\zeta \right) \varepsilon \qquad (17)$$

Shear rate is found as a function of coordinates and gap size as

$$\dot{\Gamma} = \frac{3}{2} \sqrt{ \xi^2 \zeta^2 + \left(\zeta^2 - 1 \right)^2 \varepsilon^2 } \qquad (18)$$

Total pressure is found to be

$$P = \frac{-3}{4} (1 + BnN) \left(\xi^2 - 1 \right) \varepsilon^{-1} + P_a + \left[\frac{3}{2} (1 + BnN) \left(\zeta^2 - 1 \right) + \frac{9}{4} BnN^2 \left(\frac{1}{2} \xi \zeta^3 - \xi \zeta - \frac{3}{2} \right) \right] \varepsilon \qquad (19)$$

And, total squeeze force is obtained as

$$\hat{F} = \frac{F}{\pi R^2 \eta_0 V / h} = 2 \int_0^1 \left(P - P_a + T_{zz} \right)_{\zeta=1} \xi d\xi = \frac{3}{8} (1 + BnN) \varepsilon^{-1} - \frac{33}{8} BnN^2 \varepsilon \qquad (20)$$

A plot of velocity field, shear rate distribution, and pressure distribution is shown in Figure 3.

6. Model Results Discussion

As can be seen in Figure 3A, the r-component of velocity field is of $O(\varepsilon^0)$ order and its z-component is of $O(\varepsilon^1)$. When the gap aspect ratio is large, these two components are of the same order and hence, the velocity field consists of both z- and r- direction velocities. However, when $\varepsilon \ll 1$, the z-direction velocity is very small compared to r-direction velocity and the total velocity consists of only r-direction velocity. Figure 3B shows contour plots of shear rate in the solution domain (upper right quadrant of the total flow domain in Figure 2). As is seen, at large gaps, the regions with low shear rates (plug flow) are around the stagnation points at the centers of the pole plates (around $\xi = 0$, $\zeta = 1$). As the gap size is decreased, the plug flow region changes in shape and size and moves toward the plane and axis of symmetry. Figure 3C shows how pressure distribution is changed as the gap size is decreased. At relatively large gap sizes, the $O(\varepsilon^1)$ term in equation (19) is significant and contributes to the total pressure in the fluid. Therefore, the total pressure is a function of both radial and axial coordinates. As the gap size is decreased, the $O(\varepsilon^{-1})$ rapidly increases in magnitude and becomes the dominant factor in total pressure. As seen in equation (19), this term only is a function of the radial coordinate. As a result, the total pressure becomes a function of ξ only. Hence, the solution reduces to the lubrication approximation and assumes the pressure is only a function of the radial coordinate for small gap sizes.

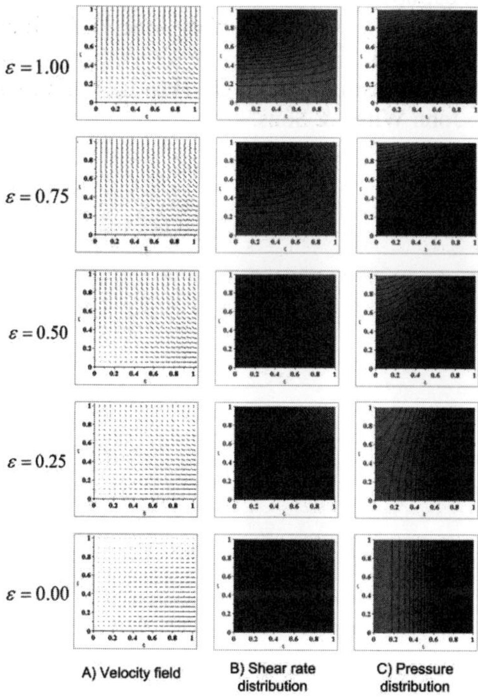

Figure 3 Model results for A) velocity field, B) shear rate, and C) pressure

484

7. Conclusion

This paper presented a novel mathematical solution for low shear rate squeezing flow of Magneto-Rheological fluids using perturbation techniques. The main focus of the model was to predict the flow behavior of MR fluids in controllable devices designed for squeeze mode. This mathematical model is capable of predicting velocity field, shear rate distribution, pressure distribution, total squeeze force, and other flow parameters. One of the advantages of this model is that the simulation results are obtained as closed-form equations. This will greatly help with design and synthesis of MR devices. Additionally, it eliminates the use of computationally expensive numerical algorithms.

References

1. Mazlan, S.A., N.B. Ekreem, and A.G. Olabi, *The Performance of Magnetorheological Fluid in Squeeze Mode.* Smart Materials and Structures, 2007. **16**(5): p. 1678-1682.
2. Farjoud, A., R. Cavey, M. Ahmadian, and M. Craft, *Magneto-Rheological Fluid Behavior in Squeeze Mode.* Journal of Smart Materials and Structures, 2009. **18**(9): p. 095001-9.
3. Papanastasiou, T.C., *Flows of Materials with Yield.* Journal of Rheology, 1987. **31**(5): p. 385-404.
4. Bird, R.B., R.C. Armstrong, and O. Hassager, *Dynamics of Polymeric Liquids.* 1987: John Wiley & Sons.

MODELING PERFLUORINATED POLYETHER BASED MR FLUIDS

K.H. GUDMUNDSSON

*School of Engineering and Natural Sciences, University of Iceland, Hjardarhaga 2-6,
Reykjavik, 107, Iceland*

F. JONSDOTTIR

*School of Engineering and Natural Sciences, University of Iceland, Hjardarhaga 2-6,
Reykjavik, 107, Iceland*

F. THORSTEINSSON

*Ossur Inc., Grjothalsi 6,
Reykjavik, 110, Iceland*

O. GUTFLEISCH

*Leibniz Institute for Solid State and Materials Research Dresden, PF 27 01 166,
Dresden, D-01171, Germany*

This study investigates the field-induced rheological characteristics of perfluorinated polyether (PFPE) based magnetorheological (MR) fluids. Four PFPE based MR fluid samples were prepared, each with a different solid loading. All samples contain a single grade of carbonyl iron powder with an average particle diameter of 2 μm. The PFPE base fluid enhances sedimentation stability and its composition and qualities are introduced in the paper. The magnetization characteristics of a fluid sample are measured and its magnetic properties are compared to known models from the literature. The shear yield stress in all fluid samples is investigated experimentally, as a function of solid loading and the magnetic flux density. The results are compared to established shear stress models for MR fluids. Model parameters are set to conform to the PFPE based MR fluids. The PFPE based MR fluids are shown to have a comparable shear yield stress, when compared to commercially available MR fluids with a comparable particle loading and particle size.

1. Introduction

Shear stress models for magnetorheological (MR) fluids are well established in the literature [1-8]. These models include both physical [1-5] and empirical models [6-8]. Based on the Maxwell's stress tensor, Ginder *et al.* [1-2] have adopted an approach to integrate the square of the magnetic field intensity to evaluate the shear stress. They employ the finite element method to estimate the variations in the magnetic field within a particle chain. Bossis *et al.* [3] have adopted a similar approach. They define the magnetostatic energy and obtain the shear yield stress as a derivative of the energy. Also, Jolly *et al.* [4] define an energy density function whose derivative, with respect to strain, equals stress. Genç and Phulé [7] confirm the work of Ginder *et al.* [2] and elaborate on this work with four empirical models.

Various empirical shear yield stress models exist in the literature [6-8]. The shear stress in established MR fluid compositions has been experimentally investigated [6]. The investigation is based on measurement data for four types of commercially available MR fluids. The measurements suggest that the shear stress, at a fixed shear-rate, can be determined by a simple model at pre-saturation levels. Furthermore, and based on experimentally investigated established MR fluid compositions, Carlson [8] presents an empirical equation relating the shear yield stress to the solid concentration and the strength of the magnetic field. This model can be used to determine the shear yield stress of MR fluids. Also, has a part of the same research, an empirical relation is presented that describes the B-H curve of MR fluids as a function of the solid loading.

The work presented in this paper applies the models from Ginder *et al.* [1,2] and from Carlson [8] to perfluorinated polyether (PFPE) [9] based MR fluids. The shear-yield stress dependency on solid loading and the strength of the magnetic field is investigated. The established models are adjusted to describe the properties of the PFPE based MR fluids. The aim is to be able to predict the characteristics of future PFPE based MR fluid compositions. The motivation for this study is an MR prosthetic knee [10,11] that employs the PFPE based MR fluid. It shares mutual goals in MR fluid design; sedimentation stability, high shear yield-stress and low off-state viscosity.

2. Perfluorinated Polyether Based MR Fluids

PFPE oil has been successfully introduced as a base fluid for MR fluids [9-11]. It is used in a commercial product, employing the MR fluid in an MR rotary brake that utilizes the fluid in shear mode. The MR rotary brake is used in an adaptive prosthetic knee [10,11].

2.1. *Magnetization Characteristics*

The magnetization characteristics of one PFPE based MR fluid sample were measured with a SQUID magnetometer. The magnetic moment was measured as a function of the applied field, at a temperature of 20°C. The sample has a solid concentration of 0.28, by volume, using a carbonyl iron powder, from BASF, with an average particle diameter of 2 μm. Particles have an iron content of 97%. Figure 2 shows the magnetization hysteresis loop for the MR suspension or the magnetic moment per unit volume. The magnetization is calculated using the sample's specific weight of 3.55 g/cm^3 and a sample size of 0.1014 g.

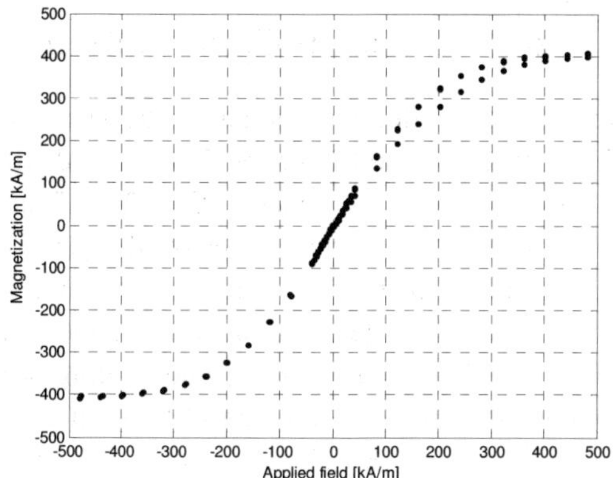

Figure 1. The magnetization hysteresis loop of a perfluorinated polyether (PFPE) based MR fluid sample with a solid loading of 0.28, by volume, and an average particle diameter of 2 microns. Particles have an iron content of 97%. Measurements are performed at a temperature of 20°C.

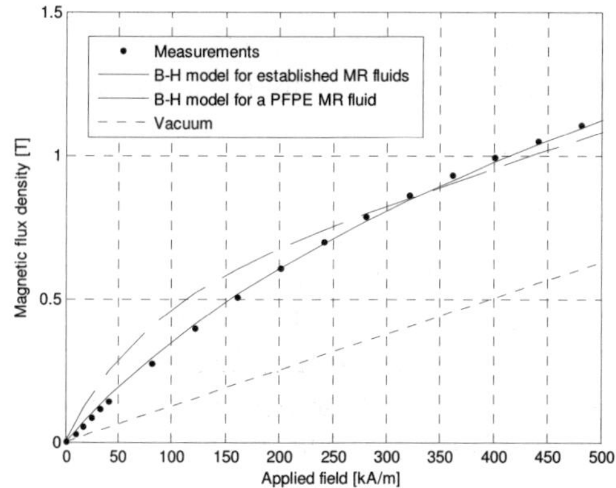

Figure 2. The B-H relation for a perfluorinated polyether (PFPE) based MR fluid with a solid concentration of 0.28, by volume, and an average particle diameter of 2 μm. Particles have an iron content of 97%. A B-H model for established MR fluids is also shown and a fine tune of that model for a FPEE based MR fluid composition.

The magnetic flux density, B, is related to the applied field, H, and the magnetization, M, by $B = \mu_0 (H + M)$. Thus, the B-H curve for a MR fluid suspension is shown in Figure 2.

The figure shows how a B-H model for established MR fluids, presented by Carlson [8], can easily be adopted to accurately describe the magnetization characteristics of the PFPE base MR fluid. The model is shown in Eq. (7).

$$B = C_1 \cdot \phi^{1.133}(1 - \exp(-C_2 \cdot \mu_0 \cdot H)) + \mu_0 \cdot H \qquad (1)$$

Carslon [8] presents this model with the parameter values $C_1 = 1.91$ T and $C_2 = 10.97$ T^{-1}. Fine tuning this model for the PFPE based MR fluid, using a least-square error fit, gives parameters values of $C_1 = 2.24$ T and $C_2 = 4.39$ T^{-1}. The induction at high applied fields can be set by the parameter C_1 and the parameter C_2 can be used to set the position the saturation point at 500 kA/m, for example. This results in an accurate relation between the induction and the applied field for PFPE based MR fluid with a solid loading of 0.28. Thus, Eq. (2), based on a model presented by Carslon [8], is an expression for the magnetization of the PFPE based MR fluid, M, as a function of solid loading and the applied field.

$$M = \frac{2.24}{\mu_0} \cdot \phi^{1.133}(1 - \exp(-4.39 \cdot \mu_0 \cdot H)) \qquad (2)$$

Eq. (1) and (2) are used to determine the magnetic characteristics of PFPE based MR fluid compositions with a solid concentration deviating from 0.28, by volume. Following the magnetic analysis, a rheological analysis is now performed.

2.2. Rheological Characteristics

The rheological characteristics of four PFPE based MR fluids samples, each with a different solid loading, were measured. Measurements are performed at shear-rates from 10^{-3} s^{-1} to 10^{3} s^{-1} for a magnetic field strength ranging from 20 kA/m to 200 kA/m. Figure 3 shows the shear-yield stress for the samples as a function of the applied magnetic field and solid loading.

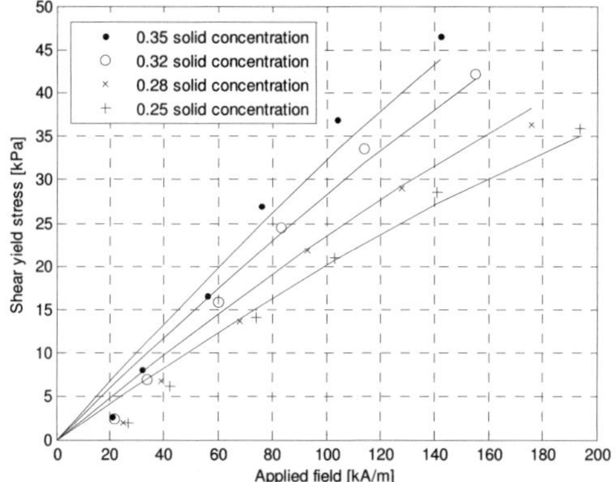

Figure 3. The shear-yield stress of perfluorinated polyether (PFPE) based MR fluid samples with a solid loading ranging from 0.25 to 0.35, by volume, and an average particle diameter of 2 μm. An empirically fitted model, based on Carlson [8], is also shown for each particle loading. Measurements are performed at a temperature of 20°C.

The figure shows also an empirically fitted model, based on Carlson [8], shown in Eq. (3).

$$\tau_y = C \cdot 271700 \cdot \phi^{C_3} Tanh(C_4 \cdot 10^{-6} \cdot H)$$
(3)

Carlson [8] suggests the C constant to be base fluid dependent having a value of $C = 0.95$, 1.0 and 1.16, depending on the base fluid. For established MR fluids, the parameters C_3 and C_4 have been found to have a value of $C_3 = 1.5239$ and $C_4 = 6.33$. This study suggests values of $C = 1.5$, $C_3 = 1.40$ and $C_4 = 3.60$ for the PFPE based MR fluids, where the squared error is used to quantify the difference between the model and the measurements.

Ginder *et al.* [1,2] suggest a shear yield stress model, below the saturation level. The model suggests a linear dependency on solid loading and a subquadratic field dependency, shown in Eq. (4),

$$\tau_{ys} = C_5 \cdot \varphi \cdot \mu_0 \cdot H^{C_6} M_s^{2-C_6}$$
(4)

where M_s is the saturation magnetization for the MR fluid. The value of M_s was measured for a MR fluid composition with solid concentration of 0.28. For other solid concentrations, Eq. (2) is used to determine the saturation magnetization.

Figure 4 shows the measured shear-yield stress, again, with the model from Ginder *et al.* [1,2] fitted with a least squared error fit.

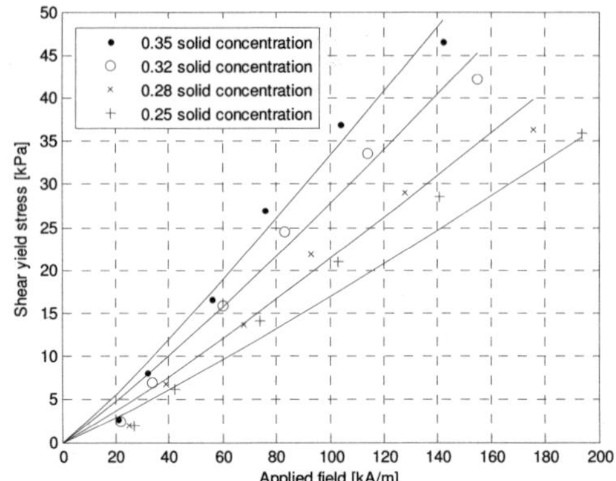

Figure 4. The shear-yield stress of perfluorinated polyether (PFPE) based MR fluids with a solid loading ranging from 0.25 to 0.35, by volume, and an average particle diameter of 2 microns. An empirically fitted model, based on Ginder *et al.* [1,2], is also shown. Measurements are performed at a temperature of 20°C.

The resulting model for the PFPE based MR fluids has a parameter value of $C_5 = 1.7$ and $C_6 = 1.1$ which compares to parameter values of $C_5 = 2.4$ and $C_6 = 0.5$ in the model from Ginder and co-workers [1,2].

The analysis has shown how a fine tune of established models for MR fluids can be used to accurately represent the characteristics of PFPE based MR fluids.

3. Conclusions

The PFPE based MR fluids exhibit field-induced characteristics that can be described with established models for MR fluids. Fine tuning the models has resulted in a more accurate description of the observed characteristics. The empirical models can potentially be used to predict the characteristics of future PFPE based MR fluid compositions.

Acknowledgments

This work is funded by the University of Iceland and Rannis, grant no. 090035021, and supported by Ossur Inc.

References

1. J.M. Ginder, and L.C. Davis, *Appl. Phys. Lett.* **65**, 3410 (1994).
2. J.M. Ginder, L.C. Davis and L.D. Elie, *Int. J. Mod. Phys. B*, **10**, 3293 (1996).
3. G. Bossis, E. Lemaire, O. Volkova and H. Clercx, *J. Rheol.* **41**, 687 (1997).

4. M.R. Jolly, J.D. Carlson and B.C. Muñoz, *Smart Mater. Struct.* **5**, 607 (1996).
5. G. Bossis, S. Lacis, A. Meunier and O. Volkova *J. Magn. Magn. Mater.*, **252**, 224 (2002).
6. M.R. Jolly, J.W. Bender and J.D. Carlson, *J. Intell. Mater. Syst. Struct.* **10**, 5 (1999).
7. S. Genç and P.P. Phulé, *Smart Mater. Struct.* **11**, 140 (2002).
8. J.D. Carlson *Int. J. Mod. Phys. B*, **19**, 1463 (2005).
9. H.H. Hsu, C.R. Bisbee III, M.L. Palmer, R.J. Lukasiewicz, M.W. Lindsay and S.W. Prince, *US Patent Spec.* 7,101,487 (2006).
10. F. Jonsdottir, E.T. Thorarinsson, H. Palsson and K.H Gudmundsson *J. Intell. Mater. Syst. Struct.* **20**, 659 (2009).
11. K.H. Gudmundsson, F. Jonsdottir and F. Thorsteinsson *Smart Mater. Struct.* **19**, 035023 (2010).

ANALYTICAL SOLUTION OF A PERISTALTIC MATERIAL TRANSPORT OF A MAGNETIZABLE FLUID

J. POPP[1*], V. BÖHM[1], V. A. NALETOWA[2], I. ZEIDIS[1] and K. ZIMMERMANN[1]

[1]*Department of Technical Mechanics, Ilmenau University of Technology,
PO Box 100565, D-98684 Ilmenau, Germany
*E-mail: jana.popp@tu-ilmenau.de
www.tu-ilmenau.de*

[2]*Department of Mechanics and Mathematics, Lomonosov Moscow State University,
Leninskiye gory, 119992 Moscow, Russia*

This article refers to the field of apedal ferrofluid based locomotion systems. An analytical solution shall be given on the matter of flow manipulation by an alternating magnetic field. The problem involves the causative magnetic field leads to the correspondingly deformed free ferrofluid surface contour and presents the evidence of an effective flow, meaning a non-zero average flow rate.

Keywords: Magnetic fluid; traveling magnetic field; perturbance; locomotion; peristaltic.

1. Introduction

Especially in miniaturized or difficult accessible areas mobile robotics applies the principles of apedal locomotion. Undulatory locomotion can be accomplished by internally driven motion resulting in repositioning of the center of mass and the outer surface.

Various publications exist on peristaltics. So investigate Friedrich et.al.[1] periodic patterns on a magnetizable fluid surface, which incorporates a contactless pumping mechanism and Park and Park[2] present a technical solution for a bifluidic peristaltic locomotion device. A related, but analytic work by Zeidis and Zimmermann[3] examines a peristaltic material transport of two fluids separated by a thin impermeable membrane. In the presents article we discuss the topic in the matter of a presumed ferrofluid flow caused by a magnetic field deforming the free fluid surface periodically. The inverse Problem (Ref. 4) is studied finding the causative magnetic field by analyzing the perturbed surface contour.

2. Setting of the Problem

We consider a magnetizable fluid in a plane flow with a velocity $\vec{v} = (u, 0, w)^T$ on a non-magnetizable rigid, impermeable base, see Fig. (1). We take the gravitational acceleration $\vec{g}$ and the atmospheric pressure p as surrounding parameters, the kinematic viscosity ν, the density ρ as the fluid properties into account. The magnetic permeability of the fluid responds to $\mu - 1 \ll 1$.

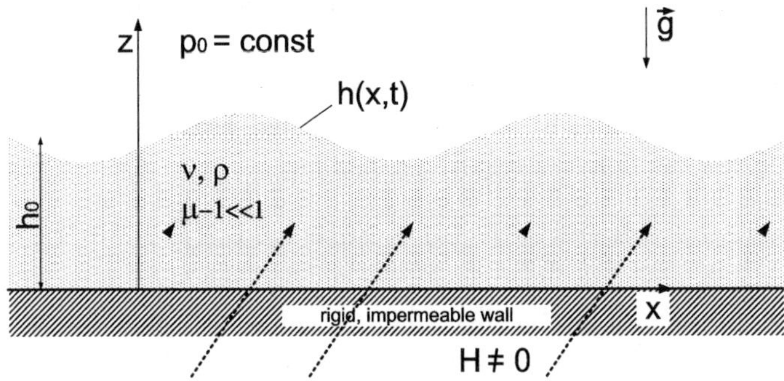

Fig. 1.   Sketch of the considered system.

An alternating magnetic field is exposed to the fluid and causes a deformation of the surface contour $h(x, t)$. At the free fluid surface at $z = h(x, t)$ the magnetic field can be described with:

$$H^2(x, t) = H_0^2 \left(1 + \varepsilon \sin(kx - \omega t)\right) \tag{1}$$

$$\text{with } \varepsilon = \frac{A^2}{H_0^2} \ll 1. \tag{2}$$

Here H_0 and A are constant. A^2 is the constant perturbance amplitude of the magnetic field strength including the demand (2). k and ω represent the wave number and the angular velocity, respectively. x is the spatial coordinate and t the time.

3. Method

The system of equations include the NAVIER-STOKES differential equation (3) and the two-dimensional continuity equation (4) for incompressible fluids with constant density

494

$$\frac{\partial \vec{v}}{\partial t} + (\vec{v} \cdot \nabla)\vec{v} = -\frac{\nabla p}{\varrho} + \nu \triangle \vec{v} + \frac{\vec{F_g}}{\varrho}, \quad (3) \qquad \nabla \vec{v} = 0. \qquad (4)$$

All equations are transferred into a dimensionless form. For this operation the following characteristic parameters are applied: The height of the unperturbed fluid layer h_0 , characteristic period time, velocity and pressure T, U_c and P_c, shear stresses τ_{ij} and characteristic weights in x and z direction F_x and F_z, respectively. The gravitational vector is $\vec{F_g} = (0, 0, -1)^T$.

The dimensionless NAVIER STOKES equation contains the characteristic numbers of REYNOLDS, EULER, STROUHAL and FROUDE, which are suitable to phrase the demand regarding the flow:

$$\text{Re} \ll 1, \quad (5) \qquad \text{Re Eu} = 1, \quad (6) \qquad \text{Sr Re} \ll 1, \quad (7) \qquad \text{Re/Fr} \sim 1. \quad (8)$$

As we consider all inertia effected forces sufficiently small in comparison to inner friction, pressure gradient and weight, we resume to the STOKES approximation for creeping flows:

$$\nabla p = \triangle \vec{v} + \vec{F_g}. \tag{9}$$

Furthermore we introduce the stream function $\Psi(x, z, t)$ and gain from the continuity equation (4) a partial differential equation of fourth order

$$u = -\frac{\partial \Psi}{\partial z}, \quad (10) \qquad w = \frac{\partial \Psi}{\partial x}, \quad (11) \qquad \triangle\triangle \Psi(x, z, t) = 0. \quad (12)$$

4. Approach to the Solution

As the external magnetic field manipulates the flow and all corresponding parameters, one can adopt its structure, see Eq. (1), to the complete system. The power series ansatz transfers the stream function $\Psi(x, z, t)$ into

$$\Psi(x, z, t) = \sum_{n=1}^{\infty} \varepsilon^n \Psi_n(x, z, t), \ \ n \in \mathbb{N}, \tag{13}$$

and secondly the periodic structure will be assigned to $\Psi(x, z, t)$

$$\Psi_n(x, z, t) = \sum_{m=1}^{\infty} \eta_{nm}(z) \cos(m\zeta) + \xi_{nm}(z) \sin(m\zeta), \ \ n, m \in \mathbb{N}, \tag{14}$$

$$\text{with } \zeta = kx - \omega t.$$

Due to the periodic character of $\Psi(x,z,t)$ Eq. (12) has to be solved merely as an ordinary differential equation. Consequently the amplitude functions $\eta(z)$ and $\xi(z)$ result in

$$\eta_1(z) = a_1 \cosh(kz) + a_2 z \cosh(kz) + a_3 \sinh(kz) + a_4 z \sinh(kz), \quad (15)$$

$$\xi_1(z) = b_1 \cosh(kz) + b_2 z \cosh(kz) + b_3 \sinh(kz) + b_4 z \sinh(kz). \quad (16)$$

Like explained in sec. 6 the order of the periodicity and the order of the power series here are already limited to $m = 1$ and $n = 1$.

5. Boundary Conditions

For the solution we take advantage of the boundary conditions. At the motionless wall at $z = 0$ the velocity of a frictional fluid equals zero: Eq. 17 and 18. Two sets of coefficients of $\eta_1(z)$ and $\xi_1(z)$ can be found from these, see Eq. 19 till 22:

$$u(x, z = 0, t) = 0, \quad (17) \qquad a_1 = 0, \quad (19) \qquad a_2 = -ka_3, \quad (21)$$

$$w(x, z = 0, t) = 0, \quad (18) \qquad b_1 = 0, \quad (20) \qquad b_2 = -kb_3. \quad (22)$$

The contour function $h(x,t)$ is also affected by the magnetic field. In this case the power series structure of Eq. (1) is adopted in the same fashion

$$h(x,t) = 1 + \varepsilon h_1 + \varepsilon^2 h_2 + o(\varepsilon^2). \tag{23}$$

The kinematic boundary condition, valid at a free surface at $z = h(x,t)$,

$$\frac{d}{dt} h(x,t) = \frac{\partial h}{\partial t} + \frac{\partial h}{\partial x} \frac{dx}{dt} \tag{24}$$

helps in context of Eq. (23) to phrase a relation between the contour function $h_1(x,t)$ and the stream function $\Psi_1(x, z = h(x,t), t)$ by concentrating on the first order of the power series

$$h_1(x,t) = -\frac{k}{\omega} \Psi_1(x, 1, t). \tag{25}$$

Another condition occurs at $z = h(x,t)$. At the intersection of the boundary layer stress is a steady function:

$$\left[-p + \frac{\gamma}{R} + \frac{B_n^2}{8\pi} \left(\frac{1}{\mu} - 1 \right) - \frac{H_\tau^2}{8\pi} (\mu - 1) \right] \vec{n} + \tau_{ij} n^j \vec{e}_i = \vec{0}. \tag{26}$$

For resolution of the remaining coefficients of $\eta_1(z)$ and $\xi_1(z)$ we dispose of the dimensions in Eq. (26). Thereby WEBER number We $= \gamma/(h_0 P_c)$ and $\delta = g h_0/(U_c \nu)$ are introduced in reference to surface tension as well as weight. With respect to the non-inductive approximation the magnetic

496

surface tension turns to $-\kappa(1 + \varepsilon \sin(kx - \omega t))$ with $\kappa = (\mu - 1)/(8\pi)$. The vector equation (26) is split into two scalar equations by multiplication with the normal vector $\vec{n} = [-h'_x/(1 + h^2_{,x})^{\frac{3}{2}}, 1/(1 + h^2_{,x})^{\frac{3}{2}}]^T$ and the function $h(x, t)$ in form of Eq. (23) is inserted. Thus the normal component is transformed into

$$-p_1(1) + \delta h_1 + We\, h_{1,xx} + 2\Psi_{1,xz}(1) - \kappa(1 + \varepsilon \sin(kx - \omega t)) = 0 \quad (27)$$

and the tangential component appears as

$$\Psi_{1,xx}(1) - \Psi_{1,zz}(1) = 0. \quad (28)$$

$p(z = 1)$ is replaced in Eq. (27) by the x-component of the continuity equation (4) and Eq. (25) is applied

$$3\Psi_{1,xxz}(1) + \Psi_{1,zzz}(1) + \delta h_{1,x} + We\, h_{1,xxx} - \kappa k \cos(kx - \omega t) = 0. \quad (29)$$

Finally we gain the missing coefficients a_3, a_4, b_3 and b_4.

6. Development of the Flow Rate

Dealing with a plane problem we attend to

$$Q = \int_0^{h(x,t)} u(x, z, t)dz. \quad (30)$$

With respect to the upper integration limit $h(x, t)$, which refers to the form of Eq. (23), finally an averaged flow rate $\bar{Q}$ can be developed

$$\bar{Q} = \tfrac{1}{T} \int_0^T Q(x, t)\, dt = C\varepsilon^2 + \mathcal{O}(\varepsilon^4) \quad (31)$$

$$\text{with } T = \tfrac{2\pi}{\omega} \text{ and } C = h_1(x, t)\Psi_1(x, z = 1, t),$$

$$\bar{Q} = -\tfrac{\varepsilon^2 k}{2\omega}\left[-(a_3^2 + b_3^2)k^2 \sinh(k)(\sinh(k) - k\cosh(k)) \quad (32)\right.$$
$$+ (a_4^2 + b_4^2)\sinh(k)(\sinh(k) + k\cosh(k))$$
$$\left. + (a_3 a_4 + b_3 b_4)(\sinh^2(k) - k^2 \cosh(2k))\right].$$

7. Discussion of the Solution - The Existence of an Averaged Flow Rate

The analytical expression of the averaged flow rate $\bar{Q}$ of Eq. (33) depends on the parameter of the traveling magnetic field and fluid parameters (viscosity, surface tension, magnetic permeability). None of the parameter is zero, so the evidence of a material locomotion is given with $\bar{Q} \neq 0$, see Eq. (33) and Fig. (2) till (4).

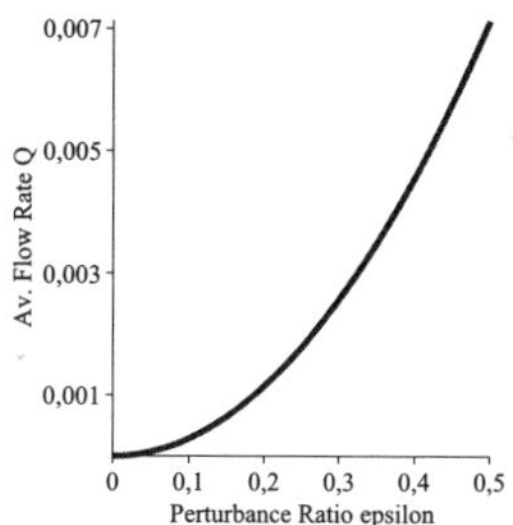

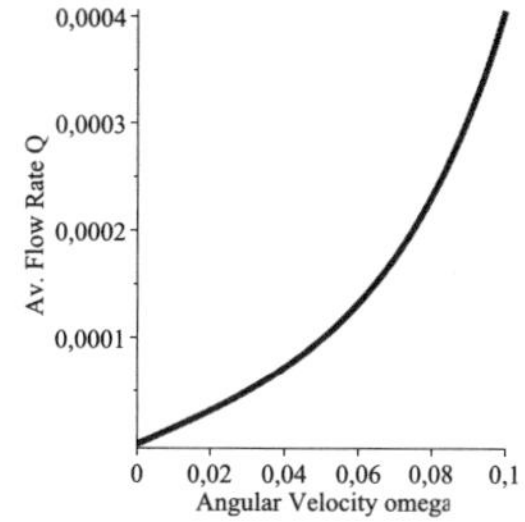

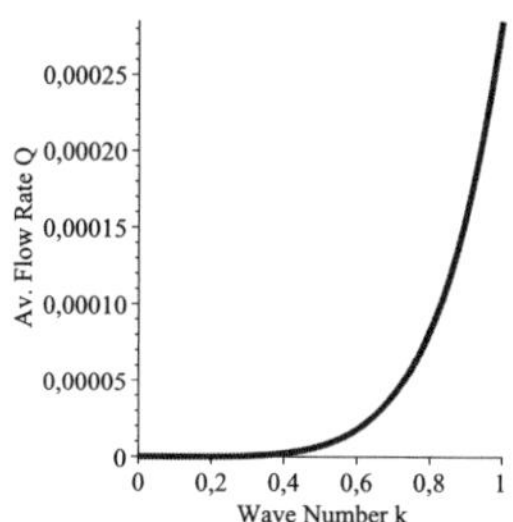

Fig. 2. Perturbance ratio ε vs. av. flow rate $\bar{Q}$.

Fig. 3. Angular velocity ω vs. av. flow rate $\bar{Q}$.

Fig. 4. Wave number k vs. av. flow rate $\bar{Q}$.

Figure (2) proves with rising ratio perturbance to constant fraction of the magnetic field increases the flow rate. This is only valid within the limitation of small ε (Eq. (2)). Similar behavior can be registered for the dependency on the wave number k and angular velocity ω. For reintroduced dimensions, one can expect in areas of higher ω a single peak at the eigen angular velocity. According to our estimations this also will lie beyond the extent of validity due to slow processes (Eq. (5)).

8. Conclusion and Prospects

The evidence of an effective mass transport depending on fluid and magnetic parameters as result of a known external alternating magnetic field is provided in analytical form.

These dependencies shall be used for the design of ferrofluid based locomotion systems, such as magnetic fluid filled envelopes either moving itself

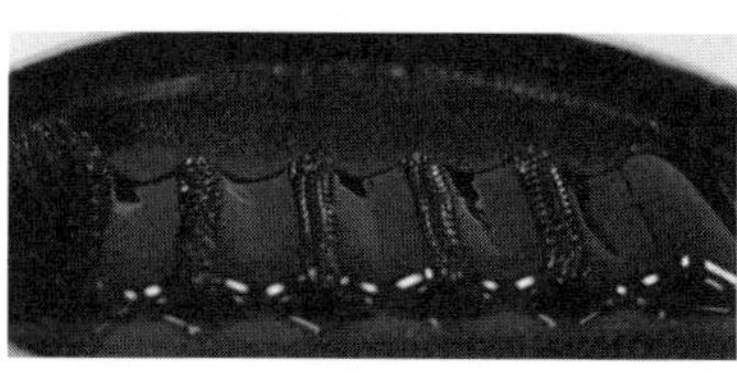

Fig. 5. Contour variable locomotion system, see Ref. 5.

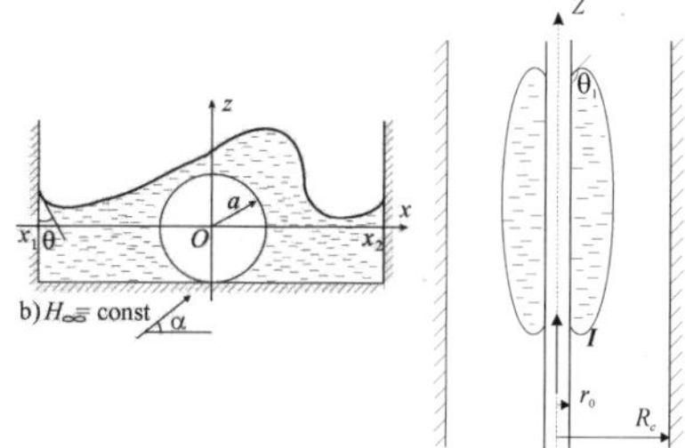

Fig. 6. Surface manipulation, see Ref. 6.

(mobile robots) or assisting secondary objects to reposition, see Fig. (5) and Ref. 5. These systems always give the problem of insufficient locomotion propulsion. For the counteracting idea the magnetic field can be selectively amplified to generate a traveling surface curvature. Investigation about the insertion of high permeable spheres or line conductors in the magnetic fluid will be continued in cooperation with Russian partners, see Ref. 6 and Fig. 6.

Acknowledgment

This work is supported by Deutsche Forschungsgemeinschaft (DFG) projects Zi 540-11/1 and Zi 540-14/1 as well as by the graduation scolarship of the Free State of Thuringia, Germany.

References

1. T. Friedrich, I. Rehberg and R. Richter, *Phys. Rev. E* **82**, p. 036304 (2010).
2. G. S. Park and S. H. Park, *IEEE Transactions on Magnetics* **36**, 3709 (2000).
3. I. Zeidis and K. Zimmermann, *Technische Mechanik. Magdeburg* **20**, 73 (2000).
4. K. Zimmermann, I. Zeidis, V. A. Naletowa and V. A. Turkow, *Journal of Magnetism and Magnetic Materials* **268**, 227 (2004).
5. K. Zimmermann, V. Böhm, I. Zeidis and J. Popp, *Sensor letters.* **3**, 351 (2009).
6. K. Zimmermann, V. A. Naletowa, I. Zeidis, V. A. Turkov, D. A. Pelevina, V. Böhm and J. Popp, *Magnetohydrodynamics* **44**, 175 (2008).

A SIMPLE SHEAR ANALYSIS OF MR FLUIDS

F. JONSDOTTIR

*School of Engineering and Natural Sciences, University of Iceland, Hjardarhaga 2-6,
Reykjavik, 107, Iceland*

K.H. GUDMUNDSSON

*School of Engineering and Natural Sciences, University of Iceland, Hjardarhaga 2-6,
Reykjavik, 107, Iceland*

E. HREINSSON

*School of Engineering and Natural Sciences, University of Iceland, Hjardarhaga 2-6,
Reykjavik, 107, Iceland*

F. THORSTEINSSON

*Ossur Inc., Grjothalsi 6,
Reykjavik, 110, Iceland*

O. GUTFLEISCH

*Leibniz Institute for Solid State and Materials Research Dresden, PF 27 01 166,
Dresden, D-01171, Germany*

The performance of most commercial MR fluid devices depends on the shear strength of the MR fluid. The field-induced yield stress can be accurately characterized experimentally by a magneto-rheometer. However, as a number of factors, for example, particle volume ratio and particle size, are known to affect the rheological properties of the fluid, such measurements can be very time consuming. Hence, computational models represent a valuable tool in the design of MR fluid devices. This study provides a computational model to quantitatively predict the shear yield stress for MR fluids. The configuration is a representative unit cell in the form of a cube of a visco-plastic material which contains a number of rigid spherical particles. The cube undergoes a simple shear deformation and the mean induced shear stress is calculated. The model is used to study the effect of particle size and volume ratio on the dynamic yield stress of MR fluids.

1. Introduction

Magnetorheological (MR) fluids consist of micron sized magnetizable particles suspended in a carrier fluid. When subjected to a magnetic field, the fluid greatly increases its apparent viscosity, to the point of becoming a plastic-like solid. A key parameter that characterizes the MR fluid is the shear yield stress. Thus, being able to predict the shear yield stress for MR fluids by a simple computational model can aid in the design of MR fluid devices. In this work a simple shear deformation model is developed and used to evaluate the field-induced shear yield stress.

Several models, both analytical and numerical, have been proposed in the past to describe the shear yield stress in MR fluids [1-9]. Most of these models are based on evaluating the magnetic interparticle forces between particles within a particle chain in the fluid. For example, Klingenberg and Zukoski [1] and Bossis and Lemaire [2], used a dipolar sphere model and calculated the restoring force which tends to align two spheres on the axis of the field. Ginder and Davis [3] and Ginder et al. [4] modeled the fluid as a collection of infinite chains of spherical magnetizable particles and used the Maxwell stress tensor to determine the shear stress needed to shear the chains. Furthermore, Jolly et al. [5] developed a quasi-static one-dimensional model. More recently, Si et al. [6] proposed a micromechanical model for MR fluids, based on theory of magnetics and statistical analysis.

In addition to the particle interaction models, continuum type models have been developed, in order to determine the yield stress. For example, Rosensweig [7] developed a continuum model of a composite based on a laminar layer structure with magnetic elements aligned in the field direction. Later, Tang and Conrad [8] extended these models and developed two- and three-dimensional laminar structure models.

This study is based on the established model of magnetic dipole-dipole interaction energy. From there, a computational model is developed to quantitatively predict the shear-yield stress for MR fluid suspension. The interaction energy is a function of the magnetic properties of the particles. However, instead of using magnetic saturation values for the particles, as many studies do, a magnetization curve for a MR suspension is obtained experimentally and used in the computational model. Hence, the model takes into account the effects of the intensity of the applied magnetic field, along with the particle size and particle fraction. The results from the computational model are compared to experimental flow curves for an MR fluid sample.

2. Shear Stresses

The analysis is based on a theoretical microscopical investigation into the interparticle forces that generate the shear-yield stress in MR fluids. Prior, to the theoretical investigation, an MR fluid composition is experimentally evaluated.

2.1. *Experimental investigation*

One MR fluid composition was prepared for experimental analysis. This sample has a solid loading of 0.28, by volume, employing carbonyl iron powder from BASF. The mean diameter of the particles is about 2 µm. Experimental data was obtained with a SEM Gemini 1530 FEG microscope, a Quantum Design MPMS-5S SQUID magnetometer and an Anton-Paar Physica MCR100 rheometer. Figure 1 shows an SEM image of a sample carbonyl iron powder.

Figure 1. An SEM image of a sample of carbonyl iron powder from BASF.

The figure shows the morphology of the sample of particles to be regular spheres. Figure 2 shows the magnetization characteristics of this sample of iron powder.

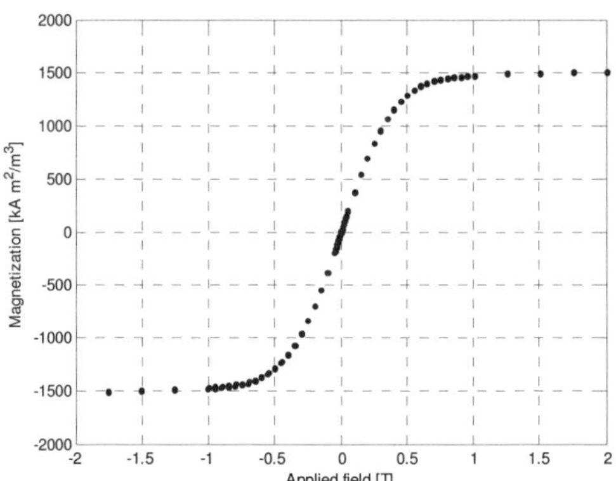

Figure 2. The magnetization characteristics of a sample of carbonyl iron powder, from BASF, with an mean particle diameter of approximately 2 µm. Measurements were performed at a temperature of 20°C.

The figure shows the particles to be saturated at a magnetic flux density of approximately 1 T. Figure 3 shows the shear stresses in the MR fluid sample.

502

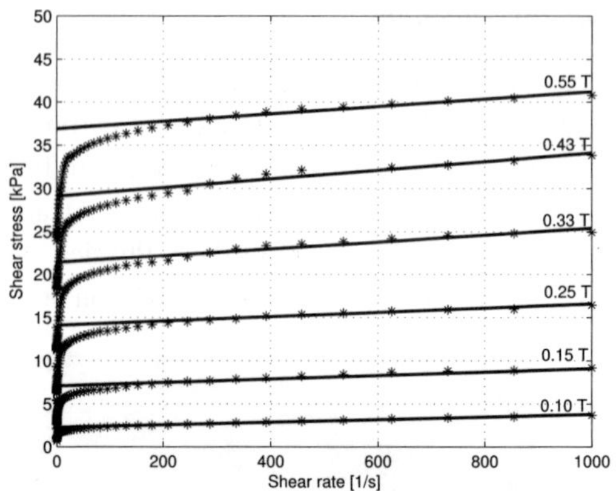

Figure 3. The field-induced rheological characteristics of an MR fluid sample with a solid loading of 0.28, by volume, using carbonyl iron powder from BASF. Measurements were performed at a temperature of 20°C.

The figure shows shear stress measurements at various field strengths along with a linear Bingham fit of the data. This concludes the experimental investigation and a theoretical investigation will now commence.

2.2. *An analytical shear stress model*

The theoretical analysis is based on established physical simple dipole model for interparticle magnetic force between two isolated dipoles. Looking at two equally sized isolated particles of radius a, each particle carries a magnetic moment shown in Eq. (1).

$$|\vec{m}_1| = M_1 \cdot 4/3 \cdot \pi \cdot a^3 \tag{1}$$

where M_1 is the magnetization of the particle. The magnetization characteristics, M, of one particular iron powder has been experimentally evaluated as a function of the applied field and is shown in Figure 2. These measurements will be used to quantify the theoretical analysis.

The interparticle magnetic potential energy between two isolated dipoles, in contact, is shown in Eq. (2) [5,9].

$$U(\vec{r},\vec{m}_1,\vec{m}_2) = \frac{\mu_0}{4\pi} \frac{\vec{m}_1 \cdot \vec{m}_2 - 3(\hat{r} \cdot \vec{m}_1)(\hat{r} \cdot \vec{m}_2)}{|\vec{r}|^3} \tag{2}$$

where $\bar{r}, \bar{m}_1, \bar{m}_2$ are the distance vector between the particles, the magnetic moment of particles 1 and 2, respectively. The dipolar interaction force is the derivative of the potential energy with respect to distance. The distance can be a measure of tension to obtain the normal force or a measure of shear to obtain the shearing force. The derivative of Eq. (2) with respect to distance, in the tension direction, is shown in Eq. (3) and, furthermore, with respect to distance, in the shearing direction, in Eq. (4).

$$F_n = \frac{3\mu_0}{2\pi d^4}(m_1 m_2) \tag{3}$$

$$F_s = \frac{\mu_0}{4\pi(d^2 + x^2)^{5/2}}(m_1 m_2) \cdot (5x\frac{d^2}{d^2 + x^2} - 3x) \tag{4}$$

where d is the distance between the dipoles or the diameter of the particles and x is the shearing distance.

To estimate the yield stress of a particular MR suspension, an imaginary unit cube is used with a defined solid concentration and a uniform particle size. Particles are assumed as perfect spheres. Thus, Eq. (5) gives the number of particles in an imaginary unit cube as a function of solid concentration, ϕ, and particle diameter, d:

$$n_V = \frac{\phi \cdot 1}{4/3 \cdot \pi \cdot (d/2)^3} \tag{5}$$

Assuming long linear chains of particles in with particles, within a chain in contact, the number of particles in a line along the cube is $1/d$ and thus the number particles in a unit area normal to the chains is shown in Eq. (6).

$$n_A = \frac{n_V}{1/d} \tag{6}$$

An estimate of the shear-yield stress in a simple shear deformation is obtained by looking at the outer-most unit area, normal to the chains. The total shearing force to shear the outer-most layer is $n_A \cdot F_s$ by which an estimate of the shear-yield stress can be obtained. The shear force has a maximum which can be found by looking at the derivative of Eq. (4) which equals the shear yield stress. This equals a shear yield stress of 13 kPa, at saturation level. This value is obtained for a particle diameter of 2 μm, a solid loading of 0.28 by volume and by shearing the outer-most layer. This gives a shear yield stress of the correct order of magnitude but small when compared to measured shear-yield stress. Research has suggested that yielding behavior is contributed only from a few of the outer-most layers [10] but the exact number is unknown. Accounting for more than one layer would increase the predicted shear yield stress.

Figure 4 shows the shear-yield stress values of the microscopic model along with the measurements from Fig. 3.

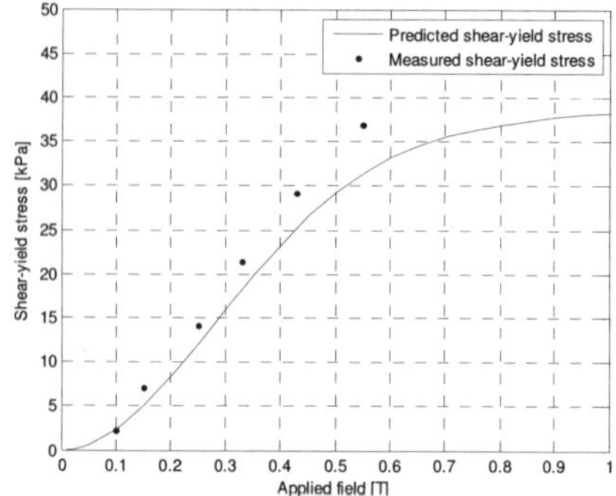

Figure 4. Shear-yield stress values from the microscopic model for a solid loading of 0.28, by volume. Measurement data for a corresponding MR fluid is also shown. The microscopic model has been scaled to fit the first data point.

The model has been scaled to fit the first data point which is the rheological measurement of the shear-yield stress at 0.1 T. An addition to the rheological data point, the model is based on measured magnetic characteristics of a sample powder, shown in Fig. 2.

The single row chain model underestimates the shear-yield stress. In addition to accounting for just one layer, this is believed to by due to the fact that particles tend to from aggregates rather than single lines of particles. Fig. 5 supports this and shows how the particles form aggregates rather than singe row chains under the influence of a constant magnetic field.

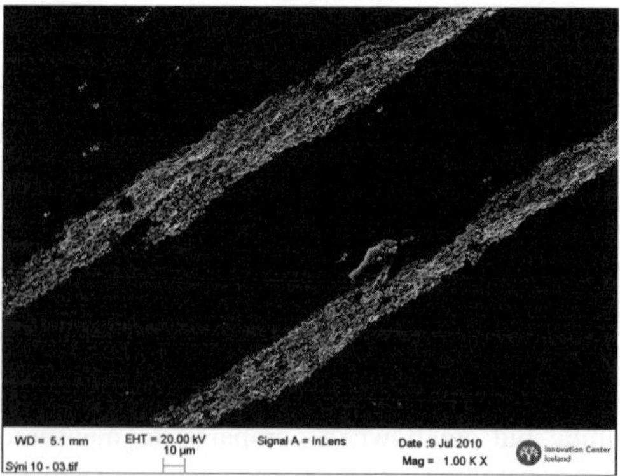

Figure 5. An SEM image of particle chains under the influence of a magnetic field.

Using this simple model to analyze the shear-yield stress gives a preliminary insight into the behavior of MR fluids. Accounting for particle aggregates will improve predictions of the model. It remains the subject of further research to investigate more thoroughly the effect of particle size and solid concentration on the field-induced shear-yield stress of MR fluids.

3. Conclusions

The microscopic simple dipole model gives an estimation of the shear-yield stress that is smaller than the measured the shear-yield stress. The model uses measurements of magnetization characteristics of carbonyl iron powder as an input data. This gives a shear-yield stress curve with a shape comparable to a measured shear-yield stress curve. By scaling the model to fit one rheological measurement point, the model accurately describes the field-induced rheological characteristics of the measured MR fluid sample.

We find the simple model presented in this study to provide a preliminary insight into the behavior of MR fluids. Or ongoing research aims to elaborate on it to be able to predict the shear-yield stress more accurately.

Acknowledgments

This work is funded by the University of Iceland and Rannis, grant no. 090035021, and supported by Ossur Inc.

References

1. D. J. Klingenberg and C. F. Zukoski, *Langmuir*, **6** 15 (1990).
2. G. Bossis and E. Lemaire, *J. of Rheol.* **35** 1345 (1991).
3. J. M. Ginder, L. C. Davis, *Appl. Phys. Letter* **65** 3410 (1994).
4. J. M. Ginder, L. C. Davis and L. D. Elie, *Int. J. Mod. Phys.*, **10** 3293 (1996).
5. M.R. Jolly, J.D. Carlson and B.C. Muñoz, *Smart Mater. Struct.* **5** 607 (1996).
6. H. Si, X. Peng and X. Li, *J. Intell. Mater. Syst. Struct.* **19** 19 (2008).
7. R. E. Rosensweig, *J. of Rheol.*, **39** 179 (1995).
8. X. Tang, H. Conrad, *J. of Phys. D: Appl. Phys.* **33** 3026 (2000).
9. P.G. de Gennes and P. A. Pincus, *Phys. kon. Mat.* **11** 189 (1970).
10. X. Tang, Y. Chen and H. Conrad, *J. Intell. Mater. Syst. Struct.* **7** 517 (1996).

IV. PROPERTIES AND MICROSTRUCTURES

POLAR-MOLECULE-DOMINATED ELECTRORHEOLOGICAL (PM-ER) FLUIDS: THE PROPERTIES AND EVALUATIONS[*]

KUNQUAN LU[1†], RONG SHEN[1], XUEZHAO WANG[1], DE WANG[1,2] and GANG SUN[1]

1Beijing National Laboratory for Condensed Matter Physics and Key Laboratory of Soft Matter Physics, Institute of Physics, Chinese Academy of Sciences, Beijing 100190, China

2 University of Science and Technology of China, Hefei 230029, China

In recent years, a new type ER fluids named as polar-molecule-dominated electrorheological (PM-ER) fluids have been developed, of which the yield stress can reach more than 100 kPa and behaves a linear dependence on the electric field. A brief description on the composition and synthesizing method for the materials is given. The main merits of PM-ER fluid are as follows: high yield stress, the shear stress increasing with shear rate up to more than $10^3 s^{-1}$, low current density, rapid electric response and anti-sedimentation. Some perspectives on PM-ER fluid and its applications are presented.

1. Introduction

Since the first discovery of Winslow on ER fluid, much effort has been paid to the study of the physical mechanism and the materials. The ER effect arising from the interaction of the polarized particles, which form chains and columns along the direction of the external electric field, has been well understood [1-4]. Typical character of this traditional dielectric ER fluid are that the yield stress is low (usually less than 10kPa) and showed a quadratic dependence on the field strength. The low yield stresses of the available traditional dielectric ER fluid can not meet the requirements for practical applications.

In recent years a serial of giant ER fluids have been developed [5-12], the yield stress of which can reach more than 100kPa and behaves a linear

---

[*] This work is supported by National Natural Science Foundation of China Grant No. 10674156; the National Basic Research Program of China Grant Nos. 2004CB619005 and 2009CB930800; the Knowledge Innovation Project of Chinese Academy of Sciences Grant No. KJCX2-YW-M07; the Instrument Developing Project of the Chinese Academy of Sciences, Grant No. YZ200758.
[†] E-mail: lukq@iphy.ac.cn

dependence on the electric field. These ER fluids are now termed as the polar-molecule dominated electrorheological (PM-ER) fluids based on a new mechanism. The polar molecules adsorbed on the particles are aligned by significantly enhanced local electric field in the gap between neighboring particles. The interaction from the aligned polar molecules is much larger than that of the polarized particles in traditional ER effect. A detail description on the mechanism has been given in our previous publications [13-15].

In this paper we are going to present the properties and perspectives of those PM-ER fluids studied in our group.

2. Materials of PM-ER Fluids

Several PM-ER fluids with high yield stress are manufactured. The solid particles based on Ti-O [7], Ca-Ti-O [8], Sr-Ti-O [9], La-Ti-O [16] were prepared with wet chemical technique. The dielectric constant of the particles should be high enough and the conductivity should be low. There must be polar molecules (O-H, C-O, C=O, N-H et al) adsorbed on the particles from the particle preparing procedure which are crucial factor for PM-ER effect differing from the traditional ER one. If the particles are heated to high temperature (about 500°C) and then all the adsorbed polar groups are released, the ER fluids consisted of the sintered particles lose their character of high yield stress showing a traditional ER effect [15].

The size of the particles is in the range of $50\text{-}10^3 \text{nm}$ with the density of 2.2-3.5g/cm^3 which depends on the species and size of the particle. For instance, the density of Ca-Ti-O particles with the size of 50-100nm is about 2.35g/cm^3. The small size and the low density of the particles are benefit against sedimentation of the suspensions. Silicone oil, mineral oil, lubricating oil or hydraulic oil can be selected as dispersion liquid.

3. Characters of PM-ER Fluids

The yield stress τ_y of PM-ER fluid can be as high as 100, 200kPa and even higher depending on the composition and concentration of the fluids. It should be emphasized that in measuring the shear stress by a conventional rheometer it is found that ER fluid slides along the electrodes and the measured values of shear stress deflect to lower side of the linear dependence on electric field when the shear stress is higher than about 100 kPa. Consequently the nominal value measured with a flat electrode is much lower than the intrinsic value of the fluid. This sliding at electrode originates in that the boundary condition at the interface of fluid-electrode is not as the same as inside of ER fluid [15,17]. Therefore, the electrodes for measuring PM-ER fluid must be treated to be rough enough. Even though the measured shear stress still is less than the intrinsic value for the PM-ER fluids with extra large yield stress. By using a cut-shearing method the measured yield stress can be close to the real value [15,18], although the measured values are not very accurate (with an error of 10 percent approximately).

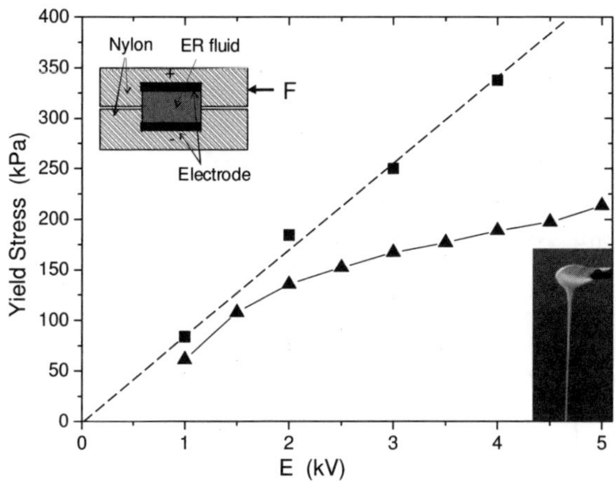

Figure 1. The yield stress vs. electric field of Ca-Ti-O particles based PM-ER fluid. The triangle symbols indicate the measured results by using parallel plate rheometer with rough electrodes. Square symbols represent the measured values with cut-shearing method as plotted in inset. Dashed line shows the ideal values of yield stress. The image at lower right shows the flowability of this ER fluid.

As an example, Figure 1 shows a measured yield stress vs. electric field for Ca-Ti-O based PM-ER fluid, which is consisted of Ca-Ti-O particles suspending in hydraulic oil with a volume fraction of 55%. It can be seen that the yield stress of this sample can reach about 400 kPa at a filed of 5 kV/mm. Obviously, the values of yield stress measured by parallel plate rheometer with rough electrodes is only about half of that measured with cut-shearing method in high τ_y region. Distinct slide on the electrodes can be observed in the measurement by using conventional rheometer.

Changing the volume fraction ϕ of the suspensions the yield stress can be adjusted. For this Ca-Ti-O based PM-ER fluid the yield stress τ_y is about 50, 100, 200 kPa at 5kV/mm in the case of ϕ=33%, 44%, 50% respectively. Also ϕ is one of the factors to influence the viscosity at zero field. In general, at a field of E=5kV/mm, the ratio of $\tau_y/\tau_y(E=0)$ can reach more than 10^4 for a fluid with $\tau_y \approx 50$ kPa while the ratio is less than 5×10^2 when $\tau_y \approx 200$ kPa.

512

The apparent shear stress $\tau_E(\dot{\gamma})$ vs. shear rate $\dot{\gamma}$ can be obtained with a clutch like apparatus by measuring the torque. Figure 2 shows a typical torque-shear rate relation for a Ca-Ti-O based dilute PM-ER fluid in which the torque $T_E(\dot{\gamma})$ increases with shear rate up to more than 4×10^3 s^{-1}. The measurement at higher shear rate is baffled by the heating effect especially for the PM-ER fluids possessing high yield stress [17]. Obviously, although $T_E(\dot{\gamma})-T_0(\dot{\gamma})$ or $\tau_E(\dot{\gamma})-\tau_0(\dot{\gamma})$ almost keeps unchanged in all region of shear rate, the ratio

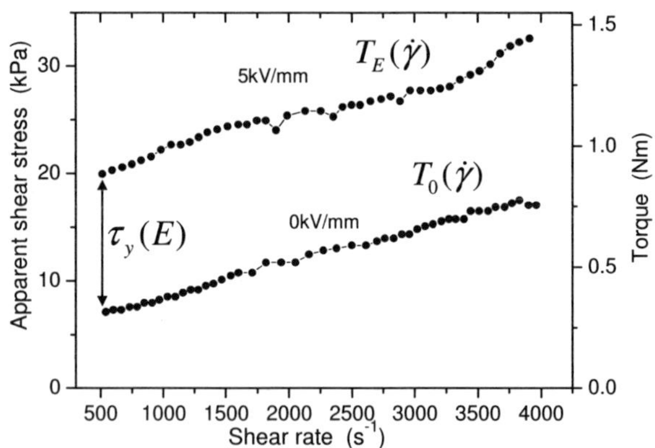

Figure 2. Relation of torque vs. shear rate for a Ca-Ti-O based dilute PM-ER fluid. The rater high torque at zero field is mainly form the sealing friction of clutch.

$$\frac{T_E(\dot{\gamma})}{T_0(\dot{\gamma})} = \frac{\tau_E(\dot{\gamma})}{\tau_0(\dot{\gamma})} \approx \tau_y(E)\alpha^{-1}\dot{\gamma}^{-1} \quad \text{(for } \alpha > 0)$$ decreases with shear rate increasing. Where α is the slope of torque increasing with shear rate, which depends on the rheological property of material and the design of clutch. The improvement in both aspects to reducing α is necessary for achieving better efficiency in the applications.

The electric response time is less than 1ms measured with a rapid square wave field. However, practically, there usually are an electric RC decays appeared in the torque measurement with a clutch. Figure 3 shows the torque responses under 0-3kV/mm square field. The time delay of the torque value happens due to the electric charging and discharging when the field turns on and off, because the dielectric constant and the resistance of the material are rather large. The RC time constant measured is about tens of millisecond, which is consistent with the calculation based on the capacity and the resistance in the arrangement of clutch. Therefore it is necessary to apply fast enough electric circuits to meet the requirement in a rapid controlling system.

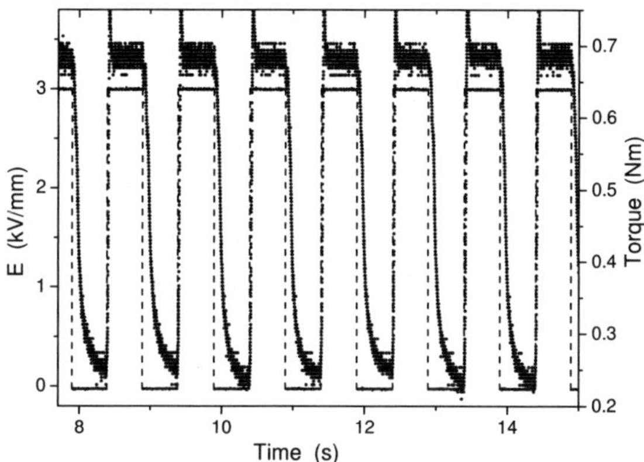

Figure 3. The torque responses under 0-3kV/mm square field at $\dot{\gamma}=50s^{-1}$. Dashed line indicates the electric field and the dots are torque values.

The electric current density of PM-ER fluids prepared in our group is normally in the range of 0.01-10µA/cm^2. Also there is almost not sedimentation observed in our PM-ER fluid because of the low density and small size of the particles.

4. Merits and Issues of PM-ER Fluid for Applications

(1) The high yield stress of PM-ER fluid being able to attain hundreds of kPa is a big advantage in technical applications, of which the maximum value is several times higher than MR fluids and hopefully to be further enhanced. To manufacture PM-ER fluids with even higher yield stress is potentially in the works. With high τ_y ER fluids smaller and more efficient devices are possibly to be fabricated by applying a relatively low electric field. Besides the flow mode and shear mode in the applications, the character of PM-ER fluids with extra high yield stress may result in a new field for applications i.e. as a smart structural material. The shear modulus of the sample shown in Figure 1 is measured to be about 60 MPa at 5 kV/mm. This modulus tunable feature by adjusting electric field is attractive in technology, for instance, in making a sandwich structure combining with metallic or other conductive materials.

(2) Low current density of the ER fluids is benefit for making the devices with low energy dissipation. Typically, the power of less than 1W is enough for

a small device, supplied by the batteries. The arbitrary shape of electrodes and high efficiency of the PM-ER fluid are favorable for designing compact or potable devices, which can be convenientlly controlled even in a multichannel operating system.

(3) There are some issues for PM-ER fluids should be further studied or improved. Firstly, the long term stability of the fluids in the applications has not been well tested. Several factors must be considered including the firmness of the polar molecules and the influence of the impurities falling out from the electrodes or other parts. Practical experiences for applications are still lacking. We also should pay attention to the surface treatment of the electrodes for the fluids with high shear stress. In respect of materials, to reduce the viscosity at zero electric field and to produce identical as well as duplicable ER fluids with high quality and a certain quantity are an ongoing process.

5. Conclusions

Several types of PM-ER fluids have been developed in our group, of which the excellent characters are described. The yield stress can be as high as 100, 200, 400kPa, respectively, by adjusting the composition and concentration of the materials. The shear stress increasing with shear rate up to more than 10^3 s^{-1}. The current density of the fluids is lower than 10 μm/cm^2 and even less than 1μm/cm^2 at a field of 5kV/mm. Moreover, almost no sedimentation of the particles in the fluids can be observed. Some perspectives and related technical issues on PM-ER are discussed. Efforts are concentrated to improve the viscosity at low zero field, long term stability and quantity production of PM-ER fluid for its practical applications.

References

1. R. Tao and J. M. Sun, *Phys. Rev. Lett.* **67**, 398 (1991).
2. L. Davis, *Phys. Rev.* **A46**, R719 (1992).
3. J. Martin, J. Odinek, T. Halsey, & R. Kamien, *Phys. Rev.* **E57**, 756 (1998).
4. H. R. Ma, W. J. Wen, W. Y. Tam, and P. Sheng, *Phys. Rev. Lett.* **77**, 2499 (1996).
5. W. J. Wen, X. X. Huang, S. H. Yang, K. Q. Lu, P. Sheng, *Nat. Mater.* **2**, 727 (2003).
6. K. Q. Lu, R. Shen, X. Z. Wang, G. Sun, W. J. Wen, Int. *J. Mod. Phys.* **B19**, 1065 (2005).
7. R. Shen, X. Z. Wang, W. J. Wen, K. Q. Lu, *Int. J. Mod.Phys.* **B19**, 1104 (2005).

8. X. Z. Wang, R. Shen, W. J. Wen, K. Q. Lu, *Int. J. Mod. Phys.* **B19**, 1110 (2005).

9. Y. Lu, R. Shen, X. Z. Wang, G. Sun, K. Q. Lu, *Smart Mater. Struct.* **18**, 025012 (2009).

10. J. B. Yin, X. P. Zhao, *Chem. Phys. Lett.* **398**, 393 (2004).

11. L. Xu, W. J. Tian, X. F. Wu, J. G. Cao, L. W. Zhou, J. P. Huang, G. Q. Gu, *J. Mater. Res.* **23**, 409 (2008).

12. Y. C. Cheng, J. J. Guo, G. J. Xu, P. Cui, X. H. Liu, F. H. Liu, J. H. Wu, *Colloid Polym. Sci.* **286**, 1493 (2008).

13. K. Q. Lu, R. Shen, X. Z. Wang, G. Sun, W. J. Wen, J. X. Liu, *Chin. Phys.* **15**, 2476 (2006).

14. K. Q. Lu ,R. Shen, X. Z. Wang, G. Sun,W. J. Wen, J. X. Liu, *Int. J. Mod. Phys.* **B21**, 4798 (2007).

15. R. Shen, X. Z. Wang, Y. Lu, D. Wang, G. Sun, Z. X. Cao, K. Q. Lu, *Advanced Materials*, **21**, 4631 (2009).

16. D. Wang, R. Shen, S. Q. Wei, K. Q. Lu, to be published in this proceeding.

17. X. Z. Wang, R. Shen, D. Wang, Y. Lu, K.Q. Lu, *Materials and Design,* **30**, 4521 (2009).

18. R. Shen, X. Z. Wang, Y. Lu, W. J. Wen, G. Sun, K. Q. Lu, *J. Appl. Phys.* **102**, 024106 (2007).

RHEOLOGY OF NOVEL FERROFLUIDS

DMITRY BORIN* and STEFAN ODENBACH

*Institute of Fluid Mechanics, Technische Universität Dresden,
Dresden, 01062, Germany*
E-mail: dmitry.borin@tu-dresden.de

The progress in the synthesis of new magnetic nanoparticles and agglomerates stimulates the development of novel ferrofluids with enhanced rheological properties. In the current work ferrofluids based on Co-nanoplatelets and clustered iron oxide nanoparticles have been considered. Steady-shear experiments and yield stress measurements of these ferrofluids have been performed using rotational rheometry.

Keywords: Ferrofluid; yield stress; flow curves; cobalt nanodiscs; clustered particles.

1. Introduction

Ferrofluids, suspensions of magnetic nanoparticles usually provided with a sterical stabilization shell of long chained polymers are commonly made using magnetite particles with a mean diameter of about 10 nm and a nearly spherical shape. The small size of the particles ensures that they are thermally agitated magnetic single domain particles in the carrier fluid, it guaranteeing that the suspensions are stable against sedimentation in the gravitational field and it keeping agglomeration due to magnetic interparticle interaction for vanishing magnetic field negligible.

From a rheological point of view fluids containing magnetite particles with this size are comparably uninteresting since interparticle interaction plays even in strong applied magnetic fields only a unimportant role and structure formation leading to significant changes of the rheological properties is negligible. Nevertheless it has been shown in the past that the small amount of larger particles usually contained in commercial ferrofluids gives rise to a variety of interesting changes of the rheological behavior in applied magnetic fields.[1,2] Moreover, first experiments containing spherical cobalt particles showed that changes in the composition of the fluid affecting the interparticle interaction lead to tremendous alterations of the fluids rheology.[3]

Recently novel ferrofluids containing particles with modified shape like e.g. Co-platelets[4] became available as well as fluids with preclustered magnetite nanoparticles[5] having low remanence but providing considerable interparticle interaction in magnetic fields.

Within the paper we present these novel fluids and discuss their rheological properties comparing with conventional ferrofluids.

2. Instrumental

For the magnetic measurements a vibrating sample magnetometer Lake Shore 7407 has been used. Field dependent investigations of the flow curves have been carried out using a shear rate controlled rheometer[1] with a cone-plate setup with an opening angle of 3° and a diameter of 76 mm combined with a Couette region. The yield stress has been measured with a specially designed stress controlled rheometer[6] using the same shear cell geometry. In the yield stress measurements we applied increasing shear stress to a fluid at rest until the first continuous measurable motion appears.

All rheological experiments were carried out at a constant temperature of 20°C.

3. Ferrofluids

The cobalt-nanodiscs were synthesized by thermal decomposition of the precursor material di-cobalt octa-carbonyl $[Co_2(CO)_8]$ as described in[4] and have been suspended in oleic acid. Characteristic parameters of the sample are collected in table 1. The modified interaction parameter λ calculated according to[2] under the assumption of a stack like particles alignment is 180.

Clustered magnetic iron oxide nanoparticles coated with a carboxymethyldextran shell were prepared for medical applications with the

Table 1. Characteristic parameters of the ferrofluid based on Co-plateles.

carrier liquid	Spontaneous magnetization kA/m	Saturation magnetization kA/m	Mean diameter nm	Mean height nm	Particle concentration vol. %
n-cetane adding oleic acid	$1{,}46{\cdot}10^3$	1,47	20	5	0,1

goal to reach a high overall magnetic moment with vanishing remanence. It was found before[5] that grains with diameters of d=10-30 nm form clusters with a size of D=50-70 nm covered with a surfactant shell. In the frame of the current work a ferrofluid based on these clusters has been prepared. The fluid was prepared from a nanocrystalline powder by use of oleic acid as surfactant and vacuum oil P3 (Pfeiffer Vacuum GmbH) as carrier liquid. The preparation followed the path described in:[7] KOH (5 molar aqueous solution, 4.3 ml), oleic acid (6.3 ml) and 10.5 ml water were added to the powder (3.0 g) to adhere to the surface of the particles in the form of oleate. The amount of oleic acid has to fit to the specific surface of the powder. After sedimentation in acetone the supernatant liquid was removed and the particles were dispersed in the oil. The characteristic parameters of this fluid are collected in table 2.

Table 2. Characteristic parameters of the ferrofluid based on clustered nanoparticles.

carrier liquid	Spontaneous magnetization kA/m	Saturation magnetization kA/m	d nm	D nm	Particles concentration vol. %
vacuum oil P3 oleic acid	$4{,}85{\cdot}10^2$	0,485	10-30	50-70	0,1

4. Flow Curves

Figure 1 shows the flow curves for the ferrofluid based on Co-platelets for various magnetic field strengths. In the absence of a magnetic field the behavior of the fluid is Newtonian. As the applied magnetic field rises, the viscosity of the fluid increases dramatically. For instance, with a magnetic field of H=2 kA/m and a stress τ=0,1 Pa applied to the fluid, the viscosity measured is about 40 times higher than the viscosity in the absence of the magnetic field. The magnetoviscous effect in this fluid is about 20 times higher compared to commercial ferrofluid samples containing magnetite particles despite the fact that such fluids contain up to 10 times more chainforming nanoparticles. This effect can be explained with the strong interparticle interaction of the magnetic particles in the fluid with nanodiscs, which is about 60 times higher than in classical magnetite ferrofluid.

It is obvious that an extrapolation of the flow curve to shear rate zero does not lead to a nameable value of yield stress, indeed the best fit of the

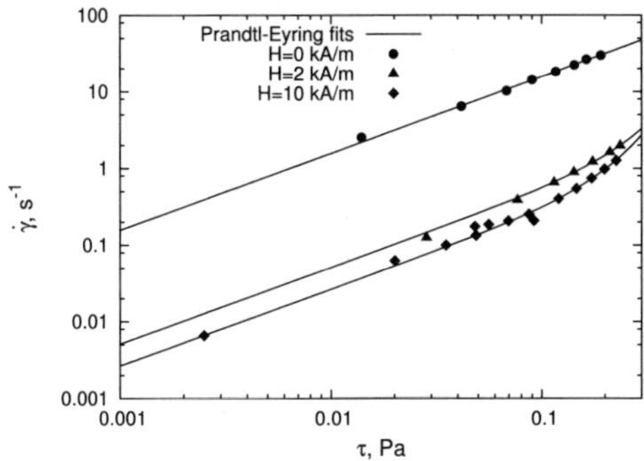

Fig. 1.   Flow curves of ferrofluid with Co-platelets for various magnetic field strengths. Solid lines are fits obtained with the Prandtl-Eyring model.

flow behavior of this fluid under the influence of a magnetic field strength is given by the Prandtl-Eyring model:

$$\dot{\gamma} = \frac{\tau^*}{\eta^*} \cdot \sinh\left(\frac{\tau}{\tau^*}\right),$$
(1)

where η^* and τ^* denot the characteristic viscosity and stress respectively.

The flow curves for the fluid with clustered iron particles measured for various magnetic field strengths are shown in figure 2. In the absence of a magnetic field the behavior of the sample is nearly Newtonian. As the applied field rises a strongly non-linear part of the flow curves appears in the range of small shear rates and the sample shows a pronounced shear-thinning behavior.

The magnetoviscous effect in this fluid, like for the sample with Co-platelets, is much stronger compared to commercial samples, which is explained with the same basic argument.

The data has been evaluated using typical models for non-Newtonian fluids and for the flow curves taken in a magnetic field strength range between 0 - 15 kA/m the best fits were obtained with a Power-Law model for which the shear stress as function of shear rate is given by

$$\tau = (K \cdot \dot{\gamma})^n,$$
(2)

where K is the flow consistency index and n is the dimensionless flow behavior index.

The best fits for flow curves corresponding to a magnetic field strength above 15 kA/m are obtained with the Herschel-Bulkley model.

$$\tau = \tau_{\mathrm{HB}} + \eta_{\mathrm{HB}} \cdot (\dot{\gamma})^n,$$
(3)

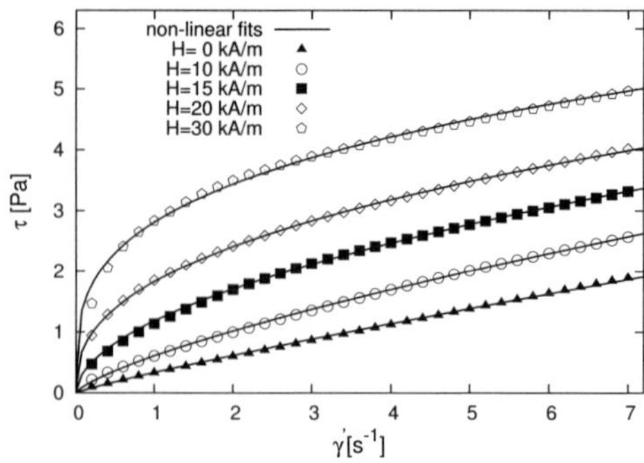

Fig. 2. Flow curves of ferrofluid with clustered particles for various magnetic field strengths. Solid lines represent non-linear fits (explanation is given in the text).

where η_{HB} is the Herschel-Bulkley or post yield viscosity and the parameter τ_{HB} denotes the yield stress.

5. Yield Stress

Using the procedure[6] in which the applied stress is increased stepwise starting from zero until a movement of the fluid is observed, the yield stress for both samples has been obtained as the maximum stress at which no motion could be measured.

The dependence of this static yield stress on magnetic field for both novel ferrofluids compared to ferrofluids based on spherical nanoparticles is presented in figure 3.

For the novel fluids the increase of yield stress with magnetic field strength is steep for small field strength and saturates afterwards. On the one hand this behavior can be explained with the magnetic interaction of the particles and on the other hand by the fact that only a small number of structures can be formed due to the low amount of magnetic material (0,1 vol. %) suspended. Clusters of the particles in these novel fluids are able to form more stable structures due to their larger size and summarized magnetization, leading to higher values of yield stress.

6. Slow Relaxation in the Fluid with Clustered Iron Particles

Time dependent rheological measurements of the fluid with clustered iron particles have shown that the time of the transient, until the shear stress in the fluid will be steady can reach several minutes and depends strongly on the strength of the applied magnetic field and shear rate. An example of

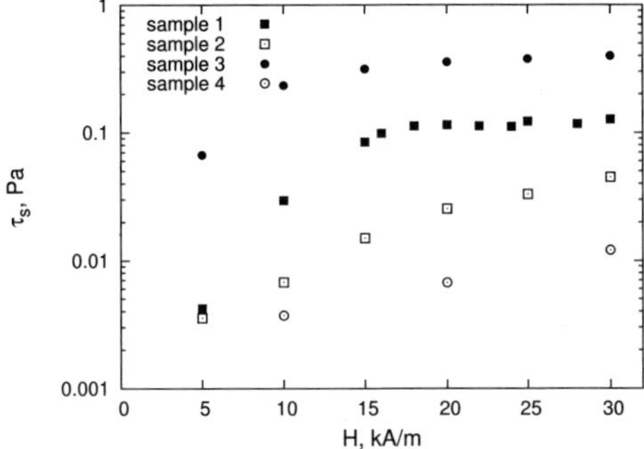

Fig. 3.   Static yield stress plotted as a function of magnetic field strength for the various ferrofluids: sample 1 is a fluid based on Co-platelets; sample 2 is a fluid based on spherical Co-particles with d=8 nm, dispersed in the oil L9 with a concentration of 2,87 vol. %; sample 3 is a fluid with clustered iron particles; sample 4 is a fluid with a spherical magnetite particles with mean diameter d=10 nm dispersed in ester with a concentration of 7,8 vol. %, containing 0,8 vol. % of particles with d >16 nm.

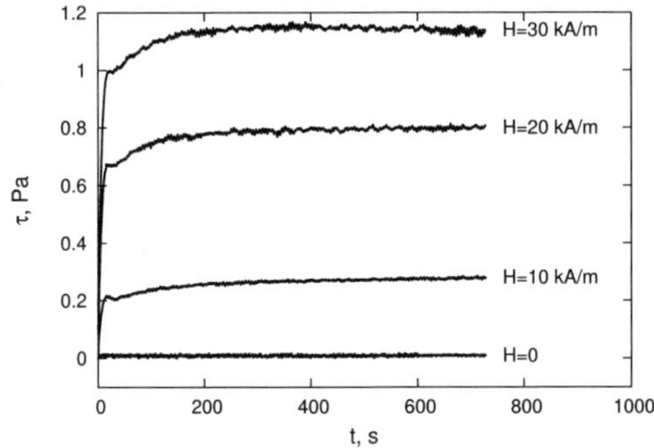

Fig. 4.   Shear stress relaxation for various magnetic field strengths after stepwise change of shear rate from 0 to 0.02 s^{-1}

the slow relaxation of the shear stress for various magnetic field strengths is presented in figure 4.

In the presence of a magnetic field the stress grows monotonically with time, reaches an overshoot, diminishes, grows again and reaches finally the steady stage. As smaller the shear rate is, as less evident is the overshoot and the stress relaxation tends to be monotonous. For large shear rate, an increasing height of the overshoot peak is found and it appears faster after the change of shear rate. Those effects can be attributed to the process of structure formation and destruction due to the simultaneous action of an applied magnetic field and shear flow. A similar behavior is known from the rheology of complex fluids like polymer melts with long chains. Calculations performed using ideas based on ref.[8] show that the chains are nearly monodisperse containing the maximal possible number of clusters for the given shear rate and almost all clusters are agglomerated in these chains. The length of the chains depends on the strength of the applied magnetic field and that leads to the increase of a polymer like behavior for increasing field.

7. Conclusions

A change of material and structural parameters of the nanoparticles has tremendous impact on the rheological properties of respective suspensions. Disc shape Co particles have a high interaction parameter resulting in strong structure formation and comparably high yield stress. Problems concerning stabilization of the disc shaped particles prevent a commercial production of such fluids. Fluids with clustered iron oxide particles have low remanence but strong magnetic particle interaction in presence of a magnetic field resulting in a higher magnetoviscous effect and enhanced yield stress compared to commercial ferrofluids. One of the features of this fluid is a slow relaxation phenomenon, resulting from the formation and destruction of agglomerates of the magnetic clusters and depending strongly on the relation between magnetic field strength and the rate of mechanical deformation. This fluid shows an increase of polymer like behavior for increasing field strength, a fact that should be taken into account for rheological measurements and possible applications of such ferrofluids with very pronounced structure formation.

Acknowledgments

The authors are very grateful to the Dr. R.Müller (Institute of Photonic Technology, Jena) for the preparation of the clustered nanoparticles and to Dr. D. Gräf (Saarland University) for the preparation of the Co-platelets.

References

1. Odenbach S., Störk H. *J. Magn. Magn. Mater.* **183**(1-2) 188 (1998).
2. Thurm S., Odenbach S. *J. Magn. Magn. Mater.* **252** 247 (2002).
3. Pop L.M., Odenbach S. *J. Phys.: Condensed Matter* **18**(38) S2785 (2006).
4. Shahnazian H. et al. *J. Phys. D: Appl. Phys.* **42**(20) 205004 (2009).
5. Dutz S. et al. J. Magn. Magn. Mater. **311**(1) 51 (2007).
6. Shahnazian H., Odenbach S. *App. Rheol.* **18** 54974 (2008).
7. Müller R. et al. *J. Magn. Magn. Mater.* **201**(1-3) 34 (1999).
8. Zubarev A.Yu., Iskakova L.Yu. *Physica A* **382** 378 (2007).

BEHAVIOR OF MAGNETO-RHEOLOGICAL FLUIDS IN MICROCHANNELS

J. WHITELEY, F. GORDANINEJAD and X. WANG

Department of Mechanical Engineering, University of Nevada, Reno, Nevada 89557
E-mail: faramarz@unr.edu

This study is focused on the behavior of magneto-rheological (MR) grease flow through microchannels with nominal internal diameters ranging from 75μm to 1mm. A PAO-based grease is synthesized and mixed with iron micro-powders with 1.1μm average particle size to create a MR suspension with 80% solids loading. A magnetic field is applied midway along the microchannel by an electromagnet, and the pressure gradient of the flow is measured across the microchannel for different channel diameters. The results show significant pressure drops for different magnetic field strengths. It is also demonstrated that there is good agreement between the experimental data and the macroscale theory, using the Hershel-Bulkley model for flow through a circular cross section. In addition, the effect of surface roughness on the pressure gradient is examined by evaluating different microchannels made of stainless steel, PEEK, and fused-silica with different values of surface roughness. It is observed that surface effect diminishes as the microchannel diameter decreases.

1. Introduction

Behavior of field-controllable fluids in fluid dampers and other devices at the macro/meso scale has been well established [1]. With the recent technological advances in micro and nanotechnology, it has become imperative to understand the behavior of MR fluid flow at the micro level. Preliminary research has produced a variety of results that deviate from the theoretical value of 64 for the product of the friction factor and the Reynolds number (f·Re), even for low Reynolds numbers. In addition, the results from the experiments that have been conducted thus far are not consistent with each other [2].

At the same time, experiments carried out by Judy and others [3] where liquid flow was studied in stainless steel and fused silica microchannels indicate that there is no deviation from conventional theory. Materials that have been studied include water, oil, alcohol, and gasses.

Since MR fluids are suspensions of magnetic particles in a liquid, such as silicon oil or water, their flow at microscale level introduces a new arduous challenge [4]. The flow behavior of MR fluid through microchannels is similar to a particle-laden flow. Studies have shown that as a rule of thumb with particle laden flows, the ratio of the particle diameter to the channel diameter must be at least 1/10 to prevent fluid from clogging. Theoretical studies of particle laden flow in microchannels do not take into account the effects of particle agglomeration, settling, and tubing diameter transitions that can result in clogging [5-6]. In this study, through the experiments, it was observed that this rule does not seem to hold true. This challenge is compounded further; as the iron particle size becomes smaller in MR fluids, the magnetic force that the fluid can develop decreases by the square of the particle radius.

2. Experimental Study

Specialized equipment for this experiment was required due to the precision of the low-volume flow rates and the high pressures that would be measured. A diagram of the experimental setup is shown in Figure 1a. The fluid delivery is operated by an 8mL stainless steel syringe driven by a high pressure programmable syringe pump that is capable of a 200lb linear force. As the fluid exits the syringe, it transitions into stainless steel tubing and components that allow for the connection of the pressure transducer. At the exit of the pressure gauge tee the microchannel test section starts. An electromagnet with a gap perpendicular to the microchannel is located in the middle of this test section and is capable of producing 0.56 Tesla when 2Amp of electric current is applied to the coil, as shown in Figure 1b. The fluid exits to atmospheric pressure at the end of the test section.

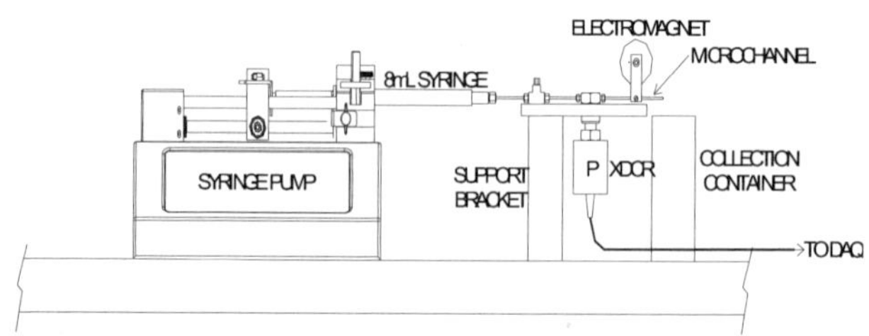

Figure 1a. Diagram of the experimental test setup.

The data acquisition (DAQ) system consists of a National Instruments (NI) SCXI-1320 controller interfaced to Labview on the computer. DC power supplies are required to create the magnetic field for the electromagnet and power the pressure transducer and NI controller.

Conventional MR fluids clog up in microchannels with IDs (Internal Diameters) below 200µm. In addition, the experimental MR fluids with nanometer sized iron powders, while able to flow through smaller microchannels, have not been able to produce a significant pressure drop from an applied magnetic field. To overcome this challenge, a special PAO-based grease is synthesized and mixed with iron micro-powders with 1.1µm average particle size. The grease maintains particle suspension and separation allowing flow through smaller microchannels down to a nominal ID of 75µm (see Figure 2). The MR grease (MRG) is pushed through various sized microchannels at different flow rates and magnetic field strengths while the pressure drop is recorded. To study the effect of different channel diameters and surface roughnesses, microchannels made of stainless steel (SS), PEEK, and fused silica materials with nominal internal diameters ranging from 75µm to 1mm are tested.

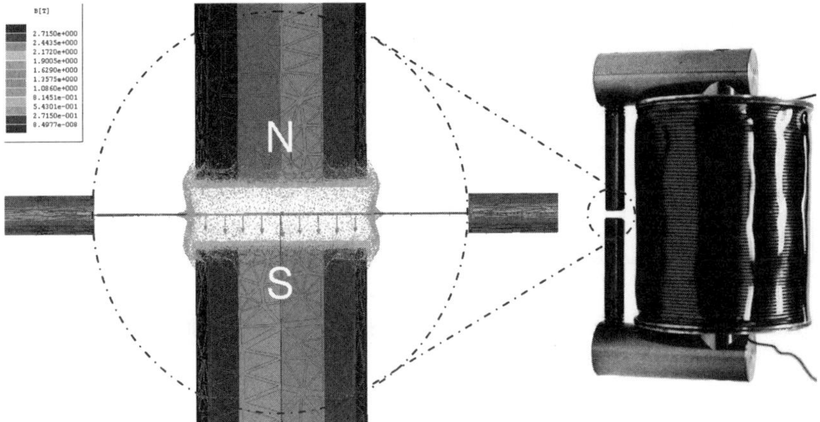

Figure 1b. Finite element analysis of electromagnet.

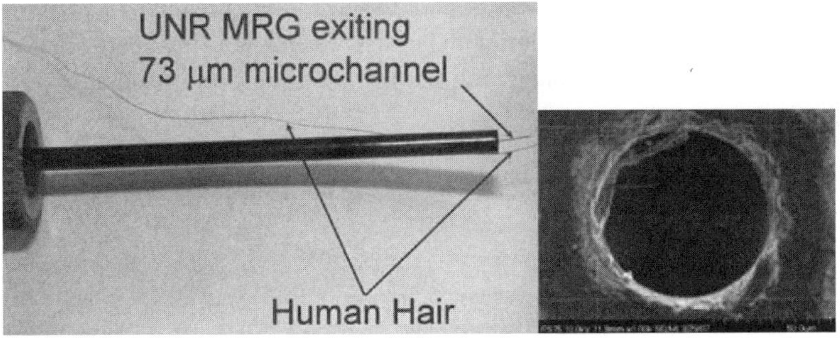

Figure 2. MRG flow through 73µm ID fused silica microchannel.

The MRG developed by the chemical engineering department at University of Nevada, Reno (UNR) that is capable to accept the addition of large amounts of carbonyl iron powders while still maintaining its base viscosity. The composition of this base grease consists of PAO as carrier medium (83.3% weight of grease), modified smectite clay (12.6% by weight) and sorbitan monooleate (4% by weight). This combination is mixed by a 500rpm mixer for

526

1 hour. The use of grease overcomes the challenge of particle settling. The effects that the weight percentage loading of ISP S-3700 micropowder has on shear stress under an applied magnetic field is shown in Figure 3.

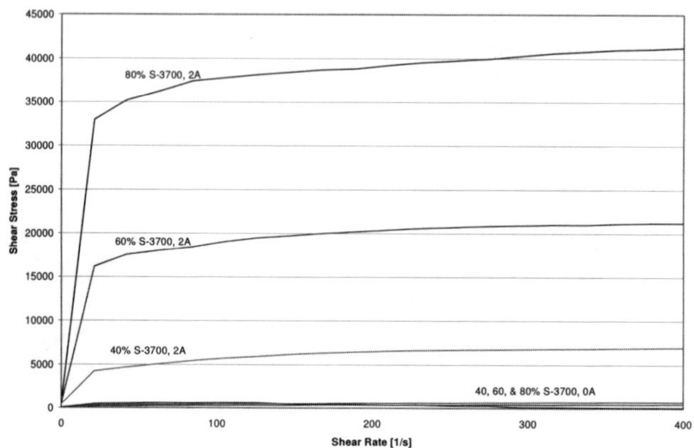

Figure 3. Flow behavior of PAO-based grease with 40, 60, & 80%wt. ISP S-3700 iron micropowder in 1mm gap of the Paar Physica MCR-300 shear rheometer when 2Amp (0.5T) field is applied.

3. Results and Discussion

Figure 4 illustrates the typical results of the raw data for pressure drop vs. time for different flow rate of UNR base grease through a microchannel of 75μm ID. Random clogging is observed during the measurements showing a large spike in pressure drop. For microchannels with IDs that are 150μm and smaller,

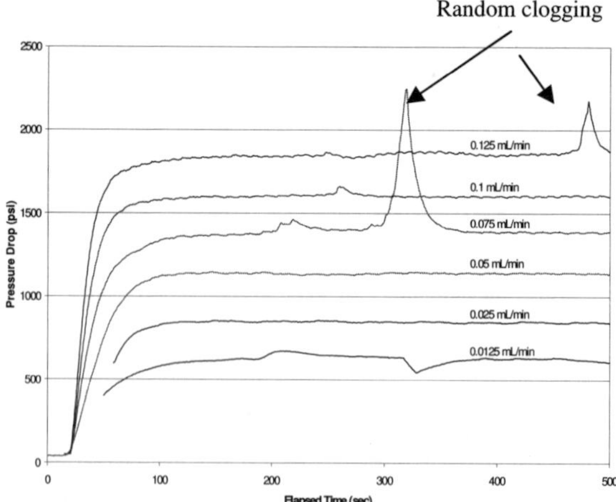

Figure 4. Raw pressure drop data for UNR base grease across 75μm nominal ID, 5cm length PEEKsil™ microchannel.

random clogging would occur. Although, not as severe, random clogging also happened for the base experimental grease with no iron micropowders. This random clogging is not necessarily due to the same mechanism as a commercial MR fluid. In the case of the latter, the MR fluid either does not flow at all through the microchannels or does consistently clog up just after flow initiation, depending on the microchannel diameter. During measurements of the pressure drop for the MRG either the off or the on (magnetic field applied) states, all of the clogging that occurred eventually broke free as the pressure increased and steady flow resumed.

Figure 5 illustrates typical experimental results compared with the Herschel-Bulkley theoretical model showing good agreement between them for this flow geometry. The detail of the Hershel-Bulkley model for flow through a circular cross section can be found in references [7,8]. In this study, all the experimental data for the pressure drop versus flow rate can be fitted by the Hershel-Bulkley flow model. This indicates that the MRG flow still follows the continuum mechanics law at least down to this microscale tested.

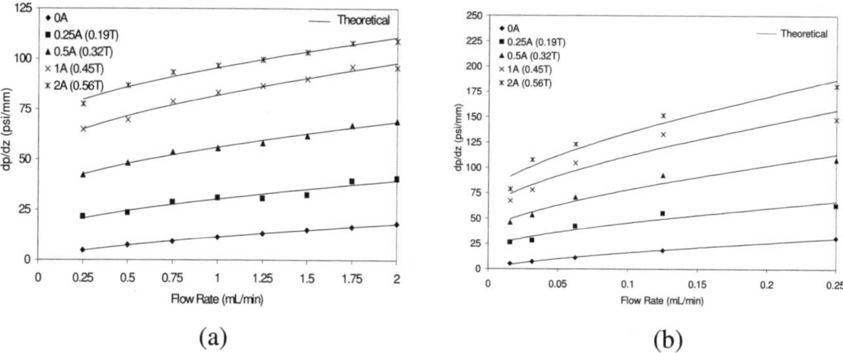

(a)   (b)

Figure 5. Pressure gradient vs. flow rate across: a) 250µm stainless steel b) 123µm PEEK microchannels. The solid lines are fitted experimental data for the MRG flow using Herschel-Bulkley model.

The dynamic force range or dynamic pressure drop range is often used to characterize the performance of the MR effect in a device. The dynamic pressure drop range is presented in Figure 5, as the area between the pressure gradient between off and maximum on state results. The dynamic pressure drop can also be measured by a ratio of the pressure drop due to the MR effect, ΔP_{MR}, to the pressure drop due to the viscous component, ΔP_{vis}. As the microchannel diameter decreases, the dynamic pressure drop $\Delta P_{MR}/\Delta P_{vis}$ also decreases, as the viscous forces dominate the flow. Differences in the effect of surface roughness between the different microchannel types are apparent at the larger diameter channels, but difficult to discern at the smallest microchannels tested.

528

It has been shown that there is a relationship between the surface roughness and the effect that the applied magnetic field has on the flow of a MR fluid. In general, by increasing the surface roughness in a channel, the measured pressure drop of MR flow increases for a given applied magnetic field [9].

Figure 6 (a,b) present the shear yield stress obtained from Hershel-Bulkley flow model vs. the applied magnetic field. Values for both stainless steel (rough) and PEEK (smooth) material types are shown for comparison. In Figure 6(a), the results show a tendency that the PEEK channels with their relatively lower surface roughness have a smaller shear stress than that of the stainless steel channels. This is consistent with the previous findings of Ref. [9].

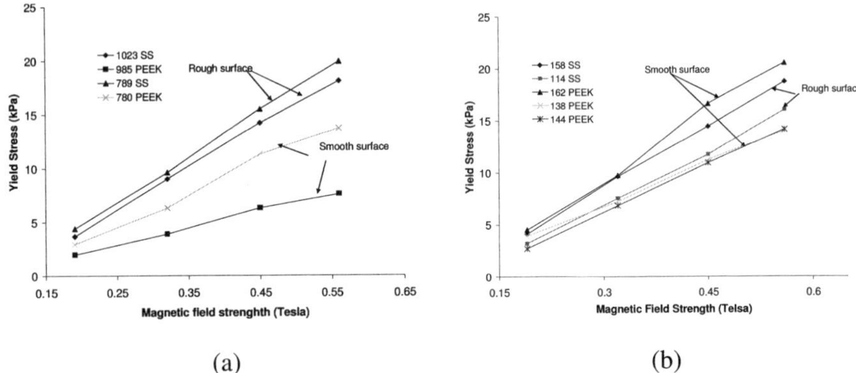

(a)   (b)

Figure 6. Measured yield stress vs. magnetic field strength for a) large stainless steel and PEEK channels, the ID values from 780µm to 1023µm. b) small stainless steel, PEEK, and PEEKsil™ channels, the ID values from 114µm to 162 µm. The surface roughness of stainless steel is 16µm RMS; while the PEEK tubing is 0.06µm RMS.

However, in Figure 6(b) the effect of the surface roughness is not easy to discern. The stainless steel and PEEK materials have a shear yield stress that is within the margin of error for the measurements. These results suggest that surface effects may diminish as the microchannel diameter deceases. This may be related to a decreasing MR effect for the smaller active fluid volume, or the increasing viscous forces acting on the fluid. Additionally, these results could be dependent on the fluid properties of the MRG. Further investigation is warranted to explore this finding in more detail.

4. Conclusions

Significant pressure drops for flow of a MR grease across microchannels with nominal IDs ranging from 75μm to 1mm to are obtained for different applied magnetic fields. The Herschel-Bulkley model for non-Newtonian fluid flow can fit all the experimental data very well. It demonstrates that the flow behavior of MR grease still follows the continuum mechanics law at least down to this microscale diameter that is tested. However, the effect of the surface roughness on the pressure gradient diminishes as the microchannel diameter decreases.

Acknowledgments

The authors are thankful for funding provided by the National Science Foundation for this project. The assistance from the Polymer Science and the Material Science and Engineering students at the University of Nevada, Reno for sample preparation, is warmly appreciated.

References

1. X. Wang and F. Gordaninejad, *Intelligent Materials*, Edited by M. Shahinpoor and H.-J. Schneider, Royal Society of Chemistry Publishing, Cambridge, United Kingdom, 339 (2007).
2. I. Papautsky, T. Ameel and A.B. Frazier, *Proc. of the ASME International Mechanical Engineering Congress and Exposition*, November 11-16, New York, NY USA., pp1-19 (2001).
3. J. Judy, D. Maynes and B.W. Webb, *Int. J. Heat Mass Trans.*, **45**, 3477 (2002).
4. Liu, J., Flores, G.A., Sheng, R., *J. Mag. Magn. Mat.*, 225, 209 (2001).
5. K. Sharp and R. Adrian, *ASME IMECE2001/MEMS*-23879, **2** (2001).
6. C. Kormann, H.M. Laun and H. J. Richter, *Int. J. Mod. Phys. B*, **10**, 3167 (1996).
7. X. Wang and F. Gordaninejad, *J. Intell. Mater., Syst. Struct.*, **10** (8),601 (1999).
8. J. Whiteley, F. Gordaninejad and X. Wang, *J. Appl. Mech.* **77**(4), 041011 (2010).
9. F. Gordaninejad, B. Kavlicoglu and X. Wang, *Int. J. Mod. Phys. B* **19** (7-9): 1297 (2005).

RELATION OF LAMELLAR STRUCTURE AND SHEAR STRESS OF DYNAMIC PM-ER FLUIDS[*]

D.K. LIU, C. LI, J. YAO, L.W. ZHOU and J.P. HUANG

State Key Laboratory of Surface Physics and Department of Physics, Fudan University, Shanghai 200433, China

To understand the dynamic rheological behavior of polar molecular electrorheological (PMER) fluids, the shear stress and viscosity of the colloids are compared with the parameters of their lamellar structures which are obtained simultaneously with the rheological characteristics using an electrorheoscope. The results of the experiments and molecular dynamics simulation indicate that the shear stress is mainly contributed by the moving particle rings, and there is an inverse correlation between the width of the moving particle rings and the shear stress.

1. Introduction

In applications of ER fluids, the shear stress at high shear rate is usually a major concern since shear thinning often observed in almost all the ER materials. To find the mechanism of shear thinning of ER fluids under high shear rate, it is necessary to understand how the lamellar structure [1,2] is formed and changed under both electric and shear fields, and how the shear stress is changed in succession. It was demonstrated there is sudden change of shear stress upon application of sudden change of shear rate, and it is proposed that it is relating to the formation of lamellar structure [3]. Two-fluid model [4] was proposed to prove that it was particle density fluctuation that was responsible to the formation of lamellar structure. Recently, Onsager principle [5,6] was employed to obtain that particle structure of dielectric ER fluid when the system is not far from an equilibrium state. Study of the dynamic structure may help us understand shear thinning of ER fluids under high shear rate. [7,8] This work concentrates on the experimental relation of the shear stress and the particle lamellar structure, and tries to understand their quantitative relation and to enrich the knowledge needed for the design of new materials that have better dynamic ER effect under high shear rate.

[*] This work is supported by grants 10974030 and 10574027 of NNSF of China, the 211 Project of MEC, the SEC "Shu Guang" project, by the STCSM-PTP 6PJ14006), and by CNKBRSF, Grant 2006CB921706.

2. Experiments

The PMER fluid consisted of TiO2 coated with 1,4-butyrolactone molecules prepared using the sol-gel method [9] and silicone oil (100cSt). The mixture ratio of the powder to silicone oil is 3g to 1.6ml.

A top-viewing electrorheoscope is home-made with a DC motor as both a rotor driving mechanism and a measure of ER shear stress which is proportional to the driving current of the motor. A silicone oil with viscosity of 20k cSt at 25°C is used to calibrate the viscosity measurement. The error of the deduced shear stress measured in this way is around 2-5%. An inverted electrorheoscope is modified from a Haake-Mars II electrorheometer by adding three microscopes with different viewing range of 35mm, 1mm and 10 μm in diameter and replacing parallel plates with two ITO glass discs as electrodes. The bottom plate rotates in the top-viewing scope, while the top one rotates in the inverted scope. In this way, lamellar structures can be recorded simultaneously when the shear stress and viscosity are measured.

2.1. *Results with the top-viewing electrorheoscope*

Using the material mentioned above, at a ratio of 3g powder to 1.6ml silicon oil (100cSt), we maintain a rotational speed of 300rpm, when E is 0v/mm, there is no obvious structure; when E reaches 500V/mm a lamellar structure emerges. Fig. 1 shows the concentration of the rings in the top view image of the lamellar structure at different field when rotational speed is 330rpm. Almost all the rings are rotating at this rotational speed. With the increasing electric field, the number of the rings increases, and so does the shear stress.

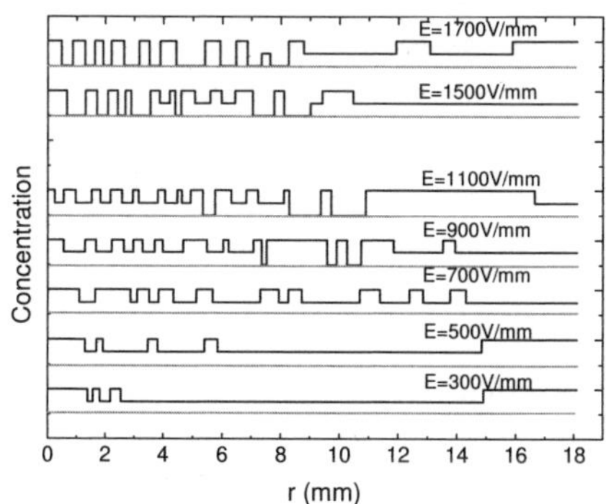

Fig. 1 The concentration of rings in the top view image of lamellar structures at different field.

When the electric field maintains at 1300V/mm, increasing the shear rate would first increases and then decreases ring numbers (Fig. 2), but its shear stress increases and viscosity decreases (Fig. 3b) monotonically.

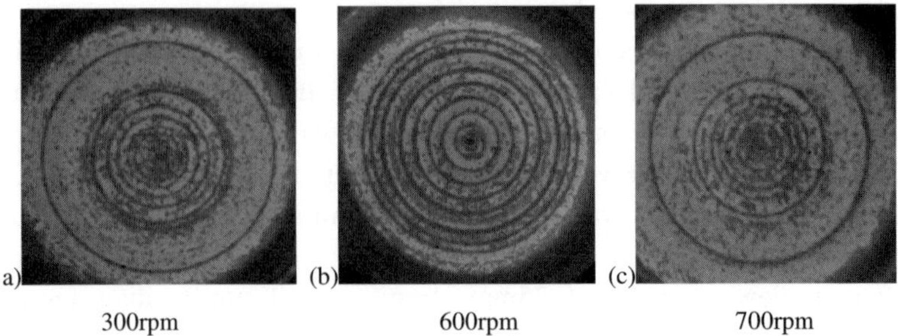

Fig. 2 The snapshots of upper lamellar pattern of the PMER fluid when E=1300V/mm at different rotational speed denoted under each pictures with driving current of the DC motor.

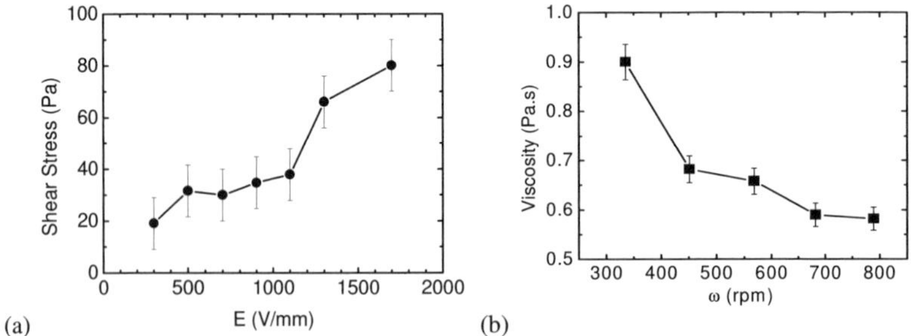

Fig. 3 (a) the shear stress versus E when ω=335prm, and (b) versus ω when E=1300V/mm.

In comparison of Figs. 1 and 3a, at a fixed shear rate and increasing electric field, ring number increases and so does shear stress. However, one can not draw a conclusion that nunber of rings is corresponding to certain shear stress since, as shown in Fig.3b, under certain field, when shear rate increases, the number of rings increases and then decreases after ω=600 rpm, while the shear stress increases and viscosity decteases monotonically. When shear rate is too high, all the rings would disappear.

2.2. *Results with the inverted electrorheoscope*

Using the inverted electrorheoscope, we recorded images viewed from the lower electrode (Fig. 4), and measured shear stress at the same time at different field while the rotational speed is kept at 300rpm. Fig. 4 shows the images of lamellar structure under different electric field. Fig. 5 is a summary of concentration, position, width and speed of particle rings at different electric fields.

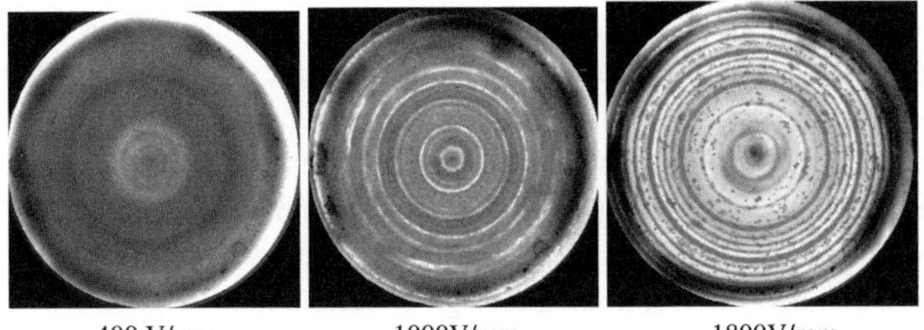

Fig. 4 The bottom view of the lamellar structure of the PMER fluid at different field when rotational speed is 300rpm.

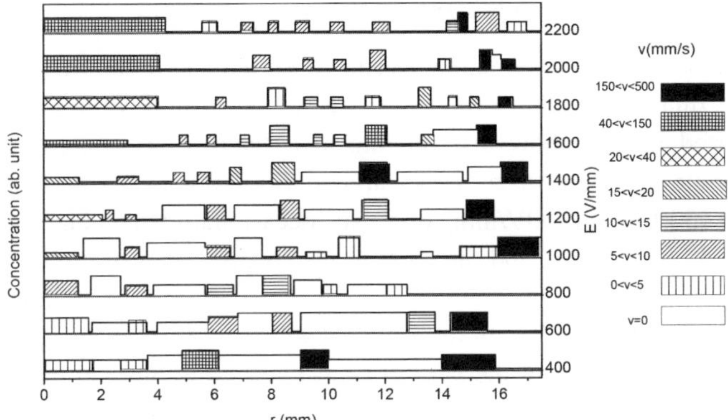

Fig. 5 Particle concentration and linear speed v of particle rings in the bottom image of lamellar structure of the PMER fluids at different field while the rotational speed is kept at 330 rpm.

We can see from Fig. 6 that, with the decrease of the electric field, the shear stress decrease linearly, also with the increase of the E, the shear stress increase linearly under 1600v/mm. So, we can see that our material is good PMER material.

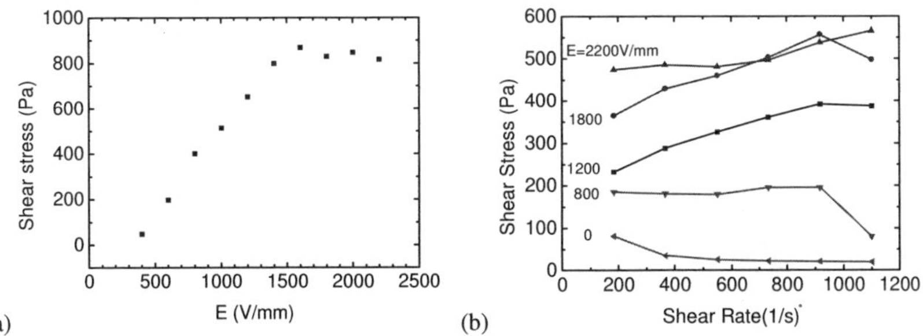

Fig. 6 (a) Shear stress of the PMER fluid at concentration of 3g/1.6ml=1.875 g/ml versus electric field when ω= 300rpm; (b) Viscosity of the fluid at different fields.

534

We want to get the quantitative comparisons between the ER effect and the parameters of lamellar structure. So we calculate the average width of the moving particle rings and find that there is apparent inverse correlation between the width of the moving particle rings and the shear stress. So, we make a guess that the shear stress is mainly contributed by the moving particle rings. To prove that we also make a comparison between shear stress and the average width of the rings that don't move, finding no sign of correlation.

3. Discussions

3.1. *Thin and moving rings contribute to shear stress*

It is found that, when the rotating speed is 300 rpm and in the field between 400 and 800V/mm, the width of rings are large (about from 0.7mm to 1.2mm), and the inverted images show that rotating and steady rings appear alternatively, corresponding to the shear stress around 300-600Pa. In the medium field between 1000 and 1200V/mm, one may occasionally find that both adjacent rings would be rotating, and the most of the rotating rings are thin while steady rings have about the same width as in the lower filed, corresponding to the shear stress around 800-1200Pa. In the higher field between 1600 and 2200V/mm, all the rings are rotating and the width of most rings are thin and only few are as large as 0.6mm, the corresponding yield stress is up to 1400 and near 2000Pa.

Fig. 7 implies when thick rings (Fig. 7a) are broken into thin ones (Fig. 7b), the corresponding shear stress is dramatically increased, and when the thin rings are merged into thick ones, the shear stress is going down. The reason why smaller walls give larger force can be drawn from the Stokes equation of viscous force for spherical particles, $F = 6\pi\eta rv$, where η is the viscosity of base liquid, r is particle radius, and v, the velocity difference between the particle and base liquid. If a large particle of radius R and volume V, is broken in to three small particles of radius r while the total volume of the three particles remains V. Thus $3F(r)/F(R) \sim 2.08$, namely, the three small particles would have shear stress as twice as much than the large one.

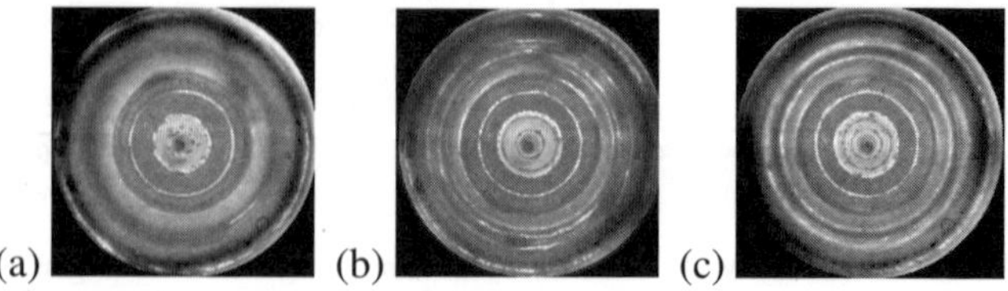

Fig. 7 Comparison of lamellar structures and their corresponding shear stress.

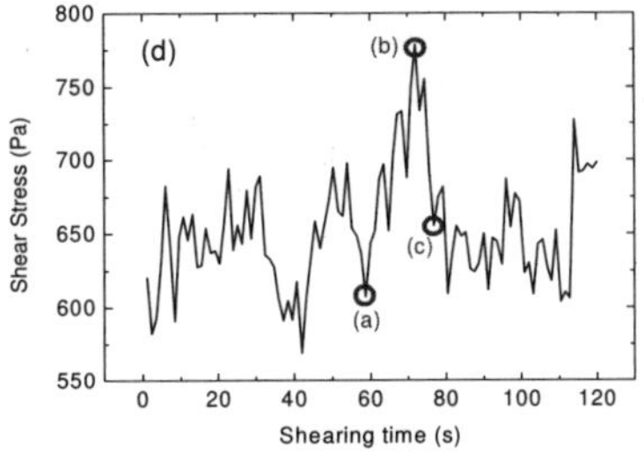

Fig. 7 (*Continued*).

3.2. *Thin and moving rings contribute more to shear stress*

It is found that parameters of lamellar structures are closely related with the rheological properties of ER fluids. These parameters of lamellar structures include the radius, width, concentration and linear speed of rings under certain electric field and rotating speed. To confirm this, a molecular dynamics simulation has been performed with the same condition as experiment (ε_p=37 ε_f=2.5, η_f=100cSt, ϕ=30% and particle size of 1μm). Since the system is symmetric, we started with a radial long box consisting of only 240 particles and linear speed of upper boundary was gradually changed. Final result is obtained by rotating such box for 13 times, and the calculation was reduced.

The result of molecular dynamics simulation is shown in Fig. 8, and the pattern of the lamellar structure has similar characteristics as seen in the experimental results. Namely, when electric field increases from 400, 800 to 2200, the number of rings increases from 6, 9 to more than 11, accordingly, and gaps between rings show quite low concentration only in high field.

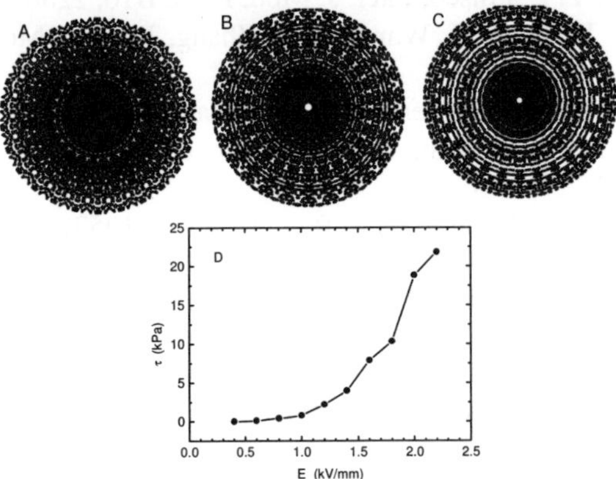

Fig. 8 The molecular dynamics simulation of lamellar structures of the PMER fluid under (a) E= 0.4, (b) 0.8, (c) 2.2 kV/mm at rotational speed of 300rpm with (d) shear stress vs. E at the solid/oil ration of 1.88 g/ml. Every state in the simulation was for 10s of equilibrium.

The Onsager principle of minimum energy dissipation rate [10] is also used to establish governing equations which are solved numerically using Comsol. Its result agrees qualitatively with the experiments reported here and will be presented elsewhere.

4. Conclusions

To understand the cause of shear thinning of polar molecular electrorheological (PMER) fluids, the shear stress and viscosity of the colloids are compared with the parameters of their lamellar structures which are obtained simultaneously with an electrorheoscope. The result of the experiments and molecular dynamics simulation indicates that the shear stress is mainly contributed by the moving particle rings, and there is an inverse correlation between the width of the moving particle rings and the shear stress. It proposed that the energy dissipation plays an important role in the shear thinning of dynamic PMER system. The undergoing work includes particle velocity measurement using Doppler effect, 3D structure of lamellar structure using a confocal microscope while sample is under electric field and shear flow.

Acknowledgments

The authors wish to thank the discussions form J.W. Zhang, W. Chen, P. Tan, J. Zheng, Y.J. Lu, Y. Jia, S. Shen, X. Zhou, D. Wu, H.Z. Zhou and S.Y. Liu.

References

1. S. Henley and F. E. Filisco, Inter. J. Mod. Phys. B16, 2286(2002).
2. X. Tang, W. H. Li, X. J. Wang, P. Q. Zhang, Inter. J. Mod. Phys. B. 13, 1806 (1999);
3. M.L. Zhang, Doctoral Dessertation, Tsinghua Univ., 2009.
4. K.V. Pfeil, M.D. Graham, and D.J. Klingenberg, Phys. Rev. Lett. 88, 188301 (2002).
5. L. Onsager and S. Machlup, Phys. Rev. 91, 1505 (1953);
6. Jianwei Zhang, Xiuqing Gong, Chun Liu, Weijia Wen, and Ping Sheng, Phys. Rev. Lett. 101, 194503 (2008);
7. J.E. Martin, J. Odinek, T.C. Halsey, R. Kamien, Phys. Rev. E 57, 756 (1998).
8. T.C. Halsey, J.E. Martin, D. Adolf, Phys. Rev. Lett. 68, 1519 (1992).
9. L. Xu, W. J. Tian, X. F. Wu, C. X. Wang, J. G. Cao, L. W. Zhou, J. P. Huang and G. Q. Gu, J. Mater. Res. 23(2) 409 (2008).
10. J. W. Zhang, X. Q. Gong, C. Liu, W. J. Wen and P. Sheng, Phys. Rev. Lett. 101, 194503 (2008).

BEHAVIOR OF THICK MAGNETORHEOLOGICAL ELASTOMERS

P. MYSORE, X. WANG and F. GORDANINEJAD

Department of Mechanical Engineering
Composite and Intelligent Materials Laboratory
University of Nevada, Reno, Nevada 89557, USA

In this study, the behavior of thick magnetorheological elastomers (MREs) has been experimentally investigated. Two types of MRE specimens of varying concentrations, with circular and rectangular shapes having thicknesses from 6.35mm to a maximum of 25.4mm are prepared. These samples are evaluated under quasi-static compression and double lap shear tests. The shear and the Young's moduli of the MREs are obtained under different applied magnetic fields. It is observed that the field induced change in modulus is independent of the thickness of the MRE and only dependent on the iron particle concentration and the magnetic field strength. With the increase in applied magnetic field, it is observed that the change in modulus from a linear trend at lower field to a non-linear trend at higher field. It is found that thick MREs are more controllable in compression mode than in shear mode.

1. Introduction

Magnetorheological elastomers are a class of materials which contain chain-like magnetizable particles embedded in a non-magnetic elastomeric matrix. They have the ability of changing their stiffness under an applied magnetic field and are hence characterized by field-dependent modulus [1]. Numerous studies have been performed on the field dependent properties of MREs in different loading conditions [2-3], but all were limited to a maximum MRE thickness of 10mm. In this work, the field-dependent compressive and shear moduli have been investigated for MREs of different thicknesses and the effect of dimension of specimen on the change in these moduli is studied.

2. Material Preparations

In the present study, carbonyl iron particles, with an average particle size of 2-8 microns are embedded in a two-part silicone elastomer to make MRE samples. Iron particle concentrations varying between 30wt.% and 70wt.% o are used. The two parts of the silicone elastomer are mixed in the ratio of 10:1. Iron particles are then introduced and a mix well. The mixture is then degassed and placed under a magnetic field of over 1 Tesla to cure. During this curing process, the iron particles aligned along the direction of the magnetic field. For

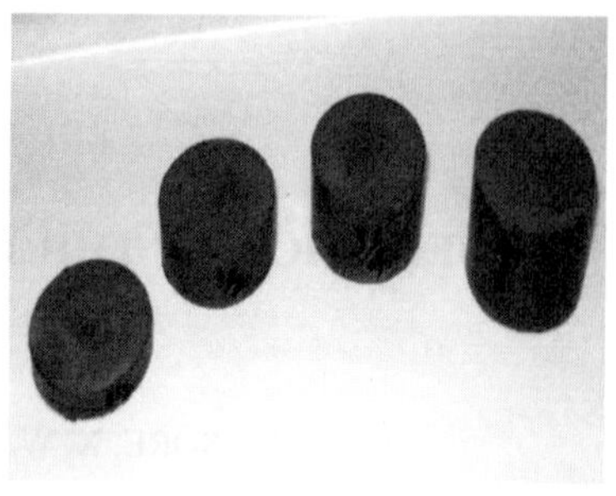

Figure 1(a). Shear test samples.

Figure 1(b). Compression test samples.

compression tests, circular cross-section samples (Figure 1(a)) and for shear tests, rectangular cross-section samples (Figure 1(b)) are prepared with thicknesses varying from 6.35mm to 25.40mm.

3. Experimental Setup

3.1. *Lap-shear test*

The lap-shear test setup (Figure 2(a)) consists of a U-shaped steel structure which houses an electromagnet to generate magnetic field and fixtures to hold the MRE in place. To ensure maximum flow of magnetic flux to the MRE, the samples are glued to 2.54mm thick steel plates which are 12.70mm wide. A commercially available adhesives is used to connet the samples to the steel plates. Two pieces of MRE are sandwiched between three thin steel bars. Shear force is applied on the MRE samples by pushing the center steel bars. Magnetic field is applied along the direction of particle alignment in the MRE. The MRE test specimen has an area of $285.16mm^2$ (22.45mm X 12.70mm) and four different thicknesses ranging from 6.35mm to 25.40mm.

3.2. *Compression test setup*

The compression test setup (Figure 2(b)) consists of a cylindrical steel casing and a T-shaped bottom steel base that rests on the bed of an Instron machine. The MRE test specimen rests on a T-shaped steel support base with a steel piston compressing it from the top. To be able to record the stiffness variation at on-state, an optimized magnetic flow path has been designed such that there is a maximum magnetic flux through the MRE specimen.

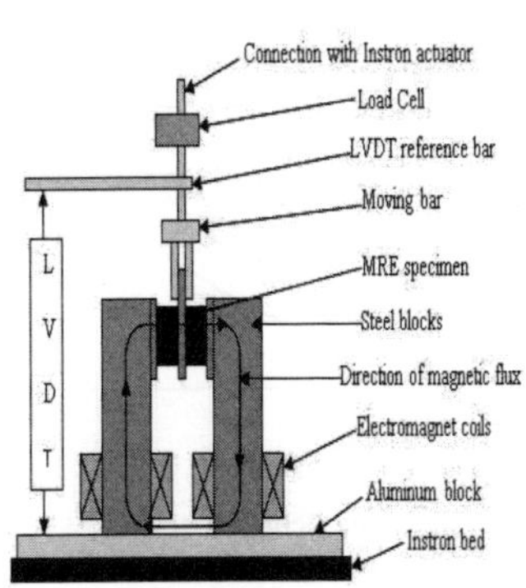
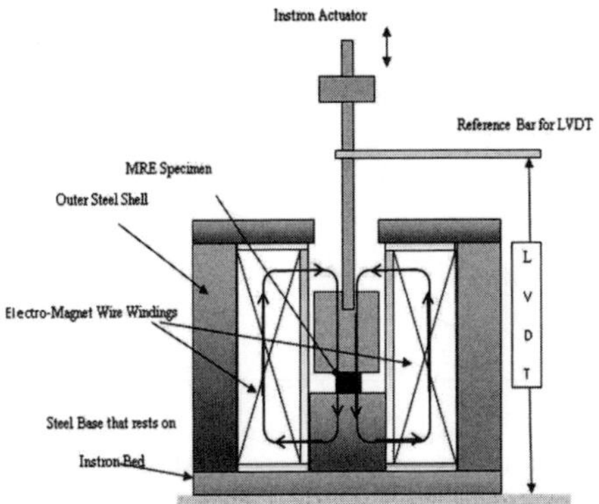

Figure 2(a). Schematic of shear test setup.　　Figure 2(b). Schematic of compression test setup.

3.3. *Test procedure*

The force on the MRE is applied by a 20-kip biaxial INSTRON system and its magnitude is recorded by a load cell. The corresponding displacement of the moving piston which represents the sample deformation is recorded through the LVDT displacement sensor. Each MRE sample is subjected to quasi-static displacement. The raw data of displacements and the corresponding force responses of the MRE samples under shear are processed to convert the force into engineering stress and the displacement into engineering strain. The measured force divided by the initial cross section area of the sample is the engineering stress. The obtained engineering strain is expressed as the ratio of total displacement to the initial thickness of the MRE samples in which the forces are being applied.

4. Results and Discussions

Figure 3(a) presents the typical stress-strain results showing the hysteresis behavior of MREs at different applied electric currents under quasi-static shear tests for a maximum strain of 5%, 10% and 15%, respectively. At small strains, it is observed that the stress-strain results follow an elliptical path which demonstrates a linear viscoelastic behavior of MREs. At higher strains it is observed that, in the presence of an applied magnetic field, the samples show signs of residual deformation at the time of unloading, in which a steep drop of stress occurs. The higher the applied magnetic field, the higher is the drop in stress value. This phenomenon may be due to the break up of particle chains, when the MRE is under post-yield deformation. It is also noticed that this

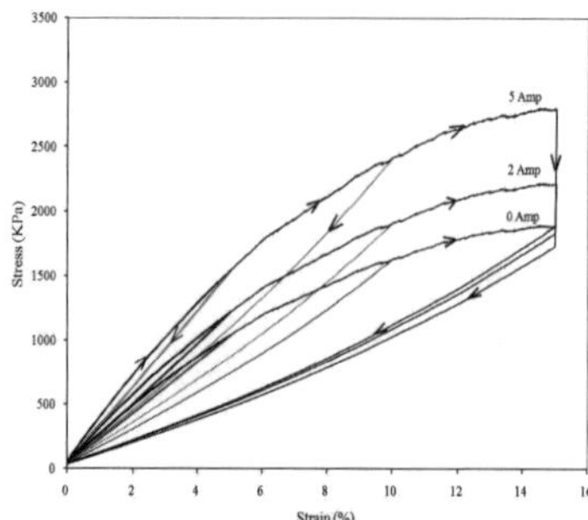

Figure 3(a). Hysteresis behavior of MREs under a quasi-static shear test with a maximum strain of 5%, 10% and 15%. The data corresponds to the 12.7mm thickness and 30wt% iron particle specimen.

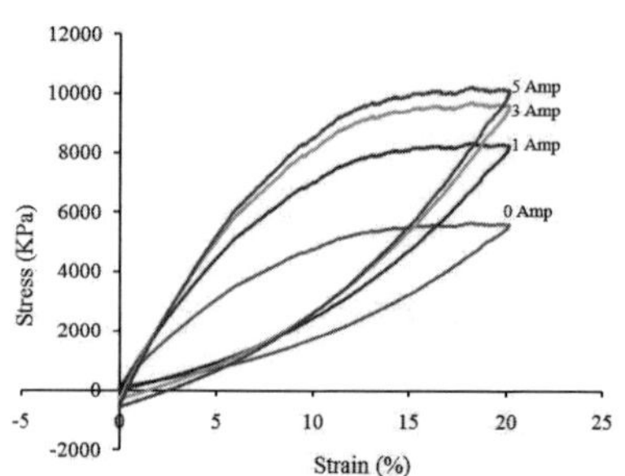

Figure 3(b). Hysteresis behavior of MREs, under a quasi-static compression test, with a maximum strain of 20%. The data corresponds to the 12.7mm thickness and 70wt% iron particle specimen subjected to different applied currents.

deformation disappears when the magnetic field is turned off. It appears that the applied magnetic field induces a pseudo-plastic effect in an MRE in shear.

Figure 3(b) presents the typical stress-strain curves showing the hysteresis behavior of MREs at different applied currents under compression tests. An offset in stress is observed for the on-state stress-strain results when compared to the off-state stress-strain data. This is due to the fact that there is an additional pull on the load cell when the electromagnet is turned on. This offset however does not affect the measurements of Young's modulus of the MRE samples. It is shown that with the applied magnetic field, the elastic modulus as well as the energy dissipated in the MRE increases.

Tangential Modulus is calculated for the stress-strain results and it is demonstrated that, even in the case of thick MRE, with increase in strain in the MRE, the shear modulus decreases. This explains the highly non-linear stress-strain behavior at high strains. Figures 4(a) and 4(b) show this phenomenon. One can also observe that both the shear and compressive moduli are almost constant when the strain is in the range of 1%. Thus, in this study, the modulus of the MRE samples are determined at strains less than 1% which is taken as a linear-elastic deformation. With the increase in the applied magnetic field, a profound increase in shear and compressive moduli is observed for all the MRE samples tested.

An electromagnetic finite element analysis is carried out to determine the imposed magnetic field strength on the MREs. Compressive and shear test setups are modeled and analyzed in three-dimensional electromagnetic analysis using ANSOFT Maxwell 3D. The magnetic permeability of MRE depends on the iron particle concentration. Carlson [4] proposed an empirical relation for the induced magnetic flux density, B, and applied magnetic field, H, for a MR fluid with various iron particle concentrations. This relation is also valid for MREs in estimation of their magnetic permeability. The following is the equation used in the electromagnetic analysis.

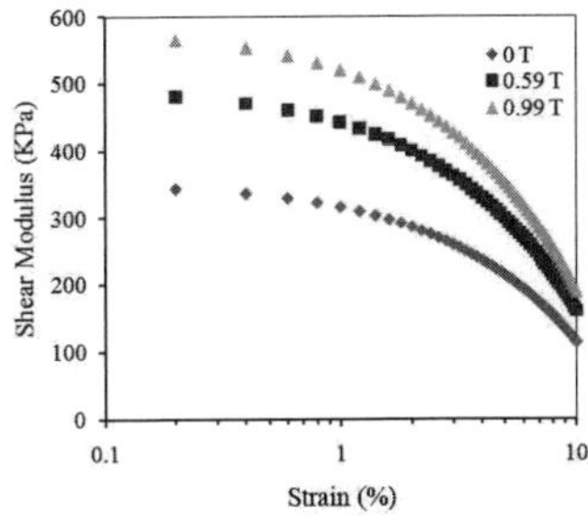

Figure 4(a). Variation of Shear modulus with strain.

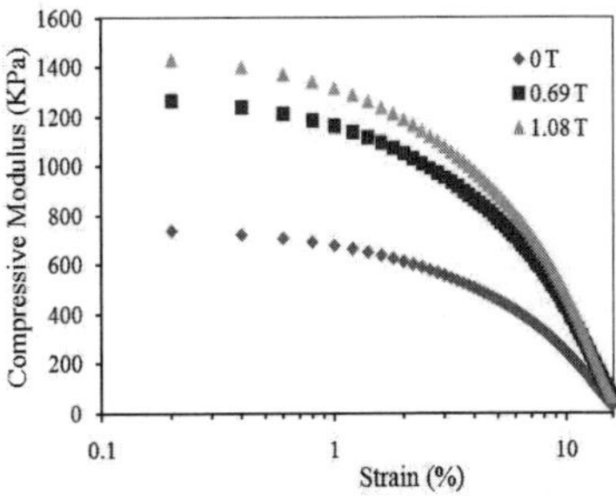

Figure 4(b). Variation of Compressive modulus with strain.

$$B = 1.91 \cdot \Phi^{1.133} \cdot [1\text{-exp} (-10.97 \cdot \mu_0 \cdot H)] + (\mu_0 \cdot H) \tag{1}$$

where Φ is the volume percentage of iron particles in the MRE mixture, and μ_0 is the permeability of free space ($4\pi \times 10^{-7}$ N/ A^2).

From this analysis, it is found that for a given sample thickness, the magnetic flux density obtained at a certain applied electric current increases with the increase in the particle concentration (Figure 5(a)) and for a given particle concentration, the obtained magnetic flux density decreases with the increase in the sample thickness (Figure 5(b)).

The magnetorheological (MR) effect, or change in modulus, is defined, as follows:

$$\Delta G = G_{on\text{-}state} - G_{off\text{-}State} \tag{2}$$

where, ΔG is the change in modulus, $G_{on\text{-}state}$ is the shear/compression modulus at on state, and $G_{off\text{-}state}$ is the shear/compression modulus at off state.

542

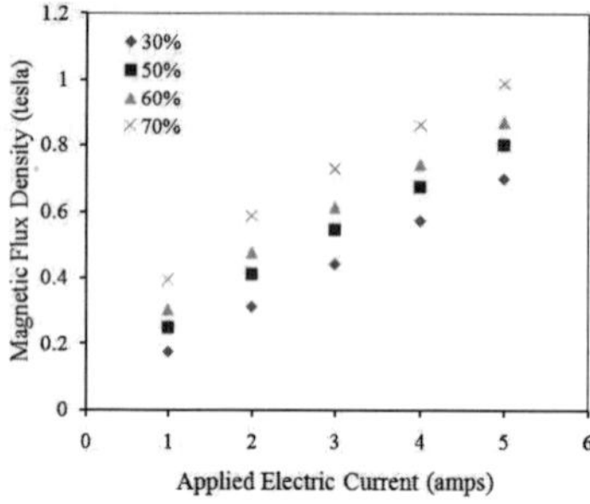

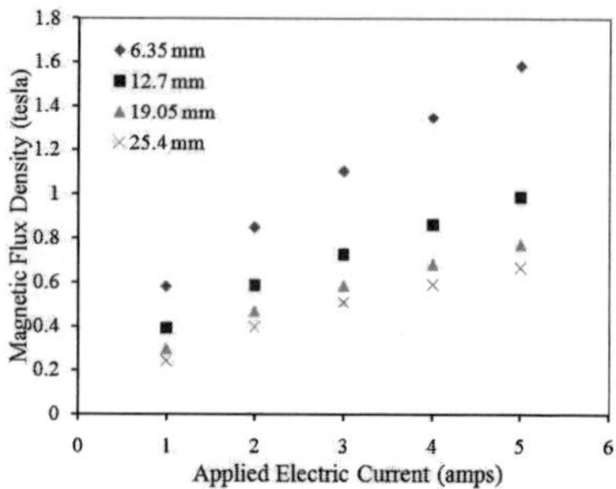

Figure 5(a). Effect of particle concentration on the magnetic field generated for a given applied current.

Figure5(b). Effect of sample thickness on the magnetic field generated for a given applied current.

At applied magnetic fields the change in the modulus is found to be independent of the sample thickness and only a function of the applied magnetic fields and the iron particle concentrations. The MR effect is shown to increase with the increase in the iron particle concentration. With the increase in applied magnetic field, it is observed that the change in modulus changed from a linear trend at lower field to a non-linear trend at higher field. Figures 6(a) and 6(b) showing typical results for the MRE specimens with 70wt.% particle concentration demonstrate these phenomenon.

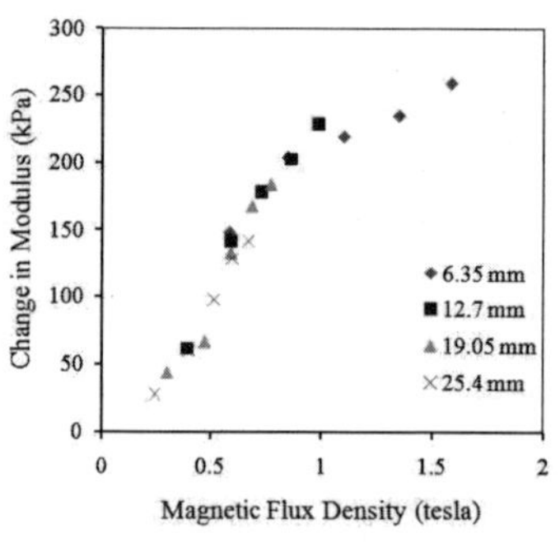

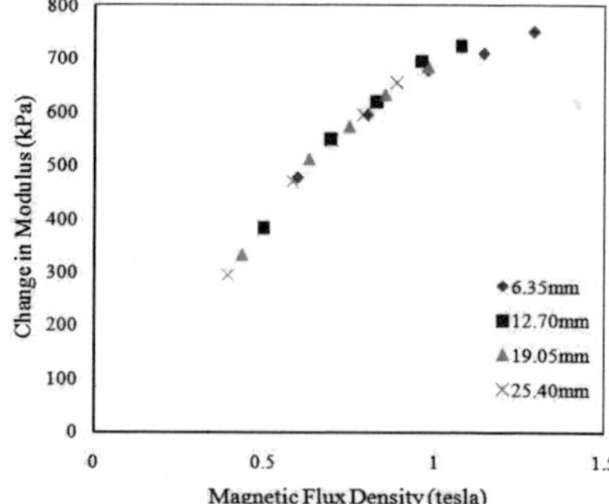

Figure 6(a). Field induced change in shear modulus for MRE with 70wt.% particle concentration.

Figure 6(b). Field induced change in compressive modulus for MRE with 70wt.% particle concentration.

The compressive Young's modulus and shear modulus of MREs as a function of applied magnetic field strength are shown in Figure 7. Both compressive and shear moduli only depend on the applied magnetic fields and are independent of the sample thickness. The maximum induced change in

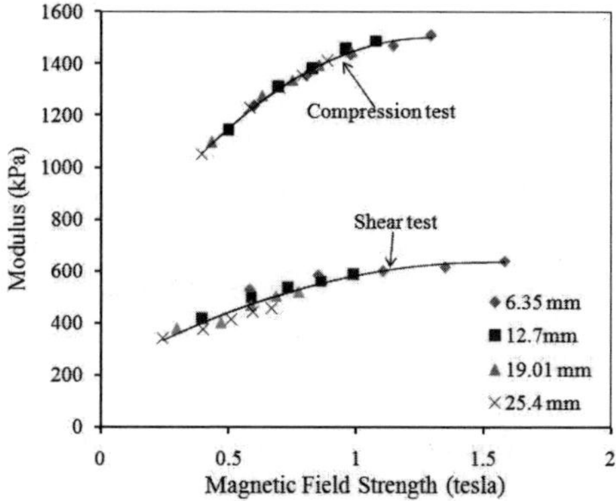

Figure 7. Measured moduli of MREs in compression and shear tests for different thicknesses and 70wt% iron particle concentration specimens as a function of magnetic field strength.

material modulus under compression is shown to be 99% whereas in shear it is found to be 68% when compared to its off-state.

5. Conclusions

Both shear and compression tests show that the MR effect of MREs increase with the iron particle concentration. The thickness of testing specimens has no effect on the measurement results. The controllability of MRE is more in compression mode than in shear mode.

Acknowledgments

The authors would like to thank the United States Navy for support of this project.

References

1. M. R. Jolly, J. D. Carlson, B. C. Munoz, and T. A. Bullions, *J. Intell. Mater., Syst. Struct.*, **7**, 613 (1996).
2. J. M. Ginder, M. E. Nicholes, and S. M. Clark, *Proceedings of SPIE – The International Society for Optical Engineering*, **3985**, 418 (2000).
3. H. X. Deng and X. L. Gong , *J. Intell. Mater., Syst. Struct.*, **18**, 1205 (2007).
4. J. D. Carlson, *Proceeding of the Ninth International ERMR Conference, Electrorheological Fluids and Magnetorheological suspension*, Edited by Lu K., Shen R., Liu J., published by World Scientific Publishing, Singapore, 531 (2004).

MODEL STUDY FOR SHEAR BEHAVIOR OF BENZENE RING OR TRIAZINE GROUP IMMOBILIZED CHITOSAN DISPERSED SUSPENSION AS ER FLUIDS

YOUNG GUN KO

Energy Mechanics Center, Korea Institute of Science and Technology, 39-1 Hawolgok-dong, Wolsong-gil 5, Seoul 136-791, Korea

UNG SU CHOI

Energy Mechanics Center, Korea Institute of Science and Technology, 39-1 Hawolgok-dong, Wolsong-gil 5, Seoul 136-791, Korea, E-mail: uschoi@kist.re.kr

YONG JIN CHUN

Department of Material Science and Applied Chemistry, Chungwoon University, San 29, Namjang, Hongsung, Chungnam 350-701, Korea

Many efforts have been spent on developing high-performance positive ER materials. In those efforts, change of functional groups of ER materials is the easiest method to approach developers' satisfaction. However, the mechanism of functional group's effect has not clearly investigated. In this study, we introduced benzene ring or triazine group into amine or carboxyl groups immobilized chitosans to observe their influence on the shear stress under an electric field. All of the curves of the shear stress plotted against shear rate were fitted well by our suggested spring-damper model.

1. Introduction

The electrorheological (ER) fluid has been one of the most promising candidate as intelligent (smart) materials for various industrial utilizations of such suspensions in active control devices such as dampers, shock, absorbers, clutches, and brakes, and in new devices include gripping devices, seismic controlling frame structures, human muscle stimulators, and spacecraft development dampers [1]. Among ER materials, polymeric materials have been studied widely due to their advantages, such as low density, easy handling, and good dispersion-property [2].

Many efforts have been spent on developing high-performance positive ER materials with merits of their fast response time and controllable shear viscosity [3]. In those efforts, the substitution of functional groups of ER materials is easiest method to increase the ER performance of materials. Chitosan shows the ER effect due to amine and hydroxyl groups [4], and these groups can be substituted to other functional groups easily. In this study, we introduced benzene ring or triazine group into amine or carboxyl groups immobilized chitosans to observe their influence on the shear stress under an electric field.

Generally, ER flow shows pseudo-Bingham flow with and without an initial decrease region of the shear stress. Therefore, the Bingham equation has been used widely as a simple and suitable model for ER fluids. However, complex fluids cannot be explained by that model. Here, our previously suggested equation using the spring-damper theory of particle-model [5,6] have treated the wide range of shear rates and specific behaviors of the shear stress in complex ER fluids.

2. Experimental

2.1. *Synthesis of Chitosan Derivatives*

Chitosan succinate was obtained by the reaction of chitosan (10 g) with succinic anhydride (80 g) in dimethyl sulfoxide (400 ml, DMSO, Aldrich) and pyridine (200 ml) solution at a 60 $^{\circ}$C oil bath for 5 h with stirring. After reaction, modified chitosan was washed with DMSO and deionized (DI) water, and dried at a temperature of 40 $^{\circ}$C in vacuum oven.

Chitosan phthalate was synthesized by the following method. Chitosan (10 g, Jakwang, Korea), phthalic acid (1 mol, Aldrich), triphenyl phospine (0.01 mol, TPP, Aldrich), diethyl azo dicarboxylate (0.01 mol, TPP, Aldrich), and dimethylformamide (500 ml, DMF, Aldrich) were put into a round flask under N_2 purging and reacted at a temperature of 35 $^{\circ}$C in an oil bath for 24 h with stirring. After reaction, modified chitosan was washed with DMF and DI water, and dried at a temperature of 40 $^{\circ}$C in vacuum oven.

Aminated chitosans were synthesized by the following method. Chitosan succinate (10 g), triethylamine (50 ml, Aldrich), amine chemicals (0.5 M, Aldrich) and DMSO (500 ml) were put into a round flask under N_2 purging and reacted at a 80 $^{\circ}$C oil bath for 24 h with stirring. After reaction, aminated chitosans were washed with DMSO and DI water and dried at a 40 $^{\circ}$C vacuum oven. Amine chemicals were diethylenetriamine, melamine, and Bismarck brown R. And these aminated chitosans were named as aminated chitosan (I), aminated chitosan (II), and aminated chitosan (III), respectively.

2.2. *Suspension Preparation and Electro-rheological Measurements*

Synthesized polymers were grinded to 5-30 μm particles using a ball mill. ER fluids were prepared by dispersing these chitosan derivatives particles into silicon oil, whose viscosity was 30 cS at 25 $^{\circ}$C. The silicon oil was dried using molecular sieves before use, and the particle concentration was fixed at 30 vol%. The rheological properties of the suspension were investigated in a static DC field using a Physica Couette-type rheometer (Physica US200) with a high-

546

voltage generator. The measuring unit was of a concentric cylindrical type, with 1 mm gap between the bob and the cup. The shear stress for the suspensions was measured under a shear rate of 1-1000 s^{-1} and the electric fields of 0-3 kV/mm.

3. Results and Discussion

The high-performance ER materials are closely related to their molecular structures. With the development of ER materials, it is clear that those materials posses branched polar group such as amine (NH_2), hydroxyl (OH), and aminocyano (NHCN), or semiconducting repeated groups. The polar groups may affect the ER behavior by playing the role of the electronic donor under the imposed electric field. Chitosan was modified with benzene ring and triazine groups to compare each functional-group. All synthesized polymers are depicted in Figure 1.

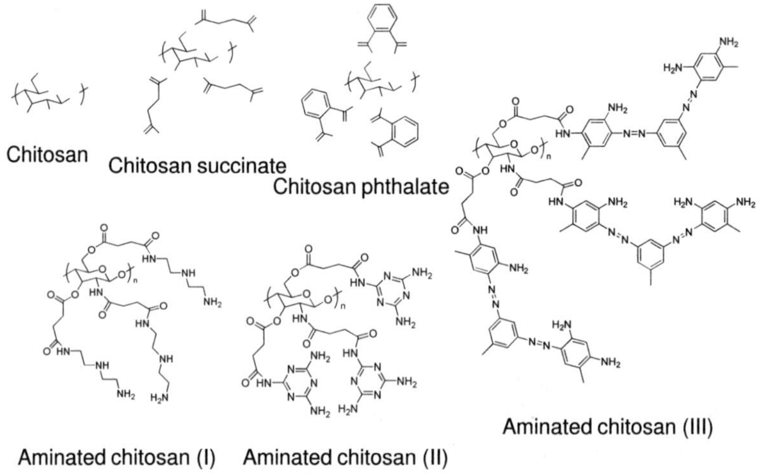

Chitosan
Chitosan succinate
Chitosan phthalate
Aminated chitosan (I)
Aminated chitosan (II)
Aminated chitosan (III)

Figure 1. Structures of chemically modified chitosans.

The Bingham equation has been used widely as a simple and suitable model for ER fluids with two parameters originating from yield stress τ_0 and Newtonian viscosity η as described below:

$$\tau = \tau_0 + \eta\dot{\gamma} \tag{1}$$

In the phenomena of ER fluids, only pre-yield behavior and Newtonian flow behavior can be explained by this equation. Therefore, this model is not sufficient to describe the plateau regime of shear stress, which is caused by competition between electrostatic interactions within the particles and hydrodynamic shear flow. ER particles are aligned under the electric field, and fibrillar shapes of the aligned ER particles flow along the fluids by shear stress. With the increase of shear rate, the aligned ER particles disassemble from long fibrillar shapes to short fibrillar shapes and particles. During this destruction, there are reforms and distortions between the particles. For our model, we consider that particles are linked to each other by springs and dampers. To provide more-effective and reasonable fitting than other model, we suggested the spring-damper model in the previous study [5].

$$\tau = \tau_0 + \eta_{app} \frac{(\dot{\gamma} - \dot{\gamma}_{slip})^2}{\dot{\gamma}} - K'_{eff}\, e^{-t_1\dot{\gamma}} \cos(-t_2\dot{\gamma} - \omega) \tag{2}$$

Here, τ_0 is the dynamic yield stress, η_{app} is the apparent viscosity, $\dot{\gamma}_{slip}$ is the slipped shear rate, K'_{eff} is the apparent-dynamic yield stress, t_1 and t_2 are time constants, and ω is the correction factor for the time constant.

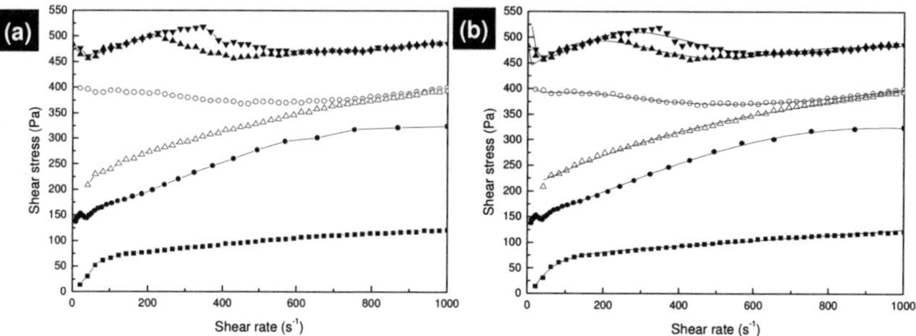

Figure 2. (a) Shear stress vs. shear rate for ■ chitosan, ○ chitosan succinate, ● chitosan phthalate, Δ aminated chitosan (I), ▲ aminated chitosan (II), and ▼ aminated chitosan (III) dispersed suspensions at an electric field strength of 3 kV/mm, and (b) fit of the suggested model to their flow curves.

According to Equation, the experimental data of shear stress on shear rate at 3 kV/mm were plotted as shown in Figure 2. In comparison with Figure 2(a), the best fits of the shear stress data against shear rate were obtained by means of the model as shown in Figure 2(b). The suggested equation treated all of the curves of shear stress against shear rate well from a simple curve to a complex curve. We divide the all complex behavior into 4 regions to analyze the curve for shear stress vs. shear rate in ER fluids [6]; (I) a region for the slow polarizations and fully-formed lamellar patterns, (II) a region for an increase of the shear stress with increase of the shear stress with increased shear rate after fully-formed fibrillar and lamellar patterns, (III) a region for an abrupt decrease of the shear stress due to the destruction of the chain and lamellar structures, and (IV) a region for the slow-destruction rate of the chain and lamellar structures due to the competition of reformation and destruction between particles. After the best fits of the shear stress data against shear rate were obtained by means of the model, obtained parameters are presented in table 1. Chitosan succinate, aminated chitosan (II), and aminated chitosan (III) showed higher values of τ_0,

Table 1. Parameters of the suggested model equation, obtained from the curve fitting of ER fluids based on chitosan and chitosan derivatives at an electric filed strength of 3 kV/mm.

samples	τ_0 [Pa]	η_{app} [Pa·s]	$\dot{\gamma}_{slip}$ [1/s]	t_1 [s]	t_2 [s]	K'_{eff} [Pa]	α
chitosan	71.47605	0.05344	0	0.02092	0.00002	92.262212	0.11299
chitosan succinate	306.21521	0.10606	64.31642	0.00450	0.00273	228.6503	-4.36159
chitosan phthalate	170.52513	0.11618	0.26422	0.00004	0.00295	59.63706	-5.22388
aminated chitosan (I)	128.80976	0.26204	0.56180	0.00034	0.00211	95.95418	-3.76170
aminated chitosan (II)	460.20153	0.02044	4.57303	0.00294	0.00913	52.12587	1.17629
aminated chitosan (III)	476.0	0.01382	288.94500	0.00258	0.00795	71.72737	0.02377

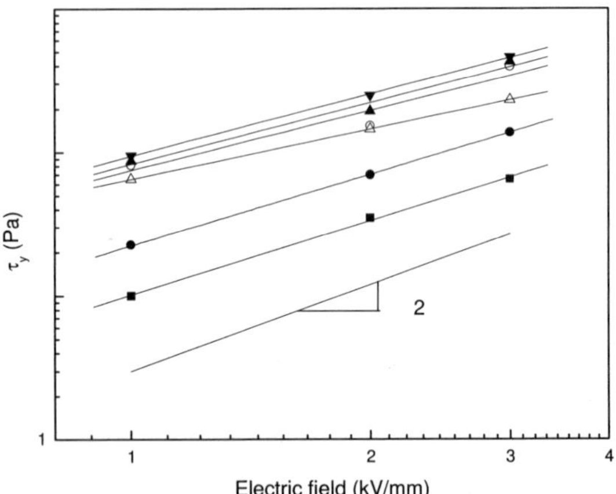

Figure 3. The yield stress for ■ chitosan, ○ chitosan succinate, ● chitosan phthalate, △ aminated chitosan (I), ▲ aminated chitosan (II), and ▼ aminated chitosan (III) dispersed suspensions under various electric field strengths.

$\dot{\gamma}_{\text{slip}}$, t_1, and t_2 than other modified chitosans. These chitosans also showed high value of shear stress against shear rate. Especially, aminated chitosan (II) and aminated chitosan (III) showed a trembling behavior of the shear stress.

The yield stresses (τ_y) obtained from various logarithmic shear stress versus shear rate curves also plotted as a function of electric strength (E) for six kinds of ER fluids as shown in Figure 3. The aminated chitosan (III) suspended ER fluid showed the highest yield stress. Especially, the aminated chitosans coupled with benzen ring or triazine group showed higher yield stress than other modified chitosans. The change of yield stress with different electric filed can be represented by the following equation $\tau_y = E^a$. The a values of chitosan, chitosan succinate, chitosan phthalate, aminated chitosan (I), aminated chitosan (II), and aminated chitosan (III) are 1.89, 1.71, 2.16, 1.15, 1.41, and 1.43, respectively. This result differ from the theoretical prediction that τ_y is proportional to the electric filed strength E^2 [7]. The difference is due to several factors, such as the degree of the polarization, the polarization rate, etc.

4. Conclusions

Benzene ring or triazine group was introduced into amine or carboxyl group immobilized chitosans to observe their influence on the shear stress under an electric field. The Bismarck brown R immobilized chitosan suspended ER fluid showed the highest yield stress. Especially, the aminated chitosans coupled with benzen ring or triazine group showed higher yield stress than other modified chitosans. According to the spring-damper model Equation, the experimental data of shear stress on shear rate were plotted. And the best fits of the shear stress data against shear rate were obtained by means of the model. Especially, chitosan succinate, aminated chitosan (II), and aminated chitosan (III) showed higher values of τ_0, $\dot{\gamma}_{\text{slip}}$, t_1, and t_2 than other modified chitosans. Our suggested model equation is very useful and meaningful for the curve fitting of the shear against shear rate, and gives a good explanation of the experimental data.

References

1. T. Hao, *Adv. Mater.* **13**, 1847 (2001).
2. B. H. Sung, Y. G. Ko and U. S. Choi, *Coll. Surf. A: Physicochem. Eng. Aspects* **292**, 217 (2007).
3. W. Wen, X. Huang, S. Yang, K. Lu and P. Sheng, *Nature Mater.* **2**, 727 (2003).
4. U. S. Choi, *Coll. Surf. A: Physicochem. Eng. Aspects* **157**, 193 (1999).

5. Y. G. Ko, U. S. Choi and Y. J. Chun, *J. Colloid Interface Sci.* **335**, 183 (2009).
6. Y. G. Ko, U. S. Choi and Y. J. Chun, *Macromol. Chem. Phys.* **209**, 890 (2008).
7. T. C. Halsey, *Science* **258**, 761 (1992).

THE EFFECT OF COMPATIBILITY OF SUSPENSION PARTICLES WITH THE OIL MEDIUM ON ELECTRORHEOLOGICAL EFFICIENCY

MARTIN STĚNIČKA, VLADIMÍR PAVLÍNEK and PETR SÁHA

*Polymer Centre, Faculty of Technology, Tomas Bata University in Zlin, TGM 275,
762 72 Zlin, Czech Republic
stenicka@ft.utb.cz*

NATALIA V. BLINOVA, JAROSLAV STEJSKAL and OTAKAR QUADRAT

*Institute of Macromolecular Chemistry, Academy of Sciences of the Czech Republic,
Heyrovsky Sq. 2, 162 06 Prague, Czech Republic*

The compatibility of solid phase – liquid medium appears to be an important factor for controlling the electrorheological efficiency of electrorheological fluids. Controlled protonation of polyaniline particles with perflourooctanesulfonic, tartaric or sulfamic acid provides particles in a broad range of hydrophobicity as a suitable model to study this phenomenon. The relation between organization of such particles in silicone oil suspensions in on/off electric field at rest and during flow at various volume concentrations was demonstrated.

1. Introduction

For more than 60 years researchers have been fascinated by a unique behaviour of smart materials which can be transformed from liquid-like to solid-like state after the application of an external electric or magnetic field [1, 2]. These intelligent fluids usually consist of polarisable solid particles homogeneously dispersed in a non-conducting liquid medium, which in the electric field organise themselves to the chain-like or columnar structures. This chaining is connected with a large change of rheological properties of the material – the viscosity of suspension steeply increases. When an external electric field disappears, the particle chains due to flow-field forces are destroyed and the suspension structure returns back to the original random state.

Since the discovery of electrorheological (ER) phenomenon, the main principles and mechanisms affecting the formation of such oriented structures have been summarized in several reviews [3–7]. Nevertheless, there are still some unexplained factors such as compatibility of the suspension particles with liquid phase that may significantly influences ER performance. In our previous paper [8], strong dependence of interfacial tension on ER efficiency in both steady shear and oscillatory mode has been described.

In the present study, the differences in formation and rheological behavior of ER structure of the polyaniline (PANI) particles with various compatibility with silicone oil medium in steady shear flow is investigated.

2. Experimental

2.1. *Synthesis of PANI particles*

Polyaniline is prepared by the oxidation of aniline in aqueous acidic medium. The standard procedure consists in the oxidation of 0.2 M aniline hydrochloride with 0.25 M ammonium peroxydisulfate [9] at room temperature. The PANI salt obtained in this manner was converted to a PANI base in excess of 1 M ammonium hydroxide, and dried. The conductivity of PANI base was 3.60×10^{-9} S cm^{-1}. The 2 g portions of PANI base were further suspended in 40 mL of aqueous acid solutions for 24 h [10]. The resulting reprotonated PANI was collected on a filter, rinsed with acetone, and dried at room temperature in air and later in a desiccator. Four suspensions of PANI particles with significant difference in hydrophobicity, and thus in compatibility to silicone oil, were employed (P1–P4; Table 1).

Table 1. The water contact angle, θ, density, ρ, and conductivity, σ, of PANI base reprotonated in the aqueous solutions of various acids.

Code	Acid	θ (deg)	ρ (g cm^{-3})	σ (S cm^{-1})
P1	5 % perfluorooctanesulfonic	102	1.448	7.49×10^{-3}
P2	None, a reference	88	1.132	3.60×10^{-9}
P3	1 M tartaric	61	1.352	1.76×10^{-2}
P4	1 M sulfamic	36	1.333	4.50×10^{-1}

2.2. *ER fluids*

ER fluids (1.2–20 vol% of solids) were prepared by mixing PANI particles with corresponding amount of silicone oil (Lukosiol M200, Chemical Works Kolin,

Czech Republic, viscosity $\eta_c = 200$ mPa s, density $\rho_c = 0.965$ g cm^{-3}, conductivity $\sigma_c \approx 10^{-11}$ S cm^{-1}). The suspensions were deposited in a desiccator and stirred at first mechanically, then in an ultrasonic bath for 30 s before each ER measurement.

2.3. *ER measurements*

Rheological measurements were carried out with a rotational rheometer Bohlin Gemini (Malvern Instruments, UK), with parallel plates 40 mm in diameter and a 0.5 mm gap, equipped with ER cell. The plates were connected to a DC high-voltage source TREK 668B (TREK, USA) providing the intensity of the electric field strength, $E = 0$–500 V mm^{-1}. Before each measurement, the suspension structure was homogenized by shearing of the suspension at the shear rate 20 s^{-1} for 100 s. The measurements were performed under room temperature.

2.4. *Monitoring of ER structure*

The formation of ER structures was monitored by optical microscope Olympus CX31 (Olympus, Japan). The electric field strength $E = 200$ V mm^{-1} was achieved by an electrometer Keithley 6517B (Keithley, USA). The micrographs of suspensions were obtained by a digital camera Olympus Camedia C-4000 Zoom (Olympus, Japan).

3. Results and Discussion

3.1. *Structure of suspensions*

The micrographs (Figures 1 and 2) demonstrate that the protonation of PANI particles with various acids significantly influence their organization in silicone oil suspension. The hydrophobic particles (Table 1; P1) treated with perfluorooctanesulfonic acid were homogeneously dispersed in continuous phase because of similar wettability. Thus, obviously only slight particle interactions set in (Figure 1a). By contrast, however, the hydrophilic particles (P4) protonated with sulfamic acid created fine chain or network structures already in the absence of the electric field (Figure 2a). When the electric field was applied, the former particles (P1) were polarized and joined the electrodes by distinct chains (Figure 1b). In the latter case, the original field-off particle arrangement changed in a dense chain-like structure oriented along the stream lines of the electric field (Figure 2b).

554

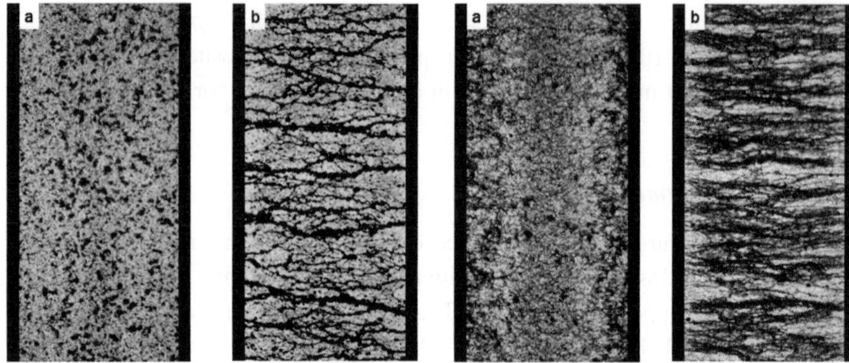

Figure 1 Micrographs of 1.2 vol% silicone oil suspensions of PANI particles protonated by perfluorooctanesulfonic acid (P1) before (a) and after (b) application of electric field $E = 200$ V mm^{-1}.

Figure 2 Micrographs of 1.2 vol% silicone oil suspensions of PANI particles protonated by sulfamic acid (P4) before (a) and after (b) application of electric field $E = 200$ V mm^{-1}.

3.2. *Steady shear*

The shear stress – shear rate dependences of the suspensions of four particle samples under study at three volume fractions in the absence (Figure 3) and in the presence (Figure 4) of the electric field showed that marked differences in suspension structure at rest are also significant at flow. The distinct effect of various particle compatibilities on flow character already appeared in the absence of electric field (Figure 3). The suspensions of hydrophobic particles (P1) were nearly Newtonian at all three particle concentrations. On the other hand, in suspensions of hydrophilic particles (P4) a significant non-Newtonian flow set in. Suspensions of rather hydrophobic (P2) and rather hydrophilic (P3) particles tended to Newtonian behavior at lower concentrations (5 and 10 vol%) and behaved as non-Newtonian at 20 vol% with marked yield-stress value. This evidently corresponds to the differences in character of particle organization as shown in Figures 1 and 2.

After the electric field application, the shear stress increases for all suspensions as a consequence of particles chaining (Figure 4). The effect of particle compatibility to silicone oil medium on ER activity at various particle concentration was demonstrated by the shear stress $\tau_{g=1}$ evaluated at the shear rate $g = 1$ s^{-1} (Figure 5). It is obvious that unlike suspensions of more hydrophobic particles P1, P2 and P3, the increase in the shear stress in the suspension of hydrophilic particles (P4) in the presence of the electric field was

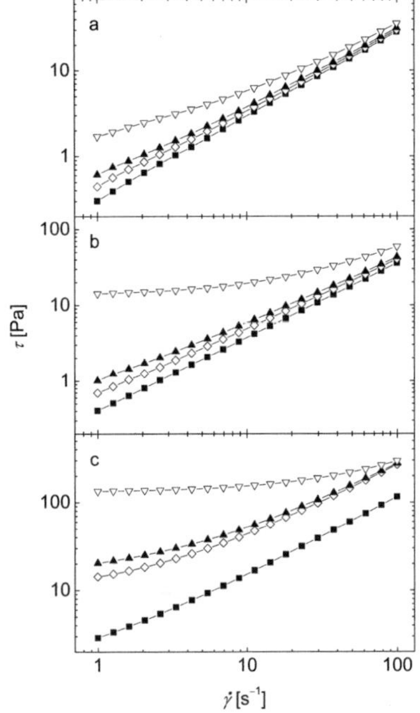

Figure 3 The shear stress, τ *vs.* shear rate, *g*, dependence of PANI suspensions in 5 vol% (a), 10 vol% (b) and 20 vol% (c) in the absence of electric field. Symbols: ■ P1, ◇ P2, ▲ P3 and ▽ P4.

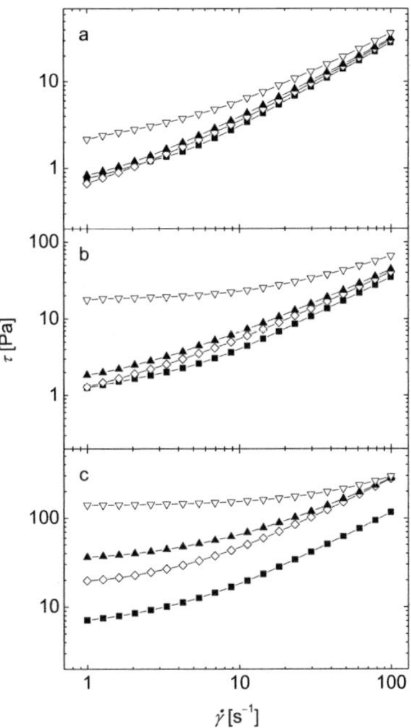

Figure 4 The shear stress, τ *vs.* shear rate, *g*, dependence of PANI suspensions in 5 vol% (a), 10 vol% (b) and 20 vol% (c) at the electric field strength, $E = 200$ V mm^{-1}. Symbols denoted as in Figure 3.

relatively low. Consequently, the ER performance, $e = (\tau_E - \tau_0)/\tau_0$, where τ_E is the shear stress under electric field application and τ_0 its field-off value, was much lower, especially at high particle concentration (Figure 6).

It is clear that the highest *e*-value in the more hydrophobic sample (P1) was a consequence above all of the low shear stress in the absence of electric field caused by good particle – oil medium compatibility. On the contrary strong interaction between hydrophilic particles (P4) resulted in high τ_0 value, the ER efficiency was low and decreased when the concentration of the particles increased (Figure 6). The ER performance of rather hydrophobic (P2) and rather hydrophilic (P3) laid between samples P1 and P4, whereas the optimal concentration of PANI particles for maximum ER response was around 10 vol%. This corresponds to observation in the earlier study [11], where the similar optimum concentration for commercially available PANI base was found.

556

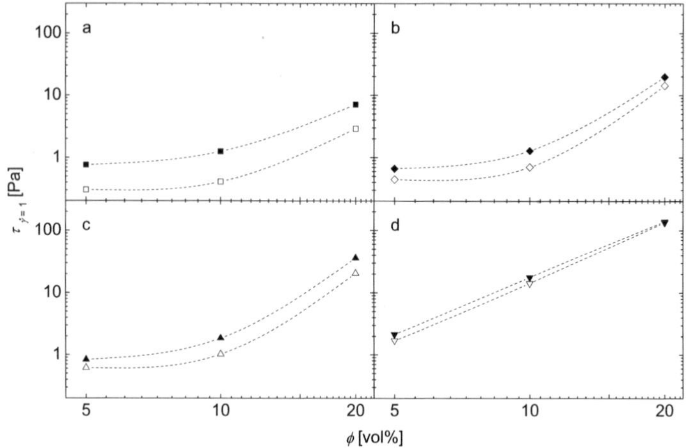

Figure 5 The shear stress, $\tau_{g=1}$ vs. particle concentration, ϕ dependence of PANI suspensions in the absence (open) and in the presence (solid) of electric field strength, $E = 200$ V mm^{-1}. Symbols: ($\square$ $\blacksquare$) P1 (a), ($\lozenge$ $\blacklozenge$) P2 (b), ($\triangle$ $\blacktriangle$) P3 (c) and ($\triangledown$ $\blacktriangledown$) P4 (d).

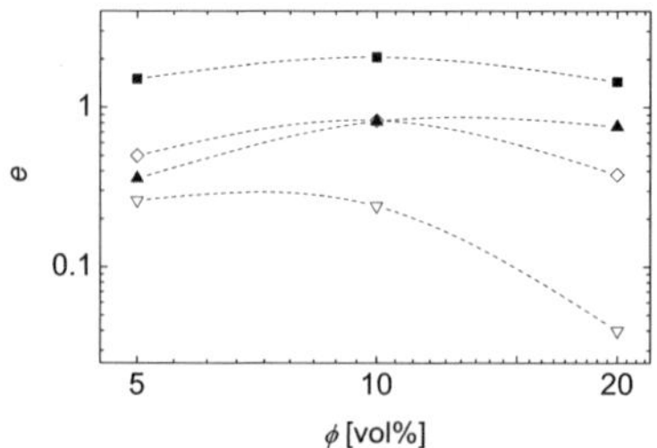

Figure 6 The dependence of the relative stress, $e = (\tau_{E=200} - \tau_{E=0})/\tau_{E=0}$, on the volume fraction of PANI particles, ϕ. Symbols denoted as in Figure 3.

4. Conclusions

The study illustrates that the compatibility between suspension particles and silicone oil crucially affects flow behavior both in the absence and in the presence of electric field. A low particle – particle interaction in suspensions of hydrophobic particles causes a low field-off shear stress and, consequently, high ER performance in a broad range of volume concentrations. In contrast, low compatibility of hydrophilic particles with silicone oil results in the formation of interparticle bonding. The shear stress of this chain-like or network suspension structure in the absence of the electric field due to limited mobility of the particles is high and, as a result, the relative shear stress is considerably lower than in suspensions of hydrophobic particles.

Acknowledgements

The authors gratefully acknowledge the financial support of the Ministry of Education, Youth and Sports of the Czech Republic (MSM 7088352101), the Czech Grant Agency (202/09/1626) and the Grant Agency of the Academy of Sciences of the Czech Republic (IAA 400500905).

References

1. W. M. Winslow, *J. Appl. Phys.* **20**, 1137 (1949).
2. J. Rabinow, *AIEE Trans.* **67**, 1308 (1948).
3. H. Block and J. P. Kelly, *J. Phys. D: Appl. Phys.* **21**, 1661 (1988).
4. T. C. Jordan and M. T. Shaw, *IEEE Trans. Electron. Insul.* **24**, 849 (1989).
5. H. Block, J. P. Kelly, A. Qin and T. Watson, *Langmuir* **6**, 6 (1990).
6. T. Hao, *Adv. Mater.* **13**, 1847 (2001).
7. O. Quadrat and J. Stejskal, *J. Ind. Eng. Chem.* **12**, 352 (2006).
8. M. Stěnička, V. Pavlínek, P. Sáha, N. V. Blinova, J. Stejskal and O. Quadrat, *J. Colloid Interface Sci.* **346**, 236 (2010).
9. J. Stejskal, J. Prokeš and M. Trchová, *React. Funct. Polym.* **68**, 1355 (2008).
10. J. Stejskal and R. G. Gilbert, *Pure Appl. Chem.* **74**, 857 (2002).
11. A. Lengálová, V. Pavlínek, P. Sáha, O. Quadrat, T. Kitano and J. Stejskal, *Eur. Polym. J.* **39**, 641 (2003).

ELECTRORHEOLOGY OF DISPERSIONS OF $Ba_xSr_{(1-x)}TiO_3$ IN SILICONE OIL UNDER AC OR DC ELECTRIC FIELD

GLAUBER M. S. LUZ and ANTONIO J. F. BOMBARD[†]

UNIFEI / ICE – Av. BPS 1303, Itajubá – MG, 37.500-903, Brasil

SILVIO L. M. BRITO

Indústria Eletromecânica Balestro LTDA, Rua Santa Cruz 1550
Mogi Mirim – SP, 13.800-911, Brasil

DOUGLAS GOUVÊA

USP / Politécnica, Av. Prof. Mello Moraes 2463, São Paulo – SP, 05.508-900, Brasil

SHEILA L. VIEIRA

sheilasama@aol.com, Villa Rica, GA 30180, USA

Electrorheology (ER) of ferroelectric materials such as nanometric $BaTiO_3$ is still not fully understood. In this work, nanoparticles of $Ba_xSr_{(1-x)}TiO_3$ (where x = 0.8, 0.9 or 1.0) were synthesized using the method of Pechini, calcinated at 950°C, and after, lixiviated under pH 1 or pH 5. A controlled stress rheometer (MCR-301) was used to make the ER characterization of dispersions made of $Ba_xTi_{1-x}O_3$ in silicone oil (30% w/w), where (a) shear stress as a function of DC electric field (under constant shear rate) or (b) shear stress as a function of shear rate (under constant AC or DC electric field) were measured. We observed that electrophoresis occurred under electric field DC, creating a concentration gradient which induced phase separation in ER fluid. On the other hand, under AC fields above 1 kV/mm, the ER effect is stronger than for DC field, and almost without electrophoresis. Furthermore, there is an AC frequency, dependent on the disperse phase, where the ER effect has a maximum.

1. Introduction

Electrorheological fluids (ERF) are composed of a continuous insulating liquid phase in which a semi conductive particulate solid phase is dispersed, or composed of an emulsion of a second semi conductive liquid phase (usually liquid crystals). The rheological properties (viscosity, shear stress, elastic and viscous modulus) of an ERF can be reversibly changed by several magnitude under an electric field (AC or DC) in the order of kV/mm. The response time of an ERF is in the order of milliseconds, which make these materials very

---

[†] Correspondence should be addressed to: *antonio.bombard@gmail.com*

attractive for mechatronics engineering and several industrial applications such as automotive industry (brakes, clutches and dampers). Ferroelectric materials such as titanates, present intrinsic polarizability and, then, are natural candidates for electrorheological studies [1].

The synthesis of $BaTiO_3$ happens traditionally by a reaction in solid state from the calcinations of a mixture of $BaCO_3$ and TiO_2. However, it is necessary temperatures on the order of 1000 to 1200°C to obtain particulates with low specific surface area and average size around or over 1μm [2-5]. It was reported that the annealing temperature also affects the ER effect under AC field [6].

In order to obtain nanometric powders, chemical methods, such as Pechini's method [7], can be used. In this method there is the formation of a chelate of mixed cations (dissolved as salt in an aqueous solution) by means of a hydroxicarboxylic acid. The solution of salt in the acid is mixed with poli-hydroxilated alcohol (usually ethylene glycol) under heat (from 70 to 110°C).

During a later moderate heating between 120°C and 250°C, the alcohol esterifies the complex and non-complex molecules of the carboxylic acid generating water which is removed by evaporation. As both the acid and the alcohol are polifunctionals, formation of a polymeric resin (polyester) takes place with the chelated cations distributed atomically along the molecular structure of the resin. This resin is then pirolized and the powder is calcinated at temperatures between 500 and 950°C.

The objective of the work was to analyze how the amount of strontium in the synthesis of nanoparticles of barium titanate, and the effect of the acid lixiviation after the calcination affect the electrorheological response of dispersions of the titanates in silicon oil (20 cSt), under AC or DC electric fields.

2. Methods and Materials

$Ba_XSr_{1-X}TiO_3$ was prepared using the polymeric precursor technique (based on Pechini's patent [6]). The molar proportion of the mixture Ba/Sr : Ti : citric acid : ethylene glycol that originated the resin was equal to 1:1:4:16. The polymeric precursor was prepared with three different compositions (**0, 1 e 2**) where the molar relation of Ba-Sr varied according to the following:

Sample 0 $\rightarrow$ 1.0 mol Ba / 1.0 mol Ti   (**$BaTiO_3$**)
Sample 1 $\rightarrow$ 0.9 mol Ba / 0.1 mol Sr / 1.0 mol Ti (**$Ba_{0.9}Sr_{0.1}TiO_3$**)
Sample 2 $\rightarrow$ 0.8 mol Ba / 0.2 mol Sr / 1.0 mol Ti (**$Ba_{0.8}Sr_{0.2}TiO_3$**)

The mixture procedure and the proportions used was similar to the one used in the work done by Cho and Hamada [8], Vinothini and Balasubramaniam [9], Ries and Varela [10]. The main difficulty observed in this phase was the solubility of the titanium isopropoxide. The total solubilization of the precursor of Ti was possible by slowly adding alcoxide to ethylene glycol under agitation at 90°C during 30 minutes, where initially a translucent solution was formed, which later became cloudy supposedly due to the precipitation of some component of Ti. At this moment citric acid was slowly added for total dissolution of the precipitates, originating again a translucent solution. Finally, barium carbonate was added under agitation until complete reaction and solubility was achieved. Agitation was then maintained for 30 minutes more. We then obtained a yellow liquid free from particulates that, when maintained under agitation for 2 hours at 140°C leads to polymerization. The resin obtained was first pre-calcinated under an oxidant atmosphere with constant air flow from a compressor and reduced heating rate of 2°C/min [11] followed by three steps of 5 hours at 200°C, 300°C and 500°C for the total decomposition of the polymer.

After being partially separated in a mortar of agate, the powders were calcinated to evaluate the continuity of the decomposition and formations of the crystalline phases. The heating rate during the calcination was maintained constant at 2°C/min and the time of treatment of 10 h was done at the called temperature of treatment which was 950°C. In order to eliminate residual barium carbonate usually found in barium titanate, we performed several acid washes using nitric acid [12].

The techniques used to characterize the powders were:

1. **Scanning Electron Microscopy (SEM)** by means of a microscope model Quanta 600 with filament FEG made by FEI Company;
2. **Specific Surface Area Analysis (BET)** using Gemini III 2375 Surface Area Analyzer (Micromeritics): a thermal treatment done before the analysis was made at 250°C for 24 hours under pressure of 100 μmHg (0.1 torr) using a joined unity of VacPrep 061 (Micromeritics) to remove possible species adsorbed to the surface;
3. **Infrared spectroscopy (FT-IR)** with an equipment made by Nicolet model Thermo-Nicolet Magna 560 capable to scan 400 to 4000 cm^{-1} (medium infrared) and resolution of 4 cm^{-1}. The analysis was conducted by diffuse reflectance (DRIFT) to characterize the surfaces of the synthesized powders;

For the electrorheological tests, we used a controlled stress rheometer (Physica MCR-301) equipped with a cell for electrorheology with Peltier temperature control for the configuration plate-plate. A plate of 25 mm of diameter and gap of 1mm was used. The electric field was applied by a high voltage power supply HCP 14-12500 (FüG). For the AC field, a function generator 33220A (Agilent) fed a high voltage supply amplifier (Trek 20/20C). This setting allowed the AC electric fields to be applied with frequency and shape of arbitrary wave. All the rheological measures were performed at 25°C.

The titanate nanoparticles and silicone oil (Dow 200, viscosity of 20 cSt at 20°C) were dried in oven during 24 h at 120°C. After they were let to cool down in a desiccator, and dispersions were prepared at a concentration of 30%. The dispersions were homogenized with the help of the tip of an ultrasound (US Horn Sonics 750 W, 20 kHz) for one minute.

3. Results and Discussion

3.1. *Electronic Scanning Microscopy and Specific Surface Area*

Figure 1 shows SEM images of particles of Sample 0, Sample 1 and Sample 2. We can observe that the titanates present high grade of aggregation and size of particles ranging from approximately 50 nm to 200 nm.

Table 1 presents the specific surface area obtained for titanates. The area of the lixiviated particles was higher than the non-lixiviated ones due to their separation during the process of lixiviation. The lixiviation was done in order to verify variations on the surface of the particles due to the solubilization of residuals carbonates.

(a)   (b)   (c)   (d)   (e)   (f)

Figure 1: SEM Images of particles: Sample 0 (a) and (d), Sample 1 (b) and (e), Sample 2 (c) and (f).

Table 1. Specific surface area (BET) for titanates.

Sample	Without Acid Wash [m²/g]	With Acid Wash [m²/g]
0	5.2	7.6
1	5.9	8.0
2	7.0	8.1

3.2. *Infrared Spectroscopy Analysis (FT-IR)*

Figure 2 shows the FTIR (DRIFT) spectrums of the sample 0 calcinated at 950°C with and without acid wash. We can observe the characteristic peaks of carbonates (which is the residual of the synthesis of barium titanate): the first one in the region of 1430 cm^{-1} that corresponds to a simple ionic carbonate CO_3^{2-} (asymmetric stretching - 1415 cm^{-1}) and a second peak of unidentate ligand carbonate (asymmetric stretching - 1480 cm^{-1}, symmetric stretching - 1370 cm^{-1}) [13]. The sharp peak in 1750 cm^{-1} is due to an organic carbonate [14]. The vibration that appears in 2450 cm^{-1} corresponds to bicarbonate. The other peaks and vibrations are not due to carbonate. Similar curves were obtained for samples 1 and 2. After lixiviation with nitric acid it was possible to remove residual barium carbonate, since vibrations at 2450, 1750, 1060 cm^{-1} were no longer observed and there was a considerable reduction of the peak at 1430 cm^{-1} for all the compositions.

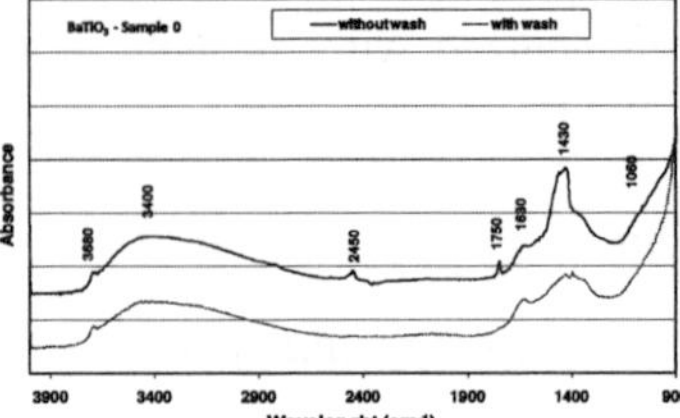

Figure 2: DRIFT spectrums of Sample 0.

3.3. *Electrorheology*

Figure 3-a, 3-b and 3-c show the ER effect on the shear stress as a function of the DC field under constant shear rate (10 s^{-1}) for Samples 0, 1 and 2, respectively. We observed voltaic arc in some samples for E > 3.5 kV/mm, showing that the conductivity of the sample had increased leading to the rupture of the dielectric. From figures 3a – 3c we can conclude that the acid lixiviation tends to increase the ER effect, and that the strontium concentration and the lixiviation pH influence the ER response in a complex way. For Sample 0 (without strontium), the curve of pH = 1 presented the strongest ER effect. However, as the concentration of strontium increased, the curve of pH = 5 had an increased ER effect presenting higher values than ph = 1 for sample 2. We observed the occurrence of electrophoresis for all samples under DC field, leading to phase separation and reduction of the ER effect after some time. Then it is difficult to guarantee reproducibility and creditability on the measurements under DC field.

In figures 4-6, under AC field, the problem due to electrophoresis was eliminated and we did not observe any phase separation between the plates of the rheometer for frequencies higher than 50 Hz. In Figure 4, for 2 kV$_{p-p}$/mm and f = 500 Hz sin, one can see that the content of strontium increases significantly the ER response for all non-lixiviated samples.

Figure 5 shows the results for Sample 2 under 2 kV$_{p-p}$/mm and f = 500 Hz sin. The treatment with acid lixiviation does not eliminate the ER effect; however it becomes much weaker under AC field. It is interesting to observe that for DC fields, the lixiviation has the opposite effect. This suggests that surface charges and different mechanisms of polarization are involved in each case, which agrees with the review done by Hao [1]. The same was observed for Samples 0 and 1: lixiviation also reduced significantly the ER effect when compared to the samples that were only calcined. Another hypothesis is that carbonate groups participate on the polarization in alternated field, in the same way oxalates and urea do in titanates (Lu *et al.* [13,14]).

Figure 6 presents the effect of the AC waveform for Sample 1, non-lixiviated. The same behavior was observed for Samples 0 and 2: the ER effect for low shear rates and a given AC frequency increases following the order:

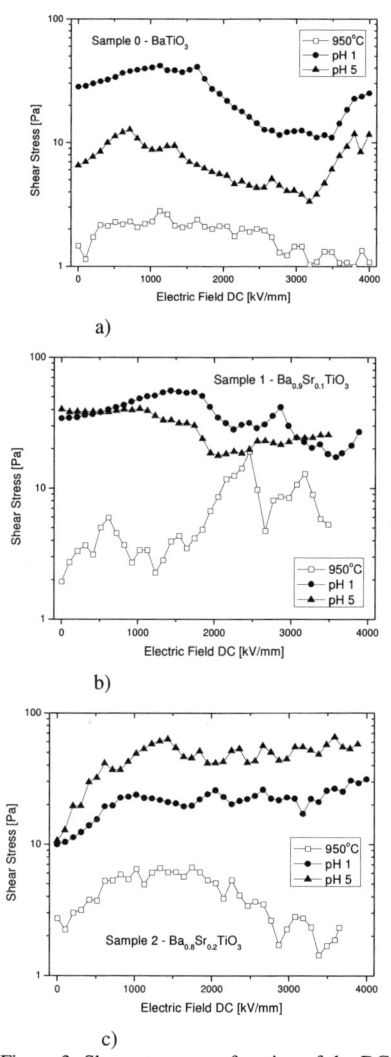

a)

b)

c)

Figure 3: Shear stress as a function of the DC electric field under constant shear rate $(10\ s^{-1})$ for Samples 0 (a), 1 (b) e 2 (c), without lixiviation and lixiviated at pH 1 or pH 5.

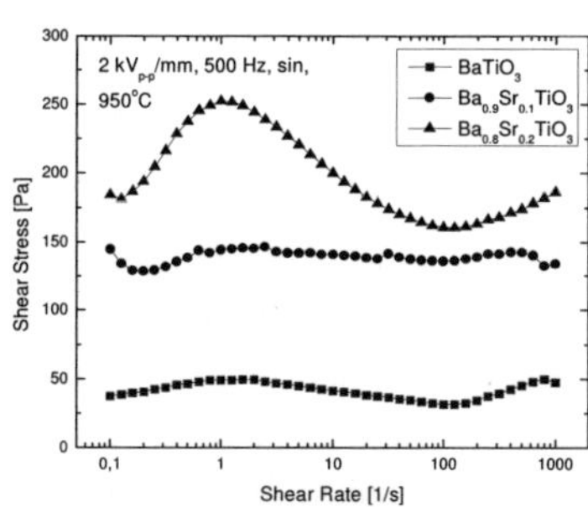

Figure 4: Effect of strontium content on flow curves under AC field for samples 0, 1 e 2, without lixiviation (E = 2 kV$_{p-p}$/mm, f = 500 Hz sin).

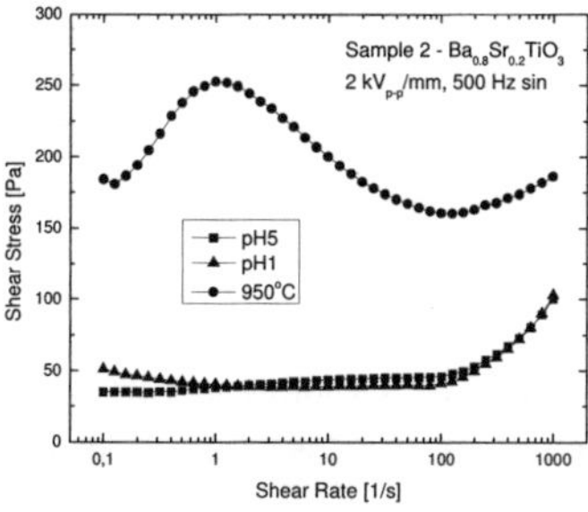

Figure 5: Effect of lixiviation on flow curves under AC field for Sample 2 (E = 2 kV$_{p-p}$/mm, f = 500 Hz sin).

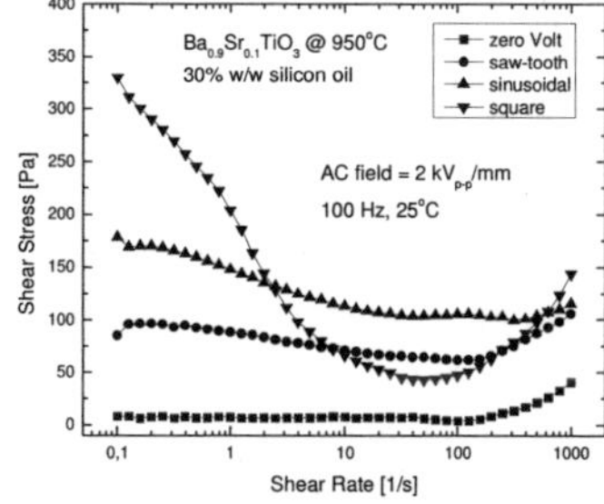

Figure 6: Effect of the waveform on flow Curves under AC field for non-lixiviated Sample 1 (E = 2 kV$_{p-p}$/mm, f = 100 Hz sin).

saw-tooth wave < sinusoidal wave < square wave [15]. The frequency of the AC field, besides the waveform, also affects the ER response. The maximum ER effect was usually observed between 100 and 500 Hz. For frequencies below 30 Hz, some electrophoresis was observed, but not as strong as for DC fields. For frequencies higher than 500 Hz or 1 kHz, the effect drops drastically, as it was reported in the literature: the particles cannot follow the rapid changes of the field.

4. Conclusions

The electrorheology of titanates nanoparticles synthesized according to Pechini's process showed to be dependent on various factors, such as: strontium content of the sample, lixiviation and the source of the electric field (AC or DC). In general, AC field showed an ER effect, if not stronger than with DC field, at least without electrophoresis. This is the main disadvantage of using DC field. Under DC field, the lixiviated samples were the ones with stronger ER effect. Under AC fields, lixiviation reduced considerably the ER effect. The frequency and the waveform of the AC field also had a significant influence on the ER response. From all nine samples studied in this work, the one that had a stronger ER effect was Sample 2, non-lixiviated under AC field and square wave. For future studies, we plan to increase the field strength and also change the concentration of the titanate in the silicone oil dispersions.

Acknowledgments

Brito and Gouvêa acknowledge Indústria Eletromecânica Balestro Ltda and FAPESP for the financial support (Proc.05/53241-9). Bombard acknowledge FAPEMIG for the grant CEX – 00531/2008 as well as the financial support to present this work on ERMR Philadelphia.

References

1. T. Hao, *Advances in Colloid and Interface Science*, **97**, 1 – 35 (2002).
2. A. Beauger, J.C. Mutin, J.C. Niepce, *Journal Materials Science*, **18**, 3041-3046 (1983).
3. P.P. Phule, S.H. Risbud, *Journal of Materials Science*, **25**, 1169-1183 (1990).
4. M.S.H. Chu, A.M. Rae, *American Ceramic Society Bulletin*, **74**, 69-72 (1995).
5. M.S. Castro, E. Brzozowski, *Journal of the European Ceramic Society*, **20**, 2347-2351 (2000).
6. Kobayashi Y, Morimoto R, Moriyama K, Satoh T, Konno M; *Nihon Reoroji Gakkaishi*, **33**, 285-287 (2005).
7. M. Pechini, U.S. Patent. #3,330,697 (1967).
8. W.S. Cho, E. Hamada, *Journal of Alloys and Compounds*, **266**, 118-122 (1998).
9. V. Vinothini, P. Singh, M. Balasubramanian, *Ceramics International*, **32**, 99-103 (2006).
10. A. Ries, A.Z. Simoes, M. Cilense, M.A. Zaghete, J.A. Varela, *Materials Characterization*, **50**, 217-221 (2003).

11. P. Durán, F. Capel, J. Tartaj, C. Gutierrez, C. Moure, *Solid State Ionics*, **141-142**, 529-539 (2001).

12. C. Hérard, A. Faivre, J. Lemaître, *Journal of the European Ceramic Society*, **15**, 135-143 (1995).

13. Y. Lu, R. Shen, X.Z. Wang, G. Sun, K.Q. Lu; *Smart Materials & Structures*, **18**, 025012 (2009).

14. W.J. Wen, X.X. Huang, S.H. Yang, K.Q. Lu, R. Sheng; *Nature Materials*, **2**, 727-730 (2003).

15. A.K. El Wahed, J.L. Sproston, E.W. Williams; *Journal of Physics D - Applied Physics*, **33**, 2995-3003 (2000).

ELECTRORHEOLOGICAL RESPONSE MEASURED WITH PECTINATED ELECTRODES[*]

RONG SHEN[1 †], DE WANG[1, 2], MINGCHUN JIAO[1], GANG SUN[1] and KUNQUAN LU[1]

[1]*Beijing National Laboratory for Condensed Matter Physics and Key Laboratory of Soft Matter Physics, Institute of Physics, Chinese Academy of Sciences, Beijing 100190, China*

[2]*University of Science and Technology of China, Hefei 230029, China*

Pectinated electrode was designed to measure the rheologcal property of PM-ER fluids. The principle of this measurement is based on the effect that the outside electric field near the edges of parallel electrode can contribute a considerable value of the field. The relation of the yield stress and the gap between pectinated electrode and workpiece, the variations of the yield stress vs the parameters of pectinated electrodes are obtained. We believe that knowledge of the pectinated electrodes and the ER behavior in between the electrode and workpiece are helpful for the application of ER fluid-assisted polishing.

1. Introduction

The viscosity of electrorheological (ER) fluid can be changed widely (from liquid-like to solid-like) under the applying electric field. Two types of traditional rheometers are generally used to measure such variation of viscosity of ER fluid, i.e. parallel-plate and inner-outer cylinder system. For traditional ER fluids whose shear stress is less than 10kPa, such rheometers can meet the requirement of measurements. As the development of ER fluids, a series of new types ER fluid with high shear tress as much as more than 100kPa have been fabricated [1-7], which is named as polar molecule dominated electrorheological (PM-ER) fluids [8,9]. For PM-ER fluids it is hardly to measured the intrinsic value of the yield stress because yield starts at the interface and the ER fluid slides along the electrodes at a large enough field strength [10,11].

To prevent such sliding of PM-ER fluids on electrodes roughening the

---

[*] This work is supported by the National Natural Science Foundation of China Grant 10674156, the National Basic Research Program of China Grant 2004CB619005 and 2009CB930800, the Knowledge Innovation Project of Chinese Academy of Sciences Grant KJCX2-YW-M07, and the Instrument Developing Project of the Chinese Academy of Sciences, Grant No. YZ200758.
[†] E-mail: rshen@aphy.iphy.ac.cn

surface of electrodes is the straightforward way [12-14]. However, it has not reached its intrinsic value since the weak bonding to the electrodes remains an unavoidable disturbing factor. In order to eliminate drawback of roughened electrode, we devised two measuring procedures inspired by the shear stress measuring equipments for solids, which are named as unilateral shearing mode and slice shearing mode by us [10,11]. With these measuring procedures the effect of fluid sliding on the electrode can be eliminated that the measured yield stresses can be taken as the intrinsic character of the ER-fluids.

In this paper, we briefly present the results of measured the yield stress of ER fluids with the designed pectinated electrode. The principle of this measurement is based on the effect that the outside electric field near the edges of parallel electrode can still contribute a certain value of the field. We believe that knowledge of the pectinated electrodes and the ER behavior in between the electrode and workpiece are helpful in the applications, for instance, using as electrodes for ER fluid-assisted polishing.

2. Experiment

The nanometer-sized calcium titanate (CTO) particles used in the experiment is synthesized by a simple co-precipitation procedure [6]. The dryed CTO particles are suspended in aviation hydraulic oil with volume fraction $\phi=40\%$, of which the yield stress is about 100 kPa at 5 kV/mm.

The fabricating method of pectinated electrodes and the arrangement for the measurement are described as follows. The copper slices with same thickness were parallelly embedded into insulating nylon, of which the intervals are ensured to be equal. Two groups consisted of alternately connected copper slices are respectively as + and − electrodes for applying the electric field as illustrated in Figure 1(b). The thickness of the copper slices is w and the width of the gapes is l. In order to detect the influence of w and l on the measured shear stress, several pectinated electrodes with different w and l are prepared. The pectinated electrode is installed in a plate rheometer and connected with torque sensor. On the other hand, a silica glass disc is glued on the lower platform driving by a step motor as shown in Figure 1(a), of which the diameter and thickness are 25mm and 3mm respectively. The distance of the pectinated electrode and silica glass disc can be adjusted and is noted as δ.

The principle of this measurement is based on the effect that the outside electric field near the edges of parallel electrode can play a considerable role. As show in Fig. 1(c), the electric field will extend outward from the surface of the pectinated electrodes according to electrostatics theory. The field distribution around the edges of electrodes depends on the geometric factors of the arrangement. The electric field strength in the gap between electrode and silica glass decreases gradually from the electrode towards the silica glass and behaves a periodic structure along transverse direction. Therefore, when ER fluid fill in the gap between the pectinated electrode and silica glass, the measured yield stress must vary with the gap distance δ and the parameter l and w of pectinated electrode.

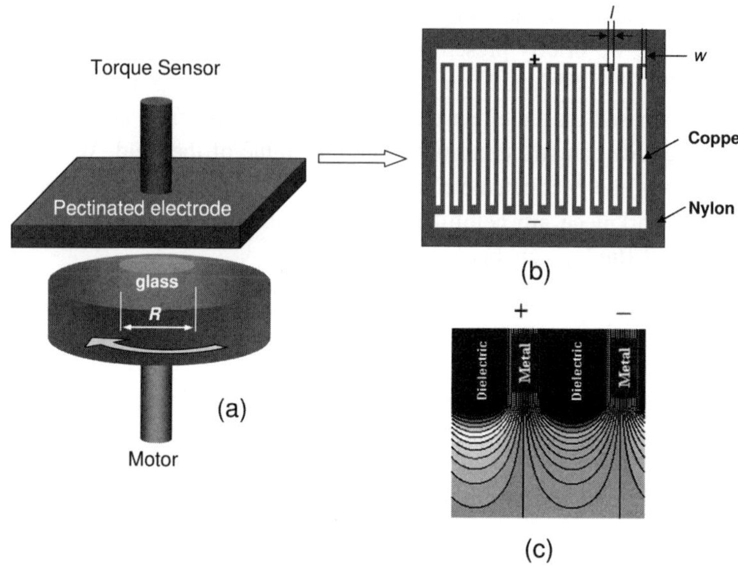

Figure 1. (a) Sketch of the arrangement for measurement. (b) pectinated electrode. (c) The field distribution near the edges of parallel electrodes.

3. Results and Analysis

We designed three pectinated electrodes with different w and l. They are nominated as electrode $1^{\#}$, $2^{\#}$, $3^{\#}$, respectively. The detailed parameter of those electrodes are shown in the caption of Figure 2.

The measured relation of ER effect versus the gap δ is shown in Figure 2. All the collected curves exhibit that the ER responses decrease with δ increasing. It is able easily to be understood and is due to the lower average field strength at larger δ. The yield stress measured by electrode $1^{\#}$ is stronger compared to that by electrode $2^{\#}$ no matter under 1000V/mm or under 1500V/mm, although the thickness of copper slices for two electrode is same (l=1.0mm). The difference is only the width w, w =0.15mm and w =0.60mm respectively. It implies that the smaller w causes the stronger strength when l is the same.

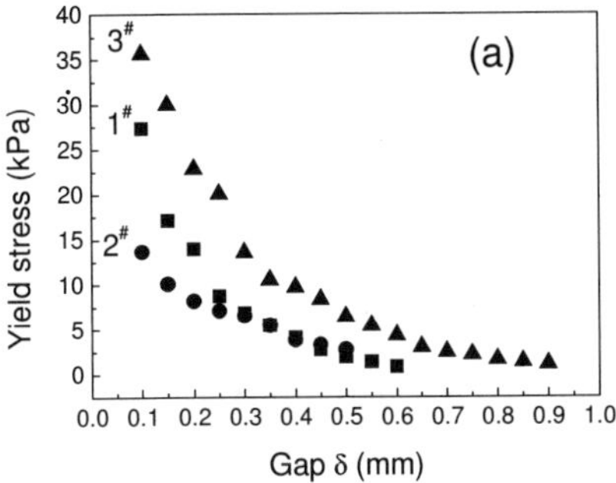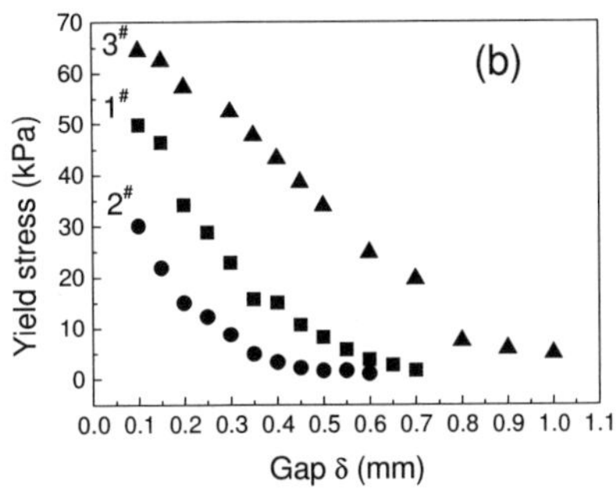

Figure 2. The dependences of the yield stress on gap δ of the pectinated electrode and silica glass disc for different w and l under different electric field (a) 1000V/mm, (b) 1500V/mm.
$1^{\#}$: w =0.15mm, l =1.0mm; $2^{\#}$: w =0.60mm, l=1.0mm; $3^{\#}$: w =0.15mm, l=1.5mm.

Figure 2 also illuminates that if the thickness w of the copper stripes is equal, pectinated electrodes with wider separation l of stripes can induce higher ER effect. The stripe separation of electrode $3^{\#}$ is l =1.5mm being wider than that of electrode $1^{\#}$ (l =1.0mm). Under same electric field, ER fluid possesses higher yield stress by using electrode $3^{\#}$ compared with electrode $1^{\#}$, although their thickness is same.

The results can be explained by electrostatics. Wider separation between stripes and thinner stripes means more the electric field will extend outward from the surface of the pectinated electrodes. The detailed calculation and discussion will be reported afterwards.

The experimental results are helpful for the application of ER fluid-assisted polishing. Pentinated electrodes can be used as polishing tool for the glass or other insulating workpieces. The reactional force between the pentinated electrode tool and the workpiece could be tuned easily by controlling the parameters of pentinated electrode, the gap between the electrode tool and workpiece, as well as the applied electric field.

4. Conclusion

We designed pectinated electrodes to measure the rheological property of ER fluids. The yield stresses of ER fluids increase with decreasing the thickness (w) of the copper stripes for same separation (l) between stripes. For the same thickness (w) between stripes, increasing the separation (l) between stripes also can increase the yield stress of ER fluids. The results are helpful for the applications of ER fluid-assisted polishing for example.

References

1. W. Wen, X. Huang, S. Yang, K. Lu, P. Sheng, *Nature Material* **2**, 727 (2003).
2. W. Wen, X. Huang, P. Sheng, *Appl. Phyi. Lett.* **85**, 299 (2004).
3. Y. Qiao, J. Yin, X. Zhao, *Smart Mater. Struct.* **16**, 332 (2007).
4. K. Lu, R. Shen, X. Wang, G. Sun and W. Wen, *Intern. J. Mod. Phys. B* **19**, 1065 (2005).
5. R. Shen, X. Wang, Z. Wang, W. Wen, K. Lu, *Intern. J. Mod. Phys. B* **18** 1104 (2005).
6. X. Wang, R. Shen, W. Wen, K. Lu, *Intern. J. Mod. Phys. B* **18**, 1110 (2005).
7. Y. Lu, R. Shen, X. Wang, G. Sun, K. Lu, Smart Mater. Struct. 18, 025012 (2009).
8. K. Lu, R. Shen, X. Wang, G. Sun, W. Wen and J. Liu, *Chin. Phys.* **15**, 2476 (2006).
9. K. Lu, R. Shen, X. Wang, G. Sun, W. Wen, *Intern. J. Mod. Phys. B* **21**, 4798 (2007).
10. R. Shen, X. Wang, Y. Lu, G. Sun, W. Wen, K. Lu, *Intern. J. Mod. Phys. B* **21**, 4813 (2007).
11. R. Shen, X. Wang, Y. Lu, G. Sun, W. Wen, K. Lu, *J. Appl. Phys.* 102, 024106 (2007).
12. X. Wang, R. Shen, D. Wang, Y. Lu, K. Lu, *Mater. & design* **30**, 4521 (2009).
13. X. Wang, R. Shen, D. Wang, Y. Lu, K. Lu, *Intern. J. Mod. Phys. B*, **21**, 4940 (2007).
14. R. Shen, X. Wang, Y. Lu, D. Wang, G. Sun, Z. Cao, K. Lu, *Adv. Mater.* **21**, 4631 (2009).

WALL SLIP EFFECTS MEASURING THE RHEOLOGICAL BEHAVIOR OF ELECTRORHEOLOGICAL (ER) SUSPENSIONS

STEFFEN SCHNEIDER

Bundeswehr Research Institute for Materials, Fuels and Lubricants
Institutsweg 1, 85435 Erding, Germany

In this work, a new method to determine the wall shear stress was developed step by step. To determine the wall shear stress, methods of the suspension rheology are being used for the first time to characterize ER fluids. This work focuses on investigations of the flow behavior of electrorheological suspensions in flow channels with different geometries at different electrical field strengths. Careful interpretation of the results with respect to different gap geometries has shown that the measured flow curves should undergo a combination of corrections. As a result it can be shown that wall slip effects can be measured under application like conditions on a hydraulic test bench.

1. Introduction

The focus of previous research activities in the field of electrorheological fluids (ERF) has mostly been on new technical applications or on fluid characterization using a single method. Comparisons between different results are somewhat hampered because standardized methods to characterize the behavior of electrorheological fluids have been lacking. This article provides a survey of applicable characterization methods of electrorheological suspensions as well as appropriate correction methods.

In 2004 a new design concept for the development of cylinder drives based on electrorheological fluids was introduced by ZAUN [1, 2]. The design of an annular gap was used in the cylinder drive itself as well as in electrorheological (ER) valves to evaluate the rheological behavior under laboratory conditions. Such an annular gap is shown in Figure 1. This type of ER valve consists of two electrodes. The annular gap between the inner and outer electrode allows for a nearly homogenous electric field. In this research the influence of gap height H and length L as well as the yield stress τ_E on the pressure difference Δp according Eq. (1) was examined. However, for simulation studies a phenomenological model was used to compare between different geometries [2, 3].

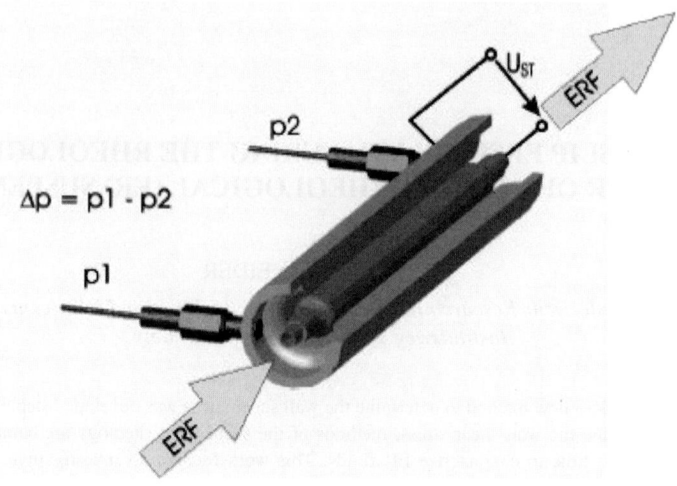

Figure 1: A sectional few of an annular ER valve [2]

$$\Delta p_E = 2 \cdot \frac{L}{H} \tau_E \qquad (1)$$

Electrorheological suspensions undergo different stresses in rotational rheometers as compared to capillary rheometers or flow channels. In general results obtained from different devices are not comparable to each other (Figure 2). One reason is that the flow behavior of electrorheological suspensions changes fundamentally in the presence of an applied electrical field. Up to now the flow has been considered Newtonian in the absence of such a field. However, in the presence of such a field, the flow approaches BINGHAM behavior (Figure 3). Another reason is that the applied field was DC for the flow channel while it was 100 Hz sine AC for the rheometer RS80 with E as the maximum value. As expected, shear rate values are at a maximum with DC and decrease with an increasing AC frequency [4, 5]. However, the discontinuity between the data sets implies additional corrections are necessary to get comparable results from different test scenarios.

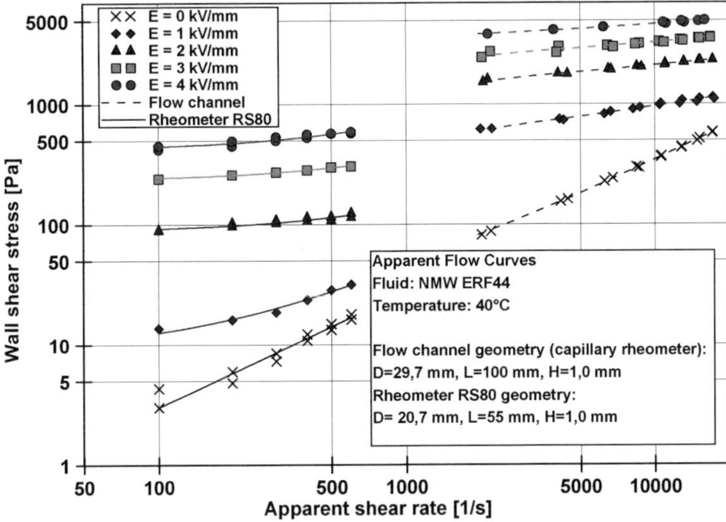

Figure 2: Differences in the results of the apparent flow curves between rheometer and flow channel (capillary rheometer)

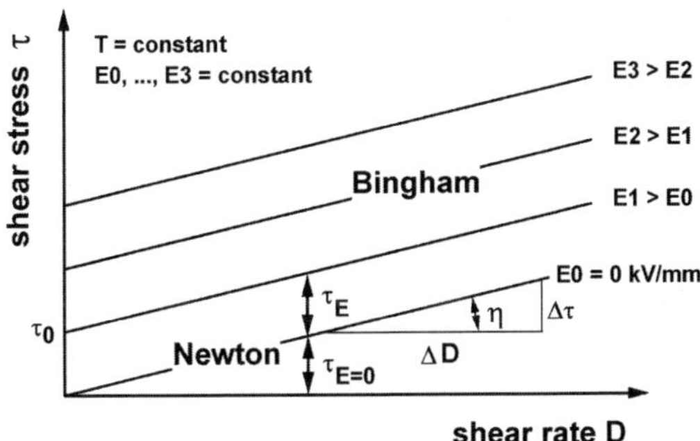

Figure 3: Idealized shear behavior of ER suspensions

2. Results and Discussion

2.1. *ER fluids and test apparatus*

The investigated electrorheological fluids, ERF44 and RheOil3.0, were developed for industrial applications based on ionically doped polyurethane particles (PUR) dispersed in silicone-oil as base fluid [6].

Measurements in shear mode were executed by a commercially available rheometer (ThermoScientific/RS80 or RS600) working with a standard Searle system in rotational mode together with a high voltage amplifier D*ASS SVU 6000/16 under AC conditions. Geometry data for bob diameter Da, rotor diameter Di, gap height H, and rotor length L are given in Table 1.

Table 1: Geometry data of sensors for rheometer measurements.

Type	Da [mm]	Di [mm]	H [mm]	L [mm]	Vsample [cm³]
Z21ER	20,7	19,7	0,5	55	2,3
Z22ER	21,7	19,7	1,0	55	4,7
Z42ER	42,4	41,4	0,5	55	5,8
Z43ER	43,4	41,4	1,0	55	11,8

Measurements in flow mode were carried out in a custom made test rig at different flow rates and defined temperature of 40°C. As a flow channel, an annular gap according to Figure 1 was used with different geometries. Geometry data for outer diameter Da, inner diameter Di, gap height H, and gap length L are given in Table 2. The gap width B is calculated according to Eq. (2) transferring the annular gap into a rectangular gap.

Table 2: Geometry data of the annular gaps for flow mode measurements.

Da [mm]	Di [mm]	Dm [mm]	H [mm]	L [mm]	B [mm]	A [mm²]
30,7	29,3	30,0	0,7	100	94,25	65,97
30,7	28,7	29,7	1,0	100	93,31	93,31
30,7	29,3	30,0	0,7	150	94,25	65,97
30,7	28,7	29,7	1,0	150	93,31	93,31
30,7	29,3	30,0	0,7	200	94,25	65,97
30,7	28,7	29,7	1,0	200	93,31	93,31
30,7	29,3	30,0	0,7	300	94,25	65,97
41	39	40	1,0	100	125,66	125,66

$$B = \pi \cdot Dm = \pi \cdot \frac{Da + Di}{2} \tag{2}$$

From the obtained data the shear stress τ_W was derived from pressure difference according to Eq. (3) and shear rate D from the volume flow according to Eq. (4). The true shear rate D_W is calculated if the fluid is a Newtonian fluid with Eq. (4). Otherwise the apparent shear rate D_S is calculated for non-Newtonian fluids with Eq. (4).

$$\tau_W = \frac{H}{2} \cdot \frac{\Delta p}{L} \tag{3}$$

$$D = \frac{6 \cdot Q}{B \cdot H^2} \tag{4}$$

The dynamic viscosity was calculated according to Eq. (5) using Hagen–Poiseuille equation.

$$\eta = \frac{B \cdot H^3}{12 \cdot L \cdot Q} \cdot \Delta p \tag{5}$$

2.2. *Results obtained by rheometer measurements*

Starting with the measurement of the base viscosity (no electrical field applied) it was found that the values vary with the gap height in a rheometer. Geometries with different diameter (D_m) but same gap height (H) show the same results especially at higher shear rates (Figure 4).

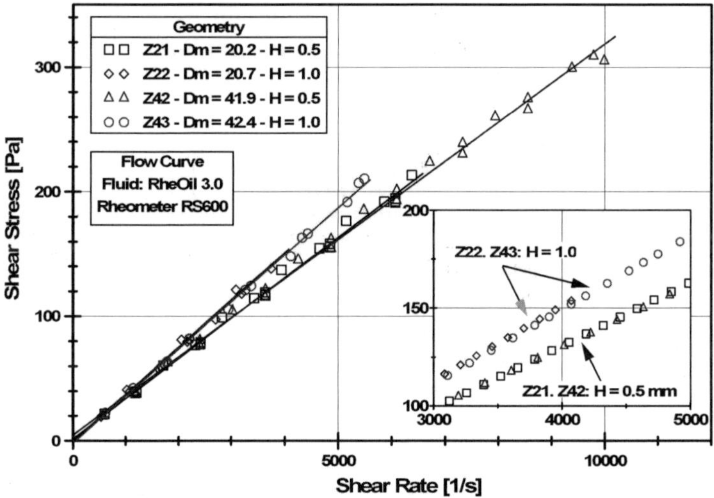

Figure 4: Differences in the results of the apparent flow curves measured in a rheometer

2.3. *Results obtained by flow channel measurements*

2.3.1. *Inlet and outlet pressure drop*

To evaluate the inlet and outlet pressure drop measurements were carried out with pressure sensors inside and outside the flow channel. For L = 100, 150, 200 mm the sensors where placed inside the channel. For L = 300 mm a sensor was placed at the inlet and outlet of the channel. From the measured values of the pressure difference for L = 100 and L = 200 mm, the pressure difference for L = 300 mm was calculated according Eq. 6.

$$\frac{\Delta p}{L} = \frac{\Delta p_1 - \Delta p_2}{L_1 - L_2} \tag{6}$$

Figure 5 shows a comparison of pressure differences measured between different lengths inside and outside of the flow channel. The calculated values for L = 300 mm correlate well with the measured values. In addition the measurements are shown in a Δp-L/D diagram as a BAGLEY plot (Figure 6). One can see that the pressure difference measured in the flow channel does not differ from the pressure difference measured outside of the homogeneous part of the flow channel. In other words, for this specific flow channel design a correction of the inlet and outlet pressure drop according to BAGLEY is not required.

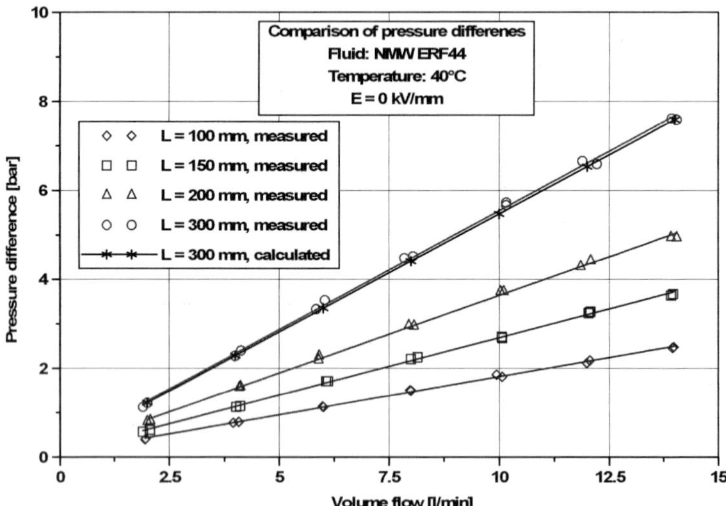

Figure 5: Comparison of pressure differences

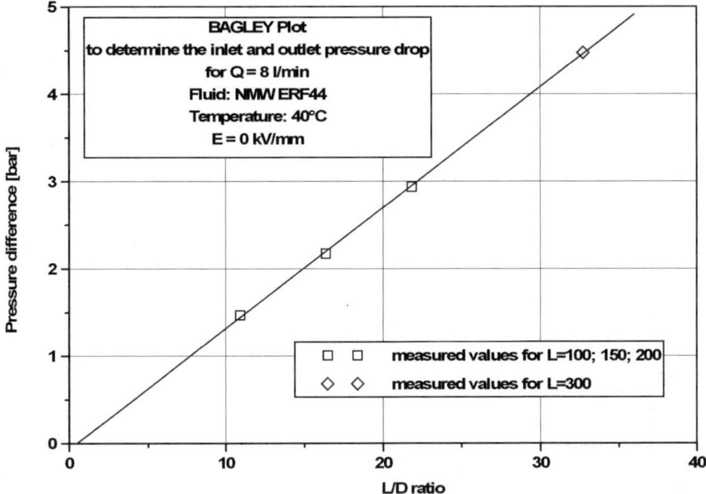

Figure 6: BAGLEY plot to determine the inlet and outlet pressure drop

2.3.2. *Non-Newtonian flow behavior*

Up to now the flow behavior of ER fluids without an applied electric field has been considered Newtonian. Considering that an ER suspension consists of polyurethane particles (PUR) dispersed in silicone-oil as base fluid, a correction of the non-Newtonian flow behavior according to WEISSENBERG / RABINOWITSCH [7] was carried out. This method takes into consideration the shear thinning behavior of non-Newtonian fluids with the condition that the velocity on the wall is zero. Apparent flow curves are transferred to true flow curves with this method by deriving the true shear rate D_W from the apparent shear rate D_S according Eq. (7).

$$D_W = \left(\frac{2}{3} + \frac{1}{3} \cdot \frac{\tau_W}{D_S} \cdot \frac{dD_S}{d\tau_W} \right) \cdot D_S \tag{7}$$

However, comparable values from different geometries could not be obtained with only the correction according to WEISSENBERG / RABINOWITSCH (Figure 7) [8].

578

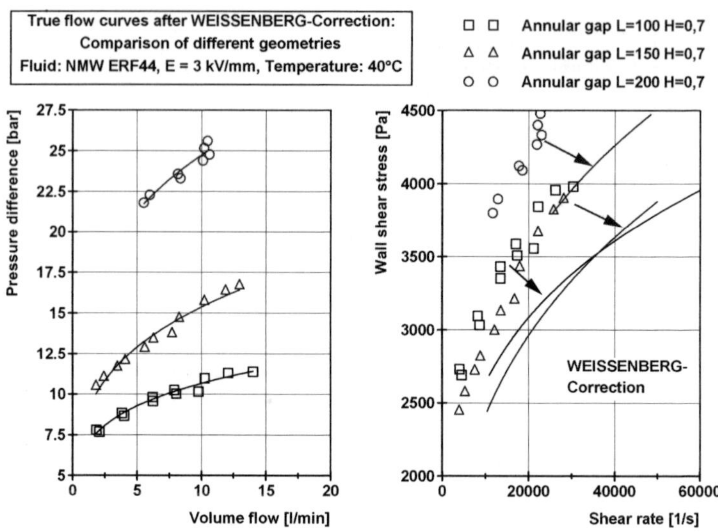

Figure 7: True flow curves after WEISSENBERG-correction for E=2 kV/mm, H=0,7 mm

2.3.3. *Wall slip*

In fluid dynamics, the no-slip condition for viscous fluids states that at a solid boundary, the fluid will have zero velocity relative to the boundary. Therefore Eq. (3) – (5) are only valid for no-slip conditions. Under slip conditions the velocity of the fluid is not zero relative to the boundary. It is known from literature that suspensions, due to their heterogeneous structure, and polymer melts, with pseudoplastic properties, tend to slip on solid boundaries [9, 10]. Even after the correction, the flow curves obtained from different geometries do not correlate to the non-slip equations. It can therefore be concluded that the non-slip condition is not a viable assumption. (Figure 8).

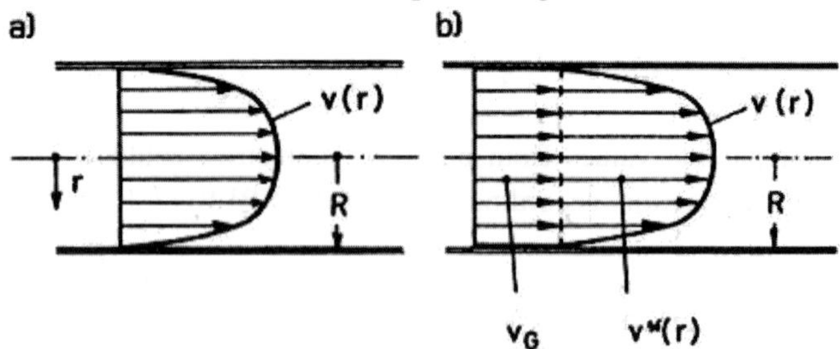

Figure 8: Velocity profiles v(r), a) without wall slip, b) with wall slip v_G [11]

An experimental procedure to investigate wall slip was published by MOONEY [12]. When applying this procedure on annular flow channels at least two different gap heights H are required. Measurements of the volume flow Q at constant wall shear stress τ_W are depicted in a so called MOONEY-Plot, in which the sliding velocity v_G corresponds with the slope $\tan\alpha$ of the resulting straight line (Eq. 8, Figure 9).

$$\left(\frac{2 \cdot Q}{B \cdot H^2}\right)_{\tau_w=konst} = \frac{2 \cdot v_G(\tau_W)}{H} + A(\tau_W) \tag{8}$$

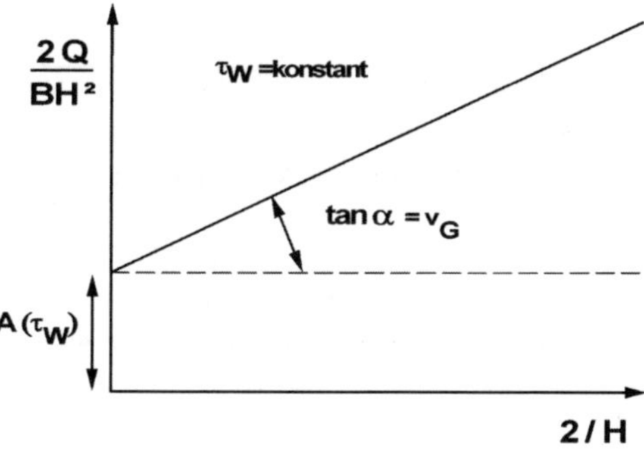

Figure 9: MOONEY-Plot to determine the sliding velocity from volume flow measurements in annular and rectangle gaps at constant shear stress

The data was examined with the MOONEY consideration to determine if there were any indications of wall slip. As can be seen in the following figures (Figure 10, Figure 11) the slope in the MOONEY-plot indicates wall slip conditions even at E = 0 kV/mm.

580

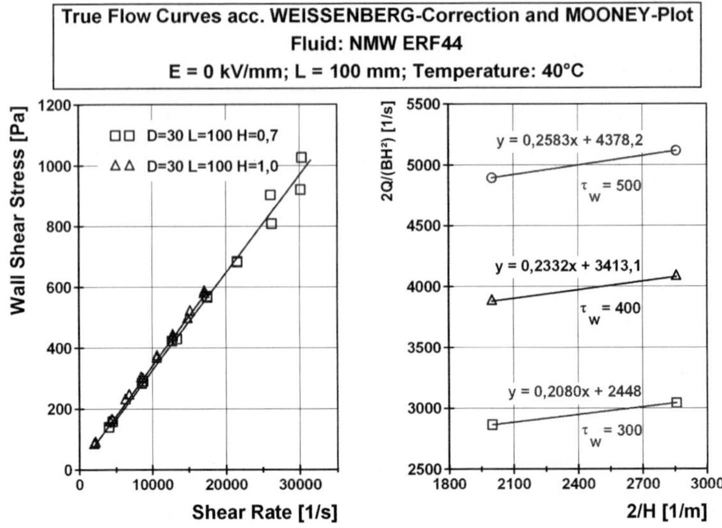

Figure 10: Comparison between true flow curve and MOONEY-Plot at E = 0 kV/mm

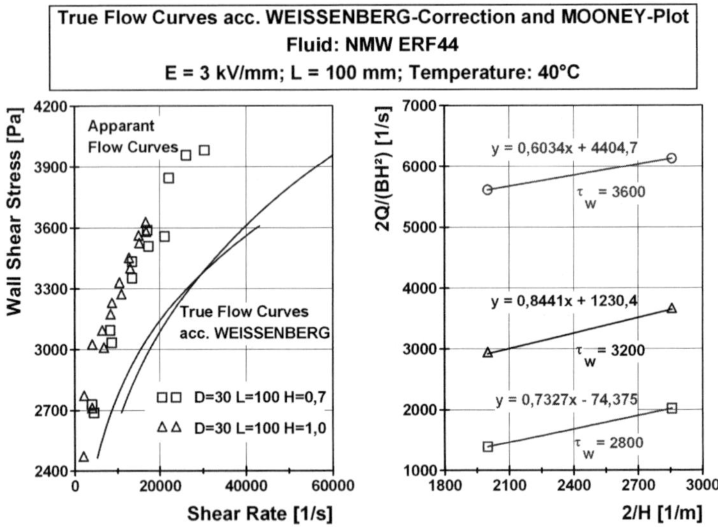

Figure 11: Comparison between true flow curve and MOONEY-Plot at E = 3 kV/mm

2.4. *Discussion*

Assuming non-slip condition and applying Hagen–Poiseuille equation on flow channel measurements will not lead to geometrically independent results. The flow behavior of ER suspensions is non-Newtonian and has to be taken into account. Therefore, additional corrections are necessary to compare results from different geometries of flow channels or rheometer measurements.

From this point of view obtaining the same results from geometries with different diameters (D_m) but same gap height (H) in rheometer measurement are also an indication of wall slip at the solid boundary of the flow channel or rheometer sensor system.

3. Summary and Conclusion

During independent research and participation in a national standardization program it was found that the viscosity obtained from different instruments and different flow geometries vary with respect to each other. Therefore, the rheological behavior was measured under laboratory conditions with a ThermoFisher RS 600 rheometer and under application like conditions on a hydraulic test bench. On the hydraulic test bench an angular flow channel was used as an ER valve.

In this work, a new method to determine the wall shear stress was developed step by step. To determine the wall shear stress, methods of the suspension rheology are being used for the first time to characterize ER fluids. This work focuses on investigations of the flow behavior of electrorheological suspensions in flow channels with different geometries at different electrical field strengths.

Careful interpretation of the results with respect to different gap geometries has shown that the measured flow curves in flow channels should undergo a combination of corrections:

- correction of the inlet and outlet pressure drop according to BAGLEY when necessary,
- correction of the non-Newtonian flow behavior according to WEISSENBERG / RABINOWITSCH,
- wall slip correction according to MOONEY.

As a result it can be shown that wall slip effects can be measured under application like conditions in a hydraulic test bench (Figure 10, Figure 11).

References

1. M Zaun, Trends in Electrorheological Valve Development. Proceedings of the 4th International Fluidpower Conference Dresden, 2004, pp 533-544.
2. M Zaun, Design Concept for the Development of Electrorheological Cylinder Drives, Int. Journal of Intelligent Materials Systems and Structures, 17, Nr. 4 2006, pp 303-307.
3. C Wolff-Jesse, G Fees: Examination of flow behaviour of electrorheological fluids in the flow mode, Proc Instn Mech Engrs, Vol. 212 Part I, 1998.

4. MJ Espin, AV Delgado, F González-Caballero, L Rejon, Rheological properties of a model colloidal suspension under large electric fields of different waveforms, J. Non-Newtonian Fluid Mechanics, Vol. 146 (2007) pp 125-135.

5. JL Sproston, AK El Wahed, R Stanway, Electrorheological Fluids in Squeeze under AC and DC Excitation, Int. Journal of Modern Physics B, Vol. 13 (1999) pp. 1861-1869.

6. M Gurka, D Adams, L Johnston, R Petricevic, New electrorheological fluids - Characteristics and implementation in industrial and mobile applications, Journal of Physics, Conference Series 149 (2009).

7. B Rabinowitsch, Über die Viskosität und Elastizität von Solen. Zeitschrift für physikalische Chemie, A 145/1929. J. Rheol., 2:210 - 222, 1931.

8. S Schneider, Methods to characterize electrorheological suspensions in consideration of the temperature influence; Helmut-Schmidt-University - University of the Federal Armed Forces Hamburg, Dept. of Mechanical Engineering, PhD. thesis, 2007.

9. FE Filisko, Alternative View for Yield Stress of ER/MR Materials. Proceedings of the 10th International Conference on ER Fluids and MR Suspensions, Lake Tahoe, USA, 18 - 22 June 2006.

10. WB Black, Wall Slip and Boundary Effects in Polymer Shear Flows, University of Wisconsin Madison, Ph.D. Thesis, 2000.

11. M Pahl, W Gleißle, HM Laun, Praktische Rheologie der Kunststoffe und Elastomere. 4. Edition, VDI-Gesellschaft Kunststofftechnik Düsseldorf, 1995.

12. M Mooney, Explicit formulas for slip and fluidity. J. Rheol., 2:210 - 222, 1931.

DYNAMIC MODELING OF MAGNETORHEOLOGICAL DAMPER USING LUMPED PARAMETER METHOD[*]

Q. H. NGUYEN[1]

[1]*Mechanical Engineering Department, Hochiminh University of Industry, Vietnam*

S. B. CHOI[2†], J. S. OH[2], M. S. SEONG[2] and S. H. HA[2]

[2]*Smart Materials and Systems Laboratory, Department of Mechanical Engineering, Inha University, Incheon 402-751, Korea*

In this work, dynamic modeling of mono-tube magnetorheological (MR) damper is performed using a lumped parameter method. After describing configuration and quasi-static modeling of the MR damper based on Bingham model, the integrated lumped parameters model of the whole damper system is obtained by taking into account the compressibility and dynamic motions of MR fluid in the annular duct, upper chamber, lower chamber and dynamics of pistons. In order to demonstrate the effectiveness of the proposed dynamic model, a comparative work between simulation and experiment is undertaken. In addition, the effect of MR fluid compressibility on the hysteresis of MR damper is investigated through computer simulations.

1. Introduction

MR and ER fluid are a class of smart materials whose rheological properties can be controlled by applying a magnetic filed and electric field, respectively. Recently, various semi-active dampers featuring magnetorheological (MR) or electrorheological (ER) fluid have been proposed and successfully applied in the real field, especially in vehicle suspension system [1-3]. It is well-known that MR/ER damper exhibits non-linear behavior with hysteresis. There have been many efforts devoted to model the non-linear behavior of ER/MR damper. These models are classified as quasi-static and dynamic models. Although the quasi-static models are capable of describing force-displacement behavior of the ER/MR damper reasonably well, they are not sufficient to describe the nonlinear force-velocity behavior of the damper [4]. In order to remedy this limitation,

---

[*] This work was supported by the National Research Foundation of Korea (NRF) grant funded by the Korea Government (MEST) (No. 2010-0015090).

[†] Corresponding author.

several dynamic models considering the hysteresis behavior of ER/MR damper have been proposed [5-7]. Essentially, these dynamic models are experiment-based models. Parameters of the models are derived based on experimental results of a real damper which results in high cost of modeling. Moreover, the identified dynamic models can provide a considerably accurate solution only at a certain value of excitation frequency and amplitude treated in experiments. Recently, Nguyen et al. [8] have proposed a lumped parameter model for a double tube ER damper. Experimental results showed that the model can well predict nonlinear hysteresis of the ER damper.

Consequently, the main contribution of this study is to apply the lumped parameter method to a mono-tube MR damper. After describing configuration and quasi-static modeling of the MR damper based on Bingham model, the integrated lumped parameters model of the whole damper system is obtained by taking into account the compressibility and dynamic motions of MR fluid in the annular duct, upper chamber, lower chamber and dynamics of pistons. In order to demonstrate the effectiveness of the proposed dynamic model, a comparative work between simulation and experiment is undertaken. In addition, the effect of MR fluid compressibility on the hysteresis of MR damper is investigated through computer simulations.

2. Configuration and Quasi-Static Modeling of MR Damper

In this study, the MR damper for vehicle suspension proposed by Sung [4] is considered which is schematically presented in Figure 1. The outer and inner pistons are combined to form a MR valve structure which divides the MR shock absorber into two chambers: the upper and lower chambers. These chambers are fully filled with the MR fluid. As the piston moves, the MR fluid flows from one chamber to the other through the annular duct (orifice) of the valve structure. The floating piston incorporated with gas chamber functions as an accumulator to accommodate the piston shaft volume as it enters and leaves the fluid chamber. By neglecting the frictional force between oil seals, the quasi-static damping force of the MR damper can be expressed as follows:

$$F_d = P_a A_s + c_{vis} \dot{x}_p + F_{MR} \operatorname{sgn}(\dot{x}_p) \tag{1}$$

where,

$$c_{vis} = \frac{12\eta L}{\pi R_d t_d^3}(A_p - A_s)^2; \quad F_{MR} = (A_p - A_s)\frac{2cL_p}{t_d}\tau_y; \quad P_a = P_0(\frac{V_0}{V_0 + A_s x_p})^\gamma \quad (2)$$

In the above, τ_y is the induced yield stress of the MR, η is the post yield viscosity of MR fluid, A_p and A_s are respectively the piston and the piston shaft effective cross-sectional areas, L is the length of the inner piston, L_p is the length of the

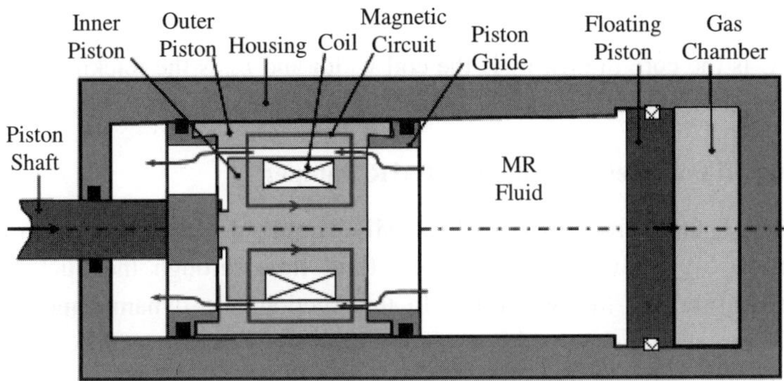

Figure 1. Configuration of the MR Damper.

magnetic pole, R_d and t_d are the average radius and gap of the MR annular duct, P_0 and V_0 are initial pressure and volume of the accumulator, P_a is pressure in the gap chamber, γ is the coefficient of thermal expansion and x_p is the piston displacement. The coefficient c depends on the MR flow velocity profile whose value is ranging from 2.07 to 3.07.

In this work, the commercial MR fluid (MRF132-DG) from Lord Corporation is used. The induced yield stress of the MR fluid as a function of the applied magnetic field intensity (H_{mr}) can be approximately expressed by [9]

$$\tau_y = C_0 + C_1 H_{mr} + C_2 H_{mr}^2 + C_3 H_{mr}^3 \quad (3)$$

In Eq. (3), the unit of the yield stress is kPa while that of the magnetic field intensity is kA/m. The coefficients C_0, C_1, C_2, and C_3 are identified based on experimental results which are 0.3, 0.42, -0.00116 and 1.05E-6, respectively.

The magnetic field intensity across the MR duct can be determined by [9]

$$H_{mr} = H_2 = \frac{N_c I \mu}{2\mu t_d + 2\mu_{mr} A_2(l_1/A_1 + l_3/A_3 + l_4/A_4 + l_8/A_8)} \quad (4)$$

where μ_{mr} and μ are the relative permeability of MR fluid and the valve core material, respectively. N_c is the number of turns of the valve coil and I is the applied current in the coil wire. H_{mr} is effective magnetic intensity. The cross-sectional areas and the lengths of magnetic circuit links are given by

$$l_1 = \frac{R_c}{2} + w_c; \; l_3 = \frac{t_{op}}{2}; \; l_4 = l_8 = L - L_p; \; A_1 = 2\pi(R_c + \frac{w_c}{2})L_p; \; A_{mr} = 2\pi R_d L_p;$$

$$A_3 = 2\pi(R - \frac{3}{4}t_{op})L_p; \quad A_4 = \pi(R^2 - (R - t_{op})^2); \; A_8 = \pi R_c^2 \tag{5}$$

where R_c is the core radius, w_c is the coil width and t_{op} is the thickness of outer piston.

3. Lumped Parameter Modeling of MR Damper

In section 2, the damping force of the MR damper was obtained based on the assumption of steady behavior of MR fluid flow through the duct and the incompressibility of the MR fluid. In this section, the dynamic modeling of the MR damper is performed considering unsteady behavior of MR fluid flows in the damper and compressibility of the MR fluid using a lumped parameter model. In the same manner of lumped parameter model for ER damper [8], the lumped parameter model of the MR damper is presented in Figure 2 and can be mathematically expressed as follows:

$$F_d = P_1 A_1 - P_2(A_2 - A_3) - m_p \ddot{x}_p - b_p \dot{x}_p \tag{6}$$

$$x_{2,1} A_2 + x_{3,1} A_3 = x_p A_2 \tag{7}$$

$$m_{2,1}\ddot{x}_{2,1} + b_{2,1}\dot{x}_{2,1} + k_{2,1}(x_{2,1} - x_{2,2}) = P_2 A_2$$
$$m_{2,2}\ddot{x}_{2,2} + b_{2,2}\dot{x}_{2,2} - k_{2,1}(x_{2,1} - x_{2,2}) + k_{2,2}(x_{2,2} - x_{2,3}) = 0 \tag{8}$$

$$\cdots\cdots\cdots\cdots\cdots\cdots\cdots\cdots\cdots\cdots\cdots\cdots\cdots\cdots$$

$$m_{2,n2}\ddot{x}_{2,n2} + b_{2,n2}\dot{x}_{2,n2} - k_{2,n2-1}(x_{2,n2-1} - x_{2,n2}) + k_{2,n2}(x_{2,n2} - x_{fp}) = 0$$

$$m_{fp}\ddot{x}_{fp} + b_{fp}\dot{x}_{fp} - k_{2,n2}(x_{2,n2} - x_{fp}) = -P_a A_2 \tag{9}$$

$$m_{3,1}\ddot{x}_{3,1} + b_{3,1}\dot{x}_{3,1} + k_{3,1}(x_{3,1} - x_{3,2}) = P_2 A_3 - F_y$$
$$m_{3,2}\ddot{x}_{3,2} + b_{3,2}\dot{x}_{3,2} - k_{3,1}(x_{3,1} - x_{3,2}) + k_{3,2}(x_{3,2} - x_{3,3}) = -F_y \tag{10}$$

$$\cdots\cdots\cdots\cdots\cdots\cdots\cdots\cdots\cdots\cdots\cdots\cdots\cdots\cdots$$

$$m_{3,n3}\ddot{x}_{3,n3} + b_{3,n3}\dot{x}_{3,n3} - k_{3,n3-1}(x_{3,n3-1} - x_{3,n3}) + k_{3,n3}(x_{3,n3} - x_{4,1}) = -F_y$$

$$m_{4,1}\ddot{x}_{4,1} + b_{4,1}\dot{x}_{4,1} - k_{3,n3}(x_{3,n3} - x_{4,1}) + k_{4,1}(x_{4,1} - x_{4,2}) = 0$$

$$\dotfill \quad (11)$$

$$m_{4,n4}\ddot{x}_{4,n4} + b_{4,n4}\dot{x}_{4,n4} - k_{4,n4-1}(x_{4,n4-1} - x_{4,n4}) + k_{4,n4}(x_{4,n4} - x_{5,1}) = 0$$

$$m_{5,1}\ddot{x}_{5,1} + b_{5,1}\dot{x}_{5,1} - k_{4,n4}(x_{4,n4} - x_{5,1}) + k_{5,1}(x_{5,1} - x_{5,2}) = -F_y$$

$$\dotfill \quad (12)$$

$$m_{5,n5}\ddot{x}_{5,n5} + b_{5,n5}\dot{x}_{5,n5} - k_{5,n5-1}(x_{5,n5-1} - x_{5,n5}) = -P_1 A_3 - F_y$$

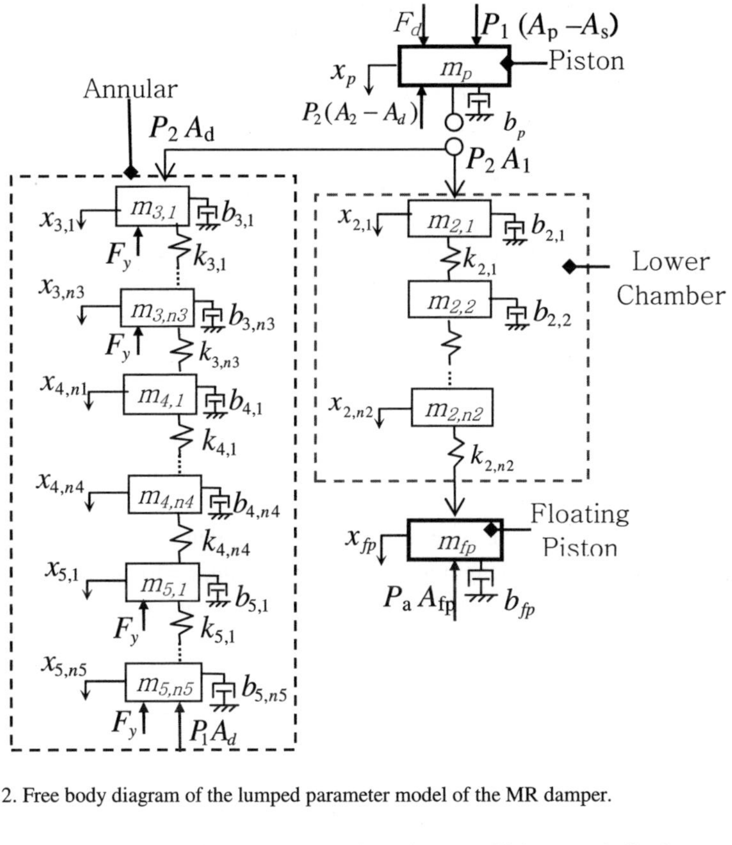

Figure 2. Free body diagram of the lumped parameter model of the MR damper.

In the above, m_p, b_p and x_p are mass, damping coefficient and displacement of the piston part, respectively. m_{fp}, b_{fp} and x_{fp} are mass, damping coefficient and displacement of the floating piston, respectively. $m_{2,i}$, $b_{2,i}$, $k_{2,i}$, $m_{3,i}$, $b_{3,i}$, $k_{3,i}$, $m_{4,i}$, $b_{4,i}$, $k_{4,i}$ $m_{5,i}$, $b_{5,i}$, $k_{5,i}$, are the analogous mass, damping and stiffness of the ith lumps of MR fluid flow in the lower chamber, the lower active MR duct, the inactive MR duct and the upper active MR duct, respectively. $x_{2,i}$, $x_{3,i}$, $x_{4,i}$ and $x_{5,i}$ are the displacements of the ith lumps of the MR fluid in the lower chamber, the lower active MR duct, the inactive MR duct and the upper active MR duct, respectively. A_{fp} and A_d are respectively the gas chamber and the annular duct effective cross-sectional areas. F_y *is* the additional force added to the ith lump of the MR fluid in the active MR duct taking into the pressure drop due to the yield stress. P_1, P_2 and P_a are pressure in the upper chamber, lower chamber and the gas chamber respectively. The pressure P_a can be determined by

$$P_a = P_0 \left(\frac{V_0}{V_0 - A_2 x_{fp}} \right)^\gamma \tag{13}$$

It is noted that in the above equations, the piston displacement xp is known. Then, Eqs. (6) – (12) can be rewritten in a matrix form as follows

$$\mathbf{M}\ddot{X} + \mathbf{B}\dot{X} + \mathbf{K}X = F \tag{14}$$

where the state vector $X = [x_{2,2}...x_{2,n2}, \ x_{3,1}...x_{3,n3}, \ x_{4,1}...x_{4,n4}, \ x_{5,1}...x_{5,n5}]^T$. $\mathbf{M}$, $\mathbf{B}$, and $\mathbf{K}$ are respectively the mass, damping and stiffness matrix. F is the resultant force vector including known displacement terms such as x_p.

4. Results and Discussion

In this study, the gas chamber is filled up with nitrogen gas whose thermal expansion coefficient is $\gamma = 1.4$, the initial volume and pressure of the gas chamber are $V_0 = 70 \times 10^{-6} m^3$ and $P_0 = 3 \times 10^5 N/m^2$, the radii of the piston part R_p, the piston shaft R_s and the annular duct R_d are 23mm, 10mm and 17.5mm, respectively, the MR duct length and gap are $L = 80mm$, $t_d = 1mm$, the pole length $L_p = 11mm$. The coil sectional area is $3.5 \times 48 mm^2$ including the nonmagnetic bobbin and 360 coil turns (wire gauge size 24). The post-yield viscosity and the bulk modulus of MR fluid are respectively 150cSt and $4 \times 10^8 N/m^2$ in this study. Figure 3 shows the simulated and measured damping force of the MR damper as function of the piston velocity when the applied current is 0.5A and 2A, respectively. The figures show a good agreement between the result obtained from the proposed lumped parameter model and experimental one. Especially, the proposed model can accurately express the hysteresis behavior of the MR damper while the quasi-static model can not, specially at the stroke extremes.

In order to evaluate the effect of MR fluid compressibility on the hysteresis simulated results with different values of MR fluid compressibility is shown in Figure 4. From the figure, it is seen that hysteresis of the damper is significantly affected by MR fluid compressibility. The smaller MR fluid compressibility (the larger bulk modulus) is the less hysteresis of the damper is.

5. Conclusion

In this research work, a new dynamic model of a mono-tube MR damper was proposed based on a lumped parameter method considering compressibility and unsteady behavior of MR fluid flows. The comparative work between the simulated results obtained from the proposed models and the measured ones have shown that the proposed lumped parameter model can accurately express the hysteresis of MR damper. Simulation results also showed that the hysteresis of the damper is significantly affected by the compressibility of the MR fluid, the greater MR fluid compressibility is the more hysteresis of the damper is.

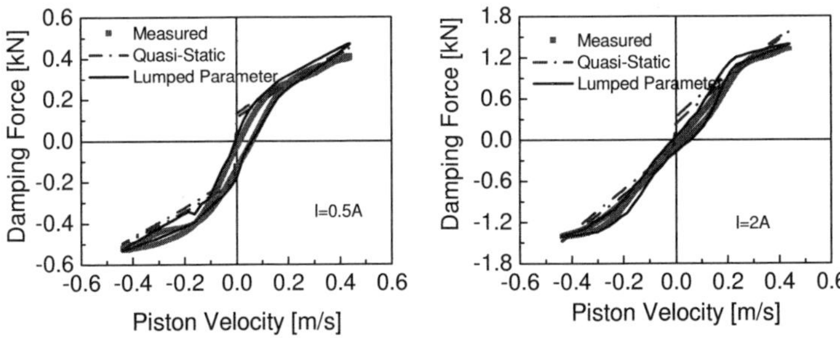

Figure 3. Damping force vs. piston velocity of the damper.

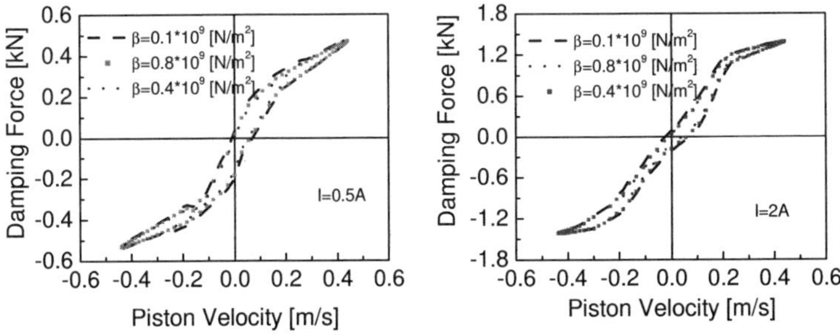

Figure 4. Dependence of damping force on the compressibility of MR fluid.

590

References

1. C. Y. Lai and W. H. Liao, *Journal of Vibration and Control* **8** 527 (2002).
2. K. G. Sung, *PhD. Dissertation*, Inha University, Incheon, Korea (2008).
3. S. B. Choi, Han Y. M. and K. G. Sung, *Int. J. Appl. Electrom.* **27,** 189 (2008).
4. G. Yang, *PhD. Dissertation*, University of Notre Dame, Indiana (2001).
5. B. F. Spencer, S. J. Dyke, M. K. Sain and J D Carlson, *J. Eng. Mech.* **123** 230 (1997).
6. S. Q. Guo, S. P. Yang and C. Z. Pan, *J. Intel. Mat. Syst. Str.* **17** (3), 3 (2006).
7. L. Bitman, Y. T. Choi, and N. M. Wereley, *Smart Mater. Struct.* **14** (11) 237 (2005).
8. Q. H. Nguyen and S. B. Choi, *Smart Mater. Struct.* **18** (11) 3 (2009).

YIELD ENRICHMENT IN MR FLUID VALVES

FERNANDO D. GONCALVES

LORD Corporation, 406 Gregson Drive
Cary, NC 27511, USA

DAVID R. DOBBS

LORD Corporation, 406 Gregson Drive
Cary, NC 27511, USA

Large, high-force, controllable MR fluid dampers often present the device designer significant challenges. This is particularly true if a significant range of force control is required at high speeds. In many instances, achieving a sufficiently high on-state force while maintaining an acceptably low off-state force can lead to device dimensions that are unreasonably large. A balance must be struck between the iron loading needed to meet the on-state force requirement and the viscosity needed to meet the off-state force requirements. A further complication is that very large diameter dampers can encounter Reynolds numbers high enough to cause significant turbulence which further compromises off-state performance. Very large, high-force dampers also present challenges relating to high static and dynamic pressures and ensuring that MR fluid in all parts of the damper does not suffer from particulate settling. In this paper we present evidence of new MR fluid valve and damper designs that allow LORD Corporation to achieve significantly enhanced performance in large, high-force dampers. We will present data that shows that an appropriately designed MR valve can appear to enrich the fluid, thus providing higher apparent yield stresses in the fluid.

1. Introduction

Magnetorheological fluid properties are normally characterized using direct-shear rheometers. These devices normally employ a constant gap, parallel plate geometry having a uniform axial magnetic field. Fluid properties such as viscosity and yield stress as a function of magnetic field strength are easily obtained. While these types of rheometers are widely used, they characterize fluid behavior over a rather narrow range of shear rates, generally less than 1000 s^{-1}. Recently, a high-shear valve-mode rheometer was built and tested to characterize MR fluids at shear rates on the order of 10^5 s^{-1}. This range of shear rates is more representative of the conditions the fluid will be exposed to in a device, particularly in a large, high-force MR fluid damper.

592

In this study the authors present the results of experiments carried out on the high-shear valve-mode rheometer. In particular, the authors will present evidence that a significant increase in the apparent yield stress is achievable with an appropriately designed MR valve. Supporting data will demonstrate that MR valves that satisfy certain conditions can appear to enrich the fluid and lead significantly higher apparent yield stresses. Employing the same design principals, the authors will demonstrate how this yield enhancement phenomenon can be realized in LORD large, high-force MR dampers.

2. Yield Stress Measurements with Direct-Shear Rheometer

To characterize the strength of MR fluids, we will often measure the yield stress as a function of the applied magnetic field. Traditionally, these measurements are carried out on a direct-shear rheometer in which one plate rotates relative to the other and the transmitted torque is recorded as a function of magnetic field strength. The resulting data is used to calculate fluid properties such as viscosity and yield stress as a function of magnetic field strength. These experiments and the resulting data have been used to develop empirical relationships for the expected yield stress as a function of magnetic field strength and iron concentration. Figure 1 shows the measured yield stress for several hydrocarbon-based MR fluids of varying iron contents and the modeled empirical relationship proposed by Carlson given by [1]

$$\tau_o = 271700 \cdot \phi^{1.5239} \cdot \tanh\left(6.33 \times 10^{-6} \cdot H\right) \tag{1}$$

where ϕ is the iron concentration by volume, and H is the magnetic field strength in A/m.

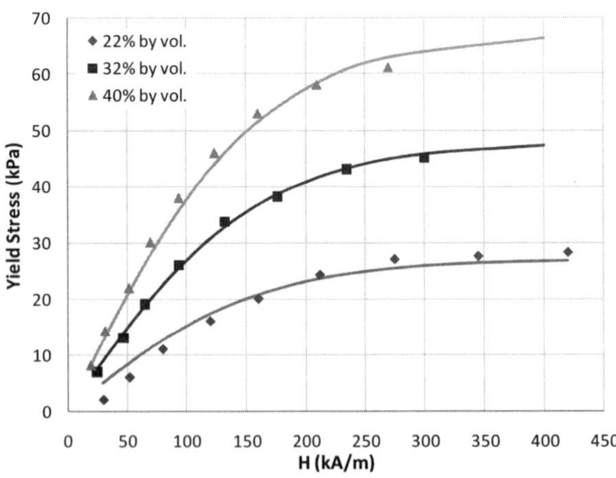

Figure 1. Yield stress measured on direct-shear rheometer

3. Yield Stress Measurements with High-Shear Rheometer

The high-shear rheometer is a pressure driven rheometer that characterizes the fluid in flow-mode rather than the direct-shear mode. The geometries of the MR cells were chosen such that the parallel plate approximation could be made [2]. Furthermore, the ratio of the length of the flow channel to the gap height is large enough that end effects can be neglected [3]. Figure 2 shows several of the valves that were used for high-shear characterization. The capillary valve is used to obtain high-shear off-state viscosity data. The other valves are used to characterize the fluid in the on-state. Each valve has a slightly different geometry to study different shear-rate regimes. Table 1 summarizes the flow geometry of each valve and the maximum shear rates achievable with each valve. A Hall probe is used to measure the field strength for each test. The Hall probe is installed in a parallel air gap having the same geometry of the MR flow channel, in this way ensuring that the measured field strength in both gaps are analogous.

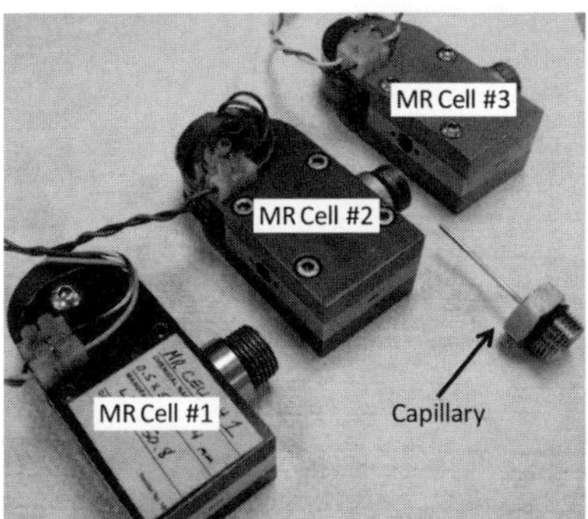

Figure 2. Modular valves used on high-shear rheometer

Table 1. MR fluid valve configurations

Valve	Gap (mm)	Channel Width (mm)	Channel Length (mm)	L/g	Shear Rate (s^{-1})
MRF#1	0.5	5	25.4	50.8	10^5
MRF#2	0.5	5	12.7	25.4	10^5
MRF#3	1.0	5	12.7	12.7	2.5×10^4

Tests are run over a range of piston velocities and are repeated at various magnetic field strengths. The result of each experiment is a force trace in which the steady state forces are calculated for each velocity. From the measured force, quantities such as pressure and shear stress are calculated. Because the valve geometry is known, the shear rate can be calculated from the piston velocity and the yield stress can be calculated at each magnetic field strength. Figure 3 shows sample data for an MR fluid with 32% iron by volume collected using the high-shear rheometer. The procedure for calculating the yield stress is well documented in the literature [4]. Careful examination of the measured yield stress indicates a substantial increase in the apparent yield values as compared with those values measured on direct-shear rheometers.

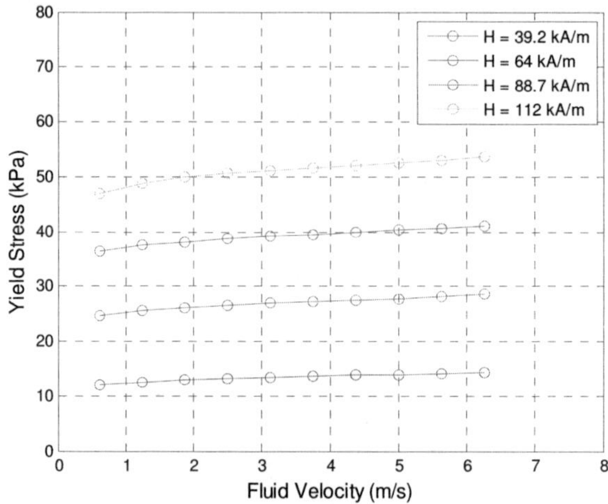

Figure 3. Yield stress measured using the high-shear rheometer

4. Yield Enrichment

The magnitude of the apparent yield stress observed in Figure 3 led to a thorough investigation comparing the yield stresses measured for several fluids on the high-shear rheometer with those predicted by the empirical relationship given in equation (1). Several fluids of varying iron concentrations were considered as were several valve configurations each with a different length to gap (L/g) ratio. The behavior under investigation is shown in Figure 4 which indicates a significant increase in the apparent yield stress as compared to the expected yield stress.

It became apparent through testing and analysis that the combination of a high pressure gradient plus a long valve length would result in dramatically increased apparent yield stress. If the MR valve is designed such that the pressure gradient is made sufficiently high and is maintained for a sufficiently long continuous flow length, the effective yield strength of the MR fluid can be dramatically increased. Under the right conditions, apparent yield strengths may be several times the expected yield strength. As indicated in Figure 5, such increases are most dramatic in fluids with low iron content. In effect, a low-iron content fluid can be made to act like a high-iron content fluid.

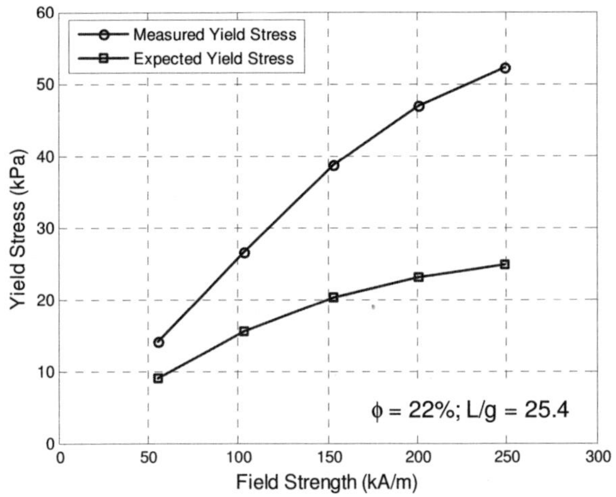

Figure 4. Yield stress comparison for MR fluid with 22% by volume iron concentration

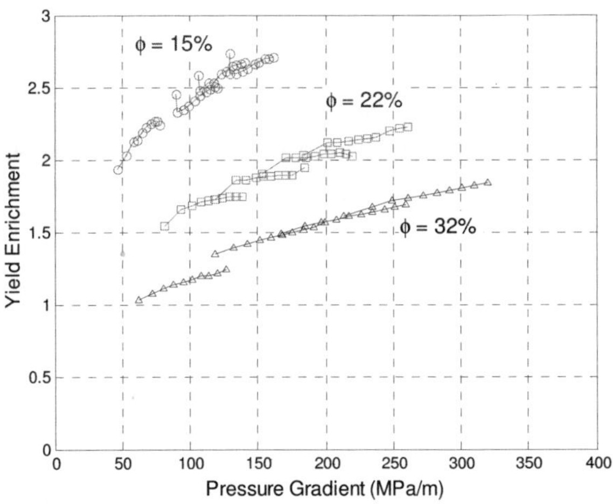

Figure 5. Yield enrichment for several MR fluids of varying iron contents (L/g = 50.8)

596

5. Design Implications

In order to realize the yield enrichment phenomenon, special care must be taken in choosing the valve geometry for the MR device. As Figure 6 indicates, the yield enrichment phenomenon is most notably observed in valves that have high L/g ratios. Perhaps a more profound observation is that the yield enrichment that occurs can be further improved by using MR fluids with low iron concentrations. The ability to use a low-iron fluid while still preserving a high on-state yield stress may help to meet certain design requirements for large high-force dampers.

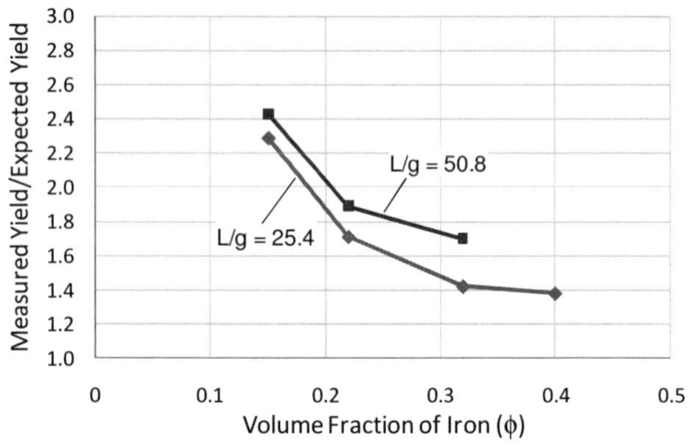

Figure 6. Yield enrichment as a function of iron concentration

Another design tool that can be useful in the design of large high-force dampers is shown in Figure 7. The horizontal axis is the L/g ratio of the valve while the vertical axis is L/g/φ where φ is the iron concentration. The valves used in existing MR dampers are captured in the small box and have L/g ratios less than 15 and L/g/φ ratios less than 50. Virtually no yield enrichment is observed in these valves. The MR fluid valves examined in this study, which have demonstrated significant yield enrichment, have L/g ratios greater than 25 and L/g/φ ratios greater than 50. MR devices that employ MR valves with geometries that fall within the larger box are likely to experience yield enrichment.

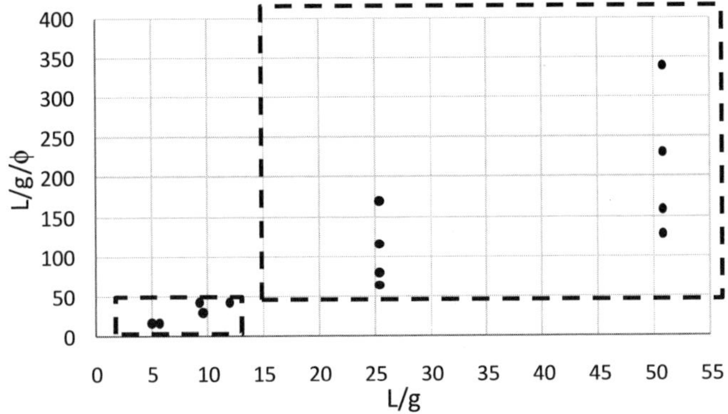

Figure 7. MR valve yield enrichment design map

6. Conclusion

Large high-force MR fluid dampers are often plagued with difficult design tradeoffs. In this study, the authors presented evidence of an MR valve design geometry that can lead to significantly increased apparent yield stresses. This phenomenon is most notably observed in fluids with low iron content. If an MR valve can be designed to capitalize on this yield enrichment phenomenon, a low iron fluid can be used to maintain an acceptably low off-state force without compromising the on-state force.

References

1. J. D. Carlson, "MR Technology Workshop – Chapter 6," LORD Corporation, Cary, NC, p. 6/9-6/10, 2004.
2. D. G. Baird, and D. I. Collias, Polymer Processing Principles and Design, Wiley Interscience, New York, NY, 1998.
3. J. R. Van Wazer, J. W. Lyons, K. Y. Kim, R. E. Colwell, Viscosity and Flow Measurement A Laboratory Handbook of Rheology, Wiley Interscience, New York, NY, 1963.
4. H. P. Gavin, "Electrorheological Dampers for Structural Vibration Suppression," Ph.D. Dissertation, University of Michigan, Ann Arbor, MI, 1994.

HIGH-SHEAR-RATE, VALVE-MODE RHEOMETER FOR MR FLUIDS

FERNANDO D. GONCALVES

LORD Corporation, 406 Gregson Drive
Cary, NC 27511, USA

J. DAVID CARLSON

LORD Corporation, 406 Gregson Drive
Cary, NC 27511, USA

RICHARD WILDER

LORD Corporation, 406 Gregson Drive
Cary, NC 27511, USA

We present details of a new, high-shear-rate, valve-mode MR fluid rheometer capable of measurement at rates appropriate to high-speed dampers and shock absorbers. The instrument is modular and allows for choosing from a number of well-defined magnetic valve geometries or a capillary tube. It requires only a modest amount of MR fluid and is easily cleaned to allow rapid testing of different fluid formulations. Of particular importance is the means for applying and accurately determining the magnetic field H applied to the fluid and provision for demagnetizing the system between measurements.

1. Introduction

Traditionally, magnetorheolgical properties of MR fluids are measured using direct-shear rheometers. Such instruments, whether commercial or custom designed, use either a constant gap, parallel plate geometry having a uniform axial magnetic field or a bob and cup geometry with an approximately uniform, radial magnetic field. Unfortunately, such instruments are only capable of measuring performance of MR fluids at relatively low shear rates, *i.e.* less than about 1000 s^{-1}. In contrast, many MR fluid damper applications such as MR fluid shock absorbers, commonly subject MR fluids to shear rates of 10^5 s^{-1} or more. Extrapolation of low shear rate measurements on such highly non-Newtonian materials as MR fluid by several orders of magnitude to predict behavior in this high rate regime is, at best, an act of faith.

Recognizing the limitations of traditional viscometers, the authors present details of a high-shear-rate rheometer that characterizes the fluid in valve-mode. The device is pressure driven and, depending on the valve geometry, can reach shear rates on the order of 10^5 s^{-1}. Characterizing the fluid in this regime is better aligned to the conditions the fluid will be exposed to in a damper for example. With special care taken to develop a magnetic circuit to uniformly energize the fluid, both the off-state and the on-state properties of the fluid can be studied. Accurate determination of the magnetic field strength is also another key feature of this rheometer.

2. High-Shear Rheometer

The high-shear-rate rheometer is shown installed on an Instron load frame in Figure 1. The load frame is fitted with a 22 kN load cell and is used to drive fluid through the valve at various speeds. The rheometer is modular in nature and can be fitted with either a 38.1 mm bore or a 57.15 mm bore. A piston drives the fluid through the valve at the commanded speed and the corresponding force is recorded. The piston has a stroke of ~200 mm and allows for several velocities to be run in a single stroke.

Figure 1. High-shear-rate rheometer installed on Instron load frame

Each of the rheometer bores has provisions for degassing the fluid while in the rheometer. This ensures that there is no entrained air in the fluid and that there is no air trapped between the piston and the fluid. Failing to degas the rheometer would introduce compliance into the data.

3. Fluid Valves

The rheometer was designed such that several valve configurations could be used. Figure 2 shows several of the valves that can be used for high-shear characterization. The capillary valve is used to obtain high-shear off-state viscosity data. The other valves are used to characterize the fluid in the on-state. Each valve has a slightly different geometry to study different shear-rate regimes. Table 1 summarizes the flow geometry of each valve and the maximum shear rates achievable with each valve.

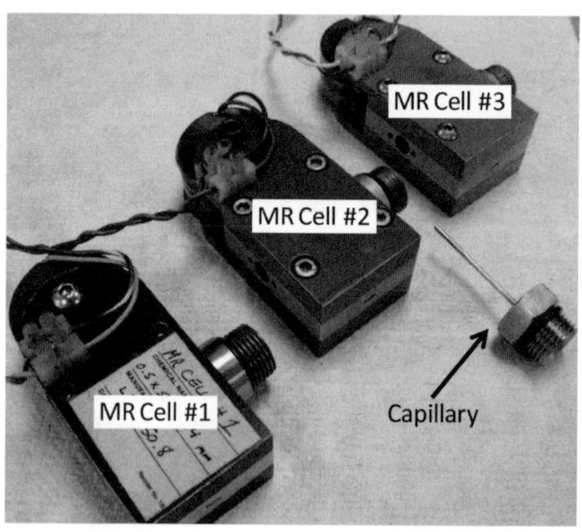

Figure 2. Modular valves used on high-shear rheometer

Table 1. MR fluid valve configurations

Valve	Gap (mm)	Channel Width (mm)	Channel Length (mm)	Shear Rate (s^{-1})
MRF#1	0.5	5	25.4	10^5
MRF#2	0.5	5	12.7	10^5
MRF#3	1.0	5	12.7	2.5×10^4
Capillary	$\phi 1.0$	-	50.0	2×10^5

A Hall probe is used to measure the field strength for each test. The measurement technique employed on the high-shear rheometer is illustrated in Figure 3. The Hall probe is installed in an air gap that is parallel to the MR flow valve. The MR fluid gap is the same as that of the air gap ensuring that the measurements are analogous. Figure 4 shows how this technique is implemented in the MR valve.

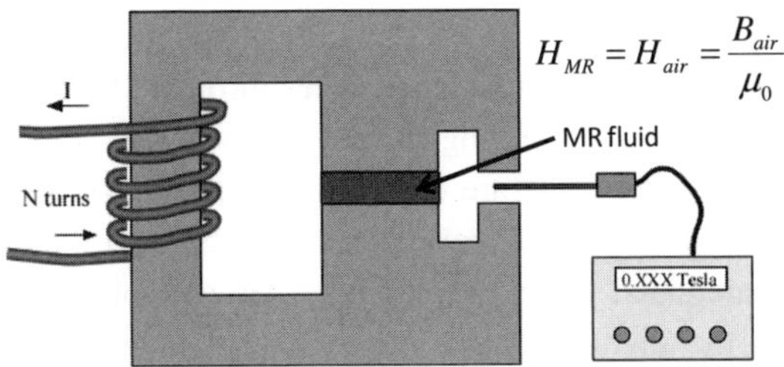

Figure 3. Measurement technique employed for measuring field strength

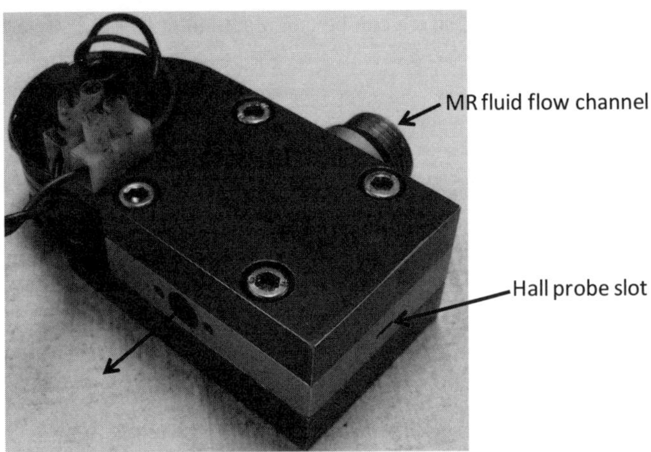

Figure 4. MR Cell showing parallel slot for Hall probe

When testing under relatively high magnetic field strengths, it may be necessary to degauss the MR valve to eliminate any residual magnetic field. Failure to eliminate even low field residuals can bias the results. One method for degaussing the MR valve is to drive the valve with an AC voltage and slowly reduce the amplitude to zero. Driving the valve with an AC voltage randomizes the field orientation and slowly reducing the amplitude of this AC signal eliminates any residual bias. This can be achieved rather easily using a Variac. This is the technique that is employed between tests in which a magnetic field is used.

4. Test Procedure

The procedure begins with sample preparation. The fluid is shaken and degassed before it is poured into the rheometer. Once the fluid is inside the rheometer, the sample is degassed again to fill any voids in the valve. Tests are run over a range of piston velocities normally ranging from ~0.8 mm/s to ~8 mm/s. In a standard test, the actuator advances through 10 velocities in a single stroke. Tests are repeated for varying current inputs to the valve.

The actuator generates pressure driven flow through the well parameterized MR valve. The result of each experiment is a force trace in which the steady state forces are calculated for each velocity. An example of the experimental data obtained on the high-shear rheometer is shown in Figure 5. From the measured force, quantities such as pressure and shear stress are calculated. Because the valve geometry is known, the shear rate can be calculated from the piston velocity and the yield stress can be calculated at each field strength.

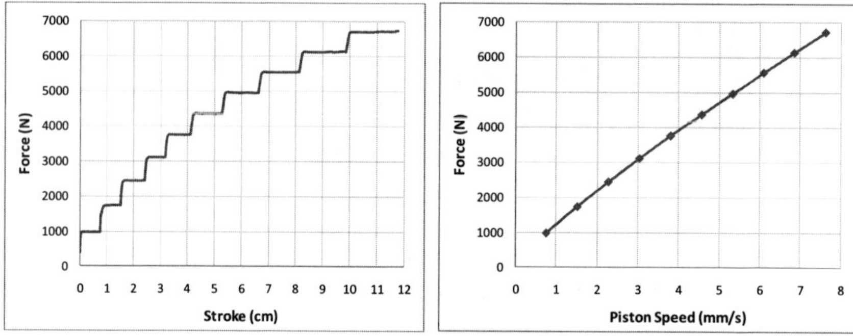

Figure 5. Sample data from high-shear rheometer

5. Data Analysis

Fluid properties such as viscosity and yield stress are extracted from the measured force for each experiment. The shear stress is computed from the mean steady-state force. In the absence of a magnetic field, the fluid is assumed to behave as a Newtonian fluid. Thus, for the capillary valve, the shear stress can be found from

$$\tau_{cap} = \frac{FD_{cap}}{\pi L_{cap}D_p^2} \tag{1}$$

where F is the measured steady-state force for a given piston speed, D_c is the capillary diameter, L_{cap} is the capillary length, and D_p is the piston diameter. The shear rate is similarly found in terms of the valve and rheometer geometry as

$$\dot{\gamma}_{cap} = \frac{8D_p^2 v_p}{D_{cap}^3} \tag{2}$$

where v_p is the piston velocity. The apparent viscosity of the fluid is then found from the measured shear stress and the calculated shear rate as

$$\eta = \frac{\tau_{cap}}{\dot{\gamma}_{cap}} \tag{3}$$

The geometries of the MR cells were chosen such that the parallel plate approximation could be made [1]. Furthermore, the ratio of the length of the flow channel to the gap height is large enough that end effects can be neglected [2]. Thus, assuming parallel plate geometry, the off-state shear stress for the MR cell is found as

$$\tau_{MRCell} = \frac{\Delta P}{L_{MRCell}} \frac{g}{2} = \frac{2Fg}{\pi D_p^2 L_{MRCell}} \tag{4}$$

where g is the gap and L_{MRCell} is the valve length. Likewise the shear rate for the MR valves is found in terms of the mean fluid velocity, u_m, in the channel.

$$\dot{\gamma}_{MRCell} = \frac{6u_m}{g} \tag{5}$$

The yield stress of the fluid is found by manipulating Phillip's cubic equation [3] and writing as a cubic in T as

$$T^3 - \frac{3}{4} P^2 T + \frac{1}{4}\left(P^3 - P^2\right) = 0 \tag{6}$$

where P and T are the non-dimensional pressure gradient and yield stress given by

$$P = -\frac{dp}{dx} \frac{g^2}{12u_m \eta} \tag{7}$$

$$T = \frac{\tau_o g}{12u_m \eta} \tag{8}$$

In equations (7) & (8), dp/dx is the pressure gradient, developed along the MR valve length, L_{MRCell}, g is the gap, u_m is the mean fluid velocity, η is the fluid viscosity, and τ_o is the yield stress. The non-dimensional pressure gradient can be written in terms of known quantities as

$$\mathcal{P} = \frac{F}{A_p} \frac{g^2}{12u_m \eta L_{MRCell}} \tag{9}$$

where A_p is the area of the piston. Substituting equation (9) into the desired root given by [4]

$$\mathcal{T}(\mathcal{P}) = \mathcal{P} \cos\left(\frac{1}{3} \mathrm{acos}\left(\frac{1-\mathcal{P}}{\mathcal{P}} \right) + \frac{4\pi}{3} \right) \tag{10}$$

the yield stress is then found using equation (8) as

$$\tau_o = \frac{12u_m \eta}{g} \mathcal{T} \tag{11}$$

The yield stress is found at each fluid velocity and each magnetic field strength.

6. Conclusion

As MR fluids continue to be considered for a wide variety of applications, their properties and characteristics must continue to be studied. While most rheometers characterize fluid behavior at relatively low shear rates, the application of these fluids often exposes them to conditions that are several orders of magnitude higher. The high-shear rheometer described in this study is capable of characterizing MR fluids at flow regimes much closer to the conditions the fluid will experience in the device.

References

1. D. G. Baird, and D. I. Collias, Polymer Processing Principles and Design, Wiley Interscience, New York, NY, 1998.
2. J. R. Van Wazer, J. W. Lyons, K. Y. Kim, R. E. Colwell, Viscosity and Flow Measurement A Laboratory Handbook of Rheology, Wiley Interscience, New York, NY, 1963.
3. R. W. Phillips, "Engineering Applications of Fluids with a Variable Yield Stress," Ph.D. Dissertation, University of California, Berkeley, CA, 1969.
4. H. P. Gavin, "Electrorheological Dampers for Structural Vibration Suppression," Ph.D. Dissertation, University of Michigan, Ann Arbor, MI, 1994.

DURABILITY TESTING ON MAGNETORHEOLOGICAL FLUIDS FOR CLUTCH AND BRAKE APPLICATIONS

CLAUS GABRIEL, GÜNTER OETTER and CHRISTOFFER KIEBURG

BASF SE,
Carl-Bosch-Strasse, 67056 Ludwigshafen/Rhein, Germany

MARTIN LAUN

Technology Consultant to BASF SE,
Robert-Blum Strasse 40, 68199 Mannheim, Germany

A twin gap plate-plate shear cell, developed by BASF, meets the requirements for reliable durability testing on magnetorheological fluids. The possibility to test the effect of life time energy input to MR-formulations within a few days is demonstrated for commercial MRF formulations (Basonetic®). MRF durability measurements as well as further analytical characterization of the sheared MRF prove that magnetorheological fluids with high content of magnetisable particles, as required for clutch/brake applications, are able to withstand specific energy intakes of more than $5 \cdot 10^{12}$ J/m³. This remarkable result should encourage designers in industry and automotive to explore new MRF applications.

1. Introduction

1.1. *Durability testing requirements*

Durability testing for clutch and brake applications requires a mimicking of typical drive cycle conditions of commercial devices. Here, the typical shear rate regime is 0 up to ~ 3000 s^{-1}. In automotive applications, the required temperature range is typically -40 – 150 °C. By magnetic flux densities between 0 and ~ 1 Tesla, the corresponding MRF shear stresses may be controlled between 0.1 and 100 kPa.

Furthermore, the MRF has to exhibit weak sedimentation and excellent redispersibility throughout the whole lifetime of a device. Since sedimentation and redispersibility of the MRF might change with energy intake, they also have to be assessed in durability testing.

1.2. *Lifetime dissipated energy intake (LDE)*

The lifetime dissipated energy intake W [J] per volume V [m³] in typical clutch and brake applications is of the order $= 10^{12}$ J/m³ and higher. The specific energy intake is calculated from the following equation:

$$\frac{W}{V} = \tau\, \dot{\gamma}\, t$$

The shear stress is denoted as τ [Pa], the shear rate $\dot{\gamma}$ [s^{-1}], and the duration of the experiment t [s]. W/V represents an integral measure of the energy dissipated in a drive cycle of a clutch or brake. It sums up all combinations of transmitted torque, differential speed and duration of the events.

2. Technical Challenges

Mimicking the energy dissipation of commercial MR clutch or brake devices by a lab-scale durability test cell results in a number of technical challenges.

2.1. *Ensure homogeneous magnetic flux density distribution*

The homogeneity of the flux density distribution in the shear gap of a durability test cell is very important. A non-homogeneous flux field would induce migration of the magnetisable particles towards flux density maxima, thus resulting in torque variations which have nothing to do with the MRF sensitivity to energy dissipation. It has been demonstrated before [1] that homogeneous flux density distributions are important for magnetorheometry or device design.

2.2. *Avoid loss of MRF base oil*

Not only the MRF itself has to withstand a high energy intake, but also the sealings of the durability test cell. Should a sealing become leaky, some loss of carrier fluid of the MRF cannot be avoided. The resulting change of particle volume concentration in a clutch type MRF may have dramatic effects on the resulting shear stress. As CIP volume concentrations are typically distinctly higher compared to damper MRF (see Fig. 1), a loss of base oil will drastically increase both off- and on-state shear stresses.

This leads to the conclusion that durability testing requires an effective sealing of the MRF, unlike open shear cells as used in typical commercial plate-plate magnetorheometers. The sealings should have a low friction to achieve a low drag torque, and the latter should not vary with time. High reproducibility of the sealing dimensions and drag are required after each replacement. In addition,

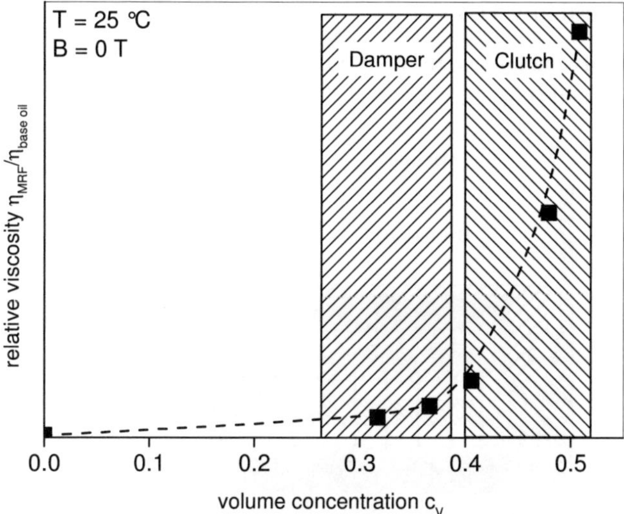

Figure 1. Relative MRF viscosity versus CIP volume concentration measured at zero flux density. The shaded areas indicate concentration regimes typical for damper and clutch type MRF.

high chemical and abrasion stability against the MRF carrier fluid and the magnetisable particles is needed to avoid swelling or degradation and wear of the sealing.

2.3. *Ensure removal of dissipated heat*

Durability testing aims at accelerating the energy dissipation encountered during the lifetime of a device. The power intake compared to the device is higher to allow for a reduction in testing time, e.g., by a factor of ten. This results in a distinct temperature increase, thus requiring effective cooling of the durability test cell.

Assuming adiabatic conditions in the MRF shear gap, the temperature rise rate may be estimated. For a power intake of 12 Watt, and an MRF volume of 2 ml, the adiabatic temperature rise rate would be as high as 2 K/s! In reality, heat conduction to the environment will reduce the temperature rise rate. This underlines the necessity of effective cooling. Yet, the temperature rise in the MRF should not be neglected.

3. BASF Durability Test Cell

Said technical challenges resulted in the design of a custom made BASF durability test cell (Fig. 2).

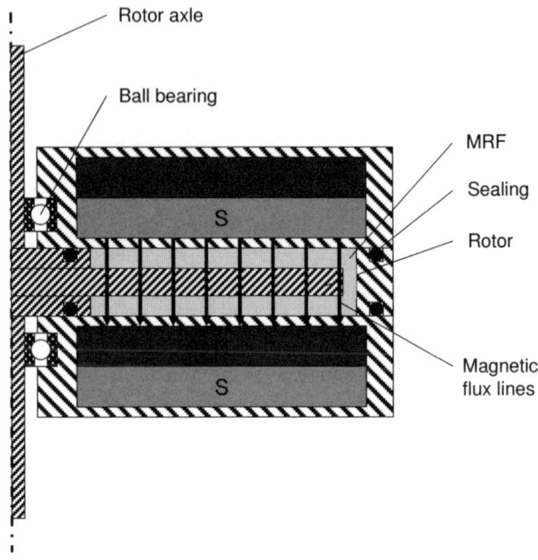

Figure 2. Schematic of the BASF durability test cell (only right hand half shown).

The durability test cell is a twin gap plate-plate rotational design, similar to the rheometer shear cell described in [2]. Advantages of this setup are the compensation of normal forces generated by the MRF in each shear gap. Moreover, an effective cooling is achieved by immersing the whole test cell into a thermostated oil bath.

A homogeneous magnetic field is achieved by placing a pair of cylindrical permanent magnets with a central bore on both sides of the shear cell. The magnetic field acts perpendicular to the plane of shear. The magnetic flux density inside the MRF gaps is around 0.5 Tesla and depends on the magnetization of the MRF itself. Stationary sealings are placed on the outer radius to achieve a leak-tight shear cell. Rotary sealings are located at an inner radius, so that their contribution to the torque measurement is only marginal.

The test cell may easily be dismantled to allow for an extraction of the MRF after a certain energy intake. The MRF volume is ~ 2 ml, the maximum torque 2 Nm, the temperature range 23 °C – 150 °C, and the shear rate regime 0 – 500 s^{-1}. The maximum allowed inside pressure of the test cell, measured by a combined temperature-pressure sensor, is approximately 10 bar. Typical test conditions are temperatures of 40 - 80 °C, shear rates between 50 and 400 s^{-1}, flux density ~ 0.5 T, and power intakes of 5 - 15 W.

4. Results

4.1. *Long-term durability testing*

The long-term durability test on a clutch and brake type Basonetic® MRF from BASF is shown in Fig. 3. The test was deliberately stopped at a specific energy intake of W/V = $6\,10^{12}$ J/m³ and lasted for about 300 hours of continuous shear!

The torque signal remains essentially constant throughout the whole test (Fig. 3). This result demonstrates the excellent long-term durability of this MRF containing 90 percent by weight CIP. Moreover, it is in excellent agreement with durability test results on a large-scale test bench using a double-gap concentric cylinder MR device as described in [3]. Here, also a high energy intake of W/V = $6\,10^{12}$ J/m³ was achieved at essentially constant torque level of ~ 100 Nm transmitted by the clutch-type MRF.

No significant difference in long-term durability was found when continuous shear was compared with interval durability testing (e.g. variation of shear rate at fixed flux density) as long as the specific power intake remained within the above mentioned range (results not shown here).

The shear stress versus flux density characteristic of the MRF after removal from the durability test cell is shown in Fig. 4 (measurement method according to [1]). Compared to a new, untreated sample, the shear stresses after such a high energy intake show only a minor increase. The rheological properties of the clutch type MRF after very high lifetime dissipated energy are very well comparable to the original properties. This again underlines the long-term durability of the Basonetic® clutch-type MRF. Scanning electron micrographs of carbonyl iron particles extracted from the MRF both before and after durability testing demonstrate no significant wear on the particles (not shown here).

4.2. *Redispersibility testing*

For redispersibility testing on magnetorheological fluids after certain energy intakes, a higher MRF quantity than the one accessible with the BASF durability test cell is required. Durability tests were therefore performed by Wiehe et al. at

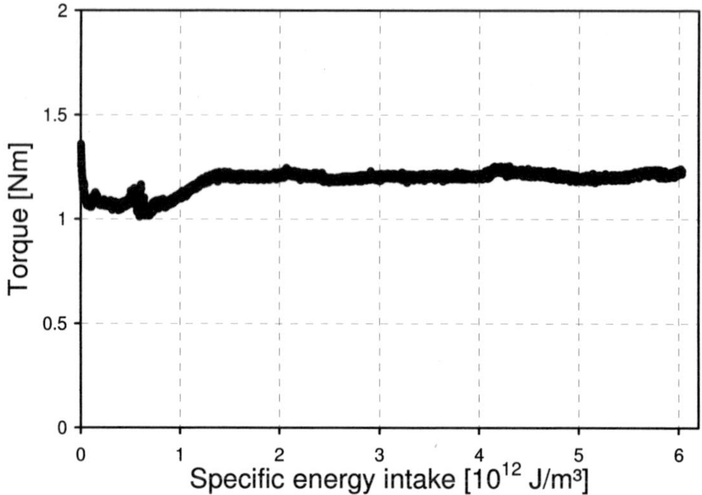

Figure 3. Long term durability test on a Basonetic® clutch-type MRF.

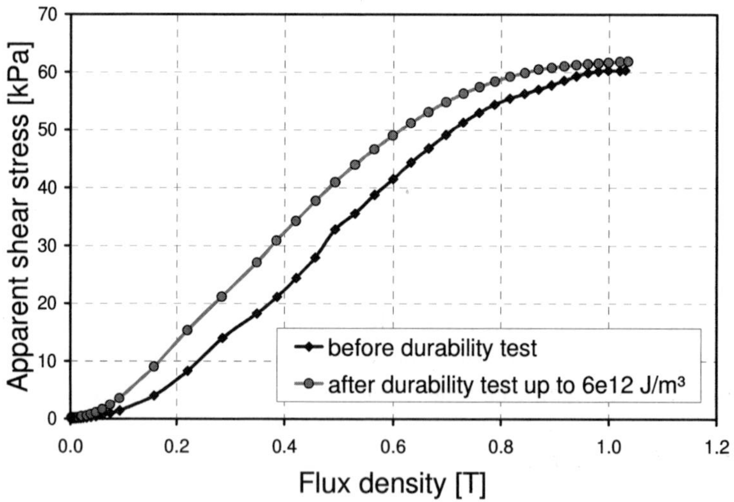

Figure 4. Shear stress versus flux density characteristic of a Basonetic® clutch-type MRF
before and after high energy intake durability testing.

W/V* [10^{12} J/m³]	W_redisp [mJ]	Evaluation
0	1	excellent
1	10	good
2	5	good - excellent
3	3	excellent

Table 1. Redispersibility test on a clutch type MRF after certain energy intakes.

Hochschule Ostwestfalen-Lippe, Germany [3] using a concentric cylinder device. The durability tests were stopped at certain energy intakes and the redispersibility work as an integral measure of the sediment hardness determined by the spatula method, see e.g. [4].

The redispersibility work changes somewhat with the energy intake the MRF has experienced. The level of redispersibility work is so low, however, that the overall evaluation of MRF redispersibility remains good to excellent. This means that the ease of redispersing the sediment of the clutch-type MRF does not change significantly with energy intake. It has also to be taken into consideration that sedimentation in narrow shear gaps, as encountered in clutches and brakes, remains weak.

5. Summary

Reliable MRF durability testing requires a leak-tight shear cell with homogeneous magnetic flux inside the MRF gap, and an effective removal of dissipated heat. These preconditions provided, clutch and brake type Basonetic® MRF are able to withstand specific energy intakes of $5 \cdot 10^{12}$ J/m³ and more. This has consistently been demonstrated by the BASF twin-gap plate-plate durability cell and the medium scale concentric cylinder testing device at Hochschule Ostwestfalen-Lippe, Germany.

References

1. H. M. Laun, G. Schmidt, C. Gabriel, C. Kieburg, *Rheol Acta.* **47**, 1049 (2008).
2. H. M. Laun, C. Gabriel, C. Kieburg, *J Rheol* **54**, 327 (2010).
3. A. Wiehe, D. Güth, J. Maas, ERMR (2010).
4. C. Kieburg, G. Ötter, R. Lochtmann, C. Gabriel, H. M. Laun, J. Pfister, G. Schober, H. Steinwender, *World Scientific,* New Jersey, 101-107 (2007).

SYNERGISTIC EFFECT IN MAGNETOELECTRORHEOLOGICAL FLUIDS WITH A COMPLEX DISPERSED PHASE

E. V. KOROBKO[*], Z. A. NOVIKOVA and M. A. ZHURAUSKI

A. V. Luikov Heat and Mass Transfer Institute, National Academy of Sciences of Belarus, 220072, 15 P. Brovka str., Minsk, Belarus
[*]E-mail: evkorobko@gmail.com

D. BORIN[†] and S. ODENBACH

Technische Universität Dresden, 01069, 10 Helmholtz str., Dresden, Germany
[†]E-mail: Dmitry.Borin@tu-dresden.de

In the paper the results of investigation of structural interactions, generated by simultaneous impact of electric and magnetic fields on fluids containing a two-component dispersed phase are presented. Dielectric, magnetic and rheological characteristics of these fluids are considered. The characteristic features of the formed structures depending on the filler type, deformation conditions and intensity of application external fields are discussed.

1. Introduction

In structure reversed fluids based on dispersed phase (magnetoelectric rheological fluids – MERF) that are sensitive to electric and magnetic fields, one can perform separate or simultaneous electrical or magnetic structurization of particles by applying independent fields. In prospect this will allow changing the character of particles interaction, controlling the structure more flexibly and broadening the range of regulation of the mechanical properties of a material. It means that the induced increase in the shear stress at two fields is higher than the sum of increments of the intensities, induced separately by the electric and magnetic fields [1 – 4]. As the index characterizing the value of the synergistic effect we will use $\xi = \dfrac{\Delta\tau_{EH}}{\Delta\tau_E + \Delta\tau_H}$ which is the ratio of the shear stress increment on simultaneous application of electric and magnetic fields to the sum of increments in the case of separate applications of the fields.

In our recent paper [5] the rheological behavior of MERF with a complex dispersed phase based on two components was investigated. It was found that on simultaneous application of two fields here also appeared the synergistic effect that by its characteristics surpassed the results obtained earlier by other authors for fluids based on a single component dispersed phase.

The aim of the present work is to perform a more detailed study of the characteristic peculiarities of a two-field impact on MERF with a complex (two-component) dispersed phase. This investigation related not only to the study of rheological response, but also to the dielectric and magnetic properties of fluids for determination of the influence of these parameters on the appearance of synergistic effect.

2. Materials and Methods

We have investigated MERF with compositions consisting of such fillers as carbonyl iron (F-CI) which is sensitive only to a magnetic field, aerosil activated by some additives (F-Si), sensitive only to an electric field, and two fillers – oxides α-Fe_2O_3 (F-α) and γ-Fe_2O_3 (F-γ), that are sensitive to both fields in different degrees. Mineral oil was used as a dispersion medium. Fluids with a concentration C of a dispersed phase of 5 vol.% and 10 vol.% with only one filler and MERF with a complex filler, consisting of 5 vol.% CI and 5 vol.% SiO_2 (F-CI/Si), or 5 vol.% α-Fe_2O_3 (F-CI/α), or 5 vol. % γ-Fe_2O_3 (F-CI/γ) have been investigated. It was earlier stated [6] that the biggest increment of rheological characteristics in magnetic field is given exactly by carbonyl iron (CI), which was chosen by us as a constant component of the dispersed phase of all complex compounds.

The rheological properties were carried out using rotational viscometer modified for measurements under two fields. The dielectric properties of MERF were determined in the range of frequencies from 10^2 to 10^5 Hz at temperature of 20 °C. The magnetic properties of MERF were determined by means of the standard technique of measuring magnetization with two Hall sensors.

3. Main Results

Dielectric properties and electrorheological effect. To evaluate the capability of dispersed phase particles to be polarized in an external electric field and to form rigid structural bridges in a dielectric dispersion medium, we have measured of the components of the complex dielectric permittivity (ε', ε'') for fluids with single and complex fillers. For the chosen fillers SiO_2, γ-Fe_2O_3, α-Fe_2O_3, like for other oxides, the interphase polarization mechanism is most characteristic. It is known [7, 8] that if for fluids with such fillers the difference between ε' at frequencies 10^2 Hz and 10^5 Hz ($\Delta\varepsilon'$) is large and the relaxation frequency f_r at

614

which the value of ε'' is maximal lies in the range $10^1 - 10^5$ Hz, then such fluids display a high electrorheological activity.

From the results shown in Fig. 1 it is seen that fluids consisting of one filler F-Si and F-γ have the strongest interphase polarization. With increase in the filler concentration of one type the values of ε', ε'' also increase without a qualitative change in the frequency dependences (Fig. 1, curves 6).

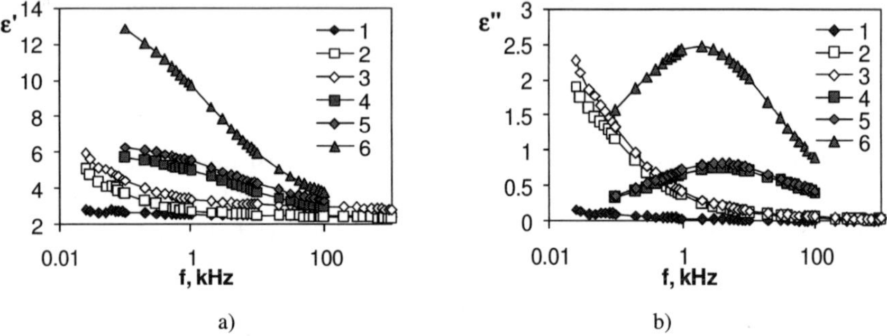

a)   b)

Figure 1. Dielectric spectra of fluids. 1 – F-α; 2 – F-Si, 3 – F-CI/Si, 4, 6 – F-γ; 5 – F-CI/γ. 1, 2, 4 – C = 5 vol.%, 6 – 10 vol.%.

The fluid F-α has a small value of $\Delta\varepsilon'$, and the value of f_r does not fall within the area of interphase polarization, thus indicating the absence of relaxational interphase polarization, and one cannot expect a high electrorheological response for such a fluid.

If we compare the electrorheological effect for these three fluids (Fig. 2b, c, d), then it is seen that the MERF F-α is least sensitive to an electric field. The function τ (E) changes a little bit more strongly in fluid with F-γ. F-Si has the strongest electrorheological response.

It is noted that the presence of carbonyl iron in MERF in addition to all investigated oxides qualitatively does not change the character of the dielectric spectra and electrorheological response τ (E), ε' (f), ε'' (f).

Magnetic properties and magnetorheological effect. In Fig. 3 the magnetization curves of the fluids under investigation are presented. For compositions with single dispersed phase the highest magnetization is displayed by the F-CI fluid (curves 5, 9). The use of only α-Fe$_2$O$_3$ as a dispersed phase provides the saturation magnetization J_s not higher than 4 kA/m (at 10 vol.%), the fluid with 10 vol.% γ-Fe$_2$O$_3$ J_s reaches 25 kA/m, the fluid with 10 vol. % CI J_s reaches 80 kA/m. The F-α and F-γ fluids reach magnetic saturation at $H_s \approx 50 \div 60$ kA/m in contrast to the F-CI fluid for $H_s > 350$ kA/m.

When using a complex dispersed phase in F-CI/α (Fig. 3, curve 7), saturation magnetization J_s is increased substantially (above 70 kA/m), and the

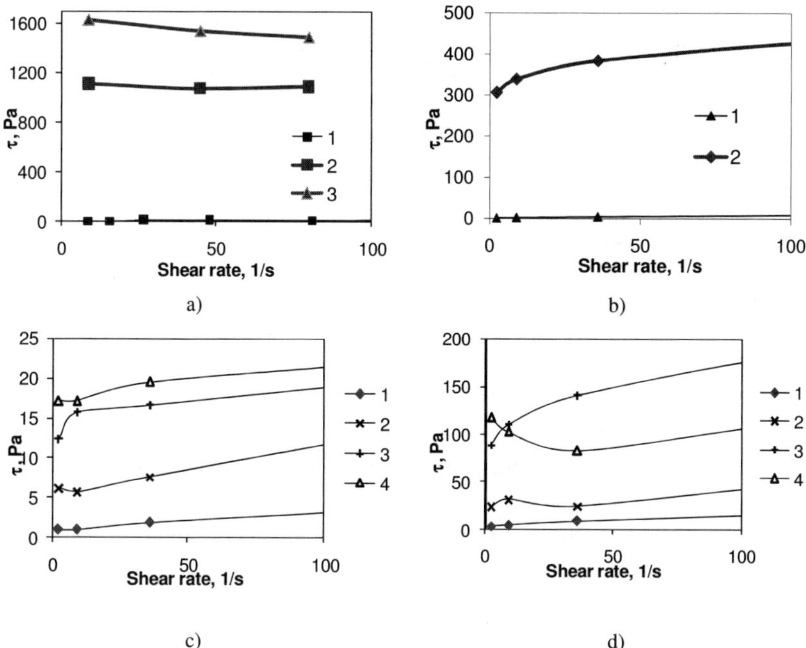

Figure 2. Flow curves for fluids with different fillers in electric and magnetic fields: a) F-CI (1 – H = 0; 2 – H= 80 kA/m; 3 – H = 100 kA/m); b) F-Si (1 – E = 0; 2 – E=1.8 кV/m); c) F-α, (1 – E=0, H=0; 2 – E=1.8 кV/m, H=0; 3 – E=0, H = 100 кA/m; 4 – E=1.8 кV/mm, H = 100 кA/m); d) F-γ (1 – E=0, H=0; 2 – E=1.8 кV/mm, H=0; 3 – E=0, H = 100 кA/m; 4 – E=1.8 кV/mm, H = 100 кA/m). C = 5 vol. %.

threshold value of the magnetic field at which the saturation occurs (H_s=380 kA/m) also increases. It should be noted, that all compositions of MERF containing 5 vol. % of carbonyl iron have similar magnetization curves that differ little when using different oxide fillers. Their saturation magnetization varies in the range $J_s = 60 \div 70$ kA/m in a magnetic field with H > 350 kA/m.

Comparing the magnetic characteristics of MERF with single dispersed phase with the magnitude of the magnetorheological effect has shown that the largest magnetorheological effect is in the fluid with CI (Fig. 2a, c, d). By increasing the intensity up to H = 60 kA/m, i.e., up to reaching the saturation magnetization (Fig. 3), one can also note the growth in the rheological characteristics of the F-α and F-γ fluids.

Comparing the values of magnetization and magnetorheological response MERF with complex dispersed phase with analogous characteristics for a fluid F-CI (Fig. 3, curve 5) allows us to assume that the growth of τ is due to the high sensitivity of carbonyl iron to the magnetic field. In using a complex dispersed

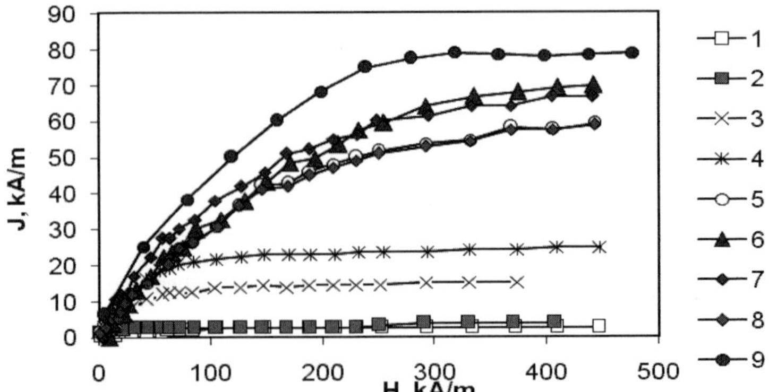

Figure 3. Magnetization curves for fluids: 1, 2 –F-α ; 3, 4 – F-γ ; 5, 9 –F-CI ; 6 – F-CI/γ; 7 – F-CI/α ; 8 – F-CI/Si. 1, 3, 5 – C = 5 vol. %; 2, 4, 9 – 10 vol. %.

phase – a maximum magnetorheological effect was revealed in F-CI/γ. The values of increments in the shear stress in a magnetic field are smaller and differ slightly for fluids F-CI/α or F-CI/Si. Thus, a magnetic field forms weaker structures in F-CI/α in contrast to its influence on F-CI/γ, since the intensity of magnetization of the α-Fe_2O_3 particles is much less.

Due to the high magnetorheological activity of CI the fluid F-CI/Si displays an increase in τ (H) in a magnetic field (Fig. 4c). However, its magnitude is not so much larger than the increment in τ (E) in contrast to other fluids in which this difference is more appreciable.

Rheological characteristics in two fields. The rheological response to the electric and magnetic fields of fluid F-α is very weak, and maximal values of τ reach only 20 Pa. For fluid with F-γ they are several times higher, moreover the shear stress grows substantially. In the case of the influence of two fields the rheological response in the range of shear rates γ >10 s^{-1} for fluid F-γ appears to be lower than that under the impact of only a magnetic field (Fig. 2d, curve 4).

Addition of 5 vol. % CI to the composition leads to a change the rheological characteristics of MERF in a magnetic field and two fields (Fig. 4a, b, c).

The presence of one component insensitive (SiO_2) or slighly sensitive (α-Fe_2O_3) to magnetic effect enhances the rheological response MERF in two fields. Possibly, thus occurs because of the action of electric field and creation of the most beneficial polarization signation for maximum structuring of each of the dispered phase components separately. The magnetosensitive γ-Fe_2O_3 and α-Fe_2O_3 particles depending on the relationship between the intensities of two fields, form agglomerates with CI particles which are stronger than those formed only by the magnetic field, since having united in structures, they are capable of being attached to the electrodes.

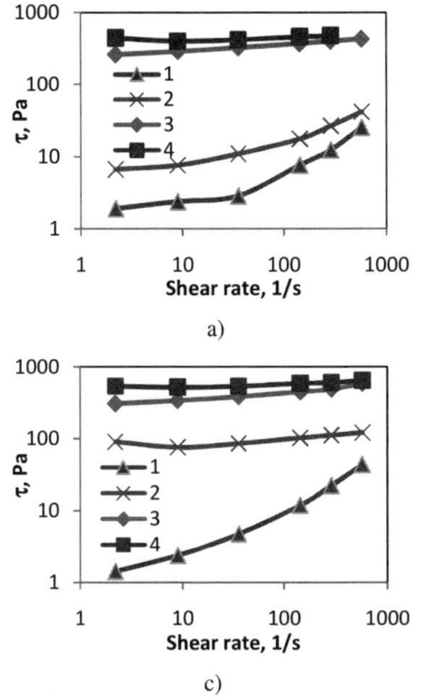

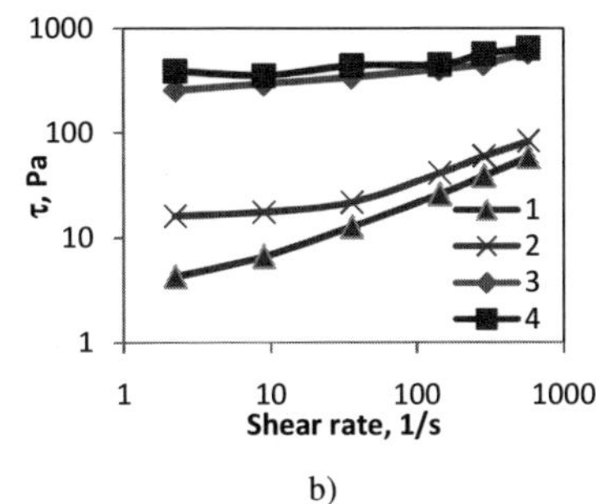

Figure 4. Flow curves: a) F-CI/α, b) F-CI/γ, c) F-CI/Si (1 – H = 0, E= 0; 2 – H = 0, E = 1.36 kV/mm; 3 – H = 100 kA/m, E = 0; 4 – H = 100 kA/m, E = 1.36 kV/mm).

These peculiarities of structuring for MERF with different fillers create a peculiar nature for every composition that displays the synergetic effect shown in Fig. 5.

For fluid F-CI/α synergetic effect is increased continuously. For F-CI/γ only when H > 60 кА/м, at H < 20 kA/m the synergetic effect is most attributable to the electric field. F-CI/Si displays (Fig. 5c) the highest synergistic effect under all intensities of electric and magnetic fields. It was found that for MERFs with α-Fe$_2$O$_3$ or γ-Fe$_2$O$_3$ the decrease in $\Delta\tau_{EH}$ in comparison with $\Delta\tau_H$ occurs at higher shear rates. On increase in the shear rate the transformations in the electrorheological structures and MR structures are intensified differently depending on the relationship between electric and magnetic forces, as well as on the number of particles involved in the aggregates. It appears possible to lower, and not only to increase, the induced stress by varying the level of intensities and hydrodynamic regimes. Probably this occurs because part of ferromagnetic particles from these structures are entrained into the common, with CI, magnetic structures increasing their strength not only due to their own magnetic properties, but also because of their attachment to the electrodes, just as in the case of the electrorheological effect.

618

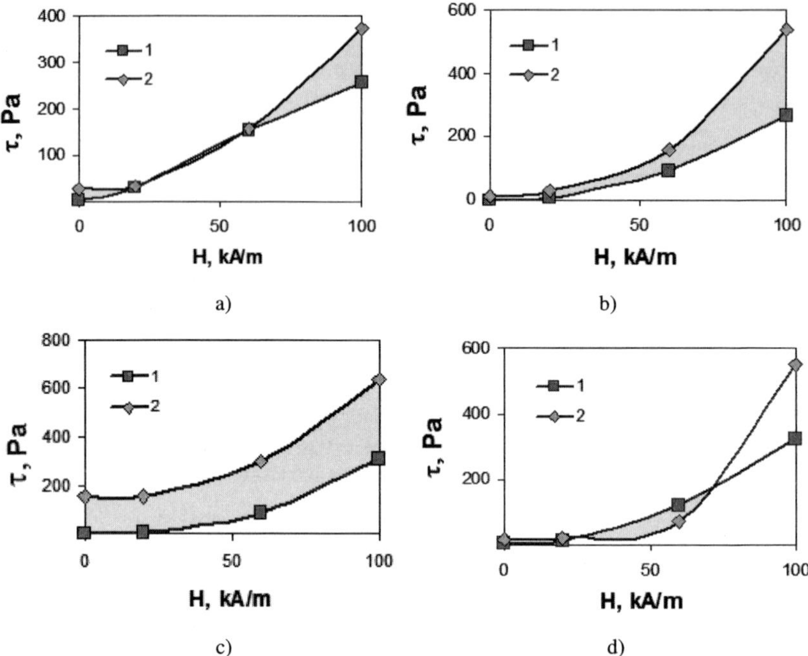

Figure 5. Shear stress of a MERF vs. the magnetic field strength: a) F-CI/γ; b), d) F-CI/α; c) F-CI/Si; a, b, c) γ = 2,2 s^{-1}, d) γ = 36 s^{-1}; 1 – E=0, 2 – E = 1.8 kV/mm.

4. Conclusion

Investigations have shown that an MERF containing a complex dispersed phase in which one of the components is highly sensitive only to the impact of an electric field (SiO_2) and the other one – carbonyl iron is highly sensitive only to the impact of a magnetic field, shows the strong synergistic effect in the entire investigated range of intensities of electric and magnetic fields. For such types of fluids the formation of separate strong electric and magnetic structures is typical.

An MERF containing ferromagnetic particles sensitive to the impact of an electric field displays an increasing synergistic effect while reaching saturation magnetization. The second region of the synergistic effect in such fluids is created by fields due to the appearance of strong electrostructures from γ-Fe_2O_3 particles in weak magnetic fields. It disappears on increase in the magnetic field strength. The MERFs containing a complex dispersed phase in which one component is weakly sensitive to the actions of both electric and magnetic fields (α-Fe_2O_3) and the other is carbonyl iron have only one zone of synergistic effect.

For such types of fluids the formation of structures identical for the particles of two fillers is also typical.

Acknowledgments

The authors acknowledge Belarusian Fund for Fundamental Research for financial support of investigations (Project No. **T10R-045**).

References

1. S. Gorodkin, W. Kordonsky, E. Medvedeva, *Problems of heat and mass transfer-91* Minsk, HMTI NAS Belarus, 48 (1991).
2. W. Kordonsky, S. Gorodkin, E. Medvedeva, *Proc. 4th Int. Conf. on ER Fluids (Feldkirch, Austria, July 20 – 23, 1993)* ed. by R. Tao, G. D. Roy, 23 (1994).
3. K. Minagawa, T. Watanabe, K. Koyama, M. Sasaki, *Langmiur* **10**, 3926 (1994).
4. K. Koyama, *Proc. 5th Int. Conf. on ER Fluids, Magneto-Rheological Suspensions & Associated Technology (Sheffield, UK, July 10 – 14, 1995)* ed. by W. Bullough, 245 (1996).
5. E. V. Korobko, M. A. Zhurauski, Z. A. Novikova, V. A. Kuzmin, *Journal of Physics: Conference Series* **149**, 012065 (2009).
6. W. I. Kordonski, S. R. Gorodkin, Z. A. Novikova, *Proc. 6th Int. Conf. on Electro-Rheological Fluids, Magneto-Rheological Suspensions and Their Applications (Yonezawa, Japan, July 22 – 25, 1997)* ed. by M. Nakano, K. Koyama, 535 (1998).
7. T. Hao, A. Kawai, F. Ikazaki, *Langmuir* **14** (5), 1256 (1998).
8. F. Ikazaki, A. Kawai, K. Uchida, T. Kawakami, K. Edamura, K. Sakurai, H. Anzai, Y. Asako, *Journal of Physics D: Applied Physics* **31**, 336 (1998).

LARGE SCALE TEST BENCH FOR THE DURABILITY ANALYSIS OF MR FLUIDS

ANSGAR WIEHE, DIRK GÜTH and JÜRGEN MAAS

Ostwestfalen-Lippe University of Applied Sciences, Control Engineering and Mechatronic Systems Liebigstraße 87, 32657 Lemgo, Germany, www.motion-ctrl.de

Durability is an essential issue of magnetorheological fluids (MR fluids) especially in brake and clutch applications. To investigate the lifetime dissipated energy (LDE) within a technical relevant scale, considering the volume of MR fluid and the transmitted torque, a long-term measuring system is developed which is based on a continuous load cell (CLC). An advantage is the possibility to vary the influencing quantities, which enables the characterization of the MR fluid during a durability experiment and application oriented load cases with changing influencing quantities. Measurements show the dependency of the attainable LDE on the power dissipation and a pressure increase, indicating wear processes of the MR fluid.

1. Introduction

In most applications MR fluids represents a novel technology. Therefore bigger uncertainties exist at the start of a product development mainly relating to the feasibility and the durability. While the feasibility can be tested by a proof-of-concept prototype with a comparable low afford, the proving of the durability is more cost intensive and often takes long periods of time. To overcome this, a model for the prediction of the durability of MR systems is needed, which should be based on reproducible and application-oriented experimental data. Since the results of durability tests, which are performed in a rheometer sized scale, probably can not be correlated to devices for different applications, it is necessary to develop a large scale, application oriented test bench. For an application oriented, representative analysis of the durability the large scale test bench has to provide a:

- shear gap volume of larger than 15 ml,
- continuously adjustable magnetic flux density of up to 1 T,
- transferable torque of larger than 100 Nm,
- continuously mechanical power dissipation of larger than 250 W,
- controllable shear gap temperature in the range from -20 °C to +150 °C,
- adjustable shear rate of up to several $1000 \ s^{-1}$.

To achieve reproducible results, the long term measuring system has to meet the following specifications:

- adequate capturing and control of the influencing and measured quantities,
- minimization of particle centrifugation, sedimentation and migration,
- easy filling and extraction of the MR fluid,
- the volume of the MR fluid outside of the energy dissipating shear gaps should be as small as possible,
- minimization of parasitic torques (i.e. caused by the sealing) to avoid affects on measurement results.

2. Description of the Long Term Measuring System

Designing the continuous load cell (figure 1) by utilizing the design methodology for rotatory MR actuators in [1], a cylindrical geometry with a horizontal oriented double shear gap is chosen to avoid the mass force induced phenomena centrifugation and sedimentation. To reduce particle migration, a homogenous magnetic field in the shear gaps is essential [2]. Therefore a particular focus is given to the distribution of the magnetic field during the design phase, which is validated afterwards by measuring the magnetic flux density. To ensure an effective cooling, a coolant duct is located near the shear gaps. The seals are located directly to the shear gaps in order to minimize the MR fluid volume, which is not involved in the energy dissipation process. The magnetic circuit is indicated by the dark grey color, while the non magnetizable areas are marked in light grey color. The distribution of the magnetic field is marked by the lines. By the design of the rotor, an easy and reproducible extraction of the MR fluid after the measurement is considered.

The durability analysis is performed on a torque test bench, which is controlled by a user programmable real-time system. Also the measurement data are recorded, pre-processed and stored for the subsequent analysis by this system. The controllable quantities are the temperature, the rotational speed and the excitation current of the continuous load cell (CLC). The measured quantity is the torque, which is transferred by the CLC. Additionally, the pressure in the area of the outer seal can be recorded.

The CLC is driven by a servo drive, while the transferred torque is measured by a torque sensor. For the cooling of the CLC, a thermostat is utilized by which tempered silicon oil is pumped through the CLC. The temperature of the CLC is controlled by an integrated closed loop controller according to the reference temperature. Since the test bench is unitized, the performance parameters, shown in table 1, can be widely adjusted.

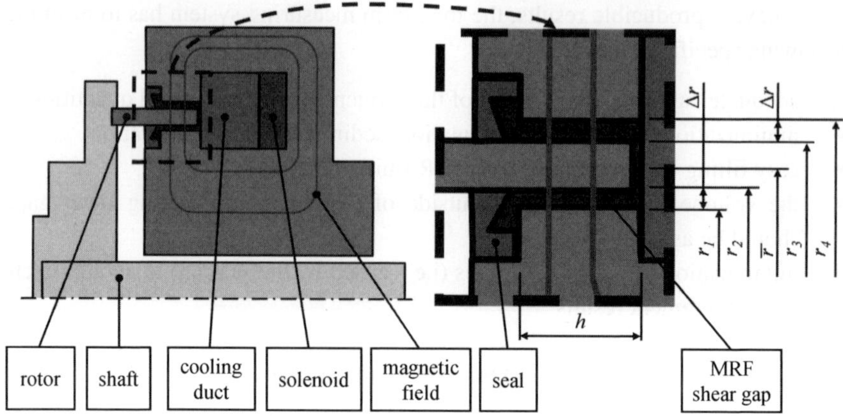

Figure 1. Schematic drawing of the continuous load cell (CLC).

Table 1. Dimensions of the continuous load cell and performance data of the long term measuring system.

description	symbol	value
shear gap radius	r_1	63.2 mm
shear gap radius	r_2	64.2 mm
shear gap radius	r_3	68.8 mm
shear gap radius	r_4	69.8 mm
mean radius	$\bar{r}$	66.5 mm
axial shear gap dimension	h	24.5 mm
shear gap volume	V_{MRF}	20.4 cm^3
measured torque	T_m	$|T_m| < 200\,\mathrm{Nm}$
rotational speed	n	$|n| < 850\,\mathrm{rpm}$
mean mechanical power dissipation	P_m	$P_m < 350\,\mathrm{W}$
excitation current	i	$|i| < 2\,\mathrm{A}$

3. Measuring Process

The aim of the measuring process is the durability analysis of MR fluids by ensuring reproducible conditions, whereas the whole characteristic regarding to the influencing quantities should be observed. As illustrated in figure 2 the beginning of a durability measurement, an initial characterization is performed, whereby the torque is measured regarding to the current, the rotational speed and the temperature. These data of the unstressed MR fluid can be used for calibration. Afterwards a load cycle is started. The main purpose is to wear the MR fluid. During the load cycle a definable load case is repeated, whereby the torque is also measured until a desired energy intake or a period of time is achieved. After every load cycle the torque behavior with respect to the influencing quantities is measured during the characterization phases. This loop is repeated until the overall energy intake is reached. At last, a final characterization is performed.

As a result of these measurements two kinds of torque data are obtained. The first one, obtained during the load cycles, is continuous by time, but typically poor in the variation of the influencing quantities. The second one, given by the characterizations, is typically more detailed, but only discrete in time.

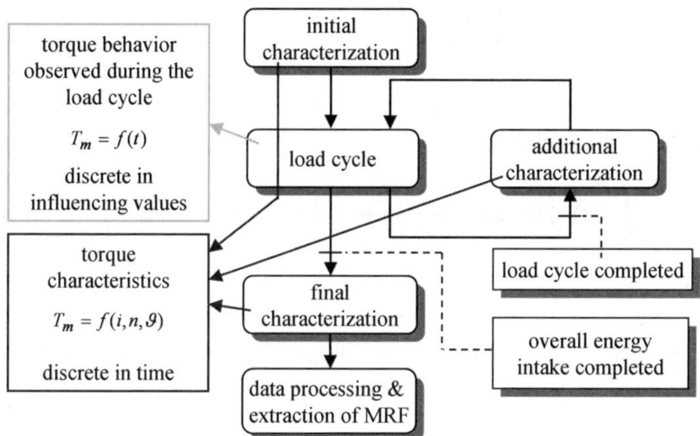

Figure 2. Measuring process with load cycles to wear the MR fluid and characterization phases to obtain the detailed behavior of the MR fluid.

Since the recorded quantities (torque vs. rotational speed, current, temperature and pressure) reflect the behavior of the CLC, these have to be converted to obtain MR fluid related quantities.

Assuming laminar flow and a small radial gap dimension Δr regarding to the mean radius $\bar{r}$, the mean shear rate $\dot{\bar{\gamma}}$ can be obtained by the rotational speed n and the dimensions, given in table 1:

$$\dot{\bar{\gamma}} = \frac{2 \cdot \pi}{60} \cdot \frac{\bar{r} \cdot n}{\Delta r}.$$

(1)

The mean shear stress $\bar{\tau}$ is given by

$$\bar{\tau} = \frac{T_m - T_p}{2 \cdot \pi \cdot h \cdot \bar{r}^2},$$

(2)

where T_m is the measured torque, which has to be corrected by parasitic torques T_p. Beside the geometric dimensions, the relation between the excitation current i and the magnetic flux density B is affected by hysteretic, nonlinear material parameters. Thus, the declaration of an explicit function is inaccurate, but this relation can for example been calculated by FE methods.

To evaluate the measurement results, the energy dissipated by the MR fluid, has to be correlated to the shear gap volume. The energy intake density w_{LDE}, also known as the life time dissipated energy (LDE, [3]), is the time integral of the power dissipation per volume:

$$w_{LDE} = \frac{1}{V_{MRF}} \int_{T_{LDE}} \int_V \tau(r,\varphi,z,t) \cdot \dot{\gamma}(r,\varphi,z,t) \cdot r \cdot d\varphi \cdot dr \cdot dz \cdot dt . \qquad (3)$$

If the assumptions of laminar flow and a small radial gap dimension are used, a simple expression is achieved:

$$w_{LDE} = \int_{T_{LDE}} \bar{\tau}(t) \cdot \bar{\dot{\gamma}}(t) \cdot dt = \frac{2 \cdot \pi}{60 \cdot V_{MRF}} \int_{T_{LDE}} \left(T_m - T_p\right) \cdot n \cdot dt = \frac{P_m}{V_{MRF}} \cdot T_{LDE} . \qquad (4)$$

Since the main topic of this contribution is the long term measuring system, the measured torque T_m in relation to the influencing quantitiess is treated as the measurement result, which represents the system behavior.

4. Results

To validate the measuring system the results are compared with those obtained on a different, rheometer sized system with a radial, double shear gap geometry [4]. The influencing quantities were chosen accordingly (table 2a). The results in figure 3a show a sufficient stable behavior for almost all known applications. The comparison of the measured torque, which is obtained during the load cycles, shows a sufficient agreement with the reference data. Only at the beginning the torque shows deviations, which is probably caused by the running-in characteristic of the different measurement systems.

The reference system [4] only operates with stationary influencing values. As an advantage of the realized long term measuring system the whole torque characteristic of the MR fluid can be observed. Exemplarily, the current depending torque behavior is depicted in figure 3b, which is obtained during the characterization phases. The measured data also indicate a stable behavior of the MR fluid.

Table 2. Parameters of a) measurement for validation and b) application oriented measurement.

parameter	value for validation	value for application oriented measurement
rotational speed n	15 rpm	40 rpm
desired temperature	20 °C	20 °C
current (characterization)	from 0 A to 1 A in steps of 0.1 A	from 0 A to 1 A in steps of 0.1 A
current (load case)	0.7 A (static)	10 min @ 0.7 A; 5 min @ 0 A (square wave)

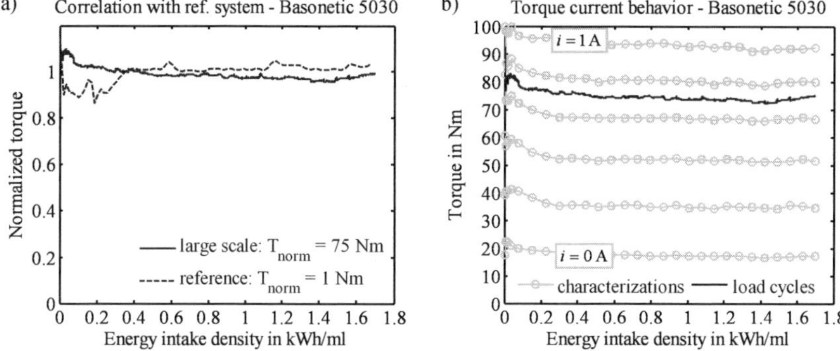

Figure 3. a) Comparison of torque data obtained from the reference system and from the long term measuring system and b) torque measured during the load cycles and during the characterization phases.

To come closer to the initial goal of an application oriented analysis, the load case, which was characterized by static conditions for the first measurement, is modified by implementing a square wave formed excitation current and by increasing the rotational speed (table 2b). As an advantage of this, it is possible to observe the on- and the off-state torque during the load cycles utilizing its pulsed behavior, which results are shown in figure 4a.

The dependency of the possible energy intake density from the power dissipation is depicted in figure 4b. The energy intake density, until an unwanted change of the measured torque is appearing (end of durability), is decreased by increasing the power dissipation. Before a change of the measured torque can be noticed an increase of the pressure is observed, which is measured in the area of the seal. This probably indicates wear effects of the MR fluid.

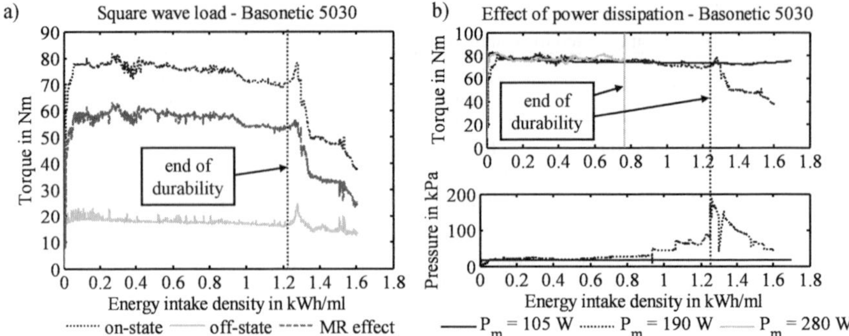

Figure 4. a) On- and off-state torque vs. energy intake density and b) effect of the power dissipation and pressure increase prior an unwanted torque change.

5. Conclusion

The development of the long term measuring system, containing the special designed continous load cell and durability experiments regarding the torque behavior of the MR fluid with respect to the energy intake density are presented in this contribution. The specifications of the long term measuring system can be summarized to an application oriented scale, indicated by shear gap volume of 20 ml and a possible continuous power dissipation of 350 W. The possibility to vary the influencing quantities enables the characterization of the MR fluid during a durability experiment and application oriented load cases. As measurements show, the examined MR fluid (Basonetic 5030) is durable for known applications. Furthermore it turns out, that the achievable lifetime dissipated energy depends on the power dissipation density.

Acknowledgments

This contribution is accomplished within the project "MRF-Bremse" (MR brake – MR based brakes and clutches), funded by the Federal Ministry of Education and Research (BMBF) of Germany under grant number 17N0307.

References

1. A. Wiehe, K. Voth and J. Maas, *JoP: Conference Series*, **149**, 2009.
2. C. Gabriel, H.M. Laun, G. Schmidt, G. Ötter, Chr. Kieburg and J. Pfister, *JoP: Conference Series*, **149**, 2009.
3. J. D. Carlson, *JIMSS*, **13**, 431, 2002.
4. C. Gabriel, G. Oetter, C. Kieburg and H. M. Laun, *Conf. proc. of ERMR 2010*, World Scientific Publishing Company.

FABRICATION AND CHARACTERIZATION OF NEW MAGNETO-RHEOLOGICAL ELASTOMERS WITH EMBEDDED HARD MAGNETIC PARTICLES

JEONG-HOI KOO

Department of Mechanical and Manufacturing, Miami University,
Oxford, Ohio 45056, USA

ALEXANDER DAWSON

Department of Mechanical and Manufacturing, Miami University,
Oxford, Ohio 45056, USA

HYUNG-JO JUNG

Civil and Environmental Engineering Department, KIAST,
Daejeon 305-701, South Korea

This study investigates a new generation of magnetorheological elastomers (MREs) based on hard magnetic materials. The type and dispersion of the filler material affects how the MREs respond to an applied magnetic field. A random dispersion of soft magnetic particles, such as iron, results in a MRE with stiffness that varies with the applied magnetic field. Unlike "soft" MREs, a dispersion of hard magnetic materials aligned in an electromagnetic field will produce an MRE with magnetic poles. When a magnetic field is applied, perpendicularly to these poles, the filer particles generate torque and cause rotational motion of the MRE blend. The primary goal of this project is to fabricate and test the properties of MREs filled with hard magnetic particles (or H-MREs). This experimental work investigated the effect of different types of filler materials by measuring the blocked force and the displacement of H-MREs with varying magnetic fields. The linear trends in the displacement and blocked force over the respective ranges give good indication of the application of H-MRE materials as bending type actuators.

1. Introduction

Magnetorheological materials are a branch of smart materials with properties and responses dependent upon the magnitude of an applied magnetic field. Since its discovery by Rabinow in 1948 the magnetorheological effect (MR) [1] has been applied to several classes of materials including fluids, foams, and most recently elastomers. The viscosity of MR fluids (MRFs) increases under applied magnetic fields, a property which lends itself to applications in the automotive industry [2-4], seismic protection [5-6], and vibration control [7].

The type and dispersion of the filler material affects the response of the MREs to an applied magnetic field and subsequently the modes of application. Conventional MREs consist of a dispersion of soft magnetic particles, such as iron, aligned in an electromagnetic field. The elastic modulus of the resulting material is magnetic field dependent. The similarity between MRFs and MREs results in interchangeability in applications. A model describing the magnetic field dependency of the stiffness of a MRE with aligned iron particles was developed by Davis [8]; a stiffness change of nearly 50% was observed. New methods of production may eliminate the need to produce MREs in the presence of a magnetic field. Lokander produced a stiffness change comparable to aligned MREs by the addition of a silicone oil additive [9].

A dispersion of hard magnetic materials aligned in an electromagnetic field produces an anisotropic, magnetically poled MRE similar to a flexible permanent magnet. The nature of MREs consisting of hard magnetic materials (H-MRE), both the permanency of embedded particles as well as the flexibility of surrounding elastomer, creates a material with applications quite different from those of conventional MREs. Since the particles are embedded in a semi-rigid median, the application of a magnetic field perpendicular to the polarity of the H-MRE will cause motion as the particles attempt to align with the applied magnetic field. The resulting motion and force can be used as a magnetic field controlled actuator.

Limited research has been performed on the new generation of MREs consisting of hard magnetic material (H-MRE), thus they will be the focus of this study. In particular this study will focus on experimental evaluation of H-MREs with different filler materials in an effort to determine the feasibility of utilizing these materials as bending type actuators as well as the effect permanent magnetic characteristics have on the response of the actuator.

2. H-MRE Sample Preparation

Volume percentages for optimal MR effect have been established for conventional MREs [9], thus, H-MRE samples were fabricated with particle volume percentages of 30% cured with a two component elastomer resin in a rectangular prism mold with dimensions of 63.5x31.8x2.5 mm^3. The silicone rubber acts as the binding agent, contributing to the stiffness and flexibility of the cured sample and the embedded magnetic particles produce the MR effect. Samples were fabricated from HS IV elastomer resin (Dow Corning Corp.).

Four types of permanent magnetic particles were studied: barium hexaferrite (BaFe12O19), strontium ferrite (SrFe12O19), samarium cobalt (SmCo5), and neodymium magnet (Nd2Fe14B). Samples were produced to resin specifications with the addition of the magnetic particles and aligned in a two Tesla magnetic for one hour. The total curing time was 24 hours at room temperature. A picture of the sample as well as a scanning electron microscope image of the anisotropic sample can be seen in Figure 1.

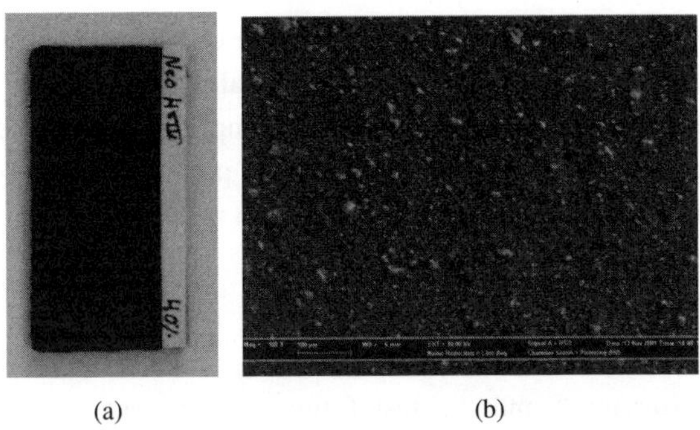

(a)　　　　　　(b)

Figure 1. (a) H-MRE Sample (b) SEM Image of Anisotropic H-MRE sample

3. Experimental Setup

In order to produce a varying magnetic field a two coil modified horseshoe electromagnet with an iron core (see Figure 2a) was constructed. The coils consist of 750 turns of 18 AWG wire. Current was supply to the coil via a Kepco Model ATE 36-15DM power supply. Figure 2b shows a finite element analysis of magnetic fields, which shows uniform magnetic fields in the gap where the MRE sample is placed. The relationship between the applied current, voltage, and resulting magnetic field can be seen in Table 1.

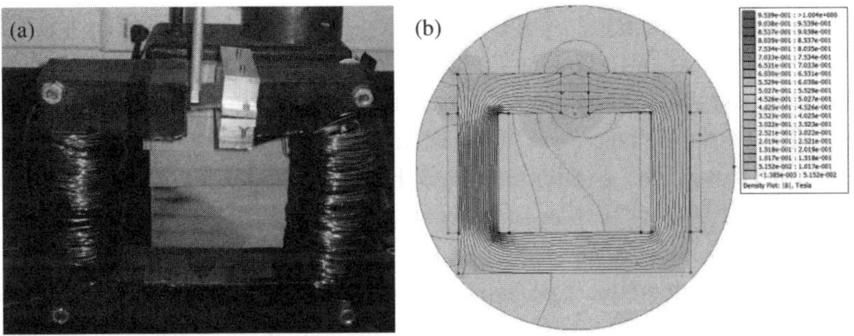

Figure 2. (a) Experimental setup (b) FE magnetic flux contour plot

Table 1. Relationship between voltage, current, and magnetic field for electromagnet configuration

Voltage (V)	Current (A)	Magnetic Field (mT)
0	0	0
5	1.28	32
10	2.59	65
15	3.85	92
20	5.09	124
25	6.25	152

The samples were positioned in a fixed-free cantilever configuration in the test region of the electromagnet via a square aluminum clamp. The samples were oriented such that the length of the sample was perpendicular to the direction of the magnetic field and the width was parallel. A high resolution load cell and a laser displacement sensor were used to measure the blocked force and the displacement of the beam, respectively.

4. Results and Discussion

Using the electromagnet and the test frames experiments were conducted to evaluate the effect of filler material on the blocked force and displacement of the response. Although the response of the different particle types is being investigated, it is important to note that the size of the particle differs (varying 40 µm – 250 µm) and thus conclusions will focus on the trends the responses and the feasibility of utilizing H-MRE materials as bending-type actuators.

4.1. *Displacement Response*

For displacement experiments, the magnetic field was varied linearly from 0 to 25 mT. This range was chosen due to the nonlinearity of the H-MRE motion at higher magnetic flux. Figure 3 gives evidence that the particle type and magnetic properties affect displacement response. For each particle type studied the displacement generated by the application of a magnetic field has a positive linear correlation over the range studied. Linearity is an important characteristic for both the use and characterization of actuators. Although the bending motion of H-MRE is only linear over a small range, considerable displacements, in excess of 10 mm for neodymium and barium hexaferrite, were able to be achieved (see Fig. 3). The non-linearity of the H-MRE displacement response at high displacement is a consequence of the elastomer median in which the particles are imbedded; as the material bends in response to the magnetic field it functions as a non-linear spring. Considering the high bending achieved H-MREs may prove to be the magnetic analog to electroactive polymers.

Moreover, comparing to other solid-type magnetic materials (such as magnetostrictive materials), the deformation of H-MREs is quite significant.

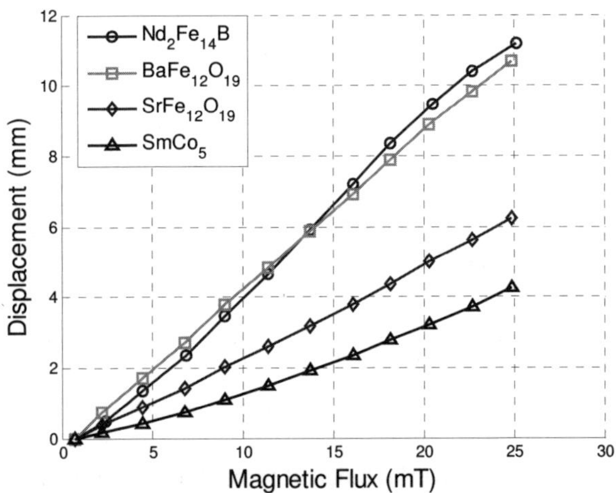

Figure 3. Displacement change of H-MREs with different particle types

4.2. *Force Response*

Similarly to the displacement response, the blocked force response exhibits a linear trend; however the blocked force maintains linearity at much higher magnetic field levels. For the blocked force experiments, the magnetic flux density varied from 0 to 152 mT. The non-linearity presence in the displacement response at high magnetic fields (> 25 mT) is not evident in the blocked force response due to the nature of the loading. Since the beam (H-MRE) is maintained horizontal during the loading (see Fig. 2a), the beam simply transmits the torque provided by the particles attempting to align within the magnetic field and does not provide restorative force due to strain in the elastomer.

Figure 4 shows the blocked force variation of H-MREs with different particle types. For the neodymium sample, the blocked force reaches to nearly 1,200 mN at the maximum magnetic flux density considered in the study (i.e. 150 mT). The magnitudes of these blocked forces are considered to be sufficient to actuate components or mechanisms in small-scale adaptive structures mechanical systems and, such as MEMS devices.

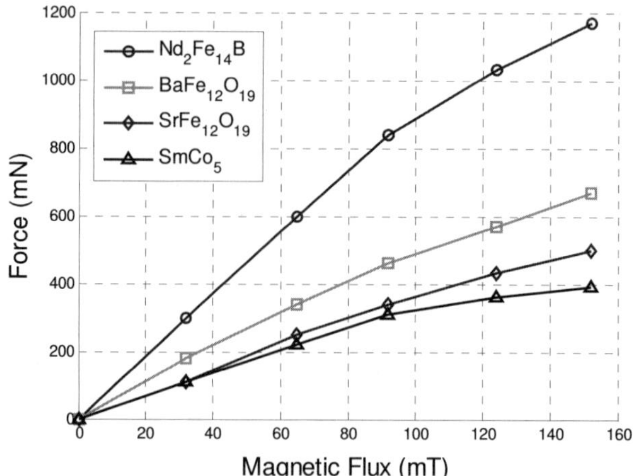

Figure 4. Blocked force change of H-MREs with different particle types

It is interesting to note that the blocked force does not follow the same trend as the displacement (see Figures 3 and 4) although the type of magnetic particle affects both responses. Neodymium based H-MRE exhibits the greatest blocked force; however, its displacement is comparable to that of the barium ferrite, suggesting that the generated responses are consequences of different properties of the embedded particles or are affected by an interaction between the filler material and embedded particles.

5. Conclusion

The focus of this paper has been the dynamic response, specifically blocked force and displacement, of H-MREs with different filler particle types under an applied magnetic field. Based on this preliminary work the properties of H-MREs, as a result of both the base elastomer and embedded particles, become evident. Trends seen by the blocked force and displacement response suggest that the displacement response is affected by an interaction between the filler material and embedded particles to which the force response is not. Overall the trends in the response of the H-MRE provide promising implications for their roles as magnetically controlled actuators. Future work will need to focus on the effect of filler percentage, base resin stiffness, and dimensionality on the dynamic response before effective actuators can be designed from H-MRE materials.

Acknowledgments

Thanks to Dr. Edelmann and the Electron Microscopy Facility at Miami University for providing the scanning electron microscope images and to Dr. M. Pechan and Dr. J. Dou and the Physics Department at Miami University for use of the electromagnet to cure the samples.

References

1. Rabinow, J. The magnetic fluid clutch. AIEE Trans 1948; 67:1308-15.
2. Phule PP, Ginder JM. Synthesis and properties of novel magnetorheological fluids having improved stability and redispersibility. Int J Mod Phys B 1999; 13(14-16):2019–27.
3. Ginder JM. Behavior of magnetorheological fluids. MRS Bull 1998; 23(8):26–9.
4. Ginder JM, Davis LC, Elie LD. Rheology of magnetorheological fluids: models and measurements. Int J Mod Phy B 1996; 10(23–24):3293–303.
5. Dyke SJ, Spencer BF, Sain MK, Carlson JD. An experimental study of MR dampers for seismic protection. Smart Mater Struct 1998; 7(5):693–703.
6. Koo, J. H., Sung, S. H., Lee, H. J., and Jung, H. J., "Seismic protection of structures using smart base isolation systems based on MR elastomers," in Proc. 4th Int. Conf. Advances in Structural Engineering and Mechanics (ASEM08), Jeju, Korea, 2008.
7. Deng, H., & Gong, X. (2008). Application of magnetorheological elastomer to vibration absorber. Communications in Nonlinear Science & Numerical Simulation, 13(9), 1938-1947. doi:10.1016/j.cnsns.2007.03.024.
8. Davis, L. (1999). Model of magnetorheological elastomers. Journal of Applied Physics, 85(6), 3348. Retrieved from Academic Search Complete database.
9. Lokander, M., & Stenberg, B. Performance of isotropic magnetorheolocial rubber materials. Polymer Testing. 22 (2003) 245-251.
10. Guth, E. Theory of Filler Reinforcement, Journal of Applied Physics, 16, 20 (1945).
11. Ginder, J. M., Schlotter, W. F., and Nichols, M.E. Magnetorheological elastomers in tunable vibration absorbers. Proc. SPIE. Vol. 4331, 103 (2001).

NUMERICAL ANALYSIS OF THE DAMPING PROPERTIES OF MAGNETORHEOLOGICAL ELASTOMERS

LIN CHEN[*] and STEVE JERRAMS

Centre for Elastomer Research, Dublin Institute of Technology, Dublin 1, Dublin, Ireland

A finite element transient analysis together with the unit cell modeling method has been employed to simulate the damping properties of magnetorheological elastomers (MREs). The influence on the damping effect of MREs of frequency and amplitude of an applied load, the shape of ferromagnetic particles, damping of the rubber matrix and properties of the interfacial layer has been investigated. The results indicated that damping in MREs is generally increased with frequency and amplitude of the external load and is more sensitive to the load at low frequencies and high amplitudes. The results also demonstrated that MREs fabricated with regular shaped particles and low damping matrix materials had the smallest damping effect. Additionally, dependence of the damping properties of MREs on the thickness and damping characteristics of the interfacial layer were also discussed.

1. Introduction

Magnetorheological elastomers (MREs) are a class of adaptive materials whose mechanical properties (such as modulus and damping) can be reversibly and instantaneously controlled by an external magnetic field [1-3]. This is achieved by the addition of microsized magnetizable particles into the polymer materials. When MREs are exposed to an applied magnetic field, the field-induced dipole magnetic forces between the particles result in the mechanical performance exhibiting field dependence.

MREs are attracted increasing attention and have been considered for wide-ranging applications over the past few years. The Ford Motor Company has patented automotive bushing employing MREs [4,5]. The stiffness of the bushing was adjusted to reduce suspension deflection and to improve passenger comfort. Ginder [6], Lerner [7] and Deng [8] have respectively developed adaptive tuned vibration absorbers based on MREs. It was noted that the absorber effect of these devices greatly depended on the damping properties of the MREs. Theoretical analysis has shown that low damping ratios lead to high

---

[*]Tel.: + 00 353 (0)1 402 3934, E-mail address: lin.chen@dit.ie

vibration reduction effects while high damping ratios result in poor vibration suppression [9]. However, minimal research to date has focused on the damping properties of MREs.

This paper investigates the damping effect of MREs using a finite element transient analysis together with the unit cell modeling method. The influence on the damping effect of MREs of amplitude and frequency of the applied load, shape of ferromagnetic particles, damping of the rubber matrix and properties of the interfacial layer are evaluated.

2. Methodology

2.1. *Principle*

Cyclic loading of polymer materials, even under small deformations, is accomplished by energy damping. The applied deformation (x) and resultant reaction load (F) form a hysteresis loop due to the phase difference between them. The area enclosed by the hysteresis loop represents the energy dissipated internally during one cycle. Specific damping capacity (a) is defined as the ratio of the dissipated energy ($? \, W$) during one complete cycle to the maximum stored energy (W) from the inception of the loading until it reaches a maximum represented by Eqn (1) [10].

$$\alpha = \frac{\Delta W}{W} \tag{1}$$

where
$$\Delta W = \oint (P_d + kx)dx = \oint F dx \tag{2}$$

$$W = \int_0^B F dx \tag{3}$$

where load F is the sum of the damping force P_d and the elastic force kx and B is the maximum displacement.

In the dynamic differential equation of a vibration system, the parameter damping ratio ($?$) is commonly used and expressed as:

$$\xi = \frac{\alpha}{4\pi} \tag{4}$$

Utilizing the concept of energy dissipation, it is possible to evaluate the damping ratio of a material by calculating the area of its force-displacement hysteresis loop.

2.2. *Finite element model*

It is assumed that the magnetic particles are uniformly distributed in the matrix, as shown in Fig. 1(a). One-quarter of the particle surrounded by its corresponding matrix was considered as a representative volume element in an axially symmetrical model. Previous research suggests that the optimum MR effect is achieved with a particle volume fraction of 33% [3]. Accordingly, the ratio of the radius of particles (R) to sides of the square unit cell (S, having any value but set to unity in this model) was determined. A typical mesh for this square unit cell is shown in Fig. 1(b). Eight-noded quadrilateral generalized axi-symmetrical elements were used. The meshing and simulation were implemented in the FEA code ANSYS 12.1.

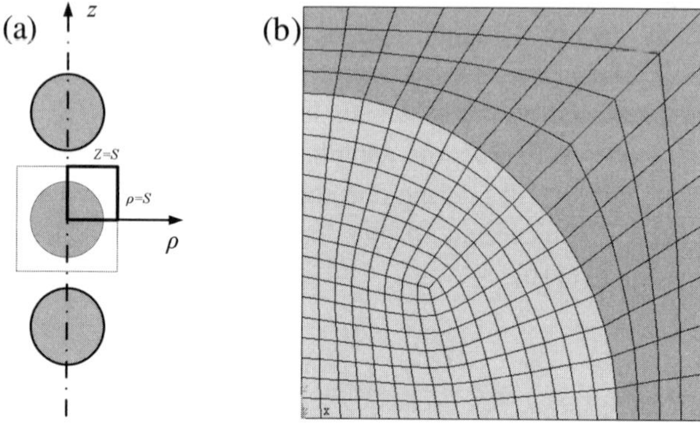

Fig. 1 (a) A uniform particle arrangement in an MRE (b) Grid mesh for the square unit cell used in the simulation.

On the axes of symmetry, the boundary conditions of the cell are

$$U_z\big|_{z=0} = 0$$
$$U_\rho\big|_{\rho=0} = 0 \tag{5}$$

where U is the displacement.

At the upper boundary ($z = s$), a harmonic displacement is applied, which can be expressed as

$$U_z\big|_{z=s} = d_0 \sin(2\pi f t) \tag{6}$$

where d_0 is the displacement amplitude, f is the external excited frequency and t represents the time. The corresponding harmonic reaction force on the upper boundary was solved by nonlinear full transient FEA. During the simulation, large deformation effects were considered.

2.3. *Material description*

The hyperelastic response of the polymer matrix can be described in terms of the principal extension ratios, λ_i, λ_2 and λ_3 which are the ratios of the lengths under load to the original lengths in the principal directions. For incompressible materials, $\lambda_1\lambda_2\lambda_3=1$. Stresses can be derived from the strain energy density function postulated by the Ogden model [11] and given by

$$W = \sum_{i=1}^{N} \frac{2\mu_i}{\alpha_i^2}(\lambda_1^{\alpha_1} + \lambda_2^{\alpha_1} + \lambda_3^{\alpha_1} - 3) \tag{7}$$

where N is the order of the Ogden model and the constants μ_i and α_i were fitted to uniaxial tension and compression experimental data as described by Johannknecht et al [12]. Using a two-term function, the fitted parameters are given by μ_1=2.106MPa, α_1=2.839; μ_2=-0.01219MPa, α_2=-6.197. The fit of nominal stress ($\partial W/\partial\lambda_1$) versus nominal strain (λ_1-1) for uniaxial deformation is compared with the measured values in Fig. 2. The hyperelastic curve fit agrees well with the experimental results.

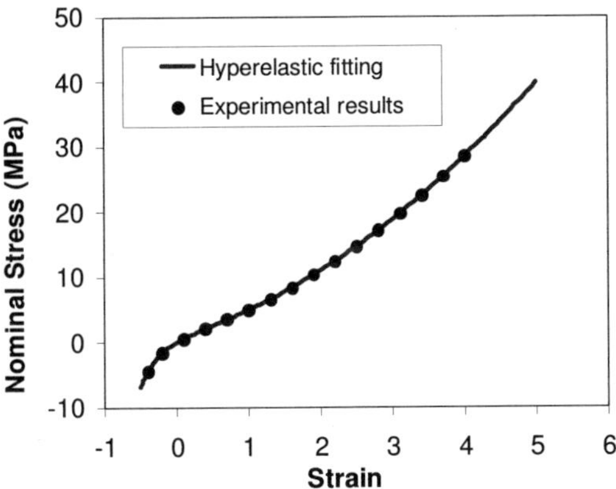

Fig. 2 A stress-strain curve fit for a polymer matrix using a 2nd order Ogden model.

Also, the elastic modulus and Poisson's ratio of the magnetic particles were considered to be constant during the simulation and were set as 200GPa and 0.25 respectively.

638

3. Simulation Results and Discussions

3.1. *Influence of amplitude and frequency of applied load*

Based on the transient FEA model, the simulations were carried out for several load cases, varying amplitudes and frequencies. The applied harmonic displacement amplitudes (d_0) were set at 0.0001, 0.001, 0.01 and 0.1m and the frequencies (f) were set at 1, 5, 10, 20, 30, 40 and 50Hz. The polymer matrix damping ratio (ξ_m) was set as 0.05 during the simulation described in section 3.1.

The variation of the force-displacement hysteresis loop for load amplitude $d_0= 0.001$m at several load frequencies is shown in the Fig. 3. Comparing the four graphs, we can see that both the maximum reaction force and area of the hysteresis loop significantly increase with the load frequency. This is because the damping force is proportional to the load velocity. So high frequency loads resulted in large damping forces. For a particular displacement amplitude, as damping force increased, more energy was dissipated and the area of the hysteresis loop increased.

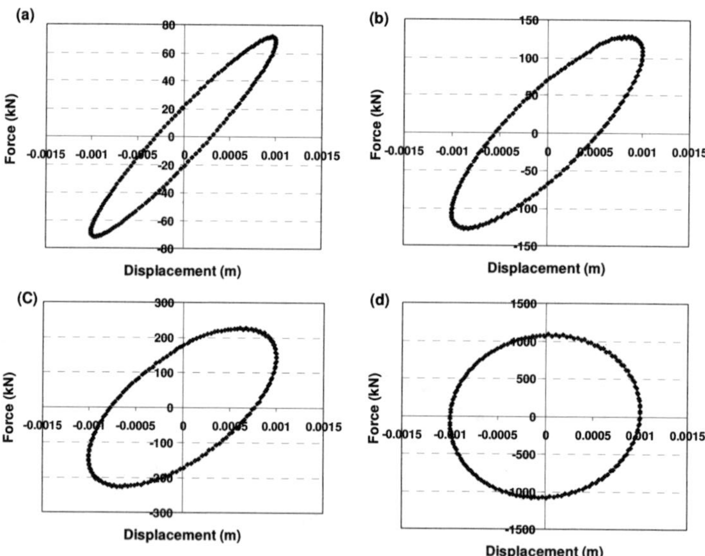

Fig. 3 The force-displacement hysteresis loop for four different load frequencies of (a) 1Hz, (b)5Hz, (c)10Hz and (d)50Hz at load amplitude $d_0= 0.001$m.

Simulations were also carried out using loads at constant frequency with varying amplitudes. It is interesting to note that the loops for the largest load amplitudes (such as $d_0= 0.1$m) are asymmetrical, as shown in Fig 4 (The result for only one frequency ($f=1$Hz) is illustrated here). The maximum tensile force was 7000kN, while the maximum compressive force reached 9000kN. This phenomenon is due to the asymmetry of the constitutive polymer matrix, as shown in Fig. 2 where the compressive modulus is higher than the tensile modulus for large deformations.

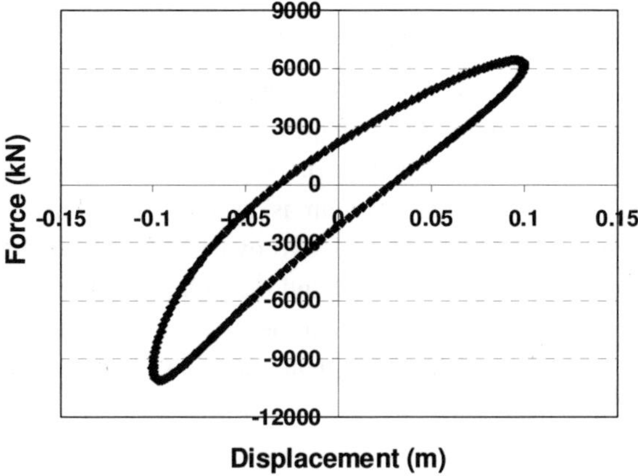

Fig. 4 The force-displacement hysteresis loop for a load amplitude of 0.1m and frequency of 1 Hz.

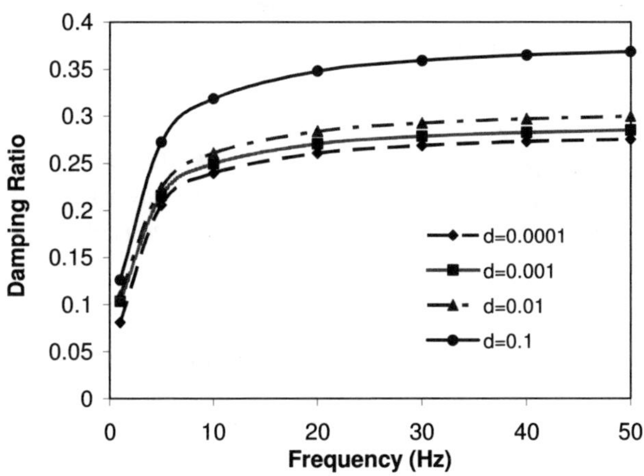

Fig. 5 Dependence of the damping ratio of an MRE on applied load frequency (f) at several values of load amplitude (d_0).

The variations of damping ratio of an MRE with load frequency for four different kinds of displacement amplitudes were obtained. They were derived from 28 hysteresis loops in total, comprising 7 frequency cases with 4 amplitude cases as shown in the Fig. 5. It can be seen that the damping ratio of an MRE

640

dramatically rises for low frequencies (less than 10Hz) and gradually increases for high frequencies. Also, damping ratio sharply increased when load amplitude reached 0.1. The results indicated that the damping ratio of an MRE was more sensitive to low frequency loads and high amplitudes.

3.2. *Influence of matrix damping and particle shape*

Simulations were carried out on different values of matrix damping (ξ_m =0.01, 0.05, 0.1, 0.15 and 0.2) and various shapes of the embedded particle (a sphere, two ellipsoids and a rectilinear column). All four shapes are axi-symmetrical about the z-axis (the coordinate system is as shown in Fig. 1(a)) and their projections in the ρ-z plane are a circle, two ovals and a rectangle, respectively. The radius in the ρ axis is denoted as R and the radius in the z axis is denoted as H. The shape of the ellipsoid particles can be defined by the value of R/H. Here the simulation parameters of load frequency and displacement amplitude were set as 5 Hz and 0.001m respectively.

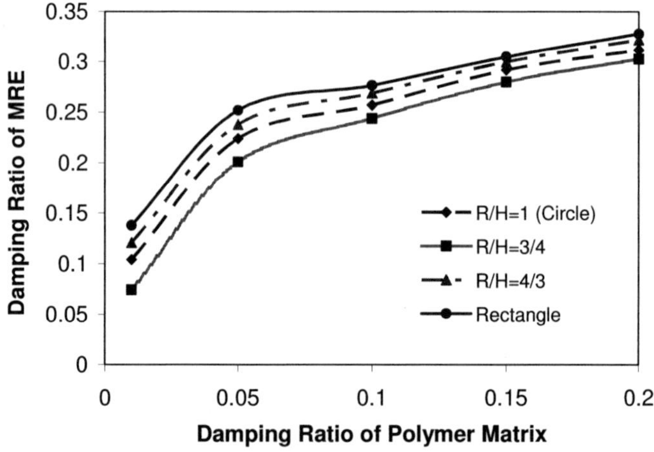

Fig. 6 Dependence of the damping ratio of an MRE on that of the polymer matrix (ξ_m) for several embedded particle shapes.

Fig. 6 shows the calculated variations in damping ratio of MREs with a polymer matrix having four different embedded particle shapes. It can be seen from the figure that damping of the MREs had a bilinear relationship with that of the matrix, which had a pronounced slope when the damping ratio of the matrix was less than 0.05 and had a smaller gradient when the damping ratio of the matrix was greater than 0.05. Comparing the damping ratio of the polymer matrix with that of the MRE for each point in Fig. 6, it was found that for an MRE damping was always larger than that of its matrix material. Hence, it was shown that the damping of an MRE is increased by inclusion of particles. Also, Fig. 6 demonstrates that rectangular shaped particles included in MREs had the largest damping effect, while the ellipsoidal particles having their major axis along the z axis (R/H=3/4) showed the smallest damping value. This is due to the influence of the shapes of the particles on stress distribution in the MRE structures. The applied load causes a high stress concentration in the MRE having rectangular shaped particles but stress is relatively uniform in the MREs having oval shaped particles of R/H=3/4. The results reveal that the regular shaped particles (such as ellipsoidal or spherical) and small damping matrices (e.g. Natural Rubber) are preferable if an MRE with low damping is required. However, if high damping is required, irregular shaped particles and large damping matrices (e.g. Butyl Rubber) are required.

3.3. *Influence of the properties of the interfacial layer*

The bound-rubber phenomenon was investigated in MREs in a previous publication [13], where an interfacial layer between an embedded magnetic particle and the matrix was observed by using a scanning electron microscope (SEM). Also, in some circumstances the particles were coated by specific polymeric materials prior to them being added to the matrix [14]. Consequently, interfacial layers are considered here. A grid mesh of a unit cell with an interfacial layer is shown in Fig. 7.

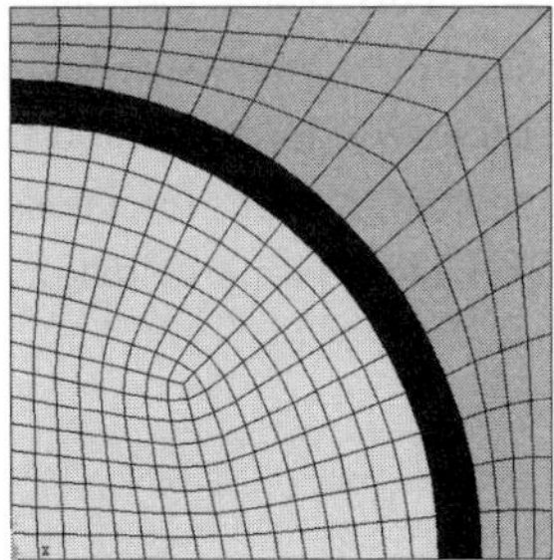

Fig. 7 Grid mesh for an interfacial layer included square unit cell.

Simulations were carried out on several interfacial layer damping ratios (ξ_i/ξ_m =0.1, 0.2, 0.5, 1, 2, 5 and 10, where ξ_i denotes damping ratio of interfacial layer and ξ_m denotes that of matrix, which was 0.05) and a range of interfacial layer thickness (t_i/R_p=0.05, 0.10, 0.15 and 0.20, where t_i denotes interfacial layer thickness and R_p denotes the mean radius of the spherical particle). Other simulation parameters load frequency and displacement amplitude were set at 5Hz and 0.001m respectively.

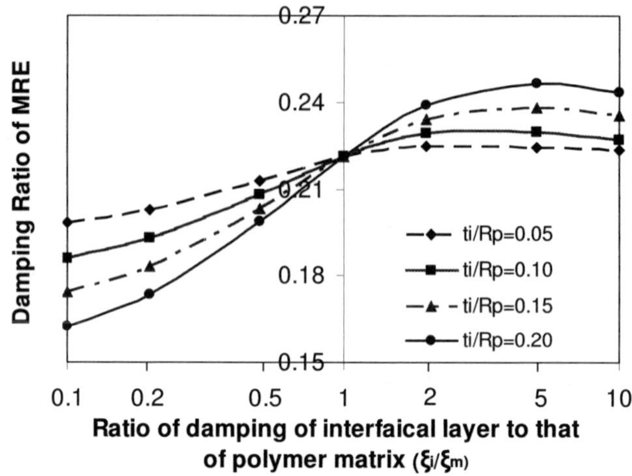

Ratio of damping of interfaical layer to that of polymer matrix (ξ_i/ξ_m)

Fig. 8 Dependence of the damping ratio of an MRE on the ratio of damping of the interfacial layer to that of the polymer matrix (ξ_i/ξ_m) under several values of interfacial layer thickness.

It can be seen from Fig. 8 that damping of the MRE generally increases with the damping of the interfacial layer up to a ratio (ξ_i/ξ_m) of 5. All the curves intersect at the axis of $\xi_i/\xi_m=1$. This result indicates that MRE damping is not affected by interfacial layer thickness when the damping of the interfacial layer equals that of the matrix. It is clearly demonstrated that the damping ratio of an MRE can be adjusted by changing the thickness of the interfacial layer. Therefore, surface modification and coating of magnetic particles can be used in MRE damping control.

4. Conclusion

The damping properties of MREs were investigated using numerical analysis techniques. The principle findings are summarized below.
- A finite element transient analysis and unit cell modeling method can be used to effectively simulate MRE dynamic properties.
- MRE damping is generally higher with increases in frequency and amplitude of an external load and is more sensitive to the load at low frequencies and

high amplitudes. Non-linearity in MRE damping is more pronounced for high deformation conditions.

- A uniform ferromagnetic particle shape and low damping characteristics in a matrix material is preferred in MREs if light damping is required.
- The damping ratio of an MRE can be adjusted by changing the thickness and damping of the interfacial layer. This can be accomplished by selection of coating materials and control of the coating procedures.

Further research is underway to determine the influence of magnetic effect and interfacial slip and these properties will be modeled in subsequent FEA simulations.

Acknowledgments

This work was made possible by funding from Enterprise Ireland under its 'Proof of Concept' programme (CAPRICE, Project Code: PC/2009/083). Additionally, the invaluable assistance provided by Prof. Xinglong Gong from the University of Science and Technology of China is gratefully acknowledged.

References

1. J. D. Carlson, M. R. Jolly, *Mechatronics* **10**, 555 (2000).
2. C. Bellan, G. Bossis, *Int. J. Mod. Phys. B* **16**, 2447 (2002).
3. L. Chen, X. L. Gong, W. Q. Jiang, J. J. Yao, H. X. Deng, W. H. Li, *J. Mater. Sci.* **42**, 5483 (2007).
4. J. R. Watson, *US Patent.* No.5609353 (1997).
5. W. M. Stewart, J. M. Ginder, L. D. Elie, Nichols and E. Mark, *US Patent.* No.5816587 (1998).
6. J. M. Ginder, W. F. Schlotter, M. E. Nichols, *In: Proceedings of SPIE 4331*, 103 (2001).
7. A. Lerner, K. A. Cunefare, *US Patent* 20050040922 (2005).
8. H. X. Deng, X. L. Gong, L. H. Wang, *Smart Mater. Struct.* **15**, N111 (2006).
9. H. L. Sun, P. Q. Zhang, X. L. Gong, H. B. Chen, *J. Sound Vib.* **300**, 117 (2007).
10. P.A. Zinoviev, Y.N. Ermakov, Energy Dissipation in Composite Materials, Technomic *Publishing Company*, p.5 (1994).
11. R. W. Ogden, *Pro. R. Soc.*, **11**, 582 (1972).
12. R. Johannknecht, PhD dissertation, VDI Verlag GmbH, Dusseldorf (1999).
13. X. Z. Zhang, X. L. Gong, P. Q. Zhang, W. H. Li, *Chin. J. Chem. Phys.* **20** 173 (2007).
14. J. F. Li, X. L. Gong, H. Zhu, W. Q. Jiang, *Polym. Test.,* **28** 331 (2009).

ON DAMPING VIBRATIONS OF THREE LAYERED BEAM CONTAINING MAGNETORHEOLOGICAL ELASTOMER

E. V. KOROBKO, Z. A. NOVIKOVA and M. A. ZHURAUSKI

A. V. Lykov Heat and Mass Transfer Institute of National Academy of Sciences of Belarus, 220072, 15 P. Brovka st., Minsk, Belarus
E-mail: evkorobko@gmail.com

G. I. MIKHASEV

Belarusian State University, 220030, 4 Nezavisimosti av., Minsk, Belarus
E-mail: mikhasev@bsu.by

In paper we present the results of investigations of viscoelastic properties of magnetorheological elastomer containing carbonyl iron particles. Frequencies of natural vibrations of three layered beam, supporting constructions of which are made from aluminum and the inner layer – from magnetorheological elastomer are calculated, the dependence of vibrations on induction of the applied magnetic field is obtained. Non stationary vibrations of the beam at pulse impact of magnetic field are found.

1. Introduction

Magnetorheological elastomers (MRE) are one of smart materials, elastic properties of which change depending on the value of the applied magnetic field. They consist of magnetic particles in deformed polymer matrix. The possibility to control viscoplastic and viscoelastic properties of MRE in a wide range allows one to use them in vibroprotecting devices.

Although rather many works are dedicated to vibroprotecting devices with the use of electrorheological and magnetorheological media [1 – 5], their calculation is a complex task for mechanical engineers that deal with development of new methods of active and semiactive damping of construction vibrations. It is explained by the fact that the response of composite construction containing MRE significantly depends on the ratio of time scale of controlling signal providing time reaction of MRE and dynamic characteristics of the controlled construction.

The principle of damping vibrations is discussed in the work on the example of three layered beam in which bearing layers are made of aluminum and the inner layer is made of the magnetorheological elastomer. Free and forced vibrations of the beam have been studied at fixed viscoelastic characteristics of MRE, and then the reaction of the beam to the nonstationary signal of magnetic field resulting in abrupt change of MRE characteristics has been investigated.

2. Experimental

As matrix for MRE a natural inorganic polymer (bentonite clay, the size of laminar particles $1 - 10$ μм) in synthetic oil was used, as a filler – particles of carbonyl iron (particle size about 20 μм). Matrix of MR elastomer was prepared by a thorough rubbing of polymer in the oil with addition of surfactant. In the prepared matrix carbonyl iron particles were introduced (about 30 wt. %).

Dependence of components of complex modulus of shear G' and G" on induction of the magnetic field B is determined by means of rheometer «Physica MCR 301» by Anton Paar.

3. Beam Vibrations at Fixed Magnetic Field Intensity

Dependencies of components of complex shear modulus on the induction of magnetic field (at the frequency of external impact 10 Hz) are presented in Fig. 1. Presented curves indicate the nonlinear dependence of shear modulus on the induction of the magnetic field of high intensity. Only at B<200 mT our results correlate with that supposition of [6] about linear dependence of imaginary shear modulus of magnetorheological material on induction of the magnetic field.

Let's consider a three layered beam of the length L (fig. 2), external layers of which are not sensitive to the magnetic field, and the inner layer represents a viscoelastic MRE. As initial equations, describing the motion of the composite beam we will use the equations obtained in work [6],

$$\rho \frac{\partial^2 w^*}{\partial t^{*2}} + 2EI \frac{\partial^4 w^*}{\partial x^4} - \frac{G^* b(h_1 + h_2)^2}{h_2}\left(\frac{\partial^2 w^*}{\partial x^2} - \frac{\partial \varphi^*}{\partial x}\right) = f \; ,$$

$$J \frac{\partial^2 \varphi^*}{\partial t^{*2}} - \frac{bh_1 E(h_1 + h_2)^2}{2} \frac{\partial^2 \varphi^*}{\partial x^2} - \frac{bG^*(h_1 + h_2)^2}{h_2}\left(\frac{\partial w^*}{\partial x} - \varphi^*\right) = 0 \; , \qquad (1)$$

where w^* - normal bending of the beam, φ^* – rotation angle of cross-section, considering tangential shears of "sandwich" layers, x - coordinate on the medium line of the beam, f – external force, t^* - time, b – beam's width, h_1 - thickness of bearing layers, h_2 – thickness of MRE, ρ – specific mass of cross-section, E – Young modulus of surface layers of the beam, G^* - complex shear

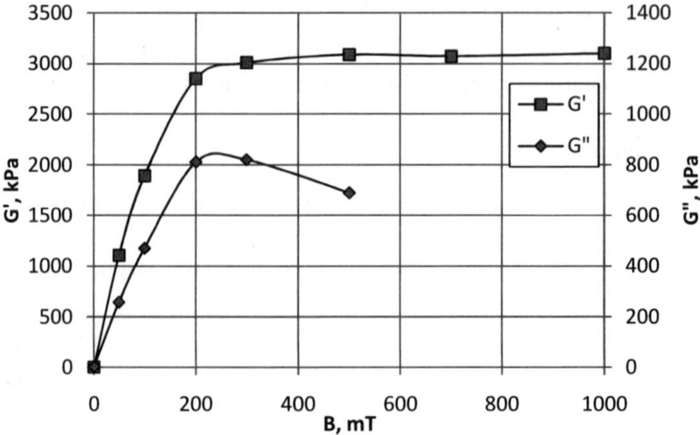

Figure 1. Dependence of real and imaginary part of shear modulus of MRE on induction of the magnetic field.

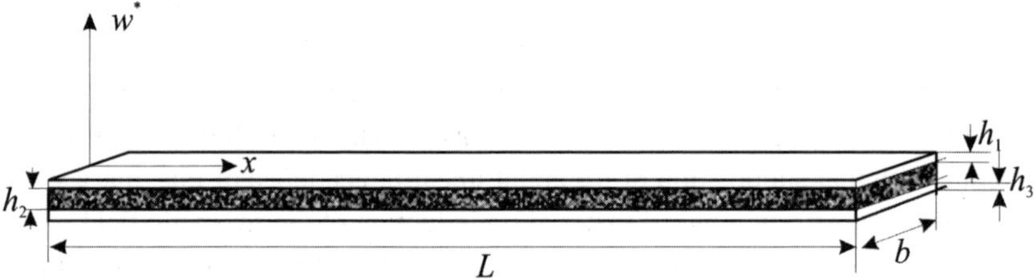

Figure 2. Three layered beam with magnetorheological elastomer.

modulus of MRE, depended on induction B of the external magnetic field, I, J – geometric and mass inertia moments of the cross-section.

On the edges let's consider conditions swiveling support

$$w^* = \partial^2 w^* / \partial x^2 = \partial \varphi^* / \partial x = 0 \quad \text{at} \quad x = 0, L. \tag{2}$$

Fur such boundary conditions natural forms of vibrations are given by functions

$$w_n^*(x, t^*) = \sin \lambda_n x \, e^{i\omega_n t^*}, \quad \varphi_n^*(x, t^*) = C_n \cos \lambda_n x \, e^{i\omega_n t^*}, \quad n = 1, 2, \ldots, \infty \tag{3}$$

$$\lambda_n = \pi n / L, \quad C_n = \frac{2\lambda_n G^*}{\lambda_n^2 h_1 h_2 E + 2G^*}, \tag{4}$$

where

$$\omega_n = \lambda_n^2 \sqrt{\frac{E}{\rho}\left[2I + \frac{G^* b(h_1 + h_2)^2}{E h_2 \lambda_n^2}\left(1 + \frac{2G^*}{\lambda_n^2 h_1 h_2 E + 2G^*}\right)\right]} \tag{5}$$

- sought imaginary frequency of natural vibrations, depending on imaginary shear modulus G^*. Moreover we should note here, that the similar formula for the frequency ω_n, deduced in [6], is erroneous.

The calculation was carried out for the beam with parameters: $h_1 = h_2 = h_3 = 0.7$ mm, $L = 390$ mm, $b = 25$ mm, bearing layers of which are prepared from aluminum, and the inner layer – from MRE with shear modulus showed in Fig. 1. The result analysis indicates that there is a weak dependence of natural vibration frequency on induction of magnetic field for all modes. In particular, by changing the induction from 0 to 400 mT the lowest natural frequency grows almost linearly from 278 Hz up to 480 Hz, and then nonlinearly decays by the following increase of the magnetic field intensity. With the increase of the mode number this dependence weakens. The dependence of imaginary part of the vibration frequency on the intensity of the magnetic field is shown in Fig. 3. As it is seen, already at $B>350$ mT the growth of the intensity of the magnetic field result in the drop in the speed of vibrations dampening, which indicates the inefficiency of the further rise of induction. Therefore, for the considered MRE the most optimal from the point of view of dampening low frequency vibrations is the value of induction $B = 350$ mT.

Let the beam undergo the impact of the external force of the intensity $f = \rho F_0(x)\mathrm{e}^{i\Omega t^*}$, where Ω - the frequency of forced vibrations. Further we do not consider the case of the resonance, taking $\Omega \neq \omega_n$ for every integer n. Solution of equations (1) with boundary conditions (2) is

$$w^*(x,t^*) = \sum_{n=1}^{\infty}\left[A_n \mathrm{e}^{i\omega_n t^*} + \frac{N_n}{\omega_n^2 - \Omega^2}\mathrm{e}^{i\Omega t^*}\right]\sin(\lambda_n x), \tag{6}$$

where $N_n = \int_0^L F_0(x)\sin\lambda_n x\, \mathrm{d}x$ – so-called generalized forces, A_n – arbitrary imaginary numbers that are found form the initial conditions of the task.

Function (9) allows us to predict the reaction of three layered beam on the applied periodical in time and arbitrary in coordinate x external load. Due to the fact that the speed of vibration dampening depends on imaginary shear modulus G^*, formula (9) could be used as the calculating one for determining the optimal regime of change the intensity of the magnetic field with the aim of the most effective damping of vibrations [6].

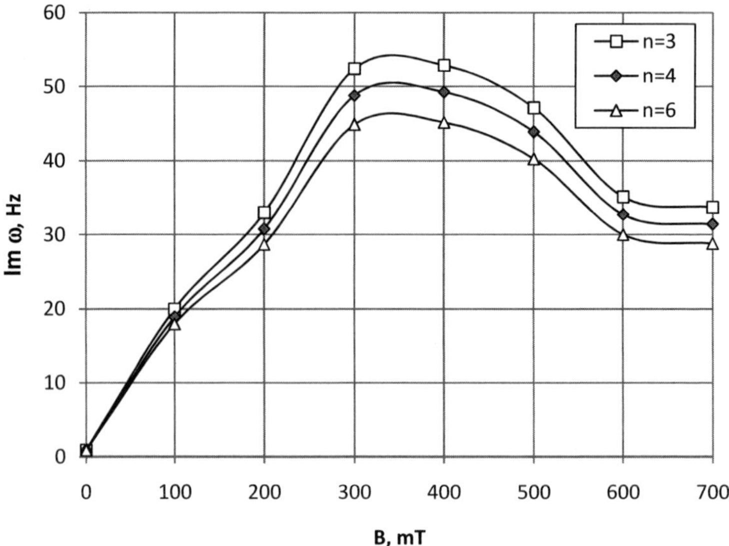

Figure 3. Dependence of the decrement of natural vibration frequency of three layered beam on the induction of the external magnetic field of the 3rd, 4th and 6th modes.

We shall note, that formulas (5), (9) are deducted at condition that shear modulus G^* is a constant value, i.e. it does not depend on time. At the same time the response of magnetorheological medium to the external magnetic pulse depends significantly on the ratio of the time scale of controlling signal and the reaction of the very medium. So, at application of the signal time of reaction of MRE is about $10^{-3} - 10^{-2}$ s. At smooth change of induction of the magnetic field time of reaction can vary substantially. In any case the obtained here relations (5), (9) do not reveal the reaction of the beam in time interval comparable with the time of MRE reaction and could be considered only at setting a certain stationary regime for viscoelastic characteristics of MRE. Therefore, relations (5), (9) could be used for solving the task of vibrations damping only on low frequencies, when the frequency of natural vibrations and also the frequency of exciting vibrations are not comparable with the speed of MRE reaction to the magnetic field. Abrupt impact of the magnetic field is a kind of "parametric blow" for a mechanical system and can excite additional high frequency modes that require further investigations.

4. The Influence of the Magnetic Field Signal on High Frequency Vibrations

For MRE the real and imaginary part of G^* at zero induction B of the magnetic field are equal G'=4,5 kPa and G''=17 kPa. Maximal values G'=3000 kPa and G''=820 kPa are reached at induction B=500 mT and B=200 mT correspondingly. At pulsed change of the induction of the magnetic field, the imaginary shear modulus $G^* = G^*(v_r\, t^*)$ – is the time function, where $v_r = 1/t_r$ – is the speed of MRE reaction on the impact of the magnetic field (and makes $\approx 10^2 - 10^3$ s^{-1}). Let $t_r = 10^{-3}$ s for the characteristic time. Let's introduce dimensionless values:

$$w^* = Lw, \quad x = Ls, \quad t^* = t_r t, \quad f(t) = 2\frac{G^*}{\varepsilon^{1/2} E}\frac{l^2}{h^2}, \quad \varepsilon = \frac{t_r}{T_v}, \quad T_v = \frac{1}{3}\sqrt{\frac{2\rho}{E}\frac{h}{b}\frac{l^2}{h^2}} \tag{7}$$

where T_v – period of low frequency vibrations of the beam. Furthermore we consider that ε – is a small parameter, characterized the speed of reaction of MRE on the magnetic field signal, relative to the lowest frequency of bending vibrations.

Using the method of multiple scale the solution will be found in the form

$$w = W(t)\sin \pi n s, \quad \varphi = \Phi(t)\cos \pi n s, \quad n = \varepsilon^{-1/2}p, \quad p \sim 1. \tag{8}$$

$$W = W_0 + \varepsilon^{1/2}W_1 + \varepsilon W_2 + \ldots, \quad \Phi = \Phi_0 + \varepsilon^{1/2}\Phi_1 + \varepsilon\Phi_2 + \ldots$$

where n - integer number of the order $\varepsilon^{-1/2}$, and W_j, Φ_j - functions of independent arguments $\tau_0 = \varepsilon^{-1/2}t$, $\tau_1 = t$, $\tau_2 = \varepsilon^{1/2}t, \ldots$ Substituting (7), (8) in (1) results in the sequence of differential equations relative to unknown functions W_k, Φ_k. Considering each of this sequences and deleting secular solutions at $\tau \to \infty$, we find the sought W_j, Φ_j. In particular, consideration being limited by first two approximations gives

$$w \approx \sin\left(\frac{\pi ps}{\varepsilon^{1/2}}\right)\left\{A_{01}\exp\left\{i\left[\frac{(\pi p)^2 t}{\varepsilon^{1/2}} + \frac{2}{9}\int_0^t f(\tau)d\tau\right]\right\}\right.$$

$$\left. + A_{02}\exp\left\{-i\left[\frac{(\pi p)^2 t}{\varepsilon^{1/2}} + \frac{2}{9}\int_0^t f(\tau)d\tau\right]\right\}\right\}, \tag{9}$$

where unknown A_{01}, B_{01} are found from initial conditions.

Relation (9) describes nonstationary high frequency bending vibrations with the frequency of the order $\varepsilon^{-1/2}$. Solution (10) could be interpreted as the high frequency reaction of the three layered beam, consisting MRE on the quik signal of the magnetic field.

5. Conclusion

Viscoelastic properties of magnetorheological elastomer containing carbonyl iron particles have been investigated. Shear modulus grow from the value about 4 kPa without magnetic field up to 3 MPa in magnetic fields with induction of 300 mT and higher. Loss modulus is maximal (about 800 kPa) at induction of the magnetic field 200 – 300 mT. Solution of motion equations for three layered beam, bearing constructions of which are made from aluminum, and the inner layer – from magnetorheological elastomer has shown that first six natural vibration frequencies of the beam grow slowly with the growth of induction of the applied magnetic field up to 400 mT and then nonlinearly decrease with the further increase of the induction of the magnetic field. It is stated that for the beam under investigation from the point of view of vibration dampening optimal is the magnetic field with induction of about 350 mT. The general solution of motion equations for the beam under forced bending vibrations has been found at stationary impact of the magnetic field. Also the analytical solution of the normal bending of the beam at pulsed impact of the magnetic field is found.

References

1. J. P. Coulter and T. G. Duclos, *Proc. 2th Int. Conf. on ER Fluids, Raleigh, North Carolina*, 300 (1989).
2. M. Yalcintas and H. Dai, *J. Smart Mater. Struct.* **8**, 60 (1999).
3. V. A. Bilyk, E. V. Korobko, G. N. Reizina, E. A. Bashtovaya, E. B. Kaberdina, V. A. Kuzmin, *Solid State Phenomena*, **144**, 220 (2009).
4. G. N. Reizina, E. V. Korobko, V. A. Bilyk, V. L. Efremov, A. E. Binshtok, *Journal of Physics: Conference Series*, **149**, 012025 (2009).
5. E. Korobko, Z. Novikova, N. Bedzik, M. Zhurauski, A. Bubulis, E Dragašius, *Proc. 8th Int. Conf. "Vibroengineering 2009" (Klaipeda, Lithuania, Sept. 16 – 18, 2009)*. Kaunas, Technologija, 27 (2009).
6. M. Yalcintas and H. Dai, *J. Smart Mater. Struct.* **13**, 1 (2004).

INFLUENCE OF FLOWING PARAMETER OF MAGNETORHEOLOGICAL POLISHING FLUIDS (MRPF) ON THE QUALITY OF PROCESSING POLYCRYSTALLINE GLASS CERAMICS

G. GORODKIN and Z. NOVIKOVA

A V Luikov Heat & Mass Transfer Institute Academy of Sciences of Belarus, 15 Brovka Str. Minsk, Belarus

The study of rheological characteristics for two types of magnetorheological polishing fluids (MRPF), MRPF-1 based on cerium oxide abrasive particles and MRPF-2 based on nano-diamond abrasive particles, was carried out. Experiments on polishing of a polycrystalline sitall laser mirror substrates with these fluids using magnetorheological finishing process were executed. The surface structure of the samples after the polishing was investigated with the atomic force microscope. The results of surface measurements for the samples under the study polished with both MRPF-1 and MRPF-2 fluids are reported in comparison with existed data obtained with regular pitch polishing method. The lowest root-mean-square roughness of 0.2 – 0.4 nm was obtained for the samples polished with MRPF-2.

1.   Introduction

A process of magnetorheological polishing of non-magnetic materials [1] is among practical applications of the magnetorheological fluids. This method is based on controlling, by means of magnetic filed, of properties of moving abrasive magnetorheological polishing fluid (MRPF) in the contact zone with a surface to be processed. Usually, the MRPF is an aqueous suspension of both ferromagnetic carbonyl iron particles and abrasive powder with addition of stabilizers [2].

We have carried out experiments to study the rheological properties of MRPFs and to analyze surface quality of the polycrystalline CO-115 sitall mirror substrates after polishing with these fluids.   Two types of MRPFs with approximately 40 vol. % of carbonyl iron were used for polishing: MRPF-1 (containing abrasive cerium oxide particles) and MRPF-2 (containing nano-diamond particles). The stabilizing additives were used in the fluids composition to provide rheological and sedimentation stability in time.   The Anton Paar «Physica MCR 301» rheometer was used to measure rheological properties of MRPFs with and without magnetic field excitation.

The surface quality for flat sitall substrates (22.5 mm in diameter and 3 mm thick) after polishing experiments with MRPF-1 and MRPF-2 were investigated in the Center of Probe Microscopy of the Riazan State Radio Engineering University (Russia) using atomic force microscope «NT MDT Solver Pro».

2. Rheological Measurements of Magnetorheological Polishing Fluids

Rheological characteristics, flow curves, static yield stress, and viscoelastic parameters such as components of complex modulus: storage modulus G' and loss modulus G'', have been measured for two MRPFs used in the study. The Figures 1 and 3 demonstrate the influence of magnetic field flux density B on viscoelastic properties of the fluids while the fluid's flow curves at different B are shown in the Figures 2 and 4.

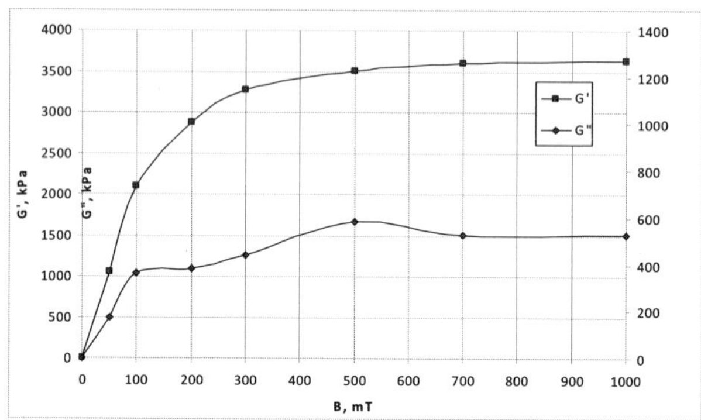

Figure 1. Complex modulus components values vs magnetic flux density for the MRPF-1.

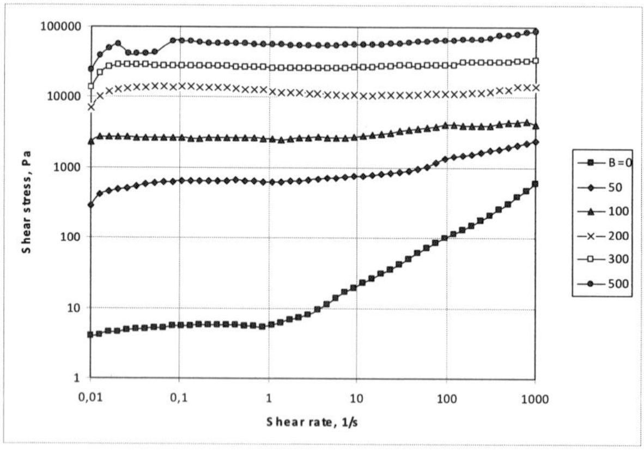

Figure 2. Flow curves of MRPF-1 for different flux densities (mT).

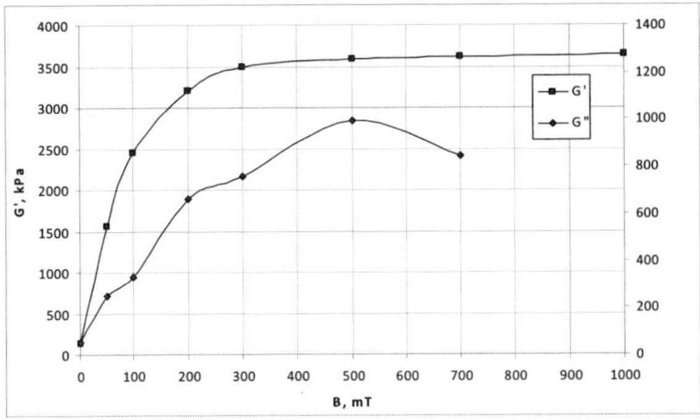

Figure 3. Complex modulus components values vs magnetic flux density for the MRPF-2.

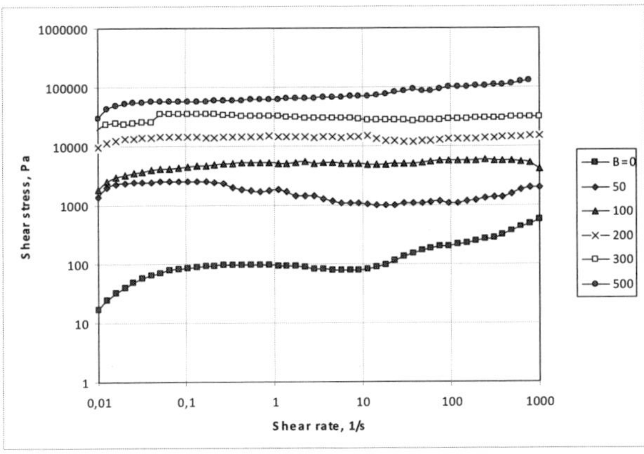

Figure 4. Flow curves of MRPF-2 for different flux densities (mT).

3. Study of the MRPF Flow Characteristics in the Delivery System of the Device for the Magnetorheological Polishing

In this study we used the polishing device similar to that reported in [2]. At first, a range of pressure drop on the section of the circulation loop after delivery pump and nozzle for the different flow rates of the MRPF and the stability of the pressure in time were defined. It was found that for the flow rate of 450 ml/min the pressure was 0.45 bar; for the flow rate 500 ml/min the pressure was 0.58 bar, and for the flow rate 600 ml/min the pressure was 0.63 bar. The deviation of the pressure from its initial value has not exceeded ~ 0.12 bar over 120 min of the fluid circulation.

Dimensions of the cross-section of the MRPF lap moving on the surface of round working tool were measured as a function of

- linear velocity of the tool (see results in the Figure 5);
- magnetic flux density in the air gap between pole pieces incorporated into the tool, as shown in the Figure 6;
- flow rate, (see Figure 7).

The fluid lap shape data were used to optimize operating mode of the polishing device.

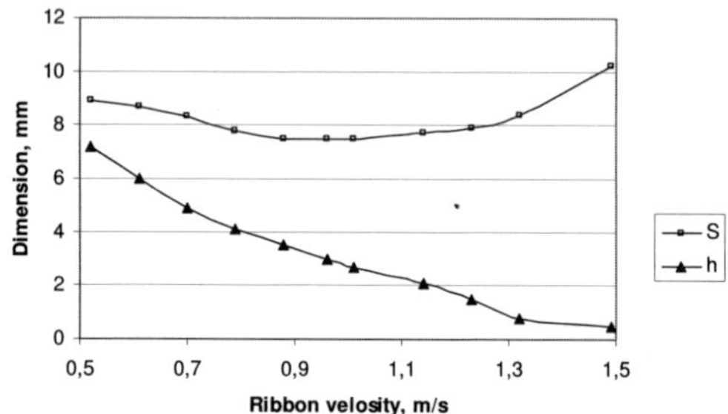

Figure 5. Cross-section dimensions of the fluid lap as a function of the tool velocity: S – width, h – height.

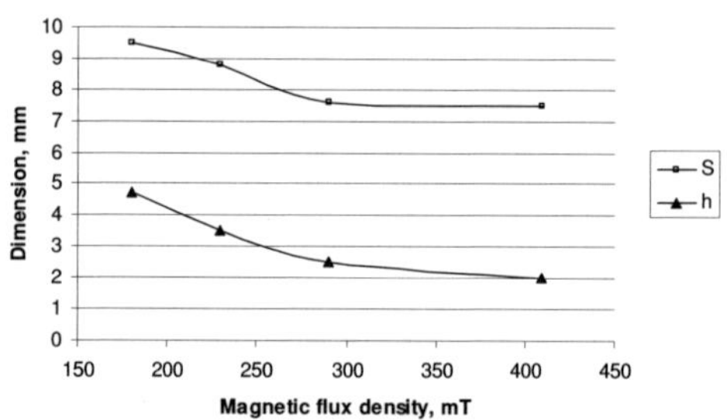

Figure 6. Cross-section dimensions of the fluid lap vs. magnetic flux density: S – width, h – height.

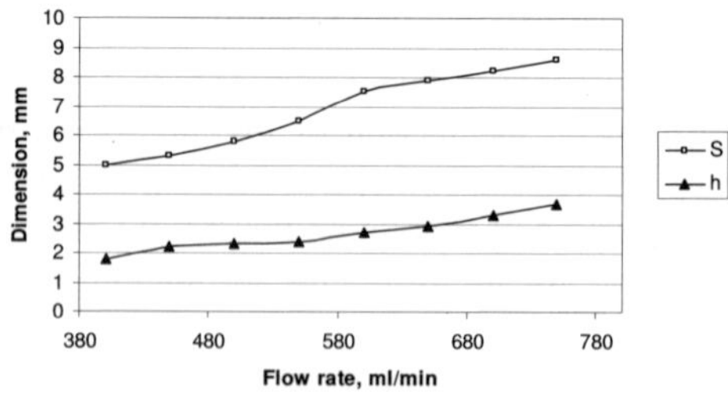

Figure 7. Cross-section dimensions of the fluid lap as a function of the flow rate: S – width, h – height.

The material removal rate, for the sitall in this particular case, was defined then varying:
-	the velocity of the lap moving with the tool surface (Figure 8);
-	the size of the gap between the tool and a polishing substrate (Figure 9);

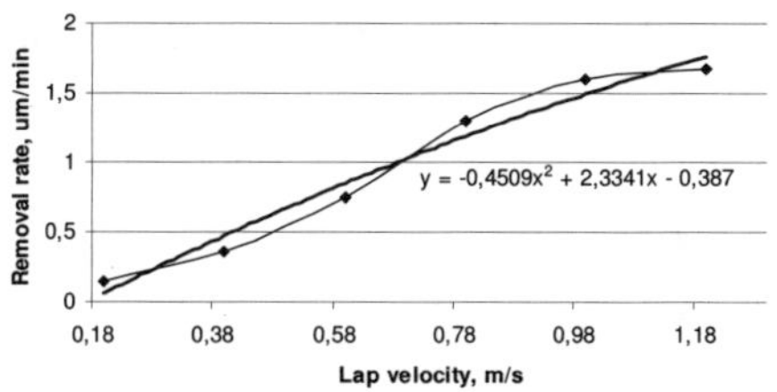

Figure 8. Removal rate for sitall vs. fluid lap velocity.

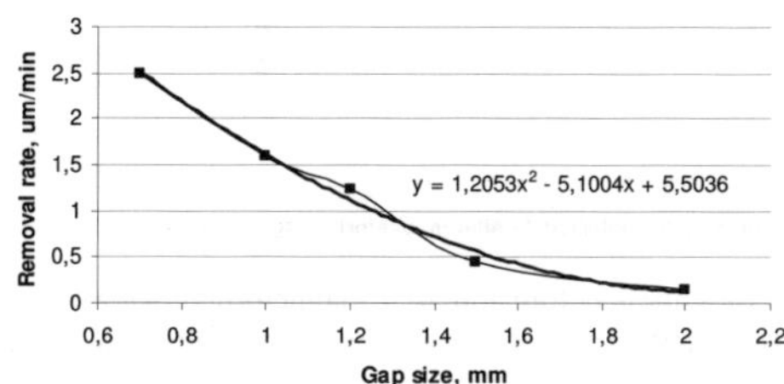

Figure 9. Removal rate for sitall vs. gap size.

Using the aforesaid data the following process parameters for the optimal polishing of the sitall has been selected:
-	flow rate of the MRPF – 600 ml/min;
-	magnetic flux density – 320 mT;
-	velocity for the MRPF lap on the working tool – 0.8 m/s;
-	gap size between the tool and sitall substrate surface – 1.32 mm.

## 4.	Polishing Results

The experimental sitall substrates were received as pre-surfaced with traditional pitch polishing method. The surface roughness of the substrates was preliminarily measured at the Center of Probe Microscopy.

All substrates were polished with the MRPF-1 or MRPF-2 using process parameters mentioned above. The substrates were sent to the Center for post-magnetorheological polishing surface analysis to compare with initial, pitch polished, results and with the industry quality etalon (keeping in the Center and using as a reference) noted as "etalon # 115" below. The surface roughness for the polished substrates was measured in seven sites around the substrates aperture using the microscope scan size of 50x50 um. The average values of both root-mean-square (RMS) and peak-to-valley (PV) roughness as well as power-spectral-density (PSD) factor for each substrate before and after magnetorheological polishing are shown in the Figures 10-12, and the average data are demonstrated in the Table 1. The PSD factor here represents correlation between lateral and vertical sizes of the surface relief irregularities.

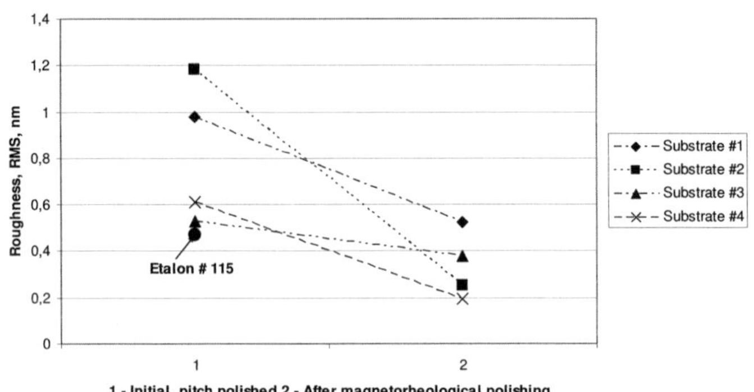

Figure 10. Surface RMS roughness before and after magnetorheological polishing. Substrate #1 was polished with the MRPF-1; substrates #2, #3 and #4 were polished with MRPF-2.

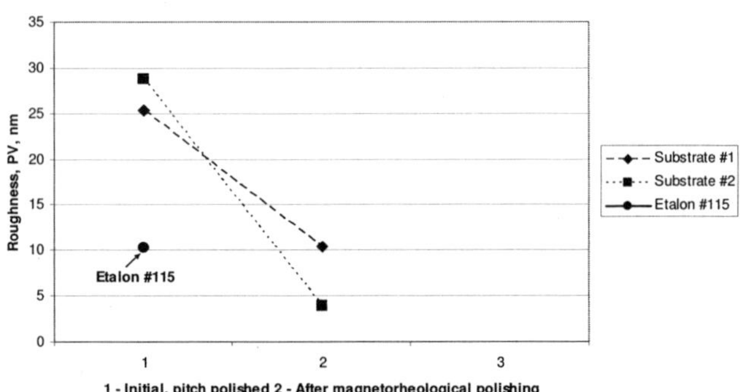

Figure 11. Surface PV roughness before and after magnetorheological polishing. Substrate #1 was polished with the MRPF-1; substrate #2 was polished with MRPF-2.

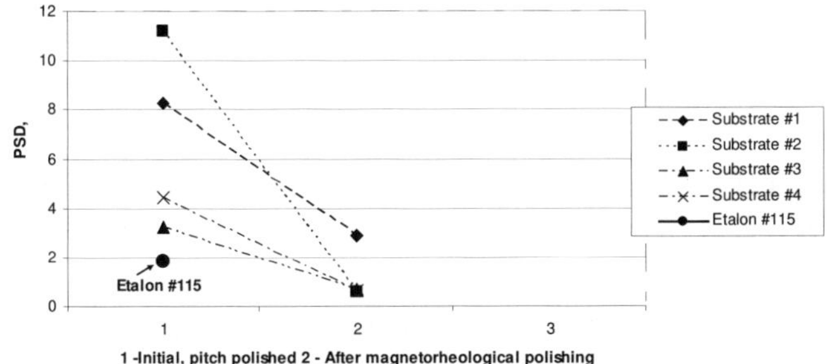

Figure 12. PSD factor before and after magnetorheological polishing. Substrate #1 was polished with the MRPFD-1; substrates #2, #3 and #4 were polished with MRPF-2 (PSD, 10^9 um^4).

Table 1. Polishing results: average data.

	RMS, nm	PV, nm	PSD, 10^9um^4
Magnetorheological polishing with the MRPF-1	0.523	10.36	2.895
Magnetorheological polishing with the MRPF-2	0.254	3.97	0.584
Pitch polishing: Etalon №115	0.469	10.33	1.827

5. Conclusion

In our experiments using magnetorheological polishing method, the abrasive magnetorheological fluid MRPF-2 with nano-diamond particles demonstrated better polishing results compared to cerium oxide based MRPF-1. The initial pitch-polished surface roughness of the sitall substrates was greatly reduced after processing with the MRPF-2 and reached in average 2.5 times lower in RMS value (1.5 times lower compared to the etalon) and 7 times lower in PV value (2.5 times lower compared to etalon). The PSD factor after polishing with the MRPF-2 was improved by 10 times compared to the initial value (3 times better then etalon data).

References

1.   W Kordonski and S Jacobs. Progress Update in Magnetorheological Finishing. Int. J. Mod. Phys, B10, 2337 (1996)

2   LLE Review #72, http://www.lle.rochester.edu/publications/lle_review/

STIFFNESS AND DAMPING IN FE SPHERICAL MICROPARTICLE MAGNETORHEOLOGICAL ELASTOMERS

A. CHAUDHURI[1], E.M. PETRO[1], N.M. WERELEY[1], R.C. BELL[2] and Y.-T. CHOI[1]

[1]*Dept. of Aerospace Engineering, University of Maryland, College Park MD 20742, USA*
[2]*Dept. of Chemistry, Pennsylvania State University, Altoona PA 16601 USA*

The study addresses experimental characterization of magnetorheological elastomers (MRE) with six different particle loadings under various magnetic field intensities. The MREs consisted of spherical iron particles, 6 to 10 μm in diameter, ranging from 0.25 to 30% weight fraction, cured in a nonmagnetic silicone rubber host elastomer. Mechanical properties of MREs such as equivalent viscous damping and complex stiffness were experimentally determined by using an Instron material testing machine with a magnetic cell. The effects of weight fraction of particles, displacement excitation frequencies, and pre-strain on the field-dependent properties of MREs were experimentally evaluated.

1. Introduction

Magnetorheological elastomers (MRE) are a class of smart materials which properties can be significantly changed by external stimuli. Typically MREs are composed of micro- or nano-sized magnetic particles, such as Fe, Co, or Ni, suspended in a nonmagnetic host material, such as silicone rubber [1,2]. The magnetic particles of MREs are aligned in response to applied magnetic field and thus MREs can control their damping and stiffness by just adjusting current input. Such advantages make MREs to be distinguished from conventional passive elastomers. Thus, recently, application of MREs to various engineering fields has widely investigated by many researchers.

Generally, the mechanical properties of MREs strongly depend on the intensity of the applied magnetic field. But, from the standpoint of MREs design, the volume fraction of magnetic particle is also an important design factor to maximize the MR effect of MREs. Thus, the primary goal of this study is to experimentally assess the influence of particle loading (i.e., weight fraction of spherical iron particles), displacement excitation frequency, and pre-strain on field-dependent properties of MREs.

2. Experimental Setup and Results

2.1. *Magnetorheological Elastomer (MRE) Samples*

Six different MRE samples consisting of 6 to 10 μm diameter Fe particles of varying weight fractions (wt%) of magnetic particles were tested. The weight fractions tested were: 0.25, 0.5, 2.0, 4.0, 6.0, and 10.0 wt%. Each sample consisted of spherical Fe particles cured in a silicone rubber host elastomer. The MRE samples were formed into cylinders of the height (h = 25.4 mm) and the radius (r = 4.75 mm).

The Fe particles in the MRE samples tested were aligned along the longitudinal axis of the sample. To align the particles, the samples were cured in an oven under a magnetic field induced by a permanent magnet. Therefore, the direction of the field applied during testing and the direction of loading should both be parallel to the longitudinal axis to maximize the MR effects.

2.2. *Testing Setup*

An Instron (Model 8841) material testing machine (refer to Figure 1) was used to deform the MRE samples and measure the resulting forces. The MRE sample sits securely on a static fixture that is connected to a 1 kN load cell. The MRE sample is compressed by a hydraulically activated piston from above. The load cell measures the forces corresponding to the piston displacement. A magnetic

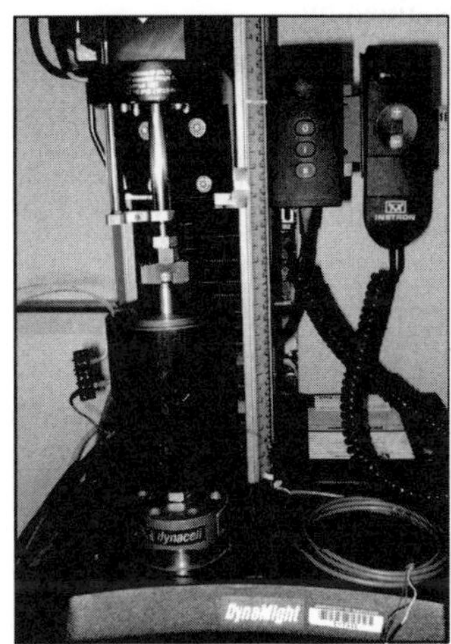

Figure 1. Instron material test machine with magnetic cell.

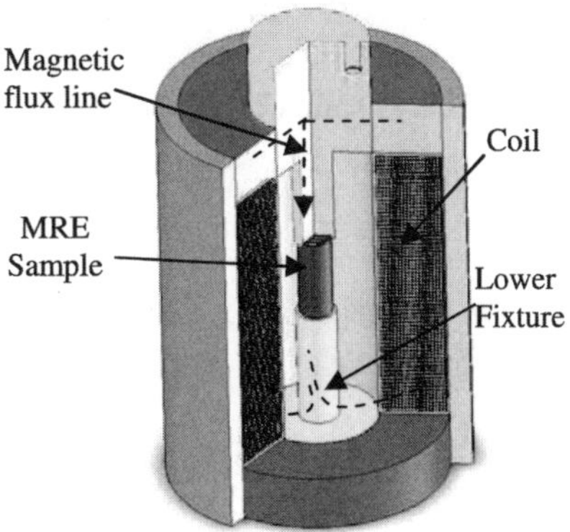

Figure 2. Schematic diagram of magnetic cell.

cell (refer to Figure 2) is configured so as to magnetically activate the MRE sample between the load cell and the piston rod. The magnetic cell consists of a bobbin wound with copper wire which fits into an outer cylindrical shell. The electromagnetic coil has an inductance of 90 mH and a resistance of 40.85 Ω at 120 Hz and DC resistance of 15 Ω. The magnetic cell produces magnetic field parallel to the direction of compressive loading, along the longitudinal axis of the MRE sample. Currents (i.e., 0.25, 0.5, 0.75, 1.0, 1.5, 2.0, and 2.5 A) were fed into the electromagnetic coil with a current amplifier so as to activate the MRE sample. A sinusoidal displacement over a range of frequencies (i.e., 1, 2, 5, 10, 15, and 20 Hz) with a constant amplitude of ±0.5 mm was used. The MRE samples were initially compressed by 2% of its nominal length. The required force to displace the MRE sample was measured by the load cell and recorded for each test.

3. Results

The performance of MRE samples was characterized in terms of equivalent viscous damping and complex stiffness.

The force (F) of the MRE under a sinusoidal displacement (x) can be expressed by an equivalent damping (C_{eq}) as follows:

$$F = C_{eq}\dot{x} = C_{eq}A\omega \cdot \cos(\omega t) \tag{1}$$

Here $\dot{x}$ is the velocity, A is the displacement amplitude, and ω is the frequency.

The energy dissipated (E) over one cycle is given by

$$E = \oint F dx = \int_0^{2\pi/\omega} F(t)\dot{x}dt \tag{2}$$

Using Eqs. (1) and (2), the equivalent damping can be expressed by

$$C_{eq} = \frac{E}{\pi A^2 \omega} = \frac{E}{2\pi^2 fA^2} \tag{3}$$

On the other hand, a complex stiffness (K^*) can be also used to characterize the performance of the MREs. First, the force of MREs is given by,

$$F = K'x + \frac{K''}{\omega}\dot{x} \tag{4}$$

where K' is the in-phase or storage stiffness which is a measure of the energy stored over a cycle. K'' is the loss stiffness which is a measure of the energy dissipated over a period. The complex stiffness, K^* is defined as:

$$K^* = \frac{F}{x} \tag{5}$$

Substituting Eq. (4) into Eq. (5) yields

$$K^* = K' + jK'' = K'(1 + j\eta) \tag{6}$$

where η is the loss factor and defined as K''/K'.

Figure 3 presents force of the MRE sample vs. displacement at 10 Hz displacement excitation frequency. The force of the MRE increases as the applied current and the particle loading increase. The average slope of the force vs. displacement plot or energy loop diagram is a measure of the storage stiffness, K', of the MRE. As shown in Figure 3(a), the slope of the energy loop diagram is nearly constant as the current input is varied. For the particle loading case as shown in Figure 3(b), the slope of the energy loop diagram becomes steeper as the particle loading increases. This implies that the MRE becomes stiffer as the particle loading increases. Below 10 wt% particle loading, the slope of the energy loop diagram barely increases. At 10 wt% particle loading, the slope of the energy loop diagram increases markedly.

Figure 4 presents the equivalent damping and complex stiffness of the MRE samples. As the applied current increases, the equivalent damping and the complex stiffness significantly increase. In addition, the particle loading also makes the equivalent damping and the complex stiffness to increase. As seen in Figure 4(a), the particle loading make greater effect to the equivalent damping of the MREs at lower displacement excitation frequency than the equivalent damping at relative higher displacement excitation frequency. But, as shown in

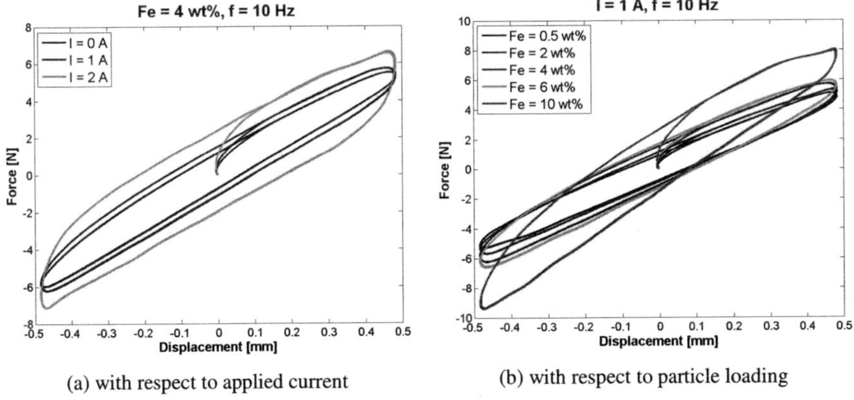

(a) with respect to applied current   (b) with respect to particle loading

Figure 3. Force of the MRE sample versus displacement at 10 Hz displacement excitation frequency.

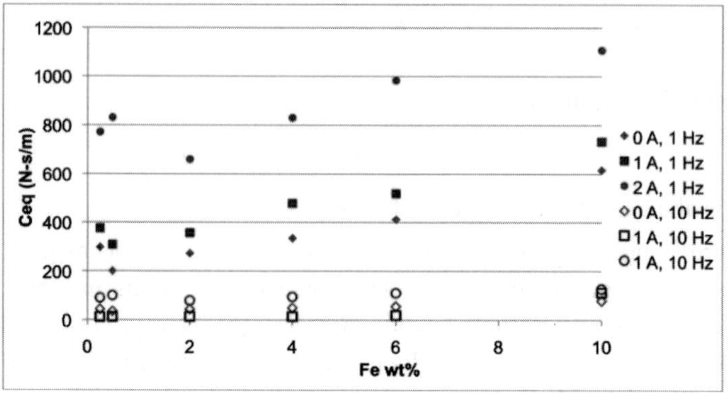

(a) equivalent viscous damping

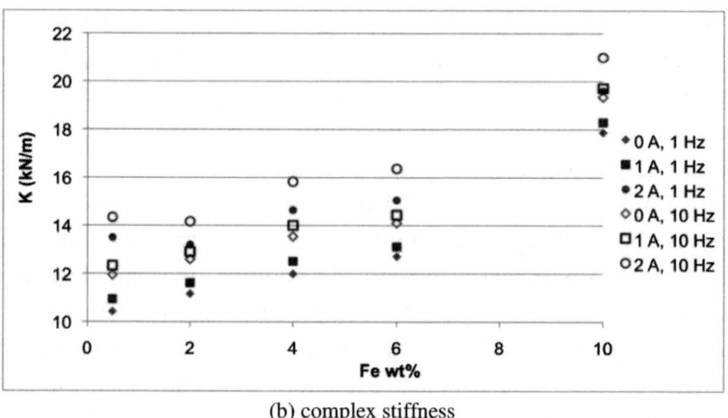

(b) complex stiffness

Figure 4. Equivalent damping and complex stiffness of the MREs with respect to particle loading.

Figure 4(b), the complex stiffness of the MREs is greatly affected by the particle loading regardless of the displacement excitation frequency.

Figure 5 presents the percentage change of the damping and the complex stiffness of the MREs with respect to the displacement excitation frequency. Here, the MRE sample had 4 wt% particle loading. As the displacement excitation frequency increases, the equivalent damping of the MRE exponentially decreases and the complex stiffness almost linearly decrease. This implies that the MR effect of the MREs dominates at lower displacement excitation frequency. But, at relatively higher displacement excitation frequency, the MR effect of the MREs becomes less dominant. The reason is that the passive (i.e., viscous) damping of the MREs becomes stronger at higher displacement excitation frequency.

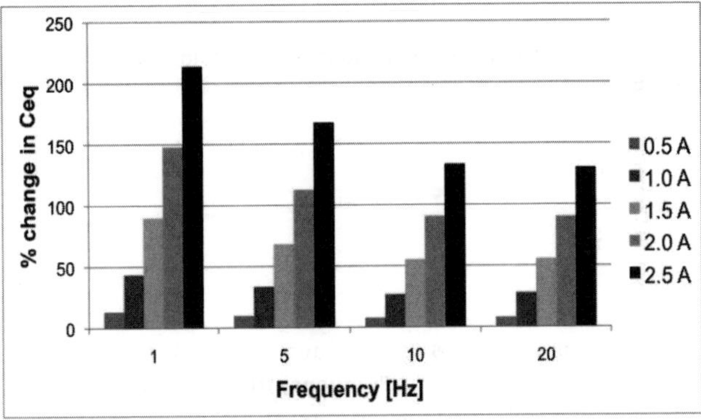

(a) equivalent damping

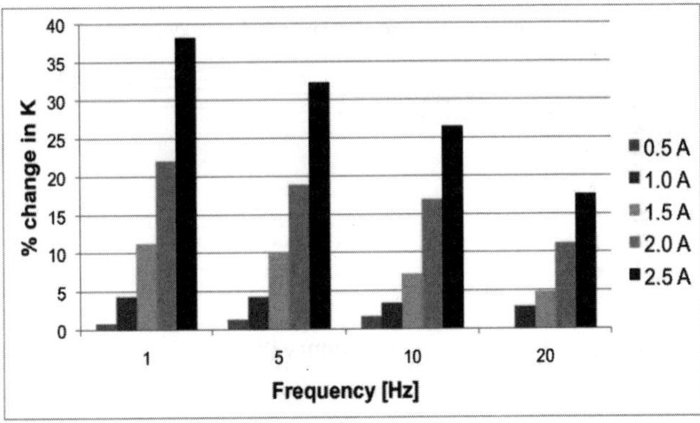

(b) complex stiffness

Figure 5. Percentage change of equivalent damping and complex stiffness of the MREs with respect to displacement excitation frequency (Fe= 4 wt% particle loading).

Figure 6 presents the equivalent damping and the complex stiffness of the MREs with respect to pre-strain. In this case, the MRE sample with 0.25 wt% particle loading was used and applied current was 2A. As seen in this figure, the equivalent damping of the MRE is not strongly affected by the pre-strain. For the complex stiffness case, the storage stiffness (K') of the MRE slightly increases as the pre-strain increases. But, the loss stiffness (K'') of the MRE does not change as the pre-strain increases.

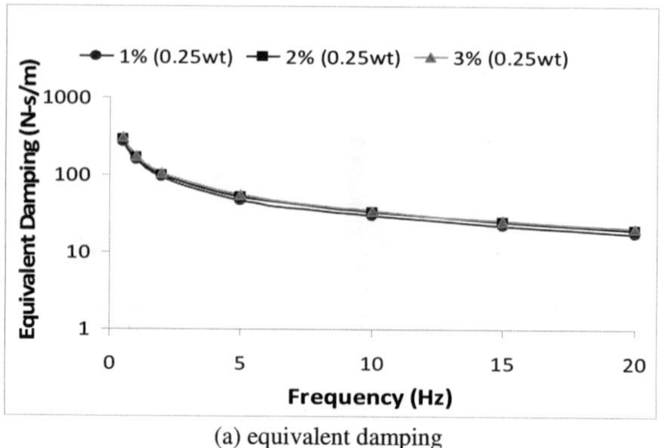

(a) equivalent damping

(b) complex stiffness

Figure 6. Equivalent damping and complex stiffness of the MREs with respect to pre-strain (0.25 wt% particle loading).

4. Conclusions

In this study, characteristics of magnetorheological elastomers (MREs) with respect to particle loading, displacement excitation frequency, and pre-strain were experimentally evaluated in terms of equivalent damping and complex stiffness.

References

1. H. J. Song, O. Padalka, N. M. Wereley and R. C. Bell, *50th Struc., Struct. Dyn., and Mater. Conf.*, 4 - 7 May 2009, Palm Springs, California.
2. H. J. Song, N. M. Wereley, R. C. Bell, J. L. Planinsek and J. A. Filer II, *J. Phys.: Conf. Series*, 11th Conf. on Electrorheological Fluids and Magnetorheological Suspensions, 149:012097. Aug. 25-28, 2008.

SENSITIVITY OF MAGNETORHEOLOGICAL DAMPER BEHAVIOR TO PERTURBATIONS IN TEMPERATURE VIA BOUC - WEN MODEL[*]

ISMAIL SAHIN[1], SEVKI CESMECI[2] and NORMAN M. WERELEY[2,†]

[1]*Akyazý Vocational School, University of Sakarya, Sakarya, 54400, TURKEY*
[2]*Dept. of Aerospace Engineering, University of Maryland,*
College Park, MD, 20742, USA
[†]*Email: wereley@umd.edu*

In this study, the temperature dependent dynamic behavior of a magnetorheological (MR) damper was characterized. To this end, an MR damper, which was designed and fabricated for a ground vehicle seat suspension application, was tested over temperatures ranging from 0 °C to 100 °C at a constant frequency of 4 Hz and a constant amplitude of 7.62 mm on an MTS-810 material testing system equipped with a temperature-controlled environmental chamber. And, the widely adopted Bouc-Wen model was assessed to characterize the temperature dependency of the MR damper through examining the trends of the model parameters. It was observed that although mBW model could capture the MR damper behavior well, some of the model parameters did not represent the physical realization of the damper based on the physical structure of the model. This is attributed to the fact that mBW has differential terms and thus, an infinite solution space and different combinations of the model parameters may yield similar results. Therefore, it was concluded that mBW model was not successful to model the temperature dependency of MR damper behavior.

1. Introduction

Magnetorheological (MR) dampers can experience large variations in temperature due to self-heating as the damper strokes in response to excitations. Energy is dissipated from the system by transforming mechanical energy into thermal energy via viscous and field dependent effects by shearing the fluid as it flows through the damper body. Self-heating may cause significant changes in the viscous damping and yield force of the MR fluid, as well as in the stiffness of the pneumatic accumulator. As the operating temperature increases, the

---

[*] Research was supported by a Fulbright Scholarship to SC, The Scientific and Technological Research Council of Turkey (TUBITAK) grant (108M635) to IS, and a MARCOM Phase 2 SBIR subcontract to NW.

viscous damping and yield force of the fluid decrease and also the stiffness of the accumulator increases due to gas law effects. For base-excited suspensions, low viscous (off-state) damping improves isolation at high frequency excitation (above system resonance); however, at system resonance, a loss in viscous damping adversely affects isolation for the passive (off-state control) MR damper. A decrease in MR yield force reduces the maximum achievable total damping force, equating to a loss in control authority. Finally, increase in the stiffness of the MR damper can shift the system resonance to a higher frequency thereby degrading isolation.

Perturbations in the operating temperature of MR dampers affect system behavior and many have studied this phenomenon. Gordaninejad and Breese [1] have modeled and predicted self-heating of an MR damper during operation, and Dogruoz *et al.* [2] have attempted to augment heat transfer and improve damper performance with the use of heat-sink fins. Batterbee and Sims [3] have developed a temperature dependent model and shown that proportional and PID feedback controllers for MR shock absorption applications can exhibit reduced force-tracking performance as temperature increases as a result of decreased fluid viscosity and yield force. Liu *et al.* [4] have investigated semi-active skyhook control for an MR vibration isolation application and demonstrated improved isolation performance using temperature compensation over uncompensated control. However, there is till much effort should be devoted to understand the effects of temperature variation on the dynamic behavior of the MR damper.

In this study, the Bouc-Wen model suggested by Spencer *et al.* [5] was assessed to characterize the temperature dependent behavior of the MR damper. To do this, an MR damper, which was designed by the team of University of Maryland and Techno-Sciences Inc. for occupant seat isolation onboard the US Marine Corps's amphibious Expeditionary Fighting Vehicle (EFV), was used. And, the temperature dependent model parameters are identified from experimental data via Matlab/Simulink®.

2. Experimental Study

The MR seat isolator (Fig. 1) uses a commercially available fluid (Lord MRF 132) to achieve the necessary maximum

Figure 1. Magnetorheological seat damper.

yield force (4300 N) and field-off viscous damping (1000 N·s/m) for the EFV seat application at an operating temperature of 50 °C. The pneumatic accumulator is filled with nitrogen gas and pressurized to 350 psi. Damper characterization was performed on an MTS 810 material testing system with an installed temperature-controlled environmental chamber (Fig. 2). The MR damper was characterized at temperatures ranging from 0 °C to 100 °C. A maximum temperature of 100 °C is chosen to stay below the rated temperature for safe operation of the damper's polyurethane rod seals; however, under severe continuous operation the MR damper can easily achieve temperatures above 100 °C.

Liquid nitrogen was fed into the environmental chamber to achieve low temperatures near 0 °C. The damper was first cooled to 0 °C and then excited, allowing the damper to self-heat up to 100 °C. An excitation of 4 Hz and 7.62 mm displacement amplitude was chosen as representative of the resonant condition of the EFV suspension system. This procedure was repeated at applied current levels of 0, 0.25, 0.5, 1.0, 1.5, 2.0, and 2.5 Amperes to evaluate temperature effects over the full range of control. A thermocouple was installed inside the damper to directly measure the operating temperature of the fluid. A linear variable differential transformer (LVDT) sensor was used to measure displacement and a load cell was used to measure transmitted force.

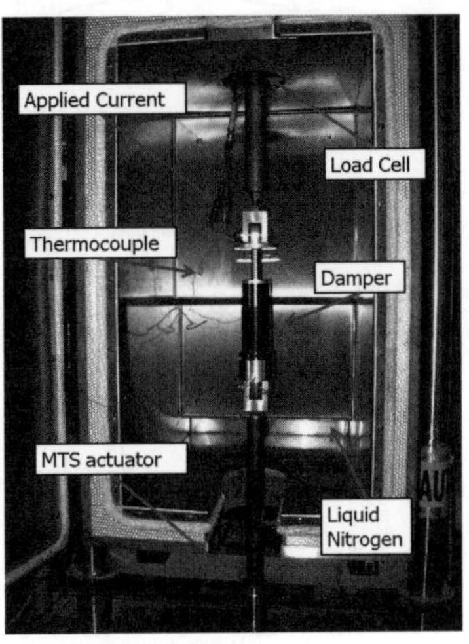

Figure 2. Temperature-controlled environmental chamber.

The measured displacement signal was Fourier filtered by choosing only the component belonging to the primary excitation frequency; however, the damper force signal was left unfiltered to capture any higher harmonics. Representative force vs. piston velocity and displacement data is shown in Fig. 3, providing the qualitative trends of the damper force behavior as a function of operating temperature. The area enclosed in a force vs. displacement curve represents the energy dissipated by the MR damper. As can be seen from Figs. 3*a* and 3*b,* the area inside each force vs. displacement curve decreases as temperature increases, indicating a decrease in the dissipated energy per cycle.

In Figs. 3*c* and 3*d*, it can be observed that as the applied current increases, both the yield force and post-yield damping increase causing high-velocity hysteresis loops that both translate outward from zero and rotate counter-clockwise. On the other hand, as the temperature increases, the high-velocity loops both rotate clockwise and translate toward zero force, illustrating the decrease in both the yield force and post-yield damping, respectively. The area within the high-velocity loops increases with increasing temperature signifying an increase in gas stiffness, which is indicated in Fig. 3*b* by the counter-clockwise rotation of the hysteresis curves.

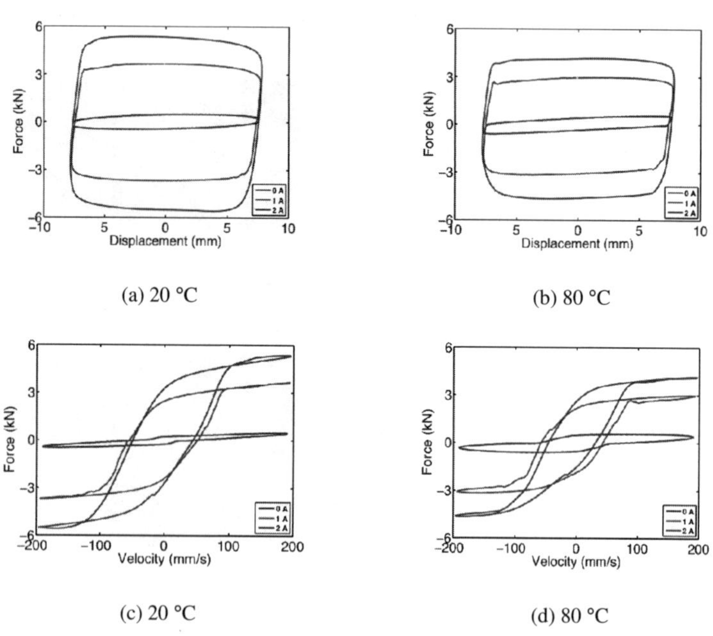

(a) 20 °C   (b) 80 °C

(c) 20 °C   (d) 80 °C

Figure 3. Measured force signal of the MR damper.

3. Modeling of the MR Damper

A modified Bouc-Wen (MBW) model [5] is shown in Fig. 4. The model is given by

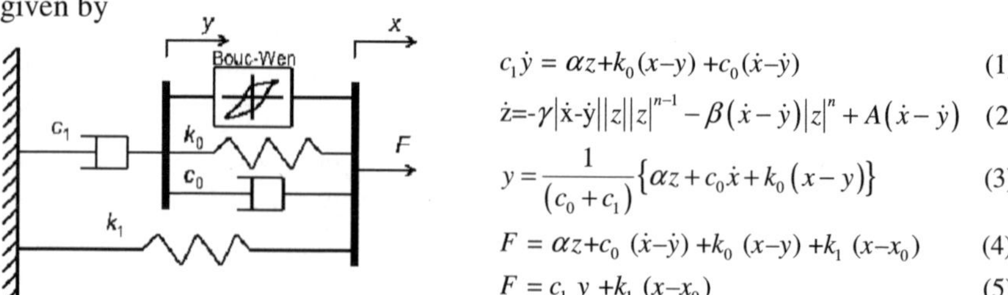

$$c_1 \dot{y} = \alpha z + k_0 (x-y) + c_0 (\dot{x}-\dot{y}) \tag{1}$$

$$\dot{z} = -\gamma |\dot{x}-\dot{y}||z||z|^{n-1} - \beta (\dot{x}-\dot{y})|z|^n + A(\dot{x}-\dot{y}) \tag{2}$$

$$y = \frac{1}{(c_0 + c_1)} \{ \alpha z + c_0 \dot{x} + k_0 (x-y) \} \tag{3}$$

$$F = \alpha z + c_0 (\dot{x}-\dot{y}) + k_0 (x-y) + k_1 (x-x_0) \tag{4}$$

$$F = c_1 y + k_1 (x-x_0) \tag{5}$$

Figure 4. Schematic of Bouc-Wen model.

where the accumulator stiffness is represented by k_1 and the viscous damping observed at larger velocities (post-yield damping) is represented by c_0. A dashpot, represented by c_1, is included in the model to produce the roll-off that was observed in the experimental data at low velocities, k_0 is present to control the stiffness at large velocities, and x_0 is the initial displacement of spring k_1 associated with the nominal damper force due to the accumulator.

To estimate the model parameters, a model was constructed in Matlab (R2010a)/Simulink and the 10 model parameters were found out by using Estimation Toolbox under Parameter Design Optimization Module for 44 experimental force vs. time data sets. In the calculations, Nonlinear Least Square method and Trust-Region-Reflective were used as an optimization technique

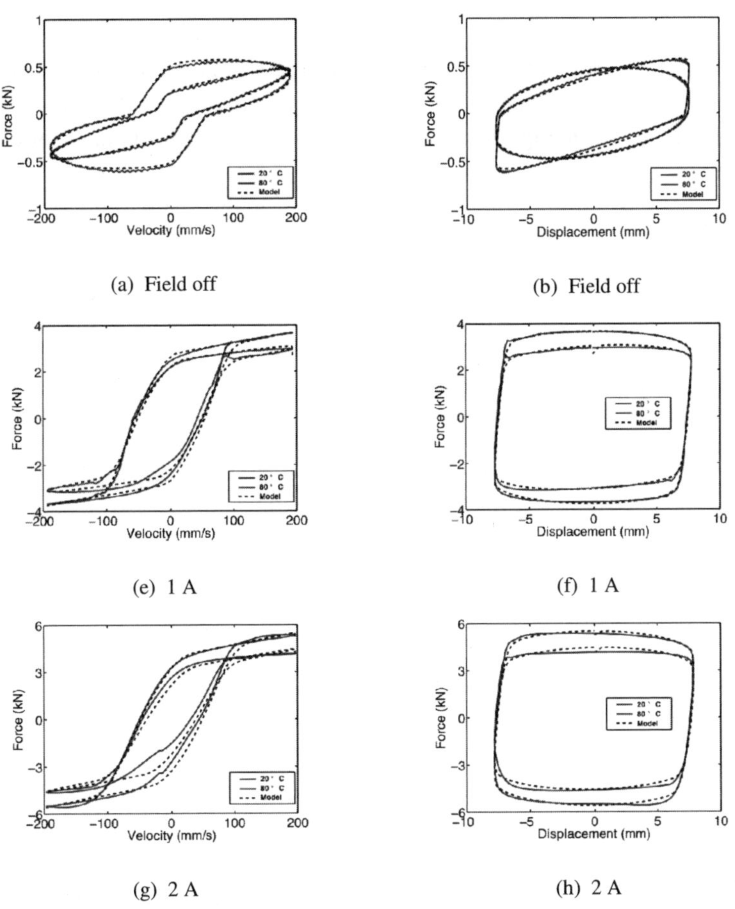

(a) Field off   (b) Field off

(e) 1 A   (f) 1 A

(g) 2 A   (h) 2 A

Figure 5. Comparisons between the model and experimental data.

670

and a solution algorithm, respectively. And, the relative error for parameters was chosen to be 1e-6. Simulations were performed on a Dell T 1500 work station (Core i7, 6 GB Ram, Quadro Fx 580).

4. Results and Discussion

Comparisons between the model predictions and experimental data are given Fig. 5 for different fields and two reference temperatures of 20 °C and 80 °C. It was observed that there is an overall good agreement between the model results and experimental data. The parameter estimates are also given in Fig. 6 as a function of temperature. In Fig. 6a, the yield force decreases as temperature increases. Similarly, the post-yield damping decreases with temperature and increases with the applied current (Fig. 6b). As the operating temperature rises from 0 °C to 100 °C, the yield force and post-yield damping decrease by 56-89% and 37-96%, respectively. In Fig. 7c, stiffness in large velocities increases along with temperature due to the increased gas pressure in the accumulator. However, as given in Fig. 7d, accumulator stiffness is fluctuating along with temperature.

Although the trends of the parameters agree with our qualitative observations from the experimental data, the decay in the yield force, α was found to be unreasonably high. And also, while k_0 yielded a physically

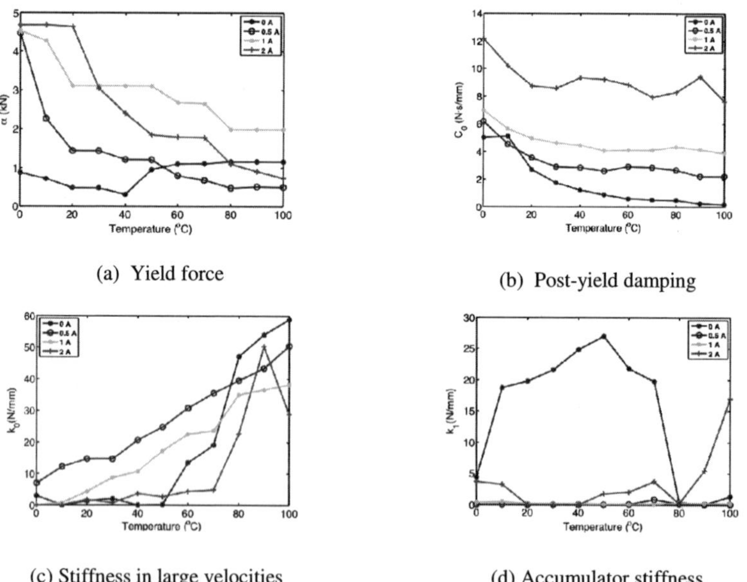

(a) Yield force

(b) Post-yield damping

(c) Stiffness in large velocities

(d) Accumulator stiffness

Figure 7. Variation of some model parameters with temperature.

reasonable trend with a continuous increase with temperature, k_1 gave a fluctuating variation with temperature having no connection to the physical realization of the damper, because stiffness was expected to increase continuously with temperature due to increased gas pressure in the accumulator. On the other hand, post-yield damping, c_0 showed an exponential decay similar to the variation of fluid viscosity with temperature as given in technical data-sheet of MRF-132. This is because the post-yield damping is directly proportional to the fluid viscosity.

5. Summary

The parameter identification results showed that, although mBW model could capture the MR damper behavior well, some of the model parameters did not represent the physical realization of the damper based on the physical structure of the model. This is because that mBW model has differential terms and thus, has an infinite solution space so that different combinations of the model parameters may yield similar results (nonunique solutions). To overcome this shortcoming, one could search for the most suitable set of starting points for each characteristic parameter and set upper and/or lower limits for each parameter. However, in this case the solution is either not converged or is likely to converge at local minima instead of at a global minimum. This intrinsic behavior of the mBW suggests that it may not be the best choice for temperature characterization of the MR damper. Apart from these shortcomings, mBW is computationally highly expensive and time consuming compared to other algebraic models studied in the literature.

References

1. F. Gordaninejad and D. G. Breese, *Journal of Intelligent Material Systems and Structures*, **10**(8), pp. 634-645 (1999).
2. M. B. Dogruoz, E. L. Wang, F. Gordaninejad, and A. J. Stipanovic, *Journal of Intelligent Material Systems and Structures*, **14**(2), pp. 79-86 (2003).
3. D. Batterbee and N. D. Sims, *Journal of Intelligent Material Systems and Structures*, **20**(3), pp. 297-309 (2009).
4. Y. Liu, F. Gordaninejad, C. A. Evrensel, U. Dogruer, M. S. Yeo, E. S. Karakas, and A. Fuchs, *Proceedings of the SPIE, Smart Structures and Materials: Industrial and Commercial Applications of Smart Structures Technologies*, SPIE, San Diego, CA, (**5054**), pp. 332-340 (2003).
5. B. F. Spencer, S. J. Dyke, M. K. Sain, and J. D. Carlson, *Journal of Engineering Mechanics*, (**123**)3, pp. 230-238 (1997).

BEHAVIOR OF MR FLUIDS AT HIGH SHEAR RATE

WEI HU and NORMAN M. WERELEY

Smart Structures Laboratory, Alfred Gessow Rotorcraft Center, Dept. of Aerospace Engineering, University of Maryland, College Park, MD 20742

The high shear rate behavior of MR fluids is investigated using a concentric rotational cylinder viscometer fabricated in-house. The rotational cylinder viscometer is designed such that a high shear rate of up to 30,000 s^{-1}can be applied to the MR fluid in a pure shear flow mode. As a comparison, the maximum shear rate of a commercially available parallel disk type rheometer is only up to 1,000 s^{-1}. To determine the shear rate of the MR fluid in the viscometer, an exact expression between torque and angular velocity is established. The yield stress and viscosity of the MR fluid is determined by fitting the expression into the measured torque and angular velocities, and the shear stress as a function of the shear rate is further derived. The magnetic filed strength across the fluid gap is determined based on an electromagnetic field analysis, and the yield stress and viscosity of the fluid as a function of the magnetic filed is established. Specifically, the stability of the MR fluid at high shear rate is also evaluated. Two commercially available MR fluids, i.e. Lord's MRF-132DG and MRF-140CG, are investigated using the rotational cylinder viscometer, and the testing results are compared to the manufacturer's data.

1. Introduction

Currently, development of magnetorheological (MR) seat suspension systems, which can provide both shock mitigation and vibration isolation, is becoming critically important for occupant protection, resulting in a broad range of potential applications for this adaptive seat suspension technology [1-3]. One key challenge in seat suspension applications involving MR energy absorbers is the generation of sufficient stroke with as low a field-off resistant force as possible, so as to effect a favorable outcome in a crash protection and vibration isolation system. A shear mode rotary vane damper is determined to be capable of providing a high yield force and unlimited stroke while maintaining a small field-off viscous force. However, a high impact velocity during a vehicle crash leads to a high shear rate applied to the magnetorheological fluid (MRF) in the shear mode damper. Due to limitations of commercially available rheometers (shear rates typically < 1000 s^{-1}), high shear rate (>> 1,000 s^{-1}) flow curves in pure shear mode is not available, so limits the design space of the rotary vane shear mode

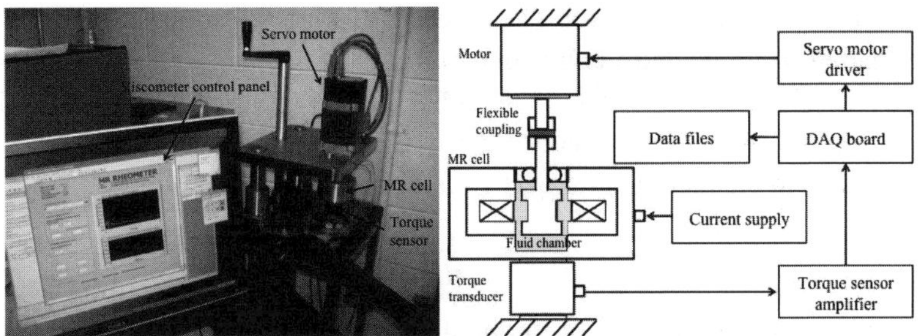

(a) Viscometer setup   (b) Schematic of the viscometer and its control system

Figure 1. Configuration of the concentric rotational cylinder viscometer

damper. Therefore, a magnetoviscometer for flow curve measurements at high shear rates is developed using a concentric rotational cylinder viscometer [4].

The MR cell of the concentric rotational cylinder viscometer is made of a stationary fluid cup and a concentric rotational bob. A fluid gap is sandwiched between the cup and bob and a uniform shear rate along the angular and axial direction can be applied to the MRF in the gap. In comparison, in a conventional parallel disk-type rheometer, the shear rate depends on the radial distance from the rotation axis, and the fluid in the gap can be expelled from between the disks even at relatively low shear rate due to centrifugal effects. Two MR fluids, i.e., Lord Corp. MRF 132DG and 140CG, are characterized using the rotational cylinder viscometer. A method to determine the shear rate is developed based on prior studies [5-6], where the flow curve is measured up to a peak shear rate of 30,000 s^{-1}. The yield stress and viscosity of the MRFs are identified using measured torque and angular velocity using a Bingham-plastic model [7], and the results are compared to the data provided by the manufacturer [8-9]. Flow curve measurements show that the MRF samples at high shear rate demonstrate stable yield stress and viscosity.

2. Viscometer Set-Up and Measurement

The configuration of the concentric rotational cylinder viscometer and its control system are given in Fig. 1. The viscometer consists of a rotating bob (spindle) attached to a servomotor, a static housing equipped with an MR cell and a fluid cup, a torque sensor, a current supply and a data acquisition system. To minimize the run-out of the shaft, the bob is supported by a precision ball bearing and is attached to the servomotor through a flexible coupling. The MRF sample is poured into the annular gap sandwiched between the concentric bob and cup. The servomotor rotates the bob with an angular velocity of up to 2,700 rpm (280 rad/s), and the transmitted torque from the bob to the housing is measured via a torque sensor. There are two bob sizes used in this investigation, i.e. a bob with a fluid gap of 0.25 mm and a bob with a fluid gap of 0.1 mm.

The rheological properties of the MR fluid is generally evaluated using shear stress and shear rate. As shown in [5], the shear stress of the MR fluid, τ, is obtained from the measured torque, T, using an equation as follows:

$$\tau = \frac{T}{2\pi L_c r^2} \tag{1}$$

where, r is the radial position of the fluid and $R \leq r \leq R+h$, R is the radius of the bob, h is the thickness of the gap, and L_c is the total gap length. The shear rate of the MR fluid, $\dot{\gamma}$, is related to the angular velocity of the bob, Ω, as follows [5]:

$$\Omega = \int_{\tau_c}^{\tau_b} \dot{\gamma} \frac{d\tau}{2\tau} \tag{2}$$

where, τ_b and τ_c are the shear stress at the bob and cup surface, respectively. In this study, the MR fluid is characterized by the Bingham-plastic model, so the shear rate of the fluid can be described as

$$\dot{\gamma} = \begin{cases} \dfrac{1}{\mu}\left(\tau - \tau_y\right) & \tau > \tau_y \\ 0 & \tau \leq \tau_y \end{cases} \tag{3}$$

where, τ_y is the field-dependent yield stress and μ is the viscosity of the fluid. Since the maximum difference between τ_b and τ_c is within 5 percent in the current viscometer configuration, the MR fluid in the gap region is assumed to be fully sheared. Thus, substituting Eq. 3 into 2 and combining Eq. 1 then gives:

$$T = \frac{4\pi L_c}{\left(\dfrac{1}{R^2} - \dfrac{1}{(R+h)^2}\right)}\left[\ln\left(1+\frac{h}{R}\right)\tau_y + \mu\Omega\right] \tag{4}$$

Apparently, the torque as a function of the angular velocity can be described by a yield component and a viscous component, which is similar to the Bingham-plastic model. Using a least squared error fitting technique, τ_y and μ can be estimated by fitting Eq. 4 to the measured torque and angular velocity. The shear rate can be further obtained using Eq. 3.

An analytical method for the rotational cylinder viscometer is also developed to estimate the magnetic field strength in the MRF across the gap as a function of the applied current. A quarter sectional view of the axial-symmetrical MR cell is shown in Fig. 2, in which close-looped magnetic flux paths are shown as dotted lines and several key dimensions for magnetic strength calculation are included, i.e. $A_g = \pi R L_c$ is the area that the magnetic flux flows through the gap, $A_b = \pi R_b^2$ is the cross-sectional area of the bob, and $A_o = \pi\left(R_o^2 - R_m^2\right)$ is the cross-sectional area of the outer magnetic tube. Because the coil number and applied current are known, N and I, the magnetic field strength at the fluid gap, H_g, can be obtained using the following equations:

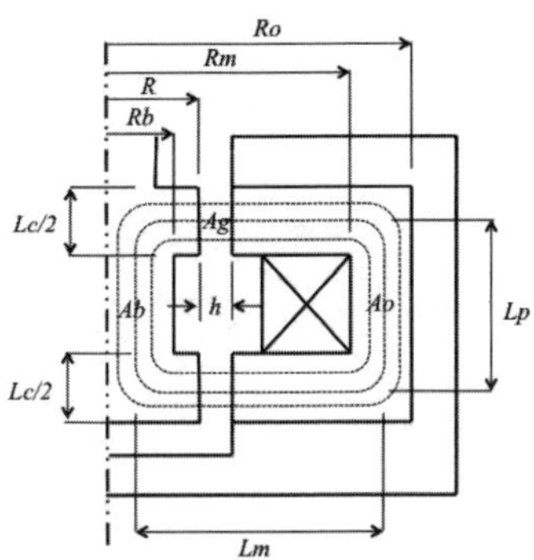

Figure 2. Schematic magnetic flux path

$$B_{mr}A_g = B_{s1}A_b = B_{s2}A_o = B_{s3}A_g$$
$$NI = 2H_g h + H_{s1}L_p + H_{s2}L_p + 2H_{s3}L_m \tag{5}$$

where, B_{mr} is the flux density at the fluid gap, B_{s1}, B_{s2} and B_{s3} are the flux density at bob, outer tube and coil armature, respectively, and H_g, H_{s1}, H_{s2} and H_{s3} are corresponding magnetic strength and are related to the flux density using MR fluid and carbon steel H-B properties.

3. Magnetorheology at High Shear Rates

Using the rotational cylinder viscometer, high shear rate measurements (up to 30,000 s^{-1}) are conducted for two MRFs, i.e. Lord's MRF-132DG and MRF-140CG, in which the volume percentage of iron particles in the fluid is 32 and 40 vol%, respectively. Both 0.25 mm and 0.1 mm gaps are introduced to apply maximum 15,000 s^{-1} and 30,000 s^{-1} shear rates to the MRFs. All measurements are performed at room temperature (25 °C).

In each testing run (with a constant applied current), eight increasing command voltages from the control system are sent to the servomotor to create eight different angular velocities. The resistant torque is measured at each velocity, and the measured torque at each angular velocity is averaged to remove signal noise in the torque measurement. The average torque and angular velocity can be used to determine the yield stress and viscosity of the MRF using Eq. 4 and the shear rate is further determined using Eq. 3. The identified MR properties, i.e. yield stress and viscosity, are compared to evaluate the effect of the gap thickness on the measured results.

The measured torque versus angular velocity of the MRF-132DG is shown in Fig. 3 and the fitted curves in the form of Eq. 4 are shown as solid lines. Notably, the MRF in the larger of the two gaps (0.25 mm) can be effectively characterized by the Bingham-plastic model, but the behavior of the MR fluid in the

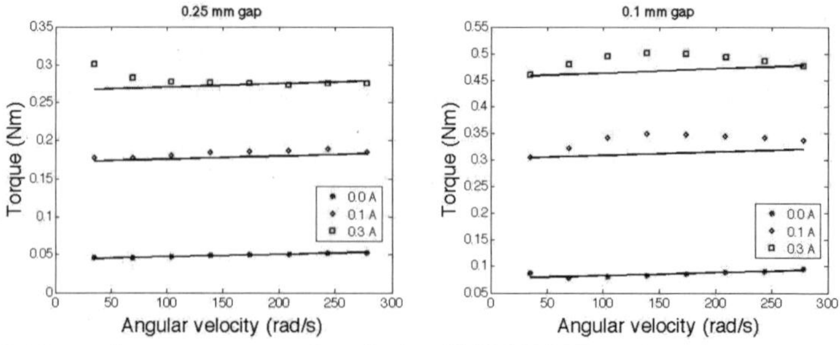

Figure 3. Measured torque versus angular velocity of MRF-132DG

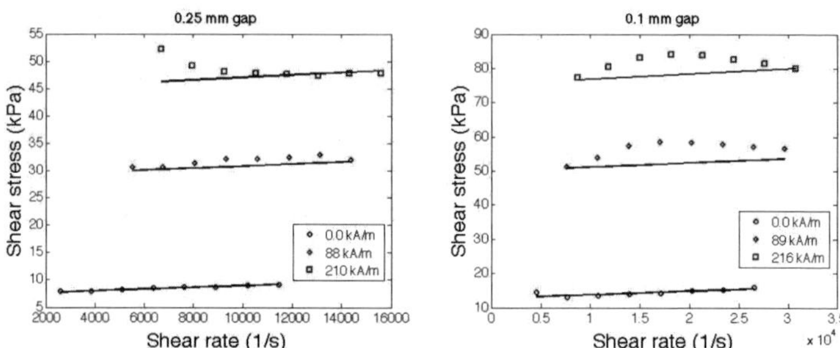

Figure 4. Shear stress vs. shear rate of MRF-132DG for gaps of 0.25 mm and 0.1 mm

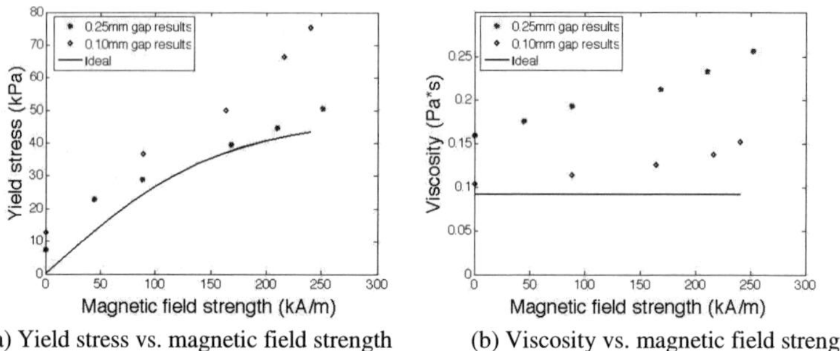

(a) Yield stress vs. magnetic field strength   (b) Viscosity vs. magnetic field strength

Figure 5. Comparison of magnetic properties of MRF-132D G

smaller gap (0.1 mm) is not sufficiently described by the Bingham-plastic model. The shear stress as a function of the shear rate at the bob surface is shown in Fig. 4, in which the magnetic field strength related to each applied current is obtained using Eq. 5. Apparently, the shear stress of the MRF does not diminish at high shear rates (up to 30,000 s^{-1}). It is observed that the zero-field shear stress of the MRF measured at 0.25 mm gap is lower than one at 0.1 mm gap, which probably results from the friction between the bob and fluid cup due to the run-out of the shaft. The identified yield stress and viscosity are shown in

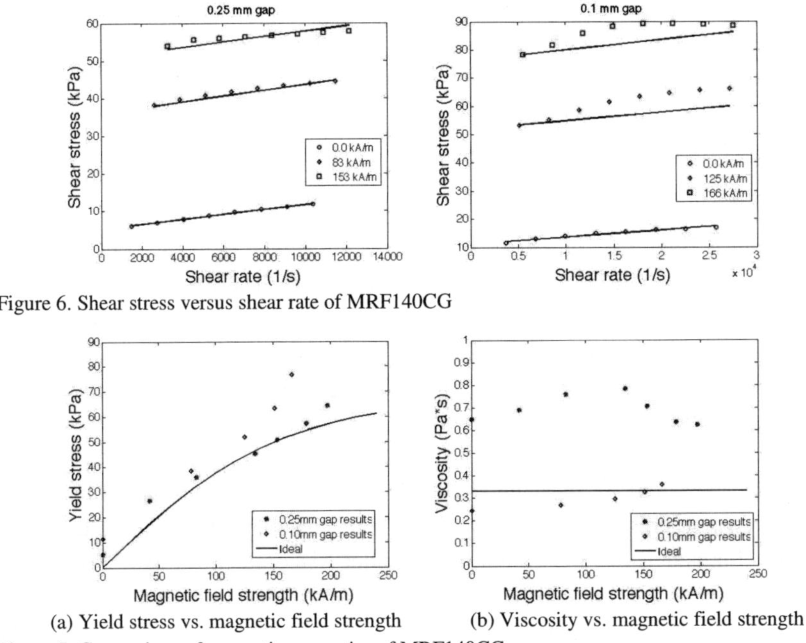

Figure 6. Shear stress versus shear rate of MRF140CG

(a) Yield stress vs. magnetic field strength (b) Viscosity vs. magnetic field strength

Figure 7. Comparison of magnetic properties of MRF140CG

Fig. 5. To verify the results of the rotational cylinder viscometer, Carlson's empirical formula [8] is used to determine the ideal yield stress, and the obtained ideal yield stress is compared to its measured quantity. As shown in Fig. 5a, if the zero-field stress is subtracted from the data, the yield stress obtained from the rotational cylinder viscometer is close to the ideal result while the magnetic field strength is below 200 kA/m. However, the obtained yield stress of the MRF in a smaller gap (0.01 mm) is much higher than the other results as the magnetic field strength is higher than 200 kA/m. The viscosity is shown in Fig. 5b and compared to the nominal viscosity provided by Lord Corp. [9]. The measured viscosity from both gap thicknesses is higher than the manufacturer's data and increases as the magnetic field increases.

Lord's MRF-140CG was also characterized using the rotational cylinder viscometer. Similar methods were used to determine yield stress and viscosity. The shear stress and shear rate at the bob surface are obtained and shown in Fig. 6. Compared to the results of MRF-132DG in Fig. 4, the achieved maximum shear rate for both 0.25 and 0.1 mm gap is lower due to the higher viscosity of MRF-140CG. The identified yield stress and viscosity is shown in Fig. 7. It is shown that the measured yield stress of the fluid in the 0.25 mm gap is close to the ideal result, and the yield stress in the 0.1 mm gap matches the ideal result at lower magnetic field strength but becomes higher as the magnetic field strength is higher than 150 kA/m. The obtained viscosity is shown in Fig. 7b and is compared to the nominal viscosity provided by Lord Corp. Notably, the measured viscosity of the fluid in the 0.1 mm gap is close to the ideal value, but the measured viscosity in the 0.25 mm gap is higher.

4. Conclusions

A concentric rotational cylinder viscometer is developed to measure the rheological properties of MRFs in a pure shear mode at high shear rate (up to 30000 s^{-1}). The MRF is characterized by the Bingham-plastic model, and an exact expression between the torque and angular velocity is established and is related to the yield stress and viscosity of the MRF. The measured torque at each angular velocity is fitted using the above expression without the need of differentiation of angular velocity with respect to the torque. The shear stress is determined using the measured torque, and the shear rate is further determined using the obtained yield stress and viscosity. The shear stress as a function of the shear rate is obtained using the rotational cylinder viscometer for two MRFs, i.e. Lord's MRF-132DG and MRF-140CG. It is noted that the shear stress of the measured fluids does not diminish at high shear rate (up to 30,000 s^{-1}) and the Bingham-plastic model describe the behavior of the MRFs more sufficiently for 0.25 mm gap than 0.1 mm gap. The experimentally identified yield stress is compared to the ideal yield stress provided by manufacturer's empirical equation. It is found the yield stress obtained from the rotational cylinder viscometer with 0.25 mm gap is close to the ideal result, but the yield stress measured from 0.1 mm gap is higher than the ideal result as the magnetic field strength goes higher (above 150 kA/m). Generally, the measured viscosity is found to be higher than the manufacturer's value. A more accurate model for the MRF may be needed to describe the fluid behavior in a smaller annular gap (0.1 mm).

References

1. M. Mao, W. Hu, Y.T. Choi and N.M. Wereley, *J. of Intel. Mater. Sys. Struct.*, 18(12), 1227 (2007).
2. G.J. Hiemenz, W. Hu and N.M. Wereley, *J. of Physics*, 149(1), (2009).
3. G.J. Hiemenz, W. Hu and N.M. Wereley, *J. of Aircraft*, 45(3), 945 (2008).
4. H.J. Song, S.R. Hong, Y.T. Choi and N.M Wereley, ERMR2008, (2008).
5. C. Huang, *Transactions of the Society of Rheology*, 15(1), 25 (1971).
6. Y.T. Choi, J.U. Cho, S.B. Choi, and N.M. Wereley, *Smart Mater. Struct.*, 14(5), 1025 (2005).
7. N.M. Wereley and L. Pang, *Smart Mater. Struct.*, 7(5), 732 (1998).
8. J.D. Carlson, ERMR2004, (2004).
9. LORD Corp, www.lord.com

AN EXPERIMENTAL INVESTIGATION INTO THE OFF-STATE VISCOSITY OF MR FLUIDS

K.H. GUDMUNDSSON

School of Engineering and Natural Sciences, University of Iceland, Hjardarhaga 2-6, Reykjavik, 107, Iceland

F. JONSDOTTIR

School of Engineering and Natural Sciences, University of Iceland, Hjardarhaga 2-6, Reykjavik, 107, Iceland

F. THORSTEINSSON

Ossur Inc., Grjothalsi 6, Reykjavik, 110, Iceland

O. GUTFLEISCH

Leibniz Institute for Solid State and Materials Research Dresden, PF 27 01 166, Dresden, D-01171, Germany

This study investigates the off-state rheological characteristics of magnetorheological (MR) fluids. Various carbonyl iron powder grades are tested, both micron- and nano-sized. The study considers both monodisperse and bidisperse MR fluid samples. The bidisperse MR fluids are both micron-sized only compositions and mixed micron- and nano-sized compositions. All fluid compositions employ a novel perfluorinated polyether (PFPE) oil as a base fluid. This base fluid is introduced and its qualities discussed. An off-state experimental investigation is conducted, for shear-rates from 0 to 200 s^{-1}. All samples exhibit the well known shear-thinning property of MR fluids but with varying proportions. The off-state viscosity is of importance in a particular application, in an MR prosthetic knee. It determines how fast the knee joint can rotate in the absence of a magnetic field. A prominent MR fluid composition is selected for the proposed application which aims to maximize to rotational speed of the MR knee joint in the absence of a magnetic field.

1. Introduction

The viscosity of non-magnetic suspensions, both colloidal and non-colloidal suspensions, has been extensively studied [1-4]. These models should be applied with care to determine the off-state viscosity of magnetorheological (MR) suspensions since they are based on submicron-sized particles. The established viscosity models [1-4] underestimate the off-state viscosity of MR suspensions [5]. Mutual design goals in MR fluid design are [6]: high field-induced shear stresses, low off-state viscosity and sedimentation stability. The research presented in this paper shares these goals but aims to explore one of them thoroughly, the off-state viscosity.

The off-state viscosity of MR fluids has not been researched to the same extend as the field-induced characteristics. However, the nature of some applications requires a low off-state viscosity. These applications include both linear dampers [7] and rotary brakes [8,9]. The work presented here is an experimental investigation into the off-state viscosity of MR fluids. Firstly, the effect of solid concentration in monodisperse MR fluids is investigated. Secondly, the effect of particle size is investigated considering five different micron-sized particles. Thirdly, the effect of nano powder is investigated and finally bidispere MR fluids with micron-sized particles are experimentally evaluated. The motivation for this study is an MR prosthetic knee [8,9] in which the off-state viscosity affects how fast the knee can rotate in the absence of a magnetic field.

2. Off-State Characteristics

The off-state characteristics of sixteen MR fluid samples where investigated. All samples employ a PFPE base fluid [10] with variable particle concentration and size. The base fluid has a relatively high viscosity [10] and facilities for sedimentation stability. Measurements where performed with an Anton-Paar Physica MCR 100 rheometer with a parallel plate measuring system. Plates with a diameter of 20 mm were used with a gap of 1 mm.

2.1. *Monodisperse with a varing solid concentration*

Four monodisperse MR fluid samples where prepared with a variable solid concentration ranging from 0.25 to 0.35 by volume. Iron powder employed is a carbonyl iron powder with an average particle diameter of 2 μm. Figure 1 shows the off-state viscosity of the four samples for shear rates up to 200 s^{-1}.

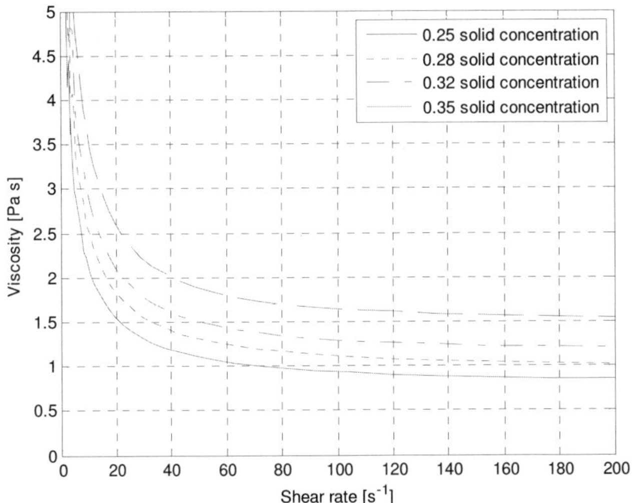

Figure 1. The off-state viscosity of monodisperse perfluorinated polyether (PFPE) based MR fluid samples with a variable solid concentration. All particles have an average diameter of 2 μm. Measurements are performed at a temperature of 20°C.

The figure exhibits the well known shear thinning property of MR fluids. At low shear-rates the viscosity is high but reaches a stable low value at higher shear-rates. At interest, in the proposed application in a prosthetic knee, is the low viscosity value at high shear-rates. This is due to a small micron-sized gap shearing the fluid in the knee resulting in high shear-rates. Figure 1 shows 50% higher off-state viscosity when a monodisperse fluid with a solid concentration of 0.25 is compared to a corresponding fluid with a solid concentration 0.35. Thus, increasing the solid concentration of a monodisperse fluid greatly deprecates the rotational speed of the prosthetic knee in the absence of a magnetic field.

2.2. *Monodisperse with a varying particle size*

Five monodisperse MR fluid samples where prepared with an iron powder with an average particle size ranging from 1 μm to 7 μm. All samples have a solid loading of 0.28. Figure 2 shows the off-state viscosity of the five samples for shear rates up to 200 s^{-1}.

682

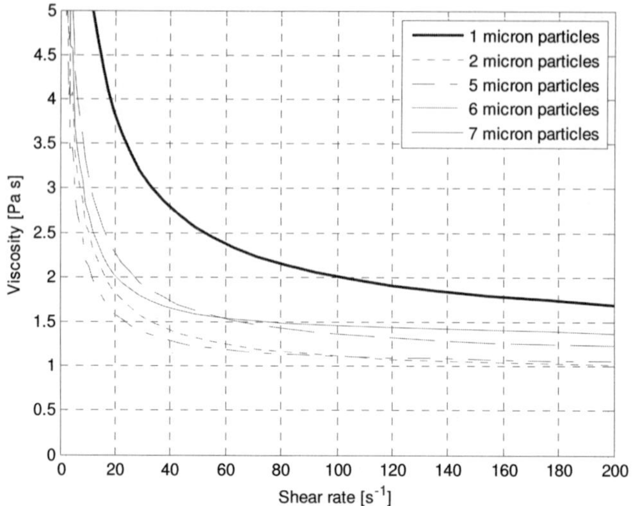

Figure 2. The off-state viscosity of monodisperse perfluorinated polyether (PFPE) based MR fluid samples with a variable solid concentration. All particles have an average diameter of 2 μm. Measurements are performed at a temperature of 20°C.

Again, the shear thinning characteristics are prominent. Of interest in the current application is the viscosity at high shear-rates. With particles in the micron size range, the off-state viscosity increases with an increasing particle size. This is true, except for the 1 μm particle composition which exhibits the highest off-state viscosity. This is believed to be due to the fact that the compositions employing the smallest particles also contain some submicron-sized particles. Submicron-sized particles will be shown to increase the off-state viscosity to a large extent in the next section. Figure 2 shows the off-state viscosity of a fluid composition employing 7 μm particles to be 40% higher than a corresponding fluid composition employing 2 μm particles.

2.3. *Bidisperse with nano-sized particles*

Four bidisperse MR fluid samples with nanoparticles where prepared, all with total solid concentration o 0.28, by volume. Average particle size for micron-sized particles is 2 μm. The nanoparticles have a size range of 100-250 nm. Figure 3 shows the off-state viscosity of the four nanoparticle samples, for shear rates up to 200 s^{-1}, along with the corresponding monodisperse sample.

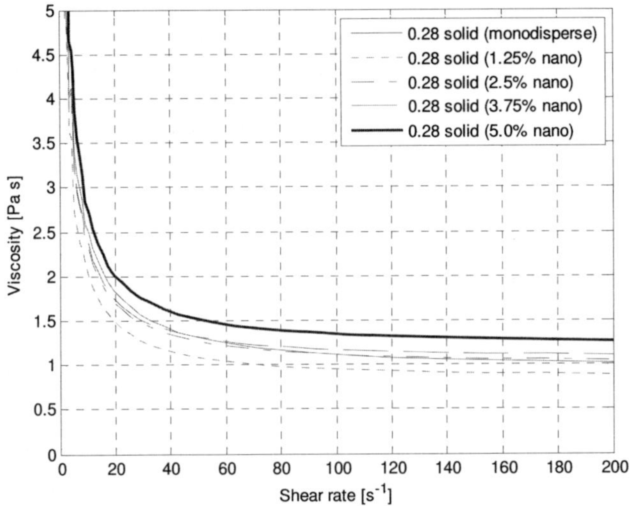

Figure 3. The off-state viscosity of bidisperse perfluorinated polyether (PFPE) based MR fluid samples with nanoparticles. Micron-sized particles have an average diameter of 2 μm and the nanoparticles have a size range of 100-250 nm. Measurements are performed at a temperature of 20°C.

It is evident from Figure 3 that nanoparticles increase the off-state viscosity although the total solid concentration is held to a constant value of 0.28. A composition with 5% of the solid medium as nanoparticles exhibits 30% higher off-state viscosity than a corresponding monodisperse fluid. Thus, replacing a small portion of the micron-sized particles with nanoparticles increases the off-state viscosity.

2.4. *Bidisperse with micron-sized particles*

Three bidisperse MR fluid samples with micron-sized particles where prepared, all with total solid concentration o 0.28, by volume. Average particle size for the smaller micron-sized particles is 1 μm and for the larger micron-sized particles is 7 μm. Figure 4 shows the off-state viscosity of the three bidisperse samples for shear rates up to 200 s^{-1} along with the two corresponding monodisperse samples.

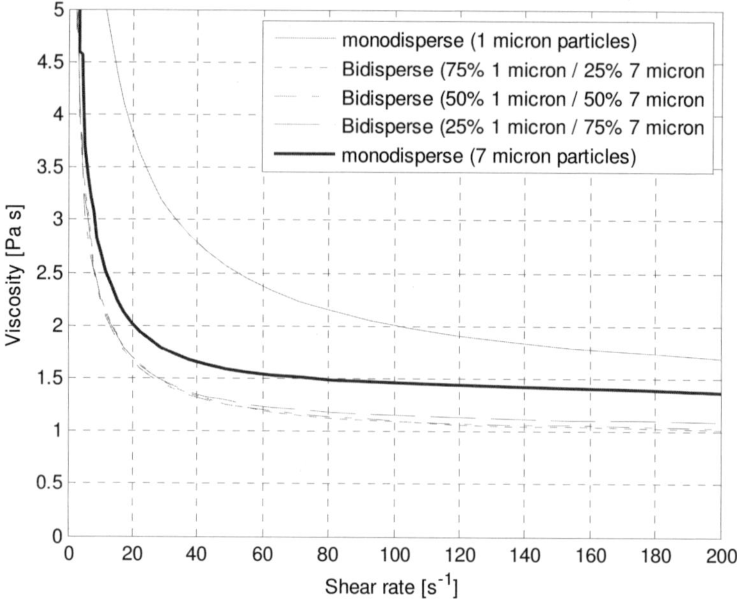

Figure 4. The off-state viscosity of bidisperse perfluorinated polyether (PFPE) based MR fluid samples with micron-sized particles. The smaller micron-sized particles have an average diameter of 1 μm while the larger have an average diameter of 7 μm. Measurements are performed at a temperature of 20°C.

Interestingly, the micron-sized bidisperse fluid compositions exhibit a notably lesser off-state viscosity when compared to the corresponding monodisperse composition. This is true for all the three bidisperse samples. They exhibit an approximately 40% lower viscosity when compared to the corresponding monodisperse composition employing the 7 μm. The bidisperse samples exhibit an 60% lower viscosity when compared to the corresponding 1 μm monodisperse composition.

3. Conclusions

The experimental investigation shows that the off-state viscosity of MR fluids is sensitive to particle size and particle concentration. This suggests that MR fluids can be tailored for the proposed application in a MR prosthetic knee.

The results show a rapid increase in off-state viscosity when increasing the solid concentration in a monodisperse composition from 0.25 to 0.35, by volume. The same holds true when increasing the average particle size from 2 μm to 7 μm, a rapid increase in off-state viscosity is observed.

The results show the nanoparticles to have a considerable negative effect on the off-state viscosity, increasing it, although only small portions of nanoparticles are used, while at the same time holding the total solid concentration constant.

The results show, favorably, how bidisperse MR fluids with micron-sized particles have a lower off-state viscosity when compared to other MR fluid samples. Our ongoing current research also, suggests bidisperse micron-sized MR fluid compositions to exhibit a high field-induced shear yield stresses. For the proposed application, in an MR prosthetic knee, the bidisperse MR fluids are an attractive option. The low off-state viscosity advantage of bidisperse MR fluid is believed to be favorably in other MR fluid applications.

Acknowledgments

This work is funded by the University of Iceland and Rannis, grant no. 090035021, and supported by Ossur Inc.

References

1. M. Mooney, *J. Colloid Sci.* **6** 162 (1951).
2. I.M. Krieger, *Adv. Colloid Interface Sci.* **3** 111 (1972).
3. G.K Batchelor, *J. Fluid Mech.*, **83** 97 (1977).
4. C.G. de Kruif, E.M.F. van lersel, A. Vrij and W.B. Russel, *J. Chem. Phys.* **83** 4717 (1985).
5. K.H. Gudmundsson, F. Jonsdottir and S. Olafsson, *New Actuators 2008 - Proceedings* (2008).
6. J.D. Carlson, *J. Intell. Mater. Syst. Struct* **13** 431 (2002).
7. M.R. Jolly, J.W. Bender and J.D. Carslon, *J. Intell. Mater. Syst. Struct* **10** 5 (1999).
8. F. Jonsdottir, E.T. Thorarinsson, H. Palsson and K.H. Gudmundsson, *J. Intell. Mater. Syst. Struct.* **20** 659 (2009).
9. K.H. Gudmundsson, F. Jonsdottir and F. Thorsteinsson, *Smart Mater. Struct.* **19** 035023 (2010).
10. H.H. Hsu, C.R. Bisbee III, M.L. Palmer, R.J. Lukasiewicz, M.W. Lindsay and S.W. Prince, *US Patent Spec.* 7,101,487 (2006).

CHARACTERISATION OF STEP RESPONSE TIME AND BANDWIDTH OF ELECTRORHEOLOGICAL FLUIDS

MARTIN GURKA and RAINO PETRIČEVIĆ

ERF Produktion Würzburg GmbH, Landwehrstrasse 1, 97249 Eisingen, Germany

STEFFEN SCHNEIDER

Bundeswehr Research Institute for Materials, Fuels and Lubricants, Institutweg 1, 85435 Erding, Germany

STEPHAN ULRICH

Helmut-Schmidt-Universität Hamburg, Holstenhofweg 85, 22043 Hamburg, Germany

The dynamic behavior of the commercially used electrorheological fluid RheOil3.0 is measured under well defined conditions. Measurements were carried out in shear mode as well as in flow mode using flow channels that provide similar flow conditions to many electrorheological applications. The response time of the change in flow behavior upon a changed electric field are measured. Measurements were carried out in time domain (single step response) as well as in frequency domain (sinusoidal frequency sweep). In most flow conditions, the observed dynamic behavior is characterized by two leading response times of which the fast one is in the millisecond range. Under stable flow or constant shear rate conditions, ionic conductivity is found to be a limiting factor for a quick step response.

1. Introduction

The combination of quick response time and high power density is a unique feature of electrorheological devices. For the practical realization of fast actuation, quickly responding ER fluids with a low base viscosity are the central element. In modern applications a bandwidth in the kHz regime could be achieved if all components (amplifiers, mechanics, fluid) operate well together. Examples for these devices are electrorheological dampers for mobile applications [1], industrial applications [2][3], or fast operating actuators [4].

The step response time of ER-fluids is dependent on fluid chemistry and in many cases on the temperature and the applied shear rate [5]. In some cases the step response time is also dependent on the applied shear mode or the way the experiment is carried out [6].

In this work the commercially available electrorheological fluid RheOil3.0 was characterized with respect to the dynamic behavior under well defined conditions and methods. Measurements were carried out in shear mode, using a standard Searle system as well as in flow-mode, utilizing flow channels permitting technical relevant conditions. Measurements were carried out in time domain (single step response) as well as in frequency domain.

2. Results and Discussion

2.1. *ER fluids and test apparatus*

The investigated electrorheological fluid RheOil3.0 is developed for industrial applications based on ionically doped polyurethane particles (PUR) dispersed in silicone-oil as base fluid.

This type of ER-Fluid exhibits outstanding electrorheological properties, a widely temperature independent yield stress, non abrasive behavior and sufficient low base viscosity. It can be synthesized in a scalable process from standard raw materials. More data on RheOil3.0 is found in [3].

Measurements in flow mode were carried out in a custom made test rig at a constant flow rate of 10000 s^{-1} and defined temperatures of 20°C, 40°C and 60°C. In this set up, two identical and flat channel valves (electrode length of 100 mm, a width of 30 mm and gap height of 1 mm) are connected in series and alternately activated by an electric field, in order to minimize the pulsation of flow rate [5]. The data were recorded with a sampling rate of 50 kS/s and a maximum phase delay between pressure transducer and high voltage amplifier of 15 μs. The response to a square wave signal as well as to a sinusoidal frequency sweep of the pressure drop is measured at field amplitudes of 5 kV/mm.

Step response measurements in shear mode were executed by a commercially available rheometer (Physica/MCR 300) working with a standard Searle system in rotational mode together with a high voltage amplifier RheCon® (Fludicon). The sampling rate was 200 S/s.

2.2. *Results and Discussion*

The temperature influence on the step response was measured under the conditions of the pulsation free flow mode rig. Figure 1 shows step responses to a square wave signal at 20°C, 40°C and 60°C at a shear rate of 10000 s^{-1}.

As visualized best in the log-log representation, a typical step response contains a more or less pronounced short response time τ_1 that affects a fast change in the pressure drop to the level p_1 and a long response time τ_2 that affects the approach to the static value by definition according to [5].

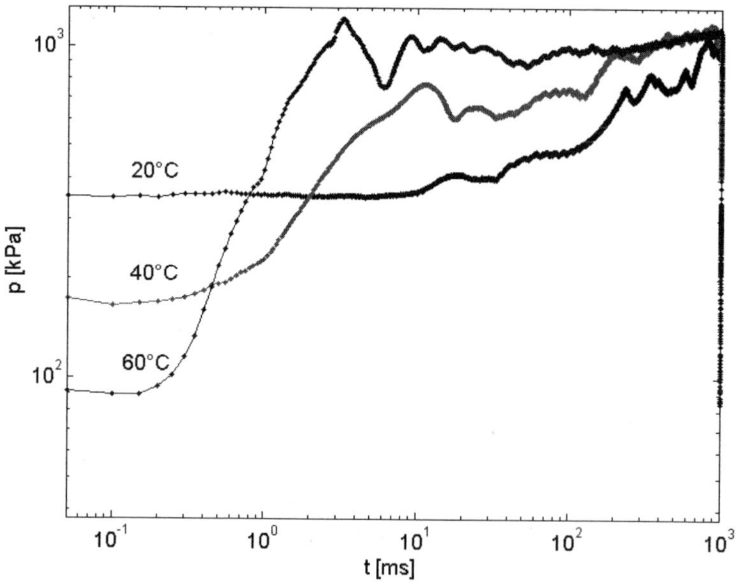

Figure 1: Log-log representation of the flow mode step response at 10000 s^{-1} and three different temperatures. A temperature dependant short response time τ_1 of e.g. 3 ms (10 ms) at 60°C (40°C) towards a pressure difference p_1 of 1200 kPa (800 kPa) is followed by a long response time τ_2 towards the static value. For 20°C, p_1 is not visible on this scale.

Qualitatively, for T < 30°C the step response is dominated by the rather long response time region, whereas for increasing temperatures above 30°C the short response time region increases together with p_1.

The results of frequency sweep measurements of the pressure difference and phase shift for 40°C and 60°C are shown in Figure 2. Corresponding to the step response measurements, the bandwidth increases significantly with temperature. Within limits, such a slope can be approximated by a second order low pass filter equivalent network.

Such a temperature activated response behavior can be explained by the ionic conductivity $\sigma(T)$ of the fluid, which in general increases exponentially with temperature due to enhanced charge mobility within the particles, according to Arrhenius' equation of the form

$$\sigma(T) = \sigma(T_0) \exp\left(\frac{b}{T+T_0}\right).$$

The dependency of step response time on conductivity will be verified in the following, by temperature independent measurements in shear mode. Preliminary, the dynamic limitations of the shear mode measurements had to be proved.

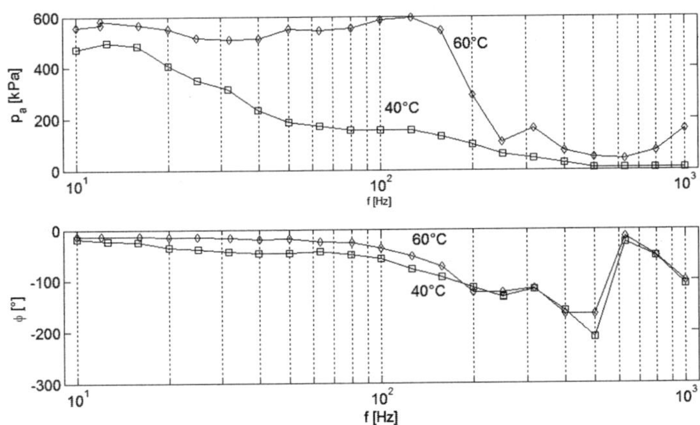

Figure 2: Frequency sweep measurements of the pressure drop amplitude and phase shift in flow mode at 10000 s^{-1} and two different temperatures.

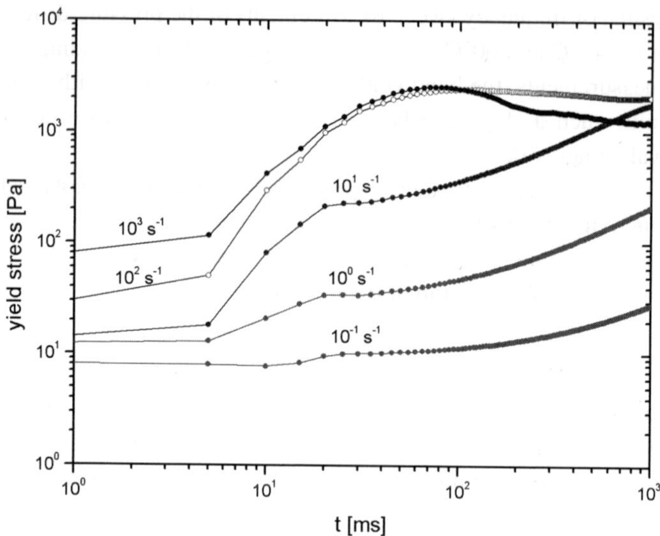

Figure 3: Log-log step response representation of RheOil3.0 yield stress in shear mode at T = 40°C and varying preset shear rates. Electric field step (5 kV/mm) is applied effectively at 5 ms.

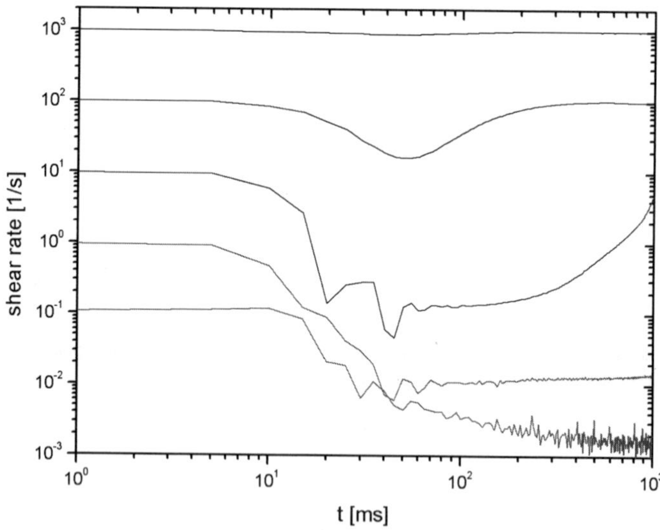

Figure 4: Shear rate instability after step response for the preset shear rates of 10^{-1} s^{-1}, 1 s^{-1}, 10 s^{-1}, 10^2 s^{-1} and 10^3 s^{-1}.

In shear mode, electric field induced changes in solid content within the cell are very limited compared to flow mode measurements, where fluid is exchanged continuously making particle accumulation upon dipole interaction possible. This generally results in a lower static yield stress of electrorheological fluids in shear mode, unless electrophoreses is dominant. Mode dependent influences on the step response behavior have not been reported yet. However, due to power limitations of the pump or motor and dynamical load changes, instabilities and pulsation of the flow or shear rate are expected. Unlike in flow mode, such instabilities are not compensable yet in shear mode. The resulting effects on a step response were studied by varying the preset shear rate in rotation. According to Figure 3 the step response seems to be shear rate limited at the first view. But due to the field enhanced thixotropic behavior of RheOil, the shear rate is not constant during the step response (Figure 4). For shear rates of 10 s^{-1} and above, a delayed relaxation to the preset shear value is observed. Below, the turning moment of the motor cannot overcome the thixotropic interaction forces since the apparent viscosity scales reciprocally with the shear rate. Conclusively, the shear rate up- and downturns seem to have retarding effects which probably dominate the step response behavior.

Despite this drawback, it is possible to get some qualitative information on the step response behavior at shear rate levels between $10^2...10^3 s^{-1}$, where a sudden shear rate drop relaxes back to the initial value within response time.

In order to verify the above conclusion about the conductivity influence on dynamic behavior, a systematic variation on ionic conductivity was performed by mixing two differently doped RheOil samples X and Y by the ratio %X/%Y, starting with 100 % pure RheOil X (100/0) with conductivity 10^{-9} S/m and ending with 100 % pure RheOil Y (0/100) with a conductivity of 10^{-8} S/m. Figure 5 shows the results of the step response at 20°C and 150 s^{-1}. In flow mode, the optimum ionic conductivity measured with respect to a fast step response is ~$5x10^{-8}$ S/m, which is in a good accordance to this result.

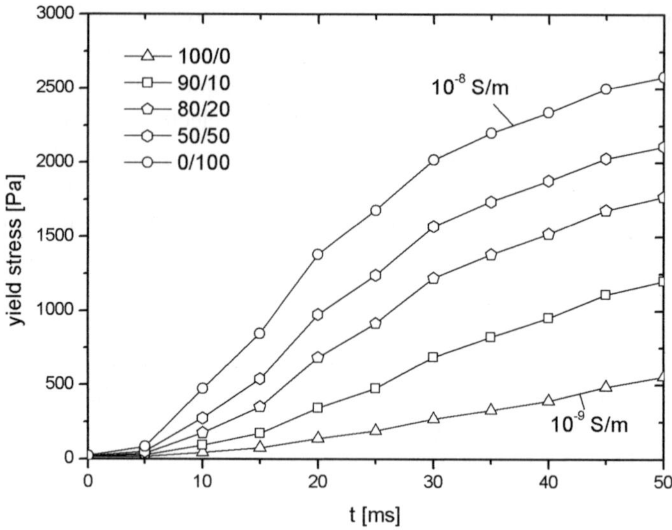

Figure 5: Fast step response region of a low ionic conductivity RheOil X (10^{-9} S/m) mixed with a high ionic conductivity RheOil Y (10^{-8} S/m). The numbers of the mixing ratio X/Y are in %. All values are measured at 20°C and 150 s^{-1}.

3. Summary and Conclusion

A step response can be characterized by an initial steep slope to a certain yield stress, usually within a few milliseconds, followed by a flat slope towards the final static value.

It was found, that step response time and bandwidth of RheOil are depending primarily on conductivity and as a consequence of Arrhenius' correlation on temperature. Faster response at constant temperature can be achieved via higher ionic conductivity (ion mobility).

One main obstacle in measuring an accurate step response time of electrorheological fluids is the pulsation of shear or flow rate after switching the electric field on and off. Pulsation can be suppressed in flow mode but not in shear mode yet, leading to response retardation as well as to yield stress peaks. The short step response time in shear mode seems to be one order of magnitude larger than in flow mode. However, qualitative conclusions from step response measurements in shear mode on a practical step response performance in e.g. flow mode seem to be possible.

References

[1] Holzmann K, Kemmetmüller W, Kugi A, Stork M, Rosenfeldt H and Schneider S, 2008 Modeling and control of an off-road truck using electrorheological dampers, Journal of Physics Conference Series, Volume 149 2009, 11th International Conference on Electrorheological Fluids and Magnetorheological Suspensions Dresden, Germany.

[2] Johnston L, 2008 Electrorheological Dampers for Industrial and Mobile Applications – an Overview of Design Variations, Product Realisation and Performance Actuator 2008, 11th International Conference on New Actuators, Proceedings 499.

[3] Gurka M, Johnston L, Petricevic R, New Electrorheological Fluids - Characteristics and Implementation in Industrial and Mobile Applications Journal of Intelligent Material Systems and Structures 2010 0: 1045389X09357974.

[4] Ulrich S, Bruns R, Böhme G, A New Electrorheological Actuator for Vibration Decoupling, will be published in: Actuator 2010, 12th International Conference on New Actuator, Bremen, Germany.

[5] Ulrich S, Böhme G, Bruns R, Measuring the Response Time and Static Rheological Properties of Electrorheological Fluids with Regard to the Design of Valves and their Controllers, Journal of Physics Conference Series, Volume 149 2009, 11th International Conference on Electrorheological Fluids and Magnetorheological Suspensions Dresden, Germany.

[6] Choi B H, Nam Y J, Park M K, Effect of Cluster Formation on Dynamic Response of an Electrorheological Fluid in Shear Flow, Journal of Physics Conference Series, Volume 149, 2009 11th International Conference on Electrorheological Fluids and Magnetorheological Suspensions Dresden, Germany.

A NOTE TO SECONDARY ELECTRORHEOLOGICAL PATTERNS

MARTIN STĚNIČKA, VLADIMÍR PAVLÍNEK, PETR SÁHA
and TAKESHI KITANO

*Polymer Centre, Faculty of Technology, Tomas Bata University in Zlin, TGM 275,
762 72 Zlin, Czech Republic
stenicka@ft.utb.cz*

PETR PONÍŽIL

*Institute of Physics and Materials Engineering, Faculty of Technology,
Tomas Bata University in Zlin, TGM 275
762 72 Zlin, Czech Republic*

The development and character of secondary electrorheological structures in the flow field under various conditions were investigated. For that purpose, suspensions of glass beads in two mediums significantly differing in viscosity have been used The influence of sequences of electric field and shear rate applications, as well as the intensity of the shearing with respect to development of electrorheological patterns is discussed.

1. Introduction

In principle, the electrorheological (ER) effect in suspensions manifests in a rapid organization of polarized particles into the chain-like or columnar structures in a direction of streamlines of external electric field [1]. When the flow is applied, superposition of shear and polarization forces sets in. Henley and Filisko [2] reported that in a flowing paraffin wax suspension of Amberlite cation exchange particles between parallel plates under electric field application a secondary structures of a series of black and white regularly spaced concentric rings were formed. The importance of the particle concentration and intensity of the electric field on the morphology of these structures was evident [3]. An evolution of ER structures under different flow conditions has been studied [4] and an analysis of ER patterns of modified materials employed as ER fluids has been performed [5, 6]. It was clear that the diameter of these ring structures, their thickness or spacing between them are complex functions of particle concentration, shear rate, electric field strength, dimensions of experimental geometry, duration of shearing, and probably many others, which have not been clarified so far.

In our previous works [7, 8], the mathematical analysis of ER patterns created in flowing suspension was presented. The current study was focused on an online monitoring of ER structures under various shearing conditions and applied electric field and their development in time. At the same time, the various sequences of the application of the electric field and shearing, as well as role of viscosity of continuous phase were analyzed.

2. Experimental

2.1. *Suspension preparation*

Suspensions (40 wt%) of glass beads (average diameter of 40 μm in two kinds of liquids namely silicone oil (Lukosiol, Chemical Works Kolín, Czech Republic, $\eta = 200$ mPa s) and liquid silicone elastomer base (SYLGARD, Dow Corning, USA, $\eta = 5\,500$ mPa s) were prepared (Table 1). After mechanical stirring, air bubbles were removed by vacuuming.

Table 1. Composition of suspensions.

code	solid phase	liquid phase	η (mPa s)
S1	glass beads	silicone oil	200
S2	(40 wt%)	silicone elastomer base	5 500

2.2. *ER measurement*

Rotational rheometer Bohlin Gemini (Malvern Instruments, UK) with parallel plates in diameter of 40 mm separated by 1 mm gap was used. Instrument was equipped with cell for ER experiments and modified for direct observation of generated structures. Geometry was connected to a DC high-voltage source TREK 668B (TREK, USA) providing the intensity of electric field strength $E = 1$ kV mm^{-1}. For the purpose of the online monitoring of the development of the ER structures, the original lower plate was replaced by transparent conducting glass sheet. Structure development was monitored via a USB microscope camera. Investigation was carried out under single shear rates ($g = 5$, 10, 20 and 40 s^{-1}) in controlled shear rate mode.

696

3. Results and Discussion

3.1. *Sequences of the electric field and shearing application*

For presentation of suspension structure development in time in a single picture collections of radial sectors corresponding to snapshots of captured video sequence has been used. When the time increases, the sectors follow anti-clockwise direction (Figure 1).

Fine structure of numerous rings appeared, when shearing started after electric field application (Figure 1a). On the other hand, lower number of wider rings joining together in time appeared while electric field was applied after shear beginning (Figure 1b).

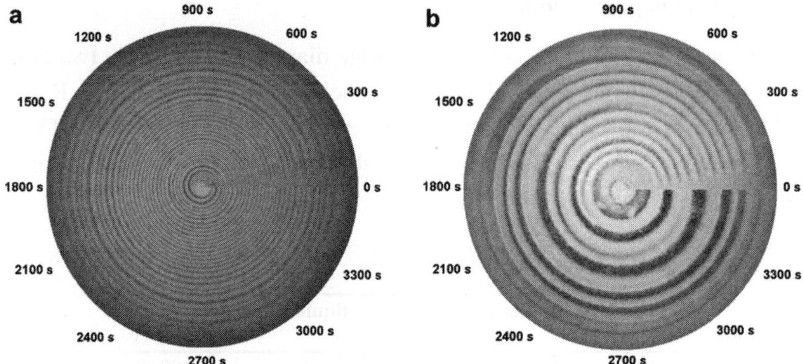

Figure 1 The radial sector artworks for suspension S2, where the electric field was applied for 60 s prior shearing ($E = 1\ \mathrm{kV\ mm^{-1}}$, $g = 10\ \mathrm{s^{-1}}$) (a) and in opposite way ($g = 10\ \mathrm{s^{-1}}$, $E = 1\ \mathrm{kV\ mm^{-1}}$) (b).

The development and consequent differences in particle organization in patterns was naturally reflected in different rheological behavior of suspensions in time (Figure 2). When the electric field was applied before shearing, the viscosity initially steeply increased and then slightly declined in time up to the stable state. In contrast, the application of the electric field to pre-sheared suspensions caused a smooth viscosity increase virtually from the very beginning. This increase probably results from transformation and merging of the ring structures together (Figure 1b).

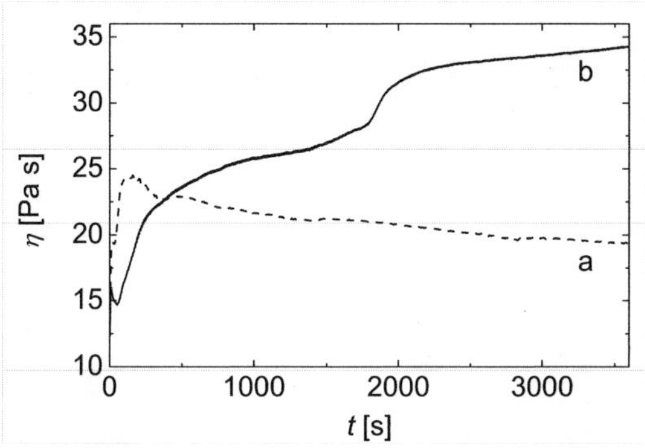

Figure 2 The time, t, dependence of the shear viscosity, η, where the electric field was applied for 60 s prior shearing ($E = 1$ kV mm^{-1}, $g = 10$ s^{-1}) (a) and in opposite sequence ($g = 10$ s^{-1}, $E = 1$ kV mm^{-1}) (b).

3.2. *Viscosity of liquid phase*

It appeared that the viscosity of liquid phase plays an important role in the development of secondary ER patterns as shown in Figures 3 and 4. In the case of low viscosity medium (suspension S1), the structures developed almost immediately throughout the sample after the shearing application. In high viscosity liquid phase (suspension S2), where the friction forces affecting the motion of the particles were higher, the development of the circles occurred much slower.

3.3. *Intensity of the shearing*

The important role of the shear rate in the development of ER patterns can be observed in Figures 5 and 6. It can be stressed out that the higher shear rate causes merging of the circles; it means increase in the ring width and decrease of their number. Also, the disappearance of the circle structures in outer part of the sample prepared from the low viscosity medium (suspension S1) sets in shorter times as the intensity of shearing increases (Figure 5). In contrast, circles in sample with high viscosity medium (suspension S2) are stable in time in all range of the shear rates used (Figure 6).

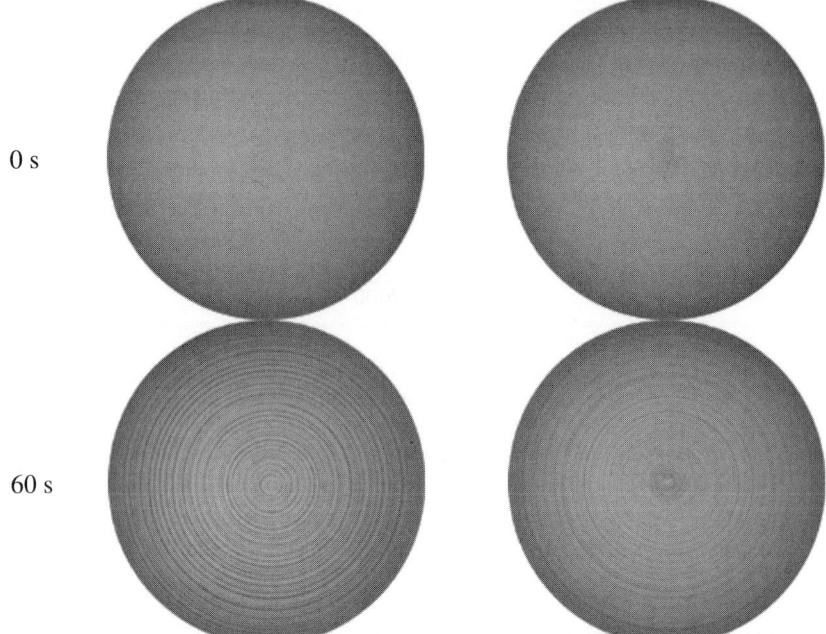

0 s

60 s

Figure 3 Development of circle patterns in time for suspension S1. Electric field was applied for 60 s prior shearing ($E = 1$ kV mm^{-1}, $g = 10$ s^{-1}).

Figure 4 Development of circle patterns in time for suspension S2. Electric field was applied for 60 s prior shearing ($E = 1$ kV mm^{-1}, $g = 10$ s^{-1}).

3.4. *Steady shear*

The ER behavior of suspensions presented as the shear rate dependence of the shear viscosity in the absence and in the presence of electric field ($E = 1$ kV mm^{-1}) is shown in Figure 7. Both suspensions demonstrated nearly Newtonian behavior in the absence of the electric field. After the application of the electric field, the flow curves of suspensions dramatically changed and showed strong non-Newtonian character.

The relative increase in viscosity of suspension (S1) in low viscosity medium was higher as a consequence of easier mobility of particles in the surrounding medium. In contrast in the case of suspension S2, the ER efficiency was markedly reduced at higher shear rates due to disintegration of ER structures. Despite this fact, in this shear rate range ring pattern structures are still developed as shown in Figure 6b.

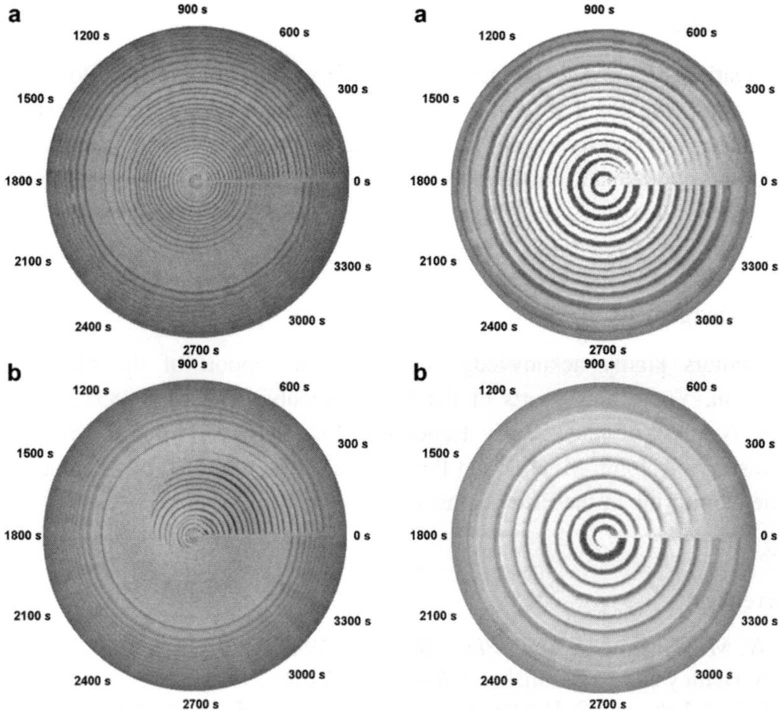

Figure 5 The radial sector artworks in time progress for suspension S1 at various shear rates: $g = 5$ s^{-1} (a) and 40 s^{-1} (b).

Figure 6 The radial sector artworks in time progress for suspension S2 at various shear rates: $g = 5$ s^{-1} (a) and 40 s^{-1} (b).

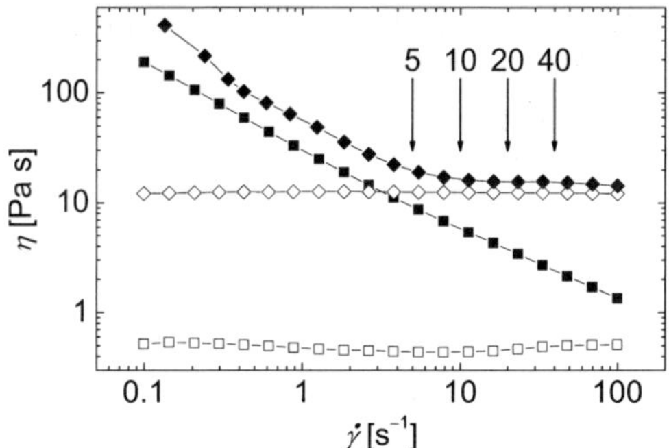

Figure 7 The dependence of the shear viscosity, η on the shear rate, g in the absence, $E = 0$ kV mm^{-1} (open) and in presence of electric field $E = 1$ kV mm^{-1} (solid). Symbols: ■□ S1 and ◆◇ S2.

4. Conclusion

Investigation of the development of secondary ring structures in flowing ER suspensions revealed a very important effect the sequence of the application of the electric field and the shear rate. Also viscosity of liquid medium significantly affects the number and thickness of the rings in developed structures. Finally, higher shear rate causes merging of the rings to the thicker structures.

Acknowledgement

The authors kindly acknowledge the financial support of the Ministry of Education, Youth and Sports of the Czech Republic (MSM 7088352101), the Grant Agency of the Czech Republic (202/09/1626). Special thanks are addressed to Antonín Minařík and Petr Smolka, who modified the rheometer for online monitoring of the ER structures.

References

1. W. M. Winslow, *J. Appl. Phys.* **20**, 1137 (1949).
2. S. Henley and F. E. Filisko, *J. Rheol.* **43**, 1323 (1999).
3. F. E. Filisko and S. Henley, *Int. J. Mod. Phys. B* **15**, 686 (2001).
4. X. Tang, W. H. Li, X. J. Wang and P. Q. Zhang, *Int. J. Mod. Phys. B* **13**, 1806 (1999).
5. F. E. Filisko, S. Henley and G. Quist, *J. Intell. Mater. Syst. Struct.* **10**, 476 (1999).
6. S. Henley and F. E. Filisko, *Int. J. Mod. Phys. B* **16**, 2286 (2002).
7. P. Ponížil, V. Pavlínek, T. Kitano and T. Dřímal, *Int. J. Math. Models Meth. Appl. Sci.* **1**, 239 (2007).
8. P. Ponížil, M. Stěnička, V. Pavlínek, T. Kitano and T. Dřímal, *J. Phys.: Conf. Ser.* **149**, 012024 (2009).

RELATION OF SLIP LENGTH AND SHEAR THINNING OF PM-ER FLUIDS[*]

J. ZHENG, C. LI, Q.G. TANG, R.D. ZHANG and L.W. ZHOU

State Key Laboratory of Surface Physics and Department of Physics, Fudan University, Shanghai 200433, China

J.H. WU, G.J. XU

Ningbo Institute of Material Technology and Engineering, Chinese Academy of Sciences, Ningbo 315201, China

Particle slip on boundaries would strongly affect the strength of electrorheological (ER) effect leading to shear thinning phenomenon under high shear rate. We modified a confocal microscope allowing samples to experience high electric field and strong shearing. This white light scattering confocal microscopy makes possible to view three dimensional lamellar structures of polar molecular electrorheological (PMER) fluids under both electric and shear fields. The particle slip length on boundaries is obtained and compared with the viscosity change measured separately to find out their relation which is further compared with the corresponding result of molecular dynamics simulation.

1. Introduction

Under electric field, the viscosity of almost all ER fluids decreases with increasing shear rate, and this effect of shear thinning become severe when particles slip on electrodes. Shear thinning is one of major obstacles towards wide applications of ER fluids as in many industries ER fluids are requited to circulate in high shear rate. It was proposed that shear slip instability may be the cause of phase bifurcation in ER fluids. [1, 2] It was pointed out [3] that the absence of an observable yield stress suggests that the colloidal particle may slip at the plates even at quite small stresses. Precise measurement of particle linear speed is necessary to detect slip length which is important to understand the mechanism of shear thinning. To understand the mechanism of shear thinning, it is desired to study the relation between the shear thinning and particle slip on boundaries.

---

[*] This work is supported by grants 10974030 and 10574027 of NNSF of China.

In the study of ER effect, non-slip boundary conditions of both liquid and particles are usually adapted for the simplicity in numerical calculations and model simulations. It is necessary to do the simulation with different condition of particle slip on boundaries.

The viscosity of ER fluids would decrease dramatically when the fluids work under high shear rate, which known as shear thinning. The reason why shear thinning happens is still not completely understood. In our research, we use the conforcal microscope under application of electric field and shear flow to observe the three dimensional lamellar structures [4,5,6,7] of the ER fluids, and this will enable us to explore the mechanism of shear thinning of ER fluids.

The columnar structure will break when sheared. However, it is not very clear where the columnar structure breaks from the experiment as they might break at the boundaries or somewhere in the middle of column. As we observed, there are both moving rings and steady rings in lamellar structure. One of the reasons may be that the columnar structure in the steady rings break in the middle (non-slip on boundaries), and the columnar structure over the rotating rings break near the bottom plate (particles slip on boundaries).

The degree of the slip is also unknown, which is related to the interaction between the fluid and the wall. We can use slip length [8,9] to describe the degree of particle slip on boundaries. When the interaction between the fluid and the wall is less than certain value, the liquid would slip on the solid wall, namely the fluid speed on the solid-fluid boundary is not zero. The virtual point where fluid speed is zero would be found by extending the line of the speed profile. The intersection is defined as a slip length, b.

2. Experiments

The PMER fluid consisted of TiO_2 coated with 1,4-butyrolactone molecules prepared using the sol-gel method [10] and silicone oil (100cSt). The mixture ratio of the powder to silicone oil is 1.88 g/ml. The viscosity is characterized with a Haake-Mars II electrorheometer with two ITO glass discs as electrodes.

The three dimensional photos of lamellar structure of the PMER fluid are obtained with an Olympus IX71 conforcal microscope under application of electric field and shear flow. A rotary disk has a pattern of linear lines that provides the effect equivalent to pin hole by high speed rotation. Light that passes through the pattern further passes through objective lens and goes to the specimen. The both disk electrodes are made of ITO glass. An LED ring light is used as a light source and another LED focal light functions as an auxiliary light source. Using a $4\times$ object lens, we are able to obtain white light scattering conforcal microscopic images to study the three dimensional particle structure of PMER fluids under both electric field and shear rate.

2.1. *White light scattering confocal microscope*

An attention was paid on the refraction index match. It was proved that when the volume fractions of particles of polar molecules coated TiO_2 particles was about 1%, clear images of particles at different depth were able to be taken while particles in other layers were out of focus. Images of well index matched samples of polar molecules modified SiO_2 particles with silicone oil or UV glue were also tested. A white light, rather than monochromic light is used as power source, and it is proved that sharp images of different depth can be taken. When the sample thickness is 0.8mm, images of the ER fluid in different layers were recorded by the CCD.

2.2. *Images of the PMER fluid at different depth*

Using the inverted electrorheoscope, we recorded images viewed from the lower electrode, and measured shear stress at the same time at different field while the rotational speed is kept at 300rpm. Fig. 1 is a series of photos of ER fluid under an electric field of 1kV/mm and a rotating speed of 313rpm. The number under each picture is the height of the layer where the object lens focuses. These photos are taken under an exposal time of 300ms.

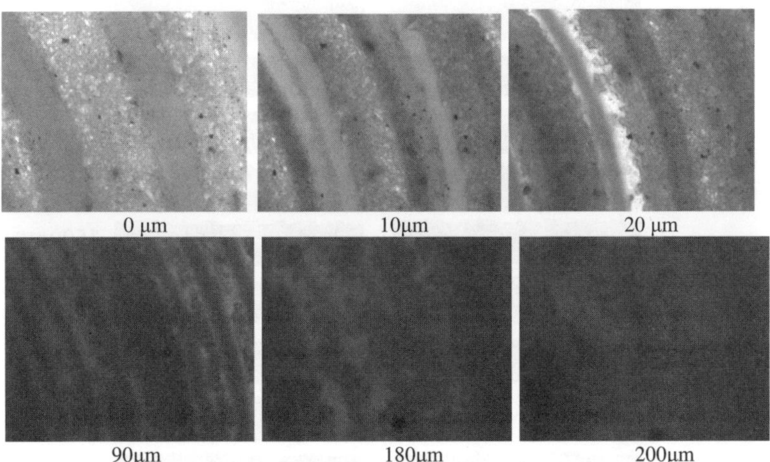

Fig. 1   White light scattering confocal microscope images of different layers of the lamellar structure of the PMER fluid when E=1000 kV/mm and rotational speed is 313rpm. The value in micron under each photo is the distance from the bottom layer.

We have also taken the photos of ER fluids under an electric field of 1kV/mm and a rotating speed of 100 and 200 rpm. Obviously, the ER fluid under a higher rotating speed has a more complex 3D structure. The reason may be that the columnar structure will not be broken under a low rotating speed, which means it may slide on the bottom.

3. Discussions

3.1. *Layers of 3D lamellar structure are grouped*

In order to view clearly the change of particle structure, the pictures in Fig. 1 are trimmed and arranged layer by layer from the bottom, as shown in Fig. 2. The red bars above each picture mean the rings below are moving, and particle structures without bars mean steady rings. In each picture, the left side is nearer to the center of rings.

We can see that the particle structures are grouped, and there are five groups in the layer ranges of 0~10, 20~30, 40~90, 110~130 and 140~170μm. Within each layer group, the structure character of moving and steady rings looks the same. The ring structure changes dramatically cross the thickness of 30~40μm. From such information we hypothesize that in this experiment, the columnar structure breaks there, and quite near to the bottom.

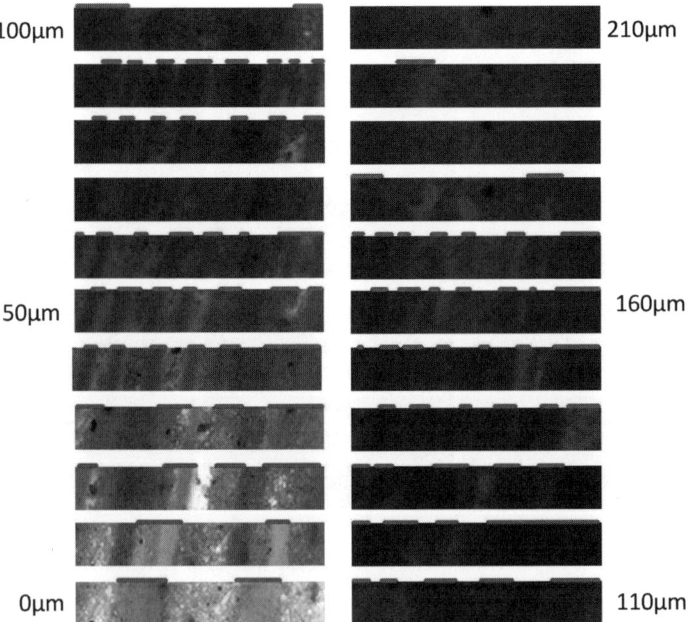

Fig. 2 A stack of white light scattering confocal microscopic images of 3D lamellar structure. Numbers on both sides indicate the height counting from the lower electrode. The applied electric field is 1000V/mm and the rotational speed is 300 rpm. In the image the width of the real sample is 1.969mm.

We took one movie of each group to measure the average rotational speed of moving rings in that movie, and we use this to represent the rotational speed of that group.

The material used is not thin enough so that the ring light cannot illuminate it well. As a result, we have to take photos in a long exposal time (about 200~300ms) and the definition of the photos decreased. That may influence the date processing and analyses since a single particle would look as a short line rather than a sharp spot in such long exposure time.

It takes some time before we can proceed to the next layer (~10s), so we cannot take all layers' movies at the same time. The lens amplification of $4\times$ is still too large and very little sample area of about $1.56\text{mm}\times1.95\text{mm}$ is captured by CCD, and a lens with less amplification factor is desired.

However, the rotating speed of the rotor is 313 rpm or 32.78 rad/s, which is a hundred times larger than that of particles in rotating rings. The error may come from the low frame speed of the CCD we used in the experiment. Through these work, we can infer that the ER fluid actually have a 3D structure and every layers in the lamellar structure is not always the same.

3.2. *Slip lengths and their comparison with shear thinning*

We separately measured viscosity (Fig.3) and the speed of some of the particles in the lowest few layers. From the three dimensional lamellar structure, one can detect moving speed of particles on each layer from bottom and up. After doing a linear fit of the particle speed in first few layers, one may find out the particle slip length b (Fig. 4) which is the intercept of the line on the depth axis z when the particle speed is zero.

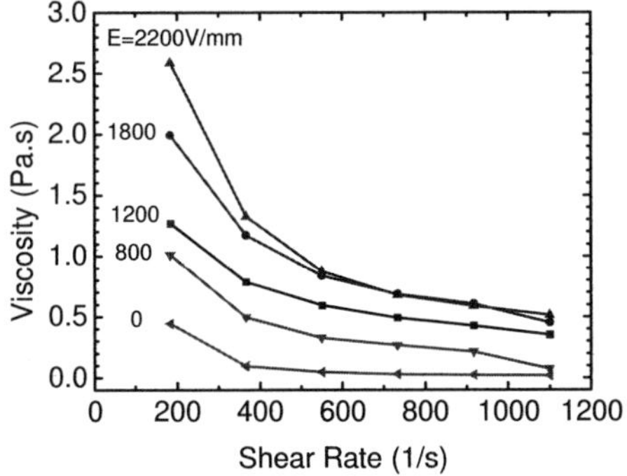

Fig. 3   Viscosity versus average shear rate of the PMER fluid of concentration of 1.88 g/ml at different fields.

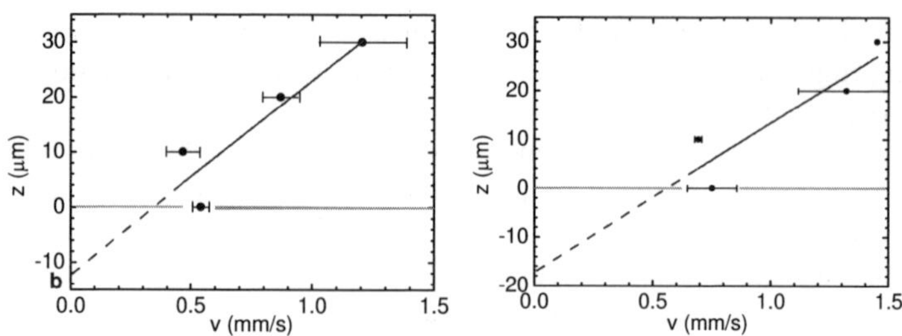

Fig. 4   Slip lengths of the PMER sample under a field of 1000V/mm at a rotational speed of 300 rpm.

It is found that the particle slip length is 0, 0 and 14 μm when rotational speed is 100, 200 and 300 rpm, respectively. From Fig. 3 one may found that the corresponding slope of viscosity versus shear rate is -5.2×10^{-3}, -1.4×10^{-3} or -0.7×10^{-3} Pa/s^2 accordingly. The viscosity slope indicates how fast the shear thinning approaches when shear rate increases.

3.3. *Simulation on relation of slip lengths and shear stress*

We have done a molecular dynamics simulation to find out the relation of slip length and corresponding shear stress using the same condition as in the experiment. The model PMER system consists of 240 interacting spherical particles of size of 1μm, dielectric constants ε_p=37, ε_f=2.5 in a liquid with η_f = 100cSt, while the volume fraction is 30%. Since our system is symmetrical, we have simulated the radial long box which greatly reduced our calculation. The external electric field is applied along the z direction. Every state was calculated for 8s from equilibrium. In our simulation, slip lengths $b = v_0/(dv/dz)$, where v_0 is the velocity of particles relative to electrode plate, and dv/dz is the velocity gradient along the field. b=5, 11 and 17μm were used. The shear stress is proportional to the energy density - $p_0 \cdot E$, where p_0 is a saturated polarization. [11] The shear stress has a similar form as in Ref. 12.

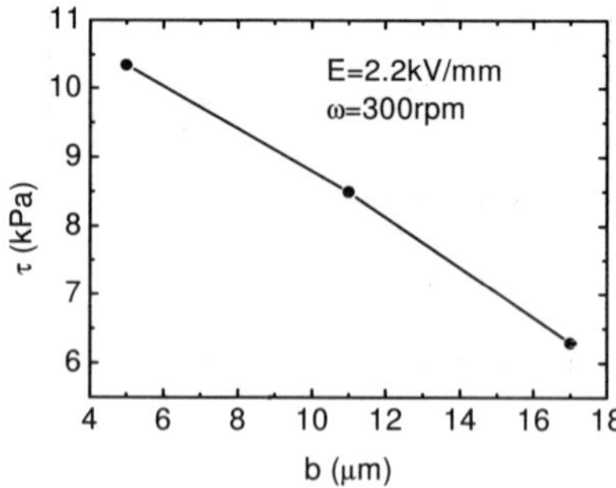

Fig. 5   The result of numerical calculation of shear stress for different slip length b.

We calculated the evolution of lamellar structure with the external electric field and shear field. The results are shown in Fig.5. It is found that, under the same shear and electric fields, the shear stress decreases as the slip length increases. In the frame of Binghan model, the relationship between viscosity and b can also be found.

4. Conclusions

Three dimensional lamellar structures of PMER fluids under both electric and shear fields are obtained via white light scattering confocal microscopy. Particle slip lengths on boundaries are deduced after the particle moving speed in the first few layers near the stator electrode is detected. Simulation of molecular dynamics simulation has also performed to draw a relation the rheological properties and particle slip lengths on boundaries. It is found that increasing shear rate would increase slip lengths, and it would decrease the viscosity of PMER fluids leading to shear thinning. Further work, both experimental and theoretical, is urged to find out a quantitative relation between the slip length and viscosity, and propose suggestions improving the functional materials.

Acknowledgments

The authors wish to thank the stimulating discussions and fruitful help form J.P. Huang, J.Y. Chen, W. Chen, P. Tan, H.Z. Zhou.

References

1. J.E. Martin, J. Odinek, T.C. Halsey, R. Kamien, Phys. Rev. E 57, 756 (1998).
2. D.F. Klingenberg and C. F. Zukoski IV, Langmuir 6, 15 (1990).
3. T.C. Halsey, J.E. Martin and D. Adolf, Phys. Rec. Lett. 68, 1519 (1992).
4. S. Henley and F. E. Filisco, Inter. J. Mod. Phys. B 16, 2286(2002).
5. X. Tang, W. H. Li, X. J. Wang, P. Q. Zhang, Inter. J. Mod. Phys. B 13, 1806 (1999).
6. K.V. Pfeil, M.D. Graham, and D.J. Klingenberg, Phys. Rev. Lett. 88, 188301 (2002).
7. R. Tao, Chem. Eng. Sci. 61, 2186 (2006).
8. L. Ren and K.Q. Zhu, Smart Mater. Struct. 15, 1794 (2006).
9. P. Sheng, W.J. Wen, Solid State Comm. 150, 1023 (2010).
10. L. Xu, W. J. Tian, X. F. Wu, C. X. Wang, J. G. Cao, L. W. Zhou, J. P. Huang and G. Q. Gu, J. Mater. Res. 23(2) 409 (2008).
11. W. J. Wen, X.X. Huang, S. H. Yang, K. Q. Lu and P. Sheng, Nature Materials, **2**, 1 (2003).
12. J.G. Cao, J.P. Huang, L.W. Zhou, J. Phys. Chem. B 110, 11635 (2006).

THE CAVITY GENERATED WITH WATER ENTRY OF A MAGNETIC FLUID COATED CYLINDRICAL MAGNET[*]

SEIICHI SUDO[†], HIRAYOSHI TAKAYANAGI and TETSUYA YANO

*Department of Machine Intelligence and Systems Engineering,
Akita Prefectural University, Ebinokuchi 84-4, Tsuchiya
Yurihonjoh, Akita Prefecture 015-0055, Japan*

HIDEYA NISHIYAMA

*Institute of Fluid Science, Tohoku University, Katahira 2-1-1, Aobaku
Sendai, Miyagi Prefecture 980-8577, Japan*

This paper presents an experimental study on the cavity generated with water entry of a magnetic fluid coated cylindrical magnet. The permanent magnet is cylindrical NdFeB magnet with the magnetic flux density B=390mT. The kerosene-based magnetic fluid is adsorbed to the permanent magnet. This projectile collided with the surface of water in the rectangular container made of the transparent acrylic plastic. The cavity of air in water created by the impact of the magnetic fluid coated cylindrical magnet was observed with high-speed video camera system. The experiments were performed under non-magnetic and alternating magnetic fields. The effect of the hydrodynamic forces on the cavity formation was clarified. The effect of the alternating magnetic fields on the pinch-off time of an air cavity was also revealed.

1. Introduction

The impact problem of a solid object or liquid droplet on a free surface of water has received a considerable amount of attention in a wide variety of field. The first systematic study of drop impacts was investigated by Worthington [1]. After his pioneering work, extensive investigations on the impact phenomena of liquid drops with solid and liquid surfaces have been conducted by a number of researchers. Reviews of the liquid drop impact problem on a liquid surface and on a solid surface can be found in Prosperetti & Oguz [3] and Yarin [3]. Authors also reported on the drop impact phenomena of magnetic fluid with a liquid surface [4] and solid surfaces [5,6] under magnetic fields. In recent years, a lot of studies on the air cavity created by water entry of solid body were reported.

---

[*] This work is partially supported by Grant-in-Aid for Scientific Research (C) (22560173).

[†] Work partially supported by grant of Collaborative Research Project, The Institute of Fluid Science, Tohoku University.

Duclaux et al. studied the collapse of a transit cavity of air in water created by the impact of a solid body [7]. Aristoff & Bush extend the work of Duclaux et al. [7] by including the effects of surface tension and aerodynamic pressure on the cavity evolution [8]. Truscott & Techet studies the complex hydrodynamics of water entry by a spinning sphere for low Froude number [9]. Grumstrup et al. reported the experimental observation of a well-defined rippling of the air cavity entering the free surface of a liquid (water or ethanol) [10]. Akers & Belmonte reported an experimental study of the impact of a solid sphere on the free surface of a viscoelastic wormlike micellar fluid [11]. However, no study has been published on the impact phenomena of a magnetic fluid-solid compound projectile on a water surface except for our previous paper [12]. In our previous paper, impact phenomena of a spherical permanent magnet coated with magnetic fluid has been investigated.

In the present paper, some aspects of the cavity created by the impact of the magnet-magnetic fluid projectile on the free surface of water were studied with a digital high-speed video camera system. The projectile is composed of a cylindrical permanent magnet and kerosene-based magnetic fluid. The effect of the magnetic fluid adsorption on the cavity was clarified. The effect of the alternating magnetic field on the projectile was also revealed.

2. Experimental Apparatus and Procedures

A schematic diagram of the experimental apparatus is shown in Fig.1. Experiments on water entry of magnetic fluid coated cylinder were performed using NdFeB permanent magnet and kerosene-based magnetic fluid. The surface of the magnet is smooth by nickel plating. The diameter of the cylindrical magnet is d_c=5mm and the length is h_c=3mm. The magnetic flux density of the magnet is B=390mT. Figure 2 shows the size and poles of two kinds of projectiles used in the experiments. In Fig.2, the cylindrical magnet (a) was used to compare the effect of magnetic fluid adsorption. The volume of magnetic fluid adsorbed to the cylindrical permanent magnet is V_m=1×10^{-7} m^3(100 μ l). This volume can completely cover the cylindrical magnet with magnetic fluid. The physical properties of magnetic fluid used in the experiment are shown in Table 1. This magnetic fluid is immiscible with water.

The magnet-magnetic fluid projectile was released out of the saucer towards the water in the rectangular container made of the transparent acrylic plastic with dimension 113mm in height and 76mm×76mm wide. The depth of the target liquid is 60mm. The impact velocity of the projectile was varied by the change of fall height H. Free fall of the projectile was performed by suddenly

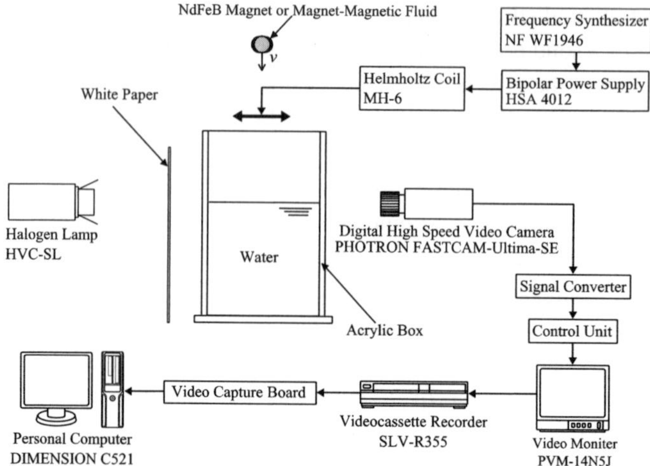

Figure 1. Schematic diagram of the experimental apparatus.

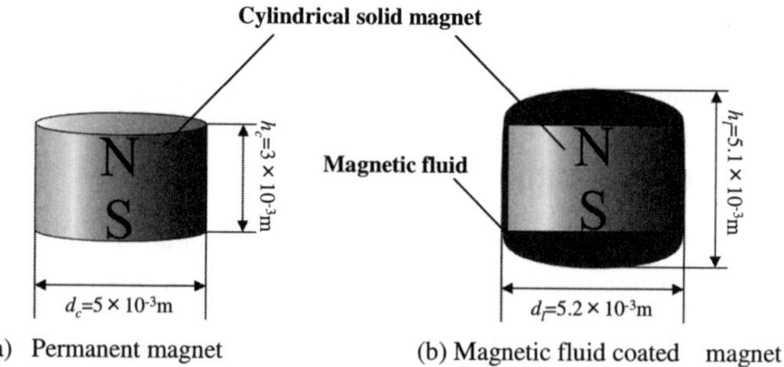

(a) Permanent magnet (b) Magnetic fluid coated magnet

Figure 2. Two kinds of projectiles used in the experiments.

Table 1. Physical properties of kerosene-based magnetic fluid.

Magnetic fluid	Ferricolloid HC-5
Density	1250 kg/m^3
Viscosity	9 mPa·s
Saturation magnetization	32 kA/m
Surface tension	0.0277 N/m

removal of the saucer. The cavity created by water entry was recorded with the high–speed video camera system. The impact phenomena of magnetic fluid coated cylinder were recorded at 4500 frames per second. A series of frames of the cavity behavior was analyzed by the personal computer. The effect of alternating magnetic fields on the water entry phenomena was also examined. The alternating magnetic field was generated by applying alternating current voltage to the Helmholtz coil. The direction of the magnetic field was parallel to the water surface. In this case, the magnetic torque T acting to the permanent magnet with magnetic moment m in the external magnetic field H_{mag} is given as follows;

$$T = m \times H_{mag} \tag{1}$$

The permanent magnet shows the rotational oscillation according to the direction of the alternating magnetic field. To compare the effects of magnetic fluid adsorption, the water entry phenomena created by the impact of dry permanent magnet were also examined.

3. Experimental Results and Discussion

3.1. *Cavity Created by Water Entry of Cylindrical Permanent Magnet*

First of all, the impact of the dry cylindrical magnet on the water surface was examined. Three typical impact phenomena of the dry cylindrical magnet on the water surface are presented in Fig.3. In Fig.3, the solid cylindrical projectiles were released from the fall height H=0.2m above the water surface. The condition of fall height H=0.2m gives the impact velocity of v_0=1.8m/s to the projectile (for Froude number F_r=v_0^2/gd_c=66; where g is the gravitational acceleration). The time step between images is δ_t =2ms. These photographs show the evolution of the cavity. In Fig.3, (a) shows the impact under the non-magnetic field, (b) shows the impact under the alternating magnetic field with the frequency f_0=45Hz, and (c) shows the impact at f_0=90Hz. The cavity formed by the water entry of cylindrical magnet under non-magnetic field shows the asymmetric growth (Fig.3 (a)). The asymmetric growth of cavity is caused by the asymmetric impact of projectile and the unsteady hydrodynamic drag acting to the projectile afterwards. Under the alternating field at f_0=45Hz, the cavity wall shows the rough texture (Fig.3 (b)). This kind of cavity growth is brought by the rotational oscillation of the projectile. The rotational oscillation of the projectile is generated by the magnetic torque described by Eq. (1). Therefore, the cavity is divided intricately as compared with the case of non-magnetic field. Under the alternating field at f_0=90Hz, the cavity growth is narrower compared with the above-mentioned conditions. In this case, the volume of air entrainment is decreased. A fast rotation of the projectile brings the symmetry of the cavity.

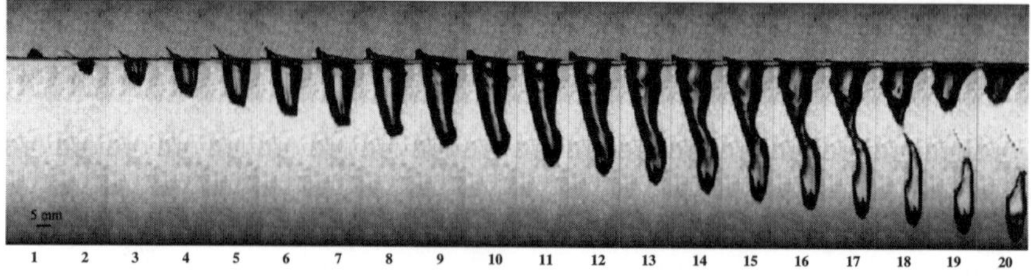

(a) Non-magnetic field (f_0=0Hz)

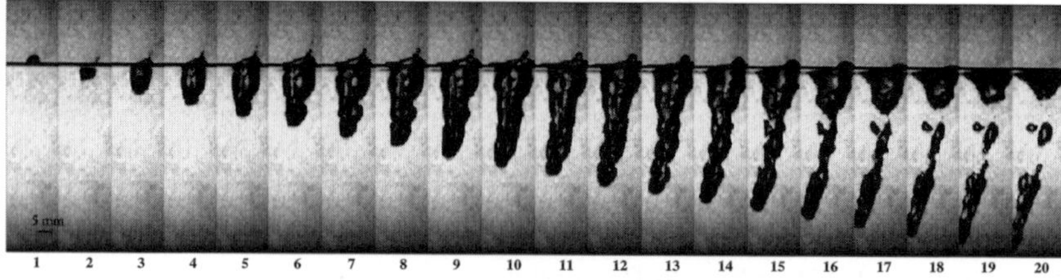

(b) Alternating magnetic field (f_0=45Hz)

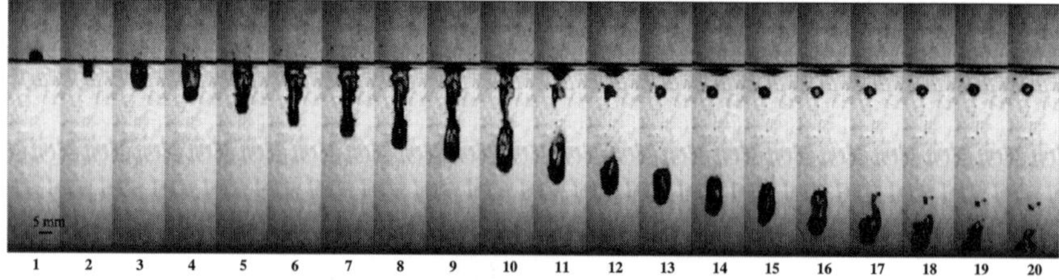

(c) Alternating magnetic field (f_0=90Hz)

Figure 3. Influence of magnetic fields for the impact phenomena of a cylindrical permanent magnet.

3.2. *Cavity Created by Water Entry of Magnetic Fluid Coated Cylindrical Magnet*

In this paragraph, the cavities created by water entry of magnetic fluid coated cylindrical permanent magnet were examined. Figure 4 shows a sequence of photographs of the water entry phenomena created by the magnetic fluid coated cylindrical magnet in each magnetic field condition. In Fig.4, the solid-liquid compound projectiles were released from the fall height H=0.2m above the water surface. The time step between images is δ_t =2ms. In Fig.4, (a) shows the impact under the non-magnetic field, (b) shows the impact under the alternating magnetic field with the frequency f_0=45Hz, and (c) shows the impact at f_0=90Hz.

The cavity formed by the water entry of magnetic fluid coated projectile under non-magnetic field shows the wider and longer growth (Fig.4 (a)) compared with Fig.3 (a). It can be seen from Fig.4 that the cavity growth is promoted very much by the adsorption of magnetic fluid. The impression of the alternating magnetic field causes the rough texture to the cavity wall. Especially, a lot of minute droplets of magnetic fluid are generated from the cavity wall into the water at f_0=45Hz. In general, the normal-field instability occurs when a horizontally extended layer of magnetic fluid is placed in a magnetic field oriented normally to the flat fluid surface [13]. Under the alternating field at f_0=90Hz, a netted shade texture is observed on the cavity wall. Furthermore, the alternating magnetic field promotes the pinch-off of the cavity (Fig.4 (c)).

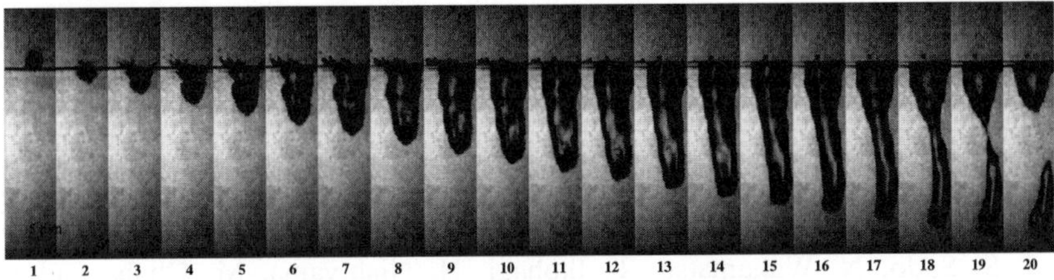

(a) Non-magnetic field (f_0=0Hz)

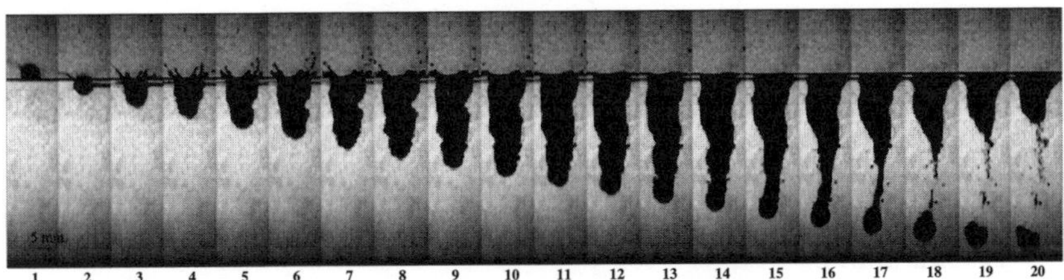

(b) Alternating magnetic field (f_0=45Hz)

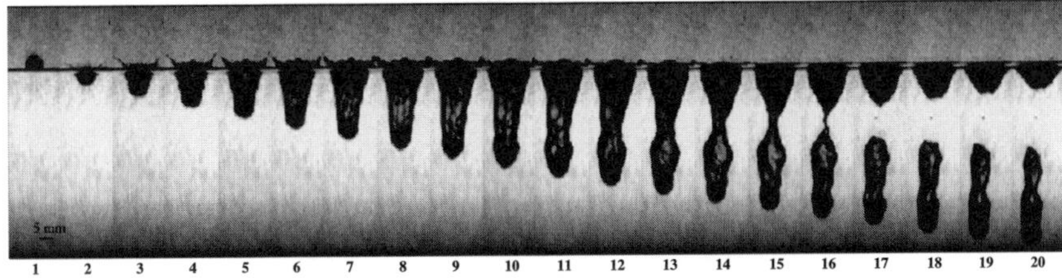

(c) Alternating magnetic field (f_0=90Hz)

Figure 4. Influence of magnetic fields for the impact phenomena of the magnetic fluid coated cylindrical permanent magnet.

4. Conclusions

The cavity generated with water entry of the magnetic fluid coated cylindrical magnet was investigated with high-speed video camera system experimentally. The results obtained are summarized as follows;

(1) The impact of the magnetic fluid coated cylindrical magnet on the free surface of water promotes the cavity growth compared with the impact only of the dry magnet.

(2) The alternating magnetic field applied to the magnetic fluid coated cylindrical magnet causes the rough texture to the cavity wall.

(3) The alternating magnetic field applied to the magnetic fluid coated cylindrical magnet promotes the pinch-off of the cavity.

References

1. A. M. Worthington, *A Study of Splashes*, Longmans & Green, London, 1908.
2. A. Prosperetti and H. N. Oguz, *Ann. Rev. Fluid Mech.* **25**, 577 (1993).
3. A. L. Yarin, *Ann. Rev. Fluid Mech.* **38**, 159 (2006).
4. S. Sudo, M. Yamabe, H. Hashimoto and K. Katagiri, *Magnetohydrodynamics* **32**, 494 (1996).
5. S. Sudo, N. Wakamatsu, T. Ikohagi, H. Nishiyama, M. Ohaba and K. Katagiri, *J. Magn. Magn. Mater.* **201**, 285(1999).
6. S. Sudo, M. Funaoka, H. Nishiyama and K. Katagiri, *Ener. Conver. Manag.* **43**, 289(2002).
7. V. Duclaux, F. Caille, C. Ybert, L. Bocquet and C. Clanet, *J. Fluid Mech.* **591**, 1(2007).
8. J. M. Aristoff and J. W. M. Bush, *J. Fluid Mech.* **619**, 45(2009).
9. T. T. Truscott and A. H. Techet, *J. Fluid Mech.* **625**, 135(2009).
10. T. Grumstrup, J. B. Keller and A. Belmonte, *Phys. Rev. Lett.* **99**, 114502(2007).
11. B. Akers and A. Belmonte, *J. Non-Newtonian Fluid Mech.* **135**, 97(2006).
12. S. Sudo, H. Takayanagi and S.Kamiyama, *Abstracts of 12th International Conference on Magnetic Fluids*, 38(2010).
13. C. Gollwitzer, G. Matthies, R.Richter, I. Rehberg and L.Tobiska, *J. Fluid Mech.* **571**, 455(2007).

ELECTRORHEOLOGICAL PROPERTIES OF AMORPHOUS TITANIUM OXIDE PARTICLES WITH DIFFERENT SIZES

FENGHUA LIU, GAOJIE XU, JINGHUA WU, YUCHUAN CHENG,
JIANJUN GUO and PING CUI

*Ningbo Institute of Material Technology and Engineering (NIMTE), Chinese Academy of
Sciences (CAS), Ningbo 315201, People's Republic of China*

A kind of amorphous titanium oxide particles with different sizes (30nm, 80nm, 240nm and 640nm) was prepared using a hydrolysis method. The microstructure, dielectric properties and ER performance were investigated. The results show that the suspensions composed of smaller particles have higher yield stress at DC electric field, but also possess higher zero field viscosity. The maximal ER efficiency of the ER fluids with the larger particle size is higher than that for the smaller particles. The particles with the sizes of 30nm, 80nm, 240nm possess an excellent antisedimentation stability, while the particle with the size of 640nm shows a poor antisedimentation stability. The particle with size of 240nm has the best comprehensive performance.

1. Introduction

Electrorheological (ER) fluid is a suspension made of micrometer- or nanometer-sized dielectric particles dispersing in an insulating liquid. Under an applied electric field, the dispersed dielectric particles will be polarized and attracted to each other to form chain or column structures. These chains and columns enable ER fluid to suddenly increase its viscosity and even change from a liquid-like state to a solid-like state. The viscosity change or liquid-solid state transition of ER fluid is rapid and reversible as the change of applied electric field [1-4]. Because of its controllable viscosity and short response time, ER fluid has attracted much interest in various areas, such as clutches, damping devices, display, human muscle simulator, and so on [5-7]. Most of these applications need ER fluid possessing large shear stress, low zero field viscosity, high ER efficiency and excellent antisedimentation stability.

Titanium oxide was considered as a promising ER material for its high dielectric constant. Many investigations were done on the ER fluids consisted of titanium oxide materials. The shear stress of pure crystalloid titanium oxide is rather low. In recent years, several authors reported that the high yield stress can be achieved by coating polar molecules on the particle surface or doping metal elements [8-10]. Even some amorphous titanium oxide based ER fluids without designed coating or doping exhibit high ER performance with high yield stress up to more than 100kPa [11]. Most of the yields stresses are beyond the required value (30kPa) for most industrial applications.

However, beside the yield stress, the zero field viscosity, electric current density, ER efficiency and antisedimentation stability of ER fluids are also the key parameters for industrial applications. Up to now, much work was devoted to the enhancement of yield stress, less more detailed study of comprehensive performance was reported. In this paper, a kind of amorphous titanium oxide particles with different sizes was prepared using a hydrolysis method. The microstructure, dielectric properties and comprehensive performance were investigated. The results show that the particle size had obvious effect on the yield stress, zero field viscosity, electric current density, ER efficiency and antisedimentation stability. And the solid particle concentration also played an important role in the rheological and ER activities of the suspensions.

2. Experiments

2.1. *Preparation of titanium oxide particles with different sizes*

All the chemical reagents in this study were of analytical grade. Titanium butoxide $(Ti(C_4H_9O)_4)$, deionized water and anhydrous ethanol were used as starting materials. Firstly, 10 ml titanium butoxide was dissolved in 100 ml anhydrous ethanol and stirred for 3 h to get a uniform solution A. A certain amount (x) of deionized water and $(200- x)$ ml anhydrous ethanol were mixed to form another homogeneous solution B. The added amount of deionized water is 160, 80, 20 and 5 ml for different samples (S1-4), respectively. Secondly, freshly prepared solution (A) was dropped into solution (B) with vigorous stirring condition at the temperature of 25 $^{\circ}$C to form amorphous titanium oxide particles with different sizes. Thirdly, the white precipitates so produced were filtered and washed several times with anhydrous ethanol, then dried in vacuum oven at 100 $^{\circ}$C to get samples 1-4.

2.2. *Preparation of ER fluids*

The samples were milled for 10 h in a mortar and dried in vacuum oven at 110°C for 5 h. Subsequently, the dried samples were mixed uniformly with silicone oil. The viscosity, density and dielectric constant of the silicone oil used in this ER fluid is 50mPas, $0.973g/cm^3$ and 2.5, respectively. A bit of oleic acid (around 0.5 wt% of the solid particles) was added to the suspensions to improve the wettability between the solid particles and silicone oil.

2.3. *Characterization and measurements of ER activity*

The morphology and grain size of the samples were examined by a Hitachi S4800 field emission scanning election microscope (FESEM) and Particle Size - Zeta Potential Analyzer (Nano ZS, Malvern). The X-ray diffraction patterns of the particles were recorded on a Rigaku D/Max-A diffractometer with CuKá radiation. The dielectric properties of the ER fluids were measured with a LCR Tester (Hioki 3532, Japan). The electrorheology of the ER fluids at different DC electric field was determined by a circular rheometer (Haake RS6000, Germany).

3. Results and Discussion

The SEM image of the samples 1-4 synthesized through hydrolysis method are expressed in Fig 1. The results show that the titanium oxide particles with different sizes were synthesized by controlling the added amount of the deionized water. The obvious distinction of the shape among the samples 1-4 is not observed. The Particle Size analysis indicates the average particle sizes of the samples 1-4 are around 30, 80, 240 and 640nm, respectively. The XRD analysis shows that all of the samples are in an amorphous state.

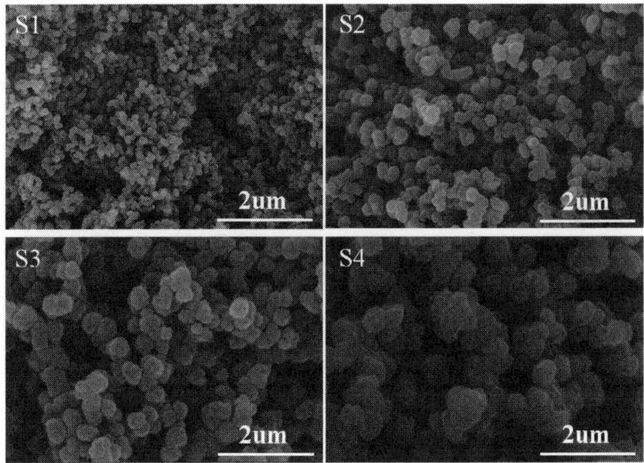

Fig. 1. The SEM image of samples 1-4 synthesized through hydrolysis method.

The dielectric properties of the titanium oxide based ER fluids with different particle sizes are expressed in Table 1. The concentration of particles in ER fluid is 10 vol%. The result indicates that the ER fluids with the smaller particle size have a higher dielectric constant and dielectric loss at low frequency.

The yield stress and the corresponding current density (inset) of the titanium oxide particles with different sizes under different DC electric field are expressed in Fig.2. The result indicates that the ER fluids with the smaller particle size have a higher yield stress and current density than that for the larger particles. The maximum yield stress of the samples with the sizes of 30nm, 80nm, 240nm and 640nm are 1.5, 1.8, 2.0 and 2.4 kPa, respectively, at E = 4 kV/mm.

Table 1. The dielectric properties of the titanium oxide based ER fluids with different particle sizes

Sample	ε_p			$tag\theta$		
	100Hz	1kHz	10kHz	100Hz	1kHz	10kHz
30nm	9.32	7.64	4.68	0.072	0.102	0.061
80nm	8.02	6.24	4.48	0.054	0.088	0.054
240nm	6.44	5.13	3.90	0.048	0.061	0.515
640nm	5.89	4.76	3.84	0.044	0.055	0.049

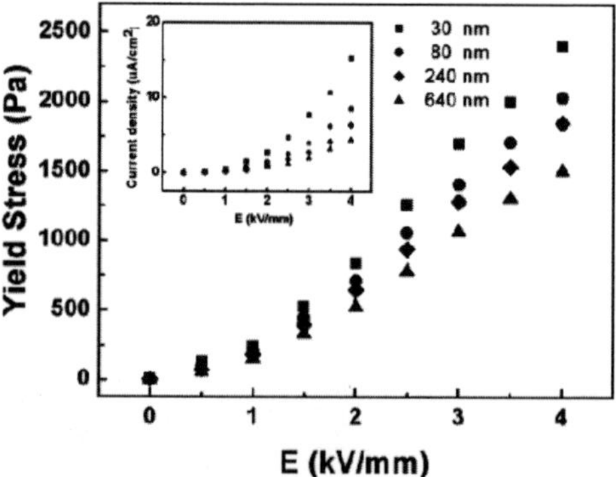

Fig. 2. The yield stress and the corresponding current density (inset) of the titanium oxide particles with different sizes. The concentration of particles in ER fluid is 10 vol%.

The dynamic shear stress of the samples with the particle sizes of 30nm, 80nm, 240nm and 640nm as a function of shear rate is shown in Fig. 3. The result indicates that the ER fluids with the smaller larger particle size have a higher dynamic shear stress than that for the larger particles. However, the ER fluids with the smaller larger particle size also have a higher zero field viscosity. The zero field viscosity of the samples with the particle sizes of 30nm, 80nm, 240nm and 640nm are 1.48, 1.25, 1.02 and 0.91 Pas, respectively.

The zero field viscosity of the samples with the particle sizes of 30nm, 80nm, 240nm and 640nm as a function of solid particle concentration is shown in Fig. 4. Fig. 4 indicates that the solid particle concentration has obvious effect on the zero field viscosity of the suspensions. At low volume fraction, the viscosity of the suspensions linearly increases with the increase of solid content.

However, when the volume fraction exceeds a critical value, a sharp increment of viscosity occurs. The suspensions composed of smaller particles have lower critical volume fraction value.

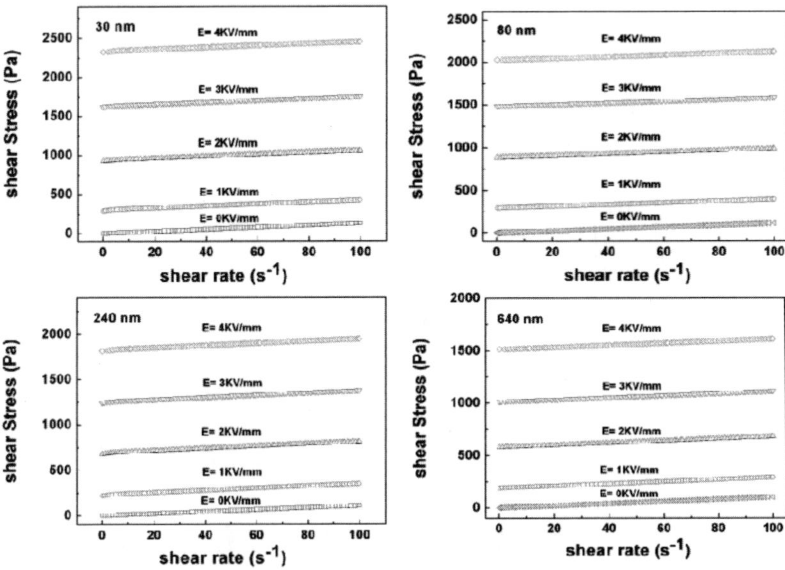

Fig. 3 The dynamic shear stress of the titanium oxide particles with the sizes of 30nm, 80nm, 240nm and 640nm under different DC electric field. The concentration of particles in ER fluid is 10 vol%.

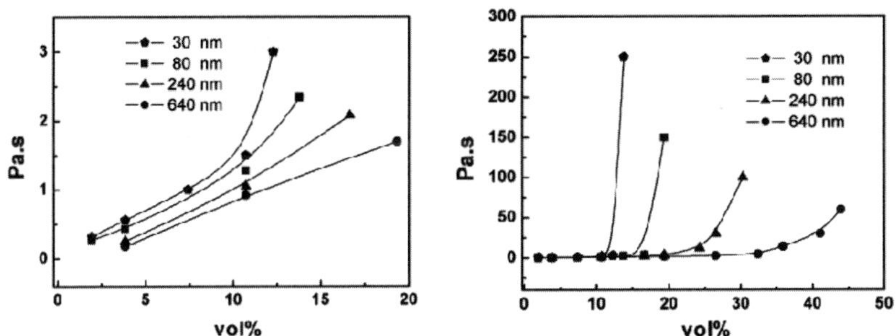

Fig. 4 The zero field viscosity of the samples with the sizes of 30nm, 80nm, 240nm and 640nm as a function of solid particle concentration.

The ER efficiency (defined by $(\tau_E-\tau_0)/\tau_0$, where τ_E is the shear stress with an electric field, and τ_0 is the shear stress at zero field) as a function of solid particle concentration is shown in Fig. 5. The ER efficiency of the suspensions increased with the increase of volume fraction, then decreased while the amounts of added particles beyond the critical values. At the lower solid particle concentrations range, the ER fluids with the smaller particle size show a

720

stronger ER efficiency than that for the larger particles. However, the maximal ER efficiency of the ER fluids with the larger particle size is higher than that of the smaller particles. The maximal ER efficiencies of the samples with the sizes of 30nm, 80nm, 240nm and 640nm at shear rate of 10 S^{-1} are 119, 210, 215, 260, respectively.

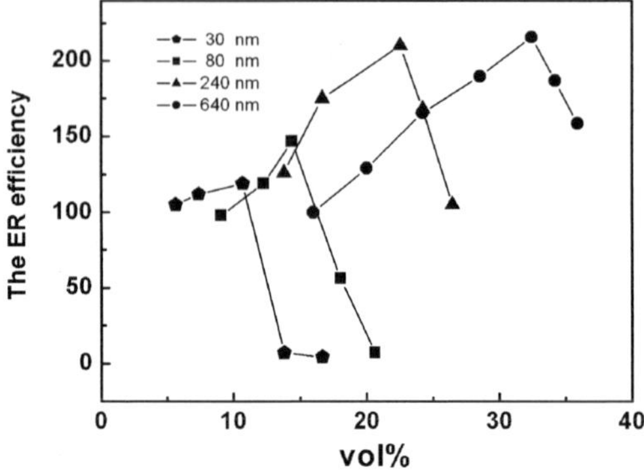

Fig. 5 The ER efficiency of the titanium oxide particles with different sizes as a function of solid particle concentration. The shear rate is 10 s^{-1} and the DC electric field is 4kV/mm.

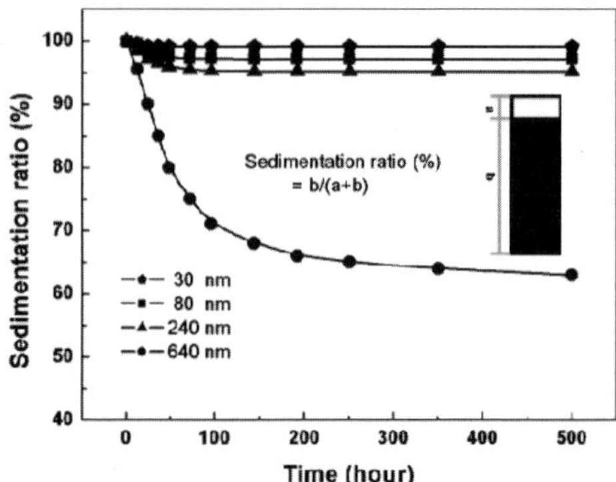

Fig. 6. Sedimentation rate of the titanium oxide based ER fluids with different particle sizes versus the static time. The concentration of particles in ER fluid is 10 vol%.

Fig. 6 shows the sedimentation rate of the titanium oxide based ER fluids with the particle sizes of 30nm, 80nm, 240nm and 640nm versus the static time. According to Fig. 6, the titanium oxide based suspensions with the particle sizes of 30nm, 80nm, 240nm possess an excellent antisedimentation stability. However, the antisedimentation stability of the titanium oxide based suspensions with the particle size of 640 nm is poor, the sedimentation ratio reaches 62% within 20 days. From the above results, we propose that the titanium oxide particle with size of 240 nm possesses the best comprehensive performance.

4. Conclusions

A kind of amorphous titanium oxide particles with different sizes (30nm, 80nm, 200nm and 600nm) was prepared using a hydrolysis method. The results show that the suspensions composed of smaller particles have higher yield stress at DC electric field, but also possess higher zero field viscosity at the same volume fraction. The maximal ER efficiency of the ER fluids with the larger particle size is higher than that for the smaller particles. The particles with the sizes of 30nm, 80nm, 240nm possess an excellent antisedimentation stability, while the particle with the size of 640nm shows a poor antisedimentation stability. The particle with size of 240nm expresses the best comprehensive performance.

Acknowledgments

This research is supported by the National Basic Research Program of China (Grant No. 2009CB930801), the National Natural Science Foundation of China (Grant No. 10904155 and 21003145), the Knowledge Innovation Project of Chinese Academy of Sciences (Grant No. KJCX2.YW.M07), the CAS/SAFEA International Partnership Program for Creative Research Teams, and by Zhejiang Provincial Natural Science Foundation of China (Grant No.D4080489 and Y4090044), Ningbo Natural Science Foundation (2009A610031, 2010A610170). We also express our gratitude to the aided program for Science and Technology Innovative Research Team of zhejiang Province and Ningbo Municipality (2009B21005).

References

1. W. M. Winslow, J. Appl. Phys. **20**, 1137 (1949).
2. H. Block and J.P. Kelly, J. Phys. D: Appl. Phys. **21**, 1661 (1988).
3. T.C. Halsey, Science **258**, 761 (1992).
4. T. Hao, Adv. Mater. **13**, 1847 (2001).
5. B.X. Wang and X.P. Zhao, J. Mater. Chem. **3**, 2248 (2003).
6. J.W. Kim, Y.H. Cho, H.G. Lee and S.B. Choi, J. Intell. Mater. Syst. Struct. **13**, 509 (2002).
7. M.S. Cho, Y.H. Cho, H.J. Choi, M.S. Jhon , Langmuir **19**, 5875 (2003).
8. J.G. Cao, M. Shen and L.W. Zhou, J. Solid State Chem. **179**, 1565 (2006).
9. J.H. Wei, L.H. Zhao, S.L. Peng, J. Shi, Z.Y. Liu and W.J. Wen, J. Sol-Gel Sci. Tech., **47**, 311 (2008).
10. J. B. Yin and X. P. Zhao, Chem. Mater. **16**, 321 (2004).
11. K.Q. Lu, R. Shen, X.Z. Wang, G. Sun, W.J. Wen and J.X. Liu, Chin. Phys. **15**, 2476 (2006).

RHEOLOGY AND MICROSTRUCTURAL EVOLUTION IN PRESSURE-DRIVEN FLOW OF A MAGNETORHEOLOGICAL FLUID WITH STRONG PARTICLE-WALL INTERACTIONS

MURAT OCALAN

Hatsopoulos Microfluids Laboratory, Department of Mechanical Engineering, Massachusetts Institute of Technology, 77 Massachusetts Av. Cambridge, Mass. 02139

Schlumberger-Doll Research, 1 Hampshire St. Cambridge, Mass. 02139

GARETH H. MCKINLEY

Hatsopoulos Microfluids Laboratory, Department of Mechanical Engineering, Massachusetts Institute of Technology, 77 Massachusetts Av. Cambridge, Mass. 02139

The interaction between magnetorheological (MR) fluid particles and the device walls that retain the fluid is critical as this interaction provides the means for coupling the physical device to the controllable properties of the fluid. This interaction is often enhanced in actuators by the use of ferromagnetic walls which generate an attractive force on the particles in the field-on state. The aggregation dynamics of MR fluid particles and the evolution of microstructure in pressure-driven flow through ferromagnetic channels were studied for the first time using custom-built microfluidic devices. The aggregation of the particles is studied in rectilinear, expansion and contraction channel geometries. With the results of this study, methods for improving MR actuator design and performance are also identified.

1. Introduction

Studies into the microstructural evolution of electrorheological (ER) and magnetorheological (MR) fluids have provided significant insight into the processes that result in their controllable bulk properties. The focus of these studies has been in the areas of chain deformation and breakup [1] and in particle aggregation & disaggregation [2-6]. Some examples of bulk rheological properties that are intimately linked to these microstructural processes include field-dependent yield stress, viscoelastic modulus and material response time.

In a quiescent MR suspension, aggregation and disaggregation phenomena are governed solely by the dimensionless ratio of magnetic force to Brownian force, $\lambda = \pi\mu_0 a^3 M_P^2 / 6k_B T$, where μ_0 is the permeability of free space, a is the particle radius, M_P is the particle magnetization, k_B is the Boltzmann constant and T is the absolute temperature [3].

In the presence of hydrodynamic forces additional dimensionless parameters are important. In this case, the aggregation dynamics are often parameterized by the Mason number, $Mn = 36\eta_0\dot{\gamma}/\mu_0 M_P^2$, which is the ratio of viscous forces to magnetic forces [4, 6]. Here η_0 is the suspension viscosity and $\dot{\gamma}$ is the shear rate. Deshmukh [6] studied aggregation phenomena in pressure-driven flow of MR fluids in microfluidic channels. The channels used in this study were manufactured from glass and poly-dimethylsiloxane (PDMS) with relatively smooth surface finishes and significant slip of the particles and aggregates along the walls of the microfluidic channels was observed.

Methods commonly used to enhance the interaction of MR fluid particles with channel walls are control of the surface roughness and manufacturing the channel from ferromagnetic materials. The latter approach is often utilized in MR fluid actuators [7-8] and therefore has more engineering significance.

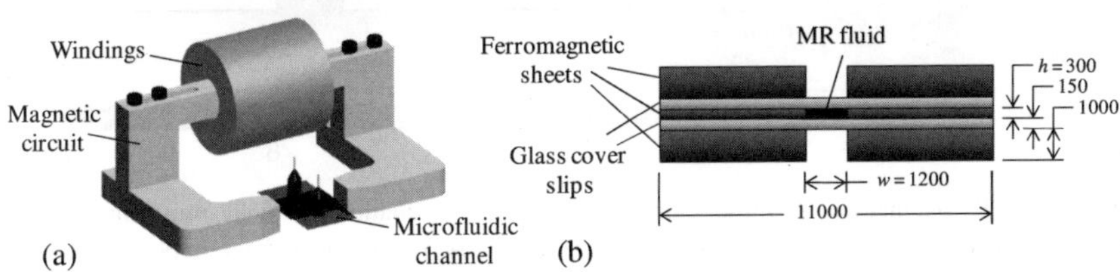

Figure 1: Experimental setup and channel construction. (a) The applied magnetic field is generated using a custom-built electromagnet. (b) Cross-sectional view of the microchannel. The channels used in this study are manufactured using a multilayered construction to provide a thin plane of fluid flow while maintaining the magnetic field uniformity. The channel dimensions (in μm) shown in the figure are utilized in all flow channels unless otherwise noted.

In the present study, we investigate the microstructural evolution of MR fluids in ferromagnetic microchannels. The strong interaction between the MR fluid particles and the channel walls creates the means to study flow phenomena in these fluids with boundary conditions replicating those realized in actuators. Furthermore, the flow of MR fluid through contraction and expansion geometries with inhomogeneous magnetic fields caused by the ferromagnetic channel walls is also studied.

2. Experimental Methods

The rectilinear channels used in this study were manufactured by adhering two ferromagnetic sheets with a separation of $h = 1.2mm$ between two glass microscope cover slips as illustrated in Figure 1(b). In order to generate a uniform field in the flow path, additional external plates were assembled on each side of the microfluidic channel.

The field uniformity in the flow channel was evaluated by a two dimensional finite element model. Results of the first simulation, presented in Figure 2(a), show that in the absence of the external plates the field uniformity is very poor with a strong gradient towards the channel walls. MR fluid flowing in this type of channel would experience an undesirable magnetophoretic force towards the channel walls. In the second model (Figure 2(b)) the external plates are also included in the model. In this case the field uniformity is significantly improved.

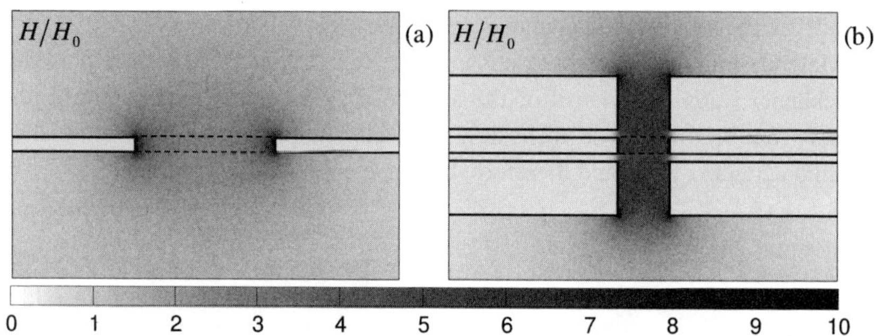

Figure 2: 2D magnetostatic finite element model for evaluation of field uniformity in the microchannels. Contours plotted for a zoomed section of the model to show the important features. (a) Without the external plates there is focusing of the field at the channel walls that would produce an undesirable magnetophoretic force on the particles. (b) The field uniformity in the microfluidic channel (shown by dashed lines) can be significantly improved with the addition of external plates.

During the experiments the channel is placed between the poles of a custom built electromagnet as illustrated in Figure 1(a). The field applied by the magnet was calibrated using a Gauss probe placed in the mid-point between the poles of the magnet. Due to dimensional constraints, it was not possible to conduct this measurement within the microchannels after assembly. The amplification of the field in the narrow gap, caused by the channel walls was estimated using the results of the finite element model presented in Figure 2(b).

Prior to each experiment the flow channel was demagnetized using a time-varying magnetic field in the form

$$B(t) = B_0 \cos(2\pi f t) e^{-\alpha t}. \tag{1}$$

Here B_0 is the initial amplitude of the signal which is normally set as the largest field applied after the preceding demagnetization cycle. The constants f and α are normally selected in the ranges $1-10 Hz$ and $0.3-1 s^{-1}$ respectively. The

MR fluid used in the experiments was prepared by suspending 1% v/v carbonyl iron particles (BASF® CR-grade) with a mean diameter of $7\mu m$ in a mixture of PDMS fluid and 10% w/w surfactant (Gelest™ DMS-T31 and DMS-S31 respectively). The fluid was pumped through the channel using a syringe pump (Harvard Apparatus® PHD 4400) with a constant flow rate, Q. The pressure drop across the channel was measured with a differential pressure transducer. Fluid flow in the microchannel was imaged using an inverted transmitted light microscope (Nikon® TE-2000S) with a 2x 0.06NA objective and a CCD camera (Matrix Vision BlueFOX™). The depth of field of this imaging system was $240\mu m$. Two image analysis techniques were utilized within this study: two-dimensional discrete Fourier transform and cluster identification. The methods and results from these techniques will be presented in a future publication; however, the descriptions of microstructure evolution presented in the following sections were quantified and validated using these methods.

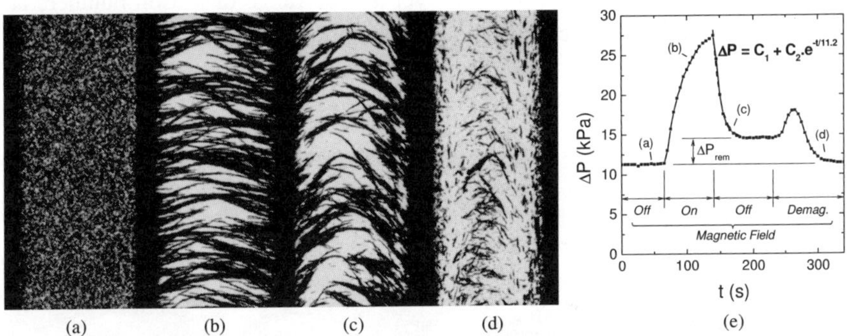

Figure 3: Aggregation and disaggregation of MR fluid flow in a ferromagnetic microchannel ($w = 1.2mm$, $h = 0.3mm$) during a typical experiment cycle. Flow is in the upward direction and the applied field is in the horizontal orientation. (a) Flow in a demagnetized channel without applied field. (b) Aggregated structures with $Mn = 6.3\times10^{-4}$. (c) Field turned off. (d) After demagnetizing field (e) Differential pressure across the flow channel during a typical experiment. The dashed line is an exponential decay curve fitted to the field-off region of the curve.

3. Results

Images recorded during a typical experiment are presented in Figure 3. In a demagnetized channel, the flow of the MR fluid occurs without aggregation of the particles. When the field is applied, the aggregates that form can be categorized into three groups by the number of channel walls the aggregate is in contact with: 0, 1 or 2 (channel-spanning). The aggregates that belong to the final group come to a complete stop within the flow channel under the influence of strong particle-wall interactions. The aggregates contacting only one of the channels also come to a stop at the channel wall; however, they rotate under the influence of the flow until they contact another aggregate or until the magnetic torque on the aggregate becomes equal to the hydrodynamic torque. Finally, the aggregates that do not have any contact with the channel walls flow continuously under the influence of the hydrodynamic forces; however, they eventually merge with a wall-contacting aggregate and stop flowing. Upon removal of the magnetic field, most of the particles in the channel freely flow once more. However, the particles close to the walls remain adhered to the interface (Figure 3(c)), forming a particle-rich layer on both sides of the channel. These layers contribute to the extra pressure drop across the channel as seen in Figure 3(e). Finally, when a demagnetizing field is applied, the remnant field in the channel walls is effectively eliminated and all the particles and aggregates flow out of the channel without the presence of any magnetic forces. The pressure drop across the channel also returns to its original value at the end of the demagnetizing step.

The aggregate structures formed over a wide range of Mason numbers are presented in Figure 4. The magnetic field is applied rapidly on a demagnetized flow channel and the images shown are recorded when the field-induced structure formation in the flow has reached a steady state.

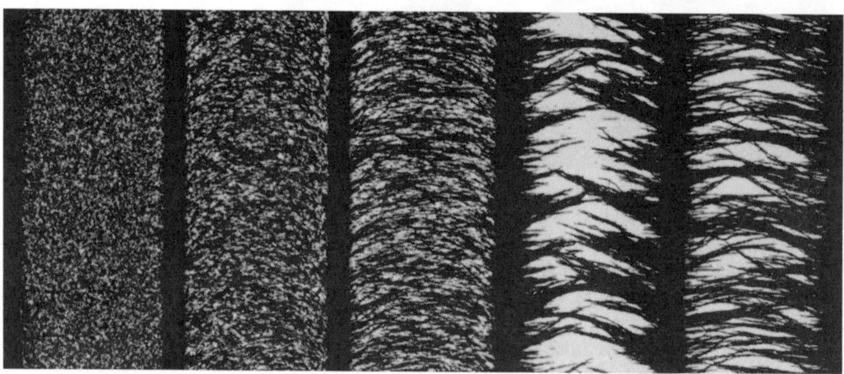

Figure 4: Evolution in aggregate microstructure of MR fluid flow. The Mason numbers in the images are (left to right) ∞ (field-off), 0.16, 0.018, 0.0016 and 0.00034. Channel dimensions are $w = 1.2mm$, $h = 0.3mm$.

The aggregation and flow phenomena in contraction and expansion flow conditions were also studied under a low-frequency $(0.1Hz)$ periodic applied field. The field was constantly on for the first half of the cycle followed by a demagnetizing field $\left(f = 10Hz, \alpha = 0.44 \text{ in Eq. (1)}\right)$ for the second half. With this selection of parameters, the decay in the demagnetizing field is rapid enough that the field is effectively removed entirely for last 40% of the cycle.

Representative images obtained in these flow conditions are illustrated in Figure 5. The aggregation in the inhomogeneous field sections outside the rectilinear channel shows a preferred direction aligned with the local field. Furthermore, an increase or a decrease in the convective velocity of the particle clusters is also observed because of the magnetic field gradients in these converging/diverging flows. A reversal in flow direction of the particulate phase was also observed in the expansion channel under high fields corresponding to $Mn = 1.6 \times 10^{-3}$.

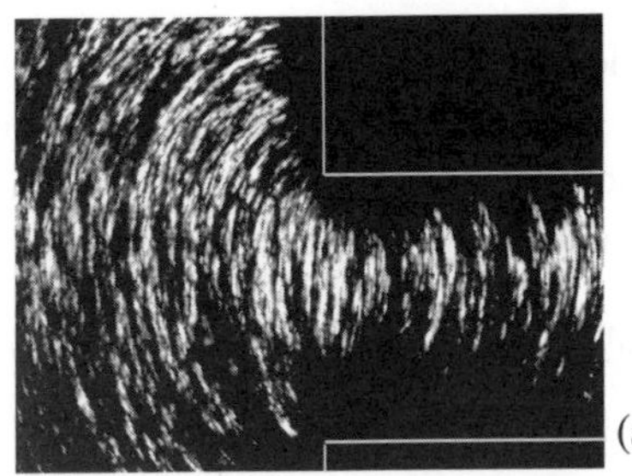
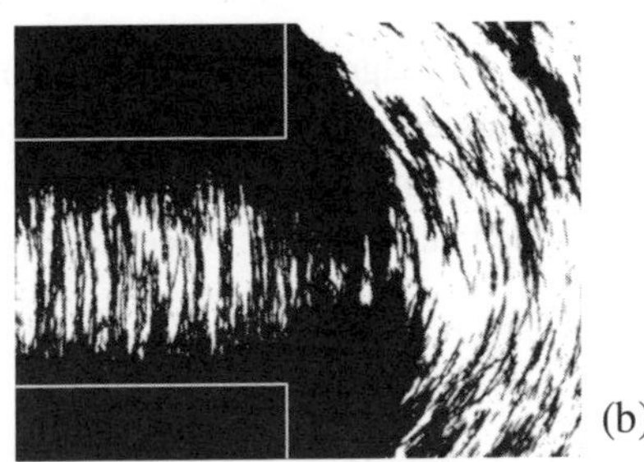

Figure 5: Contraction and expansion flow of MR fluid (width ratio, $w_1/w_2 = 3$). Square wave magnetic field with a rapid demagnetization cycle field-off state. (a) Contraction flow with $Mn = 3.2 \times 10^{-3}$, $\Phi \approx 80$. (b) Expansion flow with $Mn = 3.4 \times 10^{-4}$, $\Phi \approx 8$. The magnetophoretic force caused by the field gradient at the exit of the channel is large enough to reverse the flow of particles back into the channel.

4. Conclusions and Discussion

Flow of an MR fluid with strong particle-wall interaction was studied using a unique microfluidic channel setup with ferromagnetic walls. The forces caused by this strong interaction were demonstrated to be large enough to effectively eliminate the slip of the particulate phase against the channel walls. Under these boundary conditions the aggregation phenomena in rectilinear flow channels were studied by varying the Mason number. Our key findings include:

- Remnant magnetization of the ferromagnetic flow channels contribute to an additional field-off pressure differential across the rectilinear flow channel.
- The magnetic field gradient in the entry and exit regions of the converging/diverging flow channels can generate additional forces on the particles that may be significant as compared with the hydrodynamic force and this can drive secondary flows.
- With the combination of the magnetophoretic force in the entry & exit regions and the no-slip condition on the particulate phase, a slow increase in particle volume fraction in the converging/diverging channels was observed.

The Mason number commonly used to parameterize such flows is a ratio of viscous force and magnetic force in which the latter is obtained from the interaction of two particles in a homogeneous applied field. In the expansion and contraction flow condition, however, a significant magnetophoretic force, caused by the applied magnetic field *gradient* is generated on the particle. In the low volume fraction limit this force can be estimated by the relation (a detailed derivation can be found in [9])

$$F_{mag} = \left| \nabla\left(\mathbf{m}.\mathbf{B}\right) \right| \sim 4\pi\mu_0 a^3 \beta H_0 \left(\frac{\partial H_0}{\partial x}\right) \sim \frac{4\pi\mu_0 a^3 \beta H_0^2}{l_e}. \tag{2}$$

Here χ is the magnetic susceptibility of particle material, $\beta = \chi/(\chi+3)$ is the effective susceptibility, H_0 is the applied field and l_e is the length scale characterizing the field decay in the contraction/expansion. The ratio of this force to the hydrodynamic force scale results in the dimensionless parameter:

$$\Phi = \frac{F_{hyd}}{F_{mag}} = \frac{3\eta_0 v_b l_e}{2\mu_0 a^2 \beta H_0^2}, \tag{3}$$

where $v_b = Q/hw$ is the bulk velocity of flow. The visualizations from the experiments conducted so far suggest that this parameter is the relevant quantity to report important in the contraction and expansion flow geometries. However, future work in this area is needed to quantify the relation between the value of this parameter and the flow characteristics of MR fluids under these conditions.

References

1. Klingenberg, D.J. and Zukoski, C.F., *Langmuir*, **6**(1) 15-24 (1990)
2. Shulman, Z.P., Kordonsky, V.I., Zaltsgendler, E.A., Prokhorov, I.V., Khusid, B.M., and Demchuk, S.A., *Int. J. Multiphase Flow*, **12**(6) 935-955 (1986)
3. Fermigier, M. and Gast, A.P., *J. Colloid. Interf. Sci.*, **154**(2) 522-539 (1992)
4. Melle, S., Calderon, O.G., Rubio, M.A., and Fuller, G.G., *Phys. Rev. E*, **68**(4) (2003)
5. Dominguez-Garcia, P., Melle, S., Calderon, O.G., and Rubio, M.A., *Colloid. Surface. A*, **270**270-276 (2005)
6. Deshmukh, S.S., Massachusetts Institute of Technology, Dept. of Mech. Eng. Doctoral Thesis (2007)
7. Jolly, M.R., Bender, J.W., and Carlson, J.D., *J. Intel. Mat. Syst. Str.*, **10**(1) 5-13 (1999)
8. Herr, H. and Wilkenfeld, A., *Ind. Robot.*, **30**(1) 42-55 (2003)
9. Jones, T.B., *Electromechanics of Particles*. Cambridge: Cambridge University Press (1995)